Climbing and Walking Robots

M.O. Tokhi G.S. Virk M.A. Hossain (Eds.)

Climbing and Walking Robots

Proceedings of the 8th International Conference on Climbing and Walking Robots and the Support Technologies for Mobile Machines (CLAWAR 2005)

Dr. M.O. Tokhi
Department of Automatic Control
and, System Engineering
University Sheffield
Mappin Street
S1 3JD Sheffield
United Kingdom
E-mail: O.Tokhi@sheffield.ac.uk

Dr. M.A. Hossain
Department of Computing
University of Bradford
Richmond Road
BD7 1DP Bradford
United Kingdom
E-mail: M.A.Hossain1@Bradford.ac.uk

Professor G.S. Virk
Department of Mechanical Engineering
University Leeds
LS2 9JT Leeds
United Kingdom
E-mail: g.s.virk@leeds.ac.uk

Library of Congress Control Number: 2005934454

ISBN-10 3-540-26413-2 Springer Berlin Heidelberg New York
ISBN-13 978-3-540-26413-2 Springer Berlin Heidelberg New York

This work is subject to copyright. All rights are reserved, whether the whole or part of the material is concerned, specifically the rights of translation, reprinting, reuse of illustrations, recitation, broadcasting, reproduction on microfilm or in any other way, and storage in data banks. Duplication of this publication or parts thereof is permitted only under the provisions of the German Copyright Law of September 9, 1965, in its current version, and permission for use must always be obtained from Springer. Violations are liable for prosecution under the German Copyright Law.

Springer is a part of Springer Science+Business Media
springer.com
© Springer-Verlag Berlin Heidelberg 2006
Printed in The Netherlands

The use of general descriptive names, registered names, trademarks, etc. in this publication does not imply, even in the absence of a specific statement, that such names are exempt from the relevant protective laws and regulations and therefore free for general use.

Typesetting: by the authors and TechBooks using a Springer LATEX macro package

Cover design: *design & production* GmbH, Heidelberg

Printed on acid-free paper   SPIN: 11502869   89/TechBooks   5 4 3 2 1 0

Preface

The interest in climbing and walking robots (CLAWAR) has intensified in recent years, and novel solutions for complex and very diverse applications have been anticipated by means of significant progress in this area of robotics. Moreover, the amalgamation of original ideas and related innovations, search for new potential applications and the use of state of the art support technologies permit to foresee an important step forward and a significant socio-economic impact of advanced robot technology in the future. This is leading to the creation and consolidation of a mobile service robotics sector where most of the robotics activities are foreseen in the future. The technology is now maturing to become of real benefit to society and methods of realizing this potential quickly are being eagerly explored. Robot standards and modularity are key to this and form key components of the research presented here.

CLAWAR 2005 is the eighth in a series of international conferences organised annually since 1998 with the aim to report on latest research and development findings and to provide a forum for scientific discussion and debate within the mobile service robotics community. The series has grown in its popularity significantly over the years, and has attracted researchers and developers from across the globe. The CLAWAR 2005 proceedings reports state of the art scientific and developmental findings presented during the CLAWAR 2005 conference in 131 technical presentations by authors from 27 countries covering the five continents.

The editors would like to thank members of the International Programme Committee, International Technical Advisory/Organising Committee and National Organising Committee for their efforts in reviewing the submissions, and to the authors in responding to comments and suggestions of the reviewers. It is hoped that this edition of the CLAWAR conference proceedings forms a valuable addition to the scientific and developmental knowledge in mobile robotics.

M. O. Tokhi
G. S. Virk
M. A. Hossain

Editors

M. Osman Tokhi
Osman Tokhi received his BSc in Electrical Engineering from Kabul University, Afghanistan in 1978 and PhD from Heriot-Watt University, UK in 1988. He has worked at various academic and industrial establishments, and is currently employed as Reader at the Department of Automatic Control and Systems Engineering, the University of Sheffield, UK.

His research interests include active control of noise and vibration, adaptive/intelligent and soft-computing modeling and control of dynamic systems, high-performance real-time signal processing and control, and biomedical applications of robotics and control.

Gurvinder S. Virk
Gurvinder Virk received 1st Class Honours in Electrical and Electronic Engineering from the University of Manchester in 1977 and PhD from Imperial College, London in 1982. Since then he has followed an academic career working at Sheffield City Polytechnic, Universities of Southampton, Sheffield, Bradford and Portsmouth before joining Leeds as Professor of Robotics and Control in April 2002.

His main research interests include mobile robotics, building management and renewable energy systems, and the use of advanced model-based control to a variety of applications.

M. Almgir Hossain
Alamgir Hossain received his MSc in Applied Physics and Electronics from the University of Dhaka, Bangladesh, 1983 and PhD from the University of Sheffield, UK in 1995. He has served at various academic institutions, and is currently employed as Lecturer at the Department of Computing, University of Bradford, UK. His research interests include real-time computing, intelligent control, parallel, grid and ubiquitous computing.

Contents

Plenary Papers   1

Common situation awareness as basis for human-robot interface   3
    A. Halme

Gait restoration by functional electrical stimulation   19
    W. K. Durfee

Space robotics   27
    G. Visentin

Bio-Engineering and Biological Inspired Systems   39

ASYSTENT - Control system of a manipulator for keyhole surgery   41
    P. Sauer, K. Kozlowski, W. Waliszewski and P. Jeziorek

A biologically inspired model for quadruped locomotion   49
    B. Hennion, J. Pill and J-C. Guinot

Fuzzy logic control strategy for functional electrical stimulation in bipedal cycling   57
    R. Massoud, M. O. Tokhi and C.S. Gharooni

Insect-inspired, actively compliant hexapod capable of object manipulation   65
    W. A. Lewinger, M. S. Branicky and R. D. Quinn

Modeling and simulation of humanoid stair climbing   73
    L. Wang, Y. Zhao, M. O. Tokhi and S. C. Gharooni

Design issues of spring brake orthosis: evolutionary algorithm approach   81
    M. S. Huq, M. S. Alam, S. Gharooni and M. O. Tokhi

Recent developments in implantable and surface based dropped foot functional electrical stimulators   89
    L. Kenney, P. Taylor, G. Mann , G. Bultstra, H. Buschman, H. Hermens, P. Slycke, J. Hobby, N. van der Aa, B. Heller, A. Barker, D. Howard and N. Sha

Fluidically driven robots with biologically inspired actuators   97
    S. Schulz, C. Pylatiuk, A. Kargov, R. Oberle, H. Klosek, T. Werner, W. Rößler, H. Breitwieser, G. Bretthauer

Climbing, Navigation and Path Planning   105

Concept for energy-autarkic, autonomous climbing robots   107
    W. Brockmann

Navigation of walking robots: path planning   115
    B. Gaßmann, M. Huber, J. M. Zöllner and R. Dillmann

Study on mobility of connected crawler robotby using GA   123
    S. Yokota, K. Kawabata, P. Blazevic and H. Kobayashi

A robot that climbs walls using micro-structured polymer feet   131
    K. A. Daltorio, S. Gorb, A. Peressadko, A. D. Horchler, R. E. Ritzmann and R. D. Quinn

Novel solutions to design problems of industrial climbing robots   139
    S. Chen, J. Shang, Z. Zhao, T. Sattar and B. Bridge

Fast pointing and tracking system for mobile robot short range control via free space optical laser line of sight communication link   147
    M. H. Ahmad, D. Kerr and K. Bouazza-Marouf

Control of CLAWAR   155

4 legs positioning control of the quadruped robot by linear visual servoing using stereo omnidirectional camera   157
    N. Maru and Y. Inoue

Control of a 3D quadruped trot   165
    L. R. Palmer III and D. E. Orin

Research on obstacle-navigation control of a mobile robot for inspection of the power transmission lines based on expert system   173
    W. Ludan, W. Hongguang, F. Lijin and Z. Mingyang

Measure of the propulsion dynamic capability of a walking system A. David, O. Bruneau and J.-G. Fontaine	181
Experimental walking results of LUCY, a biped with pneumatic artificial muscles B. Vanderborght, B. Verrelst, R. Van Ham, M. Van Damme and D. Lefeber	189
Hip joint control of a legged robot for walking uniformly and the self-lock mechanism for compensating torque caused by weight T. Okada, T. Sakai, K. Shibuya and T. Shimizu	197
Controlling dynamic stability and active compliance to improve quadrupedal walking E. Garcia and P. Gonzalez de Santos	205
Complete stability analysis of a control law for walking robots with non-permanent contacts S. Chareyron and P.B. Wieber	213
Vision-based stabilization of the IDP flat output L. Beji, A. Abichou and M. A. El Kamel	221
Bus communication in robot system control G. Mahalu, A. Graur and V. Popa	229
A hybrid locomotion control approach D. Spenneberg	237
Model and Control of joints driven by fluidic muscles with the help of advanced automatic algorithm generation software T. Kerscher, J. M. Zoellner, R. Dillmann, A. Stella and G. Caporaletti	245
Modelling and control of a X4 bidirectional rotors R. Slim, A. Abichou and L. Beji	253
Stability measure comparison for the design of a dynamic running robot J. E. Clark and M. R. Cutkosky	261
Control architecture and walking strategy for a pneumatic biped robot G. Spampinato and G. Muscato	269
Time-scaling control of a compass type biped robot P. Fauteux, P. Micheau and P. Bourassa	277

Design Methodology and Gait Generation   285

Mechanical design of step-climbing vehicle with passive linkages   287
    D. Chugo , K. Kawabata , H. Kaetsu , H. Asama and T. Mishima

Integrated structure-control design of dynamically walking robots   295
    P. Kiriazov

Intuitive design and gait analysis for a closed loop leg mechanism of a quadruped with single actuator   303
    Vinayak

Design of a cockroach-like running robot for the 2004 SAE walking machine challenge   311
    M. A. Lavoie, A. L. Desbiens, M. A. Roux, P. Fauteux and É. Lespérance

Finding adequate optimization criteria to solve inverse kinematics of redundant bird leg mechanism   319
    L. Mederreg, V. Hugel, A. Abourachid, P. Blazevic and R. Hackert

Integrated system of assisted mechatronic design for oriented computer to automatic optimising of structure of service robots (SIDEMAR)   327
    C. Castejón, A. Gimenez, A. Jardón, H. Rubio, J. C. Garcia-Prada and C. Balaguer

The construction of the four legged prototype robot ARAMIES   335
    J. Hilljegerdes, D. Spenneberg and F. Kirchner

Application of waves displacement algorithms for the generation of gaits in an all terrain hexapod   343
    A. Alonso-Puig

Extensive modeling of a 3 DOF passive dynamic walker   349
    M. A. Roa, C. A. Villegas and R. E. Ramirez

Development of biped robots at the national university of Colombia   357
    M. A. Roa, R. E. Ramirez and D.A. Garzón

Design of a low cost platform for unmanned aerial vehicle mathematical modeling   365
    D. M. Alba, H. Montes, G. Bacallado, R. Ponticelli and M. Armada

Hopping and Legged Robots 373

Observer backstepping for height control of a resonance hopping robot 375
R. Fernández, T. Akinfiev and M. Armada

Standing up with motor primitives 383
V. Hamburger, K. Berns, F. Iida and R. Pfeifer

Multiple terrain adaptation approach using ultrasonic sensors for legged robots 391
S. Nabulsi, M. Armada and H. Montes

Sliding mode observer with no orientation measurement for a walking biped 399
V. Lebastard, Y. Aoustin and F. Plestan

Humanoid Robots 407

Detection and classification of posture instabilities of bipedal robots 409
O. Höhn, J. Gačnik and W. Gerth

Development of a low-cost humanoid robot: components and technological solutions 417
V. M. F. Santos and F. M. T. Silva

Analysis of humanoid robot lower extremities force distribution in standing position 425
H. Montes, P. Alarcon, R. Ponticelli and M. Armada

ZMP human measure system 433
M. Arbulú, F. Prieto, L. Cabás, P. Staroverov, D. Kaynov and C. Balaguer

Mechanical design and dynamic analysis of the humanoid robot RH-0 441
L. Cabás, S. de Torre, R. Cabás, D. Kaynov, M. Arbulú, P. Staroverov and C. Balaguer

Advanced motion control system for the humanoid robot Rh-0 449
D. Kaynov, M. A. Rodríguez, P. Staroverov, M. Arbulú, L. Cabás and C. Balaguer

Humanoid vertical jump with compliant contact 457
V. Nunez and N. Nadjar-Gauthier

Locomotion 465

A 3D galloping quadruped robot 467
D. P. Krasny and D. E. Orin

Kineto-static analysis of an articulated six-wheel rover 475
P. Bidaud, F. Benamar and T. Poulain

Momentum compensation for the dynamic walk of humanoids based on the optimal pelvic rotation 485
H. Takemura, A. Matsuyama, J. Ueda, Y. Matsumoto, H. Mizoguchi and T. Ogasawarahi

Walk calibration in a four-legged robot 493
B. Bonev, M. Cazorla and H. Martínez

Peristaltic locomotion: application to a worm-like robot 501
F. Cotta, F. Icardi, G. T. Zurlo and R. M. Molfino

Impact shaping for double support walk: from the rocking block to the biped robot 509
J.-M. Bourgeot, C. Canudas-de-Wit and B. Brogliato

Proposal of 4-leg locomotion by phase change 517
K. Morita and H. Ishihara

Introducing the hex-a-ball, a hybrid locomotion terrain adaptive walking and rolling robot 525
C. C. Phipps and M. A. Minor

Stability control of an hybrid wheel-legged robot 533
G. Besseron, C. Grand, F. Ben Amar, F. Plumet and P. Bidaud

Manipulation and Flexible Manipulators 541

Hybrid control scheme for tracking performance of a flexible system 543
F. M. Aldebrez, M. S. Alam and M. O. Tokhi

Predesign of an anthropomorphic lightweight manipulator 551
C. Castejón, D. Blanco, S. H. Kadhim and L. Moreno

Design of a "soft" 2-DOF planar pneumatic manipulator 559
M. Van Damme, R. Van Ham, B. Vanderborght, F. Daerden and D. Lefeber

Simulation and experimental studies of hybrid learning control with acceleration feedback for flexible manipulators 567
M. Z. Md Zain, M. S. Alam, M. O. Tokhi and Z. Mohamed

BNN-based fuzzy logic controller for flexible-link manipulator 575
    M.N.H. Siddique, M.A. Hossain and M.O. Tokhi

Design constraints in implementing real-time algorithms for a 583
flexible manipulator system
    M. A. Hossain, M.N.H. Siddique, M.O. Tokhi and M.S. Alam

Pay-load estimation of a 2-DOF flexible link robot 591
    N. K. Poulsen and O. Ravn

Design of hybrid learning control for flexible manipulators: a 599
multi-objective optimisation approach
    M. S. Alam, M. Z. Md Zain, M. O. Tokhi and F. Aldebrez

Intelligent modelling of flexible manipulator systems 607
    M. H. Shaheed, A. K. M. Azad and M. O. Tokhi

Wafer handling demo by SERPC 615
    N. Abbate, A. Basile, S. Ciardo, A. Faulisi, C. Guastella, M. L. Presti, G. Macina and N. Testa

Vision control for artificial hand 623
    M. Kaczmarski and D. Zarychta

Robotic finger that imitates the human index finger in the number and distribution of its tendons 631
    D. M. Alba, G. Bacallado, H. Montes, R. Ponticelli, T. Akinfiev and M. Armada

Modular, Reconfigurable Robots 639

Methods for collective displacement of modular self-reconfigurable robots 641
    E. Carrillo and D. Duhaut

Suboptimal system recovery from communication loss in a 649
multi-robot localization scenario using EKF algorithms
    P. Kondaxakis, V. F. Ruiz and W. S. Harwin

ORTHO-BOT: A modular reconfigurable space robot concept 659
    V. Ramchurn, R. C. Richardson and P. Nutter

Motion of minimal configurations of a modular robot: sinusoidal, lateral rolling and lateral shift 667
    J. Gonzalez-Gomez and E. Boemo

Modularity and System Architecture   675

The modular walking machine, platform for technological equipments   677
  I. Ion, I. Simionescu, A. Curaj, L. Dulgheru1 and A. Vasile

YaMoR and Bluemove – an autonomous modular robot with Bluetooth interface for exploring adaptive locomotion   685
  R. Moeckel, C. Jaquier, K. Drapel, E. Dittrich, A. Upegui and A. Ijspeert

On the development of a modular external-pipe crawling omni-directional mobile robot   693
  P. Chatzakos, Y. P. Markopoulos, K. Hrissagis and A. Khalid

Modularity and component reuse at the shadow robot company   701
  The Shadow Robot Company Ltd

CLAWAR design tools to support modular robot design   709
  G. S. Virk

Powering, Actuation, Efficiency   717

Pneumatic actuators for serpentine robot   719
  G. Granosik and J. Borenstein

Nontraditional drives for walking robots   727
  T. Akinfiev, R. Fernández and M. Armada

Energy efficiency of quadruped gaits   735
  M. F. Silva and J. A. T. Machado

Bellows driven, muscle steered caterpillar robot   743
  G. Granosik and M. Kaczmarski

On the application of impedance control to a non-linear actuator   751
  H. Montes, M. Armada and T. Akinfiev

MACCEPA: the Actuator with adaptable compliance for dynamic walking bipeds   759
  R. Van Ham, B. Vanderborght, M. Van Damme, B. Verrelst and D. Lefeber

A design of a walking robot with hybrid actuation system   767
  K. Inagaki and H. Mitsuhashi

Manipulators driven by pneumatic muscles 775
   K. Feja, M. Kaczmarski and P. Riabcew

Sensing and Sensor Fusion 783

New advances on speckle-velocimeter for robotized vehicles . . . 785
   A. Aliverdiev, M. Caponero, C. Moriconi, P. A. Fichera and G. Sagratella

Information processing in reactive navigation and fault detection of walking robot 793
   A. Vitko, M. Šavel, D. Kameniar and L. Jurišica

Intelligent sensor system and flexible gripper for security robots 801
   R.D. Schraft, K. Wegener, F. Simons and K. Pfeiffer

Search performance of a multi-robot team in odour source localisation 809
   C. Lytridis, G. S. Virk and E. E. Kadar

A "T-shirt based" image recognition system 817
   P. Staroverov, C. Chicharro, D. Kaynov, M. Arbulú, L. Cabás and C. Balaguer

Object shape characterisation using a haptic probe 825
   O. P. Odiase and R.C. Richardson

Detection of landmines using nuclear quadrupole resonance (NQR): signal processing to aid classification 833
   S. D. Somasundaram, K. Althoefer, J. A. S. Smith and L. D. Seneviratne

Software and Computer-aided Environments 841

Simulator for locomotion control of the Alicia^3 climbing robot . . 843
   D. Longo, G. Muscato, S. Sessa

A general platform for robot navigation in natural environments 851
   E. Celaya, T. Creemers and J. L. Albarral

Simulations of the dynamic behavior of a bipedal robot with torso subjected to external disturbances 859
   C. Zaoui, O.Bruneau, F.B. Ouezdou and A. Maalej

System Analysis, Modelling and Simulation — 867

Analysis of the direct and inverse kinematics of ROMA II robot — 869
J. C. Resino, A. Jardón, A. Gimenez and C. Balaguer

Simulation of a novel snake-like robot — 875
R. Aubin, P. Blazevic and J. P. Guyvarch

An actuated horizontal plane model for insect locomotion — 883
J. Schmitt

Industrial Applications — 891

Machine vision guidance system for a modular climbing robot used in shipbuilding — 893
J. Sánchez, F. Vázquez and E. Paz

A locomotion robot for heavy load transportation — 901
H. Ishihara and K. Kuroi

Using signs for configuring work tasks of service robots — 909
M. Heikkilä, S. Terho, M. Hirsi, A. Halme and P. Forsman

A system for monitoring and controlling a climbing and walking robot for landslide consolidation — 917
L. Steinicke, D. Dal Zot and T. Fautré

Non-destructive Testing Applications — 925

Small inspection vehicles for non-destructive testing applications — 927
M. Friedrich, L. Gatzoulis, G. Hayward and W. Galbraith

Automated NDT of floating production storage oil tanks with a swimming and climbing robot — 935
T. P. Sattar, H. E. Leon Rodriguez, J. Shang and B. Bridge

7-axis arm for NDT of surfaces with complex & unknown geometry — 943
T. P. Sattar and A. A. Brenner

Personal Assistance Applications — 951

Elderly people sit to stand transfer experimental analysis — 953
P. Médéric, V. Pasqui, F. Plumet and P. Bidaud

A portable light-weight climbing robot for personal assistance 961
applications
  A. Gimenez, A. Jardon, R. Correal, R. Cabás and C. Balaguer

Modeling and control of upright lifting wheelchair 969
  S. C. Gharooni, B. Awada and M. O. Tokhi

A humanoid head for assistance robots 977
  K. Berns and T. Braun

An application of the AIGM algorithm to hand-posture recog- 985
nition in manipulation
  D. Garcia, M. Pinzolas, J. L. Coronado and P. Martínez

Security and Surveillance Applications 993

  AirEOD: a robot for on-board airplanes security 995
    S. Costo, F. Cepolina, M. Zoppi and R. Molfino

  AIMEE: a four-legged robot for RoboCup rescue 1003
    M. Albrecht, T. Backhaus, S. Planthaber, H. Stöppler, D. Spenneberg and F. Kirchner

  Modular situational awareness for CLAWAR robots 1011
    Y. Gatsoulis and G. S. Virk

Space Applications 1021

  Design drivers for robotics systems in space 1023
    L. Steinicke

  A robotics task scheduler – TAPAS 1031
    G. Focant, B. Fontaine, L. Steinicke and L. Joudrier

  Mobile mini-robots for space application 1037
    M.-W. Han

  Teleagents for space exploration and exploitation in future hu- 1045
  man planetary missions
    G. Genta

  An expandable mechanism for deployment and contact surface 1053
  adaptation of rover wheels
    P. Bidaud, F. Benamar and S. Poirier

  A New traction control architecture for planetary exploration 1061
  robots
    D. Caltabiano, D. Longo and G. Muscato

The lemur II-class robots for inspection and maintenance of orbital structures: a system description   1069
   B. Kennedy, A. Okon, H. Aghazarian, M. Garrett, T. Huntsberger, L. Magnone, M. Robinson and J. Townsend

Lemur IIb: a robotic system for steep terrain access   1077
   B. Kennedy, A. Okon, H. Aghazarian, M. Badescu, X. Bao, Y. Bar-Cohen, Z. Chang, B. E. Dabiri, M. Garrett, L. Magnone and S. Sherrit

Tele-operation, Social and Economic Aspects   1085

Virtual immersion for tele-controlling a hexapod robot   1087
   J. Albiez, B. Giesler, J. Lellmann, J. M. Zöllner and R. Dillmann

Economic prospects for mobile robotic systems, new modular components   1095
   N. J. Heyes and H. A. Warren

Appendix A: CLAWAR-2005 Organisation   1103

Appendix B: CLAWAR-2005 Reviewers   1105

Appendix C: CLAWAR-2005 Sponsors and Co-sponsors   1106

Author Index   1107

Plenary Papers

Common Situation Awareness as Basis for Human-Robot Interface

A. Halme

Automation Technology Department, Helsinki University of Technology, Espoo, Finland, aarne.halme@tkk.fi

Abstract

One of the fundamental questions in service robotics is how we can get those machines to work efficiently in the same environment with and under the command of the average user. The answer – not an easy one – is in the development of human-robot interfaces (HRI), which should be able to transmit the will of the user to the robot in a simple but effective way preferably using means that are natural to humans. The paper introduces a generic interface concept based on exchanging cognitive information between the robot and the user, both present in the same environment. The information is exchanged through a virtual world called "common situation awareness" or "common presence" of both entities. The virtual world is a simple map augmented by database of semantic information. Functions of the HRI are based on utilizing interactively the senses of both the human and machine entities, the human when perceiving the environment and commanding the machine and the machine when looking for working targets or moving in the environment. The syntax of the communication language utilizes the objects and work targets currently existing in the common presence. A multi-tasking humanoid robot, called WorkPartner, is used to demonstrate the interface principles. The interface is designed to support interaction, including teaching and learning, with the human user in various outdoors work tasks. Natural to human communication methods, such as speech and gestures, are promoted in order to ease the user's load. Methods and hardware presented have been designed for the WorkPartner robot, but many of them are generic in nature and not dependent upon the robot physics.

Keywords: human-robot interface, service robotics, cognition, situation awareness.

1 Introduction

There are two challenging technological steps for robots in their way from factories to among people. The first to be taken is to obtain fluent mobility in unstructured, changing environments, and the second is to obtain the capability for intelligent communication with humans together with a fast, effective learning/adaptation to new work tasks. The first step has almost been taken today. The rapid development of sensor technology – especially inertial sensors and laser scanners – with constantly increasing processing power, which allows heavy image processing and techniques for simultaneous localization and mapping (SLAM), have made it possible to allow slowly moving robots to enter in the same areas with humans. However, if we compare the present capability of robots to animals, like our pets in homes, it can be said without a doubt that improvements are still much preferred.

The second step is still far away. Traditional industrial robots are mechanically capable to change a tool and perform different work tasks, but due to the nature of factory work need for reprogramming is relatively minor and therefore interactive communication and continuous learning are not needed. The most sophisticated programming methods allow task design, testing, and programming off-line in a simulation tool without any contact to the robot itself. Today's commercial mobile service robots, like vacuum cleaners and lawn mowers, are limited to a single task by their mechanical construction. A multi-task service robot needs both mechanical flexibility and a high level of "intelligence" in order to carry out and learn several different tasks in continuous interaction with the user. Instead of being a "multi-tool" the robot should be capable of using different kinds of tools designed for humans. Due to fast development in mechatronics, hardware is not any more the main problem although the prices can be high. The bottlenecks are the human – robot interface (HRI) and the robot intelligence, which are strongly limiting both the information transfer from the user to the robot as well as the learning of new tasks.

Despite huge efforts in AI and robotics research, the word "intelligence" has to be written today in quotes. Researchers have not been able to either model or imitate the complex functions of human brains or the human communication, thus today's robots hardly have either the creativity or the capacity to think.

The main requirement for a service robot HRI is to provide easy humanlike interaction, which on the one hand does not load the user too much and on the other hand is effective in the sense that the robot can be kept in useful work as much as possible. Note that learning of new tasks is not counted as useful work! The interface should be natural for human cognition based on speech and gestures. Because the robot cognition and learning capabilities are still very limited the interface should be optimized between these limits by dividing the cognitive tasks between the human brains and robot "intelligence" in an appropriate way.

The user effort needed for interactive use of robotic machines varies much. Teleoperators need much user effort, because the user controls them directly. Single tasks service robots, like autonomous vacuum cleaners, do not demand too much effort, but complexity and effort needed increase rabidly when general purpose machines are put to work.

It is quite clear today that without noticeable progress in the matter, the effort needed for operation of multi-tasking service robots will be even higher than in the case of classical teleoperators, because more data is needed to define the details of the work. The classical teleoperators have evolved greatly since the 1950's and today have reached a high standard of development especially through the development of tele-existence methods and technologies [10]. Teleoperators may be classified into four classes [2] due to their complexity, sensing, and the operator supervision status, but altogether the way these systems are predicted to develop, they will lead to user information loads too large to be practical as a human-robot interface concept for interactive service robots. Thus new concepts are needed.

Intuitively it is clear that such concepts must utilize the superior cognition and reasoning capacity of human brains, which allow for fusing different perception information and making conclusions on the basis of insufficient information. This means that controlling the robot must be based mainly on semantic or symbolic information instead of copying motions or following numeric models, which is the case even in teleoperation commanded in task level. The problem is how to get the exchange of information in such a way that the robot "understand orders" and actively interacts while performing them rather than follow them slavishly. Only in this way is there a possibility of keeping the user load within acceptable levels. The ultimate goal is to let the robots do skilled work, thought at first by the user, but gradually learned by the robot as an independent performance [8].

The approach proposed in this paper is based on the idea that as much of the information needed to make the robot to perform a work task as possible is given in symbolic or schematic form instead of numeric data. The environment itself should be utilized when possible. The human user can create relatively easily such information when using his/her cognition in the work place. The cognitive capacity of the robot is programmed so that transferring the user's will to the robot happens mostly in the form of dialogue, which makes sure that the task is uniquely defined and can be executed by the robot's internal commanding system. The dialogue is based on concepts and objects in the virtual situated awareness called "common presence". The common presence is a model of the working environment, which both entities can understand in the same way through their cognition system. The detailed meaning of common presence will be explained later, but it should be noted here that we use the term here in a different meaning than psychologists, who often include feelings and imagination in this concept. In our pragmatic presence concept the physical working place is the place where both the human user and the robot try to be "as present as possible". The human user might be also telepresent there, but the robot is supposed to be physically present.

2 Robot used to illustrate the concept

The research platform Workpartner, shown in **Fig. 1**, is a humanoid service robot, which is designed for light outdoor applications, like property maintenance, gardening or watching [1].

Fig. 1. WorkPartner service robot

Workpartner was designed as a multi-purpose service robot, which can carry on many different tasks. As the partner to the user it should be capable of performing tasks either alone or in cooperation with its master. In **Fig. 1** WorkPartner is cleaning snow on a yard – a very common task in Finland in winter time. The colored beacon shown side of the robot is a part of the user interface equipment by the aid of which he/she can easily crop the area to be cleaned.

Skills for new tasks are initially taught interactively by the operator in the form of state diagrams, which include motion and perception control actions necessary to perform the task. Design of the user interface has been done so that most of the interaction can be done (not ought to be done) with human like conversation by speech and gestures to minimize the wearable operator hardware. Different interface devices have also been developed to help the mutual understanding between robot and operator, especially in teaching and teleoperation situations.

Although the WorkPartner robot is used here as a reference robot, it should be noticed that most of the ideas and results presented are generic in nature and do not depend on the specific robot. Almost any mobile service robot with manipulation capability and with similar subsystem infrastructure could be used as the test robot as well.

2.1 HRI equipment

The main functions of WorkPartner's HRI are

- communication with the robot in all modes of operation
- task supervision, assistance and collaboration
- task definition/teaching
- direct teleoperation
- environment understanding through common presence
- information management in the home-base (Internet server)

The HRI consists of three main hardware components: operator hardware, robot hardware and home base. The home base component provides additional computing power and a connection to external databases (internet).

The core of the operator hardware is a portable PC including multimodal control interfaces, a map interface and a wireless connection to interface devices as illustrated in **Figs. 2 - 3**. The whole hardware is wearable and designed in such a way that the user can move easily in the same environment with the robot. The hardware is relatively complicated because it is designed not only for normal commanding and supervision, but also for teaching and teleoperation. Due to the nature of the work these functions might be needed without knowing it beforehand, so it is practical to make the whole system wearable at the same time. In certain cases, however, it is appropriate that the user can control the robot without wearing any operator hardware. Commanding by speech and gestures from close distances is used for this purpose.

The important interface components on board of the robot are camera, laser pointer, microphone, loudspeaker, head LEDs, and the arms. Besides for working, the arms can also be used for communicating in a dialogue mode, like a human uses his/her hands. The communication network with the operator and the home-base server is based on WLAN.

2.2 Principle of interaction

The core of WorkPartner's interaction and cognition system is a software the main parts of which are the interpreter, planner, manager, and internal executable language (see **Fig. 4**). The *interpreter* takes care of the communication and receives the commands/information from the user. Information can be a spoken command, a gesture, or data from any interface equipment. The data from interface devices and detected gestures are unambiguous and can be forwarded to the manager. Speech is broken into primitives and the syntax of the command sentence is checked. If the syntax is accepted and all the other parameters of the command - such as the objects and their locations - are known, the command is transferred to the *manager*. In the event of shortcomings in the command, the interpreter starts asking questions from the user until the information needed to plan the mission is complete.

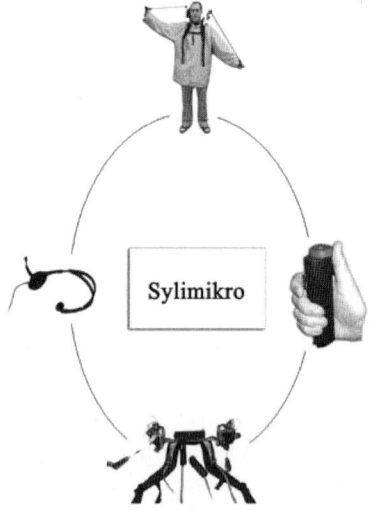

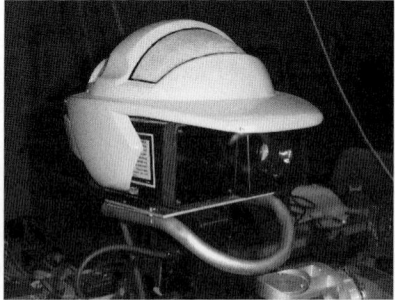

Fig. 2. User wearable hardware

Fig. 3. Robot head includes camera, laser pointer and five LEDs.

The *manager* forwards the interpreted command to the *planner*, which plans execution of the task as a set of subtasks The planner writes the plan automatically in the form of *internal executable language* (ILMR) [7], which controls the different subsystems of the robot during execution. ILMR is a XML type language acting as an intermediate link from the user, an intelligent planner, or a HRI to a robot's actions and behaviors. It provides ready features containing sequential and concurrent task execution and response to exceptions.

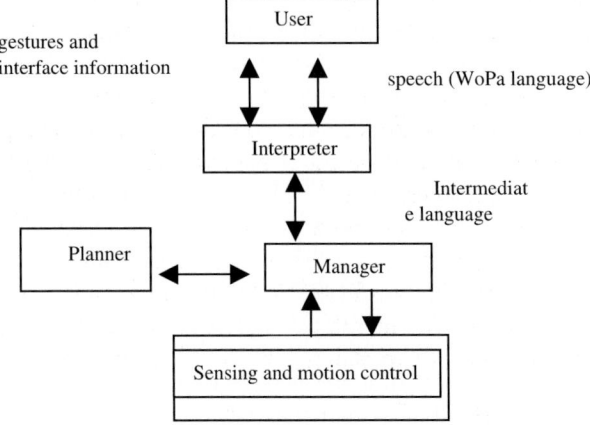

Fig. 4. Interaction principle of WorkPartner

The use of ILMR makes the software development much easier and has an important role when implementing learning capabilities for the robot. The following lines illustrates intermediate language commands

    obsavoid(on)
    speed(0.4)
    createpath(myroute,1,1,5,1,5,10,8,13,15,20)
    followpath(myroute)
    sign1a=findtarget(camera,sign)

Due to the poor performance of the commercial speech processing software and the limited speech processing capabilities on board of the robot, the commanding language between the user and the robot is formulated currently very strictly and the vocabulary is minimized. Language is based on commands starting with an imperative. For example "Partner, bring box from hall". Command processing is executed interactively. The questions to the user are formulated so that they can be answered with one or a maximum of two words. "Partner" is the prefix that starts the command, "bring" is an action verb (go somewhere, take something and bring it back), "box" is an object and "hall" is a location attribute. The object "box" may be a unique object in the common presence or it may be one of the many boxes. In the latter case the robot asks more information. Anyway, the name indicates also form of the object, which is important for gripping process. "Hall" is a known location in common presence, but the location of "box" inside it may be not known. If not, it may be given by the operator by using a more specific definition (e.g. "near door") or the robot may start searching the hall to find the object.

5. Spatial awareness

In order to cooperate, the operator and robot must have similar understanding of the environment – at least of those elements of it, which are interesting in their mutual division of work. WorkPartner has a set of traditional robotic navigation tools (GPS, dead-reckoning and laser-based map matching), which it uses depending on the present situation. The pose (position and heading) is known all the time. Therefore the spatial cognition of the robot can be constructed in its simplest way as a 2D map with fixed local coordinates and an object database that represents different properties of the environment or tasks to be performed in it. This combination of the map and objects can be visualized together as an occupancy grid to the robot and a clear object-oriented topographic map to the operator. Humans do not perceive the environment as numerical coordinates, but they can perceive and understand it well without this information.

Coordinates are, however, important for robots. A general problem is how to make both entities to understand the world with semantic information in a similar way. A classical approach to this problem is to let the robot to recognize objects named by the user using a camera or other perception sensors, and put them on their map. As well known, the difficulty of this approach is the automatic recognition of objects, which can be mostly overcome in the interactive mode of operation by letting the user recognize the objects and place them on the map by showing them to the robot in a way it understands. There are many ways how to do this. If the user has geometrically correct map available and he/she knows the position the object can be just be placed on the map. The robot is not necessarily needed in this operation. Alternatively, when moving with the robot he/she points the object in a way, which enables the robot to put it on the map. WorkPartner robot has two such pointing devices available, one is the laser pointer in its turning head and the other is "sceptre", which is a stick with colored head used by the user (see **Fig. 9**). A third way is to relate a new object to an already known object, e.g. "close to object A", in which case the known object is used as a rough position reference.

6. Common presence and how to build it

The common presence is a virtual model of the environment, which fuses the geometrical map and the data included in the different objects. The objects with approximately known location are represented by "boxes" (mini worlds) inside which they exist on the map. The boxes carry the names of the objects shown to the user so that he/she can outline the world graphically. The data base is object oriented so that details of the objects can be obtained by clicking the corresponding boxes on the map. The principle is the same used in modern object based digital maritime maps (so called ECDIS maps). Not all objects, however, have boxes. Such objects exist in the database, but have no physical location (or the present location is not know). The objects may exist also without identity, say "ball", in which case it refers to all ball objects in the common presence before identifying more.

The underlying idea for using the common presence is to transfer the task related knowledge between the user and the robot. Essential information is usually related to the boxes that include the object information and define their approximate physical location. Robot perception is supposed to be able to find and recognize the object when it is close to the corresponding box.

The methods to create a common presence should be easy, quick, and reliable. Using existing maps is not usually practical, because their validity may be a problem, and even if valid, fixing the local co-ordinate system in the right way and positioning the objects might take much time and effort. Using 3D- or 2D- laser scanners for mapping is a potential method, which is shortly described in the following.

Fig. 5 represents a laser range camera view from a parking place in the Helsinki University of Technology campus. Ranges are color coded and objects, like cars, building walls, etc. are easily recognizable for human cognition. In the next picture the user has cropped a car by mouse forming a box around it. He/she can immediate transfer this information to the presence model by giving a name (like "my car") for the box. If the object "my car" is needed during a mission of the robot – e.g. when commanding to wash it - the robot knows that it can be found inside that box in the presence model (provided of course that it has not been moved). The box may include the accurate model of the car as illustrated in the last picture in **Fig. 5**. Such information has usually no use for the user when commanding the robot but the robot may need it when doing its job.

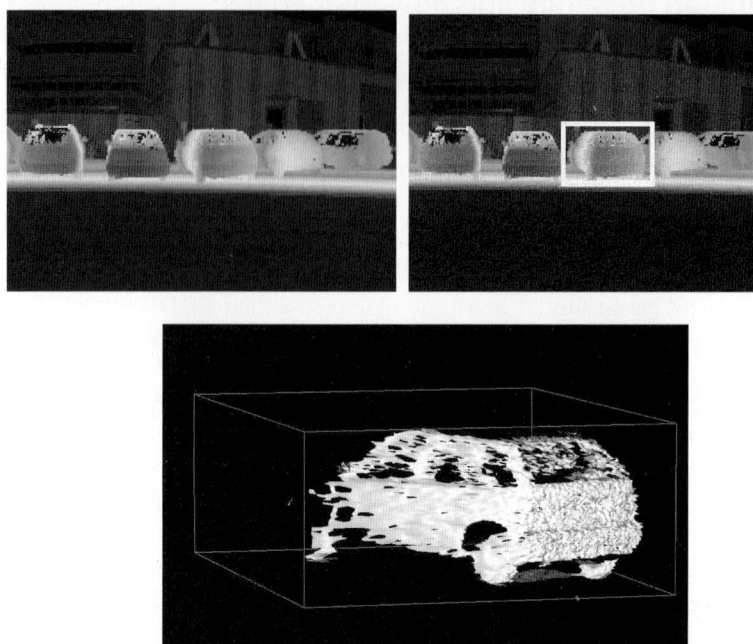

Fig. 5. Building up common presence using 3D-range laser camera.

This is a process, where human cognition can be used very effectively to create the common presence in a semi-automated way rapidly, reliably and a natural way including only the essential object information. After creation, only a simplified representation of the common presence is usually enough for operational purposes in the HRI.

Mapping of the basic geometry of the common presence can be done by many ways and means. Another possibility is illustrated in **Fig. 6**. It is a wearable SLAM system, which is based on a personal navigation system and 2D – laser scanner. The personal navigation system (PeNa), developed originally in a European Union PeLoTe project [11] for use in rescue operations, uses only dead-reckoning instruments, like a stepping odometer and heading gyro, because it is designed to operate without support from beacon systems. The stepping odometer is, however, fused with the laser-based odometer obtained by algorithmic processing of laser range data.

Fig. 6 illustrates also the result when mapping of an office corridor environment. The map made by the aid of PeNa is quite correct in proportioning, but the long corridors are slightly bending. The bending effect is due to the dead-reckoning navigation error, mainly caused by gyro drifting and odometric errors. When used as the basic map of a common virtual presence such distortion has no meaning, because a human looks more at the topology of the map when using it and the robot relies on its sensors when moving and working in the environment. Mediating the semantic information between the entities is possible in spite of geometric errors as long as the human entity can understand the main features of the map and their correspondence in the real world.

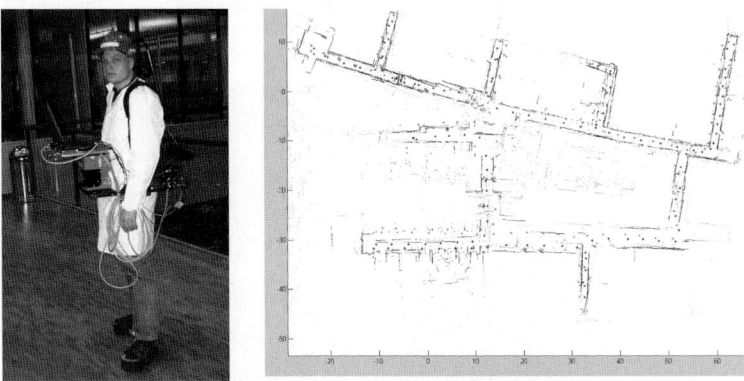

Fig. 6. Personal SLAM system developed in EU PeLoTe project.

7. Devices and means for interaction

When working with a highly interactive robot, like WorkPartner, the human user needs means to transfer his/her wish or specify information to the robot. From robot's point of view the user is only a special type of object that interacts with it. Interaction can be done directly or indirectly. The direct transfer means that the information goes directly to the awareness of the robot. In the indirect transfer the user leaves some kind of mark and related information to the real environment, where the robot can obtain it when needed. The idea of the common virtual presence is best utilized when the indirect methods are used as much as possible, because the user uses then maximally his cognition and ability to perceive entireties. In this case the user has already been able to complete the geometric information with task related conceptual information and the autonomy of the robot actions can be probably increased during the work period.

7.1 Direct teleoperation

Direct teleoperation is needed to test and move the robot when an intelligent part of the software cannot take over for one reason or another. This is the primary direct means to interact. In the case of the WorkPartner robot direct teleoperation is used e.g. to teach skilled tasks before they can be done autonomously. Another situation where direct teleoperation is used is when driving the robot from one place to another or testing its functionality. Because of the large number of degree of freedoms the robot has, a joystick alone is not a practical device for teleoperation. A wearable shoulder mounted device, called "Torso controller" was developed (see **Fig. 7**).

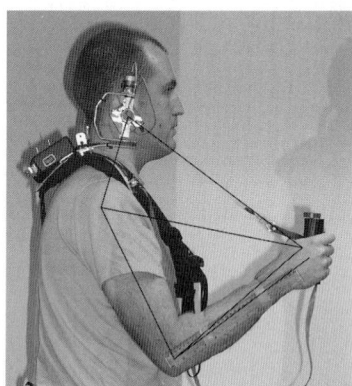

Fig. 7. Torso controller

7.2 Gestures

Gestures (and expressions) are very typical and natural way of human communication. They are used both with the speech and without it [3,4,5]. It is fairly easy to develop a sign language which is simple but rich enough in meaning for interacting with a robot. The problem, as in the case of speech, is reliable recognition of gestures used in the language. In the case of the WorkPartner robot, two different ways to detect gestures have been developed and tested. One way is to use the camera head of the robot to track the operator, who uses a colored jacket as illustrated in Figure 8. The gestures are recognized by a feature extraction algorithm, which first extracts the jacket color from the picture. This method works fairly well within short distances and in moderate illumination conditions. Another method is to use the torso controller explained above. Hand gestures can be recognized on the bases of the wrist positions. This method is not limited by the distance to the robot, but the controller is needed. Using camera based recognition allows communication with the robot without any wearable equipment.

Fig. 8. Use of gestures in commanding. Gestures are recognized by the robot's camera head, which tracks the colored jacket of the user.

Rich enough sign language can be constructed in most cases by the aid of simple static gestures, but by adding dynamic features to the gestures the language can be made more natural to use. Dynamic features are included in most sign languages used in human to human communication.

7.3 Pointing interfaces

Pointing is an important part of human communication. The purpose is to relate certain special objects with semantic or symbolic information or to give for an object a spatial meaning. Humans naturally use their hands for pointing but also technical means like pointers when the "line of the hand" is not accurate enough. In the case of human to robot communication there are several ways pointing can be realized. Pointing can be done through the virtual common presence by using a normal computer interface provided the accuracy obtained is good enough and spatial association with the object can be done easily (like in **Fig. 5**).

When pointing in the real environment a pointing device may be used. The main problem with handheld pointing devices, like laser pointers, is how to bring the pointed location to the virtual common presence geometry, usually defined in a local coordinate system. In the case of the WorkPartner robot this problem has been solved by using the robot itself as the reference point. The navigation system of the robot knows all the time the robot pose, i.e. 2-D position and heading angle, in the local coordinate system. Two different systems are in use. In one of them the user uses the laser range finder assembled in the other eye of the camera head on same optical axes as the camera. The user points through the camera image by teleoperating the camera head. The pointed locations can be transferred to the robot base coordinate system immediately and then to the common presence coordinate system if needed. Objects up to 10-15 m distances can be pointed.

The other system is called "scepter". As illustrated in **Fig. 9** the scepter is a stick with a colored ball at the tip. The visual perception with color tracking follows the tip and measures the coordinates from the image when the operator indicates that the position is right. This system is applicable within close distances only, but it does not require any active devices be with the operator. The same principle is applied when the user uses himself as the pointer. By tracking the colored jacket of the user the robot can measure his/her position and also follow him/her when moving. For example by letting the robot record the trajectory of the motion the user can show the way to travel in a work task at the same time as he/she evaluates if the road is passable.

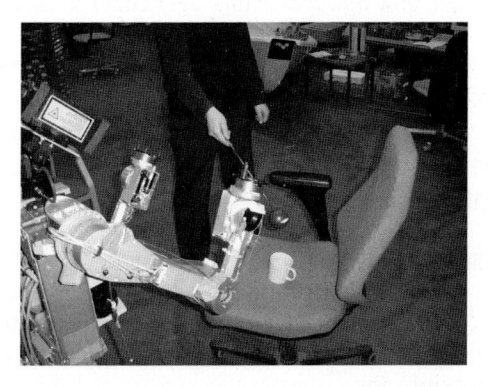

Fig. 9. Use of "scepter" for pointing.

7.4 Using signs and marks in the real environment

One very old mean of indirect communication between humans is by signs or marks left to the environment. The same principle is also applicable in human to robot communication or even robot to robot communication [9]. The signs can be passive but also active, so that they indicate their presence and information actively. Good examples of passive signs are traffic signs. Similar signs can also be used to conduct robot tasks in the working area, as illustrated in **Fig. 10**. The figure presents a hypothetical case, where a user has marked the home yard for a robot that helps in cleaning and carrying things in this environment. By simple color-coded signs it is easy to mark routes to travel, areas forbidden to cross, dangerous areas (like ditches), areas to collect litter, etc

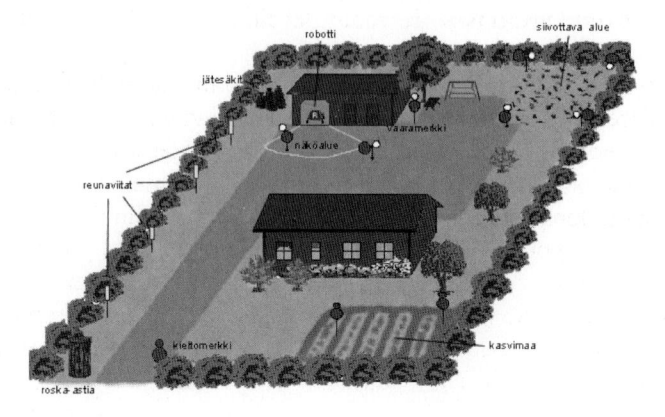

Fig. 10. Illustration how task conducting signs could be used in a home yard

One may of course mark them through the virtual common presence provided such a model is available and accurate enough, but in many cases it is easier to just mark this information in the real environment and allow the robot to read it when close enough.

8. Conclusions and discussion

Communication by the aid of symbolic or semantic information between the user and the robot is essential in effective use of future collaborative service robots. This is possibly the only way to avoid the robots becoming masters for the users as long as the autonomy of the robots can be developed highly enough. Really skilled tasks reaching a sufficient level of autonomy is still far away. This way of building up a new type of HRI technology allows the user the possibility of using his/her superior cognition capacity to load the machine instead of it loads him/her.

The underlying idea in the previous presentation is that this can be realized through "common presence" – a concept of utilizing robot cognition together with human cognition. Common presence is a virtual world presentation, which is understood in a similar way by all entities, robots and users, involved in the task. Semantic information related to the task and the real environment can be exchanged trough this world. The essential problems are how to build this world effectively, represent it for the user, and how to associate objects and concepts with it when the real world and /or tasks are changing.

In the presentation the idea behind this has been studied and demonstrated by the aid of the WorkPartner service robot developed at TKK Automation Technology Laboratory. The humanoid-type robot, having a simple command language, and the wearable user interface equipment allow the user to work with the robot in the same outdoor environment. Spatially bound information is essential in cooperative tasks. Various interactive methods and equipment have been developed and demonstrated to bring this information as a functional part of common presence.

References

1. Halme, Aarne; Leppänen, Ilkka; Suomela, Jussi; Ylönen, Sami; Kettunen, Ilkka, "WorkPartner: Interactive Human-like Service Robot for Outdoor Applications", The International Journal of Robotics Research, 2003. Vol. 22, nro 7-8, pp. 627-640.
2. Fong T., Thorpe C., "Vehicle teleoperation interfaces", Autonomous Robots, Vol. 11, No. 1, July 2001
3. Amai W., and Fahrenholtz J., Leger C. (2001). "Hands-Free Operation of a Small Mobile Robot", Autonomous Robots, Vol. 11, No. 1, July 2001
4. Fong T., Conti F., Grange S. and Baur C (2000). "Novel Interfaces for Remote Driving: Gesture, Haptic and PDA", SPIE Telemanipulator and Telepresence Technologies VII, Boston, MA, November 2000
5. Heinzmann J., Zelinsky A., "Visual human-robot interaction", 3rd conference on Field and Service Robotics (FSR2001), June 11-13, 2001, Helsinki, Finland
6. Ylönen, S., Halme, A., "Further development and testing of the hybrid locomotion of WorkPartner robot", 5th International Conference on Climbing and Walking Robots, Paris, France, September 25-27, 2002.
7. Kauppi I., "Intermediate language for mobile robots – A link between the high – level planner and low-level services in robot", Doctoral thesis, VTT Publications 510, Espoo, Finland, 2003
8. Halme A., Sievilä J., Kauppi I., Ylönen S., "Performing skilled work with an interactively operated service robot", The 4th International Conference on Field and Service Robotics, Mt. Fuji, Japan 2003.
9. D. Kurabayashi, K. Konishi, H. Asama: "Distributed Guidance Knowledge Management by Intelligent Data Carriers", Int. J. of Robotics and Automation, vol.16, no. 4, pp. 207-216, 2001
10. S.Tachi: Augmented Telexistence, Mixed Reality, pp.251-260, Published by Springer-Verlag, 1999.
11. Saarinen, Jari; Suomela, Jussi; Heikkilä, Seppo; Elomaa, Mikko; Halme, Aarne: "Personal navigation system". The IEEE/RSJ International Conference on Intelligent Robots and Systems, Sendai, Japan 2004. pp. 212-217.

Gait Restoration by Functional Electrical Stimulation

W. K. Durfee

Department of Mechanical Engineering, University of Minnesota, Minneapolis, USA   wkdurfee@umn.edu

Abstract

For over 40 years, research labs have been developing gait assist systems that use electrical stimulation of muscle to restore function to limbs paralyzed as a result of spinal cord injury. The concept of functional electrical stimulation (FES) is simple, but realization is challenging. While there are some similarities between FES-aided gait and a bipedal robot walking machine, there are also significant differences in actuators (muscles versus motors), sensors, and control strategy. Hybrid approaches that combine FES with a mechanical orthosis show promise for overcoming some of the limitations of FES-aided gait.

Keywords: Walking, electrical stimulation, spinal cord injury, FES.

1 Basics of FES

Because the human nervous system is mediated by electrical events in the form of ionic currents, artificial application of electricity to the body can be used for many applications. For example, electrical stimulation is used for therapeutic purposes such as pain suppression, muscle conditioning or wound healing. The term "Functional electrical stimulation" or FES is used when the application of low level electrical excitation is for restoration or aiding of a function that is normally under central nervous system (CNS) control, but is missing or impaired because of disease, trauma, or developmental complications. There are many applications for FES. For example, nerves in the peritoneum can be stimulated to activate bladder sphincters to cure incontinence. Cochlear implants are small implanted devices that process sounds from an external microphone and stimulate the auditory nerve in the inner ear to restore rudimentary hearing. Visual prostheses take the form of an artificial retina that stimulates receptor cells at the rear of the eye or a matrix of electrodes that stimulate the visual cortex based on signals from an external camera. There has been an explosion of implantable stimulation products from medical device companies that target deep structures in the brain for curing movement disorders. The Medtronic Activa cite for tremor suppression in Parkinson's disease patients is one example. By far the largest FES application is cardiac rhythm management with hundreds of pacemakers and ICDs implanted worldwide every day.

The application of FES most closely related to mobile robots is restoring motion to limbs paralyzed following stroke or spinal cord injury (SCI). FES for paralyzed limb control was first proposed by Liberson [1] who in 1961 used a stimulator to correct hemiplegic drop foot, a common gait disability resulting from stroke. While there has been additional FES work for stroke, systems for individuals with SCI are far more advanced.

Spinal cord injury results from disease or trauma to the spinal cord. There are approximately 250,000 in the U.S. living with SCI and approximately 11,000 new cases each year [2]. One half of those with SCI have cervical injuries leading to partial or complete quadriplegia while the other half have thoracic or sacral injuries resulting in paraplegia. The leading causes of SCI are automobiles accidents, gunshot wounds, sporting accidents (particularly diving) and falls. Because of the active nature of the events that result in spinal trauma, the mean age at the time of SCI is 29 with most falling in the 16-30 age group.

SCI damages the communication pathway between the voluntary movement centers in the brain and the muscles. The machinery below the injury, including lower motor neurons and the muscles themselves, remains unaffected by the injury. In fact, those with SCI often have sensitized reflexes (spasticity) and can have massive withdrawal reflexes in both legs when the bottom of the foot is stroked.

The basis of FES for limb control is to activate the branches of the lower motor neurons by pulses of electricity passed through surface electrodes placed on the skin over the muscle, or through implanted electrodes placed in or on the muscle, or around the nerve that supplies the muscle, to cause muscle contraction (**Fig. 1**). A multi-channel stimulator connected to a controller that can read sensors, interpret voluntary commands, and output muscle activation strength results in a system that, in theory, can restore full function to the limb. For quadriplegics, hand and forearm muscles can be activated to restore rudimentary grasp [3, 4] while for paraplegics, leg muscles can be activated to restore rudimentary gait in the vicinity of a wheelchair. The remainder of this paper is limited to the latter application.

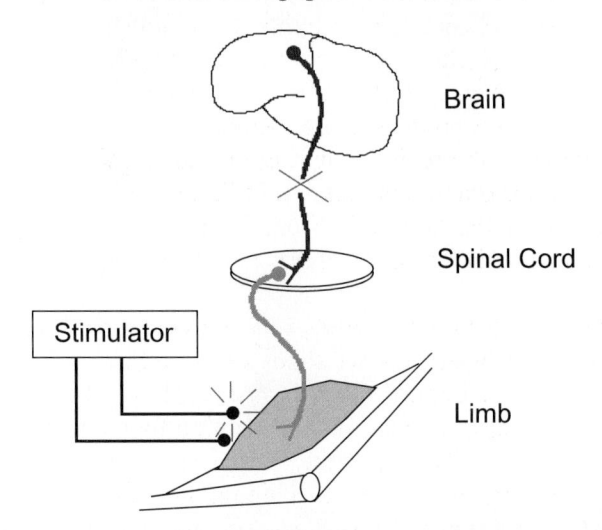

Fig. 1. Basics of FES. Spinal cord injury interrupts communication pathway between brain and muscles, but the lower motor neurons and muscles are intact. Application of current pulses through surface or implanted electrodes causes the muscle to contract and the paralyzed limb to move.

2 FES-aided gait

The inability to walk due to lower limb paralysis is a common result of thoracic level SCI. FES which uses electrical stimulation of motor nerves trigger muscle contractions, is one means for restoring rudimentary standing and limited mobility in the vicinity of a wheelchair to some individuals with SCI [5-11]. The user must have good trunk control and a strong upper body as considerable effort is required from the arms engaging parallel bars, walker or crutches for support. Despite these restrictions, successful FES users are able to ambulate for hundreds of meters with many years of use from their system [12, 13]. For example, the user shown in **Fig. 2** is T10 complete with total sensory and motor paralysis from the waist down. She is able to walk slowly, using parallel bars for support because her lower limb muscles are being activated with a rudimentary FES system [14, 15].

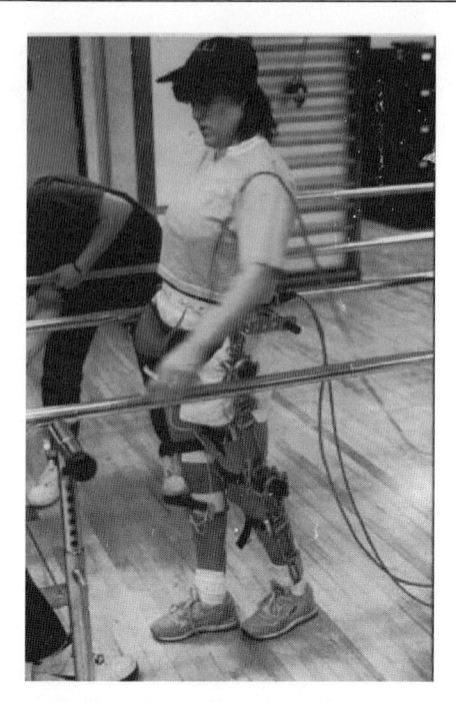

Fig. 2. User is paralyzed from the waist down following a T10 complete spinal cord injury. Rudimentary gait, using parallel bars for balance, is realized by electrical stimulation of the paralyzed muscles. This hybrid system combines the stimulation with a mechanical orthosis.

Several factors separate the performance of FES gait from normal gait under control of the central nervous system (CNS). First, while the CNS has access to all of the 60 lower limb muscles involved in gait, practical FES systems are limited to just a few channels of stimulation. Second, muscle activated by external electrical pulses fatigues far faster than muscle activated by the CNS. Third, artificial controllers must rely on a limited set of external sensors while the CNS has access to thousands of internal muscle stretch and tension sensors. Forth, the artificial system has no say over the upper body which remains entirely under voluntary control.

While FES-aided gait is similar in principle to a bipedal walking robot, there are many distinctions which separate the design and performance of FES systems from robot walking machines. Insufficient actuators and sensors is one difference, but more critical is the difference in actuators. Walking machines typically use DC servomotor actuators that have well-behaved, linear torque-current and torque-speed properties. In contrast, FES systems have muscle actuators. Muscles have exceedingly high strength-to-weight and power-to-weight characteristics, but are nonlinear, time-varying and saturate at a low peak force. The latter two properties are most critical during the rapid fatigue that greatly restricts the ability of the FES approach to restore meaningful gait for long periods. The former property means that, despite considerable efforts to model muscle input-output properties [16, 17], stimulated muscle output can be bounded but is never known with precision, making tight control of stepping motions challenging if not impossible.

To tackle the problem of fatigue that limits gait time and the problem of poor control that leads to non-repeatable stepping motions, several research groups have turned to hybrid systems that combine FES with a mechanical orthosis [18-33]. The FES system developed in our lab and shown in **Fig. 2** [14, 15] is one example of such a hybrid system. As illustrated in **Fig. 3**, the controlled brake orthosis (CBO) uses the stimulated muscle as a power source, but regulates swing phase limb motion by continuously controlling the action of orthosis-mounted magnetic particle brakes, much the same way as one controls the speed of a bicycle going downhill by manipulating the hand brakes. In addition, during the double-support phase of the gait cycle, the brace joints lock and the muscles can be turned off, delaying the time to fatigue. Users of the CBO are able to ambulate twice as far compared to when they have just the FES because of the reduced muscle stimulation [15].

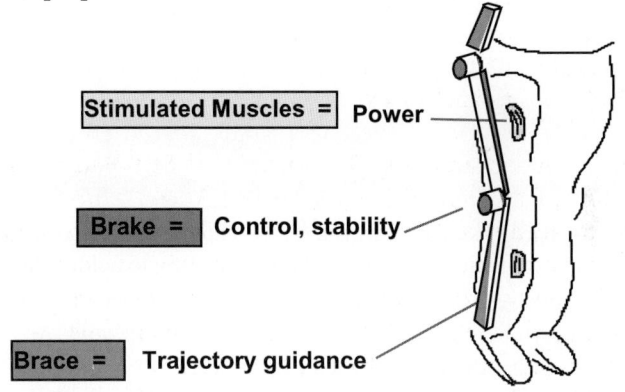

Fig. 3. The controlled brake orthosis (CBO) uses stimulated muscle for power and controls swing phase trajectory by modulating continuous brakes that act at the hip and knee. The brakes also lock the joints during double-support so that muscles are used less often thereby increasing usage time before fatigue sets in.

While the orthosis does increase function, it comes at the cost of requiring the user to don and doff a substantial piece of hardware. It is difficult to predict whether this shortcoming will ultimately be accepted by users and even more difficult to predict whether any form of FES gait assist technology will be sufficiently cost effective to merit reimbursement or individual payment.

Nevertheless, work continues on hybrid and other FES systems. For example, in our lab we are developing the energy storing orthosis (ESO) [34], another form of FES hybrid. Here, excess energy is extracted from the stimulated quadriceps muscle, stored, and piped to other joints for use later in the gait cycle (**Fig. 4**). The ESO is realized with a gas spring system to hold the leg in a flexed configuration at equilibrium, and a pneumatic system with air cylinders to convert excess extension energy at the knee to compressed air that is stored in a tubing accumulator then released into a second piston to actuate motion about the hip that is difficult to achieve with FES. The ESO makes possible a surface electrode FES gait system that uses just a single channel of muscle stimulation.

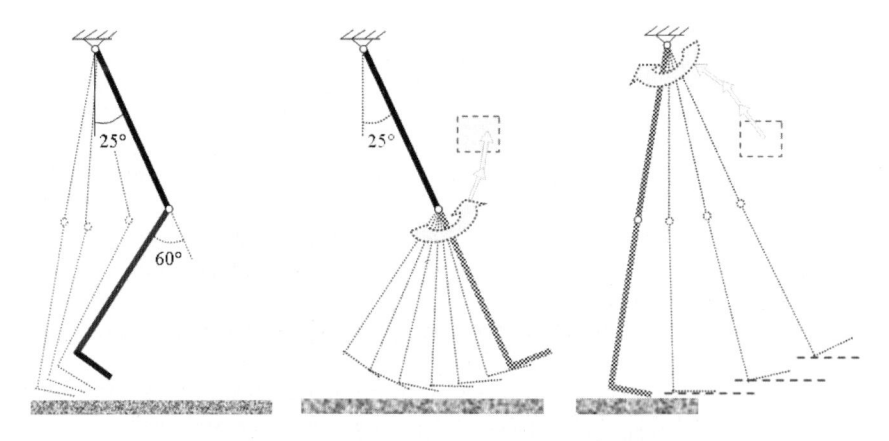

Fig. 4. The energy storing orthosis (ESO) captures and stores excess energy from quadriceps activation, and uses the stored energy to actuate hip motion later in the gait cycle.

References

1. Liberson, W.T., et al., Functional electrotherapy: stimulation of the peroneal nerve synchronized with the swing phase of the gait of hemiplegic patients. Arch Phys Med Rehabil, 1961. 42: p. 101-5.
2. SCIIN, Spinal cord injury: facts and figures at a glance - June 2005. 2005, Spinal Cord Injury Information Network.
3. Peckham, P.H., et al., Efficacy of an implanted neuroprosthesis for restoring hand grasp in tetraplegia: a multicenter study. Arch Phys Med Rehabil, 2001. 82(10): p. 1380-8.

4. Snoek, G.J., et al., Use of the NESS handmaster to restore handfunction in tetraplegia: clinical experiences in ten patients. Spinal Cord, 2000. 38(4): p. 244-9.
5. Marsolais, E.B. and R. Kobetic, Functional electrical stimulation for walking in paraplegia. J Bone Joint Surg Am., 1987. 69A: p. 728-733.
6. Kobetic, R., R.J. Triolo, and E.B. Marsolais, Muscle selection and walking performance of multichannel FES systems for ambulation in paraplegia. IEEE Trans Rehabil Eng, 1997. 5(1): p. 23-29.
7. Cybulski, G., R. Penn, and R. Jaeger, Lower extremity functional neuromuscular stimulation in cases of spinal cord injury. Neurosurgery, 1984. 15(1): p. 132-146.
8. Graupe, D. and K. Kohn, Functional Electrical Stimulation for Ambulation by Paraplegics. 1994: Krieger.
9. Kralj, A. and T. Bajd, Functional Electrical Stimulation: Standing and Walking After Spinal Cord Injury. 1989: CRC Press.
10. Marsolais, E.B. and R. Kobetic, Functional walking in paralyzed patients by means of electrical stimulation. Clin Orthop, 1983. 175(30-36).
11. Creasey, G.H., et al., Clinical applications of electrical stimulation after spinal cord injury. J Spinal Cord Med, 2004. 27(4): p. 365-75.
12. Kobetic, R., et al., Implanted functional electrical stimulation system for mobility in paraplegia: a follow-up case report. IEEE Trans Rehabil Eng, 1999. 7(4): p. 390-8.
13. Kralj, A., T. Bajd, and R. Turk, Enhancement of gait restoration in spinal injured patients by functional electrical stimulation. Clin Orthop Relat Res, 1988(233): p. 34-43.
14. Goldfarb, M. and W.K. Durfee, Design of a controlled-brake orthosis for FES-aided gait. IEEE Trans Rehabil Eng, 1996. 4(1): p. 13-24.
15. Goldfarb, M., et al., Preliminary evaluation of a controlled-brake orthosis for FES-aided gait. IEEE Trans Neural Syst Rehabil Eng, 2003. 11(3): p. 241-8.
16. Durfee, W.K., Muscle model identification in neural prosthesis systems, in Neural Prostheses: Replacing Motor Function After Disease or Disability, R. Stein and H. Peckham, Editors. 1992, Oxford University Press.
17. Durfee, W.K. and K.I. Palmer, Estimation of force-activation, force-length, and force-velocity properties in isolated, electrically stimulated muscle. IEEE Trans Biomed Eng, 1994. 41(3): p. 205-16.
18. Andrews, B., et al., Hybrid FES orthosis incorporating closed loop control and sensory feedback. J Biomed Eng, 1988. 10: p. 189-195.
19. Ferguson, K.A., et al., Walking with a hybrid orthosis system. Spinal Cord, 1999. 37(11): p. 800-4.
20. Kobetic, R., et al., Development of a hybrid gait orthosis: a case report. J Spinal Cord Med, 2003. 26(3): p. 254-8.
21. Major, R., J. Stallard, and G. Rose, The dynamics of walking using the hip guidance orthosis (HGO) with crutches. Prosthet Orthot Int, 1981. 5(1): p. 19-22.
22. Marsolais, E.B., et al., The Case Western Reserve University hybrid gait orthosis. J Spinal Cord Med, 2000. 23(2): p. 100-8.

23. McClelland, M., et al., Augmentation of the Oswestry Parawalker orthosis by means of surface electrical stimulation: gait analysis of three patients. Paraplegia, 1987. 25: p. 32-38.
24. Nene, A. and S. Jennings, Hybrid paraplegic locomotion with the Parawalker using intramuscular stimulation: a single subject study. Paraplegia, 1989. 27: p. 125-132.
25. Petrofsky, J. and J. Smith, Physiologic costs of computer-controlled walking in persons with paraplegia using a reciprocating-gait orthosis. Arch Phys Med Rehabil, 1991. 72(11): p. 890-896.
26. Popovic, D., Dynamics of the self-fitting modular orthosis. IEEE Trans Robotics Automat, 1990. 6(2): p. 200-207.
27. Popovic, D., R. Tomovic, and L. Schwirtlich, Hybrid assistive system - the motor neuroprosthesis. IEEE Trans Biomed Eng, 1989. 36(7): p. 729-737.
28. Solomonow, M., et al., Reciprocating gait orthosis powered with electrical muscle stimulation(RGO II). Part I: performance evaluation of 70 paraplegic patients. Orthopedics, 1997. 20(4): p. 315-324.
29. Solomonow, M., et al., The RGO generation II: muscle stimulation powered orthosis as a practical walking system for thoracic paraplegics. Orthopedics, 1989. 12(10): p. 1309-1315.
30. Stallard, J., R. Major, and J. Patrick, A review of the fundamental design problems of providing ambulation for paraplegic patients. Paraplegia, 1989. 27: p. 70-75.
31. Tomovic, R., M. Vukobrativic, and L. Vodovnik, Hybrid actuators for orthotic systems: hybrid assistive systems, in Adv. External Contr. Human Extremities IV. 1973. p. 73-80.
32. Yang, L., et al., Further development of hybrid functional electrical stimulation orthoses. Artif Organs, 1997. 21(3): p. 183-7.
33. Butler, P., R. Major, and J. Patrick, The technique of reciprocal walking using the hip guidance orthosis (HGO) with crutches. Prosthet Orthot Int, 1984. 8(1): p. 33-38.
34. Durfee, W.K. and A. Rivard. Preliminary design and simulation of a pneumatic, stored-energy, hybrid orthosis for gait restoration. 2004.

Space Robotics

G. Visentin

Head, Automation & Robotics Section, European Space Agency, The Netherlands, Gianfranco.Visentin@esa.int

Abstract

This presentation reviews the topic of space robotics.

In the space jargon, often any unmanned space probe is called a robotic probe. This acknowledges the challenges of largely autonomous operation in a complex mission.

In the following, however, the focus will be on space robots: systems involving arms for manipulation or some kind of locomotion device for mobility, having the flexibility to perform varying tasks.

As an introduction, some definitions and the rationale for space robotics are given. The main differences between space and terrestrial robots are highlighted, and it is shown that they are driven by the peculiar environmental, system and programmatic constraints of space missions.

A second part describes the typical architecture, sub-systems, and some key technologies of robot systems. This distinguishes between manipulator arm and rover type robots. The interdisciplinary system character of space robotics is emphasised.

Finally the two main fields of orbital robotics and planetary robotics are introduced, together with the currently perceived application scenarios (i.e. system servicing and payload tending in Low-Earth-Orbit, satellite servicing in Geostationary Earth Orbit, the assembly of large orbiting structures, and applications in exploration missions to the Moon, Mars, Mercury, comets, asteroids, and other celestial bodies).

For each application scenario, the main robotic functions are presented and some developed/in-development robotic systems, most relevant to the CLAWAR conference, are described.

Keywords: Space robotics, planetary exploration, rover, orbital robotics, servicer.

1 Rationale for Space Robotics

The reasons why robots are used in space are very similar to those that make robots used on Earth. In fact it is the relative importance of the reasons that changes. In order of importance, Space robots are adopted for their:
1. Safety: some tasks are too dangerous (e.g. because of the hostile environment) for astronauts
1. Performance: the given tasks are too difficult or impossible (e.g. because of large masses involved, high precision and repeatability required, long duration) for astronauts
1. Cost: astronauts in space require a very expensive life support infrastructure and eventually they must return to Earth, robots just need power and they can be disposed of after they have attained their goal
Differences between Space and Terrestrial Robots

In many ways, space robots are very different from terrestrial robots, be they factory-plant robots or the service robots often subject of the CLAWAR conference.

Table 1:Environmental constraints and resulting design impacts

Constraint	Typical design impact
Survive launch and landing loads (planetary landing !)	Support structures, holddown / release mechanisms, specially mounted electronic components, expensive test facilities
Function in vacuum	Careful materials selection, special lubrication, brushless motors preferred, certain sensing principles not applicable (e.g. ultrasonic), clean room integration and testing
Function under "weightlessness" (orbital applications)	Everything has to be fastened, altered dynamic effects (highly non-linear), very low backlash gearing
Function under extreme radiation exposure	Limited materials life time, shielded and hardened electronics, outdated computer performance (state-of-the-art computers not space compatible)
Function under extreme temp. and temp. variations, possibly in vacuum	Multi-layer insulation, radiators with heat pipes, distributed electric heaters, radioisotope heating units (planetary applications)
Function under extreme lighting and contrast conditions	Difficulty for vision and image processing
Function in extremely remote environment	Comprehensive testing before launch, essentially maintenance-free systems, adequate level of autonomy, in-orbit calibration and sensor-based control, effective ground operator interfaces

The particular requirements and constraints that drive the designs of space robots in a special direction can be classified into two groups. Table 1 summarises major space environmental constraints and typical design impacts, while Table 2 compiles key space system and programmatic constraints and typical design impacts.

However, environmental constraints may also have positive design impacts such us allowing robot systems physically impossible on Earth (e.g. small robots with the ability to manipulate enormous payloads, thanks to reduced gravity).

Table 2: System or programmatic constraints and resulting design impacts

Constraint	Typical design impact
High system complexity	Professional system engineering and project management needed
Long life time	Maintenance-free design desired, built-in growth potential and upgradeability (e.g. re-programmability), "orbital replaceable units" for maintenance
High reliability and safety	Product assurance measures, space system engineering standards, high documentation effort, "inherently safe" design preferred, built-in failure tolerance (redundancy) and diagnostics, problems with "non-deterministic" approaches (e.g. AI)
On-board mass very limited and expensive (esp. planetary!)	Extremely lightweight designs, arms with noticeable elastic effects (control problem), high payload mass fraction needed, slow motions
On-board power / energy very limited and expensive	Low-power electronics, very high efficiency, limited computing resources, batteries always critical for rovers, slow motions
Communications with Earth very limited and expensive	Adequate degree of autonomy, built-in checking, sophisticated ground operator interfaces to cope with signal transmission delays
Preserve micro-gravity conditions	Smooth acceleration and low speeds, high actuator motion smoothness, high gear ratios, weak joints
Limited testability on Earth	High effort for thermal vacuum and launch loads testing, approximations for 0 g motion, sophisticated simulations to verify system behaviour in 0 g
Long planning and development	Problems of staff continuity and morale, technology in space often obsolete
Development in international co-operation	Often sub-optimal project efficiency from artificial work distribution, communications and logistics problems

2 Architecture of a Space Robot System

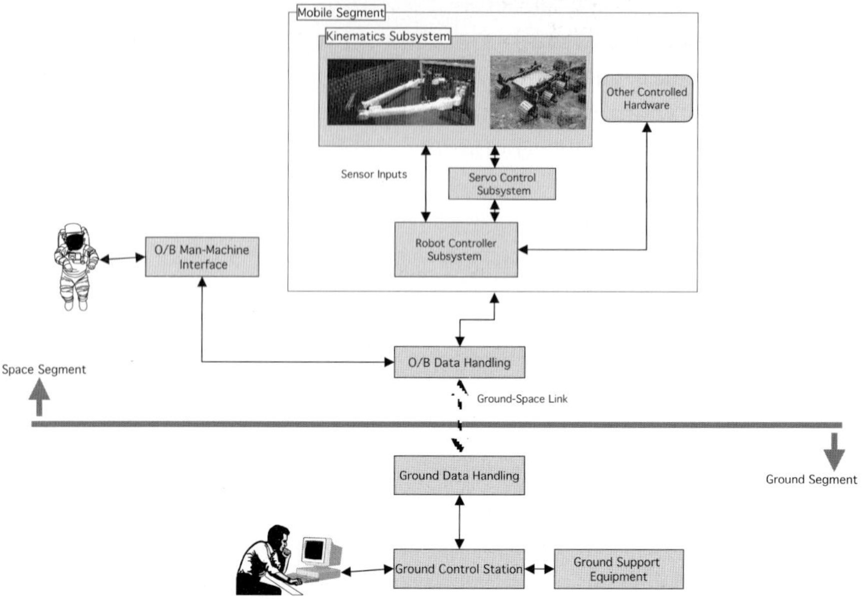

Figure 1: Typical architecture of a space robot system

A general breakdown of a typical space robot system is shown in Figure 1. The following main subsystems are introduced:

- The *Kinematics Structure Subsystem*: which is what implements the physical motion of the whole robotic system, This typically takes the shape of a [multi]arm robot for Orbital Applications, and of a rover (wheeled, tracked or legged) for Planetary Applications, although more exotic incarnations may exist (e.g. blimps, moles).
- The *Robot Controller* subsystem, which implements the autonomy of the robot, navigation, path planning and digital control of the motion.
- The *Servo Control* subsystem that takes care of controlling the electric motors, This unlike in terrestrial systems is often tightly integrated into the Kinematics Subsystem and linked with the Robot Controller by means of a digital serial bus
- The combined Robot Controller and Servo control subsystems are often referred as *Robot Avionics*. They are frequently tightly integrated on the Kinematics Subsystem to form, together with

other non-robotics subsystems (e.g. thermal control, power generation/storage) the *Mobile Segment*
- The *On-Board Data Handling* (O/B DH) subsystem is a common element of space systems and it is in charge of data storage and exchange between the *Flight Segment* and the *Ground Segment*. In case of manned missions the O/B Data Handling subsystem, may also interface with an *On-board Man-Machine Interface* (O/B-MMI) which allows astronauts to control the Mobile System.
- The *On-ground Data handling* subsystem is the ground counterpart of the O/B DH, This is often made of many elements distributed in various geographical locations to enable optimal connectivity with the space segment despite the orbital motion of this last.
- The *Ground Control Station* is the subsystem that allows programming the robot behavior, uploading the programmed sequences, and monitoring the execution of them.
- To validate the quality of the programming and to troubleshoot anomalies, the ground control station is always paired with some Ground *Support Equipment*. This is usually made of a high fidelity replica of subsystems in the Flight Segment linked with a simulated/emulated replica of the environment where the robot system is.

These subsystems may have very different appearance according to the particular robot system and its application scenario. In the following chapters a description of the application scenarios and of robotics systems ESA is developing for them is provided.

3 Orbital Robotics

Although use of robots is technically possible in the main orbital domains, such as the *Geostationary Earth Orbit* (GEO), base of the commercial/military communications and Earth observation satellites, or the Medium Earth Orbit (MEO), base of the Global Positioning System satellites, robots are solely used in the *Low Earth Orbit* (LEO).

The LEO at 300 to 700 km altitude, is the domain of manned space missions. The current Space Shuttle, the International Space Station (ISS) and the past US Skylab and Russian MIR Station are all LEO systems. All robotic applications in LEO therefore have an important element of crew interaction or operation. The main European contribution to robotics on the ISS is the *European Robot Arm (ERA)*, responsible for servicing the Russian segment of the ISS.

ESA is also preparing another robot system, EUROBOT, which will further increase the automation of the ISS.

1.1 ERA

ERA will transport large Station elements (such as folded solar array or radiator packages) and exchange Orbital Replaceable Units (ORUs). ERA is a completely symmetric 7-axes robot arm with an extended length of 11.3 m. Both ends of ERA are equipped with Basic End Effectors (BEEs) able to grasps objects equipped with suitable grapsping fixtures.

ERA features avionics integrated on the kinematics structure and it is normally anchored to a *base point*, by means of one of its BEEs. The base point also connects ERA to power and data channels. A peculiar feature ERA, interesting for the CLAWAR conference, is that ERA can relocate itself along a system of base points on the Russian part of the ISS by alternately switching the roles of end wrist and shoulder base, hence performing a sort of "walk".

ERA can be operated from either inside the Station or from a cosmonaut performing an EVA, but not from Earth. It shall be launched in the summer of 2007.

Figure 2: The European Robot Arm, undergoing EMC tests in an anechoic chamber

1.2 Eurobot

ESA is developing a new robotic system for the ISS. This system named EUROBOT [1], is intended to help or even replace EVA crew. The first concept of EUROBOT has been simulated and prototyped in a testbed to determine its characteristics and detail its operations (see Figure 3).

Figure 3: The EUROBOT testbed

In the meantime the EUROBOT has been subject of further study work (Phase-A) and some of its characteristics have changed since the initial concept.

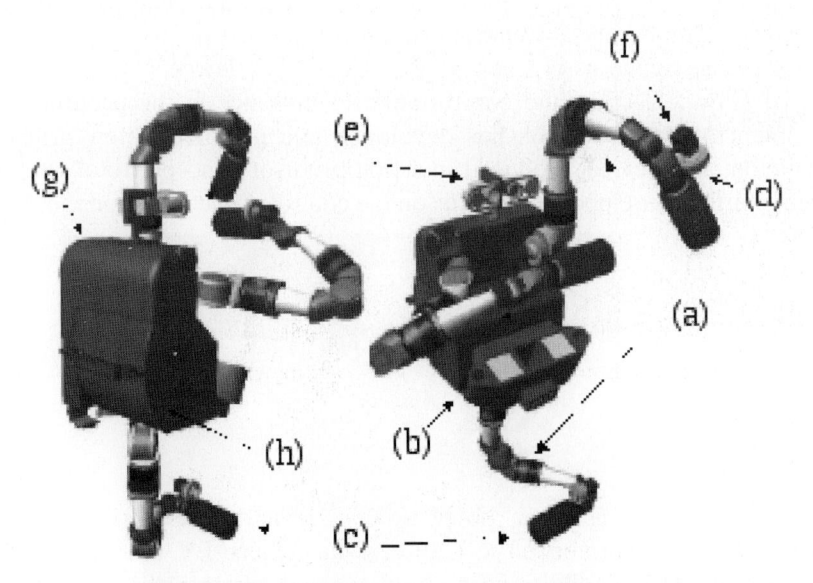

Figure 4: The Eurobot system as presently envisaged

Still the robot does not to pretend to emulate human features but tries to exploit all possible robotic advantages. The system (see Fig 1 for reference) features three 7-dof identical arms (a) arranged around a body. The arms are multi-functional and may be used as "arm" or "leg".

The EUROBOT carries a tool rack (b). Each arm may pick-up/release wrist mounted tools (c showed as cylindrical volumes) at selected locations on the tool rack by means of tool exchange devices (d). These tools may be specialised (e.g. wrench) or general purpose (a hand tool comparable to the DLR hand 2). EUROBOT is equipped with a lighting and imaging head (e), allowing human compatible stereoscopic vision, as well as optional lighting and imaging units at the wrists (f). The robot controller is housed in the body (g) and it is powered by a large replaceable battery (h). An IEEE 802.11g transceiver guarantees communication with a control station inside the space station. EUROBOT features two main modes of operation: *Programmed mode* and *telemanipulation*. The first is used when EUROBOT has to perform routine tasks that do not require involvement of a human operator, such as relocation from one side of the ISS to another. This task, relevant for the CLAWAR conference, is performed by walking attached to handrails.

Whenever the task contains elements of unpredictability or of high dexterity, the second mode is used.

Since EUROBOT is designed with dimensions and kinematics compatible with human ones, it enables to use the most effective and intuitive telepresence by means of a newly developed haptic device.

To allow effective and comfortable telepresence in the peculiar environment of the ISS, ESA has developed and patented a new arm exoskeleton. The device, produced in a first proof-of-concept prototype [2] is being further developed with focus on the control of the actuation means.

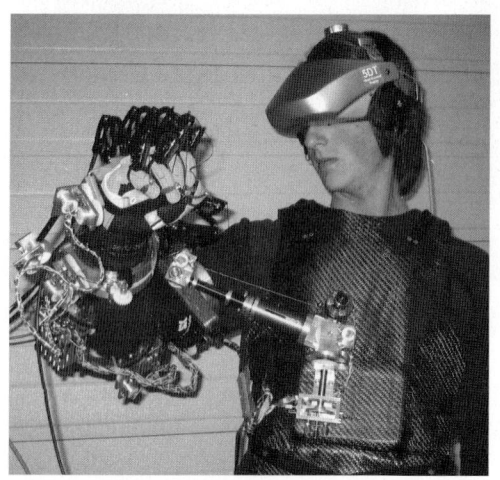

Figure 5: Telepresence MMI for EUROBOT (includes single arm exoskeleton)

4 Planetary Robotics

The other main use of Robotics is in support of scientific investigation, exploration and, in the future, in-situ resource utilization of planets.

Despite the great variety of the robotics systems (e.g. aerobots, moles, hoppers) possible in planetary environment, present space missions are only using rovers. Furthermore, thanks to the fairly traversable terrains currently being explored, and to the inherent simplicity of wheeled locomotion, only wheeled rovers are being developed. ESA, in particular is developing the EXOMARS mission.

However ESA is also studying future mission scenarios where extremely rugged terrains (e.g. cliffs, craters) need to be negotiated. For these scenarios legged rovers outperform wheeled ones. For this reason ESA has initiated a R&D activity to develop an Ambulating Robot for Autonomous Martian Investigation, Exploration and Science (ARAMIES).

1.3 ExoMars

The ExoMars mission [3], to be launched in 2011, features a descent module that will land a large (200 kg), high-mobility 6-wheels rover (see Figure 6) on the surface of Mars.

The primary objective of the ExoMars rover will be to search for signs of life, past or present. Additional measurements will be taken to identify potential surface hazards for future human missions, to determine the distribution of water on Mars, to measure the chemical composition of the surface rocks and to deploy seismic instruments.

Figure 6: Artist impression of the ExoMars rover (left) and 1/2 scale prototype of it (right)

A demonstrator of the rover has been build by ESA contractors (RCL) based on an innovative chassis design, which allows overcoming isolated obstacles twice the diameter of a wheel. The final chassis configuration is not yet frozen, however it will certainly feature the so called wheel-walking mode. This mode uses the ability of each wheel axis to move forward-backward independently from the chassis. The mode allows to negotiate very soft terrains where otherwise wheels would dig themselves down.

1.4 ARAMIES

The aim of the ARAMIES project is to develop a walking system for extremely difficult terrain, especially steep and uneven slopes. The mission scenario in which an ARAMIES rover would be used is the exploration of cliffs.

The surface of certain cliffs on Earth provides immediate access to the layers, stratified over time, which make the crust. Hence by exploring certain cliffs on Mars it is believed that a great deal of information on the history of Mars could be gathered.

To explore the close-to-vertical walls of the cliffs ARAMIES will be walk down them attached by means of a tether to a large rover (similar to the EXOMARS) stationed on top of the cliff.

The ARAMIES, activity [4] funded by ESA and DLR and run by the Bremen University has produced so far a prototype (see Figure 7), which will enable testing and tuning of bio-inspired walking behaviors.

Figure 7: The prototype of the ARAMIES rover

5 Conclusions

This paper has introduced the subject of Space Robotics with its terminology, its principal application scenarios and examples of systems being readied for them.

The paper has also shown that even in the small number of robotics missions developed at ESA the capability of walking is present in most of the systems.

6 References

1. P. Schoonejans, R. Stott, F. Didot, A. Allegra, E. Pensavalle, C. Heemskerk, (2004) *Eurobot: EVA-assistant robot for ISS, Moon and Mars*, 8th ESA Workshop on Advanced Space Technologies for Robotics and Automation,
   http://robotics.estec.esa.int/AUTOLINKS/Publishing,%20Conferences%20and%20Symposia/Astra/Astra2004/astra2004_Proceedings.pdf
2. A. Schiele, G. Visentin, (2003) *The ESA Human Arm Exoskeleton for Space Robotics Telepresence*, 7th International Symposium on Artificial Intelligence Robotics and Automation in Space,
   http://robotics.estec.esa.int/AUTOLINKS/Publishing,%20Conferences%20and%20Symposia/I-sairas/isairas2003/data/pdf/EU15paper.pdf
3. J. Vago, *(2004) Overview of ExoMars mission preparation*, 8th ESA Workshop on Advanced Space Technologies for Robotics and Automation,
   http://robotics.estec.esa.int/AUTOLINKS/Publishing,%20Conferences%20and%20Symposia/Astra/Astra2004/astra2004_Proceedings.pdf
4. D. Spenneberg, F. Kirchner, J. De Gea, (2004) *Ambulating Robots for Exploration in Rough Terrain on Future Extraterrestrial Missions*, 7th International Symposium on Artificial Intelligence Robotics and Automation in Space,
   http://robotics.estec.esa.int/AUTOLINKS/Publishing,%20Conferences%20and%20Symposia/I-sairas/isairas2003/data/pdf/EU15paper.pdf

Bio-Engineering and Biological Inspired Systems

ASYSTENT - Control System of a Manipulator for Keyhole Surgery[1]

P. Sauer[1], K. Kozlowski[1], W. Waliszewski[2] and P. Jeziorek[1]

[1]The Institute of Control and Systems Engineering, Poznan University of Technology, ul. Piotrowo 3a, 60-965 Poznan, Poland; [2]The Department of General and Laparoscopic Surgery, Hospital J. Strusia, Poznan, Poland
{piotr.sauer, krzysztof.kozlowski}@put.poznan.pl,

Abstract

In this paper, we present control system which is named „Multi-level Control System of a Manipulator for Minimally Invasive Surgery (MIS)". The overall system is based on the Stäubli robot RX60 which plays a role of a robotic assistant for surgeons in laparoscopic cholecystectomy. The desired displacements of the end of the laparoscope is obtained by a movement of a joystick or commends from the speech recognition system. In this paper simulation applications and experimental results are presented.

Keywords: robot, control system, keyhole surgery, laparoscope, force feed-back.

1. Introduction

Minimally Invasive Surgery (MIS), also well known as laparoscopy or keyhole surgery is dedicated to telepresence because the surgeon is physically separated from the workspace. The laparoscope and instruments used for the operation are inserted though trocars placed at the incisions of the abdomen. Surgical telepresence system helps the surgeon to overcome barriers, such as the patient's chest or distance if surgeon and patient are lo-

---

[1] This work was supported by the Ministry of Science and Informatics under grant no. 4 T11A 026 22, entitled " Multi-level Control System of a Manipulator for Laparoscopy in Surgery"

cated in different rooms or even hospitals. The keyhole surgery is one of the field where telerobotics enlarges the human possibilities. Surgical robots allow more precise movements of surgical instruments. The surgical robot isn't automated system but teleoperated system under direct control of the surgeon. Consequently, surgeon executes operation but his or her work is assisted by robot. The telerobotic assistant for laparoscopic surgery was developed by Taylor et al. [1]. In September 2001, Marescaux et al. successfully carried out a remote laparoscopic cholecystectomy on a 68-years old female. They used telerobotic system named ZEUS [2]. The surgeon who executes this operation was in New York while the patient was in Strasbourg [3].

In this paper, we present the results of the project „Multi-level Control System of a Manipulator for Minimally Invasive Surgery (MIS)". The goal of this project was to develop robotics tools to replace manual control of laparoscope in surgery. It was realized in the cooperation between the Institute of Control and Systems Engineering, Poznan University of Technology, and the Department of General and Laparoscopic Surgery, Hospital J. Strusia in Poznan. Presented system replaces the surgical assistant during laparoscope cholecystectomies. In keyhole surgery, surgeon views the anatomy from inside with the help of video system. The video system consists of high resolution monitor and laparoscope with camera. The laparoscope itself consists of a chain of lens optics to transmit the image of the operation site to the CCD camera connected to its outer end and optical fibers to carry light to illuminate inside the patient's body. An image of the operation site is displayed on a high resolution CRT screen. In traditional laparoscopic operation, the camera is controlled by surgical assistant (another surgeon). Consequently, the laparoscopic image often is unstable because of tremor and sudden movements of surgical assistant. To overcome these drawbacks, we proposed the multi-level control system which was named ASYSTANT. In this project we made use of Stäubli robot RX60 which is equipped with force and torque sensor JR3 and laparoscope (see **Fig. 1.**).

2. Telerobotic system - ASYSTANT

The proposed system operates under surgeon's direct teleoperation of the precision movements needed to maneuvre the laparoscope within the patien's body. It will replace the surgical assistant (another surgeon who controls laparoscope movement) during cholecystectomies.

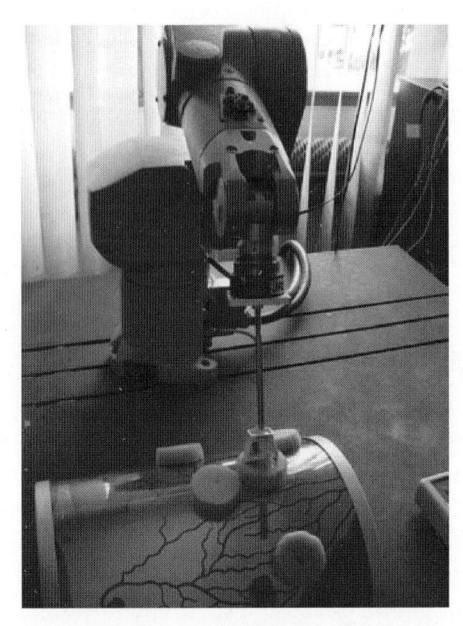

Fig. 1. View of the robot Stäubli RX60 with JR3 sensor and model of abdominal

2.1. The teleoperation task

The input device translates the joystick movements, commands from speech recognition module or commands from visual module to robot Cartesian coordinates [4]. For this action the telerobotic system must know model of the task which depends of the solution of the following problems:
- the inverse and forward kinematics of robot assistant,
- the distance along the optic axis between the camera and the insertion point.

This model is determined by the homogeneous transformation $^{B}T_{O}$ (see **Fig. 2.**) which describes relationship between the base of manipulator (frame B in **Fig. 2.**) and the insertion point where the surgery tool goes into the patient's body (frame O in **Fig. 2**). Transformation $^{B}T_{O}$ is defined as:

$$^{B}T_{O} = {}^{B}T_{E} \times {}^{E}T_{H} \times {}^{H}T_{O} \qquad (1)$$

where

- $^{B}T_{E}$ - robot forward kinematic model which defines the relationship between the base frame B and the end effector frame E fixed to the last link,
- $^{E}T_{H}$ - the relationship between the end effector frame E and the holder frame H,
- $^{H}T_{O}$ - defines the relative location of the insertion point frame O from the holder frame H.

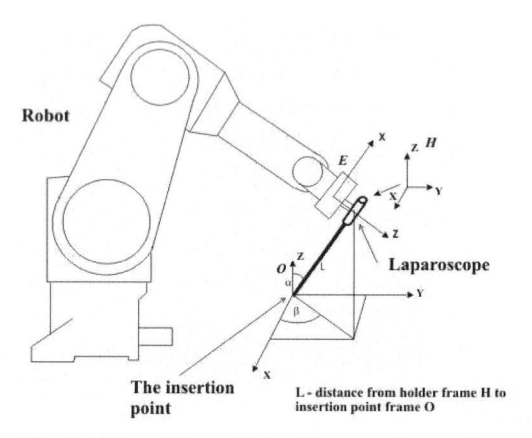

Fig. 2. The scheme of the teleoperation task

The distance L (see **Fig. 2**) along the optic axis between the insertion point (frame O) and the holder of laparoscope (frame H) is defined by infrared distance sensor GP2D12. This sensor is mounted at the end effector of the robot. The GP2D12 is connected to computer PC through MSP430 microcontroller by RS232 interface. Forward and inverse kinematics problems were solved using modified Denavit-Hartenberg notation for Stäubli RX60 manipulator with laparoscope. Kinematics solution was computed using MATLAB program and Robotics Toolbox. We tested area of the manipulator task space where the volume of the manipulability ellipsoid is big enough to assure comfortable manipulation of the laparoscope in the whole operation area [5].

2.2. Telerobotic system

The new structure of ASYSTENT system consists of the following units (see **Fig. 3**):
- **the main unit**, which allows communication between all elements of the system.
   The main unit is made up of the following blocks:

- the block of threads service which collects information from all input units, it chooses source of control based on assumed priorities,
- the teleoperation block that allows to control the laparoscope and robot movement on the bases of the teleoperation task (see subsection 2.1). This block translates the commands from joystick unit, the unit of speech recognition or vision system to robot Cartesian coordinates respectively,
- the archives block records states of the system
- **the joystick control unit**, which is used to control the robot movement by means of joystick movement. This units reads in data from joystick and sends to main unit.
- **the unit of hybrid control** which is used as force feedback control law with estimation of the environment stiffness. Force feedback is realized based of data from force and torque sensor JR3 which is mounted between the Stäubli RX60 robot and laparoscope (see Section 3).
- **the unit of the speech recognition** which is used to control the robot movement. This unit is an important interface between a surgeon and a robot because it allows the surgeon for communication with telerobotic system using natural commands. These commands are quite similar to those used during normal laparoscopic surgery without robotic system. We proposed the following set of commands: "robot left", "robot right", "robot up", "robot down", "robot closer", and "robot further"[6]. In order to recognize the commands we used a method based on pattern matching known as Dynamic Time Warping (DTW) [7].
- **the vision system,** this system recognizes the surgical tool and sends back information about position of tool's tip to the main unit of the system. It tries to manipulate the robot with laparoscope to keep the tool in the center part of the camera image [8].
- **the robot system** which consists of the Stäubli RX60 robot, its controller, force and torque sensor and laparoscope. Controller of the robot is set up in V+ environment. The communication between the robot system and main unit is executed by means of RS232 interface and Ethernet.

These units are implemented as separate applications. Communication between units is realized with the help of TCP/IP protocol.

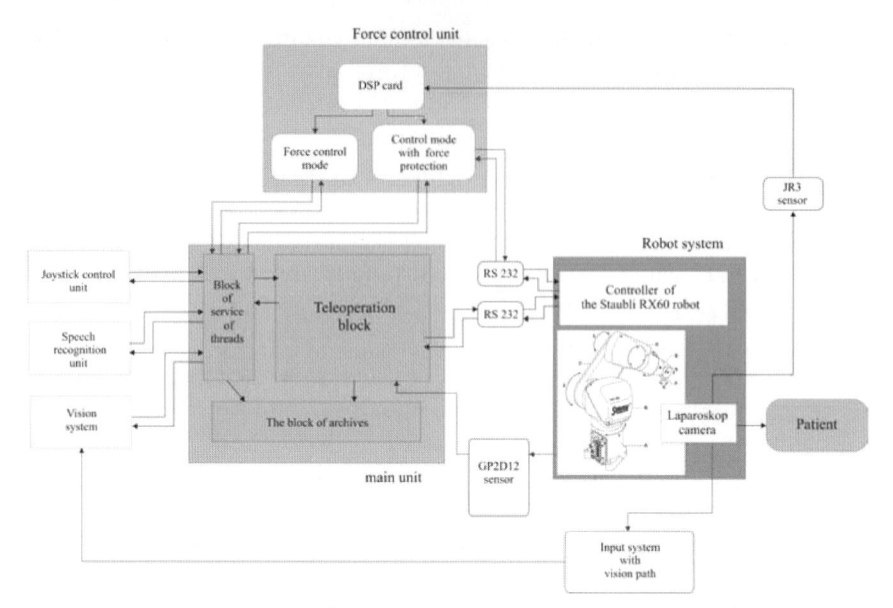

Fig. 3. The scheme of multi-level system ASYSTENT.

3. Experimental results

First we implemented the hybrid control algorithm and joystick control unit. Hybrid control includes parameter adaptation in the force feedback control law with the estimation of the environment stiffness. In Keyhole Surgery force feedback it is important to give the human operator as much intuitive information as possible about the remote environment, from which he is separated by a distance. We built three-dimensional image of abdominal with points network which are characterized by assigned values of force and torque. These values cannot be exceeded during the surgery instrument manipulation. Critical values of forces and torques are very important owing to health and patient's safety. Therefore we investigated some tissues properties. To measure in vitro properties of tissue in extension we designed and built a measurement system which consists of the Stäubli RX60 robot and a force sensor mounted between the robot and the pointer [9]. We performed experiments on the liver of pig. We measured force response between the pointer and the soft tissue in vitro conditions [10]. Results of this research were used to build the hybrid control algorithm which consists of two algorithms:
- force control algorithm which is used to insert laparoscope into patient's body by hand. In this step of control, the robot is placed in ini-

tial point under patient. Next, surgeon takes up the end of manipulator and puts pressure on it. When the force exceeds the force threshold laparoscope moves in the direction of force action. The operator can change force threshold in the program.
- Control algorithm with force protection which fulfills role of protection. It controls values of forces and torques during robot movement. The system measures forces which act at the end of laparoscope. When the values of force exceed their threshold the system is stopped. In this moment the robot can move in the other direction.

Two software systems were designed:
- First system consists of all units of the system (for example: in program *LSM_Module_Control.exe* the main unit was implemented). This software system has been written in Borland C++.
- Second system, named LAP, has been realized in V+ environment which receive data from programs in C++ through RS232 interface. This program is executed by the controller of the robot.

We tested system making use of animal soft tissue: liver of pig, liver of turkey and chicken breast. During tests we observed that the system oscillaties because force sensor doesn't measure force response between laparoscopic tool and the soft tissue because it measures friction between the laparoscopic tool and the trocar.

4. Conclusions

The robotic assistant has been proposed to help the surgeons in laparoscopic cholecystectomy. System does not require any modification of a standard operation procedure: furniture or surgery tools for its installation and putting into operation. The surgeon can control this system with the help of joystick or speech commands. For testing the control algorithms we have built experimental system with animal model which may become filled gas CO_2. Communication between computer with joystick and robot serial interface introduces delay. This delay is about 500ms. We plan testing this system on animals (pigs) in the near future. **This system can be used in endoscopy.**

References

1. Taylor R. H. et al. (1995) A telerobotics assistant for laparoscopic surgery. *IEEE Engineering in Medicine and Biology Magazine*, No 14(3), pp. 279-288, June.
2. Butner S. E., Ghodoussi M. (2003) Transforming a surgical robot for human telesurgery. *IEEE Transactions on Robotics and Automation*, No 19(5), pp. 818-824, October.
3. Marescaux J. et al (2001) Transatlantic robot-assisted telesurgery. *Nature*, No 413, pp. 379-380, September.
4. Sauer P., Kozlowski K., Waliszewski W. and Hildebrant T (2002) Telerobotic Simulator In Minimal Invasive Surgery. *Proceedings of 3rd International Workshop on robot Motion and Control RoMoCo'02*, Bukowy Dworek, Poland, pp. 39-44.
5. Kowalczyk W., Lawniczak M. (2005) Study on opportunity to use industrial manipulator in surgical-laparoscopic application, *Proceedings of 5th International Workshop on Robot Motion and Control RoMoCo'05*, Dymaczewo, Poland pp. 63-68, June.
6. Sauer P., Kozlowski K., Pazderski D., Waliszewski W., Jeziorek P.(2005) The Robot Assistant System for Surgeon in Laparoscopic interventions, *Proceedings of 5th International Workshop on Robot Motion and Control RoMoCo'05*, Dymaczewo, Poland, pp. 55-62, June.
7. Keogh E., Pazzani M. (2001) Derivative Dynamic Time Warping, *SIAM International Conference on Data Mining*, Chicago Il.
8. Dutkiewicz P., Kielczewski M., Kowalski M., Wroblewski W. (2005) Experimental Verification of Visual Tracking of Surgical Tools, *Proceedings of 5th International Workshop on Robot Motion and Control*, Dymaczewo, Poland, pp. 237-242, June.
9. Sauer P., Kozlowski K., Waliszewski W. (2003) Measurement System of Force Response of Intra-abdominal for Laparoskopy, *Proceedings of 9th IEEE International Conference on Methods and Models In Automation and Robotics*, Międzyzdroje, Poland, pp. 1037-1042.
10. Sauer P., Kozlowski K., Majchrzak J., Waliszewski W. (2005) Mesurement System of Force Response for Minimal Invasive Surgery, *Computer-Aided Medical Intervantions: tools and applications SURGETICA'2005*, Chambery, France, pp. 139-147, January.

A Biologically Inspired Model For Quadruped Locomotion

B. Hennion, J. Pill and J-C. Guinot

Laboratoire de Robotique de Paris (LRP)
CNRS FRE 2507 - Université Pierre et Marie Curie, Paris 6
18, route du Panorama, BP 61, 92265 Fontenay-aux-Roses, France
{hennion, guinot}@robot.jussieu.fr

Abstract

The biomimetism is based on the observation and imitation of nature. It is expected, by studying live walking animals, to find interesting parameters that could be used in various fields of robotics [1], and to find parameters that don't vary from one specie to another [2]. The steps described here are the different experiments and the conception of a computer model. Different animals have been studied in the past like the hedgehog, the tenrec and the shrew, and a first model has been achieved. A new model, which is based on the tortoise of Hermann, consists in a 15 links, 22 degrees of freedom system, animated with Proportional-Derivative control, which references are actual recordings of live animals.

Keywords: Dynamic simulation, Biomimetism, Quadruped locomotion, 3D modelling

1 Introduction

The basic principle is shown in **Fig. 1**. From studies of animals, a passive mechanical structure (a) is built. The experiments described in the following chapter on walking animals will give the articular dynamic data (b) needed as references to animate this model. After the choice of an adapted foot-ground interaction model (c), the model can then be animated with a dynamic simulation software (d). The measures available after the simulation will then be used to check the realism of this model by comparing it to real values, and to refine and validate the model. The types of outputs available are: the contact forces under the feet of the virtual animals (e), all the spatial and temporal characteristics of any part of the model's body (f), the torques needed to actuate the joints (g).

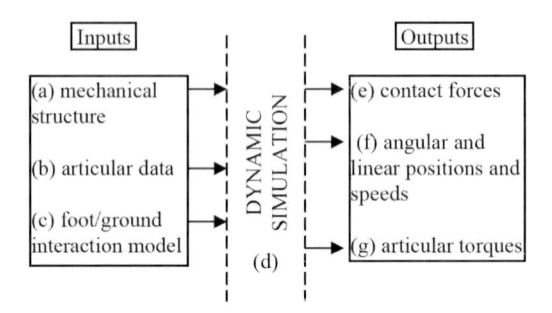

Fig. 1. Basic summary of the study

2 Experiments

The experiments took place at Museum National d'Histoire Naturelle, in a laboratory (USM 302) specialized in animal locomotion and morphology.

2.1 cineradiography

The system used is a cineradiography device, with the animals walking between a source of X-rays, and a digital camera. The bones of the animals can then be seen while moving during a locomotion cycle. The data are directly treated on a computer.

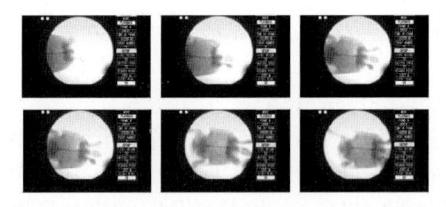

Fig. 2. X-ray video shoot of the tortoise - USM302/MNHN

An example of a cineradiography experiment is shown on **Fig. 2**. It is the top view of a walking tortoise. The same process is done in the lateral view to get the 3D record needed to reconstruct the virtual model. The time line is of course very important to synchronize temporally and spatially the different views. As one can see on the figure, it is a hard task to visualize the interesting points because of the small size of the bones and the opacity of the shell. These data will give the reference angular trajectories.

2.2 Dissection

A complementary way of getting the geometric informations was to make a dissection of a dead tortoise. All the bones have been measured, weighted, and even "digitized" to have the best knowledge of the inertia matrix of each segment.

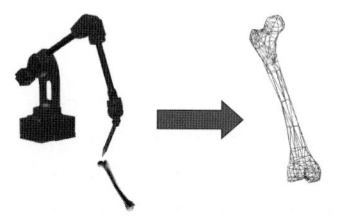

Fig. 3. Bones digitizing instrument : the *microscribe*©

3 Computer Model

The first computer model was based on the hedgehog [3]. Each of its leg were in 2D parallel plans. it has a steady walk on flat surfaces, but imperfections have emerged, like slipping of the legs. This maybe due to the incomplete modelling of the hip movements, or the displacements of the legs outside the sagittal plan.

The new model (**Fig. 4**) based on the tortoise [4] includes all the legs 3D movements of the real animal. An advantage of the tortoise is that its shell is a rigid part, so there is no complex movements due to rotations of the hip or the vertebral column. The model is closer to reality than the first one, and simulation data will be easier to compare with real data. Other studies like the effect of mass and length variations on the locomotion are also to be investigated [5].

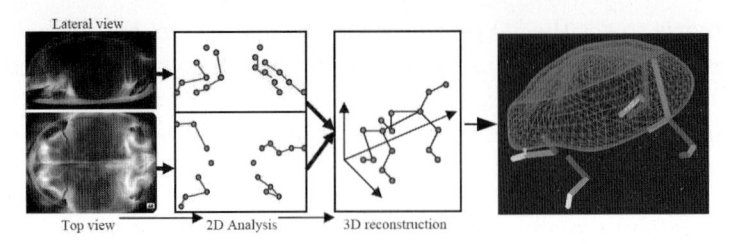

Fig. 4. 3D model from 2D images

4 Results

4.1 Angle trajectory

The references of the PD controller for all the joints of the model are real recordings. The model is also based on real measures of the different segments of the tortoise. But there are of course some inaccuracies in these measures, and some differences between the real tortoise and the model. Anyway, the angles follow well the references. To test the robustness of the whole system (geometry and control), a peek force has been put in the middle of the locomotion cycle representing half of the total weight of the model at the extremities of the legs. The angular position record of one joint (reference and measures), and the force peek are shown on **Fig. 5**. With different tries for the proportional gain of the control, the difference between the angle with and without the force was always less than 10^{-2} rad, which is very low. The zoom on **Fig. 5** shows this difference.

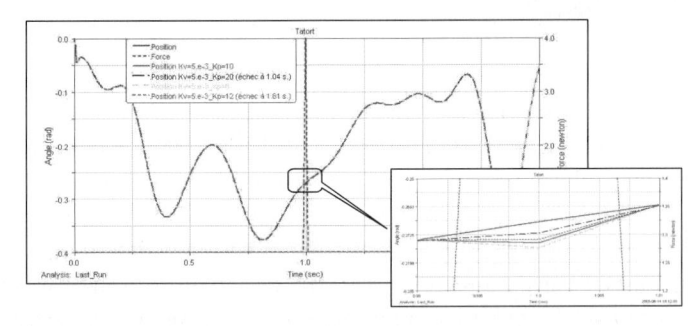

Fig. 5. Robustness test - 4N force at the extremity

4.2 Slipping study

The first improvement that has been studied was about the sliding of the model during its walk. Indeed, in this field, the model only walked half the distance covered by the real animal in the same number of locomotion cycle. The contact parameters (static and dynamic friction coefficients, and static and dynamic transition velocities coefficients) can be tuned. After an iteration work with simulation parameters, this has been greatly improved, especially on the slipping problem, and in consequence the distance covered.

The optimization on the contact parameters which gave results very close to the real animal were very small and at first view far from reality. Not satisfied with these results, it has been decided to construct a bench mark to get the real friction coefficient of the materials involved. Indeed, as we have here live materials involved (claws, skin, scales), no data have been found about this. The principle of the bench is shown on **Fig. 6**

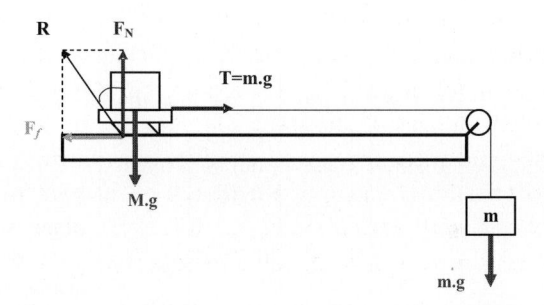

Fig. 6. Bench mark

It consists in a mobile part, where 3 claws are in contact with the ground. A mass is put on this part so M is the total mass and F_N the normal force on this part. A constant lateral force is applied through a pulley system, with a mass m. The resisting force due to friction on the mobile part is noted F_f. we want to determine the static and dynamic friction coefficient μ_{stat} and μ_{dyn}. For the first one, it is just necessary to increase the mass m until the limit where the claws are slipping. in this case $m \cdot g = F_{stat.max}$ and μ_{stat} is obtained with equation (1)

$$\mu_{stat} = \frac{F_f}{F_N} = \frac{F_{stat.max}}{F_N} = \frac{m}{M} \qquad (1)$$

To get the dynamic friction, we have to increase the mass m more than what is necessary to obtain the static coefficient. furthermore, it is important that when the mobile part slips, its acceleration remains constant along its linear trajectory. From the time and the distance covered, this acceleration a can be calculated, and the dynamic friction coefficient is obtained through equation (2)

$$\mu_{dyn} = \frac{F_f}{F_N} = \frac{m \cdot g - (M+m) \cdot a}{M \cdot g} \qquad (2)$$

During the cineradiography recording, the tortoise was walking on a paper sheet fixed over glass. As it is the references, we put paper over glass on our bench. The glass/claws contact has also been tested to compare the results. **Table 1** and **Table 2** show the results for the static friction coefficient for these two material pair.

Table 1. Claw/Paper contact - Static friction coefficient

Weight M (g)	50 (+56.9)	100 (+56.9)	200 (+56.9)	1000 (+56.9)	Test n°
μ_{stat}	0.267	0.315	0.375	0.327	1
	0.272	0.309	0.317	0.462	2
	0.283	0.323	0.335	0.519	3
	0.296	0.296	0.358	0.396	4
			0.349		5
					(paper at 90°)
		[0.64]*	[0.575]*		6
					(grainy paper)
Average	**0.2795**	**0.3107**	**0.347**	**0.426**	* : not used

Table 2. Claw/Glass contact - Static friction coefficient

Weight M (g)	100 (+56.9)	200 (+56.9)	Test n°
μ_{stat}	0.145	0.163	1
	0.156	0.142	2
	0.138	0.156	3
	0.168	0.168	4
Average	**0.1517**	**0.1572**	

One can see that the results are quite dependent on the weight put both on the mobile part (ie on the claws), and the weight necessary to pull it. Of course, as the weight increase, we can imagine that the claws enter a little inside the paper and this makes another phenomenon that prevents the claws from slipping. Anyway 0.3 can be considered as a good average static coefficient for the claws-paper contact, and expectedly the claws-glass contact is smaller (nearly half of it) at 0.15. These experiments give us small value for the claws contact. For comparison, the glass/glass static friction coefficient is more around 0.5.

For the dynamic friction coefficient, we have been unable to obtain a constant acceleration on the paper. Whatever the weight, the mobile part started slipping, then stopped, to slip again etc. On the glass, everything worked fine. this is probably due to the very low penetration of the two materials. the results are shown in **Table 3**.

Table 3. Claw/Glass contact - Dynamic

	Acceleration $(m \cdot s^{-2})$	μ_{dyn}	Speed $(m \cdot s^{-1})$	Test N°
3 tests with :	0.467	0.144	0.686	5
M=200g (+56.9g)	0.4806	0.142	0.7	6
m=50g (+2.3g)	0.448	0.147	0.32	7
Average	**0.465**	**0.144**	**0.569**	

5 Conclusion

After a long work to obtain a biomimetic model of a real animal which follow actual trajectories, many studies remain to see the importance of copying nature in the field of quadruped locomotion. the main interests seems to be in the geometry of such a model and its constituting segments. The studies so far consist mainly in the ground/leg contact problem as it remains a weak point in all modelling and dynamic simulation problems. The next step involves the type of control used, still biologically inspired, whether it consists in direct responses to sensors in the legs [6], or a more high-level scheme like learning methods [7] [8] [9].

Acknowledgement. We would like to thank the laboratory where the experiments took place and especially Marion Depecker and Sabine Renous from

USM 302 (MNHN - CNRS - UPMC - Collège de France).
The dynamic simulation software used is ADAMS, MSC SOFTWARE, from which images and curves from **Fig. 4** and **Fig. 5** are taken.

References

1. A. C. Smith, M. D. Berkemeier (1990) Passive dynamic quadrupedal walking. In *IEEE, Proceedings of International Conference on Robotics and Automation*.

2. Alexander, R. M. (1984) The gaits of bipedal and quadrupedal animals. *International Journal of Robotics Research*, vol. 3, pp. 49–59.

3. Villanova, J. (2003) *Simulation dynamique de quadrupèdes inspirés d'animaux reels*. Ph.D. thesis, Université Paris VI.

4. W. F. Walker, Jr. (1971) Structural and functional analysis of walking of the turtle, chrysemys picta mariginata. *Journal of Morphology*, vol. 134, pp. 195–214.

5. J. Schmiedeler, R. Siston, K. J. Waldron (2002) The significance of leg mass in modelling quadrupedal running gaits. In *CISM/IFToMM*, Udine, Italy.

6. A. Prochazka, S. Yakovenko (2001) *Locomotor Control: From Spring-Like Reactions Of Muscles To Neural Prediction*. .

7. Marhefka, D. W. (2000) *Fuzzy Control and Dynamic Simulation of a Quadruped Galloping Machine*. Ph.D. thesis, Ohio State University, Columbus, USA.

8. J. S. R. Jang, E. Mizutani (1996) Levenberg marquardt learning for anfis learning. In *Biennal Conf. of the N. American Fuzzy Information Processing Society*, Berkeley, CA, pp. 87–91.

9. L. Palmer, D. Orin, D. Marhefka J. Schmiedeler K. Waldron (2003) Intelligent control of an experimental articulated leg for a galloping machine. In *IEEE Proceedings of the International Conference on Robotics and Automation*, Taipei, Taiwan.

Fuzzy Logic Control Strategy for Functional Electrical Stimulation in Bipedal Cycling

R. Massoud, M. O. Tokhi and C.S. Gharooni

Department of Automatic Control and Systems Engineering, University of Sheffield, UK. R.Massoud@sheffield.ac.uk

Abstract

This paper presents the development of a fuzzy logic control (FLC) strategy for rehabilitation cycling by using functional electrical stimulation (FES) applied on disabled knee extensors, to achieve a steady crank cadence. Two fuzzy logic (FL) controllers are used to apply stimulation to both left and right knee extensors to achieve the desired knee joint motion. To guarantee the smoothness of the cycling motion and also to enhance the performance of the cycling process, an assistant actuation is applied on the crank to obtain optimum stability of the crank cadence.

Two types of actuators are tested, one is an electrical motor actuator and the other is dependent on changing the load at the crank. Both are controlled using a proportional derivative (PD) controller.

The controllers are developed and tested, and analyses are carried out for both types of actuators, and simulation results verifying the control strategy are presented and discussed.

Keywords: Functional electrical stimulation, fuzzy logic control, cycling, crank cadence, disabled.

1 Introduction

An important goal of rehabilitation exercises for the disabled is to improve their health after spinal cord injury (SCI) by exercising their large leg muscles. Functional electrical stimulation (FES) cycling is one of the most promising exercises, where electrical stimulation is applied to the paralyzed leg muscles of SCI patients to drive an exercise bicycle. FES cycling is used to reduce spasticity, reverse muscle atrophy, as well as increase cardio vascular fitness, muscle bulk, blood flow and bone density in the legs, and to reduce the chance of developing pressure sores and cardiovascular disorders such as orthopaedic intolerance [1].

The success of any FES system depends on its control system which qualifies the requirements of the control problem addressed by the disability. User supervision for the system may be sufficient in some applications; but in other applications, other levels of control could be desirable.

For FES control electrical current pulses are used to excite muscle nerves to cause the muscle contract which in turn produces muscle force that leads to a joint torque allowing the body to move.

Many studies have used different control strategies to control the rehabilitation exercise cycling by using different types of FES stimulation [2] [3] [4] [5]. Most of the researchers have used the motor to assist the pedals over dead spots in the cycling motion and as a generator to provide resistance [2] [3] [4] [6] [11]. [7] used FLC to control cycling movement induced by FES through stimulating both quadriceps and hamstrings muscle groups. Previous studies have used two groups of muscles at least. The approach presented in this paper, however, uses model-free fuzzy logic (FL) controller to control cycling motion through stimulating the quadriceps only, and to compare the results between using the electrical motor and the changing of loads methods.

A humanoid model is developed in this study and used as a test-bed to develop a suitable strategy for motion control of leg joints to result in bipedal motion in the lower extremity.

2 Methods

2.1 The humanoid model

The humanoid model was developed using Visual Nastran (VN) software. The physical dimensions of the humanoid segment, were calculated according to [8]. The humanoid is assumed to be 180 cm in height and 70 kg in weight.

The controllable joints in this paper are the knee joints, and the gear joint. Upper body joints do not have significant role in the cycling process. To make the analysis easier, the joints are constructed with one degree of freedom. This means that they are allowed to rotate around the sagittal

Fig. 1. The humanoid with bicycle

plane only. The complete model for the humanoid and the bicycle is shown in Figure 1.

2.2 Muscle model

The muscle model used is described in [9]. The model is a transfer function between electrical stimulation and the resultant knee torque. The muscle model is a first order model, given as

$$H(s) = \frac{G}{1+s\tau}$$

where:
  τ is time constant.
  G is static gain.

The muscle model is to mimic the quadriceps muscle group in the thigh, which straightens the knee joint. Thus, the muscle model works as a knee extensor.

3 Control strategy

In this paper, a stationary bicycle is driven by the VN humanoid and Matlab/Simulink environment. The controllable joints of the humanoid are the right and left knee joints, by applying stimulation pulses to the knee extensors muscle group. Since only the knee extensors are used to control the knee joint movement and cadence of the crank, other types of actuators are used to help in the control process, such as electrical motor [3], changing loads, or using mechanical storage actuator such as mechanical spring [10].

To control the knee joint movement during cycling, a reference knee trajectory is obtained by using reversed dynamic movement method. The error in angular velocity of the knee joint and the change of this error are implemented as inputs to an (FL) controller , while the output of this controller is the stimulation signal which in turn is applied to the muscle model to produce the appropriate torque that can move the knee joint in the desired trajectory.

Proportional derivative (PD) controller is also utilized in the control strategy, to control the crank cadence. The input of the PD controller is the error in the crank cadence and the output is the torque applied to the crank to keep the cadence constant at 50 rpm. Figure 2 shows the control design, where TL, TR, TM are the torque applied to the left knee joint, right knee joint and crank motor respectively, ΔVL and ΔVR are the errors in the angular velocity of the left and right knee joints respectively, Acc_L and Acc_R are the derivative of ΔVL and ΔVR respectively.

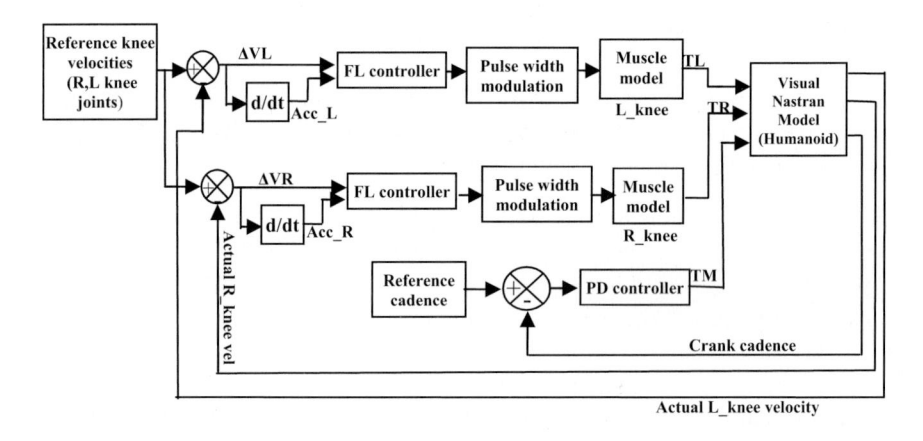

Fig. 2. Control system structure

3.1 Fuzzy logic controller

Two FL controllers were developed to control the knee joint movement for both legs, and to force the leg to follow a reference cyclical signal. Choosing fuzzy controller inputs and outputs is a very critical process, because it is important to be sure that all the information needed about the plant is available through the controller inputs.

Each of the FL controller inputs were designed with five membership functions (MFs). This results in 25 fuzzy rules described in Table 1, where ΔV and Acc are the error in angular velocity of the knee joint and change of error respectively.

Table 1. The FL rule base.

Acc / ΔV	NB	NS	Z	PS	PB
NB	NB	NS	Z	PS	PS
NS	NB	NS	Z	PS	PS
Z	NB	NS	Z	PS	PB
PS	NS	Z	PS	PB	PB
PB	Z	PS	PB	PB	PB

NB=Negative big   NS=Negative small   Z=Zero PS=Positive small
PB=Positive big

4 Results

Two control strategies are investigated with the system; using the motor to stabilize the cadence at a speed of 50 rpm, and replacing the motor with changing loads. Both designs are good in terms of controlling the knee trajectory; Figure 3 describes the reference and actual knee trajectories for both approaches.

The motor design is superior for cadence stability as compared with the changing loads method, since the motor can move in two directions, while the load resists against the movement direction. Figure 4 shows the errors in crank cadence for both motor and load designs.

Figures 5 and 6 show the errors of angular velocity and knee trajectory for both actuator designs respectively. It is clear that both designs provide satisfactory results and the error is very small, despite that the error in the motor design is notably smaller.

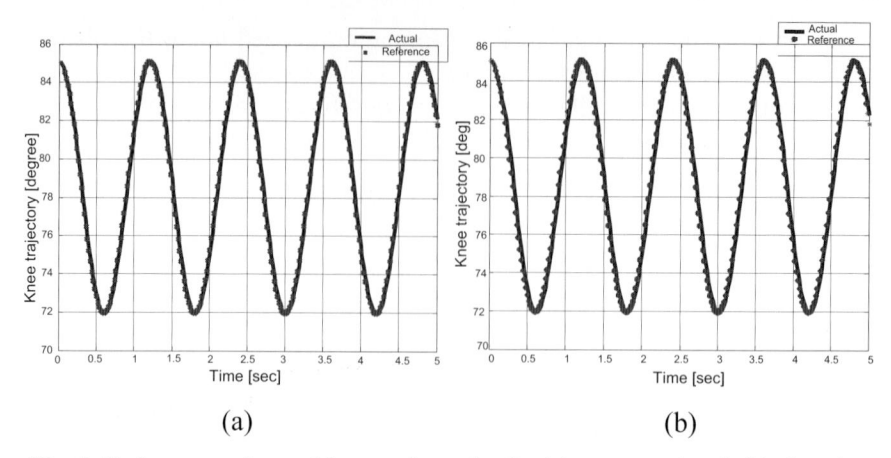

Fig. 3. Reference and actual knee trajectories for (a) motor assisted, (b) changing load approach

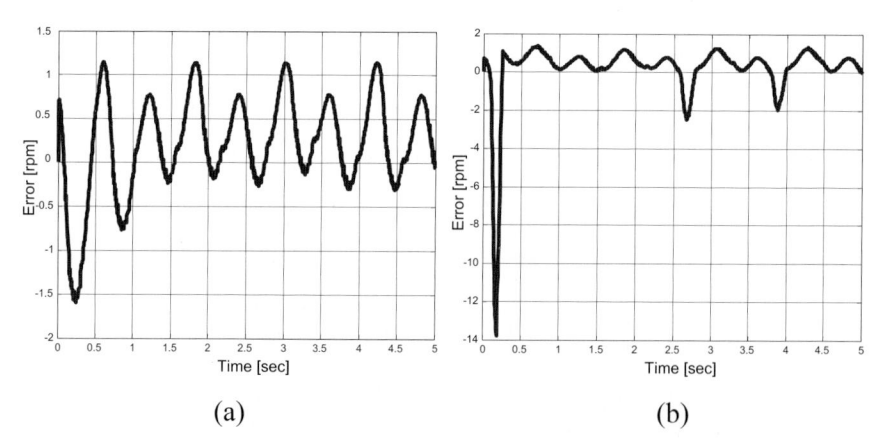

Fig. 4. The error in crank cadence for (a) motor assisted (b) changing load method

5 Conclusion

In conclusion, it is possible to drive stationary bicycle by stimulating quadriceps muscles (knee extensors) only. However, the result can be improved further with using mechanical actuators to assist the muscle stimulation.

Using FL control to position knee joint angle in the required position through quadriceps stimulation is a very effective control strategy that may help to enhance the performance of rehabilitation machines for the disabled.

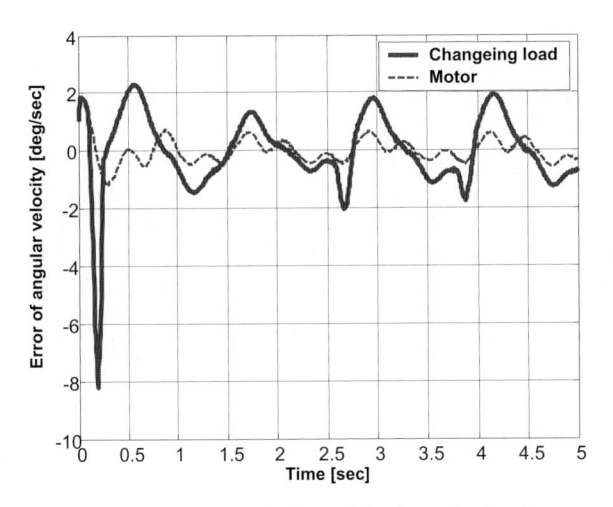

Fig. 5. Error of the angular velocity of the knee for both actuator options

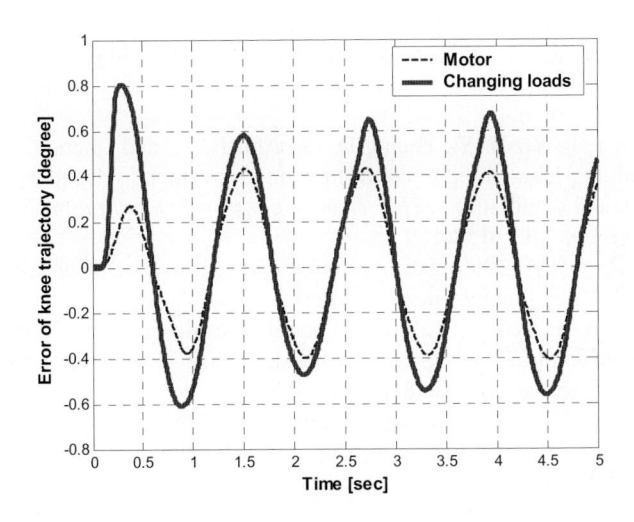

Fig. 6. Error of the knee trajectory for both actuator options

References

1. Petrofsky J. S., Laymon M. (2004) The effect of previous weight training and concurrent weight training on endurance for functional electrical stimulation cycle ergometry. *Eur J Appl Physiol,* vol. 91, pp. 392-398.
2. Pons D. J. Vaughan C. L. and Jaros G. G. (1989) Cycling device powered by the electrically stimulated muscles of paraplegics. *Medical & Biological Engineering & Computing,* vol. 27, pp. 1-7.
3. Hunt K. J., Stone B., et al. (2004) Control strategies for integration of electric motor assist and functional electrical stimulation in paraplegic cycling: utility for exercise testing and mobile cycling. *IEEE Transactions on Neural systems and rehabilitation engineering,* vol. 12, pp.89-101.
4. Fornusek C., Davis G. M., Sinclair P. J., and Milthorpe B. (2004) Development of an isokinetic functional electrical stimulation cycle ergometre. *Neuromodulation,* vol. 7, no.1, pp. 56-64.
5. Perkins T. A., Donaldson N., Hatcher N. A. C., Swain I. D., and Wood D. E. (2002) Control of leg-powered paraplegic cycling using stimulation of the lumbo-sacral anterior spinal nerve roots. *IEEE Transactions on Neural systems and rehabilitation engineering,* vol. 10, no. 3, pp. 158-164.
6. Eichhorn K. F., Schubert W., and David E. (1984) Maintenance, training and functional use of denervated muscles. *Journal of Biomedical Engineering,* vol. 6, pp. 205-211.
7. Chen J. J., Yu N. Y., Huang D. G., Ann B. T., and Chang G. C. (1997) Applying fuzzy logic to control cycling movement induced by functional electrical stimulation. *IEEE Transactions on Rehabilitation Engineering,* vol. 5, no. 2, PP. 158-169.
8. Winter D. A. (1990) *Biomechanics and motor control of human movement.* Wiley_Interscience New York.
9. Ferrarin M., Pedotti A. (2000) The relationship between electrical stimulus and joint torque: A dynamic model. *IEEE Transactions on Rehabilitation Engineering,* vol. 8, pp.342-352.
10. Gharooni S., Heller B., and Tokhi M. O. (2001) A new hybrid spring brake orthosis for controlling hip and knee flexion in the swing phase. *IEEE Transactions on Neural systems and rehabilitation engineering,* vol. 9, pp. 106-107.
11. Petrofsky J. S., Smith J. (1992) Three-wheel cycle ergometer for use by men and women with paralysis. *Medical & Biological Engineering & Computing,* vol. 30, pp. 364-369.

Insect-inspired, Actively Compliant Hexapod Capable of Object Manipulation

W. A. Lewinger[1], M. S. Branicky[1], and R. D. Quinn[2]

[1]Department of Electrical Engineering and Computer Science, Case Western Reserve University, Cleveland, OH, USA, wal4@case.edu, msb11@case.edu
[2]Department of Mechanical and Aerospace Engineering, Case Western Reserve University, Cleveland, OH, USA, rdq@case.edu

Abstract

Insects, in general, are agile creatures capable of navigating uneven and difficult terrain with ease. The leaf-cutter ants (*Atta*), specifically, are agile, social insects capable of navigating uneven and difficult terrain, manipulating objects in their environment, broadcasting messages to other leaf-cutter ants, performing collective tasks, and operating in cooperative manners with others of their kind [9][12]. These traits are desirable in a mobile robot. However, no robots have been developed that encompass all of these capabilities. As such, this research developed the Biologically-Inspired Legged-Locomotion Ant prototype (BILL-ANT-p) to fill the void. This paper discusses the features, development, and implementation of the BILL-Ant-p robot, quantifies its capabilities for use as a compliant mobile platform that is capable of object manipulation.

Keywords: Biologically-inspired, Actively-compliant, legged hexapod, object manipulation

1 Introduction

The goal of this research [10] was to develop a robot that is power and control autonomous; capable of navigating uneven terrain, manipulating objects within its environment, and active compliance with its environment; very strong for its size; and relatively inexpensive compared to other similar robots. An investigation into existing hexapod robots [2] was conducted, such as Tarry I and Tarry II [4], MAX [1], Robot I [7] and Robot II [6], the TUM Walking Machine [11], the LAURON series of robots [8], and Genghis [3]. While each robot exhibited one or more of the above-mentioned traits, none were found that encompassed all of the desired aspects. As such, the BILL-Ant-p was designed, constructed, and evaluated.

Fig. 1. BILL-Ant-p robot (left) and Acromyrmex versicolor (Leafcutter ant found in Arizona, USA, ©Dale Ward) (right)

Based on abstracted anatomy from ants [9][12] and leg coordination from stick insects [5][6][7], the BILL-Ant-p robot (Fig. 1, left) is an actively compliant 18-DOF hexapod robot with six force-sensing feet, a 3-DOF neck and head, and actuated mandibles with force-sensing pincers for a total of 28 degrees of freedom [10]. The robot actively moves in a planar motion away from external perturbations applied to the body by measuring the shift in load on the six foot-mounted force sensors. Similarly, changes sensed by the pincer-mounted force sensors while grasping an object cause actively compliant movements in the neck and the body.

2 Mechanical System

The body is constructed from 6061 aluminum and carbon fiber sheets (Fig.1, left). During a neutral stance, the BILL-Ant-p is 47cm long, 33cm wide, and 16cm and 26cm tall to the top of the body and top of the head, respectively. It weighs 2.85kg.

Layout of the body and orientation of the body-coxa (BC) joints was based as closely as possible on the body segments of various ants (Fig. 2). While the ant body has a much more compact configuration, the mechanical design was limited by the constraints of function (housing batteries and servo controller) and the connecting elements (legs and head/neck). Leg placement and orientation were designed to accommodate 90 degrees of rotation for each BC joint (maximum range of motion for the joint motors) without interfering with other legs throughout the range of motion. Front and rear BC servos are splayed 60 degrees from the medial plane. The middle BC joint motors are perpendicular to the medial plane. This pattern is similar to the ant for the middle and rear legs; however, it is not biologically accurate for the front legs. While the front BC servo orientations were chosen to produce axially-symmetric body plates, the front legs are attached to their respective servos to roughly conform to the ant's anatomy with a starting position of 15 degrees from the medial plane. All legs have ±45 degrees of motion; however the front legs have +0/-90 degrees of forward/rearward motion from starting positions of 15 degrees off the medial plane.

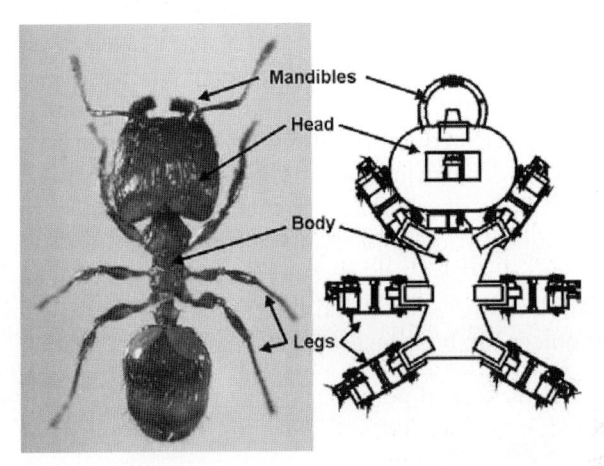

Fig. 2. Top-view body layout comparison of Pheidole fervida (left, found in Japan, ©Japanese Ant Database Group) and the BILL-Ant-p robot (right)

Each leg has three active degrees of freedom: a body-coxa (BC) joint, a coxa-femur (CF) joint, and a femur-tibia (FT) joint. MPI MX-450HP hobby motors (Maxx Products, Inc., Lake Zurich, IL, USA) are used for the joints. These servos were chosen for reliability, high torque, and affordability. The MPI servos have 8.37kg-cm of torque, can rotate through a 60° arc in 0.18sec, and the small internal dc motor consumes 1125mW of power at stall torque.

Attached directly to the ends of the tibiae are the feet. The feet provide traction and measure the load along each leg. Each foot is comprised of an Interlink Electronics, Inc. (Camarillo, CA, USA) FSR 402 force-resistive sensor sandwiched between two flat plates, which are 2.06cm square. A simple voltage divider with a 10kΩ resistor and the force sensor in series is used to measure force at the foot. Signals for each foot are connected to the IsoPod™ microcontroller ADC inputs.

The neck has three degrees-of-freedom, which allows for nimble manipulation of objects. Each degree is actuated by an MPI MX-450HP servo. At the base of the neck is the yaw servo, which is attached to the robot body. The pitch assembly is connected to the output of the yaw servo. The roll servo attaches to an aluminum plate that is connected to the underside of the carbon fiber head. This plate is also connected to the mandibles servo housing to give the mandibles assembly a strong connection to the neck.

The oval-shaped carbon fiber head is 18.54cm wide and 12.19cm long at the extremes. Attached to the neck by the roll servo mounting plate, the head is not part of the load-bearing link between the mandibles and the neck. It supports the two BrainStem microcontrollers that are used to actuate the neck and mandibles. Additional space is available for placement of future sensors, such as a miniature video camera.

Object manipulation is achieved by the twin pincer mandibles (Fig. 1, left). They are fabricated from aluminum and actuated by a single MPI MX-450HP servo. The mandibles are kept open by a lightweight spring and closed by Kevlar fiber cables attached to a pulley on the servo. The tips of the mandibles each hold twin Interlink Electronics FSR 401 force transducers. By using four sensors and having the head at an angle to the ground, mandible closing force and horizontal and vertical forces exerted by a grasped object can be measured.

3 Electrical System

The electrical system has two major components: control and power. Control consists of motor controllers (IsoPod™ and BrainStem microcontrollers) and a System Controller. Power is supplied by on-board Li-ion batteries.

A New Micros, Inc. (Dallas, TX, USA) IsoPod™ V2 SR microcontroller is used to translate System Controller commands into leg joint servo signals and return foot-mounted force sensor values. This microcontroller was chosen for several reasons: programmability, the availability of floating-point math, the ability to control up to 26 R/C servo motors (with the attached daughter board), small footprint, two serial interface ports, and low cost. There are eight ADC inputs on the IsoPod™, six of which are used to measure foot-mounted sensor forces.

Two Acroname BrainStem GP 1.0 microcontrollers (Acroname, Inc., Boulder, CO, USA) are used in the head. These PIC-based controllers have four R/C servo outputs, five 10-bit ADC inputs, five digital I/O ports, an RS-232 serial interface, I^{2}C interface bus, and a digital IR range finder input. One of the BrainStem units controls the 3-DOF neck. Three servo output ports and three ADC input ports are used to actuate and sense the status of the neck servos. The additional two ADC inputs can be used for future expansion. The second BrainStem unit controls the mandibles. The mandible servo is actuated through a servo output port and the four force transducer voltage dividers values are fed into four ADC input ports. Since these controllers have limited processing power and no capacity for floating-point math, they were not selected for use in controlling the legs.

The System Controller is a laptop computer (2.8MHz P4, 1 GB RAM, 60 GB HD) running custom software that was written in Microsoft Visual Basic 6.0 (Microsoft Inc., Redmond, WA, USA). The System Controller has a user interface to dictate commands to the robot and remotely to view robot posture and status. It is connected to both the IsoPod™ microcontroller and the router BrainStem microcontroller by two RS-232 serial ports.

The BILL-Ant-p robot is power autonomous, with four on-board 2400mAh, 7.4vdc Li-ion batteries from Maxx Products, Inc. (MPI, Lake Zurich, IL, USA). To limit the voltage to 6.0vdc, each of the batteries is connected to an MPI ACC134 6-volt Regulator. Three of the batteries are connected in parallel to provide power for the servo motors. The fourth battery supplies power to the microcontrollers.

4 Software System

A Software Interface was created using Microsoft Visual Basic 6.0 (Microsoft Inc., Redmond, WA, USA). The interface allows the operator to command robot actions and view robot status. Basic commands on the interface allow the operator to: manipulate each leg joint; set foot position in body-centric x-, y-, and z- coordinates; initiate a standing routine; adopt a standing posture; adjust body height from the ground; adjust body roll and pitch; drive the robot using speed, heading, and rotation values; manipulate roll, pitch, and yaw of the neck; and adjust position and closing force of the mandibles.

Strafing (moving in one direction while facing another) (Fig. 3, center), walking (a type of strafing where the robot is facing the direction of motion) (Fig. 3, left), rotating (zero radius turn about the body center) (Fig. 3, right), or a combination of strafing/walking and rotating movements are possible. Each foot heading during stance and swing is calculated based on the heading and rotation values. For strafing/walking all feet move a uniform direction. Rotating assumes a zero radius turn and has each foot move tangentially to the body center at a speed proportional to the distance from the foot to the body center. Vector sums of each foot's path for strafing/walking and rotating are normalized to the nominal step length to create combinations of movements (e.g. turning in an arc or rotating while moving along a straight line). An implementation of Cruse control for leg coordination is used to adjust stance length and set transition points between stance-swing and swing-stance phases [5][7]. A continuum of gaits from wave to quadruped to alternating tripod is achieved as speed is increased.

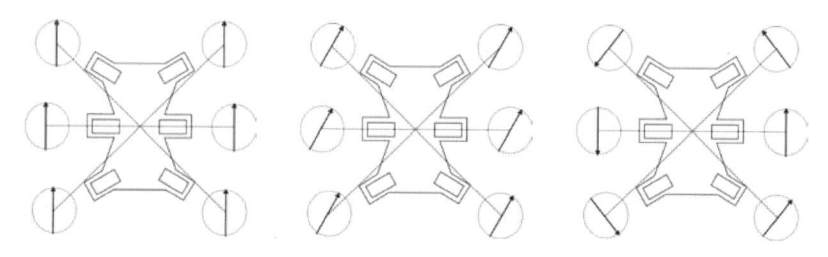

Fig. 3. Foot movements during the swing phase for walking (left), strafing at 330° (center), and rotating counter clockwise (right).

When employing an active compliance behavior, the BILL-Ant-p robot is programmed to maintain a neutral stance, where the legs are axially symmetric about the medial plane, and the neck motors are at the center of their ranges of motion. Perturbations applied to the body are measured by foot-mounted force sensitive resistors. External force magnitude and direction are calculated by summing the force sensor values in body-centric x- and y-coordinates based on the position of each foot relative to the body. Measured forces at each foot are converted to vertical forces (positive z) based on the angle of the foot with respect to the ground. The robot moves in a planar motion in the direction of, and proportional to, the force, as though attached to a virtual attractor by a virtual spring and damper.

The gripping force for the pincers is user-defined and is essentially a stiffness setting. The pincers open and close to maintain the desired gripping force. Similar to the body movements initiated from the foot-mounted sensors, the neck and body respond to forces measured by the pincers. The neck yaw motor moves within its range of motion to balance the forces measured by the pincer sensors such that the sensors values are all equal and at the desired gripping force. The yaw axis is divided into three 30° areas. When the neck is outside the central 30° area, movement commands are sent to the body to bring the neck toward a more neutral position. Body movements (planar and rotational) are initiated in response to lateral, longitudinal, and rotational forces. These commands continue until the neck is within the "body motion deadband", near the center of the ranges of motion.

5 System Performance

With fully charged batteries, the robot is able to stand and walk at 0.72cm./sec (two body lengths per minute) with its 2.85kg body weight and an additional 3.18kg of payload, and can rotate at a rate of 6.9°/sec. While standing in a neutral posture, the BILL-Ant-p is able to support its body weight and a payload of up to 8.64kg. Motor batteries allow for approximately 36min of normal operation, or about 25min of heavy lifting. Weight lifting performance declines by up to 30% over 25min as battery power is consumed.

When the body is perturbed, the BILL-Ant-p quantifies the amount and location of the force and moves away from the perturbation at a speed that is proportion to the force. This allows for robot strafing and walking movements to be initiated by pushing or pulling, rather than through the user interface. Speeds up to 0.72cm/sec were observed during several experiments including: pushing from behind, pushing from the side, pushing rearward on the head, pulling toward the head, pulling toward the rear, and pulling toward the side. Since the operator end of the attached string for pulling was raised above the robot body, the robot responded more quickly and smoothly during the pulling experiments. During pushing experiments the feet had a slight tendency to get caught on the ground.

Reactions while grasping an object were conducted by closing the mandibles on a solid block of Delrin® held by an operator. Movements of forward and backward walking, left and right strafing, and clockwise and counter-clockwise rotation were initiated by manipulating the grasped block. As expected, lateral movements and rotational movements of the block caused the neck to rotate in the direction of the applied force. Once the neck was outside the central 30° area, the legs began moving in the appropriate manner for the applied force: strafing for lateral forces and rotating for rotational forces. Forward and backward movements were observed when pulling and pushing the block. The neck rotated to equalize the forces sensed by the pincer contact plates; however the walking response was initiated regardless of the yaw motor position, as programmed.

Acknowledgements

Funding was provided in part by NSF IGERT Training Grant DGE 9972747

References

1. Barnes, D. P., "Hexapodal Robot Locomotion over Uneven Terrain," in Proc. IEEE Conf. on Control Applications. Trieste, Italy, pp. 441 – 445, September 1998.
2. Berns, K., "The Walking Machine Catalog: Walking Machine Catalog," World Wide Web, http://www.walking-machines.org/, 2005.
3. Brooks, R. A., "A Robot That Walks; Emergent Behaviors from a Carefully Evolved Network" Technical Report, MIT AI Lab Memo 1091, Cambridge, MA, USA February 1989.
4. Buschmann, A., "Home of Tarry I & II: Design of the Walking Machine Tarry II," World Wide Web, http://www.tarry.de, March 2000.
5. Cruse, H., "What Mechanisms Coordinate Leg Movement in Walking Arthropods?" Trends in Neurosciences, Vol. 13, pp. 15 – 21, 1990.
6. Espenschied, K. S., Quinn, R. D., Chiel, H. J., and Beer, R. D., "Biologically Based Distributed Control and Local Reflexes Improve Rough Terrain Locomotion in a Hexapod Robot," Robotics and Autonomous Systems, Vol. 18, pp. 59 – 64, 1996.
7. Espenschied, K. S., Quinn, R. D., Chiel, H. J., and Beer, R. D., "Leg Coordination Mechanisms in Stick Insect Applied to Hexapod Robot Locomotion," Adaptive Behavior, Vol. 1, No. 4, pp. 455 – 468, 1993.
8. Gaßmann, B., Scholl, K.-U., Berns, K., "Behavior Control of LAURON III for Walking in Unstructured Terrain," in Proc. Intl. Conference on Climbing and Walking Robots (CLAWAR '01), pp. 651 – 658, Karlsruhe, Germany, September 2001.
9. Hölldobler, B. and Wilson, E., The Ants, The Belknap Press of Harvard University Press, Cambridge, Massachusetts, 1990.
10. Lewinger, W. A., "Insect-inspired, Actively Compliant Robotic Hexapod," M. S. Thesis, Department of Electrical Engineering and Computer Science, Case Western Reserve University, Cleveland, OH, USA, 2005. (http://biorobots.cwru.edu/publications/Thesis05_Lewinger_BILL-Ant-p.pdf)
11. Pfeiffer, F., Weidemann, H. J., Eltze, J., "The TUM Walking Machine. - In: Intelligent Automation and Soft Computing," Trends in Research, Development and Applications, TSI Press, Vol. 2, pp. 167 – 174, 1994.
12. Yahya, H., The Miracle in the Ant, Ta-Ha Publishers, Inc., London, United Kingdom, 2000.

Modeling and Simulation of Humanoid Stair Climbing

L. Wang[1], Y. Zhao[2], M. O. Tokhi[1] and S. C. Gharooni[1]

[1]Department of Automatic Control and Systems Engineering, [2]Department of Applied Mathematics, The University of Sheffield, UK

Abstract

In this paper, a dynamic humanoid with a stair case model is designed to achieve a dynamically stable stair climbing gait pattern, and sequences of stair climbing locomotion analysis (weight acceptance, pull up, forward continuance, foot clearance and foot placement) are presented. A suitable trajectory control strategy is developed for the lower limb's joints (hips, knees, ankles) during stair climbing, and the strategy is tested successfully in simulation. Climbing speed of 0.5m/s has been achieved.

Keywords: Control, humanoid, modelling and simulation, stair climbing.

1 Introduction

Stair climbing (ascending and descending) is a common functional activity of daily life. A recent review of literature has revealed that only a limited number of scientific studies on the biomechanics of normal stair climbing is available [2-5]. It has been reported that stair climbing places a higher demand on the knee when compared to level walking as demonstrated by increased knee extensor moments and greater ranges of knee motion [3],[6]. In 1969 Morrison [7] presented the joint contact force at the knee for level walking, walking up and down a ramp, and walking up and down stairs. The flexor moments for stair ascent and descent were much higher than those during level walking [8].

Basically, there are three main tasks involved in this research: (i) develop a humanoid model with standard human height, weight and density within the Visual-Nastran (VN) software; (ii) gait cycle analysis in stair climbing locomotion, which separates the locomotion into several phases and demonstrates the inter-relation between them so as to produce stair climbing movement; (iii) to build a Simulink model in order to control the humanoid.

2 Model development

A humanoid model of 175 cm height and 65 kg weight was developed according to the standard measurement data from Winter et al [1]. Table 1 shows the corresponding lengths and weights obtained for the segments involved.

Table 1. Body segment lengths and weights

	Head	Upper	Fore-arm	Torso	Foot	Shank	Thigh
Weight (kg)	5.265	1.82	1.82	32.305	0.943	3.023	6.5
Length (m)	0.228	0.316	0.248	0.5661	0.039	0.418	0.417

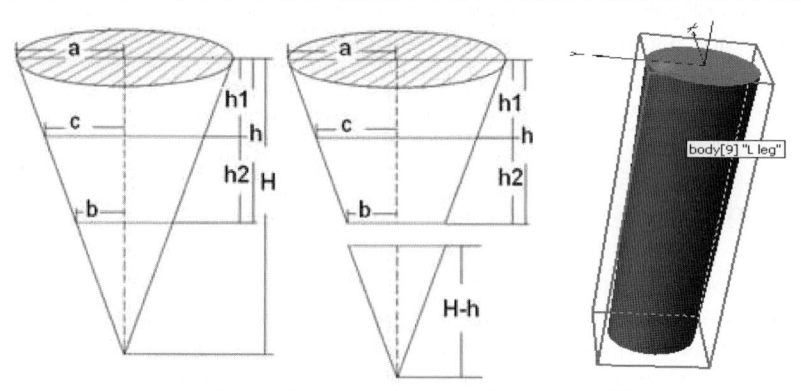

Fig. 1. Segment developed in Auto CAD

Most of the segments, such as arms and legs were developed using cylinder shapes, as shown in Figure 1, using Auto CAD, where

$$H = \frac{3(h^2 - 2h_1^2) + \sqrt{(24hh_1(h^2 + h_1^2 - hh_1) - 3(h^2 + 2h_1^2))}}{6(h - 2h_1)},$$

$$a = \sqrt{\frac{3VH^2}{\pi[H^3 - (H-h)^2 * h]}}, \quad b = \frac{a(H-h)}{H}, \quad c = \frac{a(H-h_1)}{H}$$

The parameters h, V, h_1 in the above are known from which the unknown parameters H, a, b, c can be obtained and used to develop segments in Auto CAD and import to VN.

The humanoid model is equipped with 12 joints: 1 neck, 1 head, 2 shoulder, 2 elbow, 2 hip, 2 knee, and 2 ankle joints. Each leg is driven with 3 joints, namely hip, knee and ankle joints. The upper body has one degree of freedom about the vertical axis of the pelvis. A revolute motor can generate relative force or torque to control the joints to achieve the desired position. In this study the motor is controlled based on orientation, but it can also be controlled on the basis of torque, velocity or acceleration. Six motors and six revolute joints are used to connect the segments, although some motors are inactive at certain times. It is known that the rotation for the ankle joints should be between –25 and 40 degrees, rotation for hip joints should be between –50 and 120 degrees, and for the knee joints should be between –100 and 0 degrees [12]. The stairs are defined as boxes anchored during the simulation with 0.14 m rise and 0.3 m run.

3 Gait analysis

A unique feature of stair climbing locomotion analysis is that unpredictable human behaviour must be taken into account when the control system is designed [10]. The main difficulty lies in balancing the body, reducing the contact force between the feet and steps and avoiding "slipping" of the feet in the support phase. For simplicity, in this paper the analysis of motion is concentrated on the sagital plane. The gait cycle is the period of time between any two identical events in the stair climbing, which can mainly be divided in two phases: the stance phase and the swing phase. A complete gait cycle of stair climbing includes three functional tasks, namely *weight acceptance, single-limb support* and *limb advancement*.

To balance the body, the *weight acceptance* (moving central of mass forward) is essential. To avoid foot "slipping", two linear actuators are used between each foot with ground by controlling the length of the linear actuator, which might produce large contact force leading to jumping of the model. There are several ways to produce contact force, and these will cause the system to become unstable when the contact force is extremely large.

Table 2 gives details of the phases involved in the gait cycle in each step. During stair climbing the first step and the last one are special.

Table 2. Gait cycle of stair climbing

	First Step	Second Step	Third Step	Last Step
Left Leg	FC FP	WA PU FCO	P FC EP	*WA PU S*
Right Leg	S S	P FC FP	WA PU FCO	*P FC FP*

FC = Foot Clearance phase; S = Stance phase; FP = Foot Placement phase; PU = Pull up phase; FCO = Forward Continuous phase; P = Pendulum phase; WA = Weight Acceptance phase

4 Control strategy

The integration of Matlab/Simulink with VN makes it possible for Simulink to send signals to VN and to receive signals from it in real time. In this way, the two software environments work together. In this manner Matlab can be used to develop suitable controllers while VN can be used to give a characterisation of the humanoid behaviour. This makes the simulation less complicated to understand. Hence, a control system in Simulink is essential to enable the joints' angular displacements track desired trajectories and the humanoid to perform stair climbing in VN. As various components of the system are dynamically strongly coupled, slight change in any joint causes variation in the trajectories of all the segments. Hence, it is not easy to construct a precise mathematical model that describes the dynamic behaviour of the humanoid. An open-loop control system is developed to achieve the desired trajectory. This is shown in Figure 2, where the first five inputs are used to control the state of the motors (active '1' or inactive '0') during the simulation, and the remaining six inputs control orientation of the six motors (ankle, knee and hip joint), the 'VN plant' is the block linking Simulink with VN, the output signal is the orientation of each joint during simulation as a function of time, which is exported to the workspace for later use.

Note that the reference inputs are not know prior to the simulation. Thus, the initial parts of the reference inputs are developed according to common knowledge [9], [11] and the remaining parts are completed gradually along with the simulation. As noted, the swing phase lasts for about 0.5 seconds during the first step, the peak angles of hip and knee flexion are reached almost simultaneously (at about 0.2 second), then the hip keeps the maximum value while the knee extends like a pendulum, until the desired position is achieved, after which the *foot placement* phase commences. The right leg moves like a pendulum while the motors in the left leg are inactive, so the left leg moves forward naturally due to the force of body's weight to accomplish *weight acceptance* and prepare for the next step (at about 0.8 second). After the *weight acceptance*, three joints of the left leg begin to extend, at about 1.4 seconds, the knee and hip joints are fully extended, but there is still some angular displacement in the ankle joints to prepare for the pendulum movement. At the same time, the left leg performs *foot clearance* and *foot placement*. The above analysis of the first two steps describes the procedure of developing the reference input trajectory and how different phases work together to accomplish the climbing task. The three joint trajectories of right side are 0.5 second behind those of the left side. However, ignoring this delay the trajectories for the left and right joints are exactly the same shape, e.g. the left knee (Lknee) and right knee (Rknee) trajectories are basically the same during a whole step. Even though the shapes are alike, a step is not exactly a copy of the previous step due to different circumstances affecting each step, which include the velocity of the body, the vertical position of each segment and the interaction between segments. Table 3 shows the interaction between the inputs for achieving the climbing task.

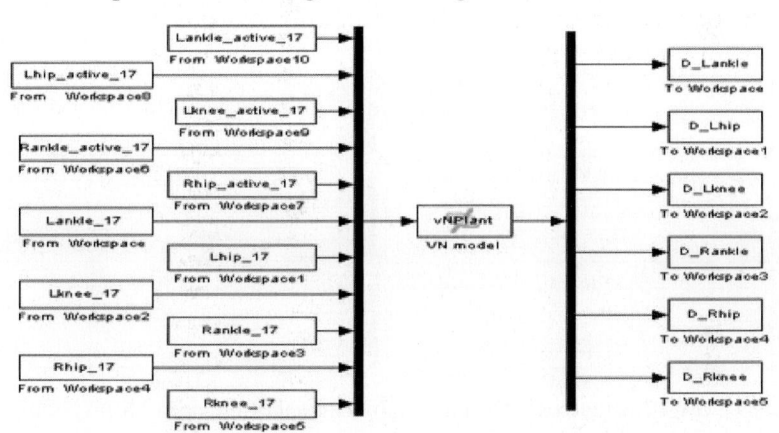

Fig. 2. Simulink representation of the open-loop control system

The input and output for each joint is thus achieved is shown in Figure 3. As noted actual output trajectory is smoother than the reference input signal. However, the output is delayed by about 0.06 seconds and the output cannot reach the desired peak angular value. Accordingly, for the peak output to reach the desired level, the peak value of the reference trajectory has to be increased. The peak error signals indicated by arrows are caused by the inactive relative joints during that time.

Following application of the open-loop control, the humanoid performed basic stair climbing locomotion, which showed some shortcomings. Thus, the left hip (Lhip) joint motor was put into a closed feedback loop with PID controllers. Consequently one more revolute motor was needed for Lhip joint which worked with the previous motor. Another input to control the states of the two motors was used (these were active one at a time). Figure 4 shows the Simulink representation of the control system thus adopted.

Table 3. Open-loop control stratege

	Hip	Ankle	Knee
Foot clearance	Open-loop orientation	Open-loop orientation	Open-loop orientation
Foot placement	Inactive	Inactive	Inactive
Weight acceptance	Inactive	Inactive	Inactive
Pull up	Open-loop orientation	Open-loop orientation	Open-loop orientation
Forward continuous	Orientation=0	Orientation=0	Inactive

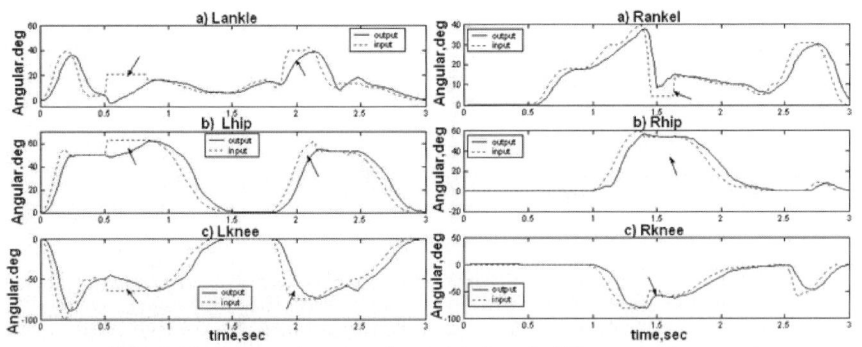

Fig. 3. Reference input and output signal of the open-loop system

Figure 5 shows the closed-loop input and output of Lhip joint, the error signal and the torque produced at this joint during the simulation. Since there is not much offset, the integral component of the PID controller was not used here, and since variation in the error signal was noticeable, the derivative term was used to improve the situation. As noted in Figure 5 the error between the actual and desired trajectory is not significant and the behaviour of the closed-loop system is not much different from the open-loop one. Accordingly, the advantage of the closed-loop control system over open-loop control system is not clearly apparent from this trial. The system is still sensitive to disturbances and parameter variations.

Figure 6 shows the sequence of simulation in VN during one gait cycle. The first step takes 0.56 seconds, whereas the first normal step takes 1 seconds, the second normal step takes 0.8 seconds, and the final step lasts 0.52 seconds.

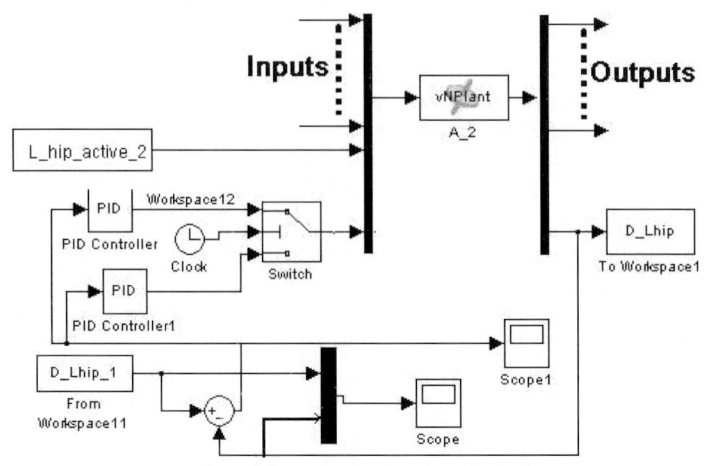

Fig. 4. Closed-loop controls Simulink diagram

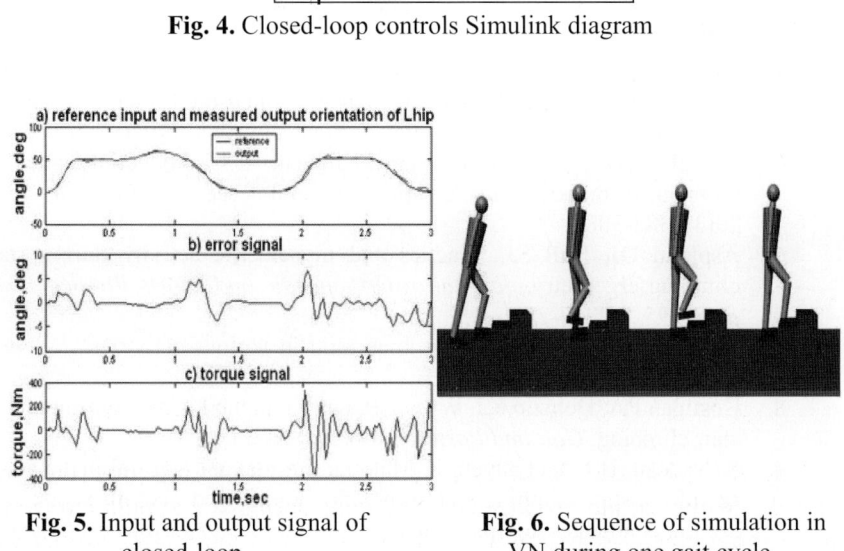

Fig. 5. Input and output signal of closed-loop

Fig. 6. Sequence of simulation in VN during one gait cycle

5 Conclusion

The development of a humanoid model, gait analysis of stair climbing locomotion and simulation and control with VN and Matlab/Simulink have been presented. Some important results have emerged from this study: Firstly, the most important phase of the locomotion is weight acceptance phase, which transfers the central of gravity forward to balance the body and to prepare for the next step, for which it is simple to apply orientation-based motor control on the joints using intuitive strategies; Secondly, the reference input trajectories vary with the rise and the run of the step, but the basic shape will not change significantly. Finally, there is limitation in using a linear actuator, which fixes an object, but also makes the system sensitive to disturbances, so it has to be used carefully.

References

1. Winter D A, *Biomechanics and Motor Control of Human Movement*, 2nd Edition. Wiley-Interscience, New York, 1990.
2. B Yu, M J Stuart, T Kienbacher, E S Growney, K-N An, Valgus-varus motion of the knee in normal level walking and stair climbing, Clinical Biomechanics Vol. 12, No. 5, pp. 286-293, 1997.
3. Andriacchi, T.P., Anderson, G.B.L., Femier, R.W., Stern, D.P.M. and Galante, J.O.A study of lower-limb biomechanics during stair-climbing. *J.Bone Joint Surg.*, 1980, 62A, 749-757.
4. McFadyen, B.J. and Winter, D.A.An integrated biomechanical analysis of normal stair ascent and descent. *J. Biomech.* 1988, 21(9), 733-744.
5. Kowalk, D.L., Duncan, J.A. and Vaughan, C.L. Abduction-adduction moments at the knee during stair ascent and descent. *J. Biomech.*, 1996, 29(3), 383-388.
6. Asplund DJ, Hall SJ. Kinetics and myoelectric activity during stair-climbing ergometry. *Journal of Orthopaedic and Sports Physical Therapy* 1995; 22:247-253.
7. Morrison JB. Function of the knee joint in various activities. *J Biomed Eng* 1969: 573-80.
8. Costigan PA, Deluzio KJ, Wyss UP, Knee and hip kinetics during normal stair climbing, *Gait and Posture* 16 (2002) 31-37.
9. S. Nadeau, B.J. McFadyen, F. Malouin, Frontal and sagittal plane analyses of the stair climbing task in healthy adults aged over 40 years: what are the challenges compared to level walking, *Clinical Biomechanics* 18 (2003) 950-959.
10. J. Shields, R.Horowitz (1998). Adaptive step rate control of a stair stepper exercise machine. Department of Mchanical Engineering, U.C. Berkeley, CA 94720-1740.
11. T. Fuhr, D. Ing, J. Quintern, Dr. med. G. Schmidt, *Stair ascending and descending with the cooperative neuroprosthesis walk*, Institute of Automatic Control Engineering, Munich and Neurological Hospital Bad Aibling, Germany.
12. Costigan PA, Deluzio KJ, Wyss UP, Knee and hip kinetics during normal stair climbing, *Gait and Posture* 16 (2002) 31-37.

Design Issues of Spring Brake Orthosis: Evolutionary Algorithm Approach

M S Huq, M S Alam, S Gharooni and M O Tokhi

Department of Automatic Control and Systems Engineering, University of Sheffield, Mappin Street, Sheffield S1 3JD, United Kingdom.
cop02msh@sheffield.ac.uk

Abstract

Spring Brake Orthosis (SBO) generates the swing phase of gait by employing a spring at the knee joint to store energy during the knee extension through quadriceps stimulation, which is then released to produce knee flexion. Spring parameters (for the knee flexion part) and the stimulus signal parameters (for the knee extension part) are the only optimizable quantities amongst the factors that determine the SBO generated knee joint trajectory. In this work, subject specific optimum spring parameters (spring constant, spring rest angle) for SBO purposes are obtained using genetic algorithms (GA). The integral of time-weighted absolute error (ITAE) between the reference and actual trajectory is defined as the cost function.

The later part of the optimization procedure (second half of the swing phase) identified two potential objective functions: (*i*) the ITAE between the reference (natural) and actual trajectory and (*ii*) the final angular velocity attained by the knee joint at the end of the excursion, which should be as low as possible to avoid (*a*) excessive stimulation, caused by the trajectory requirement, which causes fatigue, (*b*) knee damage. Multi-objective GA (MOGA) is used for this purpose.

Finally, the stimulus signal parameters are optimized for the functional electrical stimulation (FES) driven extending knee for two objectives: (*i*) square of the knee joint orientation nearest to the full extension (0°) during the whole FES assisted excursion and any instant t_c and (*ii*) square of the knee joint angular velocity at t_c, resulting in optimal knee joint trajectory.

Keywords: FES, functional electrical stimulation, hybrid orthosis, SBO, spring brake orthosis

1 Introduction

Functional electrical stimulation (FES) induces muscle contraction through electrical stimulation of the constituent motor neurons [3]. Nowadays, it appears to be a promising means of restoring limited useful movements in paralyzed people, who have lost mobility due to spinal cord injury (SCI), while having their muscles and nerves of the extremities distal to the injured region capable of generating force. But the FES stimulated muscle fatigues very quickly because of the reversed recruitment order of artificially stimulated motoneurons [4]. The consequences are twofold: (*i*) limiting the duration of the FES assisted movement, especially standing and walking, (*ii*) drastic changes in the plant (muscle) properties which pose challenges to the control task.

One of the major approaches to overcome these limitations is to reduce the use of active muscle, where possible, through the use of passive braces [5]. This is classified as 'hybrid orthosis' and combines FES with a lower limb orthotic brace.

1.1 The spring brake orthosis concept

The quadriceps muscles, when artificially stimulated, could produce much more torque than is required just to extend the knee. During knee extension of swing phase, spring brake orthosis (SBO) exploits this feature of the quadriceps through partially storing FES generated quadriceps force as potential energy in a torsion spring attached to the knee joint. A brake is then employed to maintain the knee extension without any muscle contraction, thus reducing fatigue. Then knee flexion is achieved by releasing the brake and letting the spring to return to its resting position (approx $70^0 - 80^0$). The hip flexion is simultaneously produced as a result of consequent shift in the centre of mass (CoM) of the overall leg segment during this knee flexion and is maintained throughout the required duration by applying a brake/ratchet at the hip joint. This results in a hybrid orthosis combining electrical stimulation of quadriceps muscle, spring and brake at the knee joint and brake/ratchet at the hip joint, with the activation of each of them at appropriate instants and for appropriate periods of time.

1.2 Genetic algorithms

Genetic algorithms (GAs), first proposed by Holland in 1975 [6], constitute a class of computational models that mimic natural evolution to solve problems in a wide variety of domains. A GA is formed by a set of individual elements (the population) and a set of biological inspired operators that can change these individuals. According to evolutionary theory only the individuals that are the more suited in the population are likely to survive and to generate off-springs, thus transmitting their biological heredity to new generations.

GAs are a suitable for multi-objective optimization. Due to their population-based nature, they are capable of supporting several different solutions simultaneously. The robustness of the GA in the face of ill-behaved problem landscapes increases the value of their utility. One of the first approaches to utilize the concept of Pareto optimality was reported by Fonseca and Fleming [2], and referred to as multi-objective genetic algorithm (MOGA).

2 Method

2.1 The SBO equipped leg model

A human leg consisting of the following components was modeled through a combination of visualNastran® (VN) software and Simulink®

The body segments viz. trunk, thigh, shank and foot were developed in VN software. The thigh and shank were developed as simple cylinders and the foot as rectangular box. Anthropometric data (mass, dimension, position of CoM, radius of gyration etc.) were obtained using [7].

The hip and knee joints were realized as hinge joints with single degree of freedom (DOF) within the same software environment while the ankle joint was simplified as rigid joint with no DOF.

The spring and the brake were implemented within the VN software environment. Linear torsion spring, modeled as described below, was tested at the knee joint.

The model incorporates the model of a biarticular muscle rectus femoris. It has been realized as an electrically stimulated muscle model which includes the stimulation parameters (frequency, pulse amplitude/width) and length as input and returns the resultant muscle force. This part has been facilitated by 'Virtual Muscle 3.1.5 for Matlab [8] which is embodied as a software package for use with Matlab and Simulink. The instantaneous length of the muscle which is one of the inputs to the muscle model has been fed according to [9].

2.2 The genetic algorithm optimization process

2.2.1 Spring parameters optimization

At the first stage, GA was employed to search for the optimum spring parameters. Reference trajectory for all lower limb joints for normal human gait was obtained from [7] for integral of time-weighted absolute error (ITAE) calculation, used as the objective function.

The GA optimization procedure was initialized with the following parameters:

Number of individuals	40
Number of generations	100
Generation gap	0.8
Crossover rate	0.8
Precision	20 Bit
Mutation rate	0.001

2.2.2 Stimulus signal optimization

In the initial approach, the rectus femoris stimulus signal, i.e. the position of the stimulus burst in time scale and its duration was optimized, for minimum ITAE between the reference and actual trajectory (objective1), and for minimum knee angular velocity at the end of the trajectory (objective2). Similar to the first stage, the normal gait trajectory was used as the reference for ITAE calculation. The second cost function was defined with an idea that the duration of the stimulation should be just enough to drive the knee to its full extension, which suggests the knee joint to reach ideally zero velocity when it reaches maximum extension.

MOGA [2] was utilized to optimize these two parameters. The result is shown in **Fig. 1**. As can be readily seen, the optimization did not achieve a satisfactory result. It was concluded from this that the natural trajectory is probably not the optimal trajectory for SBO operation and thus the need to search for the optimal trajectory.

2.2.3 Knee trajectory optimization

As the variable to be optimized was the trajectory, no reference trajectory was used at this stage to define any of the cost functions. The basic idea was to let the shank move freely under the influence of the stimulating pulse (and gravity, and against the spring) and keeping track of the orientation of the knee joint nearest to full extension during this excursion and the time instant corresponding to that particular orientation, t_c. Thus, the two objectives were defined as:

(*a*) the orientation of the knee joint nearest to full extension during the externally stimulated excursion that occurs at any certain time instant t_c,

(*b*) the squared angular velocity of the knee joint at t_c.

This approach in effect results in a knee joint trajectory that is characterized by these two specified object values, and thus supposed to be the optimal trajectory for this purpose.

3 Results

3.1 Optimization of spring parameters

With the ITAE between the reference and actual trajectory as the cost function to be minimized, the optimization procedure achieved the following optimum values for the spring parameters (corresponding to the minimum ITAE): spring constant (k) = 1.0278 Nm/° and spring rest angle (θ_r) = 76.75°.

3.2 Optimization of stimulus signal

Fig. 2 shows the Pareto-optimal solutions for the stimulation signal optimization process. As can be seen, although it was a minimization problem, both objective functions retain a very high value.

3.3 Optimization of knee joint trajectory

Very high (poor) values of the objective functions in **Fig. 1** led to the idea of trajectory optimization. The Pareto-optimal solutions for this search operation are shown in **Fig. 2**. It can be readily seen that the optimization procedure could achieve useful results. The knee trajectory corresponding to any pair of solutions could be used as the optimal reference trajectory for control operations with SBO.

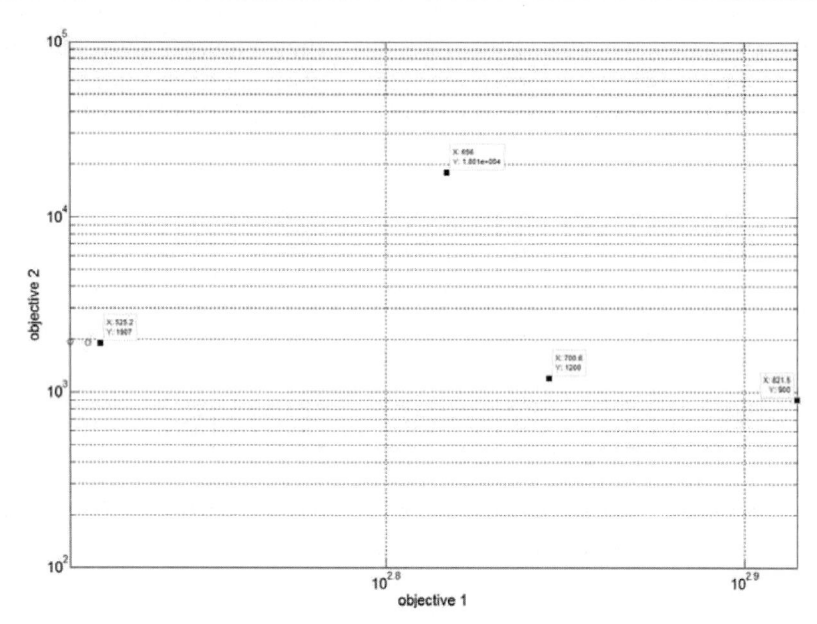

Fig. 1. Pareto-optimal solutions for stimulus signal optimization: Objective1 = ITEA between reference (natural) and actual trajectory, Objective2=knee joint angular velocity at the end of simulation (0.5sec)

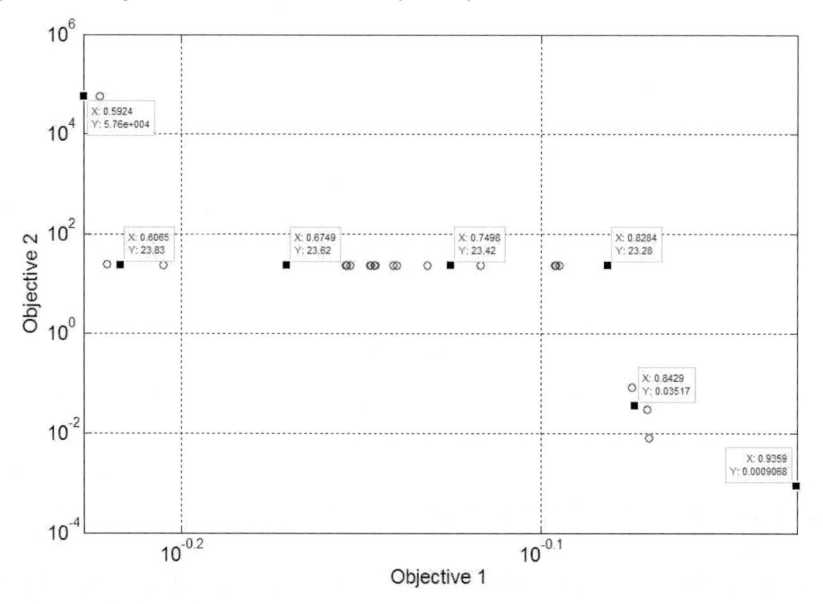

Fig. 2. Pareto-optimal solutions for knee joint trajectory optimization: Objective1=square of the knee joint orientation nearest to full extension during the whole FES assisted excursion and any instant t_c; Objective2=square of the knee joint angular velocity at t_c.

Design issues of spring brake orthosis: evolutionary algorithm approach 87

The stimulus burst defined by one of the solutions in **Fig. 2** along with the optimum spring parameters obtained in Section 3.1 were applied to the leg model. The corresponding results obtained are shown in **Fig. 3**.

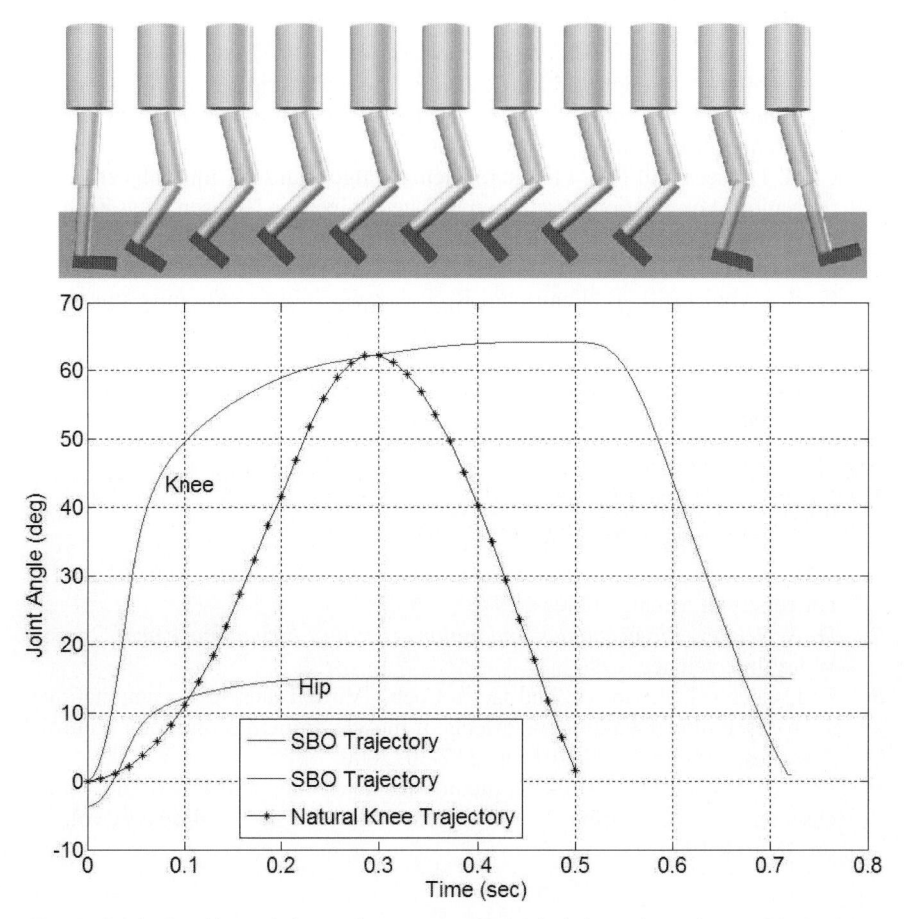

Fig. 3. Optimized knee joint and corresponding hip joint orientations with SBO

4 Discussion and conclusion

An evolutionary algorithm approach for optimization and design of subject specific SBO has been presented. In addition to optimizing the spring, the importance of a suitable optimal trajectory has been revealed and realized. This optimal trajectory is supposed to play an important role in closed-loop

SBO control operations, while the energy usage is a very important issue besides the challenges of control operation.

Reference

1. S. Gharooni, B. Heller, and M. O. Tokhi, "A new hybrid spring brake orthosis for controlling hip and knee flexion in the swing phase," *IEEE Trans Neural Syst Rehabil Eng*, vol. 9, pp. 106-7, 2001.
2. C. M. Fonseca and P. J. Fleming, "Genetic algorithms for multiobjective optimisation: Formulation, discussion and generalization," presented at Fifth International Conference on Genetic Algorithms, Morgan Kaufmann, San Mateo, CA, 1993.
3. G. R. Cybulski, R. D. Penn, and R. J. Jaeger, "Lower extremity functional neuromuscular stimulation in cases of spinal cord injury," *Neurosurgery*, vol. 15, pp. 132-46, 1984.
4. E. Rabischong and D. Guiraud, "Determination of fatigue in the electrically stimulated quadriceps muscle and relative effect of ischaemia," *J Biomed Eng*, vol. 15, pp. 443-50, 1993.
5. M. Goldfarb, K. Korkowski, B. Harrold, and W. Durfee, "Preliminary evaluation of a controlled-brake orthosis for FES-aided gait," *IEEE Trans Neural Syst Rehabil Eng*, vol. 11, pp. 241-8, 2003.
6. J. H. Holland, *Adaptation in Natural and Artificial Systems*. Ann Arbor: The University of Michigan Press, 1975.
7. D. A. Winter, *Biomechanics and motor control of human movement*, 2nd ed: Wiley-Interscience, 1990.
8. E. J. Cheng, I. E. Brown, and G. E. Loeb, "Virtual muscle: a computational approach to understanding the effects of muscle properties on motor control," *J Neurosci Methods*, vol. 101, pp. 117-30, 2000.
9. D. Hawkins and M. L. Hull, "A method for determining lower extremity muscle-tendon lengths during flexion/extension movements," *J Biomech*, vol. 23, pp. 487-94, 1990.

Recent Developments in Implantable and Surface Based Dropped Foot Functional Electrical Stimulators

L. Kenney[1], P. Taylor[2], G. Mann[2], G. Bultstra[3], H. Buschman[4,5], H. Hermens[4], P. Slycke[6], J. Hobby[2], N. van der Aa[5], B. Heller[7], A. Barker[7], D. Howard[1], N. Sha[1]

[1]University of Salford, UK; [2]Salisbury District Hospital, UK; [3]University of Twente, the Netherlands; [4]RRD bv, the Netherlands; [5]TWiN, the Netherlands; [6]Xsens Motion Technologies bv, the Netherlands; [7]University of Sheffield/Sheffield Teaching Hospitals Trust, UK.
l.p.j.kenney@salford.ac.uk

Abstract

One approach to improving the gait of patients with foot drop is the use of functional electrical stimulation (FES) as a neural prosthesis. However, there remain limitations with the current clinically used technology and the paper describes some recent developments addressing some of these problems. The paper describes initial work on an alternative surface-based solution and recent developments of an implantable two channel stimulator.

Keywords: Functional electrical stimulation, drop foot.

1 Introduction

Drop foot is a condition found in significant numbers amongst stroke, cerebral palsy, partial spinal cord injury, multiple sclerosis and other populations. The condition comes about through a loss of central control over movement of the foot and is characterized by weakness in the muscles that dorsiflex (lift) the foot, and/or spasticity in the plantarflexors (muscles that act to lower the foot). This results in reduced dorsiflexion during the swing phase of gait and an inability to achieve heel strike at the beginning of stance phase. Typically at the end of swing phase in drop foot gait the foot lands plantarflexed and sometimes inverted, whereas in healthy gait the foot lands dorsiflexed and slightly everted. Compensatory strategies are required to clear the ground during the swing phase of gait and the net result of these impairments and compensatory mechanisms is gait that is tiring, unsteady and slow.

There are a number of treatment options for drop foot patients, including physiotherapy, use of an ankle foot orthosis (AFO) and Botulinum Toxin (BoTox). The evidence base for the long-term effectiveness of any of these treatments is variable in both quantity and quality and in certain cases, rather poor. An alternative approach, first demonstrated in the early 1960s, is the use of artificially generated electrical pulses to stimulate activity in peripheral muscles, in this case muscles that control the ankle [1].

Functional electrical stimulation is the electrical stimulation of a peripheral nerve to generate nerve impulses (or action potentials) sufficient to produce functional activity in a muscle [2]. The effect of FES using conventional surface mounted electrodes is complex. Stimulation acts not only on the muscles directly supplied by the stimulated nerve, but also, via reflex action, to inhibit or stimulate other muscles. Secondly, when stimulation is delivered via a pair of conventional surface electrodes, the effective stimulation field, the volume within which stimulation levels are sufficient to generate action potentials, is difficult to control in a systematic way by adjusting electrode placement on the skin. This results in difficulties with only recruiting the targeted muscle groups. This problem is termed selectivity and is a well established limitation with surface stimulation.

There are a number of clinical FES systems currently available for patients with drop foot. The vast majority of these are based on the principle originally described by Liberson in 1961. In order to understand the principles of drop foot FES, it is useful to consider the relevant anatomy.

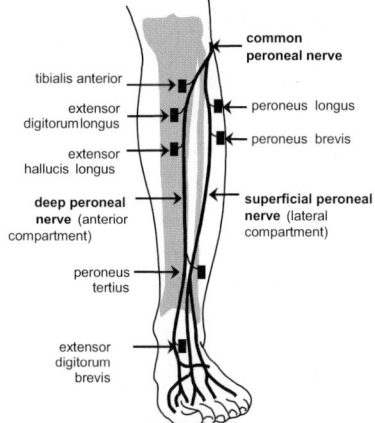

Fig. 1. The common peroneal nerve, showing its major branches and the muscles supplied

The common peroneal nerve divides at the head of the fibula into the deep and the superficial peroneal nerves. These supply both the muscle group that dorsiflexes (lifts) and inverts the foot at the ankle, and the group that everts the foot (see figure 1). Electrodes stimulating the common peroneal nerve are attached to the skin, usually one over the nerve as it passes behind the fibula head, and one over the body of the major dorsiflexor muscle (tibialis anterior).

As drop foot presents most problems during the swing phase of gait, stimulation is triggered at, or near heel off and terminated at, or near heel strike. These events are detected by a footswitch located under the heel of the affected leg. Stimulation is provided by a control box, usually situated on the waist of the user, with the stimulation level user-determined. Other, more complex arrangements are also used in certain instances.

A recent systematic review effectively summarizes the evidence base in stroke patients [3]. FES was shown to improve walking speed and also there is some evidence to show that it reduces physiological cost index (a measure of walking energy expenditure). The treatment is now being used by several thousand patients and compliance rates are high.

Nevertheless, there remain limitations with the single channel surface stimulator. Amongst these are difficulties with electrode placement, muscle selectivity and failures of leads and footswitches. Whilst the problems of external equipment have been reduced through careful design, some limitations are inherent to the surface stimulation approach itself. For example, the common peroneal nerve supplies both the anterior and lateral compartment muscles. Although a limited degree of selectivity is possible by small adjustments to the location of the electrode(s), practically this is difficult to achieve, particularly for stroke patients who often have limited upper limb function on the affected side. For this reason two alternative approaches are discussed in this paper. The two approaches are: to improve the selectivity of surface-based systems; or to implant the electrodes on the nerve(s) themselves. Before describing progress on the implant technology, a brief overview will be given on a project that addresses improving the surface-based approach.

2 Selective surface stimulator

The work described in this section is aimed at producing an intelligent, steerable electrode array-based surface FES system for drop foot patients [4,5] and is based on the idea of a virtual electrode that uses a selectable number of smaller electrodes to create a single larger, reconfigurable and steerable source.

An experiment was conducted at the University of Sheffield to investigate the possibility of using a surface, electrode array-based system to independently stimulate the deep and superficial peroneal nerves to control foot movement in healthy volunteers. The experiments were conducted with the subjects seated with the knee flexed. An array of 64, 6mm electrodes was used. As this was a preliminary study aimed at investigating the potential for the approach, the stimulation was manually controlled. A 16 channel stimulator was built that delivered constant current pulses of 0.2ms at a frequency of 25Hz. Each of the 16 channels could be manually switched to any of the electrodes. Response to stimulation was measured from a magnetic movement tracker located on the dorsal aspect of the subject's foot that measured 3d orientation relative to a nominal reference coordinate system.

A systematic search of combinations of 4 electrodes within the array was carried out. With each configuration, stimulation was increased until movement was detected and then further increased until the subject found the sensation to be intolerable. The set of 4 electrodes that were associated with the minimum current to produce dorsiflexion movement ("dorsiflexion set") was identified, as well as the sets producing inversion and eversion ("eversion set" and "inversion set"). Stimulation was then applied to the "dorsiflexion set" of electrodes until 15 degrees of dorsiflexion was achieved. The foot dorsiflexion response at this level of stimulation was usually accompanied by an unwanted degree of either inversion or eversion. While maintaining stimulation to the "dorsiflexion" electrode set, the stimulation amplitude on either the "inversion set" or "eversion set" of electrodes was increased until a balanced foot response was found, or the experiment was terminated. In 9 out of 12 subjects it was possible to find a satisfactory array configuration to produce a balanced foot response.

Work at Sheffield University has demonstrated the feasibility of the approach and began to explore the design parameters, such as electrode size and electrode-skin interface properties [5] and this is being further developed in a collaboration between the Universities of Sheffield and Salford.

3 Implantable stimulators

Since the early 1970s researchers have also attempted to address the problems inherent to the surface approach through implanting the electrodes directly onto the nerve. As the proximity of the electrode to the nerve is a major factor in determining the response to any given stimulation, this approach clearly holds potential advantage. However, early attempts at implanting a single channel stimulator directly to the common peroneal nerve lacked the selectivity to compensate for changes in the balance response of the muscles [6]. One approach to overcoming the lack of selectivity problem is to introduce a second stimulation site to allow selective stimulation of the 2 branches of the Common peroneal nerve (figure 1). It was postulated that this would allow the inversion associated with stimulation of the deep peroneal nerve to be offset by stimulation of the superificial peroneal nerve which supplies the muscles that evert the foot. In such a way, a more controllable foot motion during the swing phase should be achievable.

3.1 Development of a two channel implantable stimulator

This approach was taken up by a group in the Netherlands who developed a prototype two channel implantable stimulator in the early 1990s and the first prototype was implanted in July 2000 [7]. The stimulator consists of an external transmitter unit and an internal, silicone rubber-encapsulated receiver/electrode unit (figure 2). The transmitter unit that supplies and controls power to the receiver via an inductive link is attached over the the implanted receiver on the lateral side of the leg slightly distal to the knee. Stimulation is triggered and terminated in the same manner as the surface stimulator, using a footswitch under the heel. Amplitude of stimulation to each channel is controlled via a panel on the front of the transmitter.

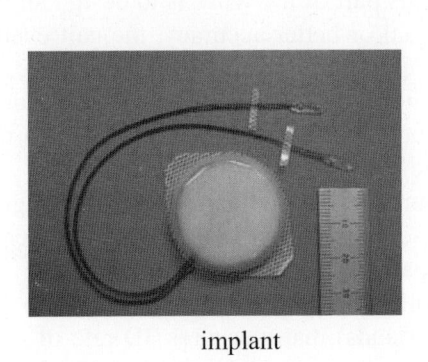

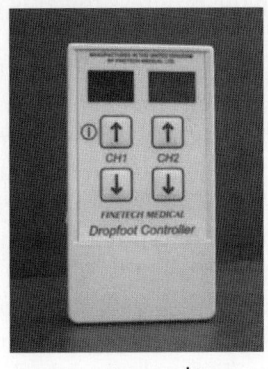

implant   transmitter

Fig. 2. Implantable stimulator system

The first subject to receive the implant (aged 37) had suffered a stroke seventeen years previously that resulted in a dropped foot. Following implantation this subject showed a much improved foot motion during the swing phase and an immediate increase in walking speed and endurance when compared with walking with/without his previous walking aid. The patient continues to use the implant daily. A high degree of selectivity over foot response has also been reported [8].

Although the pilot trials results are encouraging, a positive result from a controlled evaluation is required before the device is widely accepted. A randomized controlled clinical trial of the implantable stimulator is underway in the Netherlands to address this question (http://www.eureka.be/ifs/files/ifs/jsp-bin/eureka/ifs/jsps/projectForm.jsp?enumber=2526). These results will be reported in due course.

3.2 Improvement in functionality and usability

Two projects are underway that aim to further improve the functionality and usability of the current system and this section we will briefly review the progress in both. The first, the TUBA (Transceiver Unit for Biomedical Applications) project is concerned with three areas: The replacement of the conventional copper wound coils in the receiver with planar coils; the replacement of the present footswitch with an inertial system, mounted in the transmitter box and; the development of a more robust transmit-receive system that increases the tolerance of the system to misalignment of the transmitter relative to the receiver.

Progress is being made on all three areas, but in this paper, we will focus on the progress towards the replacement of the footswitch with an inertial sensor system. The goal of this part of the work is to design an inertial system that can register with equal or better accuracy, the gait events detected by the footswitch. These events are heel off and heel strike, corresponding to the start and termination of stimulation. By mounting the inertial system in the transmitter box, the need for additional cabling and complexity for the patient will be eliminated.

In order to define the minimum data set and hence sensor configuration required to reliably identify these events from motion of the lower leg, a series of trials with drop foot patients has been carried out. The inertial sensor system being used to collect the motion data is the MT6 (Xsens Motion Technologies, the Netherlands) that measures 3D rate of turn and 3D linear acceleration. Concurrent footswitch data and stereophotogrammetric data was also collected, to allow the processed inertial data to be compared with "gold standard" gait events [9].

The second project, EUREKA IMPULSE (Improved Mobility through imPlanted fUnctional eLectrical Stimulation of nErves), is concerned with two aspects. The first is the clinical trial of the present stimulator, referred to earlier in this paper, and the second is the development of a system for automatic tuning of the two channel stimulator.

The primary objectives of the new stimulator are to provide adequate dorsiflexion during the swing phase and to orient the foot optimally at the beginning of stance phase. Secondary objectives include the need to control the rate of dorsiflexion to avoid exceeding the stretch reflex threshold, which is often lower in stroke patients than in healthy individuals. Although the current system is open loop controlled, it would be possible to introduce feedback. In order to tune the stimulator to achieve the objectives described earlier, a method of measuring 3D foot motion is required. Accurate measurement of 3D motion of the foot using inertial sensors has now been demonstrated and progress is now being made towards using this data in a closed loop automatic tuning system to optimally set the stimulation parameters on the two channels [10].

4 Discussion and Conclusions

Drop foot is a significant impairment for many thousands of people, leading to reduced mobility and quality of life. The use of surface FES to restore movement to the affected foot has proven of benefit to many patients. The systems currently available are well tolerated and have been shown to increase both walking speed and endurance of patients. Nevertheless, there are certain limitations inherent to the current systems and in order to address some of these, two approaches are being investigated; the use of surface-based array systems to improve selectivity and implantable systems. The surface electrode-array based system has shown promise as a non-invasive approach to the problem, but further work is required to optimise the array design and array search strategy. Another alternative approach is to implant the system. In contrast to earlier implantable systems, the two-channel implantable system described here provides a certain degree of selectivity over the muscles stimulated. The paper has described the design and initial pilot trials of the stimulator. Initial results suggest that the system is of significant benefit to the patients that were implanted and hence further evaluation and development is justified. The paper concludes by describing two projects that will extend this technology further. Further results from these projects will be published in due course.

Acknowledgements

The support of the EU (TMR networks NEUROS and Neural Pro, FPV project TUBA), UK DTI (Medlink project IMPULSE) and Dutch Ministry of Economic Affairs (Eureka project IMPULSE) and the St Hubertus Foundation are gratefully acknowledged. Also, the authors would like to acknowledge the major contribution of Ir Gerrit Bultstra, who sadly died shortly before this paper was submitted.

References

1. Liberson, W.T., et al., Functional Electrotherapy: stimulation of the peroneal nerve synchronized with the swing phase of the gait of hemiplegic patients. *Arch Phys Med Rehab,* 1961. 42: p. 101-105.
2. Rushton, D.N., Functional electrical stimulation. *Physiol Meas.*, 1997. 18(4): p. 241-275.
3. Kottink, A.I., et al., The orthotic effect of functional electrical stimulation on the improvement of walking in stroke patients with a dropped foot: a systematic review. *Artif Organs*, 2004. 28(6): p. 577-86.
4. Sha, N., Development of a steerable electrode array for functional electrical stimulation, submitted in partial fulfilment of MSc, Dept Medical Physics and Clinical Engineering. 2003, University of Sheffield.
5. Sha, N., B.W. Heller, and A.T. Barker. 3D modelling of a hydrogel sheet - electrode array combination for surface functional electrical stimulation. in *Proc 9th IFESS Conf.* 2004. Bournemouth, UK.
6. Waters, R.L., et al., Functional electrical stimulation of the peroneal nerve for hemiplegia. Long-term clinical follow-up. *J Bone Joint Surg.Am,* 1985. 67(5): p. 792-793.
7. Kenney, L., et al., An implantable two channel drop foot stimulator: initial clinical results. *Artif Organs*, 2002. 26(3): p. 267-70.
8. Kottink, A.I., et al., The sensitivity and selectivity of an implantable two-channel peroneal nerve stimulator system for the restoration of dropped foot. *Neuromodulation*, 2004. 7(4): p. 277-283.
9. Findlow, A.H., Kenney,L., Howard,D. Can alternatives to the forceplate be used for accurate detection of key gait events. in *Proc 9th IFESS Conf. 2004*. Bournemouth, UK.
10. Veltink, P.H., et al., Three dimensional inertial sensing of foot movements for automatic tuning of a two-channel implantable drop-foot stimulator. *Med Eng Phys*, 2003. 25(1): p. 21-28.

Fluidically Driven Robots with Biologically Inspired Actuators

S. Schulz, C. Pylatiuk, A. Kargov, R. Oberle, H. Klosek, T. Werner, W. Rößler, H. Breitwieser, G. Bretthauer

Institute of Applied Computer Sciences, Forschungszentrum Karlsruhe GmbH, PO Box 3640, 76021 Karlsruhe, Germany. schulz@iai.fzk.de

Abstract

In this paper different robot applications are presented that are driven by flexible fluidic actuators. These pneumatically driven actuators are biologically inspired. The first robot presented is an eight legged walking machine with 48 compliant joints. Then an auto propulsive flexible endoscope will be presented, followed by a serpent and an elephant trunk.

Keywords: Robot, Bionics, fluidic actuator.

1 Introduction

The idea of using fluidic energy for actuation is a very old one and can be dated back to the Alexandrine Ktesbios (285-247 AC) [1]. For many decades pneumatic and hydraulic actuators are wide spread in industrial applications, such as heavy industries, mechanical engineering, transportation systems and medical engineering.

Fluidic actuators are characterized by a high force to weight ratio, however the efficiency is restricted by frictional loss. Therefore, pneumatic actuators for application in rehabilitation "McKibben artificial muscle" [2] and robotics „Rubbertuator" [3] were presented. These actuators are made from an inflatable inner bladder sheathed with a double helical braid which contracts lengthwise when expanded radially. Thus a contraction principle is the basis for actuation, which can also be achieved by inflating a cascade of chambers made from two or more layers of sealed plastic film [4]. Some more advantages are: elastic energy can be stored in the actuator and inherent compliancy of the joints. However, the robot design possibilities are limited by the length of the linear actuator necessary to perform a required adjustment travel. Typically they contract by 10-20% of their initial length. A different type of fluidic actuators, "pouch type actuators" [5] and "flexible fluidic actuators" [4] are integrated to the joint. Joint actuation is performed by an expansion principle, as an elastic chamber moves the levers of a joint apart, when inflated. Different applications driven by biologically inspired flexible fluidic actuators designed at the "fluidic robot lab" of the Forschungszentrum Karlsruhe, Germany will be presented in three groups in the following chapter.

2 Applications

2.1 Pneumatic spider

The aim of designing an eight-legged walking machine was transferring a biological actuation principle into a mechanical design [6]. The extension movement of the femoro-patellar and the tibio-metatarsal joint in the chilenean red tarantula (Grammostola spatulata) is performed by a hydraulic driving mechanism: A liquid (hemolymph) is pressed from cavities of the body (lacunae) into the joint [7]. It was not intended to copy nature as closely as possible, because some spiders have 30 muscles and 12 DOF in each leg [8]. Therefore the number of actuators has been reduced to three pairs of antagonistically working actuators (Fig.1) in each leg which makes a total of 48 actuators.

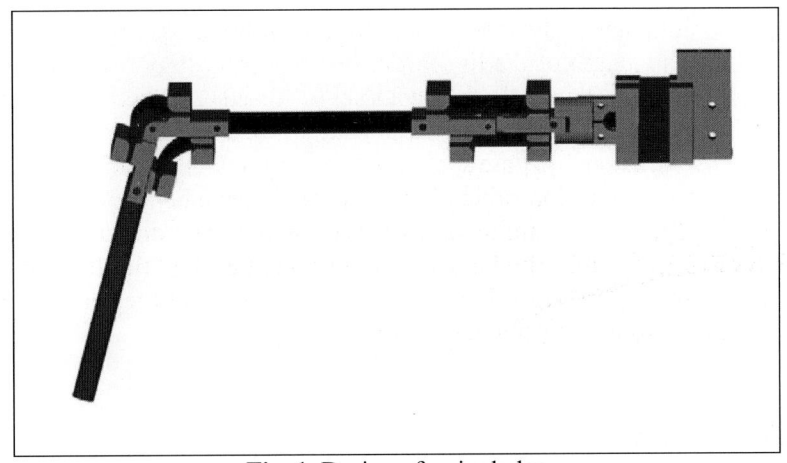

Fig. 1. Design of a single leg

All eight legs are designed with the same geometrical parameters. The proximal joint enabled the leg to move 30° forward and 30° backward from starting position, which was vertical to the bodies' longitudinal axis. The range of motion of the middle and distal joint was 50° and 70° flexion and 20° extension. Stable gait was obtained using a tetrapod gait pattern, that can be observed in a slow walking spider [8]. Due to the compliant joints of the spider's legs forces are distributed resulting in self adaptability to uneven terrain. Stable walking without any sensory information is possible up to a rise of 5% (Fig. 2). Like natural spiders the fluidic walking machine is able to continue walking, even with one or two disabled legs. Pressurized air from a gas tank has been favoured as energy source because of its lighter weight compared to a hydraulic system. The weight of the spider is 12 kg and the body length is 40cm and the span width from the tip of one leg to the contra lateral leg is 160cm when lying flat on the ground. Without external power supply the pneumatic robot-spider is able to walk 15 minutes.

Fig. 2. Pneumatic Spider

2.2 Inch worm colonoscope, serpent and elephant trunk.

Three different robot applications were designed based on the same longitudinal actuation principle. Three actuators were arranged symmetrically around the centre axis. Each actuator consists of a cascade of flexible chambers that can be inflated with pressurized air. When all three actuators are inflated simultaneously the robot elongates. By applying pressure only to one or two of the longitudinally orientated actuators a flexion movement to the opposite direction is performed. It has to be noted that in this case actuation was different to a real inch worm, serpent or elephant trunk, where muscles contract at the longitudinal axis. A prototype of a self propelled colonoscope was designed (Fig. 3) using a combination of this actuation principle and inflatable elements on either end working as „anchors".

Fig. 3. Inch worm Colonoscope

The locomotion principle is inspired by inch worms. The movement sequence starts by inflating the proximal element, resulting in contacting the colon wall and holding the colonoscope in place (Fig. 4, upper row). Then all three of the middle section actuators are inflated, elongating the body and pushing the head along the colon. Next the distal actuator is inflated, holding the head of the colonoscope in place (Fig. 4, middle rows). At the end of the sequence the distal element is inflated connecting the colon wall and the proximal and intersection elements are deflated (Fig. 4, bottom row). Compared to conventional colonoscopy the examiner does not need to manually push the device. Instead the movements are controlled via a joystick. It is expected that the examination will be less painful for the patient and analgesic drugs can be abandoned. The new colonoscope has a diameter of 18 mm.

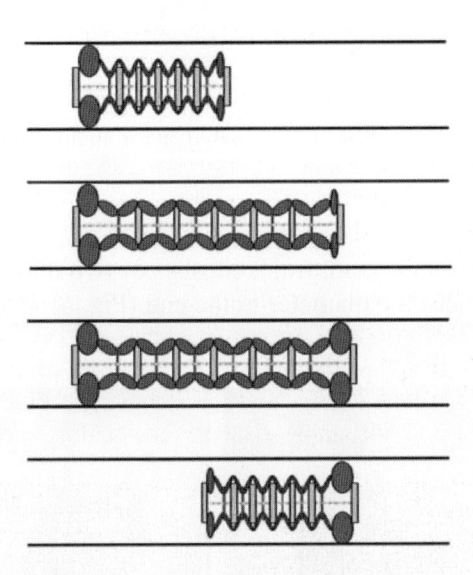

Fig. 4. Inch worm actuation principle

Two larger robots were designed using the longitudinal extension principle. The first one is a serpent with a "neck" that is animated by 9 actuators (Fig. 5). Each actuator can be controlled individually resulting in very realistic movements. The head of the serpent weighs 3kg, so the stiffness and operating pressure (8 bar) of the actuators are much higher than the actuators used in the second robot (elephant trunk).

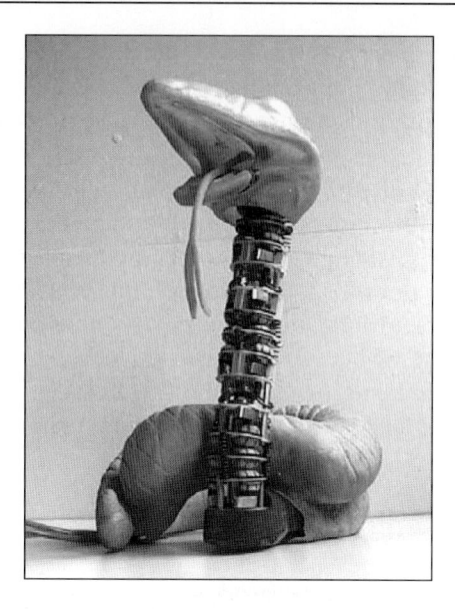

Fig. 5. Serpent with actuated "neck" before colouring.

The design of an elephant trunk consists of two different actuator sizes, as the trunk reduces the diameter to the end (Fig. 6). Flexion up to 360° is possible.

Fig. 6. Two different views of a moving elephant trunk robot.

3 Conclusions

Different robotic applications driven by biologically inspired actuators are presented. For each application different actuators have been designed in order to provide sufficient forces, stiffness, moments and stroke. All robots presented are controlled with open loop. Due to inherent compliance of the actuators the spider robot can walk over smaller obstacles and the inch worm colonoscope can move in a mechanically sophisticated surrounding. This reduces manufacturing costs and accounts for robustness. As positioning of flexible fluidic actuators is difficult, suitable angular sensors and closed loop control are under development. However, the robots presented are designed to be controllable without sensor integration.

Acknowledgements

The authors want to thank the German Federal Ministry of Education and Research for their financial support (BMBF framework programme Biotechnology).

References

1. Maskrey RH, Thayer WJ (1978) *A brief history of electrohydraulic servomechanisms.* ASME J. Dynamic Systems Meas. Control, pp. 110-116.
2. Schulte H F Jr (1961) *The characteristics of the McKibben artificial muscle.* In: *The Application of external power in prosthetics and orthotics.* National Academy of Sciences-National Research Council, Washington D. C., Appendix H, pp. 94-115.
3. Inoue K (1988) *Rubbertuators and applications for robots.* Robotics Research: The 4th International Symposium, Bolles R, Roth B (eds). MIT Press, Cambridge, Mass. p.57-63, May 1988, Univ. of California, Santa Clara, California, United States.
4. Schulz, S.; Pylatiuk, C.; Bretthauer, G.(1999) *A New Class of Flexible Fluidic Actuators and their Applications in Medical Engineering.* Automatisierungstechnik, Vol. 47, No. 8, pp. 390-395.
5. Kato I. (1985) Development of Waseda Robots - The Study of Biomechanisms at Kato Laboratory, Tokyo, Waseda University publication dedicated to I. Kato 60th anniversary jubilee.
6. S. Schulz, C. Pylatiuk, G. Bretthauer: *Walking Machine with compliant joints.* 4th International Conference of Climbing and Walking Robots (CLAWAR 2001), 24th-26th September 2001, Karlsruhe, Germany.

7. Parry, D.A.; Brown, R.H.J. (1959) *The hydraulic mechanism of the spider leg.* J. Exp. Biol. 36, pp. 423-433.
8. Sens, J. (1996) *Funktionelle Anatomie und Biomechanik der Laufbeine einer Vogelspinne.* Master Thesis, (in German), University of Saarland, Saarbrücken, Germany.

Climbing, Navigation and Path Planning

Concept for Energy-autarkic, Autonomous Climbing Robots

Werner Brockmann

University of Lübeck, Institute of Computer Engineering, Ratzeburger Allee 160, 23562 Lübeck, Germany, brockman@iti.uni-luebeck.de

Abstract

In order to build fully autarkic climbing robots, the energy consumption is of great importance. While the energy needed for locomotion is dictated by accelerating masses, the energy needed for adhesion depends on the design of the robot. Passive suction cups are a promising approach here because they are low cost, simple and robust and allow a light-weight construction of climbing robots. But in order to construct a proper system, the behaviour of passive suction cups has to be understood. This concerns the analysis of holding forces versus pressing forces as well as forces to pull the strap. Another important aspect is the height change of a suction cup caused by pulling forces. This introduces elasticity into the system in a non-linear manner. Out of this, some design considerations are derived for climbing robots with passive suction cups which are a step towards energy-autarkic climbers.

Keywords: climbing robots, energy-autarky, passive suction.

1. Introduction

In many practical applications, autonomy of a mobile robot means that the robot not only acts autonomously, i.e. it navigates and fulfils its task without any human intervention. But the robot should also carry its own power supply in order to operate without an umbilical cable. While this energy-autarky is not a severe problem for wheeled mobile robots, it is a real challenge for climbing robots because the weight of a climbing robot is of tremendous importance, and hence also the weight of the power supply. For energy-autarkic climbing robots it is thus essential to keep the weight and the energy consumption as low as possible.

One substantial part of energy consumption is required for holding a climbing robot at the surface, e.g. a wall. Besides magnetic adhesion and van-der-Walls forces, suction cups are most commonly used which are evacuated actively by a vacuum pump. Most machines use either a single large suction cup [1-3] or multiple small suction cups on each foot [4,5]. Alternatively, the body of a climbing robot is used as a large suction cup with a wheeled drive-system underneath [6,7]. In general, these suction cups are evacuated actively by at least one vacuum pump which is mounted on the robot. This is called "active suction" or "active suction cups", respectively, hereinafter.

Vacuum pumps make climbing robots not only noisy. They increase the weight and the costs of a robot, also due to additional vacuum tubes, muffles, valves, and so forth. But what is more important for energy-autarkic robots, this solution causes a more or less steady and not negligible energy consumption. Hence, it is desirable to avoid an active vacuum generation and a separate installation for vacuum transportation.

For sufficiently flat surfaces like windows, a few number of climbing robots use "passive suction cups" made of elastic material, i.e. vacuum cups which are evacuated simply by pressing them to the surface (see for instance Fig. 1a). In this way, the vacuum is generated indirectly by utilizing the robot´s locomotion system without a specific vacuum pump, and no energy is consumed for adhesion. On more or less clean surfaces, the vacuum is kept for a longer period of time. Freeing a passive suction cup by tearing it off the surface needs large forces which cause mechanical stress and thus finally a rigid and heavy mechanical construction. A better solution is to release the suction cups in a controlled manner by a specific mechanical construction which opens small (passive) valves to inflate the cups [8]. But this construction needed to handle the suction cups tends to get complicated and not only increases the weight of the climbing robot. It also leads to severe kinematic restrictions.

2. Climbing with passive suction cups

In [9] a way of passive suction was introduced for the first time which causes nearly no kinematic restrictions and is flexible and light-weight. Therefore passive suction cups with a strap are used, as Fig. 1b shows. The suction cup is pressed to the surface without pulling the strap. By that, it is evacuated and produces hence an adhesive force. The strap is pulled to release the vacuum before a suction cup is to be lifted, in order to avoid any remaining adhesive force. This pulling may be done e.g. by a small and light-weight RC-servo or solenoid.

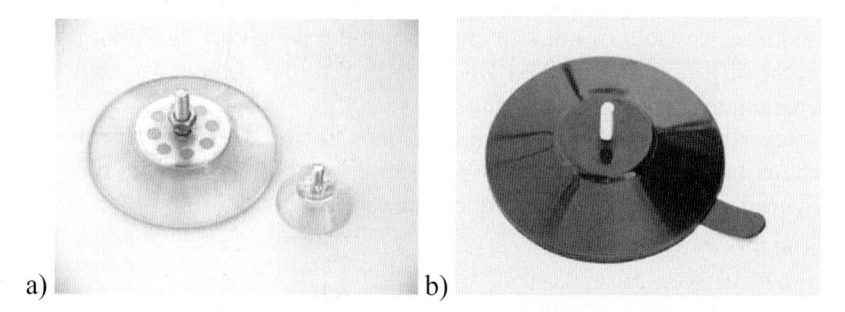

Fig. 1. Passive suction cups a) without and b) with a strap

Because passive suction cups do not limit the dexterity of climbing robots, conventional kinematics can be used and only electrical power is needed which is easily delivered by accumulators for fully autarkic climbing robots.

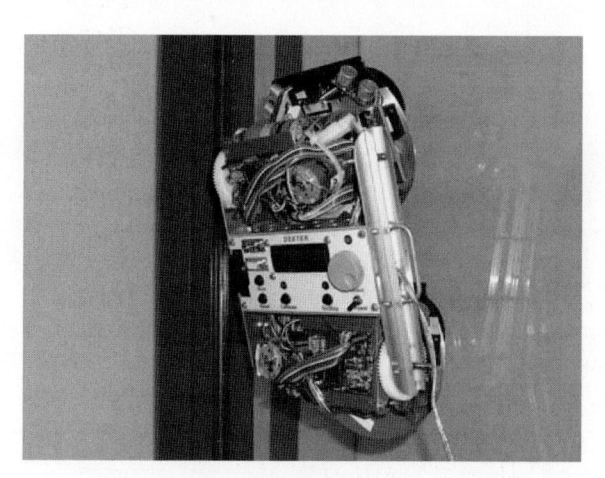

Fig. 2. DEXTER climbing at a window

The climbing robot *DEXTER* (*DEXTerous Electric Rover*), see Fig. 2, built at our institute is as a first practical example showing that this kind of passive suction can be directly combined with kinematic structures and the same dexterity known from climbing robots using active suction. Its locomotion principle is similar to e.g. [3], except that the body is not articulated. It has four degrees of freedom, i.e. each foot is capable of rotating around two axes. This enables DEXTER to move on an arbitrarily inclined surfaces in two modes, i.e. by turning over and by swiveling around the holding foot. By that, it is also capable of crossing from the floor to a wall and from a wall to another wall or to the ceiling. In total, it has a size of 36,5×22×13 cm^3 and a weight of about 3 kg. For this robot, a single suction cup is capable of holding eight times the weight of the robot and has a weight of a few grams. And DEXTER has three of them on each foot (in an arrangement according to Fig. 5a).

Nevertheless, passive suction cups show a specific behaviour which has to be considered when designing a climbing robot. Their main characteristics are thus described in the following.

3 Behavior of passive suction cups

3.1 Acting forces

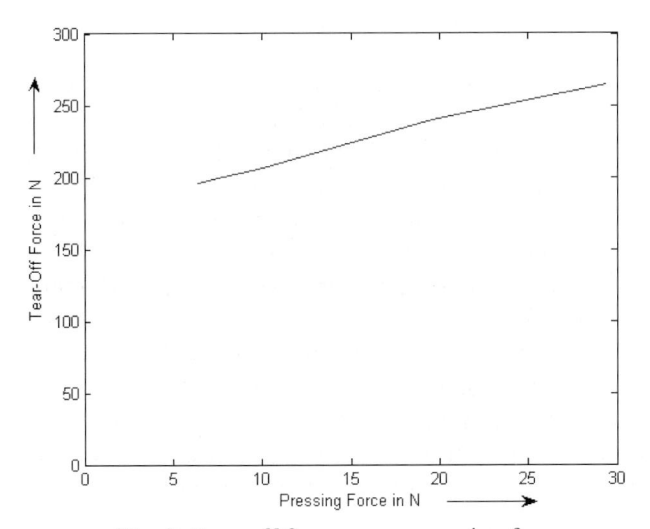

Fig. 3. Tear-off force versus pressing force

The passive suction cup of Fig. 1b has an outer diameter of 75 mm and a weight of 23 g. It is mounted simply by a threaded pin. The achievable holding force (measured perpendicular to the surface) is given by the vacuum and the effective area underneath the suction cup. In theory it is 433 N at maximum. In practice it depends on how strong a cup has been pressed to the surface, i.e. the pressing force, and hence on the remaining amount of air underneath the cup. Fig. 3 shows this dependency and gives the maximal pulling force, called the tear-off force hereinafter, with respect to the pressing force.

These measurements were taken for a clean glass surface which is the best case, in order to figure out the potential of passive suction cups. A pressing force of 5 N is needed to bend the rubber material of the suction cup such that it sucks to the surface. But even with this small pressing force, the cup is able to hold up to 200 N. Increasing the pressing force increases also the tear-off force. A limit of 270 N is reached when the pressing force is about 30 N because the top of the cup touches the surface and the remaining air underneath it cannot be evacuated further. Overall a strong amplification of the pressing force is achieved in this way and no energy is needed to hold a cup and a robot, respectively, at a surface. Furthermore, no specific energy is needed to press a cup to the surface if a rest of kinetic energy is used to push a foot to the surface. Such a pushing movement is dampened by the elasticity of the cup itself.

Adhesion also leads to a friction force which prevents the robot from slipping (down). It depends primarily on the vacuum underneath the cup, but also on the ground material, especially its friction coefficient. But more importantly, the vacuum is also depending on the pulling force. A pulling force results in an increased volume and consequently in an increased vacuum. Thus the adhesive force is increased also. It further seems as if there is a larger indenting of the cup material with the ground due to an increased vacuum. This enlarges the friction force additionally. In practice, this effect depends on the initial volume and thus also on the pressing force. Nevertheless, the robot may slip, although a suction cup is not torn off. The safety factor, i.e. the number of suction cups, should hence be designed such that slippage does not occur even in the worst case.

To release a suction cup, the strap has to be pulled. The force needed to pull it depends on the vacuum and is in the range of 0.4 N. It may be done by an ordinary RC-servo. In case of the first climbing robot DEXTER, the weight of the RC-servo to pull three straps at a time is only 7g. Passive suction cups are hence very attractive for climbing robots.

3.2 Height Change

The volume underneath a suction cup depends on the acting forces as discussed before. The height of the cup thus varies, especially depending on the pulling as well as pressing force. The effect of the pulling force on the height change outweighs the effect of the pressing force as Fig. 4 shows. While the height of the suction cup is about 10 mm for any pressing force

of time. Additionally, requirements concerning flatness and cleanness of the surface, on which the robot is climbing, may be a bit harder than for active suction. Possible applications hence range from cleaning and inspection of windows and facades or even of whole greenhouses to climbing robots for education, entertainment, and hobby [10], probably with a focus on the latter ones. Nevertheless, investigations have to be done concerning how to deal with elasticity and different surface conditions.

References

1. White TS, Hewer N et. al. (1998) SADIE, A climbing robot for weld inspection in hazardous environments. In: *Proc. 1st Int. Symp. Mobile, climbing and walking robots – CLAWAR'98*, Brussels: 385-389
2. Yoneda K, Ota Y et. al. (2001) Design of light-weight wall climbing quadruped with reduced degrees of freedom. In: Berns K, Dillmann R (eds.) *Proc. 4th Int. Conf. Climbing and walking robots – CLAWAR 2001*, Prof. Engineering Publ., Bury St. Edmunds, London: 907-912
3. Lal Tummala R, Mukherjee R et. al. (2002) *Climbing the walls*. IEEE Robotics & Automation Magazine, Vol. 9, No4: 10-19
4. Hirose S, Kawabe K (1998) Ceiling walk of quadruped wall climbing robot NINJA-II. In: *Proc. 1st Int. Symp. Mobile, climbing and walking robots – CLAWAR'98*, Brussels: 143-147
5. Gimenez A, Abderrahim M, Balaguer C (2001) Lessons from the ROMA I inspection robot development experience. In: Berns K, Dillmann R (eds.) *Proc. 4th Int. Conf. Climbing and walking robots – CLAWAR 2001*, Prof. Engineering Publ., Bury St. Edmunds, London: 913-920
6. Xu D, Liu S et. al. (1999) Design of a wall cleaning robot with a single suction cup. In: Virk GS, Randall M, Howard D (eds.) *Proc. 2nd Int. Conf. Climbing and walking robots – CLAWAR'99*, Prof. Engineering Publ., Bury St. Edmunds, London: 405-411
7. Weise F, Köhnen J et. al. (2001) Non-destructive sensors for inspection of concrete structures with a climbing robot: In: Berns K, Dillmann R (eds.) *Proc. 4th Int. Conf. Climbing and walking robots – CLAWAR 2001*, Prof. Engineering Publ., Bury St. Edmunds, London: 945-952
8. Schraft, R.D.; Simons, F. (2003) *Facade and Window Cleaning Robots for Cleaning and Inspection*. CLAWAR News, No. 11, 12-14
9. Brockmann, W.; Mösch, F.: Climbing Without a Vacuum Pump. To appear in: *Proc. 7th Int. Conf. on Climbing and Walking Robots – CLAWAR2004*, Madrid, 22.-24.9.2004
10. Brockmann, W.: Towards Low Cost Climbing Robots. *Proc. 1st CLAWAR/EURON Workshop on Robots in Entertainment, Leisure, and Hobby*, IHRT – Vienna University of Technology, Vienna, Austria, 2004, ISBN 3-902161-04-3, 13-18

Navigation of Walking Robots: Path Planning

B. Gaßmann, M. Huber, J. M. Zöllner, and R. Dillmann

Forschungszentrum Informatik, Haid-und-Neu-Str. 10–14, 76131 Karlsruhe, Germany
{gassmann,huber,zoellner}@fzi.de, dillmann@ira.uka.de

Abstract

Proper navigation of walking machines in unstructured terrain requires a path planning algorithm that reflects the flexibility of the robots movements. This paper discusses the problem and presents a real time capable path planning algorithm for walking machines that also makes use of their climbing abilities.

Keywords: walking robot, navigation, path planning

1 Introduction

The focus of this paper lies on the task of path planning: A collision free route of the central body has to be generated which connects a given start point with a given goal point. The base for the path planner is a geometrical 2½D model of the environment build out of a 3D model by projection [1]. The search for a path is a special planning problem which complexity scales with the degrees of freedom of the mobile system and the number of restrictions. Restrictions are: obstacles, optimality criteria, unknown terrain, computational cost.

[2] presents a path planner for walking robots which rates the robot position on the load capacity of the ground. The physical possibilities of the robot are not taken into account. A great inspiration for this work is the potential field based approach presented in [3]. They present a potential guided path planner based on a hierarchical rating approach. Their simple model of the walking robot is also not sufficient to meat the

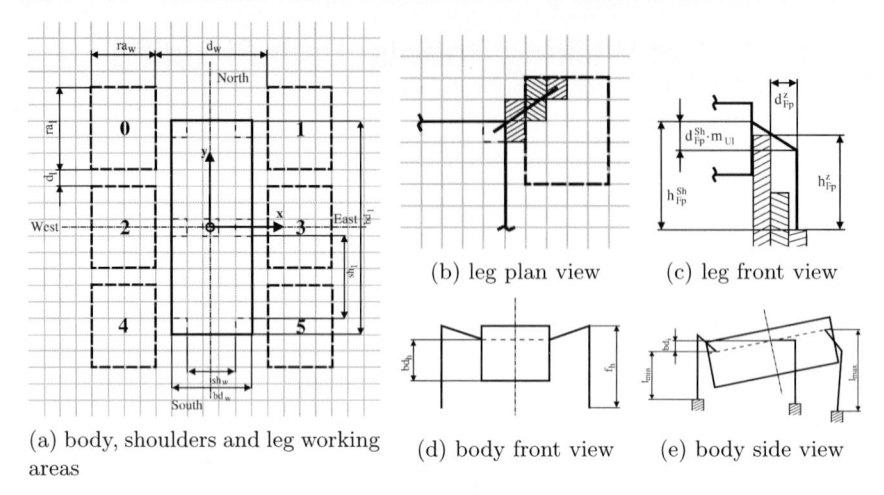

(a) body, shoulders and leg working areas
(b) leg plan view
(c) leg front view
(d) body front view
(e) body side view

Fig. 1. The walking machine model

flexibility of walking robots. The authors neglect the reciprocal action of the legs that makes it e.g. possible to climb stairs. The lifting altitude and the height of the cells have both to be taken in account like realized by [4]. There a high level planner searches a statically stable path which then is verified or rejected by a detailed foot level planner. [4] propose a complete approach which uses a "best first" rating of the path. So the resulting path surely is possible but has not to be optimal in any way. The foot level planning as verification is a desirable extension, but its computation is extremely expensive.

2 Generic Walking Machine Model

For path planning purposes an insect like walking robot could be modelled as illustrated in **Fig. 1(a)**. The robot consists of a bd_l cell long and bd_w cell wide main body, n_l equal symmetrically arranged legs with identical non-overlapping working areas of $ra_w \times ra_l$ cells with d_w respectively d_l cells between each other. The shoulders are modelled as single cells separated by sh_w respectively sh_l cells. The symmetry coordinate system is located in the middle cell of the central body with the discrete cell index (cg_x, cg_y). There are four discrete robot orientations possible, whereas the symmetry allows the consideration of only *North* and *East*.

The height bd_h of the lower part of the main body is measured from the shoulder plane. The height of the lower leg is f_h (see **Fig. 1(d)**). The legs have identical minimum and maximum support heights l_{min} and

l_{max}, which are used to determine the minimum and maximum distance between the footprint and the shoulder. The model assumes that these support heights are guaranteed all over the working area of the legs. The proposed walking machine model also respects the possibility of inclining the main body performing a *pitch* rotation which is given by bd_i mm, the maximum height difference of two successive shoulders (see **Fig. 1(e)**).

Furthermore this approach models the interaction of the legs with the environment. The lower leg is assumed to be vertical and remain in the borders of the foothold cell. The knee, a one degree of freedom (DOF) joint, connects this to the upper leg which itself is connected to the 3 DOF shoulder. **Figures 1(b)** and **1(c)** illustrate the positioning of the right front leg in plan view. The upper leg crosses some (hatched) cells which potentially could influence the minimum possible height of the corresponding shoulder. This is the cell z upon which the upper leg lies with the biggest gradient m_{ul} in respect to the foothold cell Fp adding the lower leg f_h. With the height difference h^z_{Fp} and the lateral distance d^z_{Fp} between z and Fp as well as the distance d^{Sh}_{Fp} between shoulder and foothold, the preliminary minimum shoulder height above the foothold calculates to:

$$h'^{Sh}_{Fp} = d^{Sh}_{Fp} \cdot m_{ul} + f_h \quad , \text{ where } m_{ul} = \frac{h^z_{Fp} - f_h}{d^z_{Fp}}. \tag{1}$$

Because the main body is able to incline it has to be taken into account that the position of the shoulder cell could vary. So the contact cell z could also differ. That requires to build the maximum over all cells z_q that come into question. Moreover the height of the shoulder could be restricted by the minimum support height l_{min} of the leg. Finally the minimum shoulder height h^{Sh}_{Fp} and its absolute h^{Sh}_a then compute to

$$h^{Sh}_{Fp} = \max(l_{min}, \max_{z_q}(h'^{Sh}_{Fp})) \quad \text{and} \quad h^{Sh}_a = h^{Sh}_{Fp} + h^{Fp}_a. \tag{2}$$

3 Rating function

The heart of each path planning algorithm is the base on which the decision is made if a position and orientation of the robot is valid or not. [4] guarantee at least one valid configuration of the feet but do not say anything about the quality. Following [3] this approach chooses a potential field method with the potentials representing the quality of a position in the full range of [0, 1]. But especially the inadequate rating of the 3D terrain complexity $\widetilde{A}_t$ of each position in [3] is changed completely.

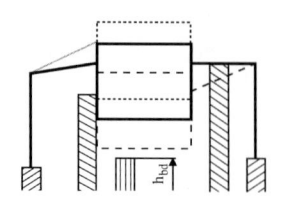

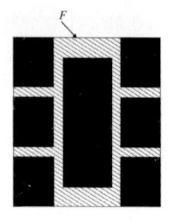

Fig. 2. (a) Shoulder height restrictions (b) Influential boundary and Free Area

3.1 Posture Availability A_p

Assuming a robot position and orientation $p = (cg_x, cg_y, cg_\gamma)$ and the positions of the legs are chosen. For all legs i the height of their foothold cell $h_a^{Fp_i}$ and their minimum shoulder height $h_a^{Sh_i}$ following the computation in equation (2) are known.

Because a *roll* rotation of the main body is not modelled, neighbouring shoulders could be handled together as *shoulder axes* $\overline{Sh_j}$ with index $j = \lfloor \frac{i}{2} \rfloor$. The heights *one* shoulder axis could adjust depend on the height of the foothold cells, the heights of the shoulders and the support heights of the legs (see **Fig. 2(a)**). The maximum height of a shoulder axis depends on the lower foothold cell and the maximum support height:

$$h_{\max}^{\overline{Sh_j}} = \min_{k \in \{2j, 2j+1\}} (h_a^{Fp_k}) + l_{\max}. \qquad (3)$$

The minimum height of a shoulder axis could either be restricted by leg obstacles and the minimum support height (both considered in the minimum shoulder height $h_a^{Sh_k}$) or by obstacles below the main body with height $h_{\max}^{bd}$:

$$h_{\min}^{\overline{Sh_j}} = \max(\max_{k \in \{2j, 2j+1\}} (h_a^{Sh_k}), h_{\max}^{bd} + bd_h) \qquad (4)$$

If $h_{\min}^{\overline{Sh_j}} > h_{\max}^{\overline{Sh_j}}$ is true the posture is physically not realizable. The **posture availability** A_p is set to zero. Otherwise with $h_{\min}^{\overline{Sh_j}} \leq h_{\max}^{\overline{Sh_j}}$ it has to be examined if the body is able to maintain the given shoulder heights. A straight line has to be found which lies in between the minimum and maximum heights of the shoulder axes not extending the maximum body inclination. Thereby it is not necessary to compute the straight line explicitly. $A_p(p) = 1$, if a corresponding straight line exists, otherwhise 0. That calculation assumes that all legs of the robot are on the ground although three legs would suffice for a statically stable stand. This form of general stability is chosen for an easier and faster calculation.

3.2 Terrain Accessibility A_t

The posture availability A_p gives a statement on the validity of a given combination of footholds. But each foothold cell could freely be chosen out of its working area which leads to $(ra_l \cdot ra_w)^{n_l}$ possible combinations. For performing a real-time online path planning where more than one main body position has to be examined, the computational cost for the **terrain accessibility** A_t has to be reduced.

Cells of one working area with similar heights and similar minimum shoulder heights could be treated all at once. Therefore the cell $p_\mathbb{P}$ is defined as **representative** of a **cell area** $\mathbb{P}$, whereas $p_\mathbb{P}$ has only few differences in its height $h_a^\mathbb{P}$ and its minimum shoulder height $h_a^{Sh_\mathbb{P}}$ in order to all cells of $\mathbb{P}$.

The second idea for reduction is given by the concept of *ordinal optimisation* [2], which states that considering a limited number of samples of footholds gives likely a "good enough" solution. Following that it is possibly enough to check the posture availability only on some of the combinations of footholds. Therefore the set $\mathbb{M}_i$ of all cell areas of a working area i is split into subsets $\mathbb{T}_{i,j}$ of size C. Herein C defines the **degree of combination**. Like that each working area consist of $n_i^T = \lceil \frac{|\mathbb{M}_i|}{C} \rceil$ subsets. A reduction of the complexity only takes place if not all subsets of different working areas are combined with each other. A working plan could follow:

1. Determine the cell areas $\mathbb{P}$ of all working areas i.
2. Split $\mathbb{M}_i$ into n_i^T subsets $\mathbb{T}_{i,j}$ according the degree of combination C.
3. For each i choose a subset $\mathbb{T}_{i,j}$.
4. Combine all cell areas of different subsets with each other and calculate for each combination c the posture availability $A_p(c)$.
5. For all working areas i with highest n_i^T: remove the subset from $\mathbb{M}_i$.
6. Back to 3 until all $\mathbb{M}_i$ are empty.

According to this each subset $\mathbb{T}_{b,j}$ of the set $\mathbb{M}_b$ with maximum cardinality $n_b^T = \max_i(n_i^T)$ takes part exactly in one combination procedure. Therefore n_b^T combination procedures take place. The amount m_j of single combinations $c_{j,l}$ (with $l \in \{1, \ldots, m_j\}$) of one combination procedure j is determined by the cardinality of the contributing subsets:

$$m_j = \prod_{i=0}^{n_l - 1} |\mathbb{T}_{i,j}|, \quad \text{with } |\mathbb{T}_{i,j}| \leq C. \tag{5}$$

One single combination $c_{j,l}$ represents the plurality of posture availability checks of all cells of the chosen cell areas $\mathbb{P}_i^{c_{j,l}}$ which counts to

$$n_{c_{j,l}} = \prod_{i=0}^{n_l-1} |\mathbb{P}_i^{c_{j,l}}| \quad , \text{ whereas } \mathbb{P}_i^{c_{j,l}} \in \mathbb{T}_{i,j}. \quad (6)$$

The terrain accessibility of a robot position p could then be calculated:

$$A_t(p) = \frac{\sum_{j=1}^{n_b^T} \sum_{l=1}^{m_j} n_{c_{j,l}} \cdot A_p(c_{j,l})}{\sum_{j=1}^{n_b^T} \sum_{l=1}^{m_j} n_{c_{j,l}}} \quad (7)$$

3.3 Terrain Complexity $\overline{A_t}$

The amount of A_t in equation (7) gives a statement on how good a place on the terrain is accessible for a walking robot. To define a repulsive potential the codomain has to be inverted. Therefore the **terrain complexity** $\overline{A_t}$ is defined as $\overline{A_t}(p) = 1 - A_t(p)$ with $\overline{A_t}(p) \in [0,1]$.

3.4 3D-Terrain Complexity $\widetilde{A_t}$

The function $\overline{A_t}(p)$ could be extended to allow in special situations a faster rating of the terrain by a precocious exclusion of extreme obstacles and the cells in their direct surrounding. To avoid a collision of the robot with a too high obstacle it has to be ensured that the obstacle does not enter the rectangle which surrounds the main body and the working areas of the legs (see **Fig. 2(b)**). The intersection of that rectangle with the cells of the working area and the central body leads to the **free area** $\mathbb{F}$. In each of these areas the highest cell p_{max}^i, $p_{max}^\mathbb{F}$ and p_{max}^{bd} with heights h_{max}^i, $h_{max}^\mathbb{F}$ and h_{max}^{bd} are determined. Then the following situations are checked:

1. $p_{max}^\mathbb{F}$ extends all cells p_{max}^i distinctly: $h_{max}^\mathbb{F} > \max_i(h_{max}^i) + l_{max}$.
2. p_{max}^{bd} extends all cells p_{max}^i distinctly: $h_{max}^{bd} > \max_i(h_{max}^i) + l_{max}$.
3. One of the cells p_{max}^i extends the cells of other working areas distinctly. Taking the inclination into account, test each two cells p_{max}^i and p_{max}^j: $\max(h_{max}^i, h_{max}^j) + l_{min} > \min(h_{max}^i, h_{max}^j) + l_{max} + bd_i \cdot |\lfloor \frac{i}{2} \rfloor - \lfloor \frac{j}{2} \rfloor|$.

The walking machine is in an influential boundary of a too high obstacle if one of these three tests evaluates positive. With that the **3D terrain complexity** $\widetilde{A_t}$ of a position $p = (cg_x, cg_y, cg_\gamma)$ gets $\widetilde{A_t}(p) = 1$, if p lies in an influential boundry, $\widetilde{A_t}(p) = \overline{A_t}(p)$, otherwise.

(a) LAURON III walking over a gap

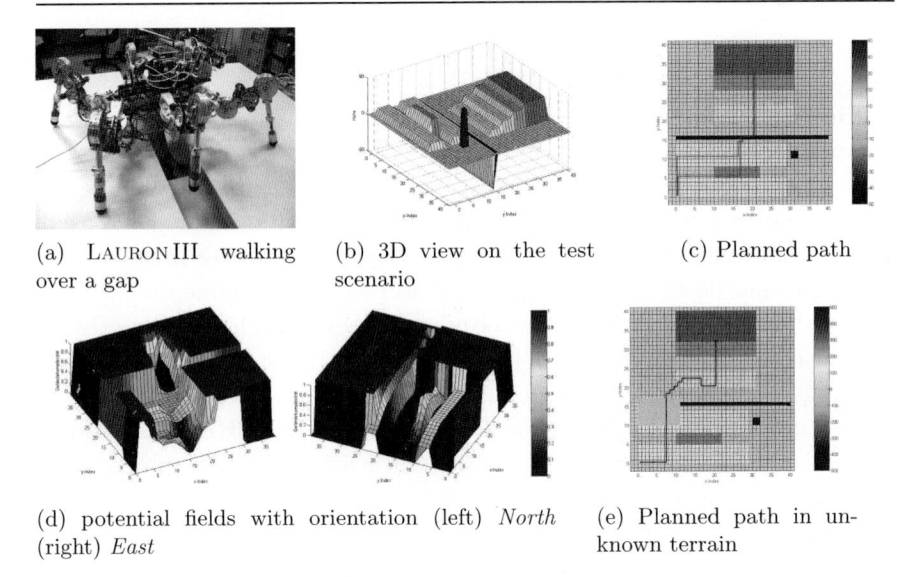

(b) 3D view on the test scenario

(c) Planned path

(d) potential fields with orientation (left) *North* (right) *East*

(e) Planned path in unknown terrain

Fig. 3. Test scenario, 3D Terrain complexity and planned path

4 Potential field

The potential function over the terrain is defined by the weighted sum of the repulsive potential $U_r(p)$ given by the rating function and the attractive potential $U_a(p)$ given by the goal. The potential is searched by a best first planning following the gradient. The basic structure of the potential field algorithm directly follows [3]. The extensions introduced by this work are explained briefly: To reflect the turning of the robot this work regards four robot orientations. The 3D terrain complexity is stored in the environment map so that later planning calls are more efficient. If the planner approaches a part of the map with only few known cells, the rating returns a constant 3D terrain complexity value e, the exploration value. By lowering that value the robot could be motivated to explore the environment, by increasing e the robot avoids the unknown.

5 Experimental Results

Several experiments have been done to prove the functionality of our described approach. For demonstrating the planning abilities the robot LAURON III had to walk over a 1 m deep gap of different sizes (see **Fig. 3(a)**), had to climb steps and to walk through doors with different openings.

Experiments covering the time consumption of this planning approach also have been done.

Figure 3(b) illustrates a complex scenario: Steps in the upper part of the map with the goal point on the little plateau on the top, a one cell wide gap, a high pillar and a view obstacles at the lower part where the starting point is located. A cell in the map has the dimension of 5×5 cm. In **Fig. 3(d)** the calculated potential fields are shown: The areas around the pillar and besides the steps are not accessible, the influential borders forbid that. The steps are rated with low values near the middle of steps with the orientation *North*. Whereas climbing the steps while sideways walking is not permitted. Depending on the factors K_r and K_a different ways from start to goal have been selected (see **Fig. 3(c)**). On adding unknown terrain to the map (the grey cells in **Fig. 3(e)**) the planner was still able to plan a way through the unknown area. The approach of [3], which was also implemented, didn't even allow the crossing of the gap.

6 Summary and Outlook

This paper presents an approach for path planning of walking robots. Based on a generic model of a walking machine the rating of the 3D terrain complexity is described. The posture availability states the validity of a given foothold combination. The terrain accessibility rates how good a position is accessible for the robot. The (3D) terrain complexity then extends this rating for the use with a potential field path search algorithm. The proposed path planner models robot inclination and is able to handle unknown environment. Above all the plan could be computed online on the walking robot and is real time capable in terms of seconds.

References

1. Gaßmann, B., Frommberger, L., Dillmann, R., and Berns, K. (2003) Improving the walking behaviour of a legged robot using a three-dimensional environment model. In *Proc. of the 6th CLAWAR*, pp. 319–326.
2. Chen, C.-H. and Kumar, V. (1996) Motion planning of walking robots in environment with uncertainty. In *Proc. of the ICRA*, pp. 3277–3282.
3. Bai, S., Low, K. H., and Teo, M. Y. (2002) Path generation of walking machines in 3d terrain. In *Proc. of the ICRA*, pp. 2216–2221.
4. Eldershaw, C. and Yim, M. (2001) Motion planning of legged vehicles in an unstructured environment. In *Proc. of the ICRA*, pp. 3383–3389.

Study on Mobility of Connected Crawler Robot by Using GA

Sho Yokota[1], Kuniaki Kawabata[2], Pierre Blazevic[3], and Hisato Kobayashi[1]

[1] Hosei University, Tokyo Japan yokota_sho@ybb.ne.jp
[2] RIKEN(The Institute of Physical and Chemical Research), Saitama Japan kuniakik@riken.jp
[3] Laboratoire de Robotique de Versailles, Versailles France

Abstract

This paper shows a relationship between the number of crawler links and its climbing ability. We consider a robot that consists of serially connected crawlers and try to let it climb up a step. To derive this relation, the robot has to climb step as high as possible. Therefore each joint require optimized motion to the step. The approach is that; each joint angle function is expanded by the Fourier series, and the coefficients and period are derived by a Genetic Algorithm. While we can get each maximum climb-able step height corresponding to the number of links, we can make clear the relation that the height of the climb-able step increases as the number of crawler links increases.

Keywords: mobility, connected crawler, rough terrain, Genetic Algorithm, Fourier Series, ODE.

1 Introduction

There are many researches for development of rough terrain mobile robot. The main theme common to these researches is to improve the mobility performance on rough terrain. In order to achieve this main theme, many research adopted connected crawler mechanism, which crawler links were connected by some joints. Lee designs two links one joint type connected crawler mechanism that uses two triangular crawlers, and shows the high mobility performance by the comparison of climb-able step height between proposed mechanism and a usual one track type[1]. "Souryu-III" is the connected crawler robots of 3 links 2 joints type, and it shows high mobility in some basic experiments such as climbing a step and stepping over a gap[2]. "MOIRA" is 4 links 3 joints type connected crawler, and the experiment reports the maximum climb-able step height.[3]. As mentioned above, the mobility performance was improved by differnet number of links in each researches. Although such research is here, there is no case which shows the relationship between the number of links and mobility performance. When a connected crawler mechanism is designed, there isn't design guidelines like whether how many links should be appropriate, no research mentioned this problems. It's a big problem that there are no discussions in spite of the important matter that the number of links is related to the mobility performance.

Therefore this research tries to demonstrate the relationship between the number of links and mobility performance which isn't cleared until now. Especially, in this paper, we set the environment as one step, and derive the relationship(**Fig. 1**). Because the climbing step ablitiy is important as one of the most fundamental mobility index [4], climbing step experiment is taken by many researches as a evaluation experiment of mobile mechanisms on rough terrain.

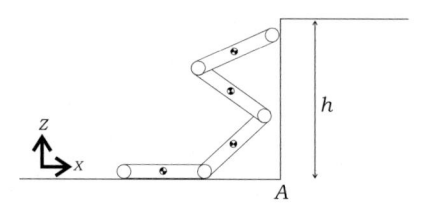

Fig. 1. The assumed model

2 Problem and approach when the maximum climb-able step height of n-links crawler is derived

There is an optimization problem of the joint motions, when the maximum climb-able step height of the n-links connected crawler is derived. Here, n means the number of links of connected crawlers. Because if the joints can't realize suitable motion for the step, the robot may not exercise climbing potential maximally. It is expected this problem is complicated caused by the increase in the number of links. For example, although it is possible to solve the optimized joint motion analytically in case of about 2 links, in case of more 3 links, it's impossible to solve it analytically. Thus the searching method is appropriate for this problem rather than analytical method. However, the round robin-like method of searching isn't so realistic, because the amount of search becomes fatness and calculation time becomes enormous. Therefore, we can think about the following idea as one of the approach to this solution. If a certain approximate function can express an optional joint angle function in a few parameters, that can get the required joint motion in a shorter time than a round robin-like search. Therefore we try to express each joint angle function by using the approximate function.

3 Proposed method

In this section, we will mention how to search that approximate function and parameters, and show how to derive maximum climb-able step height.

3.1 The approximate function

The approximate function must be possible to differentiate twice, so as to find an angular velocity and angular acceleration. It is also required that the function is periodic, and has a few parameter, and contains boundary condition. Therefore, Fourier series is useful as a function which satisfy these conditions [5]. Thus, in this paper, Fourier series approximates a joint angle functions. And the equation (1) is Fourier series for this approximation,

$$\theta_k(t) = \sum_{i=0}^{j} \alpha_i \cos \frac{i}{T} 2\pi t + \sum_{i=1}^{j} \beta_n \sin \frac{i}{T} 2\pi t \qquad (1)$$

Here, k means the number of joints, j refers to the number of oder of Fourier series, T means the period. α_i, β_i, T are parameters which are searched.

3.2 Searching for parameters of the Fourier series

Searching for each coefficient and period of the Fourier series is one of the searchig problem which derive the optimized answer in wide area. Many reseaach proposed to use GA for such this problem [6]~[7]. Because, GA is able to find a comparatively excellent answer in the utility time, and fit various problems. Therefore this paper searches the unknown coefficient of the Fourier series by GA, too. In this paper, we use simple GA [8], and set following parameters(**Table 1**).

We also set the equation (2) for evaluating the chromosomes.

Table 1. Parameters of GA

Number of chromosomes	10
Gene length for a coefficient[bit]	10
Crossover rate[%]	25
Mutation rate[%]	1

$$E = h + \frac{1}{t} + \frac{\sum_{i=1}^{n} x_i + \sum_{i=1}^{n} z_i}{1000} \quad (2)$$

Thus, h is the step height which the robot could climb up, t is the time for climbing up a step. Then, it is understood that the evaluation of the gene is high when the robot could higher step in shorter time. On the other hand, when the robot couldn't climb a step, we set $h = 0$, $t = 100$ as a penalty. However, in this conditions, the evaluation of gene which couldn't climb up a step becomes equal, and it makes difficult to execute crossover. Therefore the third clause of the equation (2) exists as the valuation item. Here, x_n, z_n are the center of gravity coordinates of each link. Thousand of the denominator is numerical value to scale it 1000 down to prevent from that the evaluation when the robot couldn't climb a step, and some links are being lifted high is higher than it when robot could climb a lower step.

3.3 The methode to derive the maximum climb-able step height

In order to evaluate gene, we have to appropriately acquire the position of center of gravity of each link and distinguish whether the robot could climb or not. The consideration of the interaction with the environment are very important, because mobility performance of the mobile mechanism concerns with topography characteristic closely [4]. Therefore we must consider dynamics and an interaction between robot and environment. Thus we adopt ODE(Open Dynamics Engine)[9] to calculate these value. In this paper, we derived maximum climb-able step height by integrating ODE and GA. And the calculation process is shown in **Fig. 2**.

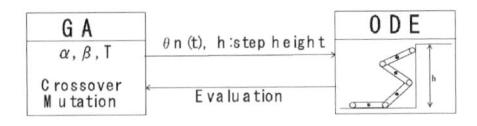

Fig. 2. Overview of the simulation system

GA gives joint angle and step height, and ODE calculate dynamics. After that, ODE distinguishes whether or not the robot could climb, and returns the evaluation to GA. GA makes a gene evolve based on the evaluation, and optimizes angle function of joint. Then the robot could become higher step in shorter time. A robot is considered to climb a step when all in the center of gravity of each link is higher than the height of the step h and it is on the right of A (in **Fig. 1**).

4 Deriving the maximum climb-able step height of n-links

In this section, we derived maximum step height of n-links based on the method mentioned in previous section. we set the conditions and assumption as follows.

- Full length robot body is $L = 2[m]$, and Total robot mass is $M = 2[kg]$ (**Fig. 3**).
- Radius of crawler sprocket is 0.1[m].
- The crawler velocity is constant 0.1[m/s].
- The actuators have enough torque for driving joints.
- Initial each joint angle is 0.0[rad].
- The range of each joint is -2.0~2.0[rad].
- The range of Fourier coefficients is -2.0~2.0.
- The range of Fourier Period is 10~60[sec].
- The height of step h is 0.5~2.0[m].

4.1 Deriving results

First, we derived maximum climb-able step height of 4-links. The number of generations is 500 and order of Fourier series is 5. The results are shown

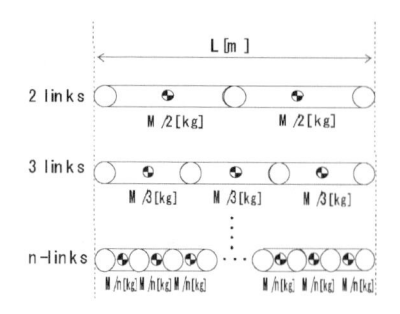

Fig. 3. Definition of the robot size

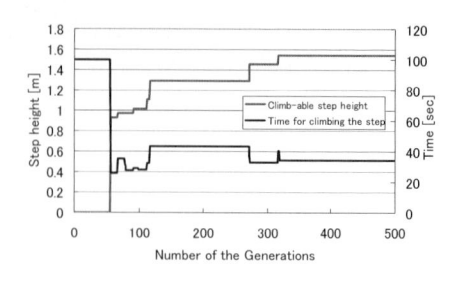

Fig. 4. Transition of the climb-able step height derived by GA (4link)

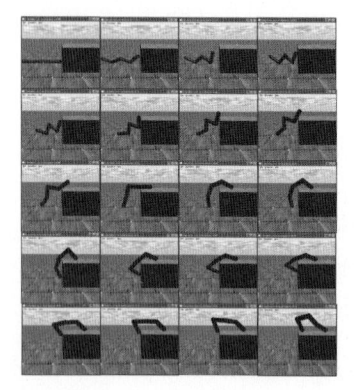

Fig. 5. Connected robot climb the step by using optmized joint motion derived by GA(4 links, h= 0.9[m], 56 generations)

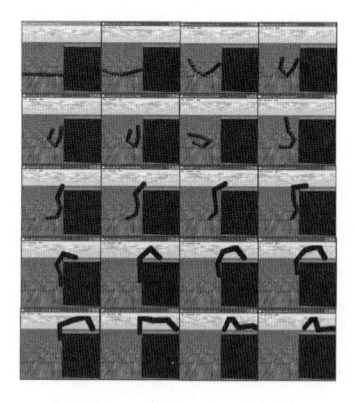

Fig. 6. Connected robot climb the step by using optmized joint motion derived by GA(4 links, h=1.5[m], 500 generations)

in **Fig. 4**~**Fig. 6**. In the **Fig. 4**, we can confirm that the robot could climb higher step when the number of generations is increased. The time required to climb the step also has been shortened. **Fig. 5** and **Fig. 6** show the robot was climbing the step. In **Fig. 5**, the height of step is 0.9[m] and the unmber of generations is 56. In **Fig. 6**, the height of step is 1.544[m](this is the maxiimu climb step height of 4 links).

We also derived the maximum step height of 2~10 links by same way. The results are shown in **Fig. 7**. It was confirmed that the robot could climb higher step when the number of generations is increased as well as the case of 4 link, and maximum climb-able step height at each link was derived.

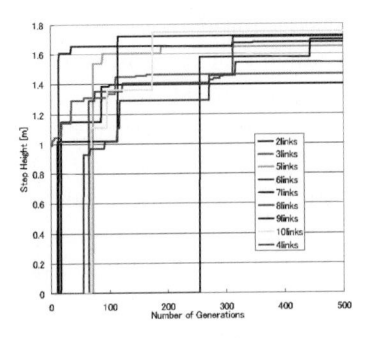

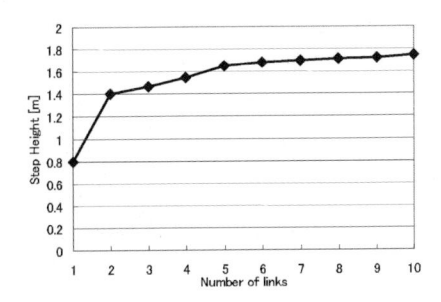

Fig. 7. Transition of the climb-able step height derived by GA (2inks 10links)

Fig. 8. Relation between the number of links and climb-able step height

4.2 The relationship betewwn the number of links and mobility

Since the maximum climb-able step height of each link was shown by **Fig. 7**, the relationship between the number of links and mobility performance of connected crawler is demonstrated. This relationship is shown in **Fig. 8**. There are a few changes in maximum climb-able step height after 6 links, and it can confirm that it reaches saturation obviously from this graph. Consequently, when the connected crawler shows high mobility performance, the appropriate number of links is more than 2 links and less than 5 links.

From the above results, **Fig. 8** of this paper showed the relationship between the number of links and mobility performance for the first time. And this new knowledge can be thought to provide important design guidelines about the connected crawler mechanism.

5 Conclusion

In this paper, we proposed the method which found the maximum climb-able step height of the connected crawler mechanism, and derived the relationship between the number of links and mobility performance which can be one of the design guidelines. The conclusions of this paper are as follows.

- The relationship between the number of crawler and mobility was cleared.

- Though mobility performance was raised caused by the increase in the number of links, too, that rate of change was small in comparison with the one before 5 link after 6 link, and confirmed that it was small as for the effect of the improvement in mobility performance to arise by increasing the number of links.
- A joint angle function was approximated by the Fourier series and the parameter was searched by GA, and it had the problem of the parameter search substituted for the problem of the orbit search.

References

1. C.H. Lee, S. H. Kim, S. C Kang, M.S.Kim, Y.K. KwakF "Double -track mobile robot for hazardous environment applications"C Advanced Robotics, Vol.17, No,5, pp.447-495, 2003
2. T. Takayama, M. Arai, S. HiroseF "Development of Connected Crawler Vehicle "Souryu-III" for Rescue Application"C Proceedings of the 22nd Annual Conference of the Robot Society of Japan, 3A16, 2004 (*in Japanese*)
3. K. Osuka, H. KitajimaF "Development of Mobile Inspection Robot for Rescue Activities:MOIRA"C Proceedings of the 2003 IEEE/RSJ Intl. Conference on Intelligent Robots and Systems, pp.3373-3377, 2003
4. T. Inoh, S. Hirose, F. MatsunoF "Moblility on the erregular terrain for rescue robots"C Proceedings of robotics symposia, 1B2, 2005 (*in Japanese*)
5. Y. YokoseCT. IzumiF "Minimization of Disspated Energy of a Manipulator with Coulomb Friction using the GA Increasing the Caluculated Genetic Information Dynamically"CTransaction of JSCES, Paper No.20040024, 2004 (*in Japanese*)
6. O.A.MohammedCG.F.UlerF "A Hybrid Technique for the Optimal Design of Electromagnetic Devices Usign Direct Search and Genetic Algorithms"CIEEE Trans. on Magnetics, 33-2, pp.1931-1937, 1997
7. Y.YokoseCV.Cingosaki, H.YamashitaF "Genetic Algorithms with Assistant Chromosomes for Inverse Shape Optimization of Electromagnetic devices"CIEEE Trans. on Magnetics, 36-4, pp.1052-1056, 2000
8. S. Kobayashi, M. YamamuraF "Search and Learning by Genetic Algorithms"C Journal of the Robotics Society of Japan, Vol.13, No,1, pp.57-62, 1995 (*in Japanese*)
9. R. Smith "Open Dynamics Engine", http://ode.org/

A Robot that Climbs Walls using Micro-structured Polymer Feet

Kathryn A. Daltorio[1], Stanislav Gorb[2], Andrei Peressadko[2], Andrew D. Horchler[1], Roy E. Ritzmann[3], and Roger D. Quinn[1]

[1] Biologically Inspired Robotics Laboratory, Department of Mechanical and Aerospace Engineering, Case Western Reserve University, 10900 Euclid Ave., Cleveland, OH 44106-7222, USA, *rdq@case.edu*;
[2] Evolutionary Biomaterials Group, Department Arzt, Max-Planck-Institute for Metals Research, 70569 Stuttgart, Germany, *s.gorb@mf.mpg.de*;
[3] Department of Biology, Case Western Reserve University, 10900 Euclid Ave., Cleveland, OH 44106, USA, *roy.ritzmann@case.edu*

Abstract

Insect-inspired foot materials can enable robots to walk on surfaces regardless of the direction of gravity, which significantly increases the functional workspace of a compact robot. Previously, Mini-Whegs™, a small robot that uses four wheel-legs for locomotion, was converted to a wall-walking robot with compliant, conventional-adhesive feet. In this work, the feet were replaced with a novel, reusable insect-inspired adhesive. The reusable structured polymer adhesive has less tenacity than the previous adhesive, resulting in less climbing capability. However, after the addition of a tail, changing to off-board power, and widening the feet, the robot is capable of ascending vertical surfaces using the novel adhesive.

Keywords: small insect-inspired adhesive wall-climbing robots

1 Introduction

Robots that could climb smooth and complex inclined terrains like insects and lizards would have many applications such as exploration, inspection, or cleaning [1]. Cockroaches climb a wide variety of substrates using their active claws, passive spines, and smooth adhesive pads [2]. Beetles and Tokay geckos adhere to surfaces using patches of microscopic hairs that provide a mechanism for dry adhesion by van der Waals forces [3]. In-

(a) (b)

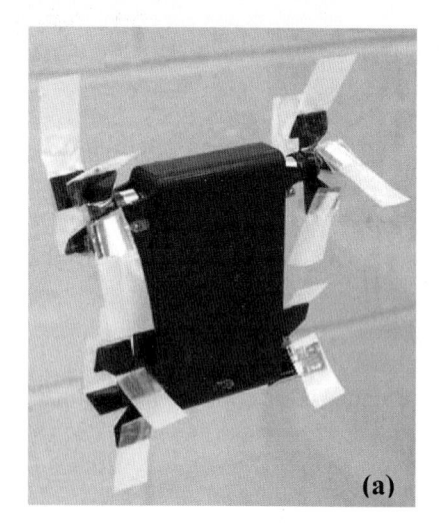

Fig. 1. Mini-Whegs™ 7 on vertical glass (a) with office tape feet and (b) with micro-structured polymer feet and 25 cm long tail (tail not shown).

spired by these animal mechanisms, new adhesives are being developed [4][5] and robots are needed to test them. Waalbots, for example, test various climbing mechanisms using traditional adhesives in preparation to testing biologically inspired adhesives [6].

Observations of insects can also provide inspiration for the kinematics of the legs of a robot. Flies make initial contact with the entire broad, flexible attachment organ (pulvilus) [7]. A slight shear component is present in the movement, which provides a preload to the surface of the attachment device. Similar shearing motion has been previously described as a part of the attachment mechanism of a single gecko seta [3]. Minimal force expenditure during detachment is also important. Disconnecting the entire attachment organ at once requires overcoming a strong adhesive force, which is energetically disadvantageous. This principle of contact formation with the entire pad surface and peeling-like detachment has been applied here to the design of a robot with climbing ability (**Fig. 1**).

2 Mini-Whegs™

Mini-Whegs™ are a series of small robots that use a single motor to drive their multi-spoke wheel-leg appendages for locomotion [8]. The spokes allow Mini-Whegs™ to climb over larger obstacles than a vehicle with similarly sized wheels. We previously developed a Mini-Whegs™ that can be used to test new bio-inspired adhesive technologies for wall climbing [9].

Mini-Whegs™ 7 (5.4 cm by 8.9 cm, 87 grams) is power-autonomous, radio-controlled, and has a total of four wheel-legs, each with four spokes. The feet are bonded to contact areas on the ends of the spokes and the flexibility of the feet acts as a hinge between the feet and spokes. The feet contact the substrate, bend as the hub turns, peel off the substrate gradually, and spring back to their initial position for the next contact. We previously reported that this robot can climb glass walls and ceilings using standard pressure sensitive adhesives [9]. This paper describes results for that robot walking on glass walls and ceilings using adhesive feet made from office tape and the adaptation of that vehicle so that it climbs walls with a biologically-inspired material.

3 Bio-inspired Materials

Two polymer samples were tested, a smooth one and an insect-inspired surface-structured one. Both samples are made of two-compound polymer polyvinylsiloxane (PVS) (President® light body, Coltene, Switzerland). The smooth samples (thickness = 0.4 mm) were molded from a clean glass surface. The structured samples made of the same polymer were obtained from the company Gottlieb Binder GmbH & Co. KG (Holzgerlingen, Germany). The base thickness of the structured sample was approximately 0.4 mm. The protrusions were about 100 µm high and about 40 µm in diameter. Young's modulus, E, of the bulk polymer is 2.5 to 3 MPa [4].

3.1 Traction Properties On Glass

The tangential forces, *i.e.* traction, of a 1.5 cm by 3.5 cm flat sample had typical stick-slip behavior with a maximum force of 0.25 N/cm^2 and zero minimal force. For the structured sample, maximal force was 0.11 N/cm^2. After series of 3–4 trials, the flat sample no longer attached to the substrate properly. After cleaning with water, the traction ability was recovered, showing that traction of the flat sample is sensitive even to slight contamination. For the structured sample, such sensitivity to contamination was not evident.

3.2 Adhesion Properties On Glass

Peeling testing was performed for characterisation of adhesive properties of structured samples. Peeling is delamination of a thin film from the substrate under action of a loading force, F, acting under an angle Θ to the substrate. In the experiment, the peeling force, needed for delamination of the polymer tape (flat and structured), was measured.

PVS samples (25 mm width) were attached to a clean smooth glass surface in a horizontal position and loaded with a weight. Then, the tilt angle of the glass was increased by steps of 2.5° until peeling occurred (**Fig. 2a**). The normalized equilibrium force, F/b, is plotted versus the peeling angle, Θ, in **Fig. 2b**.

The Kendall model of peeling was applied to estimate the adhesion energy [10]:

$$\left(\frac{F}{b}\right)^2 \frac{1}{2Ed} + \left(\frac{F}{b}\right)(1 - \cos\Theta) - R = 0 \qquad (1)$$

where F is the peeling force, d is the thickness of the adhesive, b is the width of the tape, E is the elastic modulus of the film material, Θ is the peeling angle, and R is the energy required to fracture unit area of an interface. The adhesion energy, R, for the structured material was 0.90 J/m^2 and 0.49 J/m^2 for the flat material. These tests demonstrate that structuring does benefit the polymer's adhesion at this range of peeling angle. However, similar testing of new Scotch® tape yields approximately 16 times the adhesive of the structured polymer.

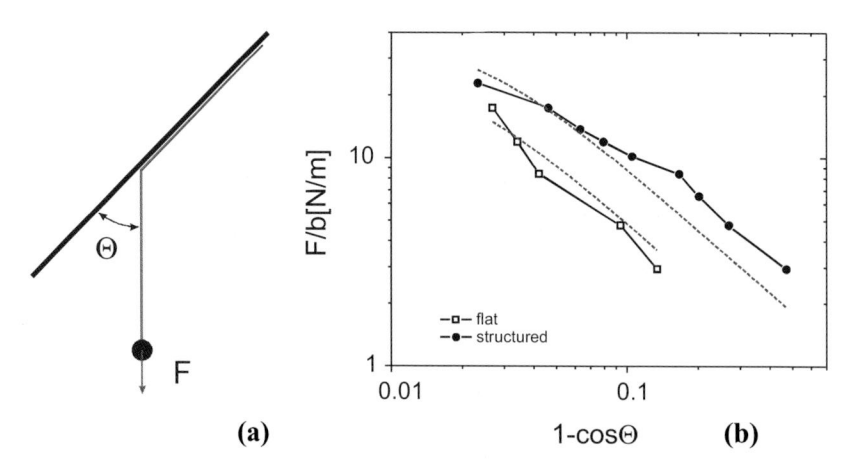

Fig. 2. (a) Diagram of the peeling experiment. (b) Normalized equilibrium force, F/b, versus peeling angle, Θ, for flat and structured materials. Dashed lines indicate fit corresponding to Kendall's model of peeling [10].

4 Robotic Climbing Failure Modes

There are two fundamental modes of failure for a surface-climbing robot. First, the robot can slip along the substrate due to insufficient tangential (traction) forces. Second, there may be insufficient normal (adhesive) forces, causing the robot to tumble away from the wall. Support behind the back axle (*e.g.* a tail) can reduce the likelihood of tumbling while increasing the tendency of slipping.

The vehicle falls backward from the wall when the feet on the front axle are not tenacious enough to support the normal force, N_1, required to balance the moment of the weight. By summing moments about the rear foot contact point (**Fig. 3a**), the magnitude of N_1 for a vehicle without a tail is

$$N_{1_{NoTail}} = \frac{hW}{a} \text{ (tensile)} \tag{2}$$

Therefore the 0.23 N of normal force is required for the 87-gram robot, with $a \approx$ wheelbase of 7 cm, and $h \approx$ leg length of 1.9 cm. Since the supportable normal force decreases with peeling angle, the critical position occurs when the wheel-hub on one side has just peeled up one foot and has not yet applied the next foot. At that instant, the foot on the other side of the axle must be capable of supporting the moment of the robot's weight otherwise the robot will rotate away from the wall, inhibiting proper foot

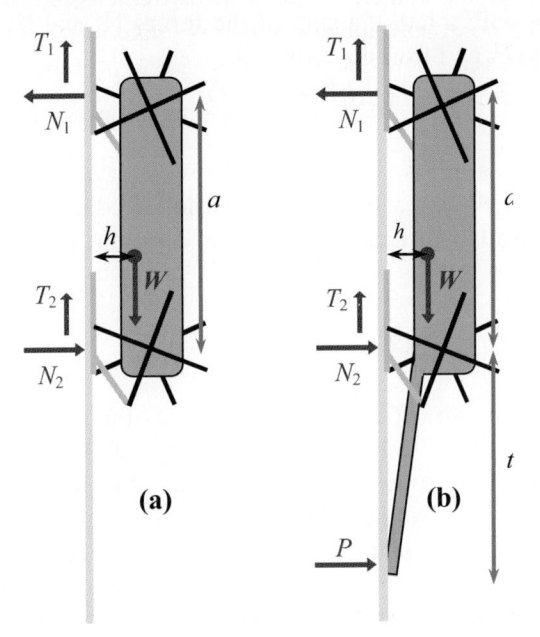

Fig. 3. Free body diagrams of (a) robot without tail and (b) robot with tail on vertical surface.

placement. Because the feet are out of phase by 45°, this should occur when the peeling foot is parallel to the substrate. From video of the robot with Scotch® tape feet, the average peeling angle at that instant is 60° (SD = 5°, n = 6). Using Equation (1), for the structured polymer the force per unit width is 0.018 N/cm. The component in the normal direction is 0.016 N/cm, which would require the feet to be at least 14 cm wide.

However, if a tail is added, the normal force, P, at the tail/wall contact point can aid the adhesive in countering the moment of the weight. By summing the moments about the rear foot contact point (**Fig. 3b**), the required adhesive force from the front feet is

$$N_{1\,WithTail} = \frac{hW - tP}{a} \quad \text{(tensile)} \tag{3}$$

Thus, the robot with a tail will require less normal force to prevent it from tumbling backwards on a substrate, assuming a lightweight tail.

The disadvantage of the tail is that it decreases the traction forces that prevent the robot from slipping. Because the adhesive materials are pressure sensitive, the tangential forces are largest when the normal forces are most compressive. For the robot without a tail, the rear normal force, N_2, will always be equal and opposite the front normal force, N_1:

$$N_{2\,NoTail} = \frac{hW}{a} \quad \text{(compressive)} \tag{4}$$

For a robot with a tail, the sum of the forces N_1 and N_2 must be equal and opposite to P, and from Equation (3):

$$N_{2\,WithTail} = \frac{hW - (a+t)P}{a} \quad \text{(compressive)} \tag{5}$$

Thus, N_2 with a tail is always less than without a tail. If the tail is long and stiff enough, N_2 can actually be in tension. Reducing the contact force decreases the available traction, which may cause the robot to slip.

5 Robot Performance

The performance with office tape demonstrates the potential of future adhesive climbing robots. With office tape, no tail was needed and the robot (87 grams) was able to climb reliably enough to test steering, obstacle climbing, and ceiling walking. The vehicle walked up, down, and sideways on vertical planes of glass using Scotch® tape feet. Further, the robot walked inverted all the way across the underside of a 30-cm-long horizontal surface. The vehicle also demonstrated successful transitions from the floor to a vertical wall and from a wall to the floor. Gradual steering was accomplished. In a test to demonstrate climbing distance, the robot ascended a 70 cm vertical surface four consecutive times at a speed of 5.8 cm/sec, without falling, a total of 280 cm. Afterwards, the robot fell with increasing frequency as the tape became dirty or damaged [9].

When the 1.6 cm wide tape was replaced with the same size pieces of polymer and the batteries were removed, the 76-gram robot was able to climb an incline of 50°, but fell backwards from the substrate at higher angles. By adding a 6.6 cm tail and widening the front feet to 2.6 cm, the robot (at 110 grams) was able to scale a 60° incline reliably. It scaled the entire length of the incline (39 cm) at a speed of 8.6 cm/sec. The robot made 13 similar-length runs without requiring washing, lasting 1.8 times longer than Scotch® tape. Reversing the driving direction on the wall resulted in the robot falling and catching itself on the substrate.

By lengthening the tail to 25 cm and widening the back feet to 2.6 cm, the now 132-gram robot was able to climb a vertical glass surface (**Fig. 1b**). With the tail and widened feet, the robot can be placed on a vertical surface and rest indefinitely. Walking on the vertical surface was less reliable than with the tape: the robot would slide or lose traction on the substrate 44% of the time (n = 16). In the trials in which the robot did make progress, the robot walked an average of 18 cm. The longest walk was 58 cm long (the entire length of the surface) at 2.3 cm/sec. The polymer feet retained their traction/adhesive properties for several hours of testing and could be renewed by washing with soap and water.

6 Discussion

The ability to transition between orthogonal surfaces, steer, and overcome small obstacles is feasible for a robot with compliant adhesives, as demonstrated by the Scotch® tape feet. A lighter robot would be more stable on the substrate, allowing more complex maneuvers. In addition, a lighter robot may not need a tail, which can get in the way of transitions. With a body flexion joint, the robot may even be able to make transitions around more difficult external angles [11]. Mounting the axles farther away from the wall than the center of mass would allow more space for longer spokes and feet without losing stability. The addition of frictional material on the tail of the robot, where the normal forces are compressive, may reduce the tendency to slip down the substrate.

While the current robot only walks on clean smooth glass, a practical climbing robot would be able to traverse rougher surfaces as well. This will require the adhesives to be resistant to dust and oils. Additionally, alternative attachment mechanisms, such as insect-like claws or spines, could be added to take advantage of surface roughness.

Acknowledgements

This work was supported by NSF/IGERT grant DGE-9972747 and U.S. Air Force contract F08630-03-1-0003.

References

1. Sangbae K., Asbeck A. T., Cutkosky M. R. and Provancher W. R. (2005) SpinybotII: Climbing Hard Walls with Compliant Microspines, *Int. Conf. on Advanced Robotics (ICAR '05)*, Seattle, USA.
2. Frazier S. F., Larsen G. S., Neff D., Quimby L., Carney M., DiCaprio R. A. and Zill S. N. (1999) Elasticity and Movements of the Cockroach Tarsus in Walking, *J. Comp. Physiol. A*, vol. 185, pp. 157–172.
3. Autumn K., Liang Y. A., Hsieh S. T., Zesch W., Chan W. P., Kenny T. W., Fearing R. and Full R. J. (2000) Adhesive force of a single gecko foot-hair, *Nature*, vol. 405, pp. 681–685.
4. Peressadko A. and Gorb S. N. (2001) When Less is More: Experimental Evidence For Tenacity Enhancement by Division of Contact Area, *The Journal of Adhesion*, vol. 80, pp. 1–15.
5. Sitti M. and Fearing R. S. (2003) Synthetic Gecko Foot-Hair Micro/Nano-Structures for Future Wall-Climbing Robots, *Int. Conf. on Robotics and Automation (ICRA '03)*, Taipei, Taiwan.
6. Menon C., Murphy M. and Sitti M. (2004) Gecko Inspired Surface Climbing Robots, *IEEE ROBIO '04*, Shenyang, China.
7. Niederegger S. and Gorb S., (2003) Tarsal movements in flies during leg attachment and detachment on a smooth substrate, *J. Insect Physiol.*, vol. 49, pp. 611–620.
8. Morrey J. M., Lambrecht B. G. A., Horchler A. D., Ritzmann R. E. and Quinn R. D. (2003) Highly Mobile and Robust Small Quadruped Robots, *Int. Conf. on Intelligent Robots and Systems (IROS '03)*, pp. 82–87, Las Vegas, USA.
9. Daltorio K. A., Horchler A. D., Gorb S., Ritzmann R. E. and Quinn R. D. (2005) A Small Wall-Walking Robot with Compliant, Adhesive Feet, *Int. Conf. on Intelligent Robots and Systems (IROS '05)*, Edmonton, Canada.
10. Kendall K. (1975) Thin-film peeling – the elastic term, *Journal of Physics D, Applied Physics*, vol. 8, pp. 1449–1452.
11. Ritzmann R. E., Quinn R. D. and Fischer M. S. (2004) Convergent evolution and locomotion through complex terrain by insects vertebrates and robots, *Arth. Struct. Dev.*, vol. 33, pp. 361–379.

Novel Solutions to Design Problems of Industrial Climbing Robots

S.Chen[1], J.Shang, Z.Zhao, T.Sattar and B. Bridge

Centre for Automated and Robotic NDT, [1]Department of Electric, Computer and Communication Engineering, Faculty of Engineering, Science and Built Environments, London South Bank University, UK
chensc@lsbu.ac.uk

Abstract

This paper aims to identify some design problems that affect system performance of climbing robots and to present novel solutions to the problems. The first problem is the interference of the umbilical cord for the robot with its mobility and dynamics; the second problem is the slow traveling speed of climbing robots, affecting the overall working efficiency of the system; the third one is unforeseen variations of surface curvature of objects on which a vacuum suction type of robot climbs, which reduces suction pressures and adhesion forces pulling the robot onto the surface; and the fourth is the intrinsic safety requirements of electrical components in flammable environments. The technical solutions have been developed and tested in laboratory and industrial field trials and have been proven to be well founded and unique. It has been verified that new techniques from other engineering areas, e.g. materials and information engineering, can be used to formulate novel solutions to overcome the obstacles. The technical approaches used in these solutions will strengthen the state of the art and contribute to the advance of climbing robot applicable technology.

Keywords: climbing robot, automated NDT, safety, nuclear, design

1. Introduction

Research on climbing robots has intensified since 1990s. Most of this research is applications directed because of the invaluable potential of climbing robots for deployment as carriers to replace human operators in the accomplishment of essential safety critical tasks in hazardous environments such as nuclear reactors, petrochemical plants, fire-fighting scenes, etc. Up to date over 200 prototypes aimed at such applications have been developed in the world. However, few of them can be found in common use due to unsatisfactory performance in on-site tests. So an enormous amount of effort worldwide in research and prototyping lays fallow. This situation has to change otherwise sponsors and end users will lose confidence in this technology and its advance will be hindered.

The main reason for this situation is that researchers have not adequately addressed practical solutions to problems that largely emerge in the later phases of structural and prototype design, or rather, they put excessive resources and effort on theoretical work in the early phase of conceptual design. Design of a climbing robot is the realization of the designers' ideas and initiatives to fulfill the functionality and specifications of the entire system. The design is implemented through three phases from conceptual design, structural design to prototype design. Presently most research at the design stage has been focused on conceptual design, where kinematics is solved mathematically or theoretically, while structural and prototype design for system dynamics, synthesis and coordination are overlooked to some extent, leaving relevant problems unsolved and hence leading to a mismatching of different functions and poor performance characteristics. This is why in, general, practical deployment of climbing robots has not been welcomed by targeted end users.

This paper seeks to recognize the main design problems affecting the system performance of climbing robots and to present novel solutions to these problems. The focus is to look at some practical details which regularly present problems to designers, yet are commonly overlooked at the proposal stage of applications based climbing robot projects. The first one is the interference of umbilical services cables with robot mobility and dynamics. The second one is a low traveling speed of a climbing type of robots, affecting the overall working efficiency of the system. Thirdly there is a problem of unforeseen local variations in surface curvature of objects on which a pneumatically adhering robot climbs, which reduces the suction levels and adhesion forces pulling the robot onto the surface. The fourth difficulty for consideration is the intrinsic safety requirements of electrical components in flammable environments like inside gas storage tanks where some climbing robots are used to deliver sensors to perform inspection tasks.

2. Interference of an umbilical with robot mobility and dynamics and the solution

An umbilical normally consists of power lines for electric and pneumatic supply and signal wires for control and sensing but of course depends on what instrumentation and payload is carried on-board the robot. The umbilical for some of our climbing robots used for non-destructive testing have a mass of one kg per meter so that a robot climbing to a height of 20m has to carry an additional payload of 20 kg, which is nearly the same weight as that of the climbing robot. This results in not only a lower ratio of weight to payload but also poor motion dynamics. This problem is mainly caused by control and sensor signal wires. In order to solve this problem, a client-server serial link control module has been developed dedicated to climbing robots as shown in **Fig.1.(a)**. The module is placed next to a mouse to emphasize its small size. Using these modules the umbilical can be made significantly lighter, for example a robot umbilical that can number up to 70 parallel wires for an 11 degrees of freedom robot is reduced to an umbilical which uses only 7 wires.

So many wires can be reduced because all the control modules mounted on the robot are serially connected as shown in **Fig.1.(b)**. The host computer located on ground communicates with the modules via serial I/O. The host computer can send commands to one or a group of the modules, or read information from one module. Thus, the communication cable between the host computer and the on board modules is just a serial cable, and the weight of the umbilical is significantly reduced. A swimming robot developed in the centre has reduced its umbilical weight by up to 5 folds due to using this novel control strategy.

(a) (b)

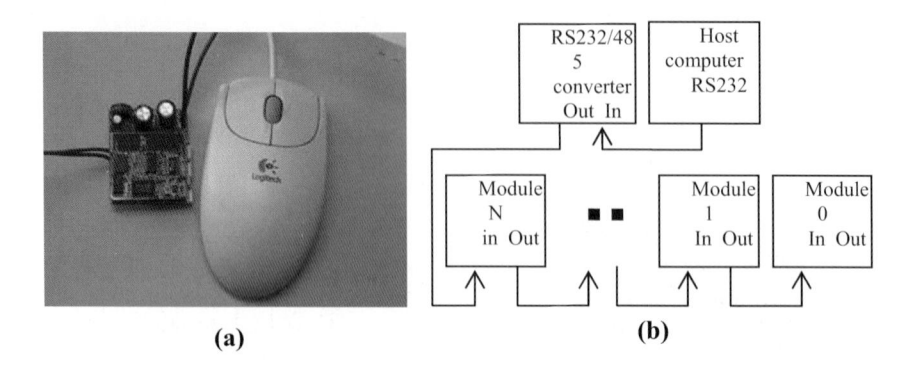

Fig.1. Client-server featured control module and method

3. A low traveling speed of a robot and the solution

One design consideration for climbing inspection vehicles is to increase travel speed to achieve cost effective inspection times. Existing pneumatically powered vehicles have a speed of less than 1 m/min, which does not satisfy targeted end-users. In order to solve this problem, a new mechanism to increase the traveling speed of the vehicle has been invented as shown in **Fig.2. (a)**. It employs one more pair of side actuating cylinders than existing climbing robots. The prototypes equipped with this mechanism are able to operate at nearly double the speed of existing vehicles of similar actuator stroke length [1][2].

As shown in **Fig.2. (b)**, a mechanism with two pairs (an inside pair and an outside pair) of actuating cylinders climbs up from the initial position 0 to the first position I and finally to the second position II to complete a cycle of movement. In the first step (from 0 to I), the outside actuating cylinders are stuck to the surface, the platform moves forward by a stroke length. The inside pair of the cylinders also move forward a stroke length relative to the platform. In the next step (from I to II), the inside pair of cylinders are attached to the surface. Unlike the mechanism with one pair of cylinders, at this time the platform is not stuck to the surface but moves forward by a stroke length through support of the inside cylinders. Meanwhile, the outside cylinders move forward by one stroke length relative to the platform. In this way, the vehicle moves forward by two stroke lengths in a full cycle of movement of two stroke lengths. Thus, in a given time span, the mechanism with two pairs of cylinders can move forward by one more stroke length than the mechanism with only one pair of cylinders. As a result, its travel speed is doubled.

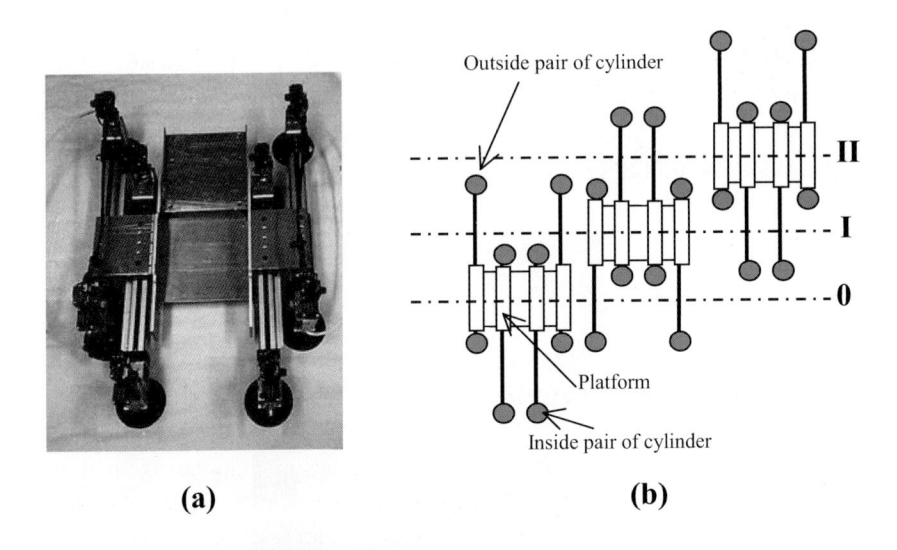

(a) (b)

Fig.2. Working principle of the mechanism

4. Unforeseen variations of surface curvature of objects and the solution

Climbing robots moving on a large surface may encounter variations in surface curvature, e.g. when traversing the length of an aircraft fuselage. The robot should be able to adapt to these curvatures. A new design of robot feet is introduced that imitate octopus tentacles [3]. This design is aimed at gaining excellent adaptation of the feet on terrains with different curvatures. Technically, the foot design should be flexible to adapt to various curvatures and it should have many small suction pads for robust adhesion. The flexible feet should be stable when the climbing robot adheres to the surface in order to obtain a stable platform so that vibration does not affect task performance e.g. data acquisition by the deployment of NDT sensors. To achieve the best stability of the vehicle structure, we have developed a new method to make the flexible multi joint mechanism as stable as three non-co-linear points which are intrinsically stable.

As shown in **Fig.3.**, the mechanism consists of two sets of supports, the outside set and inside set, 4 legs each support. For the outside support, each leg is mounted with a foot with 4 suction pads located at four corners while for inside support, each leg has two feet, each with 4 pads, located around the centre. 4 pads are mounted on a foot with 4 joints and further the foot on the support with a joint, and finally the support mounted on the body of the vehicle with a joint. Totally 64 joints are used in this structure. Therefore the maximum flexibility up to 16 degrees of freedom (DOFs) is created.

Two of the feet on the inside support

One of the feet on the outside support

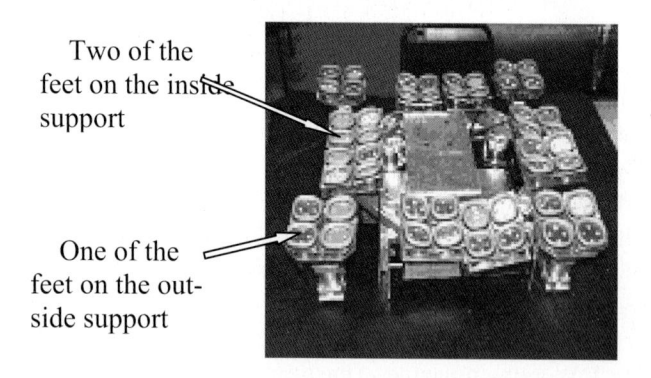

Fig.3. Octopus tentacle imitated robot feet

Obviously, the structure is flexible to adapt curvatures while the feet approaching the surface. But once the feet are adhere to the surface, all the degrees of freedom are restrained. The robot works like a rigid body so that NDT inspection without vibration is achieved.

5. Intrinsic safety requirement in flammable environments and the solution

A new pneumatic servo-motor has been especially designed and developed in-house (see **Fig.4**.) to replace the need for an electric motor. All presently existing climbing robots use electric motors and thus have difficulty in conforming to the legislation for intrinsic safety. A fully pneumatic approach cannot be used for precision positioning of the vehicle due to the compressibility of air. The solution adopted was to use solenoid-operated control valves which were intrinsically safe and that could be used in potential explosive atmospheres.

Fig. 4. Electro-pneumatic servomotor mechanism

Resolution of the servomotor is determined by the tooth number of the indexing wheel. For the mechanism discussed, the resolution is up to 1.0°. Rotation accuracy of the mechanism is affected by factors associated with electric, pneumatic and mechanical components. Whether systematic errors or stochastic errors are generated depends on their characteristics [4]. The factors affecting the systematic error includes a response time of control valves and a backlash of gears; while the factors affecting the stochastic errors consist of compressed air disturbance and irregularity of the tooth profile of the indexing wheel. The total rotation error is the sum of the systematic error and the stochastic error. For the developed model, the maximum rotation error of the servomotor is 0.6° and the relative rotation error is 4 %.

6. Conclusion

This paper recognizes some design problems of climbing robots and presents novel solutions to these problems. These technical solutions have been tested in laboratory and industrial field trials and have been proven to be well founded and unique. It has been verified that new techniques from other engineering areas, e.g. materials, information engineering and communications, can be used to formulate novel solutions to overcome the obstacles which hinder the progress of climbing robot applications. The technical approaches used in these solutions will strengthen the state of the art and contribute to the advance of climbing robot applicable technology.

Acknowledgements

This paper reflects the work of the research projects funded by the European Community under FP4 and FP5 (1994-1998, 1998-2002)

References

1. Chen S. (1999) *Wall-climbing Robotic Non-Destructive Testing System for Remote Inspection*, PhD Thesis, London South Bank University Library.
2. Sattar T., Chen S., Bridge B. and Shang J. (2003) *Robair: Mobile Robotic System to Inspect Aircraft Wings and fuselage*, CARs & FOF, Kuala Lumpur, Malaysia.
3. Chen S., Sattar T. and Bridge B. (2001) Analysis and Calculation of Payload Carrying Capacity for Climbing Robotic Non-destructive Testing System, *IEEE International Conference on Mechatronics and Machine Vision in Practice*, p. 393-398.
4. Iborra A., Pastor J.A., Álvarez B., Verná C. and Fernández J.M. (2003) Robots in radioactive environment, *IEEE Robotics & Automation Magazine*.
5. Virk G.S. (2002), Clawar modularity – the guidance principles, *Proceedings of the 6th international conference on climbing and walking robots*.

Fast Pointing and Tracking System for Mobile Robot Short Range Control via Free Space Optical Laser Line of Sight Communication Link

M. H. Ahmad, D. Kerr, K. Bouazza-Marouf

Wolfson School of Mechanical and Manufacturing Department, Loughborough University, Loughborough, LE11 3TU, United Kingdom
M.H.Ahmad@lboro.ac.uk

Abstract

In this paper, the short range control of mobile robots using an optical pointing and tracking system is described. The technique uses fast hardware based methods for tracking a laser beacon, where the focused spot is maintained in the center of a CMOS camera image. This system is designed for use with a laser line-of-sight (LOS) communication system for short range tele-operation of a mobile robot.

Keywords: Short range control, Line-of-Sight, CMOS camera.

1 Introduction

The use of radio frequency (RF) and wired control for tele-operation of mobile robots may be prohibited in certain hazardous environments such as nuclear power plants[1]. Umbilical cables reduce mobility and often become entangle and broken, rendering the robot inoperable. RF control systems can cause interference with the safe operation of the plant[2]. Under these circumstances, the use of free space LOS optical links for tele-operation would be preferable[3, 4]. Most current research and development of LOS optical links has tended to concentrate on satellite communication applications, indoor wireless local-area-network (LAN) and point-to-point links between PCs and appropriate devices, rather than on mobile robot telemetry[5, 6]. However, these are in general not suitable for short range control of mobile robots, where unexpected changes of speed and direction, ambient light conditions, and environments are common.

When using LOS optical beams as tele-operation links for the control of mobile robots, a continuous pointing and tracking system is required between the operator's command post and the mobile robot[7, 8]. The work described here concerns the design and development of a fast pointing and tracking system, as part of a LOS optical link, for mobile robot tele-operation at relatively short ranges from 1m to 20m. The optical pointing and tracking system produced as part of this research is simple and inexpensive in its design and construction.

2 The Pointing and Tracking System

The pointing and tracking is achieved using a LM9637 digital CMOS image camera from National Semiconductor to track an infrared beacon mounted close to the receiving device. The technique uses fast, hardware based methods for tracking the beacon, where the focused spot is maintained in the center of the camera image. The camera is mounted on a 2-axis DC motor driven pan and tilt unit while the laser beacon is made from a fibre coupled to an infra-red laser diode with its free end machined at a cone angle of 60° to allow even horizontal illumination. Fig. 1 shows the pointing and tracking system with LOS optical links. A pair of high power transmission LEDs are mounted coaxially with the camera optics. If the camera is made to maintain the beacon centrally within its field of view, then the transmitter LEDs and receiver remain aligned.

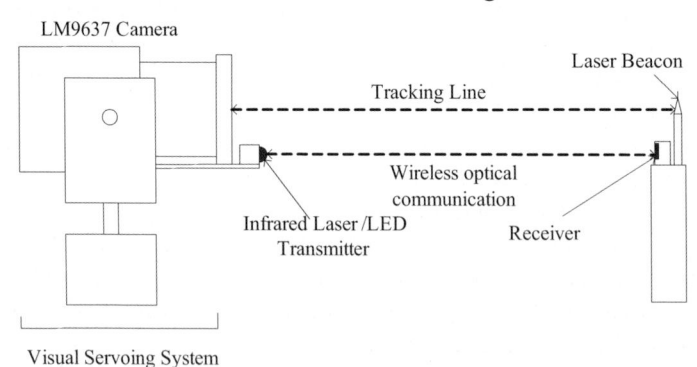

Fig. 1. Pointing and tracking system for LOS optical links

The architecture of the system is shown in Fig. 2. It consists of 3 parts, a frame grabber to digitise the camera image, the spot tracker to calculate the position of the target within the window and motor drive controls to maintain correct physical pointing and tracking.

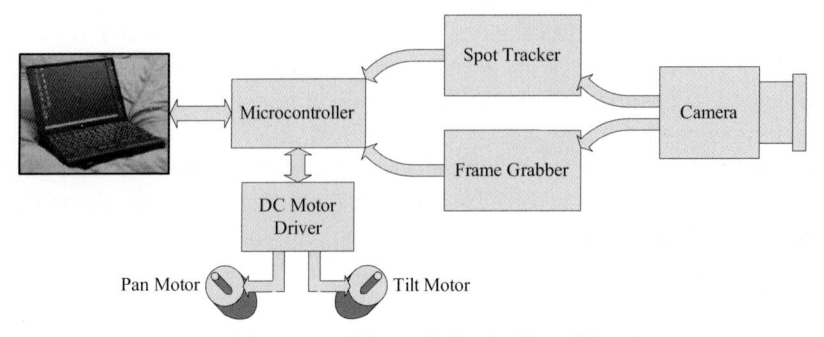

Fig. 2. Architecture of the pointing and tracking system

The advantages of using the digital CMOS image camera are its low cost and power consumption, increased frame readout, access flexibility to the pixels array and low smear and blooming.

2.1 Camera Windowing

The traditional tracking method of continuously finding the object in an entire camera image requires excessive processing time. A smaller image window increases the frame rate and reduces the processing time[9]. Therefore, the speed of the tracking beacon can be minimized by optimizing the size of the tracking image window in the CMOS camera.

The target laser beacon is focused to an image sensor plane as a small spot or blob. The estimated size of this spot for different values of lens focal length is shown in Fig. 3. This estimation is important when selecting the optimal window size for the camera. Windowing is easily achieved in the CMOS camera with the random access addressing capability. The size is carefully selected to ensure the target will always be within the window when tracking the movement of the laser beacon.

In this design, the window is selected to be 100x100 pixels; which is about 3% of the area of the full 488x648 pixels array of the camera. Fig. 4 shows the image of the projected target within the window at a range of 3m, with the camera lens aperture set to f12.

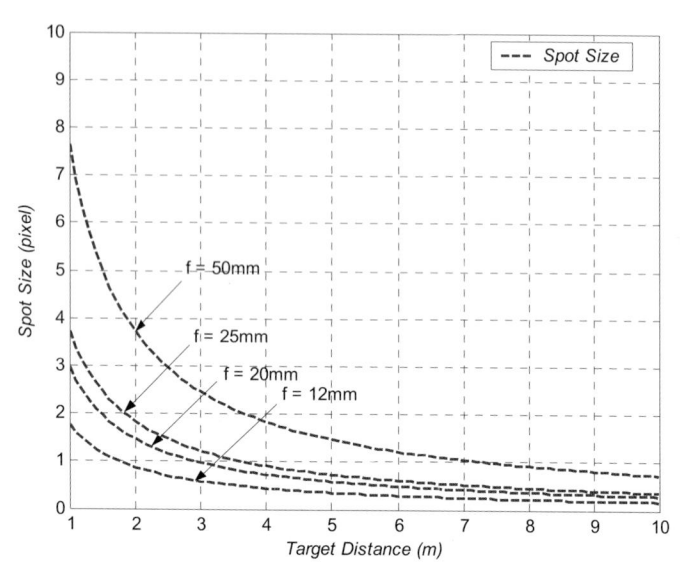

Fig. 3. Estimated spot size for 12mm, 20mm, 25mm and 50mm focal length

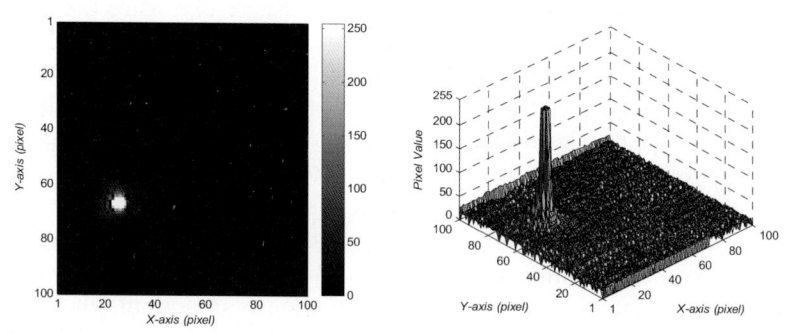

Fig. 4. Image of a target taken from 100x100 pixels window at 3m range

3 Control Design

The system uses closed-loop control to maintain the tracking camera in line with the target. Fig. 5 shows the basic control diagram. Pan angle (θ) and tilt angle (Φ) are actively controlled to keep the target spot at the center of the image plane, by controlling the angular rotation of the DC motors of the camera pan and tilt unit.

Fast pointing and tracking system for mobile robot ...

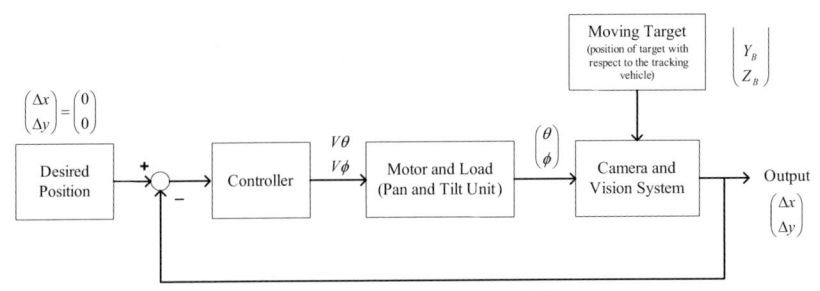

Fig. 5. Control diagram of the tracking system
($V\theta$ and $V\Phi$ are voltages corresponding to the desired pan and tilt motions)

From Fig. 5, $(X_B\ Y_B\ Z_B)$ is the position of the laser beacon with respect to the tracking camera. Δx and Δy lie on the image plane of the sensor, whilst the z axis is directed towards the target laser beacon.

The position of the laser beacon focused spot on the image plane, relative to a fixed Cartesian axis, is computed every frame. The deviation of the spot, in the x and y directions, from the center of the image sensor window is then used as an error signal.

Field Programmable Gate Array (FPGA) devices are used to control the counters of a pixel clock from the CMOS camera to deduce the Δx and Δy position of the target spot. By using a 7ns clock signal from the FPGA and 14.318 MHz to generate a pixel clock to the CMOS camera, it takes approximately 0.1ms to scan the 100x100 pixel window, from top to bottom in a raster fashion. When the target spot has been located, i.e. when a fixed brightness threshold is exceeded, the line number and pixel position for the current line are used to calculate the Δx and Δy offset from the image center. Fig. 6 shows a layout of the signal processing used to obtain the laser beacon spot position from the CMOS camera image.

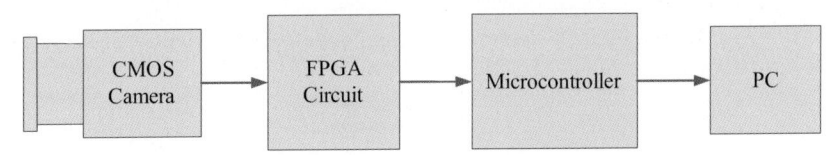

Fig. 6. Block diagram for target error detection

The FPGA circuits process out the video synchronization signals and pass the data on to the counters which determine the line number and pixel position of the target spot. The error signal is then sent to the microcontroller, in the form of two 8-bit digital signals corresponding to the offset (Δx, Δy) of the spot from the image window center.

The processor converts the two error signals to corresponding PWM values, the greater the error signal the higher the PWM value. The PWM signals are used to drive the two DC motors on the camera pan and tilt unit in order to keep the camera pointing continuously towards the beacon.

3 System Tests and Performance

A simulation test of the pointing and tracking system has been carried out using a 3D model for a 1m to 8m distance from the target laser beacon to the CMOS image camera. The target moves through the simulated environment with a maximum velocity of 300mm/s and a maximum acceleration of 150mm/s^2. The target position is defined with respect to the camera X, Y and Z axes. Fig. 7 shows the 3D model of the planned target movement and Fig. 8 shows the corresponding result from the pan and tilt units of the tracking system.

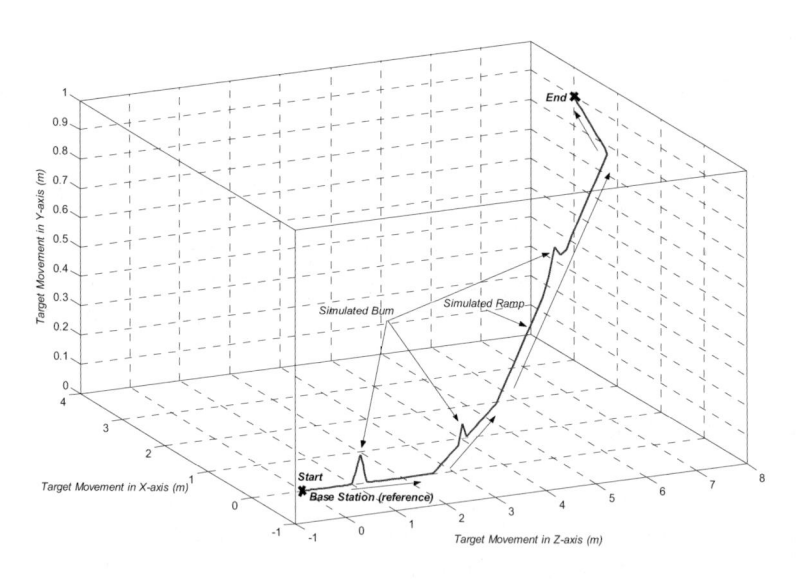

Fig. 7. Planned target motion with respect to a fixed base station.

Fast pointing and tracking system for mobile robot ...

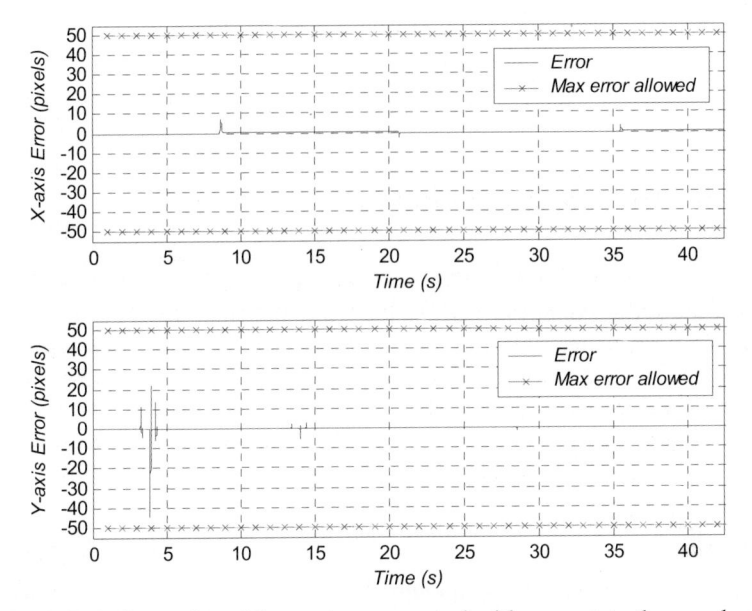

Fig. 8. Pointing and tracking errors expressed with respect to the x and y axes of the image frame of reference

"Static" tracking, which corresponds to very slow target beacon movement relative to Z axis, gave an estimated accuracy of less than 1 pixel. As the target changed direction, with an accompanying change in acceleration, this accuracy was reduced but still remained within the maximum allowed error for the system as shown in Fig. 8. The maximum error chosen insures that the target spot remains on the selected image window at all times.

4 Conclusions

A fast pointing and tracking system for short range mobile robot control using a free space optical LOS communication link has been developed. The system should be very robust and reliable, as well as simple in construction and relatively inexpensive. The system is designed to operate at only relatively short ranges from 1m to 20m. The response speed is effectively faster at greater distances since angular displacements are smaller. The static pointing accuracy can be maintained, but the dynamic accuracy is reduced when the target changes direction and acceleration.

References

1. T. Kobayashi, K. Miyajima, and S. Yanagihara, "Development of remote surveillance squads for information collection on nuclear accidents," Advanced Robotics, vol. 16, pp. 497-500, 2002.
2. H. G. Nguyen, N. Pezeshkian, A. Gupta, and N. Farrington, "Maintaining Communication Link for a Robot Operating in a Hazardous Environment," ANS 10th Int. Conf. on Robotics and Remote Systems for Hazardous Environments, Gainesville, Florida, USA, March 2004.
3. D. Kerr, K. Bouazza-Marouf, and T. C. West, "An inexpensive pointing and tracking system using TV cameras and an optical fibre beacon," Mechatronics, vol. 9, pp. 919-927, 1999.
4. D. Kerr, K. Bouazza-Marouf, K. Girach, and T. West, "Free space laser communication links for short range control of mobile robots using active pointing and tracking techniques," IEE Colloquium on Optical Free Space Communication Links, London, UK, February 1996, pp 11/11 - 11/15.
5. J. i. Kawaguchi, T. Hashimoto, T. Misu, and S. Sawai, "An autonomous optical guidance and navigation around asteroids," Acta Astronautica, vol. 44, pp. 267-280, 1999.
6. R. Ramirez-Iniguez and R. J. Green, "Indoor optical wireless communications," IEE Colloquium on Optical Wireless Communications (Ref. No. 1999/128), London, UK, June 1999, pp 14/1 - 14/7.
7. T. Motoki, Y. Sugiura, and T. Chiba, "Automatic Acquisition and Tracking System for Laser Communication," IEEE Transactions on Communications, [legacy, pre - 1988], vol. 20, pp. 847-851, 1972.
8. G. S. Mecherle, A. K. Rue, and G. T. Pope, "Automatic acquisition and tracking for laser communication using video techniques," Military Communications Conference (MILCOM 88), '21st Century Military Communications - What's Possible?', IEEE, 1988, pp 543 - 555.
9. M. Vincze, "Optimal window size for visual tracking for uniform CCDs," Proceedings of the 13th International Conference on Pattern Recognition, 1996, pp 786 - 790.

Control of CLAWAR

4 Legs Positioning Control of the Quadruped Robot by LVS using Stereo Ominidirectional Camera

Yukinari Inoue[1] and Noriaki Maru[2]

[1] Graduate School of Systems Engineering, Wakayama University,Sakaedani 930, Wakayama 640-8551, Japan
[2] Department of Systems Engineering, Wakayama University, Sakaedani 930, Wakayama 640-8551, Japan maru@sys.wakayama-u.ac.jp

Abstract

We propose a positioning control method of the 4 legs of the quadruped robot by Linear Visual Servoing(LVS) using stereo omnidirectional camera. It is based on the linear approximation of the relation between binocular visual space and joint space. Binocular visual space is defined by vergence angle and horizontal and vertical viewing directions. LVS uses the constant linear approximation matrix with neither camera angles nor joint angles to calculate feedback command. Therefore, LVS is very robust to calibration error, especially to camera angles and joint angles and the amount of caliculation is very small to compared to the conventional visual servoing scheme.

Keywords: Linear Visual Servoing, Quadrupted Robot, Omnidiredtional Stereo Camera, Binocular Visual Space

1 Introduction

Quadruped walking robot can walk in discrete environment. In quadruped walking robot, conventional control method of swing leg is either a simple feedback based on potentiometer or an only feedforward. But in such environment, the accurate positioning control of swing leg to safe landing point by 'Visual Servoing' is required. However, the conventional visual servoing methods [1],[2] have some problems. For example, they require complex non-linear calculation with the exact camera angles and joint angles. Therefore, they are not so robust to calibration errors of camera angles.

To solve these problems, 'Linear Visual Servoing'(indicated as LVS) is proposed[3]. It is based on the linear approximation of the relation between binocular visual space and joint space. Binocular visual space is defined by vergence angle and horizontal and vertical viewing directions. LVS uses the constant linear approximation matrix with neither camera angles nor joint angles to calculate feedback command. Therefore, LVS is very robust to calibration error, especially to camera angles and joint angles and the amount of caliculation is very small to compared to the conventional visual servoing scheme. We proposed a leg positioning control method of the quadruped robot by LVS using normal stereo camera. However to control 4 legs of the quadruped robot, two set of stereo camera and vergence control are needed. In this paper, we propose a positioning control method of the 4 legs of the quadruped robot by LVS using stereo omnidirectional camera.

2 LVS

2.1 Model of Quadruped Robot

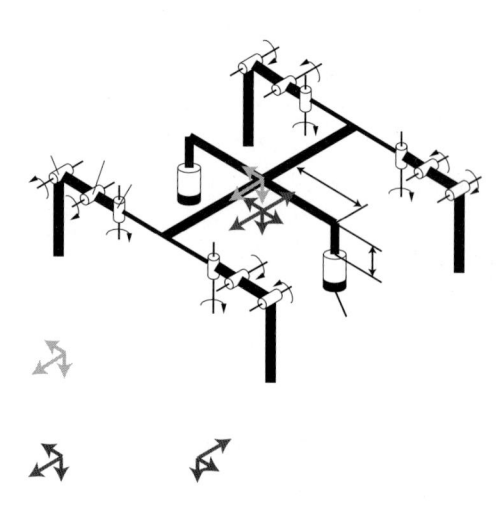

Fig. 1. Model of the quadruped robot

Fig. 1 shows the leg-eye coordination of the quadruped robot which has two omnidirectional cameras. Robot coordinate system $\sum_r (X_r, Y_r, Z_r)$ is located at the center of the quadruped robot as shown in **Fig. 1**. The origin of the Camera coordinate system $\sum_c(X_c, Y_c, Z_c)$ is located at the center of the two omnidirectional camera. We call the right forward leg Leg1. Leg1 consists of three links and three joints. The shoulder joint of the Leg1 is located at (X_{off}, Z_{off}) in the robot coordinate system. The shoulder joint has 2 d.o.f. We call these joints j_{10}, j_{11} respectively. The elbow joint has 1 d.o.f. We call this joint j_{12}. Let L_0 be the link length between j_{10} and j_{11} and L_1 be the link length between j_{11} and j_{12} and L_2 be the link length between j_{12} and end tip of the leg.

2.2 Binocular Visual Space

The binocular visual space is defined by the vergence angle γ and the viewing directions θ, δ (See **Fig. 2**). This space has been employed by psychologists and physiologists as a model of binocularly-perceived space.

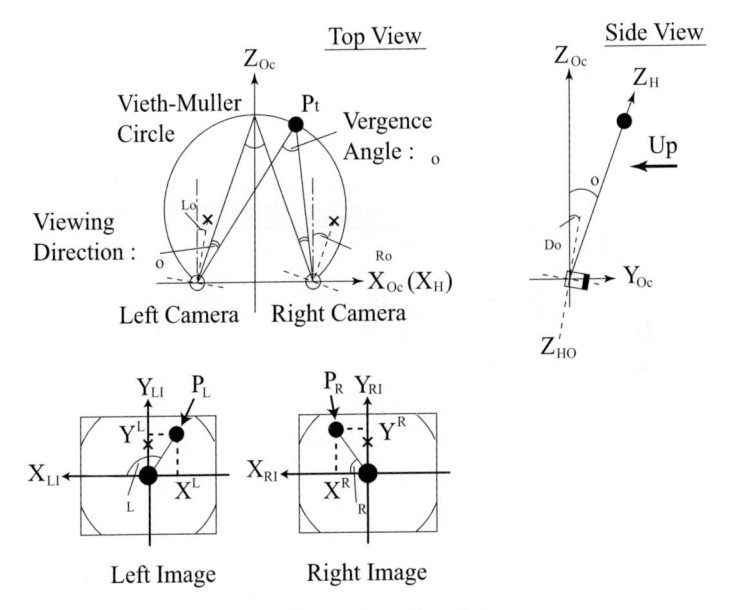

Fig. 2. Binocular Visual Space

The coordinates of a feature point projected on the camera image planes are transformed into binocular visual coordinates by

$$\mathbf{V} = \begin{bmatrix} \alpha_L - \alpha_R \\ (\pi + \alpha_L + \alpha_R)/2 \\ \alpha_D \end{bmatrix} + \begin{bmatrix} -\theta^L + \theta^R \\ (-\theta^L - \theta^R)/2 \\ \xi^L + \xi^R \end{bmatrix} \quad (1)$$

$$\xi^i = \tan^{-1} \frac{(b^2 + c^2)\sin\zeta^i - 2bc}{(b^2 - c^2)\cos\zeta^i}$$

$$\zeta^i = \tan^{-1} \frac{f}{\sqrt{x^{i2} + y^{i2}}} \quad (i = L, R)$$

where α_L, α_R is the angle between y_{RI}, y_{LI} axis in omnidirectional image and Z_H axis respectively and α_D is the vertical angle of omnidirectional camera and θ^i(i=L,R) are directional angle given by $\theta^i = \tan^{-1}(y^i/x^i)$ and a,b,c are constant values of hyperbolic mirror.(see **Fig. 3**)

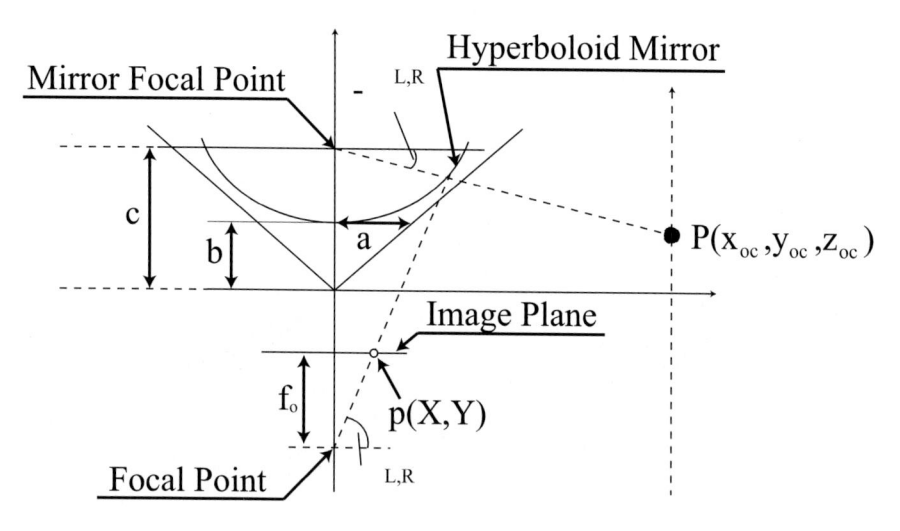

Fig. 3. Internal structure of Omnidirectional Camera

2.3 LVS

We linearize the transformation between binocular visual space and joint space of Leg1 using the least-squares approximation in the space defined in **Fig. 1**.

Then the transformation from binocular visual space $\mathbf{V} = (\gamma, \theta, \delta)^T$ to joint space $\mathbf{j} = (j_{10}, j_{11}, j_{12})^T$ is given by

Table 1. Range of Leg Joint Angle

$-80° \leq j_{10} \leq 30°$	$every 10°$
$-60° \leq j_{11} \leq 60°$	$every 10°$
$-60° \leq j_{12} \leq 40°$	$every 10°$

$$\mathbf{j} = \mathbf{RV} + \mathbf{C} \tag{2}$$

where

$$\mathbf{R} = \begin{bmatrix} A_1 & B_1 & C_1 \\ A_2 & B_2 & C_2 \\ A_3 & B_3 & C_3 \end{bmatrix} = \begin{bmatrix} 1.99 & 2.38 & -0.40 \\ 1.47 & 1.04 & 1.24 \\ 3.00 & 0.43 & -0.30 \end{bmatrix}, \tag{3}$$

$$\mathbf{C} = \begin{bmatrix} D_1 \\ D_2 \\ D_3 \end{bmatrix} = \begin{bmatrix} -174.48 \\ -50.84 \\ -113.14 \end{bmatrix}. \tag{4}$$

Fig. 4 shows the result of linear approximation.

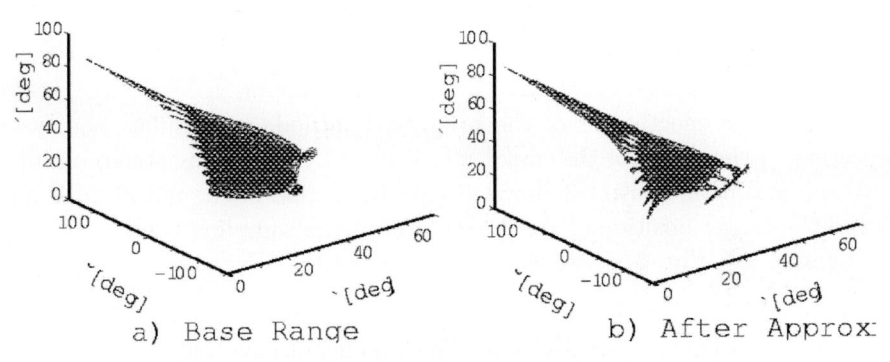

Fig. 4. Result of Approximation(Binocular Visual Space)

Using these relationship, LVS is given by

$$\begin{aligned} \mathbf{u} &= -\lambda \mathbf{R}(\mathbf{V} - \mathbf{V}_d) \\ &= -\lambda \mathbf{R} \begin{bmatrix} (-\theta^L + \theta^R) - (-\theta_d^L + \theta_d^R) \\ \{(-\theta^L - \theta^R) - (-\theta_d^L - \theta_d^R)\}/2 \\ (\xi^L + \xi^R) - (\xi_d^L + \xi_d^R) \end{bmatrix} \end{aligned} \tag{5}$$

where **u** are control signals to joint velocity controllers, **V** is the binocular visual coordinates of the leg point, $\mathbf{V_d}$ is the binocular visual coordinates of a target, λ is a scalar gain and **R** is the linear approximation matrix of the inverse kinematics obtained in the previous section. LVS is very robust to calibration error, especially to camera angle errors and joint angle errors, because the control law includes neither camera angles nor joint angles.

When controling other legs except Leg1, the sign of the element of R changes as shown in **Fig. 2**

Table 2. Sign of Constant(**R**)

	A_1	A_2	A_3	B_1	B_2	B_3	C_1	C_2	C_3
Leg2	—	—	—	—	—	—	{	—	{
Leg3	—	—	—	{	{	{	{	—	—
Leg4	{	{	{	—	—	—	—	{	—

3 Simulation

To show the effectivness of the proposed method, we simulate the trajectory of the 4 legs of the quadrupted robot by LVS using stereo ominidirectional images. **Fig. 3** shows the configuration of the initial position and the target position of the 4 legs. We set the sampling time as 66[ms] and λ as 3.0. **Fig. 5** shows the trajectories of the 4 legs.

Table 3. Configuration

	Initial Points[mm]	Target points[mm]
Leg1	(246,124,201)	(400,243,200)
Leg2	(246,124,-201)	(200,243,-500)
Leg3	(-246,124,201)	(-400,243,400)
Leg4	(-246,124,-201)	(-200,243,-400)

From **Fig. 5**, we can see that 4 legs converge to the target.

a) 3-D Trajectory

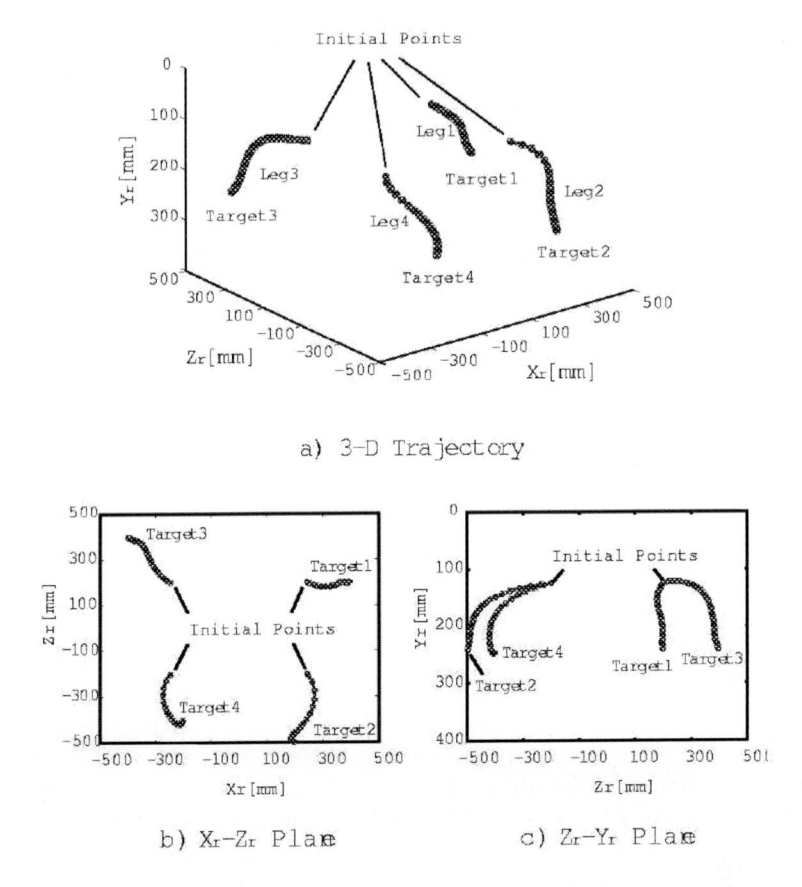

b) X_r-Z_r Plane

c) Z_r-Y_r Plane

Fig. 5. Trajectory of Each Leg Point

4 Conclusion

We have proposed positioning control method of the 4 legs of the quadruped robot by LVS using stereo omnidirectional images. We have shown the effectiveness of the proposed method by simulation. In future, we will adapt positioning control by LVS into gait.

References

1. L.E.Weiss, A.C.Sanderson and C.P.Neuman, "Dynamic sensor-based control of robots with visual feedback", *IEEE Trans. Robotics and*

Automation,Vol.RA-3, No.5, pp. 404-417,1987.
2. K.Hasimoto, T.Ebine and H.Kimura, "Visual servoing with Hand-Eye manipulator - Optical Control Approach", *IEEE Trans. Robotics and Automation*,Vol.12, No.5, pp. 766-774,1996.
3. T.Mitsuda, N.Maru, and F.Miyazaki, "Binocular visual servoing based on linear time-invariant mapping", *Advanced Robotics*, Vol.11, No.5, pp. 429-443,1997.

Control of a 3D Quadruped Trot

L. R. Palmer III and D. E. Orin

Department of Electrical and Computer Engineering
The Ohio State University
Columbus, OH USA
palmer.216@osu.edu, orin.1@osu.edu

Abstract

Legged vehicles offer several advantages over wheeled vehicles, particularly over broken terrain, but are presently too slow to be advantageous for many tasks. Trotting, the precursor to galloping for many quadrupeds, employs high-speed actuation to coordinate the intermittent ground contacts for each leg. Compliant elements and high-power actuators combine to perform a complex interchange of potential and kinetic energy during these short thrust intervals. These complexities, the frictional and contact losses that occur during normal running, plus the high number of degrees of freedom make three-dimensional (3D) dynamic quadruped motion very difficult to model accurately for control. For this reason, most of the research effort has been focused on simplified planar systems, only allowing motion in the sagittal plane. Many of these controllers only perform well around a fixed operating point and cannot regulate heading for desired changes in running direction. A 3D trotting controller which overcomes the above problems is presented here. Simulation results show the system responding appropriately to changes in the desired speed up to 3 m/s and heading up to 20 deg/s.

Keywords: 3D, control, quadruped, trot

1 Introduction

Legged vehicles offer several advantages over wheeled vehicles, particularly on broken terrain. In the future, legged vehicles will be used for military reconnaissance and time-critical search and rescue operations.

Fig. 1. Complete trotting stride [4]. Diagonal pairs are synchronized to contact and leave the ground together.

Presently, however, legged machines are too slow to be considered for these high-speed tasks. To address the issue of speed, this paper will present an effective controller for a high-speed quadruped trot.

Cursorial quadrupeds select gaits at various speeds based mainly on energy considerations. Many researchers have studied the quadruped bound[1] as a high-speed dynamic gait [1, 2, 3], but very few animals naturally bound. Instead, trotting is usually employed as the precursor to galloping, which is ultimately used at top speeds. Because trotting and galloping are commonly used together, we expect their control algorithms to share a number of similar components. Likewise then, understanding the quadruped trot will serve as a valuable foundation to understanding the quadruped gallop. **Figure 1** shows the complete trotting stride with two flight phases and two stance phases.

Research toward this goal of high-speed quadruped motion has focused on planar quadrupeds which are only allowed to move in the sagittal plane [5, 6]. Although potentially useful in the study of biped gaits, the planar trot analysis done by Berkemeier [2] and others has not yet been extended to produce satisfactory control on 3D machines. This is primarily because when motion is allowed in all three planes, the control efforts in the sagittal plane are negatively affected by the motion and corresponding control in the frontal and transverse planes. When this coupling is not actively accounted for, the accuracy of the controller suffers.

Control algorithms have also been developed for quadruped machines moving in steady state trajectories [6]. These algorithms only maintain stability when the system is well initialized and without the presence of significant disturbances. Raibert [7] tested a pacing, trotting, and bounding machine using the virtual leg concept derived from single-leg and

---

[1] In the bound, the legs are in effect linked in lateral pairs. First the front legs contact and leave the ground together, then the rear legs. In the trot, diagonal leg pairs are synchronized. In the gallop, the legs contact and leave the ground asynchronously.

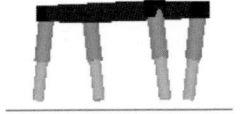

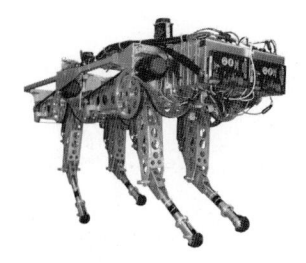

Fig. 2. System in simulation.

Fig. 3. KOLT vehicle at Stanford University [8].

biped algorithms and required leg pairs to contact and leave the ground synchronously. This method was effective for trotting but could not be extended to the asynchronous gallop gait, which is our goal.

This paper presents a control scheme which overcomes the above problems to regulate forward velocity and yaw rate in a 3D quadruped trot. Height, pitch and roll motion, and lateral velocity are of secondary importance, but need to be stabilized to maintain a stable trot. The system is controlled in a fully-dynamic simulation environment as shown in **Fig. 2**. The controller will ultimately be tested on the experimental KOLT (Kinetically Ordered Locomotion Test) Vehicle at Stanford University (**Fig. 3**).

The next section describes the system model used in this work. Section 3 discusses the control mechanisms. Section 4 presents some results of the controller, followed by a summary in Section 5.

2 Quadruped Model

A kinematic diagram of the quadruped system is shown in **Fig. 4**. The legs each have two actuators at the shoulder and hip joints, one for abduction and adduction of the leg and another to protract and retract the leg. The resultant leg angles are θ_a and θ_t respectively. Another actuator at the knee adjusts leg length, r. This motor is prismatic and acts in series with a passive spring with fixed stiffness. The spring is used to store and return energy as muscles, tendons, and ligaments do in biological systems. The quadruped weighs 60 kg and stands 60 cm high with the leg springs at nominal length. The size and mass were chosen to closely match those of a mid-sized goat.

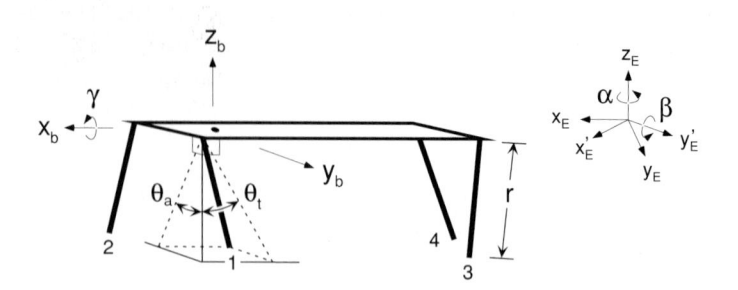

Fig. 4. Simplified kinematic model. Roll, γ, pitch, β, and yaw, α, are Z-Y-X Euler angles about x_b, y'_E, and z_E, respectively. The axis y'_E is y_E rotated by α. Forward and lateral velocity, v_{x_b} and v_{y_b}, are measured along the body fixed axes, x_b and y_b. Each leg has three degrees of freedom. There are two actuators at the shoulder and hip joints, one for abduction and adduction of the leg by varying θ_a (about x_b), and another to protract and retract the leg, by varying θ_t (about y_b). Another actuator at the knee and elbow joints changes leg length, r.

Figure 4 also shows the specific body states to be stabilized: roll, γ, pitch, β, yaw, α, forward velocity, v_{x_b}, lateral velocity, v_{y_b}, and height, p_{z_E}, at the apex of the flight phase. The peak height is recognized when the vertical body velocity is zero and occurs at a time referred to as top of flight (TOF). A cycle starts at a TOF and concludes at the next TOF, each leg having contacted the ground and exerted an appropriate impulse during that time.

3 Control Mechanisms

At contact, each leg's passive spring compresses to absorb energy from the ballistic flight. When the spring is maximally compressed, additional energy is applied to the system by further compressing the spring. The leg is then allowed to lengthen naturally, returning energy to the system. The additional thrust energy added at maximum compression is applied by all four legs to overcome frictional and contact losses during the stride. This common mode energy, η, is also adjusted to accelerate and decelerate the body.

Pitch stability is achieved by the biologically inspired method of redistributing the vertical impulses between the fore and hind limbs [9]. The differential leg thrusts, δ, are computed from a linearly-approximated model of the system. This is accomplished by testing the response of the system starting from several combinations of pitches and pitch rates, and applying a range of thrust differentials at maximum compression. These initial conditions and inputs plus the resulting pitches and pitch rates at the next TOF are approximated into a state-space model of the system using a least squares method. The matrix, $\mathbf{\Phi}$, and vector, $\mathbf{\Gamma}$, are solved for, resulting in the state-space model:

$$\mathbf{x}[k+1] = \mathbf{\Phi}\mathbf{x}[k] + \mathbf{\Gamma}u[k], \text{ where} \tag{1}$$

$$\mathbf{x} = \begin{bmatrix} \beta \\ \dot{\beta} \end{bmatrix}, \text{ and}$$

$$u = \delta.$$

From Eq. 1, we can show that

$$\mathbf{x}[2] - \mathbf{\Phi}^2 \mathbf{x}[0] = [\mathbf{\Gamma} \ \mathbf{\Phi}\mathbf{\Gamma}] \begin{bmatrix} u[1] \\ u[0] \end{bmatrix}, \text{ or} \tag{2}$$

$$\mathbf{z} = \mathbf{W}_c \mathbf{u} \tag{3}$$

after substitution. From Vaccaro [10], an nth-order controllable system can go from its current state to any other state in n time steps. If the controllability matrix, $\mathbf{W_c}$, is nonsingular, then the system is controllable and Eq. 3 can be inverted to form

$$\mathbf{u} = \mathbf{W}_c^{-1} \mathbf{z}. \tag{4}$$

Since our goal, $\mathbf{x}[2]$, is always $[0, 0]^T$, Eq. 2 can be simplified and leads to

$$\mathbf{u} = -\mathbf{W}_c^{-1} \mathbf{\Phi}^2 \mathbf{x}[0]. \tag{5}$$

Substituting again,

$$\mathbf{u} = \hat{\mathbf{W}}\mathbf{x}[0], \text{ or}$$

$$\begin{bmatrix} \delta[1] \\ \delta[0] \end{bmatrix} = \begin{bmatrix} \hat{W}_{1,1} & \hat{W}_{1,2} \\ \hat{W}_{2,1} & \hat{W}_{2,2} \end{bmatrix} \begin{bmatrix} \beta[0] \\ \dot{\beta}[0] \end{bmatrix}. \tag{6}$$

For any starting position in the state space, the algorithm produces a differential for the first step, $\delta[0]$, and a differential for the second step,

$\delta[1]$. A predicted state for the next TOF, $[\beta[1], \dot\beta[1]]^T$, can be computed from Eq. 1. If the model is exact, the machine will achieve zero pitch and zero pitch rate after the second step.

A luxury of this method is that the algorithm can be restarted at every TOF and produce the same results. This is necessary for our quadruped because the model is not exact. The true TOF state will not match the predicted intermediate step, so the corresponding second step differential is no longer valid. Instead, the algorithm can be restarted after one step using the actual TOF state and good convergence still occurs. For the front legs, the additional spring energy added at maximum compression is $\eta + \delta$, and the rear leg additional energy is computed by $\eta - \delta$.

The differential thrust used to control pitch causes an unwanted moment on the roll axis. This moment is offset by a torque on the abduction/adduction axis, τ_a, and is computed by

$$\tau_{a_i} = k_p \gamma + k_d \dot\gamma + \frac{|N_i(f_i)|}{\sum_{i=1}^{4} |N_i(f_i)|} \sum_{i=1}^{4} N_i(f_i), \text{ for } i = 1, ..., 4. \qquad (7)$$

The first two terms form a proportional-derivative (PD) controller to eliminate existing roll and roll rate. The last term is a feedforward term used to negate the roll moment being produced by uneven leg thrusting. Leg kinematics are used to compute the roll moment, N_i, being applied by foot force, f_i, and an opposing torque is applied. The total moment seen by the body is the sum of moments from all legs. Each leg is directed to oppose the percentage of the total moment that it is responsible for.

Leg touchdown angles are used to regulate velocity. The thigh angle, θ_t, in **Fig. 4** can be set to adjust stride length, which affects forward velocity, v_{x_b}. In this work, known working leg angles for a discrete number of speeds are linearly interpolated to produce leg angles for all speeds in the desired range. This method is also used to compute the common mode thrust (before applying differential) and the abduction/adduction angles, θ_a.

Foot contacts for diagonal leg pairs are synchronized by extending and shortening the leg to compensate for body roll and pitch. Yaw regulation is achieved by placing the front legs to the outside of the turn and the hind legs to the inside. This method was suggested by Raibert [7], but because the system states are highly coupled, this approach can cause instability if both roll and pitch are not adequately controlled. Results of successful control on all three axes are provided in the next section.

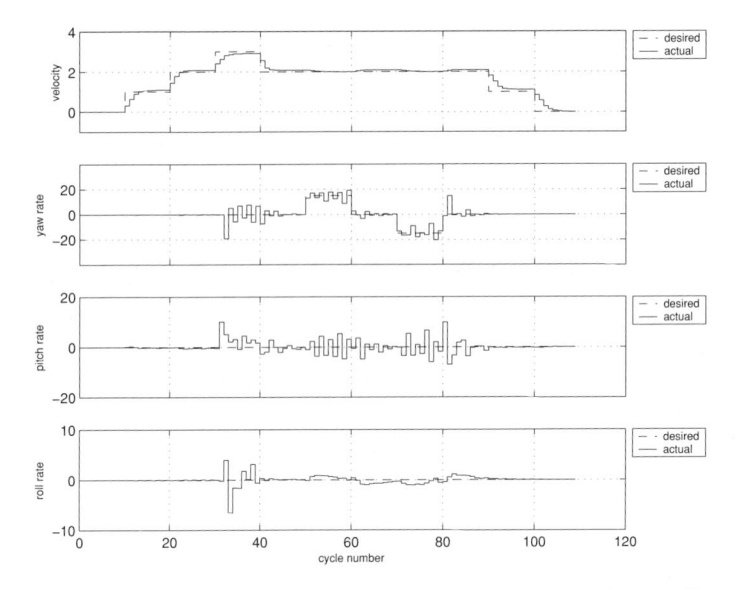

Fig. 5. 3D trotting. The velocity and yaw rate controllers perform well in tracking the commanded input. The linear pitch controller is less robust at higher speeds, suggesting that the system is more nonlinear at these speeds.

4 Results

Figure 5 shows the performance of the system in simulation responding to desired velocity and yaw rate changes. The linear pitch controller is less robust at higher speeds, suggesting that the system is more nonlinear at these speeds. The coupling that exists between the three axes of motion is evident here as changes in desired yaw rate (cycle 50) disturb the motion of both roll and pitch. The controller accounts for these disturbances and maintains the system's stability. Stability here is described as sustained regulation of system outputs and the ability to compensate for moderate disturbances.

5 Summary

The trotting controller presented here successfully regulates forward speed up to 3 m/s and turns at rates up to 20 deg/s. This appears to be the first reported regulation of quadruped heading while running at high speeds. Plans are for the robustness of this algorithm to be tested on the KOLT vehicle, and the results will also be a valuable stepping stone toward the development of a galloping controller at even higher speeds.

Acknowledgments

Support was provided by grant no. IIS-0208664 from the National Science Foundation to The Ohio State University.

References

1. M. Buehler, R. Battaglia, A. Cocosco, G. Hawker, J. Sarkis, and K. Yamazaki, "SCOUT: A simple quadruped that walks, climbs, and runs," in *Proceedings of the IEEE International Conference on Robotics and Automation*, (Leuven, Belgium), pp. 1707–1712, 1998.
2. M. D. Berkemeier, "Modeling the dynamics of quadrupedal running," *International Journal of Robotics Research*, vol. 17, pp. 971–985, September 1998.
3. A. Neishtadt and Z. Li, "Stability proof of Raibert's four -legged hopper in bounding gait," Tech. Rep. 578, New York University, September 1991.
4. P. P. Gambarian, *How Mammals Run*. New York: John Wiley & Sons, 1974.
5. P. Nanua and K. J. Waldron, "Energy comparison between trot, bound, and gallop using a simple model," *Journal of Biomechanical Engineering*, vol. 117, pp. 466–473, November 1995.
6. H. M. Herr and T. A. McMahon, "A trotting horse model," *International Journal of Robotics Research*, vol. 19, pp. 566–581, June 2000.
7. M. H. Raibert, "Trotting, pacing, and bounding by a quadruped robot," *Journal of Biomechanics*, vol. 23, suppl. 1, pp. 79–98, 1990.
8. J. G. Nichol, S. P. Singh, K. J. Waldron, L. R. Palmer III, and D. E. Orin, "System design of a quadrupedal galloping machine," *International Journal of Robotics Research*, vol. 23, no. 10-11, pp. 1013–1027, 2004.
9. D. V. Lee, J. E. A. Bertram, and R. J. Todhunter, "Acceleration and balance in trotting dogs," *Journal of Experimental Biology*, vol. 202, pp. 3565–3573, 1999.
10. R. J. Vaccaro, *Digital Control: A State-Space Approach*. New York: McGraw-Hill, Inc, 1995.

Research on Obstacle-navigation Control of a Mobile Robot for Inspection of the Power Transmission Lines Based on Expert System

Wang Ludan [1,2], Wang Hongguang [1], Fang Lijin [1], Zhao Mingyang [1]

1 Robotics Laboratory, Shenyang Institute of Automation, The Chinese Academy of Sciences, Shenyang, 110016, P.R.China
2 Graduate School of The Chinese Academy of Sciences, P.R.China
wangludan@sia.cn   hgwang@sia.cn   ljfang@sia.cn   myzhao@sia.cn

Abstract

This paper presents an obstacle-navigation control strategy for a mobile robot suspended on overhead ground wires of 500Kv extra-high voltage (EHV) power transmission lines, based on a novel movement mechanism. The control system of the mobile robot is based on a hybrid architecture expert system, which is mainly composed of a knowledge base and inference engine and can be applied to obstacle-navigation autonomously. In the control system, environment state data and fuzzified obstacle-data are input to the expert system so that the process of obstacle-navigation can be generated. Hence, joints variables can be calculated by solving the robot kinematics equations. According to the obstacle-navigation process and the calculated joints variables, the controller drives the motor rotations and the robot can overcome the obstacles.

Keywords: mobile robot; expert system; obstacle-navigation.

1 Introduction

Recent research on mobile robots that can run on high voltage power transmission lines has raised much interest as it can perform part or all the tasks of power transmission line equipment inspection and relieves humans from dangerous and tiring work. The inspection robot described here has two identical manipulators and each manipulator is composed of two arms: an 'up' arm and a 'down' arm. Each up arm consists of a wheel combined with a gripper. The wheels can rotate individually or all together and drive the robot to run on the power line. The gripper is connected to the up arm with a revolute joint, and the up arm is connected to the down arm with a revolute joint also. Each down arm is connected to the robot body with a prismatic joint. Thus the manipulators can translate on the robot body individually because the two prismatic joints share one linear guide orbit. The front manipulator cannot translate to the back and exceed the rear manipulator and vice verse.

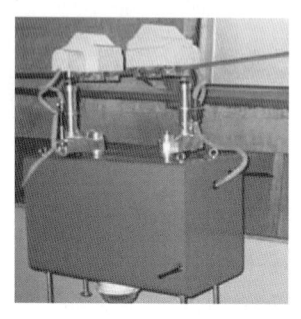

Fig .1 Photo of robot for inspection

There are many types of obstacles on the high voltage power transmission lines such as counterweight, single clamp, and dual clamps. Due to the different configuration and dimension of the obstacles it is impossible to overcome all the obstacles with one process. To solve this problem we can model the obstacle in advance and then tell the robot to "lift the front gripper "--"rotate the back wheel"……step by step the robot can overcome the obstacle. Obviously using this pattern we must know clearly what obstacle the robot encountered and we must program the process off-line. If the obstacle is not modeled or is irregular then the robot does not know how to overcome the obstacle. So the final method is that the robot can autonomously generate the locomotion process on-line according to the obstacle encountered.

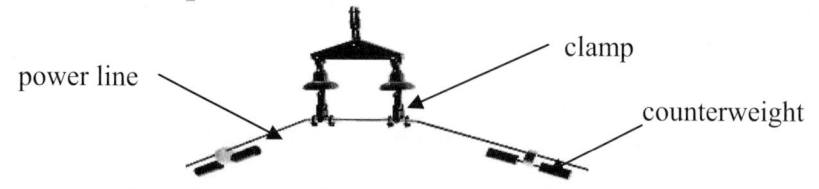

Fig.2 Obstacle environment on overhead ground wire

In the robot control system an expert system is adopted to on-line deduce the process of obstacle-navigation for the mobile robot. An expert system or knowledge based system are computer programs embodying knowledge about a narrow domain for solving problems relating to that domain. An expert system usually comprises two main elements, a knowledge base and an inference mechanism. In our expert system the knowledge base contains the locomotion rules of the robot. Inputs are the parameters of any obstacles and other factors that might affect the navigation process. The output is the process and joint variables for the robot to navigate the obstacle.

This paper is organized as follows: Following the introduction, section 2 describes the kinematics model of the robot, section 3 details the obstacle-navigation control strategy based on an expert system, 3.1 presents the inputs and outputs of the robot expert system, 3.2 details the fuzzification operation for the obstacle navigation. Section 3.3 describes the knowledge base and inference engine of the robot expert system. 3.4 is the realization of obstacle-navigation, section 4 contains the result of experiments, and section 5 presents the conclusion and the acknowledgments.

2 Kinematics Model of inspection robot

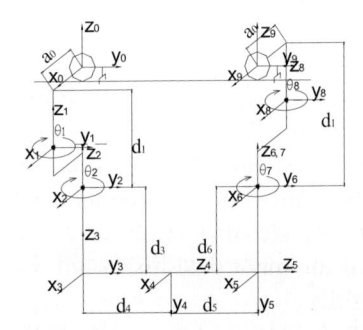

Fig . 3 Coordinate frames of inspection robot

In the robot system the control is realized in the joint space, while the task level commands are normally expressed in Cartesian space. As the mobile robot operation demands that at least one gripper grasps the overhead ground wires, the base coordinate frame can be set up for each gripper. If the base coordinate frame is installed in the rear gripper, the relative position and orientation equations of front gripper are as follows:

$$p_x = -\sin(\theta_1 + \theta_2)(d_4 + d_5) + a_0(1 - \cos\theta_1 - \cos(\theta_1 + \theta_2 + \theta_7) + \cos(\theta_1 + \theta_2 + \theta_7 + \theta_8)) \quad (1)$$

$$p_y = \cos(\theta_1 + \theta_2)(d_4 + d_5) + a_0(-\sin\theta_1 - \sin(\theta_1 + \theta_2 + \theta_7) + \sin(\theta_1 + \theta_2 + \theta_7 + \theta_8)) \quad (2)$$

$$p_z = d_6 - d_3 \quad (3)$$

$$\theta_x = 0 \quad (4)$$

$$\theta_y = 0 \quad (5)$$

$$\theta_2 = \theta_1 + \theta_2 + \theta_7 + \theta_8 \qquad (6)$$

In the equations $\theta_1, \theta_2, \theta_7, \theta_8, d_3, d_4, d_5, d_6$ are joint variables.

3 Development of the robot control system based on an expert system

3.1 Input and output of the expert system

To design an expert system there are many factors to be considered. In order to make the robot safely overcome an obstacle, the expert system should not only know the obstacle information but also the outside state and inside state of the robot. The following are the parameters that should be input into the expert system:
Configuration parameter of the obstacle: Includes geometric dimension of the obstacle, position and orientation of the obstacle.
Working state of the robot: Robot state that might influence the process of obstacle-navigation: position of the robot, state of electrical source.
The outside instruction and state: Remote control instruction and outside state that might affect obstacle-navigation.
After the processing of the input data, the robot expert system should generate the following results:
Permit the obstacle to be overcome or not: Tell the robot if it can overcome the obstacle or can't.
Available process for the robot to overcome the obstacle: Inform the robot process for obstacle-navigation.
Joints variables: Joints variables for obstacle-navigation.

3.2 Fuzzification of the obstacle figure

The input to the expert system includes the configuration parameters of the obstacle, but it is difficult for an expert system to "understand" this data and to process it. So prior to input to the expert system, we must first deal with these configuration parameters by fuzzification. Consider the configuration of the robot and the obstacle, then the fuzzification operation of the obstacle parameters can be achieved in six directions: up, down, left, right, front and rear. When the robot navigates an obstacle, whatever the configuration the obstacle is, there are 3 states. In state 1, both the front manipulator and the rear manipulator are located at the front of the obstacle; in state 2 the front gripper is located at the front of the obstacle and the rear is located at the back of the obstacle. Finally in state 3, both of the grippers are located at the back of the obstacle. So the process of obstacle-navigation is to change the robot state from state 1 to state 2 then to state 3. In other words: to move the front gripper to the back of the obstacle, and then grasp the wire, and finally to move the rear gripper.

Because the different configuration of obstacles will effect the way of moving the gripper, such as if the right of the wire has an obstacle, then the gripper must rotate 180°. It can then move forward if there is an obstacle on the wire and before moving forward, and the gripper must be lifted. The fuzzification of obstacle is performed in 6 directions. In each direction if there is no obstacle then the direction will be marked 0, if there is an obstacle but the gripper can move forward after some essential adjustment, then this direction will be marked 1, if there is an obstacle that the gripper can't pass that direction this will be marked 2. Fig .4 shows the fuzzification of a counterweight.

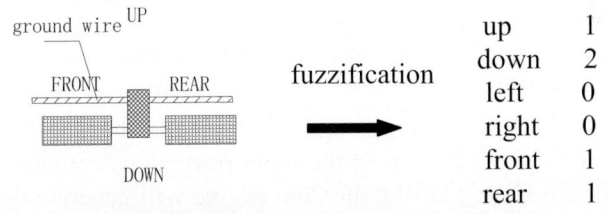

Fig.4 Fuzzification of a counterweight

For the counterweight example if there is an obstruction in the up direction, and if the gripper is lifted within the locomotion range it can pass the obstacle. If in the down direction there is an obstruction, then the gripper can't pass even if the down arm is contracted to its min, so the Up is 1 and the Down is 2. The opposite direction is judged in the same way, so after fuzzification each obstacle can be coded into 6 bit number.

3.3 Knowledge base and inference engine of the expert system

With the input to the expert system the inference engine can reason with the knowledge base and then generate the process for the robot to overcome an obstacle. The knowledge contained in the system has been compiled from two main sources: from the mobile robot designer's theory and the robot operators' experiences. In our system all the basic actions of the robot are stored into the knowledge base in the form of rules. The following are examples of the rules of our robot expert system.
Rules for obstacle-navigation:

IF *"power level low"* **THEN** *"obstacle-navigation is not permitted"*
IF *"wind power high"* **THEN** *"obstacle-navigation is not permitted"*
IF *"obstacle-navigation permitted"* **THEN** *"check the obstacle code"*;
IF *" the up is 1 "***THEN** *"front paw move forward"*
IF *"gripper move"* **THEN** *"lift the gripper"*
IF *"lift gripper"* **THEN** *"open gripper"*
; ....

With these rules stored in the knowledge base and with inputs to the inference engine we can generate the result need. Here the inference engine adopted is the 'production system' and following is the arithmetic of this system:

Step 1: put the initial fact (or data) into the dynamic data base;

Step 2: test (or match) the target with the fact in the dynamic data base. If the target is satisfied, then the reasoning is successful, return;

Step 3: match the fact in the dynamic data base with the rules in the knowledge base. The rules that have been successfully matched are added to the dynamic data base and a rules aggregate is built;

Step 4: if the rules aggregate is empty then failed; return; else go to step 2;

A case study:
Suppose the power level is high and other outside states satisfy the qualification of obstacle-navigation, if the input obstacle is counterweight whose configuration code is 121100, then the engine will generate the process for navigation: "open the front gripper", "front gripper lift", "rear wheel rotate forward", "front gripper down to the wire ","open rear gripper", ….. Thus the robot can navigate the counterweight step by step within the process.

3.4 Realization of obstacle-navigation control

In order to realize obstacle-navigation the joint's variables should be input to the robot controller. See from the kinematics equation of the mobile robot, when only one gripper grasps the wire the other gripper has 4 DOF (degree of freedom), while the joint number is 8, Thus the robot is a redundant system. It is difficult to obtain the inverse kinematics if all the joints are unlocked, while with the obstacle-navigation process some joints can be locked and the joint inverse kinematics can be obtained. For example if the robot wants to realize "front gripper move forward" this means the gripper just changes position in the direction x, then with relative posi-

tion equation (2) the unlocked joint variable is d_5 so the inverse kinematics is:
$$d_4 = \frac{p_y - a_0(-\sin\theta_1 - \sin(\theta_1 + \theta_2 + \theta_7) + \sin(\theta_1 + \theta_2 + \theta_7 + \theta_8))}{\cos(\theta_1 + \theta_2)} - d_5 \qquad (7)$$

Here $\theta_1, \theta_2, \theta_7, \theta_8, d_5$ are fixed so if the dimension of the obstacle in the y direction is l_y then the change of joint variable d_4 is given by:

$$\Delta d_4 = \frac{p_{yd} - p_{yo}}{\cos(\theta_1 + \theta_2)} = \frac{l_y}{\cos(\theta_1 + \theta_2)} \qquad (8)$$

With the joint variable the controller can control the robot and overcome the obstacle.

4 Experiment

The field experiment line is a simulated power transition line with a counterweight and is situated in the factory where the robot was manufactured. The obstacle condition was consistent with a real power transition line in China. A remote control platform communicates with the robot through an RF transmitter and receiver FC206. Following are the experimental procedures:

1. The robot runs on the line and is stopped when it encounters a counterweight.
2. The remote control platform sends the obstacle parameter to the robot.
3. The expert system generates the process and articulation trace for obstacle navigation, and the robot navigates the obstacle.

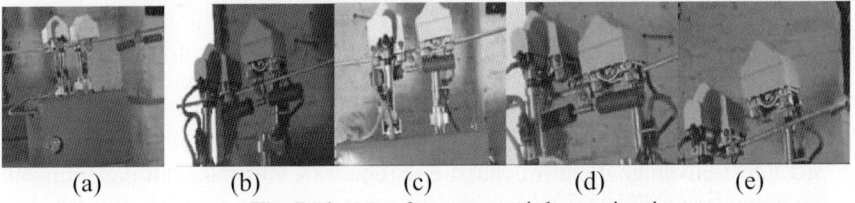

(a)　　(b)　　(c)　　(d)　　(e)

Fig.5 Photos of counterweight-navigation

The process of counterweight-navigation is: (a.) robot stop ;(b.) front gripper lift (c.) front gripper move forward (d.) front gripper down to the wire (e.) rear gripper lift and move forward.

See above the pictures of obstacle-navigation, the experiment validated the function of the robot expert system and can successfully generate the processes needed for obstacle-navigation.

5 Conclusion

Obstacle-navigation control of an inspection robot is described in this paper. With the expert system approach the mobile robot can overcome obstacles without intervention from a remote control platform and the need to previously model the obstacle. The results of the experiments performed have validated this method.

Acknowledgements

The authors acknowledge the support of Ling Lie, Li Yong engineers of the Robotics laboratory in Shenyang Institute of Automation, CAS. We would like to thank the works in the No.228 factory, without their hard work the robot could not be manufactured and tested successfully.

References

1. T. Tsujimura, T. Yabuta, T. Morimitsu, "Design of a wire-suspended mobile robot capable of avoiding path obstacles", IEE Proc.-Control Theory Appl., Vol. 143, NO. 4, July 1996, pp. 349-357
2. M.Cemal Cakir,Ozgur, Kadir Cavdar, "An expert system approach for die and mold making operations", Robotics and Computer-Integrated Manufacturing'21(2005)pp.175-183
3. Mineo Higuchi, Yoichiro Maeda, Sadahiro Tsutani, "Development of a Mobile pectionIns Robot for Power Transmission Lines", J. of the Robotics Society of Japan, Japan, Vol. 9, pp. 457-463, 1991
4. Tang Li, Fang Lijin, "Research on an inspection robot control system of power transmission lines based on a distributed expert system", J. of Robot, P.R.China, No. 3, May 2004
5. Joseph Giarratano, Gary Riley, Expert Systems Principles and Programming, CT: Thomson Learning, 2002
6. Michele Lacagnina,Giovanni Muscato,Rosario Sinatra, "Kinematics,dynamics and control of a hybrid robot Wheeleg ",Robotics and Autonomous System 45 (2003) 161-180
7. E. Garcia., P. Gonzalez de Santos , Crespo A. , "Mobile-robot navigation with complete coverage of unstructured enviroments", Robotics and Autonomous System 46 (2004) 195-204
8. Joseph Giarratano, Gary Riley, Expert Systems Principles and Programming, CT: Thomson Learning, 2002
9. Takeshi Tsujinmura,Takenori Morimitsu, "Dynamics of mobile legs suspend from wire",Robotics and Autonomous Systems 20(1997)pp85-98
10. Shinya Tanaka, Yoshinaga Maruyama, Kyoji Yano,"Work Automation with the Hot-Line Work Robot System 'Phase ☐' " ,Proceding of the 1996 IEEE International Conference on Robotics and Automation,Minncapolis.Mirmesota –April1996.

Measure of the Propulsion Dynamic Capability of a Walking System

A. David, O. Bruneau and J.-G. Fontaine

Laboratoire Vision et Robotique, ENSIB University of Orléans, France;
[anthony.david,olivier.bruneau,jean-guy.fontaine]@ensi-bourges.fr

Abstract

This paper presents criteria based on dynamics in order to control an under-actuated robot. The control strategy takes into account the intrinsic dynamics of the robot as well as the capability of its joint torques and modifies the dynamic effects generated by the robot directly.

Keywords: Dynamics criteria, walking gait, under-actuated robot.

1 Introduction

When humans walk, we actively use our own dynamic effects to ensure propulsion. Current research uses the dynamic effects to generate walking gaits. However, they use either pragmatic rules based on qualitative studies of human walking gaits ([1], [2]) or the theory of the passive walk with particular designs for the robot ([3],[4]) or reference trajectories calculated via various method ([5],[6],[7]). Our objective is to carry out a more adaptive and universal approach based on the dynamic equations. In this paper, we develop criteria based on dynamics in order to control a robot while taking into account its intrinsic dynamics as well as the capability of its joint torques and modifying its dynamic effects directly.

Even if these criteria are universal, we specify them for a planar under-actuated robot: RABBIT ([8],[9]) (Fig.1). RABBIT is composed of one trunk and two legs without a foot. The parameters used to describe its configuration are given in Fig.2. R_p is chosen as the reference point.

The organization of this paper is as follows. In part two, we introduce the dynamic propulsion criterion. In part three, the dynamic propulsion potential, and in part four, we describe the control strategy, based on these criteria, and the results obtained.

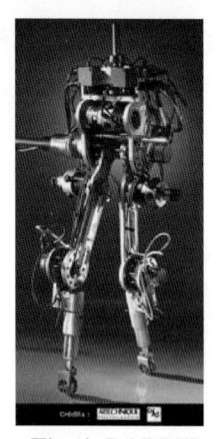

Fig. 1. RABBIT

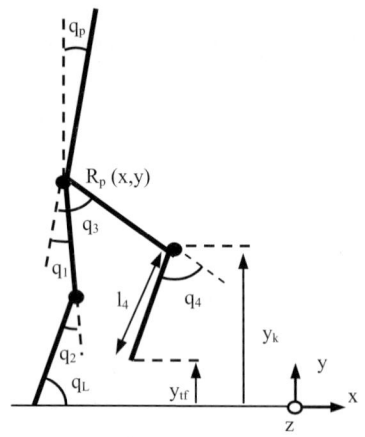

Fig. 2. Model of RABBIT

2 Dynamic propulsion criteria

To start, we define the dynamic propulsion criterion. The idea is to exactly quantify the required accelerations to generate the desired trajectory, defined at R_p. To do this, we define two generalized forces. Let $\bar{F}_r$ be the resulting generalized force applied to the robot at R_p and generating the real trajectory. Let $\bar{F}_i$ be the resulting generalized force which should be applied to the robot at R_p to generate the desired trajectory. The dynamic propulsion of the robot will be ensured if $\bar{F}_i$, the ideal resulting generalized force, is equal to $\bar{F}_r$, the real resulting generalized force.

2.1 Analytical expression of the dynamic propulsion criterion

To obtain the analytic expression of the dynamic propulsion criterion, we use the Lagrangian dynamic equations, applied at R_p:

$$\overline{\overline{M}}_p \bar{a}_p + \overline{\overline{H}}_k \bar{a}_k + \overline{C}_p + \overline{G}_p = \overline{\overline{D}}_{1p}^T \overline{F}_{c1} + \overline{\overline{D}}_{2p}^T \overline{F}_{c2} \quad (1)$$

$$\overline{\overline{H}}_k^T \bar{a}_p + \overline{\overline{M}}_k \bar{a}_k + \overline{C}_k + \overline{G}_k = \overline{\overline{D}}_{1k}^T \overline{F}_{c1} + \overline{\overline{D}}_{2k}^T \overline{F}_{c2} + \overline{\Gamma} \quad (2)$$

where $\bar{a}_k$ is the joint accelerations vector, $\bar{a}_p$ the trunk accelerations vector, $\overline{F}_{cj\,(j=1,2)}$ the contacting forces vector and $\overline{\Gamma}$ the joint torques vector. The index p refers to the trunk and the index k to the legs.

$\bar{F}_r$ is obtained while isolating $\bar{a}_k$ in (2) and introducing it in (1). The function terms unique to the trunk are grouped on the one hand, and the function terms unique to the legs and function of both on the other hand:

$$^p\overline{\overline{M}}_p \overline{a}_p + ^p\overline{C}_p + ^p\overline{G}_p = \left[\sum_{i=1}^{2}\overline{\overline{D}}_{ip}^T \overline{F}_{ci} - \overline{\overline{H}}_k \overline{\overline{M}}_k^{-1} \sum_{i=1}^{2}\overline{\overline{D}}_{ik}^T \overline{F}_{ci}\right] - \left[^k\overline{\overline{M}}_p - \overline{\overline{H}}_k \overline{\overline{M}}_k^{-1} \overline{\overline{H}}_k^T\right]\overline{a}_p + \overline{\overline{H}}_k \overline{\overline{M}}_k^{-1}\left[\overline{C}_k + \overline{G}_k - \overline{\Gamma}\right] - ^k\overline{C}_p - ^k\overline{G}_p \quad (3)$$

The left term is the trunk's dynamics and the right term is $\overline{F}_r$.

$\overline{F}_i$ is calculated with the trunk's dynamics where we replace the real values defining the real trajectory by their desired values defining the desired trajectory.

3 Dynamic propulsion potential

With the dynamic propulsion criterion, we know the required acceleration to generate the desired trajectory. However, does the robot have the capability to apply this acceleration? To answer this question, we introduce the dynamic propulsion potential. This criterion represents the capability of a system to generate a desired resulting generalized force, expressed as a range bounded by the maximum and the minimum resulting generalized force that it can apply at a reference point. This criterion integrates the constraints related to the system's design and the technology used as well as the generation of motions. In the case of RABBIT, it is possible to express one criterion for each leg and one criterion for the robot.

3.1 Analytical expression of the dynamic propulsion potentials

To obtain the expression of these potentials, we use the Lagrangian dynamic equations (1) and (2). First of all, the limits of the field of variation of the joint torques are expressed. The potentials are then calculated.

3.1.1 Field of variation of the joint torques

The idea is to limit the field of variation of the joint torques, given by the joint technology, such as the joint's actuators having the capability to stop the motion before reaching the joint stop, while taking into account the evolution of the dynamics. In spite of the coupling of the problem because of the dynamics, we decouple it in order to validate the approach. To resolve this problem, we use three main relations.

The first relation (4) integrates the joint technology. The angular velocity and the joint torque give the maximum and the minimum joint torques the joint actuator can generate to the next step of time:

$$(\dot{q}_i, \Gamma_i) \Rightarrow (\Gamma_i^{max}, \Gamma_i^{min}), \text{ where } i=1,2,3,4 \quad (4)$$

The second relation (5) links the joint torques vector $\overline{\Gamma}$ to the joint acceleration vector $\overline{a}_k$. The trunk accelerations vector $\overline{a}_p$ is isolated in (1) and introduced in (2):

$$\overline{\Gamma} = \left[\overline{\overline{M}}_k - \overline{\overline{H}}_k^T \overline{\overline{M}}_p^{-1} \overline{\overline{H}}_k\right] \overline{a}_k + \overline{C}_k + \overline{G}_k - \overline{\overline{H}}_k^T \overline{\overline{M}}_p^{-1} \left[\overline{C}_p + \overline{G}_p\right] + \overline{\overline{H}}_k^T \overline{\overline{M}}_p^{-1} \sum_{i=1}^{2} \overline{\overline{D}}_{ip}^T \overline{F}_{ci} - \sum_{i=1}^{2} \overline{\overline{D}}_{ik}^T \overline{F}_{ci} \quad (5)$$

The third relation (6) gives the new angular velocity ($\dot{q}_i^{t+\delta t}$) and position ($q_i^{t+\delta t}$), knowing the new angular acceleration ($\ddot{q}_i^{t+\delta t}$) and the old angular velocity ($\dot{q}_i^t$) and position (q_i^t):

$$\dot{q}_i^{t+\delta t} = \dot{q}_i^t + \ddot{q}_i^{t+\delta t} \delta t, \quad q_i^{t+\delta t} = q_i^t + \dot{q}_i^t \delta t + \ddot{q}_i^{t+\delta t} \frac{\delta t^2}{2}, \text{ with } \delta t \text{ the step of time} \quad (6)$$

With these relations and the maximum and minimum joint stops $\left[q_i^{min}, q_i^{max}\right]$, we calculate the limits of the field of variation of each joint torque. With (4), Γ_i^{max} and Γ_i^{min} are calculated and only bounded by actuators limits. Then, with Γ_i^{max}, Γ_i^{min} and (5), the maximum and minimum angular acceleration $\ddot{q}_i^{max}$ and $\ddot{q}_i^{min}$ are calculated. In the next step, new angular acceleration limits $\ddot{q}_i^{max(n)}$ and $\ddot{q}_i^{min(n)}$ are calculated by taking into account the joint stops q_i^{max} and q_i^{min}. Only the method to find $\ddot{q}_i^{max(n)}$ is explained in this paper. However, it is exactly the same method for $\ddot{q}_i^{min(n)}$ calculation.

With $\ddot{q}_i^{max}$ and (6), the angular velocity and position which will be generated are calculated. We are now able to determine if the joint actuator has the capability to stop the motion before reaching q_i^{max}. At each step of time the minimum joint torque which can be generated is computed in order to take into account the evolution of the dynamics. For that, we use the relations (4), (5) then (6) to have the evolution of the angular joint acceleration, velocity and position. If the joint actuator can stop the motion before the joint stop then it can generate without risk Γ_i^{max}, otherwise we calculate the new maximum joint torque which does not allow the joint stop to be exceeded.

3.1.2 Dynamic propulsion potentials

We use the expression of $\overline{F}_r$ (3), separate in the two subsystems and each related to a leg:

$$\overline{F}_{rj} = \left[\overline{\overline{D}}_{jp}^T - \overline{\overline{H}}_{kj} \overline{\overline{M}}_{kj}^{-1} \overline{\overline{D}}_{jk}^T\right] \overline{F}_{cj} - \left[^{kj}\overline{\overline{M}}_p - \overline{\overline{H}}_{kj} \overline{\overline{M}}_{kj}^{-1} \overline{\overline{H}}_{kj}^T\right] \overline{a}_p + \overline{\overline{H}}_{kj} \overline{\overline{M}}_{kj}^{-1} \left[\overline{C}_{kj} + \overline{G}_{kj} - \begin{pmatrix} \Gamma_{hj} \\ \Gamma_{gj} \end{pmatrix}\right] - ^{kj}\overline{C}_p - ^{kj}\overline{G}_p \quad (7)$$

where Γ_{hj} and Γ_{gj} are respectively the hip and knee joint torque of the leg j.

Measure of the Propulsion Dynamic Capability of a Walking System

We introduce the constraints related to the robot's design and the generation of walking gaits. For instance, concerning the dynamic propulsion along the x direction which is the most important to generate walking gaits, the joint of crucial interest is the knee joint if the leg is in contact, and the hip joint if the leg is in transfer. So, the potential P_j of the leg j is:

$$[\overline{F}_{ri}^{min}, \overline{F}_{ri}^{max}] = [\min\{\overline{F}_{ri}(\Gamma_{hj}^{min}, \Gamma_{gj}), \overline{F}_{ri}(\Gamma_{hj}^{max}, \Gamma_{gj})\}; \max\{\overline{F}_{ri}(\Gamma_{hj}^{min}, \Gamma_{gj}), \overline{F}_{ri}(\Gamma_{hj}^{max}, \Gamma_{gj})\}] \qquad (8)$$

if the leg j is in transfer, and :

$$[\overline{F}_{ri}^{min}, \overline{F}_{ri}^{max}] = [\min\{\overline{F}_{ri}(\Gamma_{hj}, \Gamma_{gj}^{min}), \overline{F}_{ri}(\Gamma_{hj}, \Gamma_{gj}^{max})\}; \max\{\overline{F}_{ri}(\Gamma_{hj}, \Gamma_{gj}^{min}), \overline{F}_{ri}(\Gamma_{hj}, \Gamma_{gj}^{max})\}] \qquad (9)$$

if the leg j is in contact. Knowing the potential of each leg, the potential P of the robot is finally:

$$[\overline{F}_r^{min}, \overline{F}_r^{max}] = [\overline{F}_{r1}^{min} + \overline{F}_{r2}^{min}, \overline{F}_{r1}^{max} + \overline{F}_{r2}^{max}] \qquad (10)$$

4 Control strategy

With these criteria, we develop a strategy to control RABBIT and we have validated it with the simulation of the first step of a walking gait. To start, we define the desired average velocity at R_p. As an illustration, a linear velocity $\dot{x}$ is imposed and the desired trunk angle q_p is equal to zero. The evolution of the height y is free. So we ensure the dynamic propulsion along the x direction and around the z direction only. The next idea is to modify, via the joint torques, the dynamic effects generated by the legs in order to make $\overline{F}_r$ converge towards $\overline{F}_i$. For that, we calculate the desired joint torques vector $\overline{\Gamma}^d$, with (3), where we replace $\overline{F}_r$ by $\overline{F}_i$ and $\overline{\Gamma}$ by $\overline{\Gamma}^d$. With the equations along the x direction and around the z direction, we can calculate $\overline{\Gamma}^d$ directly. However, we do not take into account the constraints related to the robot's design and to the generation of walking gaits. These constraints are introduced as functions of the locomotion phases.

4.1 Double support phase

During this phase, we have four joint actuators to ensure the dynamic propulsion and the closed loop constraint (Fig. 2 with $y_{tf}=0$). As we have three equations, we use a generalized inverse to calculate $\overline{\Gamma}^d$.

In order not to exceed the capability of the robot, we use the dynamic propulsion potential of each leg. We compare the force generated by the legs to their dynamic propulsion potential along the x direction. If a leg reaches its maximum capability, we decide to start the take off of the concerning leg to perform the single support phase.

We see that the robot generated the desired moving velocity at R_p (Fig.3), without exceeding its capability, or the capability of each leg along the x direction (Fig.4). After 1,3 seconds, the maximum capability of the back leg along the x direction reduces. Indeed, the knee joint of the back leg approaches its joint stop (Fig.5). So, when the back leg reaches its maximum capability, we decide to start the take off to perform the single support phase.

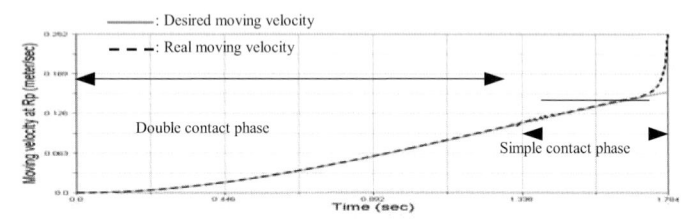

Fig. 3. Desired and real moving velocity at R_p

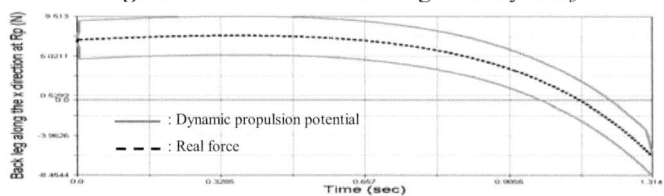

Fig. 4. Dynamic propulsion potential and real force of the back leg along the x direction at R_n

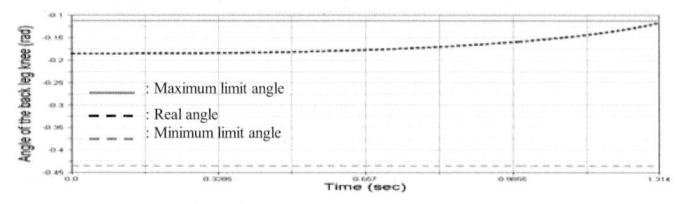

Fig. 5. Angle and limit angle of the back leg knee

4.2 Single support phase

During this phase, we have four joint actuators to ensure the dynamic propulsion and to avoid the contact between the tip of the transfer leg and the ground. For that, the transfer leg's knee is used to avoid the contact while the three other joints perform the dynamic propulsion of the robot.

The joint torque of the transfer leg's knee is calculated with a computed torque method using non linear decoupling of the dynamics. The desired joint acceleration, velocity and position of the transfer leg knee are calculated with inverse kinematics. We express the joint torque of the transfer leg's knee as a function of the three other joint torques and of the desired control vector components of the transfer leg's knee.

With the three other joints we perform the dynamic propulsion. It is possible that the robot does not have the capability to propel itself along the x direction dynamically. In this case, we limit the desired force with its dynamic propulsion potential. Then, we distribute this desired force with the dynamic propulsion potential of each leg. In order to keep a maximum capability for each leg, the desired force generated by each leg is chosen to be as further as possible from the joint actuators limits. In this case, we have three equations, one for the desired force along the x direction for each leg, and one for the desired force around the z direction. To calculate the three joint torques, the joint torque of the transfer leg's knee is replaced by its expression as a function of the three other joint torques. So, we first calculate the three joints torque performing the dynamic propulsion, then the joint torque avoiding the contact between the tip of the transfer leg and the ground.

We see that the robot can ensure the propulsion along the x direction (Fig.6) and generates the desired moving velocity (Fig.3). Moreover, the control strategy involves naturally moving of the transfer leg (Fig.7). After 1,675 seconds, the robot can not ensure exactly the desired propulsion along the x direction. Indeed, the transfer leg is passed in front of the contact leg and the system is just like an inverse pendulum under gravity and for which the rotational velocity increases quickly.

5 Conclusion and perspectives

In this paper, we presented two criteria based on dynamics: the dynamic propulsion criterion and the dynamic propulsion potential, to control an under-actuated robot. The control strategy took into account the intrinsic dynamics of the robot as well as the capability of its joint torques and modified the dynamic effects generated directly by itself. It was validated with the generation of the first step of a dynamic walking gait.

Our future work will consist of developing a new controller in order to exploit the capability of the robot during the double contact phase. Then, we will develop a control method to ensure the landing in double contact phase and thus generate dynamic walking gaits without reference trajectories.

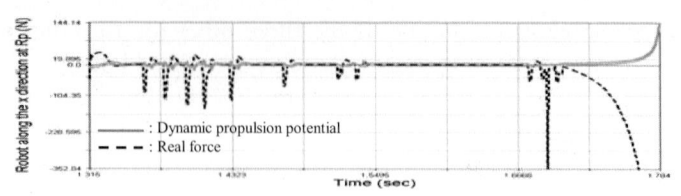

Fig. 6. Dynamic propulsion potential and real force of the robot along the x direction at R_p

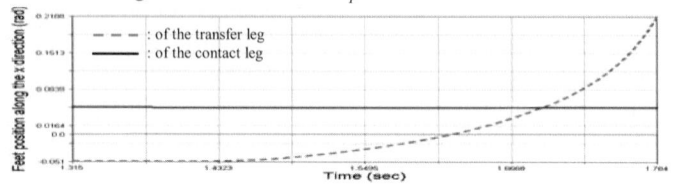

Fig. 7. Feet position along the x direction

References

1. Pratt J.E., Chew C.-M., Torres A., Dilworth P., Pratt G. (2001) Virtual Model Control: An Intuitive Aprroach for Bipedal Locomotion. *Internationl Journal of Robotics Research*, vol. 20, no. 2, pp. 129-143.
2. Sabourin C., Bruneau O., Fontaine J.G. (2004) Pragmatic Rules for Real-Time Control of the Dynamic Walking of an Under-Actuated Biped Robot. *IEEE Proc. of Int. Conf. On Robotics and Automation (ICRA)*, New Orleans, USA, 26 April-1 May.
3. Asano F., Yamakita M., Kamamichi, Luo Z. (2002) A novel gait generation for biped walking robots based on mechanical energy constraint. *IEEE Proc. Conf. on Intelligent Robots and Systems (IROS)*, Lausanne, Switzerland, 30 September-4 October.
4. Khraief N., M'Sirdi N.K., Spong M.W. (2003) Nearly passive dynamic walking of a biped robot. *Proc. European Control Conference*.
5. Chevallereau C., Djoudi D. (2003) Underactuated Planar Robot Controlled via a Set of Reference Trajectories. *International Conference on Climbing and Walking Robots (CLAWAR)*, Catania, Italy, 17-19 September.
6. Grizzle J.W., Abba G., Plestan F. (2001) Asymptotically stable walking for biped robots : analysis via systems with impulse effects. *IEEE Transaction on Automatic Control*, vol. 46, no. 1, pp 51-64.
7. Canudas-de-Wit C., Espiau B., Urrea C. (2002) Orbital Stabilization of Underactuated Mechanical Systems. *IFAC*, Barcelone.
8. Buche G., "ROBEA Home Page", http://robotrabbit.lag.ensieg.inpg.fr/English/.
9. Chevallereau C., Abba G., Aoustin Y., Plestan F., Westervelt E.R., Canudas C., Grizzle J.W. (2003) *RABBIT : a testbed for advanced control theory*. Control Sys. Mag., vol. 23, no. 5, pp. 57-59.

Experimental Walking Results of LUCY, a Biped with Pneumatic Artificial Muscles

Bram Vanderborght, Björn Verrelst, Ronald Van Ham, Michael Van Damme, and Dirk Lefeber

Vrije Universiteit Brussel, Department of Mechanical Engineering
Pleinlaan 2, B-1050 Brussels, Belgium
bram.vanderborght@vub.ac.be
http://lucy.vub.ac.be

Abstract

This paper discusses in detail the control architecture of the biped Lucy which is actuated with pleated pneumatic artificial muscles. Pleated pneumatic artificial muscles have interesting characteristics that can be exploited for legged locomotion. They have a high power to weight ratio, an adaptable compliance and they can absorb impact effects. The trajectory generator calculates trajectories represented by polynomials based on objective locomotion parameters, which are average forward speed, step length, step height and intermediate foot lift. The joint trajectory tracking controller is divided in three parts: a computed torque module, a delta-p unit and a bang-bang pressure controller. The control design is formulated for the single support and double support phase, where specifically the trajectory generator and the computed torque differs for these two phases. The first results of the incorporation of this control architecture in the real biped Lucy are given.

Keywords: Biped, Pneumatic Artificial Muscle, Dynamic Control.

1 Introduction

Bipeds can be divided into 2 main categories: on one side the completely actuated robots who don't use natural or passive dynamics at all, like Asimo [1], Qrio, HRP-2 [2] and on the other side the "passive walkers" who don't need actuation at all to walk down a sloped surface or only use a little actuation just enough to overcome friction when walking over level ground like the Cornell biped, Delft biped Denise and MIT robot Toddler [3].

The main advantage of the last group of robots is that they are highly energy efficient but unfortunately they are of little practical use. They have difficulties to start, they can't change their speed and cannot stop compared to a completely actuated robot. But these robots swallow up a lot of energy and energy consumption is an important issue for bipeds.

Our goal is to develop a robot somewhere between these two. We want to adapt the natural dynamics as a function of the imposed trajectories. For this research we need a joint with adaptable compliance. This is not possible by using electrical drives because they need a gearbox to deliver enough torque at low rotation speeds which make the joint stiff. We choose to use Pleated Pneumatic Artificial Muscles because in an antagonistic setup the compliance is controllable [4]. But the control of these actuators is not so commonly know and consequently requires an extra effort to use them. The Pleated Pneumatic Artificial Muscle is developed at our department. When inflated, the pleats of the membrane will unfold, the muscle will contract while generating high pulling forces. The muscles can only pull, in order to have a bidirectionally working revolute joint one has to couple two muscles antagonistically, where the applied pressures determine both position and stiffness. They can also be directly coupled without complex gearing mechanisms.

2 The Robot

The Robotics and Multibody Mechanics Research Group of the Vrije Universiteit Brussel has built the planar walking biped "Lucy" (**Fig. 1**). The goal of the project is to achieve a lightweight bipedal robot which is able to walk in a dynamically stable way, while exploiting the passive behaviour of the PPAM's in order to reduce energy consumption and control efforts. It weights 30kg and is 150cm tall. It uses 12 muscles to actuate 6 degrees of freedom so Lucy is able to walk in the sagittal plane. A sliding mechanism will prevent the robot from falling sideways.

3 Control Architecture

To induce fast and smooth motion a dynamic controller is required. The considered controller is given in the schematic overview of **Fig. 2** and is a combination of a global trajectory generator and a joint trajectory tracking controller.

3.1 Trajectory Generator

The trajectory generation is designed to generate walking movements based on objective locomotion parameters chosen for a specific robot step, which are average forward speed, step length, step height and intermediate foot lift. These parameters are calculated by a high level path planning control unit, which is beyond the scope of the current research. The trajectories of the leg links are represented by polynomials and is a simplified version of the method developed by Vermeulen [5] where the upper body and hip velocity is held constant. This is sufficient for slow walking speeds. In the future the method developed by Vermeulen [5] will be used to achieve faster walking. Here the trajectory planner generates motion patterns based on two specific concepts, being the use of objective locomotion parameters, and exploiting the natural upper body dynamics by manipulating the angular momentum equation. Thus taking the motion of the upper body into account and not keeping it at a fixed angle as is the case in this paper. The effectiveness of this method has been proven in simulation [6].

3.2 Joint Trajectory Tracking Controller

The joint trajectory tracking controller can be divided in three parts: the computed torque module, the delta-p unit and the bang-bang pressure controller. The trajectory planner, computed torque module and delta-p unit are implemented on a central computer; the bang-bang controller on a micro-controller (each joint has its own microcontroller). The communication system between computer and microcontrollers uses the USB 2.0 high speed protocol and works at 2000Hz. *Computed Torque* After the trajectory generator the joint drive torques are calculated. This control unit is different for the single and double support phase.

During single support the torques are calculated using the popular computed torque technique consisting of a feedforward part and a PID feedback loop. When the robot is in single support phase the robot has 6 DOF and the 6 computed torques can be calculated. During the double support phase, immediately after the impact of the swing leg, three geometrical constraints are enforced on the motion of the system. They include the step length, step height and the angular position of the foot.

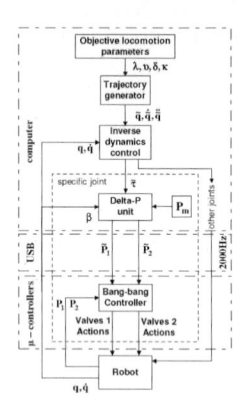

Fig. 1. Photograph of Lucy   **Fig. 2.** Control architecture

Due to these constraints, the robot's number of DOF is reduced to 3. The dynamic model has to be reformulated in terms of the reduced independent variables. The dependant variables are related to the dependant variables through the kinematical jacobian. An actuator redundancy arises because there are 6 actuators which need to control only 3 DOF. To be able to calculate the 6 joint drive torques a control law as proposed by [7] can be used, which is based on a matrix pseudoinverse. The disadvantage is that these results are discontinue when switching between single and double support phase. An alternative way to distribute torques over the actuators, is to make a linear transition of torques between the old and new single support phase, by calculating the applied torque as if the robot is in single support phase. The advantage of this strategy is that there are no torque discontinuities when switching between single and double support phase. The disadvantage is that the calculated torques are not dynamical correct values, but the double support phase is rather short and a feedback loop is implemented. Experimental results show this is a good strategy not causing serious problems for the regarded motions.

Delta-p Control For each joint an estimated computed torque is available. The computed torque is then feeded into the delta-p control unit, one for each joint, which calculates the required pressure values to be set in the muscles. The generated torque in an antagonistic setup with two muscles is given by:

$$\tau = \tau_1 - \tau_2 = p_1 l_1^2 r_1 f_1 - p_2 l_2^2 r_2 f_2 = p_1 t_1(\beta) - p_2 t_2(\beta) \qquad (1)$$

with p_1 and p_2 the applied gauge pressures in front and back muscle respectively which have lengths l_1 and l_2. The dimensionless force functions of both muscles are given by f_1 and f_2 [4]. The kinematical transformation from forces to torques are represented by r_1 and r_2 which results, together with the muscle force characteristics, in the torque functions t_1 and t_2. These functions are determined by the choices made during the design phase and depend on the joint angle β. Thus joint position is influenced by weighted differences in gauge pressures of both muscles.

The two desired pressures are generated from a mean pressure value p_m while adding and subtracting a Δp value:

$$\tilde{p}_1 = p_m + \Delta p \quad ; \quad \tilde{p}_2 = p_m - \Delta p \tag{2}$$

The mean value p_m will determine the joint stiffness and will be controlled in order to influence the natural dynamics of the system[9]. Feeding back the joint angle β and using expression (1), Δp can be determined by:

$$\Delta p = \frac{\tilde{\tau} + p_m \left[(t_2(\beta) - t_1(\beta)) \right]}{t_2(\beta) + t_1(\beta)} \tag{3}$$

The delta-p unit is thus a feed-forward calculation from torque level to pressure level and uses estimated values of the muscle force function and estimated kinematical data of the pull rod mechanism.

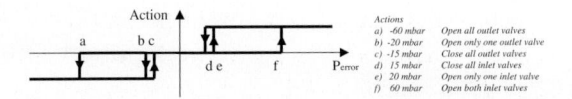

Fig. 3. Multi-level bang-bang control scheme

Bang-bang Pressure Controller In order to realize a lightweight, rapid and accurate pressure control, fast switching on-off valves are used. The pneumatic solenoid valve 821 2/2 NC made by Matrix weighs only 25g. The opening time is about 1 ms and it has a flow rate of 180 Std.l/ min. A set of 2 inlet and 4 outlet valves are used per muscle. In the last control block the desired gauge pressures are compared with the measured gauge pressure values after which appropriate valve actions are taken by the multi-level bang-bang pressure controller with dead zone (**Fig. 3**) [8].

4 Experimental Results

In this section the measurements of the walking biped are shown with the following chosen objective locomotion parameters : mean forward velocity

$\nu = 0.02$ m/s; steplength $\lambda = 0.10$ m; stepheight $\delta = 0.0$ m; footlift $\kappa = 0.04$ m. The calculated single support and double support phase durations are respectively 4 s and 1 s.

The graphs (**5-7**) depict 6 single support phases and 6 double support phases. **Fig. 5** shows the desired and real joint angles β_i of the ankle ($i = 1$), knee ($i = 2$) and the hip ($i = 3$) respectively. The definitions of the oriented relative joint angles are giving in **Fig. 4** (counterclockwise positive).Vertical lines on all graphs show the phase transition instants. Due to the nature of the bang-bang pneumatic drive units and the imperfections introduced in the control loops, tracking errors can be observed. Especially when phase transitions occur, since these introduce severe changes for the control signals. But tracking errors are not that stringent as it is for e.g. welding robots, the most important thing is that the overall dynamic robot stability is guaranteed. **Fig. 6** visualizes the

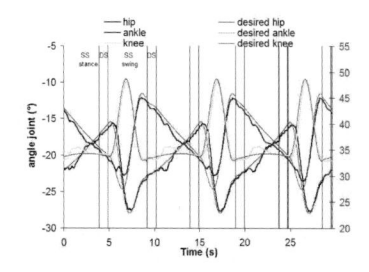

Fig. 4. Definition of the oriented relative joint angles (CCW positive)

Fig. 5. Desired and measured angle of left hip, knee and angle joint

actual applied torque for the knee of the left leg, which consists of a PID feedback part and computed torque part. The computed torque controller is working well, but the robot parameters still have to be fine-tuned to lower the action of the PID controller. The pneumatics are characterized by pressure courses in both muscles of each joint. **Fig. 8** and **Fig. 9** depict desired and measured gauge pressure for the front and rear muscle of the knee of the left leg. All these graphs additionally show the valve actions taken by the respective bang-bang pressure controller. Note that in these figures a muscle with closed valves is represented by a horizontal line depicted at the 2 bar pressure level, while a small peak upwards represents one opened inlet valves, a small peak downwards one opened exhaust valves and the larger peaks represent two opened inlet or four opened outlet valves. The desired pressures are calculated by the delta-p

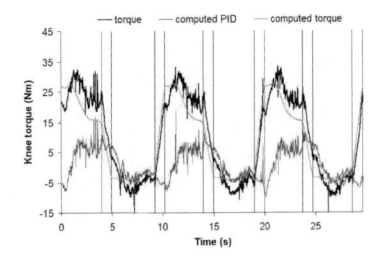

Fig. 6. Left knee torque

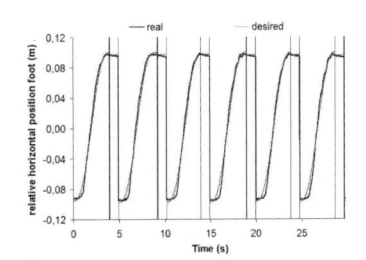

Fig. 7. Desired and real horizontal position swing foot

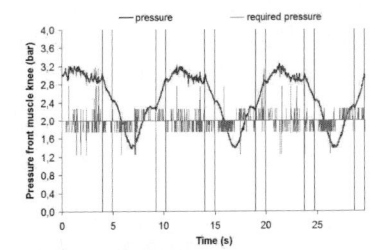

Fig. 8. pressure and valve action of front left knee muscle

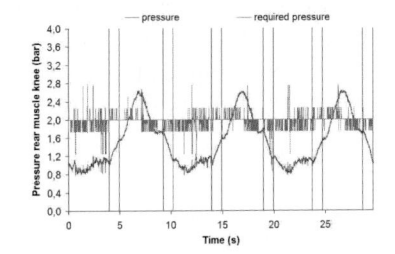

Fig. 9. pressure and valve action of back left knee muscle

unit. For this experiment the mean pressure p_m for all joints is taken at 2 bar, consequently the sum of the pressures in each pair of graphs, drawing the front and rear muscle pressures, is always 4 bar. It is observed that the bang-bang pressure controller is very adequate in tracking the desired pressure. Currently a lot of valve switching is required due to the fix compliance setting. By incorporating the natural dynamics of the system the switching will be reduced [9].

Fig. 7 depict the horizontal position of the swing foot. Note that the swing foot moves twice the step length during single support phase. Compared with the desired objective locomotion parameter only small deviations can be observed. This proofs the global performance of the proposed control strategy.

The position of the zero moment point is not measured yet and it is impossible to calculate the position out of the movements of the robot because the accelerations are not measured. In the future ground force sensors will be installed to track the ZMP. If necessary this information will be placed in an extra feedback loop to achieve postural stability during faster walking.

5 Conclusions

This paper discusses in detail the control architecture of the biped Lucy which is actuated with pleated pneumatic artificial muscles. The current control architecture focusses on the trajectory generator and the joint trajectory tracking controller. The effectiveness of the proposed controller is shown by performing walking experiments. Some videos of these movements can be seen at the website: http://lucy.vub.ac.be.

References

1. M. Hirose, Y. Haikawa, T. Takenaka, and K. Hirai, "Development of humanoid robot ASIMO," in *Proceedings of the IEEE/RSJ International Conference on Intelligent Robots and Systems*, 2001.
2. K. Yokoi and al., "Humanoid robot's applications in HRP," in *Proceedings of the IEEE International Conference on Humanoid Robots*, (Karlsruhe, Germany), 2003.
3. S. H. Collins, A. Ruina, R. Tedrake, and M. Wisse, "Efficient bipedal robots based on passive-dynamic walkers," *Science*, vol. 18, pp. 1082–1085, February 2005.
4. F. Daerden, D. Lefeber, P. Kool, and E. Faignet, "Free radial expansion pneumatic artificial muscles," in *Proceedings of the 6th International Symposium on Measurement and Control In Robotics*, pp. 44–49, 1996.
5. J. Vermeulen, B. Verrelst, D. Lefeber, P. Kool, and B. Vanderborght, "A real-time joint trajectory planner for dynamic walking bipeds in the sagittal plane," *Robotica*, in press.
6. B. Verrelst, J. Vermeulen, B. Vanderborght, R. Van Ham, J. Naudet, D. Lefeber, F. Daerden, and M. Van Damme, "Motion generation and control for the pneumatic biped lucy," *International Journal of Humanoid Robotics (IJHR)*, in press.
7. C.-L. Shih and W. Gruver, "Control of a biped robot in the double-support phase," *IEEE Transactions on Systems, Man and Cybernetics*, vol. 22, no. 4, pp. 729–735, 1992.
8. R. Van Ham, F. Daerden, B. Verrelst, D. Lefeber, and J. Vandenhoudt, "Control of pneumatic artificial muscles with enhanced speed up circuitry," in *Proceedings of the 5th International conference on Climbing and Walking Robots and the Support Technologies for Mobile Machines*, pp. 195–202, September 2002.
9. B. Verrelst, R. Van Ham, B. Vanderborght, J. Vermeulen, D. Lefeber, and F. Daerden, "Exploiting adaptable passive behaviour to influence natural dynamics applied to legged robots," *Robotica*, vol. 23, no. 2, pp. 149–158, 2005.

Hip Joint Control of a Legged Robot for Walking Uniformly and the Self-lock Mechanism for Compensating Torque Caused by Weight

T. Okada, T. Sakai, K. Shibuya and T. Shimizu

Department of Biocybernetics, Faculty of Engineering, Niigata University,
Ikarashi 2-8050, Niigata, 950-2181 JAPAN   okada@bc.niigata-u.ac.jp

Abstract

This paper describes hip joint control of the legged robot, PEOPLER (Perpendicularly Oriented Planetary Leg Robot) to make the robot walk uniformly. Equations are shown expressing robot speed while 4 legs change attitude patterns periodically as the hip joint rotates. With differentiation of each equation we can find angular velocity of the hip joint so that the robot can walk forward at a constant speed. Also, we show vertical displacements of the robot in connection with the angular velocity. Moreover, we devised a leg driving mechanism composed of worms and worm wheels for compensating torque caused by weight operating on the hip joint. Typical features for worm gearing such as generating lateral and axial forces are analyzed. In order to prevent the shaft from shifting toward the axial direction, we show how strong axial stoppers should be placed on a propeller shaft connecting the worms. These considerations are verified by experiments using PEOPLER Mk.2 and a miniature sized walker.

Keywords: Legged robot, worm gears, weight torque cancellation, stride.

1 Introduction

Legged robots are useful for moving on irregular terrain since it can step over obstacles. Biologically inspired robots have been developed [1], [2]. Crawling robots are also useful on this terrain [3].

However, some robots move too fast in order to keep their dynamic stability while others move slowly to get enough torque from the mini-motor. We must calculate kinematics or dynamics for some robots. Particularly, a robot with a multiple-jointed configuration needs complicated algorithms to find angular variables for generating a constant speed in walking, turning, and spinning. Our legged robot PEOPLER has 4 legs [4]. Each leg can kick and proceed simply by repeating a cyclic motion generated by coaxial driving mechanisms. Its leg attitude patterns and stable direction change have been discussed [5]. The robot's speed varies. Also, the robot can not stand unless extraordinary power is applied.

This paper describes hip joint control to allow the robot PEOPLER to walk at a constant speed. In fact, we introduce equations expressing robot speed while the 4 legs change attitude patterns periodically as the hip joint rotates. Differentiation of each equation tells us the angular velocity of the hip joint so that the robot can walk forward at a constant speed. We show that time dependent angular displacements of the hip joint depend on the leg attitude pattern. We also demonstrate vertical displacements of the robot. Moreover, we devised a leg driving mechanism using worm gears for compensating torque caused by weight operating on the hip joint. Lateral forces generated in these gears are analyzed. In order to reduce the affect of these forces, we show smart design using the worm gears. Experiments by PEOPLER Mk.2 and a miniature sized walker verify the effect of these considerations.

2 Hip joint control for a designated speed

To make the robot walk at a constant speed, we speculate time dependent hip joint angular parameters and extract mathematical formulations.

2.1 Robot speed when hip joint rotates at a fixed speed

Leg attitude is always vertical when the stride is controlled in our normal stride mode. However, the leg inclines forward or backward a little in other modes [1](see **Fig.1**). This makes the robot's speed change when the hip joint rotates at a fixed speed.

The value of γ_{max} is given to determine the stride. Let's suppose that γ_{max} = $\pi/6$[rad], then instantaneous attitude angle γ is calculated accordingly as θ_r increases. These angles versus θ_r are shown on the left side of **Fig.2**. In the coordinate system (x, y) in Fig.2 (right), horizontal displacement of the robot's centre G is written as

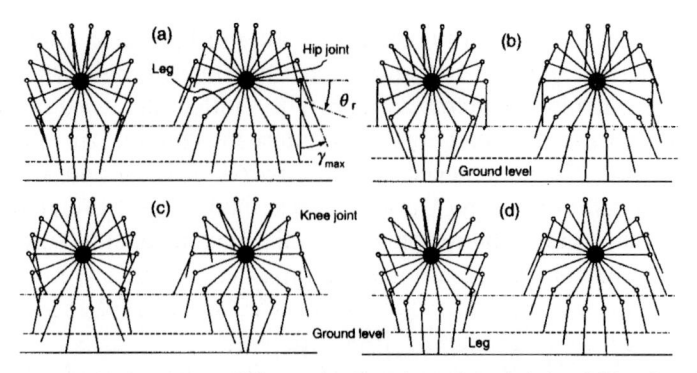

Fig. 1. Four typical patterns of leg attitude. The left and right sides of each pattern are utilized to generate short and long stride modes, respectively

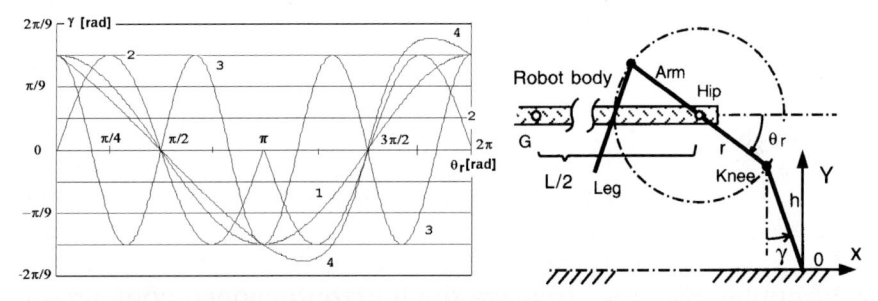

Fig. 2. Left: Characteristics of γ versus θ_r depending on the leg attitude patterns from 1 to 4. Right: Illustration of the parameters of r, h, γ, L and θ_r.

$$X_G = -L/2 - r\cos\theta_r - h\sin\gamma \quad (1)$$

Since θ_r is a function of time, differentiating Eq.(1) yields the horizontal speed of the robot's body, say V_x, then

$$V_x = (dX_G/d\theta_r)(d\theta_r/dt) \quad (2)$$

Suppose that ω_r (=$d\theta_r/dt$) is constant, then we obtain horizontal speed depending on the different patterns of the leg attitudes. Actually,

$$V_x = \begin{cases} \omega_r[\sin\theta_r\{r + h\gamma_{max}\cos(\gamma_{max}\cos\theta_r)\}], & \text{Pat.1} \quad (3a) \\ \omega_r[r\sin\theta_r - 2h\gamma_{max}\cos(2\theta_r)\cos\{\gamma_{max}\sin(2\theta_r)\}], & \text{Pat.2} \quad (3b) \\ \omega_r[r\sin\theta_r + 3h\gamma_{max}\sin(3\theta_r)\cos\{\gamma_{max}\sin(3\theta_r)\}], & \text{Pat.3} \quad (3c) \\ \omega_r[r\sin\theta_r + hK_1/K_2\sin\gamma_{max}], & \text{Pat.4} \quad (3d) \end{cases}$$

where
$$K_1 = 2K_3\sin\theta_r + \sin(2\gamma_{max})\cos^2\theta_r \quad (4)$$
$$K_2 = 2K_3^{3/2} \quad (5)$$
$$K_3 = 1 + \sin(2\gamma_{max})\sin\theta_r \quad (6)$$

These equations are utilized to obtain the data in **Fig.3** under $\omega_r=1$ [rad/s], $r=20$[cm] and $h=25$[cm]. The figure shows that the robot sways back and forth in a long stride in patterns 2 and 3. The short stride of pattern 3 makes the same sway. The short stride of pattern 2 does not cause the same sway, but makes the robot speed up and down. Patterns 1 and 4 make the robot's speed rather smooth. Therefore these patterns are preferable. Pattern 4 is the most preferable since it shows the least amount of sway.

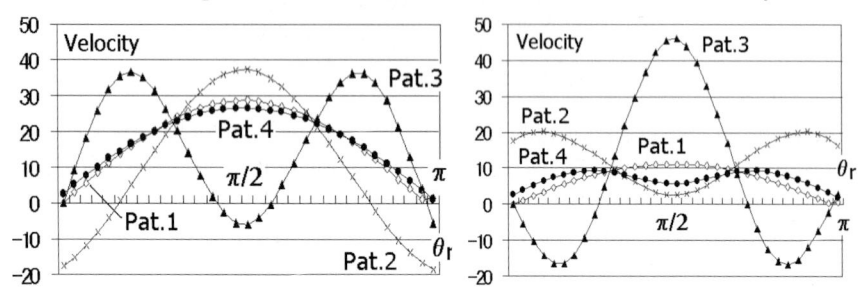

Fig. 3. V_x versus θ_r under θ=constant. The left and right views show the cases when $\gamma_{max}=\pi/9$[rad] and $-\pi/9$[rad], i.e. long and short stride modes, respectively.

2.2 Angular variables to make the horizontal speed constant

It is possible to find hip joint control variables so that the robot can walk smoothly at a constant speed V_x. Actually, one can get $\omega_r = V_x/f(i)$, where $f(i)$ is the term written in equations from (3a) to (3d), depending on the pattern number $i(=1$-$4)$.

2.3 Vertical vibration occurs when the robot walks

From Fig.2 (right), we can write the vertical position of point G as

$$Y_G = r \sin\theta_r + h \cos\gamma \quad (8)$$

Also, vertical speed, say V_y, is determined by differentiating Y_G to obtain the results

$$V_y = \begin{cases} \omega_r[r\cos\theta_r + h\gamma_{max}\sin(\theta_r)\sin(\gamma_{max}\cos\theta_r)\}], & \text{Pat.1} \quad (9a) \\ \omega_r[r\cos\theta_r - 2h\gamma_{max}\cos(2\theta_r)\sin\{\gamma_{max}\sin(2\theta_r)\}], & \text{Pat.2} \quad (9b) \\ \omega_r[r\cos\theta_r + 3h\gamma_{max}\sin(3\theta_r)\sin\{\gamma_{max}\cos(3\theta_r)\}], & \text{Pat.3} \quad (9c) \\ \omega_r\cos\theta_r[r/2 + h\{K_3 - \cos(2\gamma_{max})\}\sin\gamma_{max}/(4K_3^{3/2})], & \text{Pat.4} \quad (9d) \end{cases}$$

Simulation results of Y_G and V_y are shown on the left and right sides of **Fig. 4** when $\gamma_{max}=\pi/9$[rad], $r=20$[cm] and $h=15$[cm].

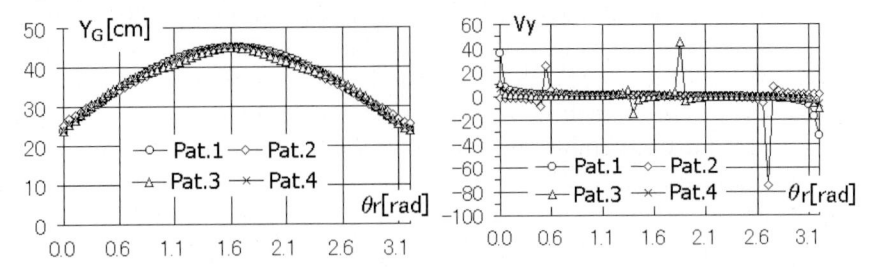

Fig. 4. Left: Displacement of the robot's centre on the road when θ_r changes. Right: Characteristics of V_y versus θ_r when the horizontal speed V_x is constant.

Fig.4 (right side) shows that vertical vibrations are almost the same in the four patterns and the sway width is about 20[cm]. The speed V_y is not smooth because $\theta_r=n\pi/2$[rad] is critical when there is a change in its sign (n; natural number). Generally, variation at $\theta_r=n\pi$[rad] is greater than that at $\theta_r=(n\pm 1/2)\pi$[rad]. It can be observed that instantaneous change appears twice in one stride in patterns 1,2 and 3. Pattern 4 is an exception in reducing the vertical displacement. These results imply that pattern 4 is better in realizing smooth motion.

3. Self-lock force by using worms and worm wheels

Worm gearing composed of a pair of worm and worm wheel is used for obtaining large speed reductions between nonintersecting shafts making an angle of $\pi/2$[rad] with each other [6]. We use the gearing not only for getting large reductions, but also for compensating hip joint torque caused by robot's weight. An input torque on the worm wheel is locked by the high reduction rate. This is called the self-lock effect.

The legs have a role in making the stance phase and swing phase cyclical. These phases are drawn on the left side of **Fig.5**. If the driving power to the hip joint is disconnected, the robot can no longer maintain its configuration because the robot's weight acts as a torque to drive the hip joint. In general, this causes the robot to subside and land on all four legs.Different handed threads cause automatic compensation of the force operating on the propeller shaft (see right of Fig.5). Pitch plane of the worm wheel in **Fig.6** is helpful in showing parameters such that F_a; axial force, F_u; lateral force, F_r; radial force, α; pressure angle, β; pitch angle, μ;

friction coefficient, Q (Q_a, Q_u); friction force and λ; pitch plane including the contacting point O. It follows that

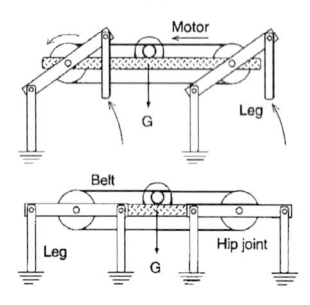

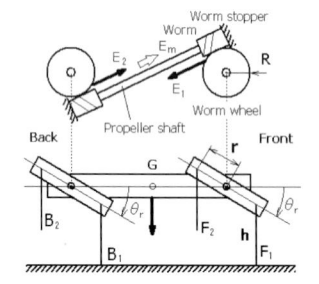

$$F_a = F_n(\cos\alpha \sin\beta - \mu\cos\beta) \tag{10}$$
$$F_u = F_n(\cos\alpha \cos\beta + \mu\sin\beta) \tag{11}$$
$$F_r = F_n \sin\alpha \tag{12}$$

Fig. 5. Use of the self-lock for compensating axial force operating on the propeller shaft(left) and diagonal alignment of the propeller shaft with two worms (right).

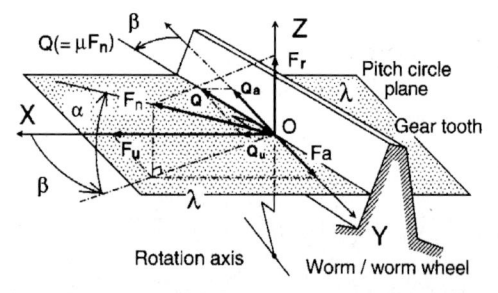

Fig. 6. Estimation of orthogonal forces F_a, F_u and F_r operating at gear contact

Eliminating F_n from Eq.(10) and combining Eqs.(11) with (12) produce

$$F_a = F_u(\cos\alpha \sin\beta - \mu\cos\beta)/(\cos\alpha \cos\beta + \mu\sin\beta) \tag{13}$$
$$F_r = F_u \sin\alpha/(\cos\alpha \cos\beta + \mu\sin\beta) \tag{14}$$

4 Smart combination of the worm gearing mechanism

Robot landing is drawn from one view as shown on the left side of **Fig.7**. The right side of the figure shows the total drawing from the top. The landing only on the legs of F_2 and B_1 or F_1 and B_2 generates the same directional force on the propeller shaft. But these landings are quite unusual. Therefore diagonal alignment of the propeller shaft makes robot design smart and enables the robot to stand by itself even when the power source to drive the hip joint is disconnected. The alignment is possible regardless of long or short strides. Merits of combination use are our aim.

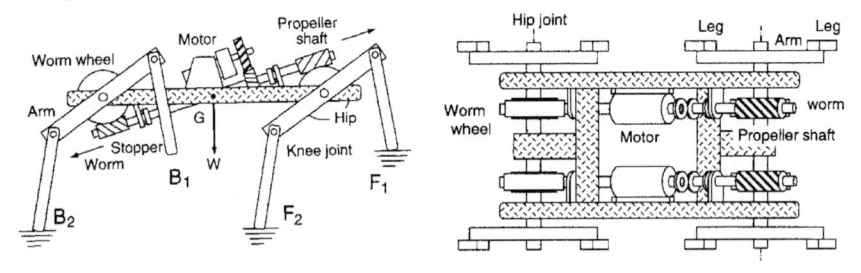

Fig. 7. Combination use of the pairs of worms and worm wheels for smart design. The top right view illustrates the total arrangement of the robot's driving parts.

5 Experiments for verifying the speed and the self-lock

The PEOPLER demonstrated walking in the four leg attitude patterns. Each of the walks was recorded on a movie [7] so that we could analyze the performance of the speed control with V_x (=0.076[km/h]). Snapshots were selected from the movie with interval of 1.7 sec. and listed in **Fig.8**.

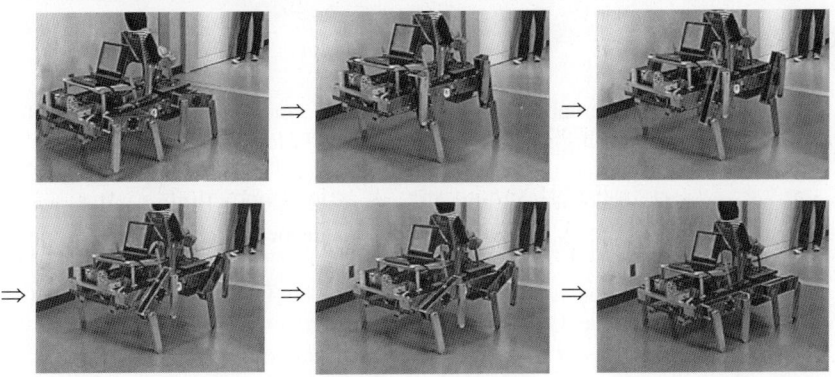

Fig. 8. Demonstration of walking at a designated speed by the PEOPLER Mk.2

The walk was performed with the leg attitude pattern 4 under γ_{max}= -10[deg], i.e. short stride of 21.1[cm]. Physical dimensions of the PEOPLER are r=16[cm], h=29.7[cm] and L=63.4[cm]. From the figure, it can be verified that the centre G goes forward as the hip joint rotates π[rad].

The miniature sized walker based on the smart design is tested for walking. In the experiment, we could observe that the walker starts and stops without deforming instantaneous configuration regardless of different weights put on its body. Thus, one can recognize the self-lock effect.

6 Concluding remarks

To make the legged robot walk at a constant speed, time dependent functions for controlling hip joints were clarified. Vertical displacements of the robot appearing in this case were also shown in mathematical equations. Snapshots selected from the movie showed that the angular variables are valid in making the robot walk at a constant speed.

Also, we introduced the driving mechanism using pairs of worms and worm wheels. The equations expressing axial and lateral forces are useful in estimating strength for designing the power transmission system. We clarified the advantages when using the worms and worm wheels.

Worm gears are not useful for enhancing driving power and extraordinary power including the weight torque is still needed when the robot begins to walk. Therefore, this technique is unsatisfactory. Our future plans are to devise a better technique that can function as a counter balance.

References

1. Moore E.Z., Campbell D., Grimminger F. and Buehler M. (2002) Reliable stair climbing in the simple hexapod RHex, *Proc. of IEEE Conf. on Robotics and Automation*, Washington DC, USA, Vol.3, pp.2222-2227.
2. Kimura H., Fukuoka Y. and Nakamura H. (2000) Biologically Inspired Adaptive Dynamic Walking of the Quadruped on Irregular Terrain. *J. Robotics Research*, vol. 9, pp.329-336.
3. Hirose S., Sensu T and Aoki S. (1992) The TAQT Carrier: A practical Terrain-Adaptive Quadru-Track Carrie Robot, *Proc. of IEEE/RSJ Int. Conf. on Intelligent Robots and Systems*, Tokyo, Japan, pp.2068-2073.
4. Okada T., Hirokawa Y and Sakai T. (2003) Development of a rotating four-legged robot, PEOPLER for walking on irregular terrain. *Proc. of Clawar Conference*, Catania, Italy, September, pp.593-600.
5. Okada T., Hirokawa Y, Sakai T., Shibuya K. (2004) Synchronous landing control of a rotating 4-legged robot, PEOPLER, for stable direction change. *ibid.*, Madrid, Spain, September, pp. - .
6. Theodore Baumeister. (1967) *Standard handbook for mechanical engineers*. McGraw-Hill Book Company, pp.142-145(section 8).
7. http://okada.eng.niigata-u.ac.jp/engstudy.html

Controlling Dynamic Stability and Active Compliance to Improve Quadrupedal Walking

E. Garcia and P. Gonzalez de Santos

Industrial Automation Institute (CSIC) 28500 Madrid, Spain
egarcia@iai.csic.es

Abstract

This paper aims to solve two main disadvantages featuring walking robots: the lack of reacting capabilities from external disturbances, and the very slow walking motion. Both problems are reduced here by combining, (1) an active compliance controller, which helps the robot react to disturbances, and (2) a dynamic energy stability margin that quantifies the impact energy that a robot can withstand. This dynamic energy stability margin is used as a new term in the active compliance equation to compensate for unstable motions. As a result, the combined controller helps the robot achieve faster and more stable compliant motions than conventional controllers. Experiments performed on the SILO4 quadruped robot show a relevant improvement on the walking gait.

Keywords: Active compliance, dynamic stability, normalized dynamic energy stability margin, quadruped robot.

1 Introduction

In spite of the well known theoretical advantages of walking robots over wheeled robots to move on very irregular terrain, legged systems are still far from being considered as a real locomotion system in industrial and service applications. The main obstacle found, apart from its mechanical complexity, is the very low velocity that walking robots feature to allow a stable motion. Moreover, walking robots easily tumble down due to unexpected external and internal disturbances. To solve these inconveniences, there is a need to generate walking gaits able to react to robot dynamics at the same time that they manage on natural environments.

Active compliance controllers have been used to control walking robot gaits successfully [1–3]. This kind of controllers produce a motion similar to that produced from a spring connecting a foot directly to the body following Hook's law. This helps the robot recover from small perturbations. However, when walking on inclined ground, or for big disturbances, the transient response of the compliant motion can itself produce instability. Our goal is to control the gait by means of an active compliance system, however we move one step forward by including in the active compliance controller a term proportional to the variation of stability margin, to improve machine speed and stability in the presence of internal perturbations (caused by robot dynamics) and external perturbations (caused by the environment). The stability margin used is of the dynamic type, the Normalized Dynamic Energy Stability Margin, NDESM [4], that quantifies the impact energy that a robot can withstand.

The outline of this paper is as follows: In Sect. 2 we revise the NDESM; then, in Sect. 3 we present the active-compliance approach that enables to improve gait speed and stability. The complete approach is validated through experimentation using a real quadruped robot in Section 4.

2 Measuring dynamic stability

In order to achieve a compliant motion able to react to the environment and to internal dynamics, a dynamic stability margin has to be used for controlling the gait. In this work the NDESM has been chosen because it measures the amount of impact energy a robot can withstand without tumbling down as a consequence of dynamic effects.

The general expression of the NDESM is:

$$S_{NDESM} = \frac{\min(E_i)}{mg}, \quad i = 1 \ldots 4 \qquad (1)$$

where E_i stands for the stability level of the ith side of the support polygon, which physically represents the increment of mechanical energy required to tumble the robot around the ith side of the support polygon, computed from (see **Fig. 1**):

$$E_i = mg|\mathbf{R}|\,(\cos\phi - \cos\varphi)\,\cos\Psi + (\mathbf{F_{RI}} \cdot \mathbf{t})|\mathbf{R}|\,\theta + (\mathbf{M_R} \cdot \mathbf{e_i})\,\theta - \frac{1}{2}I_i\omega_i^2 \qquad (2)$$

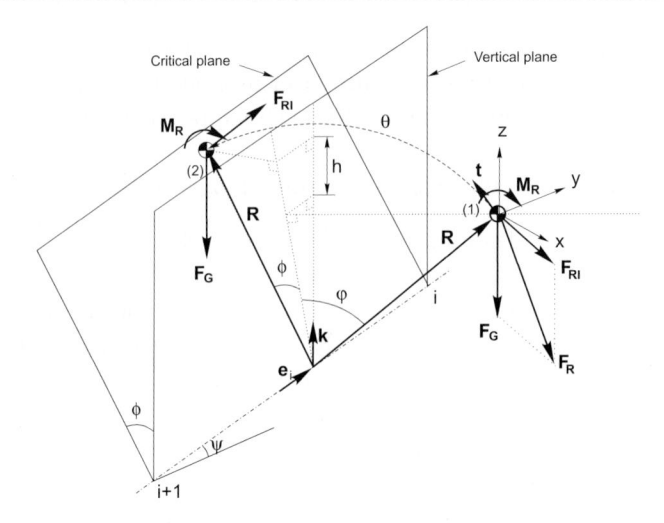

Fig. 1. Geometric outline for the computation of S_{NDESM}.

where **R** is a vector orthogonal to the ith side of the support polygon that points to the COG position, $\mathbf{F_{RI}}$ is the resultant inertial force acting on the COG, $\mathbf{M_R}$ is the resultant moment acting on the COG, I_i is the moment of inertia around the rotation axis i, ω_i is the initial angular velocity of the COG, Ψ is the terrain inclination angle, and ϕ, φ and θ are the rotation angles around the i axis. φ is the rotation angle required to position the COG inside the vertical plane (see **Fig. 1**); ϕ is the angle that the COG rotates from the vertical plane to the critical plane, where the resultant moment acting on the COG vanishes. Finally, θ is the addition of both rotations. Unitary vector **t** is tangent to the COG trajectory and $\mathbf{e_i}$ is the unitary vector that goes around the support polygon in the clockwise sense. The first three terms on the right side of Equation (2) represent the potential energy required for the tumble caused by gravitational and non-gravitational forces and moments, while the fourth term represents kinetic energy (see [4] for a more detailed explanation).

3 Active compliance with stability compensation

The principle of active compliance in a walking robot consists in controlling the motion of each leg in support phase so that steady-state force errors at the foot are considered linearly proportional to displacements errors following Hook's law [2]. However, active compliance can be also

Fig. 3. The SILO4 quadruped robot

proposed approach. The first experiment is aimed at showing the better adaptation to the environment and to disturbances that is achieved by using the proposed active compliance with stability compensation. The SILO4 robot has been placed in a 10-degree-inclined ground and commanded to start walking using first a conventional active-compliance controller ($K_p = 1.0\,s^{-1}$, $K_f = 4 \cdot 10^{-4}\,m/s/N$ for every leg) and later compared with the proposed active-compliance with stability compensation ($K_p = 0.8\,s^{-1}$, $K_f = 4 \cdot 10^{-4}\,m/s/N$, $K_{sx} = 0.5\,m^{-1}$, $K_{sy} = 1.0\,m^{-1}$ for every leg). In both cases above, at the moment of leg lift, the robot tilts downhill a little because of the perturbing impulse of the leg in transfer. As shown in **Fig. 4**(a), when the conventional active-compliance controller is used (thin line), the stability margin decreases and then it keeps at low level. However, when the proposed active-compliance with stability compensation is used (thick line), the robot recovers from the perturbation and the stability margin increases after a transient response and it stabilizes at a stability level even higher than the initial state (helped by a small decrease of robot height).

The second experiment shows the improvement on gait speed that can be achieved by using the proposed approach. In this experiment the robot walks uphill on a 5-degree slope. The gait stability margin has been plotted for different gait average speeds in **Fig. 4**(b). As a result, three curves have been obtained which show the evolution of the gait stability margin as gait speed increases for a position-controlled gait (thin, solid line), an active-compliance controlled gait ($K_p = 1.0\,s^{-1}$, $K_f = 4 \cdot 10^{-4}\,m/s/N$ for every leg in support) (dashed line) and an active compliance controlled

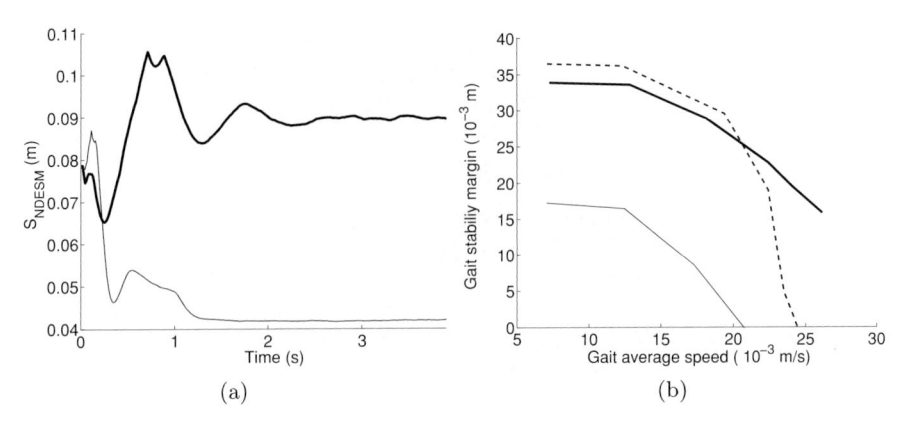

Fig. 4. (a) Result of applying a conventional active-compliance controller (*thin line*) vs. the proposed active-compliance with stability compensation controller (*thick line*) on the gait stability margin. (b) Curves of gait stability vs. average gait speed for a 5-degree inclined ground for three cases: without active compliance (*dashed line*), with conventional active compliance (*solid, thin line*), and with active compliance and stability compensation (*solid, thick line*).

gait with stability compensation ($K_p = 1.0\,s^{-1}$, $K_f = 4 \cdot 10^{-4}\,m/s/N$, $K_{sx} = 0.5\,m^{-1}$, $K_{sy} = 0.8\,m^{-1}$ for every leg in support) (thick line). As this figure shows, the active-compliance controller is not adequate for inclined ground because the compliant motion reduces gait stability, and it gets unstable for a gait average speed of 21 mm/s. The position-controlled gait is more stable than the active compliance with stability compensation at low speeds due to the slow transient response of the proposed controller, however, at high speed, the rigidity of the position-controlled gait produces large oscillations that make the gait unstable, while the proposed active compliance with stability compensation achieves a stable motion up to the highest average speed that the actuators allow. This experiment has been done for increasing gait speeds until the actuator speed limits were found, at 26.2 mm/s.

5 Conclusions

In this paper, a new active compliance controller has been proposed that improves conventional active compliance by means of a stability compensation term. This compensation term is based on a dynamic energy stability measurement that considers the impact energy a body can withstand.

Experiments have been done to show the improvement of the proposed approach. The proposed gait controller compensates for destabilizing effects that, augmented by the compliant behavior of conventional controllers causes instability, specially when the robot walks on inclined surfaces. The proposed approach also allows to increase gait average speed up to actuator limits, improving position-based gait controllers. The active compliance with stability compensation does not improve position-based controllers at low speeds due to the slow transient response of the compliant controller. However, this can be improved by using predictive filters. This will be our future work.

Acknowledgements

This work has been partially funded by CICYT (Spain) through Grant DPI2004-05824. The first author is supported by a postdoctoral CSIC-I3P contract granted by the European Social Fund.

References

1. Klein, C.A. and Briggs, R.L. (1980) Use of active compliance in the control of legged vehicles. *IEEE Transactions on Systems, Man, and Cybernetics*, vol. SMC-10, no. 7, pp. 393–400.
2. Gorinevsky, D.M. and Schneider, A.Y. (1990) Force control in locomotion of legged vehicles over rigid and soft surfaces. *The International Journal of Robotics Research*, vol. 9, no. 2, pp. 4–23.
3. Palis, F., Rusin, V., and Schneider, A. (2001) Adaptive impedance/force control of legged robot systems. In *Proc. Int. Conf. Climbing and Walking Robots*, Kalsruhe, Germany, pp. 323–329.
4. Garcia, E. and Gonzalez de Santos, P. (2005) An improved energy stability margin for walking machines subject to dynamic effects. *Robotica*, vol. 23, no. 1, pp. 13–20.
5. Waldron, K.W. (1986) Force and motion management in legged locomotion. *IEEE Journal of Robotics and Automation*, vol. RA-2, no. 4, pp. 214–220.
6. Gonzalez de Santos, P., Galvez, J.A., Estremera, J., and Garcia, E. (2003) Towards the use of common platforms in walking machine research. *IEEE Robotics and Automation Magazine*, vol. 10, no. 4, pp. 23–32.
7. Industrial Automation Institute, C.S.I.C. (2002) *The SILO4 Walking Robot*. Madrid, Spain, Available: http://www.iai.csic.es/users/silo4/.

Complete Stability Analysis of a Control Law for Walking Robots with Non-permanent Contacts

Sophie Chareyron[1] and Pierre-Brice Wieber[1]

INRIA Rhône-Alpes, ZIRST-655 avenue de l'Europe-Montbonnot-38335 Saint Ismier Cedex
`sophie.chareyron@inrialpes.fr,Pierre-Brice.Wieber@inrialpes.fr`

Abstract

We're interested here in the stability analysis of a simple control law for walking robots without any assumption on the state of the contacts. We propose to use the framework of nonsmooth dynamical systems since it provides a general formulation of the dynamics that does not depend on the contact state. Classical stability results cannot be applied in this framework, so we needed to derive a Lyapunov stability theorem for Lagrangian dynamical systems with non-permanent contacts. Based on this theorem, we will then prove, here, the stability of this control law.

Keywords: Lyapunov stability, nonsmooth Lagrangian systems, biped robots

1 Introduction

One of the main specificities of walking robots is their non-permanent contact with the ground which impairs their stability. The only stability analyses of control laws that have been proposed so far for walking robots have been distinctly focusing on each contact phases, with the strong assumption that these phases are never perturbated [1, 2]. We aim here at analysing the stability of a regulation of the position of a walking robot without any assumption on the state of these contacts. So, we proposed in [3] to work in the framework of nonsmooth dynamical systems, what provides a general formulation of the dynamics that does not depend on the contact state. Classical stability theorems cannot be applied to this framework, so we needed to derive in [3] a Lyapunov stability theorem and a Lagrange Dirichlet theorem specifically for Lagrangian dynamical systems with non-permanent contacts. Based on these theorems, we will prove here the stability of a simple control action that realizes the regulation of the position and contact forces of a walking robot.

2 Position and force regulation law for walking robots

2.1 Walking machines with non-permanent contacts

With n the number of degrees of freedom of the walking robot, let us consider a time-variation of generalized coordinates $q : \mathbb{R} \to \mathbb{R}^n$ and the related velocity $\dot{q} : \mathbb{R} \to \mathbb{R}^n$:

$$\forall t, t_0 \in \mathbb{R}, \ q(t) = q(t_0) + \int_{t_0}^{t} \dot{q}(\tau)\, d\tau.$$

We assume that the interaction between the feet of the walking machine and the ground is modelled by non-penetrating rigid bodies, which implies that the feet cannot penetrate or stick on the ground. That can be described by a set of unilateral constraints on the position of the system: $\varphi(q) \geq 0$, $\varphi : \mathbb{R}^n \to \mathbb{R}^m$. The dynamics of walking robots can be described in this case by a classical Lagrange differential equation:

$$M(q)\frac{d\dot{q}}{dt} + N(q, \dot{q})\dot{q} + g(q) = u + r + f, \qquad (1)$$

with $M(q)$ the inertia matrix, $N(q, \dot{q})\dot{q}$ the corresponding nonlinear effects, $g(q)$ the gravity forces, u the control action, r and f the normal and tangential contact forces (also referred as the reaction and friction forces). In the following, we are not going to precise here the model of these contact forces, we will only use the dissipativity property of friction: $f^T \dot{q} \leq 0$.

Note that walking robots can be seen as underactuated systems, indeed it can be shown in [4] that the vector of generalized coordinates q of a walking robot can be decomposed in q_1, a vector gathering the position of the robot articulations and q_2 a vector describing the position and orientation of one solid of the robot with respect to the environment. Since the actuators of the robot produce a torque τ that acts only on the positions of its articulations, the actuation u can be shown to have the following structure inside the robot dynamics (1)

$$M(q)\frac{d\dot{q}}{dt} + N(q,\dot{q})\dot{q} + g(q) = \begin{bmatrix} \tau \\ 0 \end{bmatrix} + r + f, \qquad (2)$$

and at equilibrium points, when $\dot{q} = 0$ and $\dfrac{d\dot{q}}{dt} = 0$, it reduces to

$$r + f = g(q) - \begin{bmatrix} \tau \\ 0 \end{bmatrix} \quad \text{or} \quad \begin{bmatrix} r_1 + f_1 \\ r_2 + f_2 \end{bmatrix} = \begin{bmatrix} g_1(q) - \tau \\ g_2(q) \end{bmatrix}. \qquad (3)$$

Relation (3) shows an equilibrium condition between the three external forces: the contact forces, the gravity forces and the actuation.

2.2 A control law through potential shaping

We want to realize a regulation of the position and contact forces of a walking robot to some desired position q_d and desired contact forces $r_d + f_d$. The torque τ has to be designed to compensate the part $g_1(q_d)$ of the gravity forces so that we can obtain the desired reaction forces $f_{1d} + r_{1d}$ at the desired position q_d: $\tau(q_d) = g_1(q_d) - r_{1d} - f_{1d}$. In order to do so, we'll consider a control law designed through potential shaping, and more precisely, following [5], we choose the following potential function

$$\tilde{P}(q_1) = \frac{1}{2}(q_1 - q_{1d})^T W(q_1 - q_{1d}) + (r_{1d} + f_{1d} - g_1(q_d))^T (q_1 - q_{1d}),$$

with W a symmetric positive definite matrix and q_{1d} the desired position of the robot articulations. We add to the derivative of this potential function a dissipative term $T\dot{q}_1$, with T a positive definite matrix, in order to obtain the following Proportionnal Derivative control law with precompensation of the gravity and desired contact forces:

$$\tau = -W(q_1 - q_{1d}) + g_1(q_d) - r_{1d} - f_{1d} - T\dot{q}_1. \qquad (4)$$

3 Nonsmooth dynamical systems

3.1 Some geometry for systems with non-permanent contacts

We have seen in section 2.1 that the non-penetration of perfectly rigid bodies can be expressed as a constraint on the robot position, a constraint that will take the form of a closed set, $\Phi = \{q \in \mathbb{R}^n | \varphi(q) \geq 0\}$, in which the generalized coordinates q are bound to stay, $\forall t \in \mathbb{R}$, $q(t) \in \Phi$. When q is in the interior of the domain Φ, there is no interaction between the walking robot and its environment. On the other hand, when the system lies on the boundary of Φ, the robot and its environment interact, what generates contacts forces $r + f$. Concerning the normal forces, this fact can be described trough the inclusion

$$-r \in \mathcal{N}(q) \tag{5}$$

involving the normal cone $\mathcal{N}(q)$ of Φ at q as defined in [6] and as illustrated in **Fig. 1**.

Now, we can observe that when the system reaches the boundary of Φ with a velocity $\dot{q}^-$ directed outside of this domain, it won't be able to continue its movement with a velocity $\dot{q}^+ = \dot{q}^-$ and still stay in Φ (**Fig. 1**). A discontinuity of the velocity will have to occur then, corresponding to an impact between contacting rigid bodies. This can be described by the fact that the velocity $\dot{q}^+$ has to belong any time to the tangent cone $\mathcal{T}(q)$ of Φ at q as defined in [6] and as illustrated in **Fig. 1**. Note that the velocity after this impact $\dot{q}^+$ can be related to the velocity before the impact $\dot{q}^-$ by modelling this impulsive behavior through a contact law. We are not going to precise here this contact law, we will only use the fact that the impact is a dissipative phenomenom, which implies that the kinetic energy $K(q, \dot{q})$ satisfies

$$K(q, \dot{q}^+) \leq K(q, \dot{q}^-). \tag{6}$$

For a more in-depth presentation of these concepts and equations which may have subtle implications, the interested reader should definitely refer to [7] or [3, 8].

3.2 Nonsmooth Lagrangian dynamical systems

Classically, solutions to the dynamics of Lagrangian systems such as (1) lead to smooth motions with locally absolutely continuous velocities. But we have seen in section 3.1 that discontinuities of the velocities may occur when the coordinates of such systems are constrained to stay inside closed sets. These classical differential equations must therefore be turned into measure differential equations [7]

$$M(q)\, d\dot{q} + N(q, \dot{q})\, \dot{q}\, dt + g(q)\, dt = u\, dt + dr + f\, dt, \tag{7}$$

where the reaction forces are now represented by an abstract measure dr which may not be Lebesgues-integrable. This way, the measure acceleration $d\dot{q}$ may not be Lebesgues-integrable either so that the velocity

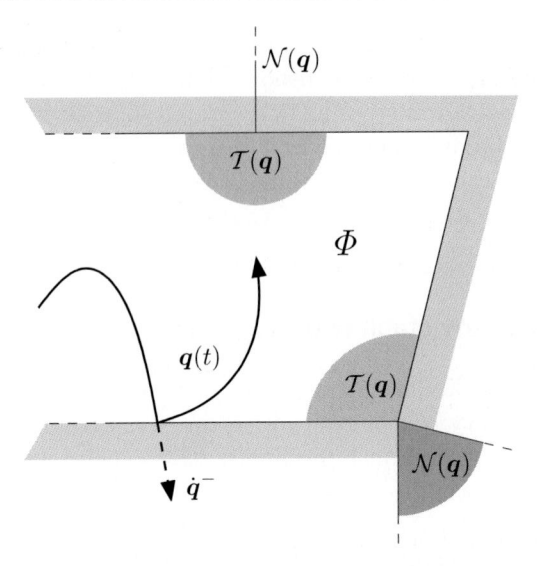

Fig. 1. Examples of tangent cones $\mathcal{T}(\boldsymbol{q})$ and normal cones $\mathcal{N}(\boldsymbol{q})$ on the boundary of the domain $\varPhi$, and example of a trajectory $\boldsymbol{q}(t) \in \varPhi$ that reaches this boundary with a velocity $\dot{\boldsymbol{q}}^- \notin \mathcal{T}(\boldsymbol{q})$.

may not be locally absolutely continuous anymore but only with locally bounded variation, $\dot{\boldsymbol{q}} \in \mathrm{lbv}(\mathbb{R}, \mathbb{R}^n)$ [7]. Functions with locally bounded variation have left and right limits at every instant, and we have for every compact subinterval $[\sigma, \tau] \subset \mathbb{R}$

$$\int_{[\sigma,\tau]} d\dot{\boldsymbol{q}} = \dot{\boldsymbol{q}}^+(\tau) - \dot{\boldsymbol{q}}^-(\sigma)..$$

A function f has a locally bounded variation on $\mathbb{R}$ if its variation on any compact interval $[t_0, t_n]$ is finite:

$$\mathrm{Var}(f; [t_0, t_n]) = \sup_{t_0 \leq \ldots \leq t_n} \sum_{i=1}^n \|f(t_i) - f(t_{i-1})\| < +\infty.$$

Rather than for this definition, it is for their properties that functions with bounded variations are useful. Notably, functions with locally bounded variation can be decomposed into the sum of a continuous function and a countable set of discontinuous step functions [9]. In specific cases, as when the definition of the dynamics (7) is piecewise analytic, its solutions can be shown to be piecewise continuous with possibly infinitely (countably) many discontinuities [10]. In this case, it is possible to focus distinctly on each continuous piece and each discontinuity as in the framework of hybrid systems [11]. But this is usually done through an ordering of the discontinuities strictly increasing with time, what is problematic when having to go through accumulations of impacts. The framework of non-smooth analysis appears therefore as more appropriate for the analysis of impacting systems, even though the calculus rules for functions with bounded variation require some care.

3.3 Some Lyapunov stability theory

The Lyapunov stability theory is usually presented for dynamical systems with states that vary continuously with time [12], [13], but we have seen that in the case of nonsmooth mechanics, the velocity and thus the state may present discontinuities. Lyapunov stability theory is hopefully not strictly bound to continuity properties, and some results can still be derived for discontinuous dynamical systems both in the usual framework of hybrid systems [11] and in the framework of nonsmooth analysis [3].

In the following we will prefer the latter for the reasons mentioned in section 3.2, for which we proposed in [3] the following corollary of a Lagrange-Dirichlet theorem:

Corollary 1. *Consider a nonsmooth Lagrangian dynamical system experiencing external forces composed of normal contact forces and other Lebesgues-integrable forces $\mathcal{F}$. If these Lebesgues-integrable forces derive from a coercive C^1 potential function $P(\boldsymbol{q})$ with a dissipative term $\boldsymbol{h}$:*

$$\mathcal{F} = -\frac{dP}{d\boldsymbol{q}}(\boldsymbol{q}) + \boldsymbol{h}, \quad \text{with } \dot{\boldsymbol{q}}^T \boldsymbol{h} \leq 0,$$

then the set $\mathcal{S} = \{\operatorname*{Arg\,min}_{\Phi} P(\boldsymbol{q})\} \times \{\boldsymbol{0}\}$ is Lyapunov stable.

In our case, the Lebesgues-integrable forces $\mathcal{F}$ acting on the dynamics (7) are composed of the gravity forces, the actuation, and the friction forces, $\mathcal{F} = -\boldsymbol{g}(\boldsymbol{q}) + \boldsymbol{u} + \boldsymbol{f}$.

With $G(\boldsymbol{q})$ the potential of the gravity forces, if we replace the control law $\boldsymbol{u}$ by its expresion (4), $\mathcal{F}$ can be expressed as the derivative of potential functions plus dissipative terms

$$\mathcal{F} = -\frac{dG}{d\boldsymbol{q}}(\boldsymbol{q}) - \frac{d\tilde{P}}{d\boldsymbol{q}}(\boldsymbol{q}) - \begin{bmatrix} T \\ 0 \end{bmatrix} \dot{\boldsymbol{q}}_1 + \boldsymbol{f},$$

deriving therefore from the potential $P(\boldsymbol{q}) = G(\boldsymbol{q}) + \tilde{\mathrm{P}}(\boldsymbol{q}_1)$, and we can conclude through corollary 1 on the stability of the set $\mathcal{S} = \{\underset{\Phi}{\mathrm{Arg\,min}}\, P(\boldsymbol{q})\} \times \{\mathbf{0}\}$ without making any assumption on the state of the contacts. Now, as shown in [3, 8], this set corresponds to the equilibrium positions of the system. The control law (4) has been designed so that the desired position $\boldsymbol{q}_d$ is an equilibrium position. We can assume under mild conditions [8] that it is the only one and it is therefore stable.

3.4 Why such a "simple" control law

First of all, note that the control action (4) can't compensate the impulsive behaviors of the contacts. Now, we have seen in section 3.1 that these impulsive behaviors originating from the discontinuities of the velocity are related to the kinetic energy through relation (6). It is therefore natural to use the energy of the system for the stability analysis, leading to the Lagrange-Dirichlet theorem for nonsmooth Lagrangian dynamical systems [3] and its corollary 1 that is used here. Since the control law (4) has been designed by potential shaping, the energy of the system appears as a natural candidate for its stability analysis. For the same reason, the choice of a control law trying to compensate completely the system dynamics such as a computed torque [1, 2], appears to be not very judicious because this energy can't be used any longer for the stability analysis. Finally, it is not possible to compensate completely the external forces in the case of walking machines because of the underactuation (2) what explains the difference with the control law proposed in [8].

4 Conclusion

Based on a Lyapunov stability theorem for nonsmooth Lagrangian dynamical systems derived in [3], we then proved the stability of a simple control law that realizes the regulation of the position and contact forces of a walking robot without any assumption on the state of the contacts. We were thus able to propose for the first time a complete stability analysis of the regulation of the position of a walking robot.

Acknowledgements

This work was supported by the European project SICONOS IST2001-37172.

References

1. Park, Jonghoon, Youm, Youngil, and Chung, Wan-Kyun (2005) Control of ground interaction at the zero-moment point for dynamic control of humanoid robots. In *Proceedings of the 2005 IEEE International Conference on Robotics and Automation*, Barcelona, Spain.
2. Westervelt, E. R., Grizzle, J. W., and Koditschek, D. E. (2003) Hybrid zero dynamics of planar biped walkers. *IEEE Transactions on Automatic Control*, vol. 48, no. 1, pp. 42–56.
3. Chareyron, S. and Wieber, P.B. (2004) Stabilization and regulation of nonsmooth lagrangian systems. Tech. Rep. 5408, INRIA Rhône-Alpes.
4. Wieber, P. B. (2000) *Modélisation et commande d'un robot marcheur anthropomorphe*. Phd thesis, Mines de Paris.
5. Wang, D. and McClamroch, H. (1993) Position and force control for constrained manipulator motion: Lyapunov's direct method. *IEEE Trans. Robot. Automat.*, vol. 9, no. 3, pp. 308–313.
6. Hiriart-Urruty, J.-B. and Lemaréchal, C. (1996) *Convex Analysis and Minimization Algorithms*. Springer Verlag, Two volumes - 2nd printing.
7. Moreau, J.-J. (1988) Unilateral contact and dry friction in finite freedom dynamics. In *Nonsmooth mechanics and Applications*, Moreau, J.-J. and Panagiotopulos, P.D., Eds., vol. 302, pp. 1–82. Springer Verlag.
8. Chareyron, S. and Wieber, P.B. (2005) Position and force control of nonsmooth lagrangian dynamical systems without friction. In *Lecture Notes in Computer Science*, Manfred Morari, Lothar Thiele, Ed. vol. 3414, p. 215, Springer-Verlag GmbH.
9. Moreau, J.-J. (1988) Bounded variation in time. In *Topics in Nonsmooth Mechanics*, Moreau, J.-J., Panagiotopulos, P.D., and Strang, G., Eds., pp. 1–74. Birkhäuser.
10. Ballard, P. (2000) The dynamics of discrete mechanical systems with perfect unilateral constraints. *Archive for Rational Mechanics and Analysis*, , no. 154, pp. 199–274.
11. Ye, H., Michel, A., and Hou, L. (1998) Stability theory for hybrid dynamical systems. *IEEE Transactions on Automatic Control*, vol. 43, no. 4, pp. 461–474.
12. Khalil, H.K. (1996) *Nonlinear systems*. Prentice-Hall.
13. Zubov, V.I. (1964) *Methods of A.M. Lyapunov and their application*. Noordhoff.

Vision-based Stabilization of the IDP Flat Output

L. Beji[1], A. Abichou[2] and M.A. El Kamel[2]

[1] LSC Laboratory, CNRS-FRE2494, Université d'Evry Val d'Essonne.
40, rue du Pelvoux, 91020, Evry Cedex, France.
beji@iup.univ-evry.fr
[2] LIM Laboratory, Ecole Polytechnique de Tunisie, BP 743, 2078 La Marsa, Tunisia azgal.abichou@ept.rnu.tn

Abstract

In this paper flatness based control is proposed and applied to stabilize the unstable equilibrium of a Inverted Double Pendulum (IDP) mounted on a cart integrating elasticity at joints. The flat output of the system serves as a point to be stabilized using vision feedback control. The proposed approach can be applied to a number of applications such as controlling a robot with feet, studying the characteristics of the human movement in the upright and modelling and controlling buildings subject to seismic excitations (movements resulting from an acceleration given to the cart). It is prove that the vision-based control is robust with a perturbing acceleration which excites the cart.

Keywords: IDP, highly perturbed cart, stabilization, vision.

1 Introduction

The classical inverted pendulum (cart and pole system) has been widely used in control laboratories to demonstrate the effectiveness of control systems in analogy with the control of many real systems [1, 2]. The double inverted pendulum is an extension of the inverted pendulum system [3, 4]. Different control approaches can be investigated to address the dynamics and high-order nonlinearities of such systems.

The double inverted pendulum belongs to the class of under-actuated mechanical systems consisting of three interconnected systems (two pendulums, one cart) with only one actuator to move the cart. Thus, it is a challenging problem for the control design [5],[6]. In Fliess et al. [3] the inverted double pendulum is treated assuming that the suspension point is moving in vertical and horizontal directions. These two directions are the control variables and the system was proved to be not flat. The same control method as the one explained in detail for the Kapitsa pendulum is used for the double pendulum. Energy and passivity based controls of the double inverted pendulum on a cart were proposed by Zhong in [4], where the swing up controller brings the pendulum from any initial position to the unstable up-up position. Double inverted pendulum was adopted by Guelton [7] to characterize human motion in the upright using fuzzy logic-like estimator to control the motion.

In this paper flat output is constructed for the unit cart-IDP (Inverted Double Pendulum). This output can be marked to the vision based stabilizing control. The dynamics of the system are transformed with the unknown control input in the image plan. The objective is to stabilize the unstable equilibrium of the IDP even with a perturbing acceleration which excites the cart.

2 Modeling of the cart-IDP system

Consider a double inverted pendulum mounted on a cart. The cart is considered mobile with respect to a reference frame. The Euler-Lagrange formulation is used to model the system.

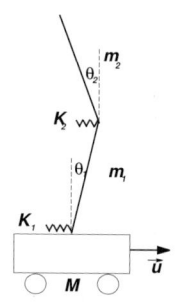

Fig. 1. The Inverted-Double-Pendulum mounted on a cart

Let x denotes the position of the cart and θ_1, θ_2 the affected angles to each pendulum. These variables are defined with respect to the reference frame. The two segments are considered with the same lengths (l), each element of inertia with respect to $y-axis$ is such that $J_1 = \frac{m_1 l^2}{3}$ and $J_2 = \frac{m_2 l^2}{3}$. $m_{1,2}$ is the segment mass.

Note that $k_{1,2}$ is the elasticity coefficients of the joints. Then a non rigid joints are considered in our problem. These joints are without control-inputs (not actuated). Then, the IDP mounted on a cart belongs to the family of under-actuated systems. The only input to the system is the force given to the cart denoted by u. In the following, the cart is considered perturbed with external acceleration $\ddot{x}_g$. Let $X = (x, \theta_1, \theta_2)^t$ denotes the state vector, the dynamic of the system can be obtained from the Euler-Lagrange formalism,

$$I(X)\ddot{X} + H(X, \dot{X}) + C(X, k_1, k_2) = U \qquad (1)$$

with

$$U = \begin{pmatrix} M + m_1 + m_2 \\ (m_1 + 2m_2)l \\ m_2 l \end{pmatrix} \ddot{x}_g + \begin{pmatrix} 1 \\ 0 \\ 0 \end{pmatrix} u$$

where M is the mass of the cart. The inertia matrix of the system is as

$$I(X) = \begin{pmatrix} M + m_1 + m_2 & (m_1 + 2m_2)l\cos\theta_1 & m_2 l\cos\theta_2 \\ (m_1 + 2m_2)l\cos\theta_1 & J_1 + l^2(m_1 + 4m_2) & 2m_2 l^2 \cos(\theta_1 - \theta_2) \\ m_2 l\cos\theta_2 & 2m_2 l^2 \cos(\theta_1 - \theta_2) & J_2 + m_2 l^2 \end{pmatrix} \qquad (2)$$

The vector which regroups the centrifugal and coriolis terms is

$$H(X, \dot{X}) = \begin{pmatrix} -(m_1 + 2m_2)l\dot{\theta}_1^2 \sin\theta_1 - m_2 l\dot{\theta}_2^2 \sin\theta_2 \\ 2m_2 l^2 \dot{\theta}_2^2 \sin(\theta_1 - \theta_2) \\ -2m_2 l^2 \dot{\theta}_1^2 \sin(\theta_1 - \theta_2) \end{pmatrix} \qquad (3)$$

The gravity and elastic forces are regrouped as follows

$$B(X, k_1, k_2) = \begin{pmatrix} 0 \\ k_1\theta_1 + k_2(\theta_1 - \theta_2) - (m_1 + 2m_2)gl \sin\theta_1 \\ -k_2(\theta_1 - \theta_2) - m_2 gl \sin\theta_2 \end{pmatrix} \qquad (4)$$

With small angle approximation, the dynamics reads

$$M\ddot{X} + KX = U \qquad (5)$$

with the appropriate constant matrices M and K.

In the following, we consider $m_1 = m_2 = m$, $k_1 = k_2 = k$ and let us regroup the state in $Z = (x, \theta_1, \theta_2, \dot{x}, \dot{\theta}_1, \dot{\theta}_2)$.

Proposition 1. *The equilibrium point $Z_e = (0_{\mathbb{R}^6}, 0)$ is unique and non stable for $k = k' \neq (\frac{5}{2} + \frac{1}{2}\sqrt{13})glm$ or $k = k'' \neq (\frac{5}{2} - \frac{1}{2}\sqrt{13})glm$.*

Proof. The proof is straightforward considering that the cart is immobile and $\ddot{x}_g = 0$. Then the problem is reduced to conditions of existence and unicity of the solution of system $0 = A'Z'_e$ with $Z'_e = (\theta_1, \theta_2, \dot{\theta}_1, \dot{\theta}_2)$. The k', k'' are solutions of the A' determinant: $d(k) = \frac{9}{28}\frac{k^2 - 5mglk + 3g^2l^2m^2}{m_1^2 l^4}$. The instability of the equilibrium Z'_e can be checked via the roots of A' characteristic equation: $P(X') = X'^4 - (b_4 + b_1)X'^2 + d(k)$.

3 Vision-based stabilization of the cart-IDP system

The objective is to develop a control approach to stabilize the system equilibrium point even in presence of acceleration given to the cart. Vision-based control will be considered. In the system coordinates $Z = (x, \theta_1, \theta_2, \dot{x}, \dot{\theta}_1, \dot{\theta}_2)$, the inversion of system (5) leads to

$$\dot{Z} = A(a_i)Z + S(s_i)\ddot{x}_g + \zeta(\zeta_i)u \qquad (6)$$

where the a_i, s_i and ζ_i can be easily obtained from the inversion of system (5) and in accordance with proposition 1. It is proved below that the system is flat by constructing its flat output. The dynamic of the system will be parametrized as function of this output. Further details about flatness can be found in [3]. Moreover, the vision-based control will be used to stabilize this output, consequently the cart-IDP unit.

3.1 Flatness-notion for the Cart-IDP

A system is flat if all states and control inputs are function only of the flat output and their time derivatives.

Lemma 1. *The IDP mounted on a cart is flat with $J = x + h_1 L \theta_1 + h_2 L \theta_2$ is its flat output. For $k \neq \frac{12mgl}{19}$ and $R = Lf_2(a_5 f_2 + a_6 f_3) - Lf_3(a_3 f_2 + a_4 f_3)$, we have*

$$\begin{aligned} h_1 &= \frac{-f_1(a_5 f_2 + a_6 f_3) + f_3(a_1 f_2 + a_2 f_3)}{R} = \frac{1}{3}\frac{35mgl - 53k}{12mgl - 19k} \\ h_2 &= \frac{f_1(a_3 f_2 + a_4 f_3) - f_2(a_1 f_2 + a_2 f_3)}{R} = \frac{1}{3}\frac{21mgl - 26k}{12mgl - 19k} \end{aligned} \qquad (7)$$

Then the system is reduced to $J^{(6)} = \nu$, with ν as a new input and $L = 2l$.

Proof. The time derivative of J is calculated until order 6, and where the first derivative is $\dot{J} = \dot{x} + h_1 L \dot{\theta}_1 + h_2 L \dot{\theta}_2$.

$$J^{(6)} = d_7 \theta_1 + d_8 \theta_2 + s_1 x_g^{(6)} + fu \tag{8}$$

h_1 and h_2 are solutions of the following system of equations where the first one is imposed at order 2 in $\ddot{J}$ and the second is imposed at order 4 in $J^{(4)}$.

$$h_1 L f_2 + h_2 L f_3 = -f_1 \tag{9}$$
$$h_1(La_3 f_2 + La_4 f_3) + h_2(La_5 f_2 + La_6 f_3) = -a_1 f_2 - a_2 f_3 \tag{10}$$

Then, the system can be reduced to

$$J^{(6)} = \nu \tag{11}$$

Remark 1. The auxiliary input ν will be subject of 2D-vision based stabilization. This problem is detailed in the following.

3.2 Stabilization with an external camera

Our objective is to stabilize the following coordinates in the vicinity of the unstable equilibrium

$$N = (x + Lh_1 \theta_1 + Lh_2 \theta_2, Lh_1 + Lh_2) \tag{12}$$

Note that the N abscissa is the system flat output, more explained in **Fig.2**

Let the camera be as in **Fig.3** such that the primitives constituting the image (Chariot-IDP) are not degenerated. Consider that the projection of N in the image plan is superimposed with the origin of the image frame at the equilibrium $(0,0,0,0,0,0)$.

Theorem 1. Let $\underline{s} = (\frac{J}{z}, \frac{Lh_1+Lh_2}{z})$ be the instantness real visual information, $\underline{s}^* = (0, \frac{Lh_1+Lh_2}{z})$ be the visual objective information (the reference) and $\underline{e} = C(\underline{s}-\underline{s}^*)$ the visual error with C is the combination matrix. Then

$$\nu = (CL_s^t)^{(-1)}(\lambda_0 \underline{e} + \lambda_1 \underline{\dot{e}} + \lambda_2 \underline{\ddot{e}} + \lambda_3 \underline{e}^{(3)} + \lambda_4 \underline{e}^{(4)} + \lambda_5 \underline{e}^{(5)})$$

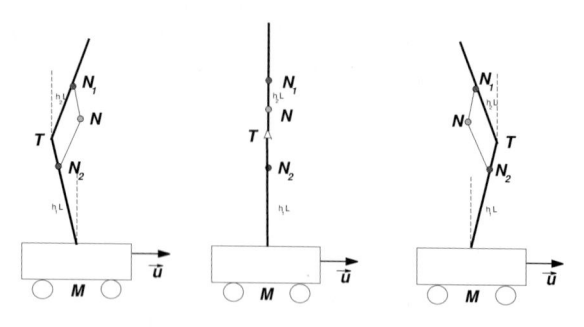

Fig. 2. Behavior of N

and the control that stabilizes the flat-output, consequently the perturbed cart-IDP, is given by

$$u = -\frac{\lambda_0}{f}x - \frac{(\lambda_0 d_1 + \lambda_2 d_3 + \lambda_4 d_5 + d_5 a_3 + d_6 a_5)}{f}\theta_1 - \frac{(\lambda_0 d_2 + \lambda_2 d_4 + \lambda_4 d_6 + d_5 a_4 + d_6 a_6)}{f}\theta_2$$
$$-\frac{\lambda_1}{f}\dot{x} - \frac{(\lambda_1 d_1 + \lambda_3 d_3 + \lambda_5 d_5)}{f}\dot{\theta}_1 - \frac{(\lambda_1 d_2 + \lambda_3 d_4 + \lambda_5 d_6)}{f}\dot{\theta}_2 - \frac{x_g^{(6)}}{f}$$

with J is the flat output, $f = d_5 f_2 + d_6 f_3 \neq 0$ and λ_i are the appropriate constant gains.

Proof. Recall that the relation between the kinematic of the camera and the visual information is

$$\underline{\dot{s}} = L^t T \qquad (13)$$

where T is the kinematic of the camera with respect to reference R_N defined in N. L is the interaction matrix.

Taking $\underline{s} = (\frac{x + h_1 L\theta_1 + h_2 L\theta_2}{z}, \frac{Lh_1 + Lh_2}{z}) = (\frac{J}{z}, \frac{Lh_1 + Lh_2}{z})$, with z is the distance between the image plan and the objective. One stabilizes ($\theta_1 = 0$, $\theta_2 = 0$, $x = 0$), then the reference visual information can be defined by $\underline{s}^* = (0, \frac{Lh_1 + Lh_2}{z})$. The interaction matrix is as $L^t = \begin{pmatrix} -1/z & 0 \\ 0 & -1/z \end{pmatrix}$ and it is shown that $T = \begin{pmatrix} -\dot{J} \\ 0 \end{pmatrix}$ which is deduced from the kinematic of N.

Without confusion, one notes $L^t = -\frac{1}{z}$, $T = -\dot{J}$, then $\underline{\dot{s}} = \frac{\dot{j}}{z}$ and so on for the needed high order derivatives of $\underline{s}$. As a result $\underline{\dot{e}} = CL_{\underline{s}}^t T$.

In the following, we look for an exponential stabilization and decoupling of visual error. So, let

$$\underline{e}^{(6)} = -\lambda_0 \underline{e} - \lambda_1 \underline{\dot{e}} - \lambda_2 \underline{\ddot{e}} - \lambda_3 \underline{e}^{(3)} - \lambda_4 \underline{e}^{(4)} - \lambda_5 \underline{e}^{(5)} \qquad (14)$$

The λ_i are stable coefficients. In order to establish a relation between T and ν then to stabilize the flat output J by the camera (visual information), we have first

$$T^{(5)} = (CL_{\underline{s}}^t)^{(-1)}(-\lambda_0 \underline{e} - \lambda_1 \underline{\dot{e}} - \lambda_2 \underline{\ddot{e}} - \lambda_3 \underline{e}^{(3)} - \lambda_4 \underline{e}^{(4)} - \lambda_5 \underline{e}^{(5)}) \quad (15)$$

however as $T^{(5)} = -J^{(6)} = -\nu$, ν can be deduced as above. It remains to calculate the force u which could be given to the cart. For this, we can refer to $J^{(6)}$ equation and deduce the control-input as given above with $C = 1$.

4 Simulation Results

The parameters used in simulation are: $g = 9.8$, $\lambda = 1$, $\lambda_1 = 6$, $\lambda_2 = 15$, $\lambda_3 = 20$, $\lambda_4 = 15$, $\lambda_5 = 6$, $l = 0.5m$, $x_0 = 0$, $M = m = 1Kg$ and $k_{1,2} = mgl$. The cart is excited with $\ddot{x}_g = ae^{-zt}\sin(w_1 t)\sin(w_2 t)$, and where $a = 0.06$, $w_1 = 0.5\pi$, $w_2 = 10\pi$ and $z = -\frac{ln(10^{-3})}{23}$ (highly seismic excitation). The flat output is $J = x + \frac{6}{7}L\theta_1 + \frac{5}{21}L\theta_2$. The results given in **Fig.3**, confirm the stabilization algorithm.

5 conclusion

The problem of stabilization of the IDP mounted on a cart has been solved via vision based feedback. the case of a debarked camera has been considered. The interest has been in the flat output of the unit which does not belong to the system. This problem should be solved in practice. Indeed, the system was highly perturbed under the effect of an acceleration given to the cart. It was considered that the acceleration was due to earth. The case of seismic phenomenon for example can be considered, which was the intention first in the study of excited buildings. Indeed, the problem of stabilizing a building is transformed to a finite number of inverted pendulums. In this paper examples of two inverted pendulums have been given. To be close to reality, future studies will consider the camera mounted on the cart.

References

1. Bradshaw A, Shao J (1996) Swingup control of inverted pendulum systems. In Robotica, V.14, 397-405

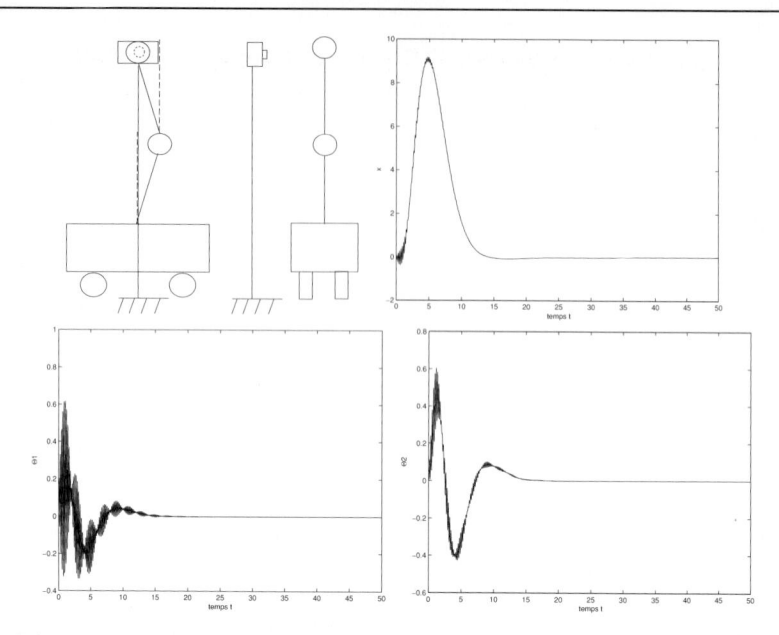

Fig. 3. Position of the camera and the stabilization results (x,θ_1,θ_2).

2. Yamakita M, Nonaka K, Sugahara Y, Furuta K (1995) Robust state transfer control of double pendulum. In IFAC Symposium Advances in Control Education, 205-208, Tokyo, Japan
3. Fliess M, Levine J, Martin P, Rouchon P (1995) Flatness and effect of nonlinear systems: Introduction theory and examples. In Int. J. of Control
4. Zhong W, Rock H (2001) Energy and passivity based control of the double inverted pendulum on a cart. In Proc. of the IEEE Int. Conf. on Control Applications, Mexico City, Mexico
5. Albouy X, Praly L The use of dynamic invariants and forwarding for swinging up a spherical inverted pendulum. http://cas.ensmp.fr/praly/Publications
6. Praly L Introduction to some Lyapunov designs of global asymptotic stabilizers. http://cas.ensmp.fr/praly/Publications
7. Guelton K (2003) Estimation des caractistiques du mouvement humain en station debout. Mise en oeuvre dobservateurs flous sous forme descripteur. Thesis of Université de Valenciennes and Hainaut Cambresis, N03-41

Bus Communication in Robot System Control

G. Mahalu, A. Graur, V. Popa

Electrical Engineering Department, The "Ştefan cel Mare" University of Suceava, Romania

Abstract

In this paper work are proposing some original solutions particular to the data transfer mean and protocols corresponding. The bus imagined has some specific features that conferring originally behaviors. The protocol involved is thinking in this application to working in logical mean with the hardware-designed structure. In this sense is performed entire hardware in relation with the signal type generated by the various kinds of sensors. Was be given attention to the adapting electronic signal structures like as an interface between sensors involved, from one the side, and the physical bus system (with all its conflict problems which can appear). The protocol will be considering these features and will be thinking in so manner to increasing the capabilities of the system.

Keywords: bus, communication, system, microcontroller, channel.

1 Introduction

The modern data process control has like necessary a high informational traffic. This is valuable both in the data quantity and the data velocity transfer and in this sense are involved some functionality requests which put in front the efficient arbiter and control protocols. All these specificity can rise an architectural problem set which resolving both hardware and software level.

A data bus communication can constitutes a challenge for system designer. A same task imposes some features which use the compromise solutions. In this mode, from hardware viewpoint, a system bus request:

- minimize the number of physical lines – responding by the simplify of the field structure request, and implicit to the simplify of the all interfaces, of the minimum transfer time and of the international electromagnetic compatibility actual norms;
- using of the intelligent circuits which work into the master-slave tandem – request of the functional compatibility with the modern microprocessor/microcontroller system structures involved in the communication and processing data;
- assure data transfer through a channel set with different physical nature and redundant functionality, that warn the informational link fault and rise the reliability of the data transfer and security mechanisms.

From software viewpoint the system bus must respond to the actual access technologies and data security adapting requests, with implementing of the protocols and compatibilities between different management structure types.

2 The approach solution

In the try of identify the bus type which responds most of the above hardware and software requests, we focused about the I^{2}C (Inter - Integrated Circuits) bus type. This bus uses three physical lines:

- a data bidirectional line, noted SDA (Serial DAta);
- a clock bidirectional line, noted SCL (Serial CLock);
- a ground line, noted GND (GrouND).

Each bus connected component is recognized by unique address and can function to the distinct times like a data transmitter/receiver in relation with the role played into system. To the I^{2}C bus can be connected more master and slave devices. The initiative to execute one data bus transfer is of the master device. Each time, only one master device has the bus control and this will access only one slave device. The master-slave pair will make a data transfer from master to slave (writing regime) or from slave to master (reading regime). An example of bus connection by more devices is shown in the figure 1.

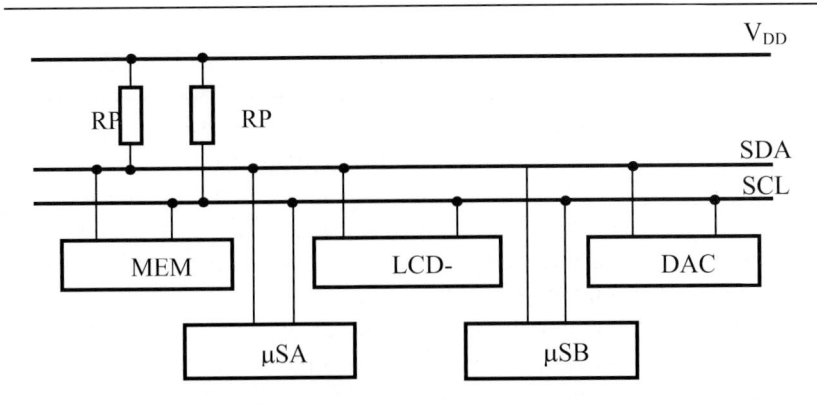

Fig. 1. More devices connected by I^{2}C bus

The data and clock are bidirectional lines. The free state of these lines is 1 logical. The traffic participant devices are connected to all these lines through logical access gates of open collector (or open drain) type which performing the AND-cabled logical operator.

The means of the notations are:

- RP – polarizing resistor
- MEM – memory block
- LCD-DRV – LCD driver
- DAC – digital-analog converter
- µSA, µSB – microsystems
- V_{DD} – supply voltage
- SDA – serial data signal
- SCL – serial clock signal

The bus control access is performed by one concurrent process. Gain of the control and set free of the bus are processes signaling on the following conditions (enabled by the 1 logical on the SCL):

- *START* – transition from 1 in 0 logical for SDA
- *STOP* – transition from 0 in 1 logical for SDA

The bus is considered busy after performed *START* condition and is considered set free after performed *STOP* condition. These signal combinations are generating by the master device (microsystem). In figure 2 is shown the above conditions.

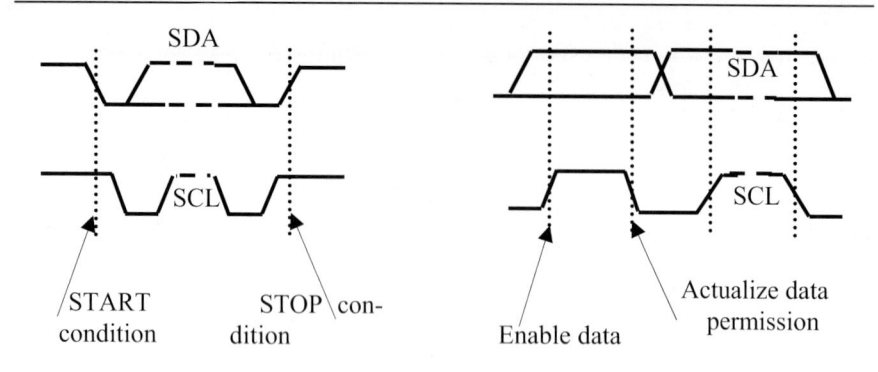

Fig. 2. START and STOP specification

Fig. 3. Transfer data on the I^{2}C bus example

The data is transferred on the bus like packets of bits which called *words*. Most usual example of *word* is the byte. Each transferred word between transmitter and receiver is accompanied by one confirmation bit. In this mode transferring data on the word constitutes an easy controllable operation. More that, can be possible finding a way to speed transfer adapting between transmitter and receiver. For this, after word complete reception, just to next word reception time, receiver forcing in 0 logical the SCL line.

Thus, the transmitter is switched into the wait state and data transfer is resumption after the SCL line clearance. The transfer confirmation is performed by the receiver device, using a clock pulse which is generated by the master, called confirmation bit. After the last bit transmitted on line, the transmitter will clearance the data line (SDA=1). If the receiver has correct received data, it will force the SDA line in 0 logical on during of the confirmation bit (figure 3). The master has duty to confirm this bit transmitting and operate consequently. Thus, if detected confirmation, it continues with a new transfer command (for next word). Contrary, and in non-existent data transferring case, the master device forcing the STOP condition and release the SDA line.

3 Bus synchronization

During data transferring bus, each master device must generating clock signal. After logical synthesis on all these signals is performed the active clock signal. This act consists into synchronization mechanism for all master devices function. Consequently, is performing finally bus access arbitration.

The arbitration mechanism can be useful understand on the figure 4 chronograms base. The clock signal that was forced in 0 logical by one master device, release a measuring process for a t_{low} length on each device, parallel with forcing SCL line in the low state. After this length exhaustion, the master devices will release one by one the SCL line. When the last device it released this line switching on high state. The switching SCL line from 0 in 1 logical conducts to releasing a new measuring process for the t_{high} length when the SCL line is released (let in high state by the master devices). The first master device which will exhausts t_{high} length measuring will command the SCL line in low state so that will be take again generating for a new clock period on the SCL line.

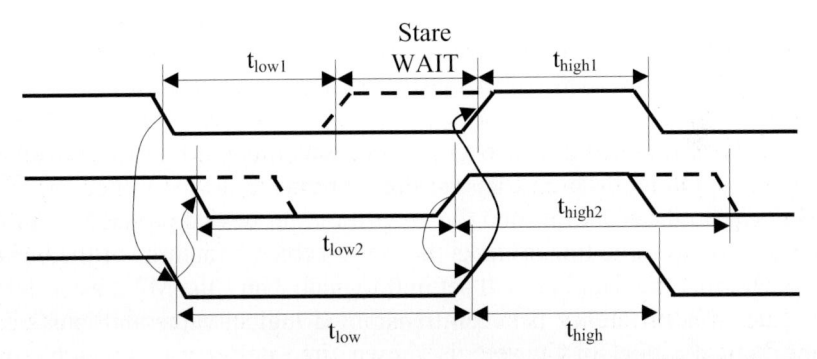

Fig. 4. The clock synchronization

In figure 4 is shown the way for access synchronization performing on the data bus. We are considered the clock signal regeneration by two master devices.

To clock signal bus generating and to synchronization performing will contribute all bus connected master devices. The clock signal obtained is characterized by:

- t_{low} – the SCL forced in low state length – is referred to most big t_{low} duration which is generated by the bus connected devices;
- t_{high} – the SCL forced in high state length – is referred to most small t_{high} duration which is generated by the bus connected devices.

4 Bus access arbitration

After detecting STOP condition, two or more connected bus master devices trying -sometimes simultaneous - to gain the bus control. In these situations is necessary to act an arbitration protocol which must solution all confliction states. The major advantages of approach solution of the I^{2}C bus consist in arbitration management parallel with data transferring.

In the above example, the two master devices, synchronized, will provide the first bit on the data line. Another hand, the master devices will compare the transmitted logical value with the SDA attached state logical value. If the two values are identically the devices will perform the next bit transmitting, following the same procedure. Results that in the limit case when more master devices will transmit simultaneously same information is not necessary inhibition in access to bus line, for some those. Unlike above case, when one or more devices transmitting a logical different value that all those from other master devices, the mechanisms are different.

We'll considerate two master devices which begin transmitting simultaneously. On the n-th bit transmit the first master device (noted M1) puts 0 logical on the SDA line and the second master device (noted M2) puts 1 logical on the same line. Thanks the AND-cabled structures of the bus access circuits, the data line will set in 0 logical. Thus, the M2 master device will detect a difference between transmitted logical value and existent on line logical value. In figure 5 is shown time delivery of the arbitration process.

5 Bus connected devices addressing

Each slave device has allocated an unique address or a group address. When a master device gain the bus access will specify the slave device's address with it whish communicate. The slave device address is specified

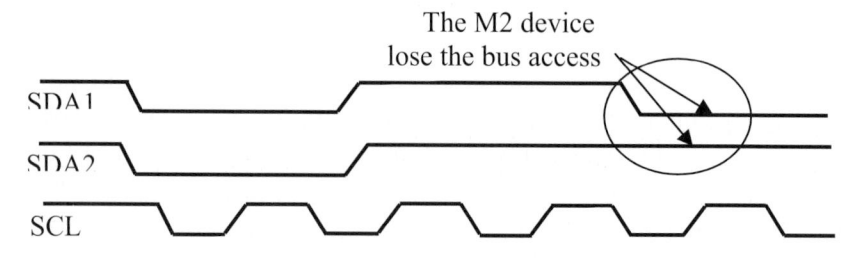

Fig. 5. The bus access arbitration

on seven bits, in standard format, or on ten bits, in extended format. The addressing process is same delivered:

- a master device whish to initiate a transfer after arise the STOP condition detected on the bus, forcing the START condition;
- the master device will provide in line, successive, the seven address bits of the slave device with it whish to change data;
- the bus connected slave devices which are in function, after the START condition detected on bus, will serial receiving the seven master's transmitted address bits;
- after reception of the seven address bits, each slave device will compare the received bits with own address and because an address is allocated to only one slave device result that only one device will recognize own address;
- the next master's transmitted bit will point the transfer direction (from master to slave in transmitting mode or from slave to master in receiving mode);
- the master device will release the data line and SDA switch in 1 logical;
- if address was be recognized by the slave device then it will force the data line in 0 logical on during of the clock pulse provided by the master device, and point so own address recognition and accepting transfer direction;
- if address was not be recognized by any slave device then SDA line stand in 1 logical on the next clock pulse length and point to master device that addressed devices is non-existent, and thus the master device will end the current transfer cycle forcing on the bus the STOP condition.

Following way of address allocation become possible to connect more similar devices to bus. Each device belong to devices group for was be allocate an address domain.

6 Conclusions

The SAUB design imposes a series of functional considerations, at physical level and at logical level respectively, specifying the conditions when a same bus can be used with a multiple physical channels support. From possible solutions was be selected I^{2}C type, extended to situation that in addition to classical used channel (electrical channel) appear new channels, with different physical nature, like as supplementary radio-channel SSS-FH with binary FSK modulation.

References

1. Mihalcea A, Şerbănescu A., Tabarcea P. (1992) *Sisteme moderne de comunicaţie*, Editura Militară, Bucureşti.
2. Creangă I, Pui C. (1997) *Microprocesoare de comenzi din receptoare TV*, Editura Teora, Bucureşti.
3. Mahalu G., Ciufudean C. (2002) *Radio-Transmitter System for PC Controlled Structures,* Proceedings of the Second European Conference on Intelligent Systems and Technologies ECIT 2002, Iaşi, July 17-20, ISBN 973-8075-20-3, CD recorded.
4. Mahalu G., Graur A., Ciufudean C. (2003) *Virtual Interface in Control and Measurement Instrumentation*, XIII-th International Symposium on Electrical Apparatus and Technologies SIELA 2003, Plovdiv, May 29-30, vol.I, pg. 147-151, ISBN 954-90209-2-4, Plovdiv, Bulgaria.
5. Mahalu G., Graur A., Ciufudean C. (2002) *PC Signal Acquisition and Control Board for Technological Processes,* ELMA, IEEE Section Bulgaria, Tenth International Conference on Electrical Machines, Drives and Power Systems, Sofia, Bulgaria, ISBN 954-902-091-6, vol.1, pp.263-268.
6. Mahalu G., Graur A., Ciufudean C., Pentiuc R., Ioachim P., Milici M., Milici D. (2002) *Intelligent Driver System for Technological Processes,* ELMA, IEEE section Bulgaria, Tenth International Conference on Electrical Machines, Drives and Power Systems, Sofia, Bulgaria, vol.1, ISBN 954-902-091-6, pp.281-285.
7. Mahalu G., Graur A., Pentiuc R., Mahalu C. (2005) *Bus Communication for Multi Sensor Operations*, Advanced Sensor Payloads for UAV, NATO Research and Technology Organisation, RTO-MP-SET-092 Reference, May 02-03 2005, Lisabon-Portugal.
8. Mahalu G., Graur A., Ciufudean C. (2003) *Virtual Video-Sensitive Interface Used in Robotic Control*, Proceedings of the 20th International Symposium on Automation and Robotics in Construction ISARC2003, Eindhoven, Netherlands, 21-24 September 2003, pg. 313-318, ISBN 90-6814-574-6.
9. Mahalu G., Pentiuc R., Mahalu C. (1999) *Intelligent Supervisor System For Energetically Nets*, 1'er Symposium International Euretech "Systemes et Technologies Modernes", Institut Euretech Settat, 16/17 Juillet 1999, SETTAT – MAROC, pg.47-48.

A Hybrid Locomotion Control Approach

Dirk Spenneberg

Robotics Group, University of Bremen, Department of Mathematics and
Computer Science, Bibliothekstr. 1, D-28359 Bremen, Germany
dspenneb@informatik.uni-bremen.de

Abstract

This paper presents the joint control part of a new hybrid locomotion control concept. The concept is based on a CPG-model, a reflex model inspired by artificial neurons, and a posture control model. It has been successfully tested on the 8-legged walking robot SCORPION.

Keywords: Hybrid Control, CPG, Reflex, Posture, Bezier Splines.

1 Introduction

A lot of different approaches for controlling locomotion in multipods already exist but most of them are focused on reflex or Central Pattern Generator (CPG) only and do not take into account a flexible posture control. Therefore, an enhanced locomotion control concept to improve the flexibility of motion production is presented. This concept is based on experiences gained in the SCORPION project [1] and has been firstly tested with the SCORPION robot.

2 The Rhythmic Motion Control

Central Pattern Generators (CPG) are the major mechanisms in animals to control and to produce rhythmic motion. They are characterized by the ability to produce rhythmic motion patterns via oscillation of neuronal activity without the need of sensory feedback. However, many CPGs get at least a sensory feedback about the load and the position on the corresponding driven joint. Thus CPGs are mostly used for closed loop control of the rhythmic motion of actuators. To modulate their behavior, their rhythmic patterns can be changed in frequency, phase, and amplitude. This functionality can not be transfered directly on a walking robot because it possesses motor-driven joints instead of muscles. From the robotics point of view it is reasonable to develop a more abstract CPG model which captures only the basic principles of the functionality of CPG motor control.

The CCCPG-approach [2] is a possible step in that direction, but its mechanisms are still tightly connected to the research on lobsters and the nitinol-actuators used in the lobster robot. Instead, we want to pursue the goal to formulate a cpg-model which has no direct biological archetype.

An important feature of CPG control is the ability to describe motion with rhythmic signals, which can be modulated in their phase, frequency, and amplitude, to change the resulting motion in its timing of execution, its duration, and its strength/speed. In our model, the closed loop control (via sensory load and position feedback) can be implemented by a PID-controller. If we use a rhythmic trajectory as a set value for a standard PID-controller, which receives the position signal from the corresponding joint, we have a closed loop control of the rhythmic motion in different load states comparable to the behavior of a real CPG.

Recapitulating this means that our abstract CPG-model consists of a controller-module and a unit to produce rhythmic trajectories in the joint angle space. For the controller unit one can use approaches already existing. For the rhythmic trajectories we describe a CPG-Pattern P as a function of part-wise fitted together polynoms of the form: $y(t) = \sum_{a=0}^{3} k_a \cdot t^a$.
A part X is described by the coefficients of the polynom $k_a(X)$, its length $l(X) \in \mathbf{N_0}$ on the x-axis, the phase offset $\theta(X) \in [0,1]$, its scalability $S(X) \in \{0,1\}$, and an optional subpart-list, if the part is constructed by subparts. A complete rhythmic pattern is then described by a list of parts with the end of the list pointing to the start of the list. To describe a part, Bezier-polynoms which are described by the following equation:

$$P0 \cdot t^3 + P1 \cdot 3 \cdot t^2 \cdot (1-t) + P2 \cdot 3 \cdot t \cdot (1-t)^2 + P3 \cdot (1-t)^3 = P \quad (1)$$

P0 and P3 are supporting points and P1 and P2 are control points of the curve. This coherence is illustrated in **Fig.1**. Bezier curves have some advantages, they are smooth (optimal for DC-motor-control) and the controllable gradients at their end-points allowing a smooth transition from one part to the next. In our special case, where we use the Bezier curves

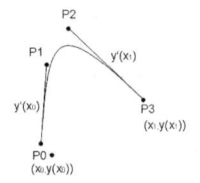

Fig. 1. A Bezier Curve 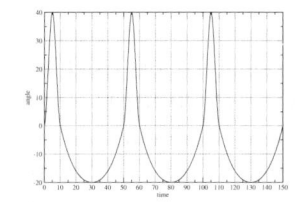 **Fig. 2.** A Rhythmic Pattern

only to describe a trajectory in the 2-dimensional joint angle space in dependence of the time t, it is possible to assign exactly one $y(t)$ to each t, thus $y(t)$ is a function. In this case we can compute the coefficients k_0 to k_3 in the following way:

$$k_0 = y_0 \; ; \; k_1 = y'(x_0) \tag{2}$$
$$k_2 = ((2y'(x_0) + y'(x_1)) \cdot x_1^2 + 3(y_0 - y_1) \cdot x_1)/-x_1^3 \tag{3}$$
$$k_3 = (2y_1 - (y'(x_0) + y'(x_1)) \cdot x_1 - 2y_0)/-x_1^3 \tag{4}$$

In this solution we have used P1 and P2 only indirectly to compute the gradients $y'(x_0)$ und $y'(x_1)$, we only need to know the chosen gradient at P0 and P3. P0 has the coordinates $(x_0, y(x_0))$ and $(x_1, y(x_1))$ are the coordinates of P3. The position-algorithm computes to every timestep t for the current Part X_a of the whole pattern P (consisting of n parts and with a offset θ) the following equation:

$$pos(t) = y(t + \theta) \pmod{l_P} \tag{5}$$

$$\text{where } y(x) = \sum_{i=0}^{3} k_i(X_a) \cdot x \; ; \; x \in X_a = [x_0(a), x_1(a)[\tag{6}$$

$$\text{and } X_a \in P = \{X_1, X_2...X_n\} \; ; \; l_P = x_1(n) - x_0(1) \tag{7}$$

For a smooth transition between parts, we define the following constraints:

$$x_1(i) = x_0(i+1) \; ; \; y'(x_1(i)) = y'(x_0(i+1)) \tag{8}$$
$$x_0(0) = x_1(n) \; ; \; y'(x_00) = y'(x_1(n)) \tag{9}$$

To get an even more compact way of describing trajectories, we distinguish two types of supporting points, (Type 0)-points where $y'(x) = 0$ (extreme points) and all other points (Type 1). The gradient of the Type 1 points is given by the gradient of the line through its direct neighbor points. The length $l(X_a)$ of a part X_a is equivalent to $x_1(a) - x_0(a)$.

activity, we can use the posture control also for direct control of each joint, e.g., for manipulating objects.

Reflexes are neuronal mechanisms which transform sensory input into motoric action - often with using one or more interneurons. The motoric action depends mostly on the strength of the stimulus and the interneuron circuit. In contrast to the commonly assumption that reflexes are fixed reactions, it is possible that the interneurons can be reprogrammed and thus change the motoric action[4]. In addition, the occurrence of reflexes is often sent to higher neuronal centers. A simple reflex model has an input-, an activation-, and a response-function.

$$f(t) = (\sum_{i=1}^{n-1} f(t-i) + \sum_{i=1}^{n_{sense}} w_i^S * S(i))/n \text{ (input function)} \quad (15)$$

$$a(f(t)) = 1/1 + e^{(-f(t)/\sigma(t))} \text{ (possible activation function)} \quad (16)$$

$$\sigma(t) = \sum_{i=1}^{n_{control}} w_i^C * C_i(t) \text{ (control of } \sigma) \quad (17)$$

The control signals C_i are given from the outside (the behavioral level). The response function $r(t)$ is activated when $a(f(t))$ is positive, and can be as complicated as needed. The response function is responsible for the reaction of the controlled motor-joint(s). An example of the reflex is the stumbling correction reflex implemented in the SCORPION and AIMEE[5] robot. The biological example is described in [6]. This reflex is only active in the first two thirds of the swing phase of a leg. If the leg hits an obstacle, a response is triggered which lifts the leg higher up to overcome the disturbance. In the stance phase this response is not triggered when the leg is disturbed. In our model this switching behavior can be controlled with the control signals C_i. The input for the input function is the current drawn by the shoulder motor which drives the leg forward (thoracic joint of the SCORPION). Because of the low load on the leg in the swing phase the threshold is chosen low, thus disturbances (blocking the motor) will activate the reflex. In the stance phase the threshold is set high, thus resulting in no activation of the reflex. The response function is a fixed function which for a certain period of time moves the leg up.

4 Integration of the Models

All components, rhythmic motion control (cpg-model), posture control, and reflex control are integrated into the motoric level as depicted in **Fig. 7**. In this approach, each joint has its own motoric level which is modulated and coordinated by a the behavioral level. The behavioral level features, for example, behavior processes responsible for obstacle avoidance, path following, etc.. The behavioral level has a multitude of interfaces to send signals to the motoric level. The behavioral level can

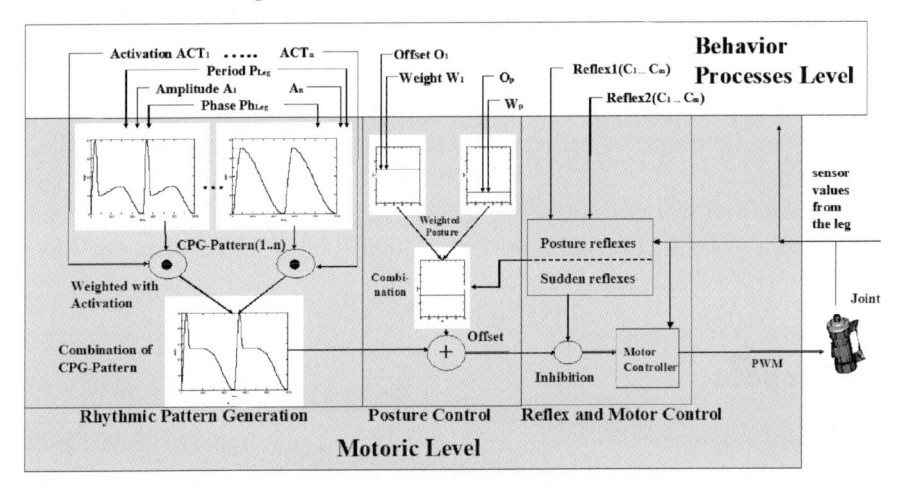

Fig. 7. The Motoric Level

choose the activity and the weights for the cpg-patterns. It can modulate the period, the amplitude, and the phase of each pattern. The period is identical for all joints, the phase for all joints of one leg. On these values, several processes can take influence at the same time. The result is computed by a merge-process which uses a weighted sum principle (for further details see [7]). This results in one final pattern. This pattern can then be offset on the y-axis by the Posture Control. Again, in the Posture Control we can have simultaneous influences which form one final offset for each joint. After the offset is applied, this pattern is given to the motor controller which moves the joint through this trajectory. Therefore, the motor controller gets the current and position values from the joint. The reflexes also get these values. We implemented two types of reflexes. Type-1: Reflexes which control the posture and are almost all the time active (low threshold). They take influence on the offset via their response function to control, for example, pitch and roll of the system or to keep the distal joint perpendicular to the ground when the SCORPION robot walks forward. Type-2: Reflexes, like the above described "'Stumbling Correction Reflex"', which inhibits the values from the rhythmic and posture control and writes it own values to the motor controller.

5 Results

The implemented mechanisms enable the SCORPION robot to walk very robust through rough terrain.

The Posture Control gives the behavior level new means of locomotion. E.g, it possible by only changing the posture of the SCORPION and using the unchanged rhythmic motion for forward walking, to brachiate upside down along a beam. In addition, first successful walking experiments have been conducted with two four-legged robots, the AIMEE[5] and the ARAMIES robot. More up-to-date information on these experiments can be found at http://www.informatik.uni-bremen.de/robotik.

The presented work is sponsored by the Deutsches Zentrum fuer Luft und Raumfahrt (DLR Foerderkennzeichen 50JR0561) and European Space Agency (ESA contract. 18116/04/NL/PA).

References

1. Kirchner, Frank, Spenneberg, Dirk, and Linnemann, Ralf (2002) A biologically inspired approach towards robust real world locomotion in an 8-legged robot. In *Neurotechnology for Biomimetic Robots*, Ayers, J., Davis, J., and Rudolph, A., Eds. MIT-Press, Cambridge, MA, USA.
2. Ayers, Joseph (2002) A conservative biomimetic control architecture for autonomous underwater robots. In *Neurotechnology for Biomimetic Robots*, Ayers, Davis, and Rudolph, Eds., pp. 241–260. MIT Press.
3. Bowerman, R. F. (1975) The control of walking in the scorpion i. leg movements during normal walking. *Journal of Comparative Physiology*, vol. 100, pp. 183–196.
4. Reichert, Heinrich (1993) *Introduction to Neurobiology*. Thieme.
5. Albrecht, M., Backhaus, T., Planthaber, S., Stoeppeler, H., Spenneberg, D., and Kirchner, F. (2005) Aimee: A four legged robot for robocup rescue. In *Proceedings of CLAWAR 2005*. Springer.
6. Forssberg, H. (1979) Stumbling corrective reaction: A phase-dependant compensatory reaction during locomotion. *Journal of Neurophysiology*, vol. 42, no. 4.
7. Spenneberg, Dirk, Albrecht, Martin, and Backhaus, Till (2005) M.o.n.s.t.e.r.: A new behavior-based microkernel for mobile robots. In *Proc. of the 2nd European Conference on Mobile Robots*.

Model and Control of Joints Driven by Fluidic Muscles with the Help of Advanced Automatic Algorithm Generation Software

T. Kerscher[1], J.M. Zoellner[1], R. Dillmann[1], A. Stella[2], and G. Caporaletti[2]

[1] Forschungszentrum Informatik, Haid-und-Neu-Str. 10-14, 76137 Karlsruhe, Germany, kerscher@fzi.de. [2] EICAS Automazione S.p.A., Via Vincenzo Vela, 27 – 10128 Torino, Italy, gabry@eicas.it

Abstract.

The content of this paper describes the model and control of a elastic joint driven by fluidic muscles including the nonlinear behavior of the fluidic muscle, the valves and the joint dynamics. Such elastic joints have a lot of advantages like passive compliance, low power to weight relation. The control of the joint is developed with the help of a professional software tool named EICASLAB which has been realized within the ACODUASIS Project founded by the European Commission in the frame of the Innovation Program aiming at transferring to the robotics sector the EICAS methodology

1 Introduction

The FZI at the University of Karlsruhe started 2000 to use fluidic muscles as actuators for robotic system [6]. This is motivated by the fact of building biologically inspired robots. In order to take advantage from nature not only the mechatronical parts, but also the actuators should be imitated.

Different groups in the robotics sector are using fluidic muscles as actuator. In all cases control is a big issue because of the elastic behaviour of these muscles [2, 3, 4]. Nevertheless in future 'soft' actuators are needed for robots which interact with humans or in human environments.

At the moment fluidic muscles have been used for walking research with the sixed legged robot AirBug [6] and the test-leg for the four legged mammal like robot PANTER (figure 1) [7]. Right now, we only use joint-controller for walking instead of controller for the whole legs. In the future we will develop a controller for the whole leg with the help of a professional software tool called EICASLAB [1].

Different assumption can be defined for leg-joints for walking-machines due to the fact, that walking is a cyclic motion. There are two different walking-phases the power phase and the return phase. During power phase the leg holds and pulls the robot. There may be interacting forces between the legs with ground contact and large disturbance torques. During the return phase the leg to perform a fast motion from the last point of the past power phase to the first point of the next power phase.

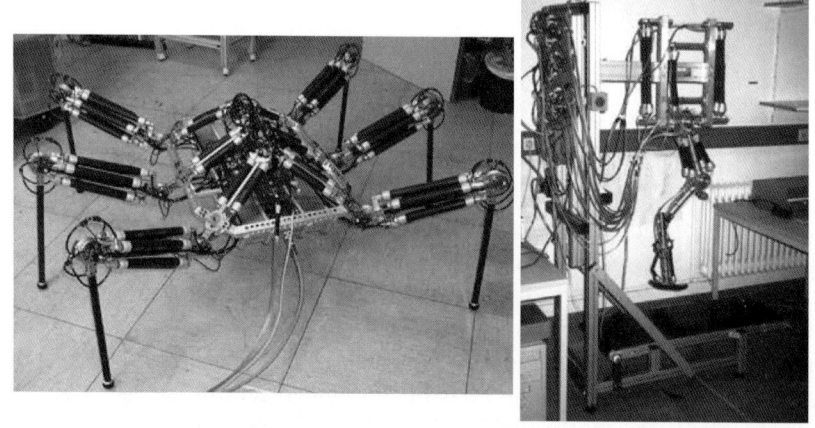

Fig. 1. (left) Six legged robot AirBug driven by fluidic muscles, (right) Test-Leg for the quadruped running PANTER-robot

The desired walking behaviour makes great demands on joint-controllers for walking machines. Due to this the software tool EICASLAB [1] was used to find and test joint-controller. With the help of EICASLAB it is possible to use automated algorithm and code generation. For the set up of the controller and for modelling the control problem in simulation it is necessary to have an accurate model of the joint which should be controlled.

2 Model of a joint driven by fluidic muscles

The fine model of a general test-rig for elastic joints (see figure 2) is introduced, so that the found controller can be easily adapted to an elastic actuated robot joint.

To find the equation of motion for the joint the Euler-equation is used:

$$J \cdot \ddot{\varphi} = -r \cdot F_{MusA}(\kappa_A, \dot{\kappa}, p_A) + r \cdot F_{MusB}(\kappa_B, \dot{\kappa}, p_B) - \cos(\varphi - \varphi_0) \cdot \frac{1}{2} \cdot F_g + f_s \cdot l \quad (1)$$

with joint inertia J, muscle forces $F_{MusA,B}$, muscle pressures $p_{A,B}$, muscle contractions $\kappa_{A,B}$, gravitation force F_g, angle between inertial position and horizontal plan φ_0, length of the joint l and disturbance force F_s (assumed that it is always orthogonal to the joint). φ should be equal zero for the joint position where both muscle have the same contraction length.

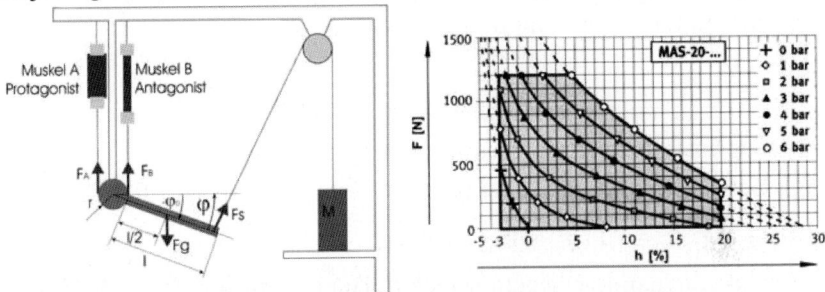

Fig. 2. (left) Schematic joint driven by pneumatic msucles. (right) Static correlation between force, pressure and contraction [5].

There is a linear correlation between the joint angle and the contraction of the muscle if the tendons between muscle and joint are tensed, so the contraction can be calculated with the help of the joint-angle:

$$\kappa_A(\varphi) = \frac{r}{l_{A0}}(\varphi_{A0} - \varphi), \quad \kappa_B(\varphi) = \frac{r}{l_{B0}}(\varphi - \varphi_{B0}) \quad (2)$$

$\varphi_{A0,B0}$ are the joint angles where the muscle have there initial length l_0.

2.1 Equation for the force and pressure of the muscle

The fluidic muscle can be modelled as a spring and a parallel damper. The force of the muscle F_{Mus} is correlated with the relative muscle pressure p, the contraction κ and the derivative of the contraction $\dot{\kappa}$ of the muscle (see figure 2).

$$F_{Mus}(\kappa, \dot{\kappa}, p) = F_{spring}(\kappa, p) + F_{damper}(\dot{\kappa}, p). \quad (3)$$

with

$$F_{damper}(\dot{\kappa}, p) = -C_D \cdot (p + P_n) \cdot \dot{\kappa}$$
$$F_{spring}(p, \kappa) = \mu \cdot (\pi \cdot r_0^2) \cdot p \cdot (a \cdot (1 - (a_\varepsilon \cdot e^{-p} + b_\varepsilon) \cdot \kappa)^2 - b) + \sigma(-\kappa) \cdot (-f_0) \cdot \kappa \quad (4)$$

The different parameters of the equations are the damping coefficient C_D, absolute ambient pressure P_N, correction value μ, muscle radius r_0, geometric muscle parameter a, b, muscle force correction parameter a_ε, b_ε and the correction parameter f_0 for $p = 0$.

The pressure in the muscle can be calculated by the following equations:

$$P_{Mus} = P_N \cdot \frac{V_{air}}{V_{Mus}} \Rightarrow \dot{P}_{Mus} = P_N \cdot \left(\frac{\dot{V}_{air}}{V_{Mus}} - V_{air} \cdot \frac{\dot{V}_{Mus}}{V_{Mus}^2} \right) \tag{5}$$

$\dot{V}_{air}$ can be found with the help of the Bernoulli-equation:

$$\dot{V}_{air} = f_V \cdot C_a \cdot A_V \cdot \sqrt{(P_0 - P)}. \tag{6}$$

Two cases must be distinguished:
- Filling: $f_V = 1$, $P_0 =$ absolute pressure of the gas storage, $P = P_{Mus}$.
- Emptying: $f_V = -1$, $P_0 = P_{Mus}$, $P = P_N$.

C_a is an aerodynamic correction factor. The area of the valve A_V is proportional to the airflow to the muscle. The Muscle volume V_{Mus} is correlated with the contraction of the muscle and the initial muscle length l_0. It can be approximated by:

$$V_{Mus}(\kappa) = a(l_0) \cdot \kappa + b(l_0) \Rightarrow \dot{V}_{Mus}(\dot{\kappa}) = a(l_0) \cdot \dot{\kappa} \tag{7}$$

The amount of air volume in the muscle under normal pressure can be calculated by:

$$V_{air} = V_{Mus}(\kappa) \cdot \frac{P_{Mus}}{P_N} \tag{8}$$

The differential equation resulting for the muscle-pressure is:

$$\dot{P}_{Mus} = P_N \frac{f_V \cdot C_a \cdot A_V \cdot \sqrt{(P_0 - P)}}{V_{Mus}} - P_{Mus} \cdot \frac{\dot{V}_{Mus}(\dot{\kappa})}{V_{Mus}(\kappa)}. \tag{9}$$

2.2 Model for High-speed switching valve

The switching valves used to control the muscle pressure have three possible states:
1. closed: no airflow;
2. opened for filling: max. opening area for airflow into the muscle;
3. opened for emptying: max. opening area for airflow out of the muscle.

The valves operate with pulse width modulation (PWM). For a given input u the valve activation can be calculated:

$$A_V = \text{sign}(u) \cdot \text{rect}\left(t, \frac{|u|}{u_{max}}\right). \tag{10}$$

u_{max} is the biggest possible input and

$$\text{rect}(t, e) = \begin{cases} \dot{V}_{max} & \text{for } k \cdot T_{PWM} \leq t < (k+e) \cdot T_{PWM} \\ 0 & \text{for } (k+e) \cdot T_{PWM} \leq t < (k+1) \cdot T_{PWM} \end{cases} \tag{11}$$

with $0 \leq e \leq 1$, the pulswidth T_{PWM}, $k = \text{floor}\left(\dfrac{t}{T_{PWM}}\right)$.

2.3 State equation

For the description of the whole dynamics the state equation for nonlinear systems is used. State variables are: $x_1 = p_A$, $x_2 = p_B$, $x_3 = \varphi$ and $x_4 = \dot{\varphi}$. The input variables are $u_1 = A_{VA}$ (opening area valve A) and $u_2 = A_{VB}$ (opening area valve B). The following state equation (shown only for the case of filling the muscle) ($\overline{\varphi}_0 = \varphi_{A0} = \varphi_{B0}$) is found:

$$\underline{\dot{x}} = \begin{pmatrix} P_N \dfrac{f_V \cdot C_a \cdot u_1 \cdot \sqrt{(P_C - x_1)}}{V_{MusA}\left(\dfrac{r}{l_0} \cdot (\overline{\varphi}_0 - x_3)\right)} - x_1 \cdot \dfrac{\dot{V}_{MusA}\left(-\dfrac{r}{l_0} \cdot x_4\right)}{V_{MusA}\left(\dfrac{r}{l_0} \cdot (\overline{\varphi}_0 - x_3)\right)} \\ P_N \dfrac{f_V \cdot C_a \cdot u_2 \cdot \sqrt{(P_C - x_2)}}{V_{MusB}\left(\dfrac{r}{l_0} \cdot (x_3 - \overline{\varphi}_0)\right)} - x_2 \cdot \dfrac{\dot{V}_{MusB}\left(\dfrac{r}{l_0} \cdot x_4\right)}{V_{MusB}\left(\dfrac{r}{l_0} \cdot (x_3 - \overline{\varphi}_0)\right)} \\ x_4 \\ \dfrac{1}{J} \cdot \left(-r \cdot F_{MusA}(\dfrac{r}{l_0} \cdot (\overline{\varphi}_0 - x_3), -\dfrac{r}{l_0} \cdot x_4, x_1) + r \cdot F_{MusB}(\dfrac{r}{l_0} \cdot (x_3 - \overline{\varphi}_0), \dfrac{r}{l_0} \cdot x_4, x_2) - \cos(x_3 - \varphi_0) \cdot \dfrac{1}{2} \cdot F_g\right) \end{pmatrix}$$

$$\underline{y} = \begin{pmatrix} x_3 & x_1 & x_2 \end{pmatrix}^T \tag{12}$$

3 Modelling and control of the elastic joint using the EICAS-Lab software tool

The control design is carried out according to the EICAS methodology that allows guaranteeing the required performance in presence of disturbances

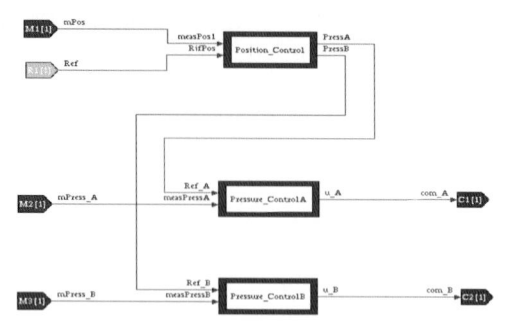

Fig. 3. Hierarchical Control

and uncertainty in the plant [8, 9]. A feedback control is designed on the basis of a simplified linear model, without considering the plant-model uncertainty. In order to get the best control performance, the plant control is typically designed according to the methodology presented in [1, 8] including:

- the estimation of future equivalent additive disturbances acting on the plant inputs so that their effect can be directly compensated. This action is computed by means of the "state and disturbance observer", together with the estimation of the state values,
- an open loop control action, which is computed by means of the "reference generator", together with the required state values,
- the feedback state control, computed by the "closed-loop control".

Then the total command is composed of three contributes: the open loop command, the compensation of disturbances and the closed-loop command. The plant control architecture related to the joint-model is realized in EICASLAB. It is a hierarchical control (see figure 3) composed of:

- one *position control*,
- two *pressure controls*.

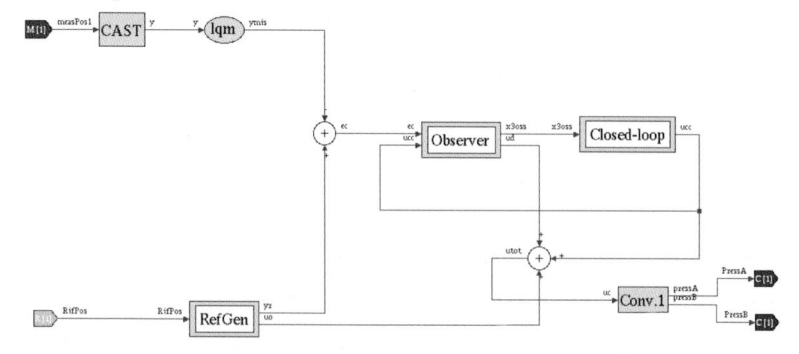

Fig.4. Position Control implemented in EICASLAB SIMBUILDER

Model and Control of joints driven by fluidic muscles with ... 251

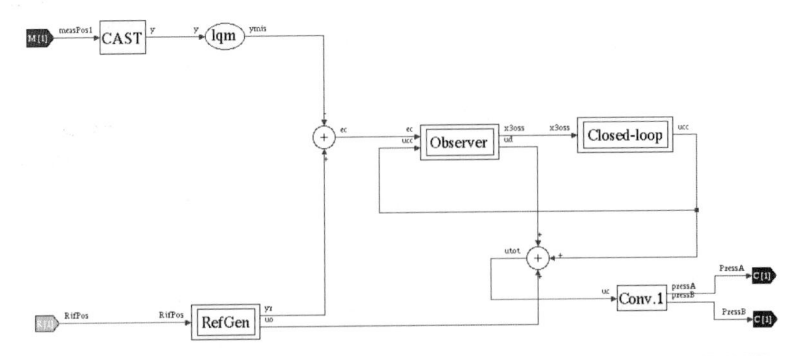

Fig. 5. Pressure Control implemented in EICASLAB SIMBUILDER

The *position control* has as input the measured position and the reference one and provides as output the reference pressures for the *pressure controls* that compute the commands for the muscles. The position control and the pressure control structures are shown respectively in figure 4 and in figure 5.

The control is tuned and its performance assessed by means of the EICASLAB simulator, where the fine model is used to simulate the plant. In this way the control immediately works well, without requiring set up in field. The simulation results are shown in figure 5.

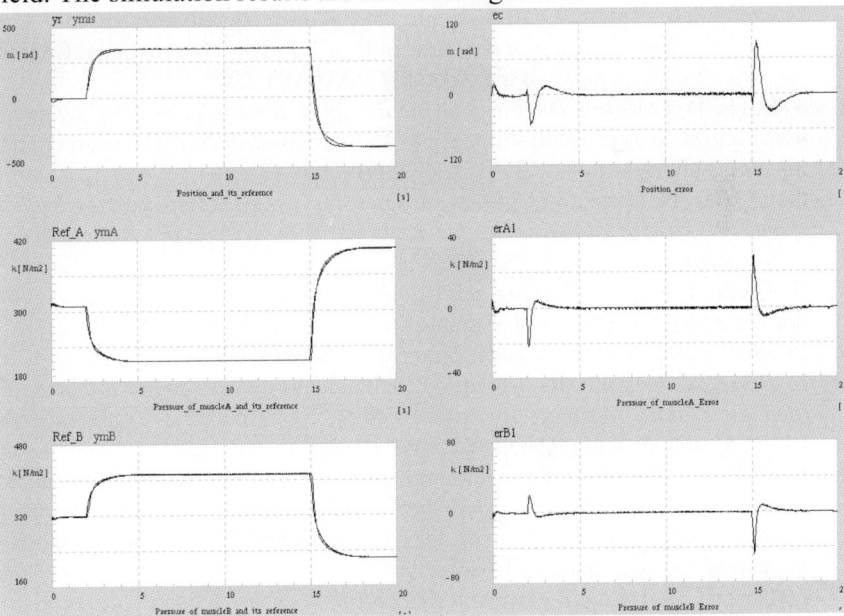

Fig. 6. Simulation results of the joint control using EICASLAB SIM

4 Conclusion and outlook

This paper presents the model and control of a joint driven by fluidic muscles. This control was implemented and simulated with the professional software tool named EICASLAB. For the implementation on the real robot PANTER the controller will be adapted for the microcontroller boards used to control the robot. This can be done also with the help of the EICASLAB software by using the automatic code generation function of the tool.

Acknowledgments

The work has been realised within the ACODUASIS Project (IPS-2001-42068), a three-year project founded by the European Commission in the frame of the Innovation Program.

References

1. G. Caporaletti, The ACODUASIS project – a professional software tool supporting the control design in robotics, In Proc. of the 6th International Conference on Climbing and Walking Robots (CLAWAR), 2003
2. S. Davis, D. Caldwell *The bio-mimetic design of a robot primate using pneumatic muscle actuators*, In Proc. of the 4th International Conference on Climbing and Walking Robots (CLAWAR), 2001
3. C.P. Chou, B. Hannaford *Static and Dynamic Characteristics of McKibben Pneumatic Artificial Muscles*, International Conference on Robotics and Automation, 1994, Vol. 1, 281-286
4. B. Tondu, P. Lopez *Modeling and Control of McKibben Artificial Muscle Robot Actuators*, IEEE Control Systems Magazine, April 2000, Vol. 20, 15-38
5. Produktkatalog 2001 -- *Antriebe "http://www.festo.com"*
6. K. Berns, V. Kepplin, R. Müller, M. Schmalenbach: *Six-legged Robot Actuated by Fluidic Muscles.* In Proc. of the 3th International Conference on Climbing and Walking Robots (CLAWAR), 2000
7. J. Albiez, T. Kerscher, F. Grimminger, U. Hochholdinger, R. Dillmann, K. Berns: *PANTER - prototype for a fast-running quadruped robot with pneumatic muscles.* In Proceedings of the 6th International Conference on Climbing and Walking Robots, 617-624, 2003.
8. F. Donati, M. Vallauri: *Guaranteed control of almost- linear plants.* IEEE Transactions on Automatic Control, vol. 29- AC, 1984, pp. 34-41
9. Website of the ACODUASIS-project funded by the European Community: http://www.fzi.de/acoduasis, June 2005.

Modelling and Control of a X4 Bidirectional Rotors

Rajia Slim[1], Azgal Abichou[1] and Lotfi Beji[2]

[1] LIM Laboratory, Ecole Polytechnique de Tunisie, BP 743, 2078 La Marsa, Tunisia. `rajia_slim@yahoo.fr` `azgal.abichou@ept.rnu.tn`
[2] LSC Laboratory, CNRS-FRE2494, Université d'Evry Val d'Essonne. 40, rue du Pelvoux, 91020, Evry Cedex, France. `beji@iup.univ-evry.fr`

Abstract

A particular structure of a four rotors mini-flying robot where two rotors are directional (X4 bidirectional rotors) is presented in the paper. The two internal degree of freedom leads to a transformation between the equivalent system of the control-inputs and the rotor force-inputs which is not a diffeomorphism. This makes our system different from that of classical flyer robot (X4 flyer). The dynamic model involves five control inputs which will be computed to stabilize the engine with predefined trajectories. For the underactuated system we show that the nonlinear model can be locally-asymptotically stabilized by a smooth time-varying control law. The simulation results are presented to illustrate the control strategies.

Keywords: Mini-UAV, stabilizing control, averying system, Lie algebra.

1 Introduction

Terrain missions control of Unmanned Areal Vehicles (UAV) is still an open and difficult problem for scientific research and in engineering design. In our laboratory *Lsc*, a mini-UAV, is under construction subject of some industrial constraints. The areal flying engine couldn't exceed $2kg$ in mass, and $50cm$ in diameter with $30mn$ as time of autonomy in flight. With these characteristics, we can presume that our system belongs to the family of mini-UAV. The idea of our group was fixed on a four-rotor flyer with protected rotor blades. It is an autonomous hovering system, capable of vertical take-off, landing and quasi-stationary (hover or near hover) flight conditions. Compared to helicopters, named quad-rotor [8], [1], the four-rotor rotorcraft have some advantages [9], [4]: given that two motors rotate counter clockwise while the other two rotate clockwise, gyroscopic effects and aerodynamic torques tend to cancel in trimmed flight, and where two rotors are directional (X4 bidirectional rotors). An X4 bidirectional rotors mini-flyer operates as an omnidirectional UAV. Vertical motion is controlled by collectively increasing or decreasing the power for all four dc-motors. Lateral motion, in x-direction or in y-direction, is achieved by differentially controlling the motors generating a pitching/rolling motion of the airframe that inclines the collective thrust (producing horizontal forces) and leads to lateral accelerations. A model for the dynamic and configuration stabilization of quasi-stationary flight conditions of a four rotor vertical take-off and landing (VTOL), named as X4-flyer vehicle, is given in [4]. Dynamic motor effects are incorporating [4] and a bound of the perturbation error was obtained for the coupled system. The X4-flyer stabilization problem is also studied and tested by Castillo [2] where the nested saturation algorithm is used. The idea is to guarantee a bound of the roll and pitch angles with a fixed bound of control inputs. Motivated by the need to stabilize aircrafts that are able to take-off vertically as helicopters, the control problem was solved for the Planar Vertical Take-Off and Landing (PVTOL): I/O linearization procedure in [7], theory of flat systems in [3] [5] and [6]. In this paper, we present the dynamic model of the X4 bidirectional rotors and we show that the nonlinear underactuated model can be locally-asymptotically stabilized by a smooth time varying control law.

2 Configuration descriptions and modelling

The X4 bidirectional rotors mini-flyer robot is a system consisting of four individual electrical fans attached to a rigid cross frame. It is an omnidirectional (vertical take-off and landing) VTOL vehicle ideally suited to stationary and quasi-stationary flight conditions. We consider a local reference airframe $\Re_G = \{G, E_1^g, E_2^g, E_3^g\}$ attached to the center of mass G of the vehicle. The center of mass is located at the intersection of the two rigid bars, each of them supports two motors. Equipment (controller cartes, sensors, etc.) onboard are placed not far from G. The inertial frame is denoted by $\Re_o = \{O, E_x, E_y, E_z\}$ such that the vertical direction E_z is upwards. Let the vector $\xi = (x, y, z)$ denote the position of the center of mass of the airframe in the frame $\Re_o$. While the rotation of the rigid

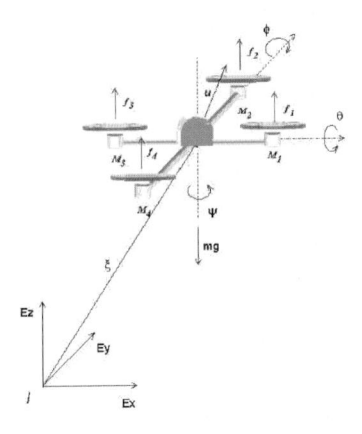

Fig. 1. Frames attached to the X4 bidirectional rotors rotorcraft

body is determined by a rotation $R : \Re_G \rightarrow \Re_o$, where $R \in SO(3)$ is an orthogonal rotation matrix. This matrix is defined by the three Euler angles ψ(yaw), θ(pitch), ϕ(roll) which are regrouped in $\eta = (\psi, \theta, \phi)$. A sketch of the X4 bidirectional rotors mini-flyer is given in figure (2) where rotors 1 and 3 are directional.

3 Dynamic of motion

We consider the translation motion of $\Re_G$ with respect to (wrt) $\Re_o$. The position of the center of mass wrt $\Re_o$ is defined by $\overrightarrow{OG} = (x\ y\ z)^t$, its time derivative gives the velocity wrt to $\Re_o$ such that $\frac{d\overrightarrow{OG}}{dt} = (\dot{x}\ \dot{y}\ \dot{z})^t$, while the second time derivative permits to get the acceleration: $\frac{d^2\overrightarrow{OG}}{dt^2} = (\ddot{x}\ \ddot{y}\ \ddot{z})^t$. then :

$$\begin{aligned} m\ddot{x} &= u_2 S_\psi C_\theta - u_3 S_\theta \\ m\ddot{y} &= u_2(S_\theta S_\psi S_\phi + C_\psi C_\phi) + u_3 C_\theta S_\phi \\ m\ddot{z} &= u_2(S_\theta S_\psi C_\phi - C_\psi S_\phi) + u_3 C_\theta C_\phi - mg \end{aligned} \quad (1)$$

where m is the total mass of the vehicle. The vector u_i, $i = 2, 3$ combines the principal non conservative forces applied to the X4 bidirectional flyer airframe including forces generated by the motors and drag terms. Drag forces and gyroscopic due to motors effects will be not considered in this work. The lift (collective) force u_3 and the direction input u_2 are such that :

$$\begin{pmatrix} 0 \\ u_2 \\ u_3 \end{pmatrix} = f_1 e'_1 + f_3 e'_3 + f_2 e_2 + f_4 e_4 \qquad (2)$$

with $f_i = k_i w_i^2$, $k_i > 0$ is a given constant and w_i is the angular speed resulting of motor i. Let :

$$e'_1 = \begin{pmatrix} 0 \\ S_{\xi_1} \\ C_{\xi_1} \end{pmatrix}_{\Re_G} ; e'_3 = \begin{pmatrix} 0 \\ S_{\xi_3} \\ C_{\xi_3} \end{pmatrix}_{\Re_G} ; e_2 = e_4 = \begin{pmatrix} 0 \\ 0 \\ 1 \end{pmatrix}_{\Re_G} \qquad (3)$$

Then we deduce

$$\begin{aligned} u_2 &= f_1 S_{\xi_1} + f_3 S_{\xi_3} \\ u_3 &= f_1 C_{\xi_1} + f_3 C_{\xi_3} + f_2 + f_4 \end{aligned} \qquad (4)$$

where ξ_1 and ξ_3 are the two internal degree of freedom of rotors 1 and 3, respectively. These variables are controlled by dc-motors and bounded $-20 \leq \xi_1, \xi_3 \leq +20$; e_2 and e_4 are the unit vectors along E_3^g which imply that rotors 2 and 4 are identical of that of a classical quadrotor (not bidirectional).

The rotational motion of the X4 bidirectional flyer will be defined *wrt* to the local frame but expressed in the inertial frame. According to Classical Mechanics, and knowing the inertia matrix $I_G = diag(I_{xx}, I_{yy}, I_{zz})$ at the center of the mass.

$$\begin{aligned} \ddot{\theta} &= \frac{1}{I_{xx} C_\phi} (\tau_\theta + I_{xx} \dot{\phi} \dot{\theta} S_\phi) \\ \ddot{\phi} &= \frac{1}{I_{xx} C_\phi C_\theta} (\tau_\phi + I_{yy} \dot{\phi}^2 S_\phi C_\theta + I_{yy} \dot{\phi} \dot{\theta} C_\phi S_\theta) \\ \ddot{\psi} &= I_{zz} \tau_\psi \end{aligned} \qquad (5)$$

with the three inputs in torque :

$$\begin{aligned} \tau_\theta &= l(f_2 - f_4) \\ \tau_\phi &= l(f_3 C_{\xi_3} - f_1 C_{\xi_1}) \\ \tau_\psi &= l(f_1 S_{\xi_1} - f_3 S_{\xi_3}) \end{aligned} \qquad (6)$$

where l is the distance from G to the rotor i, $i \in \{1, 2, 3, 4\}$.

The model given above can be input/output linearized by the following decoupling feedback laws :

$$\begin{aligned} \tau_\theta &= -I_{xx} \dot{\phi} \dot{\theta} S_\phi + I_{xx} C_\phi \tilde{\tau}_\theta \\ \tau_\phi &= -I_{yy} \dot{\phi}^2 S_\phi C_\theta - I_{yy} \dot{\theta} \dot{\phi} S_\theta C_\phi + I_{yy} C_\theta C_\phi \tilde{\tau}_\phi \\ \tau_\psi &= I_{zz} \tilde{\tau}_\psi \end{aligned} \qquad (7)$$

Then the dynamic of the system is defined by :

$$\begin{cases} m\ddot{x} = u_2 S_\psi C_\theta - u_3 S_\theta \\ m\ddot{y} = u_2(S_\theta S_\psi S_\phi + C_\psi C_\phi) + u_3 C_\theta S_\phi \\ m\ddot{z} = u_2(S_\theta S_\psi C_\phi - C_\psi S_\phi) + u_3 C_\theta C_\phi - mg \\ \ddot{\theta} = \tilde{\tau}_\theta \\ \ddot{\phi} = \tilde{\tau}_\phi \\ \ddot{\psi} = \tilde{\tau}_\psi \end{cases} \quad (8)$$

We can regroup the dynamic of y and z as follow

$$\begin{pmatrix} m\ddot{y} \\ m\ddot{z} \end{pmatrix} = M \begin{pmatrix} u_2 \\ u_3 \end{pmatrix} + \begin{pmatrix} 0 \\ mg \end{pmatrix} \quad (9)$$

So for $\theta \neq (2k+1)\frac{\pi}{2}$ and $\psi \neq (2k+1)\frac{\pi}{2}$ $\forall k \in \mathbb{Z}$, we can consider the following transformation :

$$\begin{pmatrix} u_2 \\ u_3 \end{pmatrix} = M^{-1} \left(\begin{pmatrix} v_1 \\ v_2 \end{pmatrix} - \begin{pmatrix} 0 \\ -mg \end{pmatrix} \right) \quad (10)$$

Then (8) can be written as follow :

$$\begin{cases} m\ddot{x} = v_1 f_1 - (v_2 + mg) f_2 \\ m\ddot{y} = v_1 \\ m\ddot{z} = v_2 \\ \ddot{\theta} = \tilde{\tau}_\theta \\ \ddot{\phi} = \tilde{\tau}_\phi \\ \ddot{\psi} = \tilde{\tau}_\psi \end{cases} \quad (11)$$

We assume that $(\theta, \phi, \psi) \simeq (0, 0, 0)$, then (11) can be written as :

$$\begin{cases} m\ddot{x} = v_1(\psi - \theta\phi + \psi\theta^2) - (v_2 + mg)(\psi\phi + \theta^2\psi + \theta) \\ m\ddot{y} = v_1 \\ m\ddot{z} = v_2 \\ \ddot{\theta} = \tilde{\tau}_\theta \\ \ddot{\phi} = \tilde{\tau}_\phi \\ \ddot{\psi} = \tilde{\tau}_\psi \end{cases} \quad (12)$$

4 Stabilization and tracking control

Feedback stabilization control of systems, that exhibit a number of degree of freedom greater than the number of inputs, is a challenge problem in control theory.
In this paper, time-varying control and average approach will be combined to stabilize the X4 bidirectional rotors with predefined trajectories.

4.1 Control strategy

In this section, we assume that (θ, ϕ, ψ) have a small values nearly to zero; so (12) is equivalent to the following system presented as :

$$\begin{cases} m\ddot{x} = v_1\psi - (v_2 + mg)\theta \\ m\ddot{y} = v_1 \\ m\ddot{z} = v_2 \\ \ddot{\theta} = \tilde{\tau}_\theta \\ \ddot{\phi} = \tilde{\tau}_\phi \\ \ddot{\psi} = \tilde{\tau}_\psi \end{cases} \quad (13)$$

To stabilize this system, we divide it in two subsystems as follow. The first subsystem is a total actuated system

$$\begin{cases} m\ddot{z} = v_2 \\ \ddot{\theta} = \tilde{\tau}_\theta \\ \ddot{\phi} = \tilde{\tau}_\phi \end{cases} \quad (14)$$

and the second subsystem is

$$\begin{cases} m\ddot{x} = v_1\psi - (v_2 + mg)\theta \\ m\ddot{y} = v_1 \\ \ddot{\psi} = \tilde{\tau}_\psi \end{cases} \quad (15)$$

4.1.1 Stabilization of the first subsystem

We notice that (14) is a total actuated system then it is easy to specify the expression of the control which stabilizes the system .
Theorem With the feedback laws

$$\begin{cases} v_2 = m\ddot{z}_d - mk_1^z(\dot{z} - \dot{z}_d) - k_2^z m(z - z_d) \\ \tilde{\tau}_\theta = \ddot{\theta}_d - k_1^\theta(\dot{\theta} - \dot{\theta}_d) - k_2^\theta(\theta - \theta_d) \\ \tilde{\tau}_\phi = \ddot{\phi}_d - k_1^\phi(\dot{\phi} - \dot{\phi}_d) - k_2^\phi(\phi - \phi_d) \end{cases} \quad (16)$$

the dynamic of z, θ and ϕ are exponentially-asymptotically stable with the appropriate choice of the gain controller parameters k_j^i with $i \in \{z, \theta, \phi\}$ and $j \in \{1, 2\}$.
remark : Then, with the following expression of the feedback laws v_2 and $\tilde{\tau}_\theta$, the variable input θ converges exponentially to zero juste as v_2. Thus, the system (15) is equivalent to :

$$\begin{cases} m\ddot{x} = v_1\psi \\ m\ddot{y} = v_1 \\ \ddot{\psi} = \tilde{\tau}_\psi \end{cases} \quad (17)$$

4.1.2 Stabilization of the second subsystem

In this section, we present a time-varying control law which stabilize asymptotically x and y to zero and ψ to constant.

Theorem The feedback laws

$$\begin{cases} v_1 = \nu_y + 2\sin(\frac{t}{\epsilon}) \\ \tilde{\tau}_\psi = \left[\nu_x - 2\dot{\nu}_y\dot{\psi} - \psi\ddot{\nu}_y\right]\sin(\frac{t}{\epsilon}) \end{cases} \quad (18)$$

with :

$$\begin{cases} \nu_x = -k_1^x m(\ddot{x} - \ddot{x}_d) - k_2^x m(\dot{x} - \dot{x}_d) - k_3^x m(x - x_d) + mx_d^{(3)} \\ \nu_y = -mk_1^y(\dot{y} - \dot{y}_d) - mk_2^y(y - y_d) + m\ddot{y}_d \end{cases} \quad (19)$$

stabilize asymptotically the dynamic of x, y and ψ with the appropriate choice of the gain controller parameters.

Proof:
We derive the dynamic of x, we obtain :

$$\begin{cases} mx^{(4)} = \tilde{\tau}_\psi v_1 + 2\dot{\psi}\dot{v}_1 + \psi\ddot{v}_1 \\ m\ddot{y} = v_1 \end{cases} \quad (20)$$

With the expression of v_1 and $\tilde{\tau}_\psi$ given in the theorem, the average system of (20) is :

$$\begin{cases} mx^{(4)} = \nu_x \\ m\ddot{y} = \nu_y \end{cases} \quad (21)$$

Then, with the appropriate choice of the gain controller parameters, (21) converge exponentially to zero, so x and y converge asymptotically to zero. Thus ψ converge to constant.

5 Numerical results

The X4 bidirectional rotors mini-flying machine presented by the LSC-group is tested in simulation in order to validate some motion planning algorithm considering the proposed control laws. We have considered a total mass equal to $m = 2Kg$ which is approximately near to the reality. We have considered also the following values of the coefficients of a Hurwitz polynomial (the gain controller parameters):
$k_1^z = k_1^\theta = k_1^\phi = 20$, $k_2^z = k_2^\theta = k_2^\phi = 100$
$k_1^x = 12$, $k_2^x = 52$, $k_3^x = 96$, $k_4^x = 64$, $k_1^y = 2$, $k_2^y = 1$ et $\epsilon = 10^{-4}$

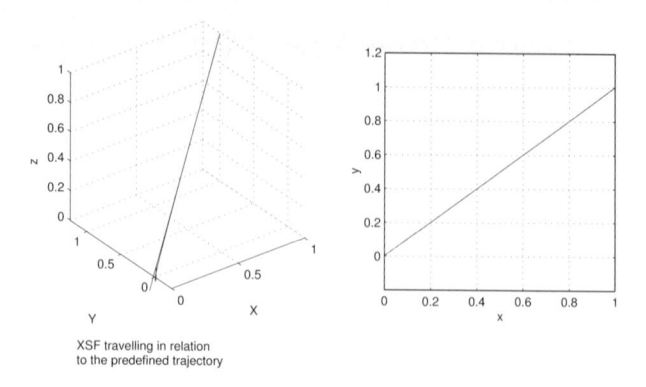

Fig. 2. XSF travelling

References

1. E. Altug, J. Ostrowski and R. Mahony, "Control of a quadrotor helicopter using visual feedback," *Proceedings of the IEEE Conference on Robotics and Automatiron*, Washington DC, Virginia, USA, 2002.
2. P. Castillo, A. Dzul, and R. Lozano, "Real-time stabilization* and tracking of a four rotor mini-rotorcraft," *IEEE Transactions on Control Systems Technology*, Accepted, 2004.
3. M. Fliess, J. Levine, P. Martin and and P. Rouchon, "Flatness and defect of nonlinear systems: Introductory theory and examples," *International Journal of Control*, 61(6), pp. 1327-1361, 1995.
4. T. Hamel, R. Mahony, R. Lozano and J. Ostrowski, "Dynamic modelling and configuration stabilization for an X4-flyer," in *IFAC, 15th Triennial World Congress*, Barcelona, Spain, 2002.
5. P. Martin, S. Devasia and B. Paden, "A different look at output tracking: Control of a VTOL aircraft," *Automatica* 32(1), pp.101-107, 1996.
6. P. Martin, R. M. Murray and P. Rouchon, "Flat systems, equivalence and trajectory generation," Ecole des Mines de Paris, *Technical report*, April 2003.
7. J. Hauser, S. Sastry and G. Meyer, "Nonlinear control design for slightly non-minimum phase systems: Application to v/stol aircraft," *Automatica* 28(4), pp. 665-679, 1992.
8. J. Leitner, A.J.Calise and J.V.R.Prasad, "Analysis of adaptive neural networks for helicopter flight controls," in *AIAA J. of Guidance, Control, and Dynamics*, (20)5, pp.972-979, sept.1997.
9. P. Pound, R. Mahony, P. Hynes and J. Roberts, "Design of a four rotor aerial robot," *Proceedings of the Australasian Conference on Robotics and Automation*, Auckland, November 27-29, 2002.

Stability Measure Comparison for the Design of a Dynamic Running Robot

Jonathan E. Clark[1] and Mark R. Cutkosky[2]

[1] University of Pennsylvania, Philadelphia, PA 19104
   jonclark@grasp.upenn.edu
[2] Stanford University, Stanford, CA 94305

Abstract

Estimating the effect of design changes on the stability of robots designed to run over rough terrain is a difficult task. No current predictor or measure is currently universally accepted. The most common techniques, including the 'stability margins' commonly used in walking machines and the return-map eigenvalue analysis used on simple models, are not fully applicable to these fast and complex systems. This paper describes a comparison of three approaches to measuring stability applied to a redesign of our hexapedal running robot, *Sprawl*.

Keywords: dyanamic, running, hexapedal, stability

1 Introduction

There have been many measures for stability created for walking and running robots over the years. None of them, however, is wholly adequate for predicting successful running for a complex system over rough terrain. Standard static and dynamic stability margins [1, 2] are commonly violated by running systems due to their high speeds and airborne phases. The other general approach to analyzing system stability, return map eigenvalue analysis, has thus far only been successfully applied to relatively simple models of robotic systems [3, 4].

The development of polypedal dynamic robots such as *Rhex*[5] and *Sprawl*[6] (see **Fig.1**(a)) which can run rapidly–over five body lengths per second–and over rough terrain, motivates an examination of the currently available tools and the development of new ones.

Fig. 1. (A) A *Sprawl* robot running over a small obstacle, (B) the dynamic simulation used in the study.

This paper discusses three versions of previously proposed measures of stability that can be applied to dynamic runners. Each of these measures is implemented on a dynamic simulation of a robot (described in [7]) which allows for efficient and repeatable tests and can be implemented in the design phase of development. Additionally, the measures can be used to test the effect of varying both physical and control parameters–an essential requirement for analyzing these robots. The measures examined in this paper are:

- The range of slopes that can be traversed in steady-state running
- An adapted version of a stability margin
- Speed of recovery from disturbances during running.

The measures are examined in the context of a parameter variation exercise on a *Sprawl* robot. Although each of these measures is applied to running on flat surfaces, in aggregate they may consider many of the effects that result from running on rough terrain.

2 Overview of Stability Comparison

While the stability measures described here can be used to inform any number of design studies or optimization efforts, the effect of changing a *Sprawl* robot's nominal leg configuration is chosen as an example since it has been shown to have a primary effect on both the robot's speed and ability to self-stabilize.

As was done in [7], *Sprawl's* nominal leg posture can effectively be paramatrized by the two-dimensional space defined by the intersection points of the line of action of each leg's piston. For each stability measure studied 121 different leg settings which span the reasonable configuration space were selected. As shown in **Fig.2**(a), a circle whose radius is proportional to the steady-state velocity is drawn for each of these configuration points. In particular, the leg configuration that results in the fastest forward velocity is designated by the intersection of the three gray dotted lines. The circles with x's inside represent configurations that resulted in the robot crashing or 'nose-diving'.

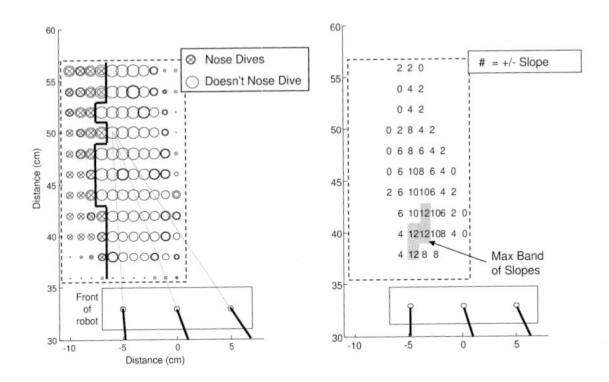

Fig. 2. (A) The effect of leg configuration on velocity and stability. The diameter of the circle at each leg configuration point is proportional to the forward velocity. The dark line denotes the boundary of configurations that crash or 'nose-dive'. (B) The effect of leg configuration on the band of traversable slopes.

2.1 Slope Invariance

A potential measure of how well a robot can run on rough terrain is its ability to traverse a range of slopes, an approach taken by Pratt et al. [8] and Taga [9].

For these tests the simulation was run on slopes up to ± 20 *degrees* with 2 *degree* gradations. The steady state velocity and presence of 'nose dives' were examined to determine which trials resulted in good running. **Fig.2**(b) shows the band of slopes that was achievable for each configuration. In each case, this is a plus/minus band. In other words, a six indicates that the simulation was able to run well for each inclination between and including ± 6 *degree* slope.

Fig.2(b) shows that, as one might expect, operation in the center of the stable parameter space increases the slope invariance. The most interesting finding, however, is that increasing the sprawl angle (moving down the line of action of the middle leg) significantly increases the band of slopes that can be handled. Although the maximum speed occurs at a narrower sprawl angle, increasing the sprawl angle significantly improves the robot's invariance to ground inclination.

A weakness of this measure is that it does not capture the effects of high-frequency terrain changes and impulses on the system.

2.2 Stability Margins

Most classical stability margins do not apply to a running robot with aerial phases. One exception is the recently proposed "Wide Stability Margin" (WSM) as used on *Tekken* [10]. This measure uses the ground projection of each foot and the center of mass, as shown in **Fig.3**(a). Although this measure was developed to analyze walking, it can also be applied to running robots with flight phases.

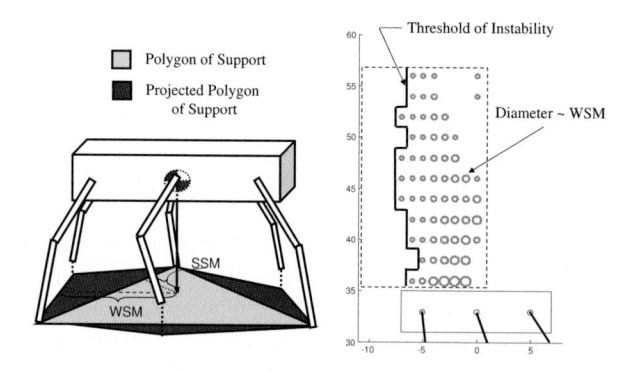

Fig. 3. (A) Diagram showing the standard and projected polygons of support, and how the sagittal projection of the Wide Stability Margin (WSM) compares to the static stability margin (SSM). (B) Effect of leg configuration on the sagittal WSM. The dark line represent the border of instability as defined the WSM. The size of the circles represents the magnitude of the WSM for each leg configuration

Each circle in **Fig.3**(b) shows the minimum WSM value over a stride for a range of leg configurations. The dark line denotes the boundary of stable values and has a good correlation with stability as defined by nose diving (see **Fig.2**(a)).

Fig.3(b) also shows how decreasing the magnitude of the mean leg angle (moving to the right of the plot) increases the size of the stability margin. This is expected as it increases the distance between the nominal location of the feet and the center of mass. Unfortunately, as far as the effect of changing posture goes, these results simply tell us that greater stability in terms of WSM accompanies going slower.

For configurations (left of the dark line) that nose dive, the WSM decreases over a number of strides until contact is made. The gradual nature of this instability suggests that if the maximum rotation of the front leg angles were sensed, the onset of instability could be detected and the nominal orientation could be altered to right the robot.

2.3 Perturbation Response

The path taken in this study to analyze the perturbation response of the simulation is two fold. First the response of a single configuration to a variety of disturbances is analyzed, and the resulting coupling between each direction of motion is observed. Secondly, the effect of changing the stance parameters on the rate of recovery for two representative disturbances is observed.

2.3.1 Perturbation Response Behavior

For the first part of the perturbation study, a nominal configuration of the simulation is chosen in the middle of the stable region of leg angles which strikes a balance between forward velocity and slope invariance.

After achieving steady-state locomotion, the simulation is disturbed with impulses equal to approximately 15% of the forward momentum of the robot. Three positive and three negative impulses are applied sequentially at the center of mass of the body in orthogonal directions aligned with the orientation of the body. In addition, disturbance torques are applied about the same axes. All twelve of the impulses are applied during an airborne phase.

The details of this study can be found in [11], the conclusions of which include the following:

- roll, pitch, and yaw motions are strongly coupled,
- disturbances in the fore-aft direction do not excite rotational motions,
- any disturbance to an angular motion will slow the forward velocity,
- pitch disturbances settle at the slowest rate.

Consequently, forward velocity and pitch were chosen as the representative responses for evaluating the effect of changing configuration.

2.3.2 Configuration Dependant Response Rate

Fig.4(a-b) compares the pitch response to the speed response for a range of leg orientations. For each configuration the settling time is plotted, with the circle size being proportional to the number of strides required before the state has settled to its nominal value. The gray areas highlight the regions that converge the most quickly.

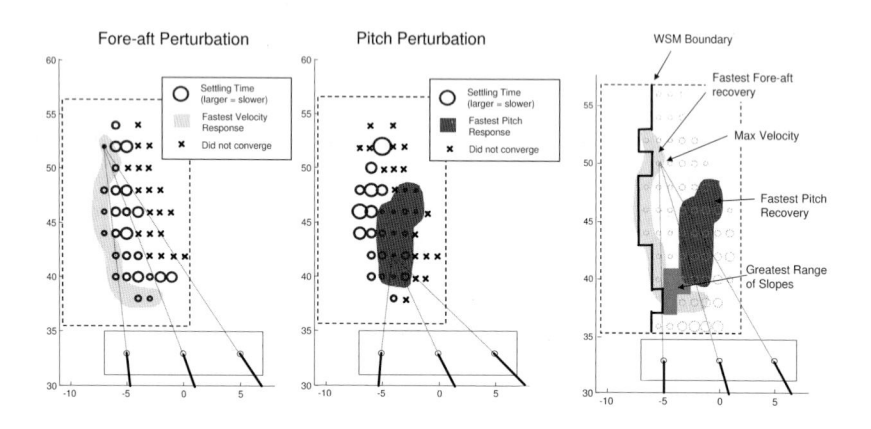

Fig. 4. Effect of leg configuration on (A) fore-aft and (B) pitch perturbation settling times. (C) Comparison of the effect of leg posture for four different stability measures considered in this chapter. The circles represent the magnitude of the WSM for each of the stable configurations.

As **Fig.4**(a) shows, the region of fastest convergence for the fore-aft perturbation response lies on the border of unstable values. The region of the configuration space that results in the fastest running overlaps with this region. This may be due to stronger locking into the limit cycle, but these configurations also pose a greater danger of crashing due to calibration or alignment errors.

The pitch response to a pitching disturbance impulse is shown in **Fig.4**(b). The region of fastest convergence occurs near the middle of the stable range. Values to the rear (right) and top of the stable region converge poorly for both types of disturbances. These values are not unstable in the sense of nose diving, but also do not show a sharp period-1 gait locking.

The slow settling time and strong dependence of the pitching response on other types of disturbances suggests that **Fig.4**(b) may be an indicator for predicting the effect of leg configuration on how rough a terrain the robot can successfully traverse.

2.4 Stability Measure Comparison

Each of the measures discussed takes a different approach to quantifying stability and, as the parameter study on leg posture shows, are dramatically different in their interpretation of the effect of such changes.

Fig.4(c) graphically overlays the results from these three measures. The size of the circles at each configuration is proportional to the WSM for that setting. The leg configurations shown by the gray dotted lines result in the maximum speed and would be a good setting for fast operation on smooth terrain. For operation on unknown or rough terrain, however, a more sprawled leg configuration coinciding with the intersection of the fastest pitch recovery and slope invariance regions might be better.

3 Conclusions and Future Work

This comparison of existing stability measures has shown that each approach only captures an aspect of stability for a running robot. While each of these measures can be independently improved (the WSM by using the center of pressure rather than center of mass, the slope tests by considering transitions, and the perturbation tests by measuring the basin of attraction rather than settling time) it is not clear that even with these improvements any one of them will be adequate to achieve the goal of predicting the effect of design changes on the stability of a running robot over rough terrain.

What is needed is experimental evidence linking these measures to how robots perform on real terrain. The characterization of rough terrain, and what stability means on such terrain, however, is a difficult and unsolved problem in its own right.

A glimmer of hope can be found in the correspondence in results between slope and pitch perturbations. If a particular area in the design space (such as increased leg sprawl) improves both the steady-state inclination measure and the transient perturbation response, then perhaps it will also improve the robot's performance in natural terrains.

Acknowledgements

We would like to thank Darryl Thelen and Trey McClung for their help in developing the simulation used in this paper. The development of the *Sprawl* robots was funded by the Office of Naval Research under grant N00014-98-1-066. The first author is currently supported by the IC Postdoctoral Fellow Program.

References

1. McGhee, R. B. and Frank, A. A. (1968) On the stability properties of quadruped creeping gaits. *Mathematical Bioscience*, vol. 3, pp. 331–351.
2. Orin, D. E. (1976) *Interactive Control of a Six-Legged Vehicle with optimization of both stability and energy*. Ph.D. thesis, The Ohio State University.
3. Seipel, J, Holmes, P., and Full, R. (2004) Dynamics and stability of insect locomotion: a hexapedal model for horizontal plane motions. *Biol. Cybern.*, vol. 91, no. 2, pp. 76–90.
4. Mombaur, K. D. (2001) *Stability Optimization of Open-loop Controlled Walking Robots*. Ph.D. thesis, Universitaet Heidelberg.
5. Altendorfer, R., Moore, N., Komsuoglu, H., Buehler, M., Brown H. B, Jr., McMordie, D., Saranli, U., Full, R., and Koditschek, D. E. (2001) Rhex: A biologically inspired hexapod runner. *Autonomous Robots*, vol. 11, no. 3, pp. 207–213.
6. Cham, J. G., Bailey, S. A., Clark, J. E., Full, R. J., and Cutkosky, M. R. (2002) Fast and robust: Hexapedal robots via shape deposition manufacturing. *International Journal of Robotics Research*, vol. 21, no. 10.
7. Clark, J. E., Thelen, D.G., and Cutkosky, M. R. (2004) Dynamic simulation and analysis of a passively self-stabilizing hexapedal running robot. In *ASME Proceedings, IMECE '04*, Anaheim, CA.
8. Pratt, Jerry, Chew, Chee-Meng, Torres, Ann, Dilworth, Peter, and Pratt, Gill (2001) Virtual model control: an intuitive approach for bipedal locomotion. *International Journal of Robotics Research*, vol. 20, no. 2, pp. 129–143.
9. Taga, G. (1995) A model of the neuro-musculo-skeletal system for human locomotion ii. real-time adaptability under various constraints. *Biological Cybernetics*, vol. 73, no. 2, pp. 113–121.
10. Fukuoka, Y., Kimuar, H., and Cohen, A. H. (2003) Adaptive dynamics walking of a quadruped robot on irregular terrain based on biological concepts. *International Journal of Robotics Research*, vol. 22, no. 3-4, pp. 187–202.
11. Clark, J. E. (2004) *Design, Simulation, and Stability of a Hexapedal Running Robot*. Ph.D. thesis, Stanford University.

Control Architecture and Walking Strategy for a Pneumatic Biped Robot

G. Spampinato, G. Muscato

Università degli studi di Catania, Dipartimento di Ingegneria Elettrica Elettronica e dei Sistemi (D.I.E.E.S.) Viale A.Doria 6 - 95125 Catania, Italy, e-mail: gspampi@diees.unict.it , gmuscato@diees.unict.it

Abstract

The paper aims to describe the design methodologies and the control architecture of a ten degrees of freedom biped robot. The prototype realized is pneumatically actuated, and the legs size and dimensions are anthropometric with the human beings. The control strategy has been specifically designed in order to ensure a proper stability during the motion. In particular, a three level feedback loop controller generates the walking trajectories in a different way for the stance and the swing leg. In particular the stance leg is actuated through a simple but efficient force control approach based on a different interpretation of the Virtual Model Control strategy, while the swing leg is controlled through a set of parabolic trajectories generated during the gait. In other words the parabolic trajectories are generated directly in the operative space in order to take into account the basic walking parameters, like the step length or the lateral balancing. A set of preliminary experimental results are also reported.

Keywords: Biped Locomotion, Posture Stability, Pneumatic Control, Trajectory Generation, Control Architecture.

1 Introduction

The study of the walking cycle in biped robot allows to understand more accurately the strategy actuated by human beings in keeping balance in order to maintain a proper stability level during the gait. Moreover the pneumatic actuators have been adopted in this work as an alternative solution to the electric motors. Due to its low weight-force developed ratio, and to the high elasticity properties, the pneumatic actuators allow the controller to adapt the stiffness property of the leg during the stance phase. In particular, new control strategies and architectures have been specifically designed in order to improve the articulations compliance during the motions. So the present work provides a biomechanical platform in order to perform research activities upon advanced humanoid walking concepts like posture control with energy preserving during the gait, skipping or running [1]. The aim of this work is also the development of new technologies for the realization of bio-mechanical limbs for motoric gait rehabilitation [2], where the low weight-force ratio provided by the pneumatic actuators plays an important role.

2 Mechanical Overview

The design of the shape and the dimensions of the single parts that compose the robotic prototype are human inspired, so the dimensions of the links and the articulation movements, are similar to the corresponding biological ones [3]. The mechanical structure of each leg is made up of four links, corresponding to the pelvis, thighbone, shinbone, and foot, jointed by five degrees of freedom. In particular the knee joint has one rotational degree of freedom, while the ankle joint and the hip joints, have two rotational degrees of freedom. In Fig.1 the prototype overall structure is shown. As discussed in the previous chapter, each articulation is actuated by pneumatic pistons. In particular two pistons actuate the ankle and the hip articulations, while the knee articulation is actuated by only one piston. In this way the hip and the ankle articulations have two degrees of freedom each, while the knee articulation is moved by one degree of freedom. The ankle and the hip articulations are made up by a cardanic joint that provides both roll and pitch degrees of freedom. In particular the two degrees of freedom are strictly coupled with the two pneumatic pistons that actuate the articulation. In other words, a single movement of a single degree of freedom involves the motions of both pneumatic pistons and vice versa. That point will be better underlined through the kinematical point of view in the next sections.

Moreover the pneumatic pistons are actuated in both directions, so the articulation structure does not require the typical antagonistic scheme typical of other bio-inspired robots using, for example, the McKibben artificial muscles.

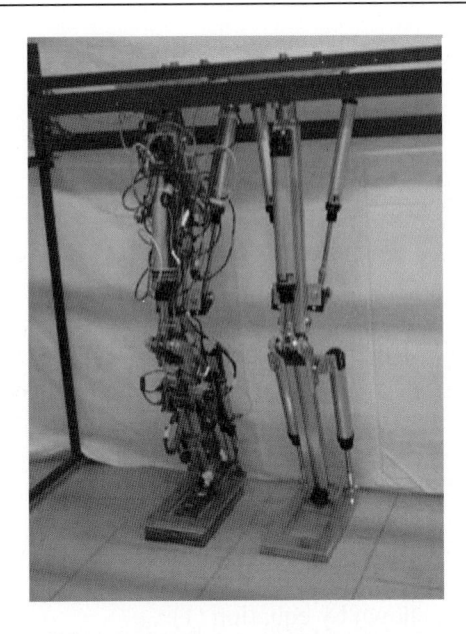

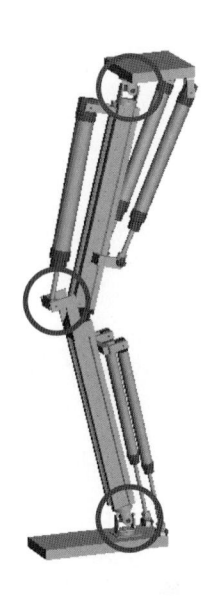

Fig. 1. The ten degrees of freedom mechanical structure of the biped robot. Five degrees of freedom move the three articulation of each leg

3 Kinematics and Kinetic Analysis

As will be described more in depth in the next sections, the control architecture has been designed to generate the joint trajectories according to the information coming by the sensors placed on board the robot. In particular, a linear potentiometer is placed behind each actuator in order to measure the piston length during the motion. So the control algorithm can deduce the articulations position during the gait through an appropriate kinematic inversion process. In order to find the best algorithm, some different approaches has been proposed and tested [4],[5]. The inversion process can be achieved once known the Jacobian matrix specifically obtained for each articulation geometrical structure. Looking at the structure of the hip and ankle joints, the joint prototype shown in Fig.2 can be taken into account as a more general geometric structure, from whose the hip and the ankle joints can be derived simply defining the a, b, c and d segments according to the articulation geometry.

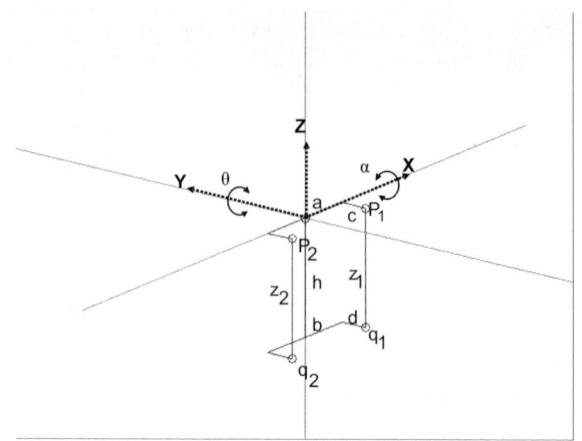

Fig. 2. General geometric structure for the hip and ankle joints

The 2x2 Jacobian matrix that has to be found in order to put into relation the time variations of the piston lengths with the time variations of the pitch-roll degrees of freedom, as shown by equation (1).

$$\begin{bmatrix} \dot{z}_1 \\ \dot{z}_2 \end{bmatrix} = J(\theta,\alpha) \cdot \begin{bmatrix} \dot{\theta} \\ \dot{\alpha} \end{bmatrix} = \begin{bmatrix} -\dfrac{P_1 - q_1}{z_1} A_1 & -\dfrac{P_1 - q_1}{z_1} B_1 \\ -\dfrac{P_2 - q_2}{z_2} A_2 & -\dfrac{P_2 - q_2}{z_2} B_2 \end{bmatrix} \cdot \begin{bmatrix} \dot{\theta} \\ \dot{\alpha} \end{bmatrix} \quad (1)$$

$$A_i = [S(-\hat{\omega}_y) \cdot R_\theta \cdot R_\alpha] \times \tilde{q}_i$$
$$B_i = [R_\theta \cdot S(\hat{\omega}_x) \cdot R_\alpha] \times \tilde{q} \quad (2)$$

In witch $\tilde{q}_1 = [b \ -d \ -h \ 1]^T$ and $\tilde{q}_2 = [-b \ -d \ -h \ 1]^T$ represents the actuators contact point with the articulation according to the main reference frame [5]. Once found the right Jacobian matrix, this can be used into a feedback integration method for inverting the kinematical problem like, for example, the inversion of the Jacobian method [6]. Also for the knee articulation the Jacobian matrix has been found according to the geometrical structure shown in Fig.3.

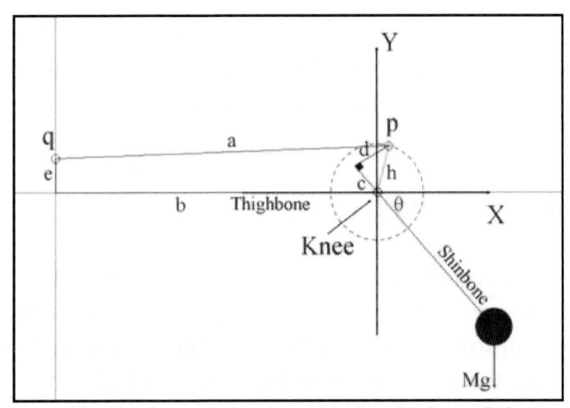

Fig. 3. Geometric structure for the knee joints

In this case the jacobian matrix found is a real number that puts into relation the time variation of the piston length with the time variation of the knee degree of freedom. It can be calculated as shown in the equation (3).

$$\dot{a} = J(\theta) \cdot \dot{\theta} = -\frac{p-q}{a} A \cdot \dot{\theta} \qquad (3)$$

Where $A = [S(-\hat{\omega}_z) \cdot R_\theta] \times \tilde{q}$ and $\tilde{q} = [-c \ d \ 0 \ 1]^T$ that represents the knee contact point with the actuator (indicated by p) in the shinbone reference frame.

Moreover, from the kinetic point of view, the articulations analysis can be managed in order to find some relations between the forces supplied by the actuators and the torques generated by the articulations. To do that, due to the virtual works principle, the jacobian matrices found for each articulation have to be transposed, like shown by the equations (4).

$$\begin{bmatrix} \tau_{A\_roll} \\ \tau_{A\_pitch} \end{bmatrix} = [J_A]^T \cdot \begin{bmatrix} F_{A1} \\ F_{A2} \end{bmatrix}, \tau_K = J_K^T \cdot F_K, \begin{bmatrix} \tau_{H\_roll} \\ \tau_{H\_pitch} \end{bmatrix} = [J_H]^T \cdot \begin{bmatrix} F_{H1} \\ F_{H2} \end{bmatrix} \qquad (4)$$

The matrices J_A J_H and J_K represents the jacobian matrices for the ankle, hip and knee articulation respectively.

$$\begin{bmatrix} F_X \\ F_Y \\ F_Z \\ M_X \\ M_Y \end{bmatrix} = J_{5Dof}^{-T} \cdot \begin{bmatrix} \tau_{H\_roll} \\ \tau_{H\_pitch} \\ \tau_K \\ \tau_{A\_roll} \\ \tau_{A\_pitch} \end{bmatrix}, J_{Art} = \begin{bmatrix} J_H^{-1} & 0 & 0 \\ 0 & J_K^{-1} & 0 \\ 0 & 0 & J_A^{-1} \end{bmatrix}, \begin{cases} F_Z = J_{Art}^T \cdot \tau \\ \tau = J_{5Dof}^T \cdot F_{Ext} \end{cases} \qquad (5)$$

Moreover, through the 5x5 matrix J_{Art} defined in the equation (5), and the 5x5 jacobian matrix J_{5Dof} calculated for each leg, it is possible to find the force needed by the actuators (indicated with F_Z) in order to generate

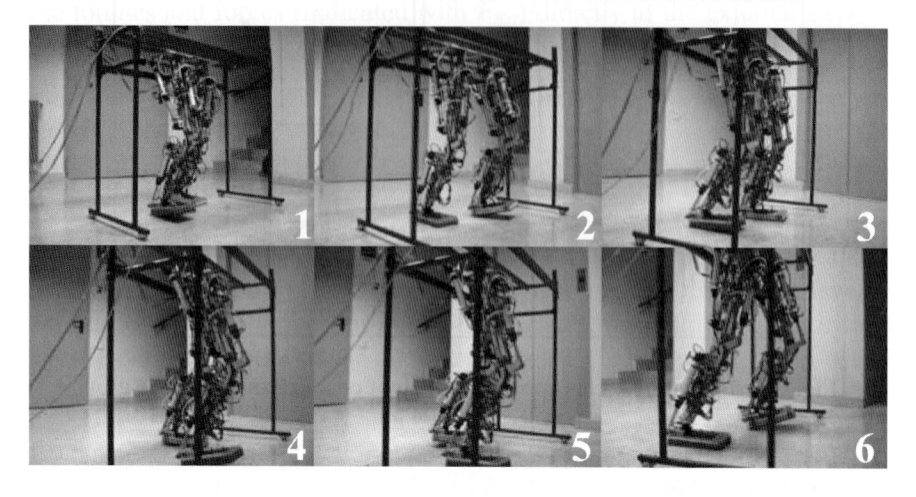

Fig. 6 A 60 cm length step cycle.

References

1. Raibert. M., *"Legged Robots that balance"*, Cambridge, MIT Press 1986.
2. H. Kazerooni, Jean-Louis Racine, Lihua Huang, and Ryan Steger. *"On the Control of the Berkeley Lower Extremity Exoskeleton (BLEEX)"*. Proceedings of the 2005 IEEE International Conference on Robotics and Automation Barcelona, Spain, April 2005.
3. G.Muscato, G. Spampinato, *"A Pneumatic Human inspired robotic Leg: Control Architecture and Kinematical Overview"*, Accepted for publication in the "International Journal of Humanoid Robotics" May 2005.
4. G.Muscato, G. Spampinato, *"A Human inspired Five Dof robotic Leg: Kinematical Model and Control Architecture Overview"*, Submitted to "Mechatronics" June 2005.
5. G. Muscato, S. Spampinato" *Kinematical behavior analysis and walking pattern generation of a five degrees of freedom pneumatic robotic leg* ", 7th International Conference on Climbing and Walking Robots CLAWAR 2004, Madrid (Spain),22-24 September 2004.
6. *"Modeling and Control of Robot Manipulators"*: Sciavicco – Siciliano, McGraw-Hill, 1996.
7. G. Muscato, S. Spampinato *"A Multilevel Control Architecture for a pneumatic Robotic Leg"* 10th IEEE International Conference on Emerging Technologies and Factory automation ETFA 2005, Catania (Italy), 19-22 September 2005.
8. J.Pratt, C-M Chew, A.Torres, P.Dilworth, G.Pratt. *"Virtual Model Control: an intuitive Approch for Bipedal Locomotion"*. The International Journal of Robotics Research. Vol. 20, February, 2001.

Time-scaling Control of a Compass Type Biped Robot

P Fauteux, P Micheau and P Bourassa

Université de Sherbrooke, Mechanical Engineering Department, Sherbrooke, QC, J1K 2R1, Canada. philippe.micheau@usherbrooke.ca

Abstract

This paper presents a robust control strategy driving an actuated compass gait robot towards steady gaits. The originality lies in the generation of the swing leg references as a simple function of the supporting leg angle. Simulations and experimentation showed that the system exhibits asymptotically stable walking cycles with large and strong basins of attraction.

Keywords: Biped, control, limit cycle, basin of attraction, virtual time

1 Introduction

Investigations involving biped robots are usually done with one the following goals in mind: creating a complementary tool for the understanding of human walking mechanisms or paving the way for future robotic developments. Either way, the complexity of the bipedal gait remains an obstacle and explains why searchers spend years understanding its subtleties.

The compass gait, as described by Goswani et al. [1] and Garcia, Ruina and al [2] is widely accepted as the simplest model of bipedal locomotion and recognized, by the biomechanists, as the most basic sub-action that explain the overall walking mechanism. This simplicity allows insight into the human gait and the drawing of strong foundations for future robotic developments. Also, this gait being dynamic, energy efficiency is possible.

The goal of this study was to create the simplest dynamic biped robot walking on level ground with a stable limit cycle and a large and strong attractor. This paper presents simulation and experimentation results of a simple controller ensuring a large basin of attraction in order to reach a dynamically stable walking mode without any complex planning of the trajectory.

2 Dynamic modelling of the compass gait

The chosen model (**Fig. 1**) walks with a compass gait similar to the one described by many authors [1-2]. In order to recover from energy losses due to friction and impacts or modify the walking cycle frequency, torque is applied at the hip and an impulsive force is generated under the post-impact swing leg. The robot possesses retractable feet of negligible mass.

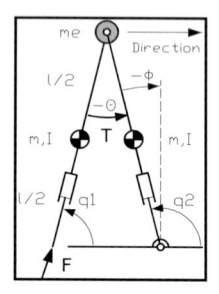

Fig. 1. Theoretical model.

The dynamics of this model is described by a set of nonlinear ordinary differential equations and instantaneous algebraic switching relations. These relations were used by the authors to study possible control strategies [3-5]. Motion equations were developed using Lagrange formalism:

$$A(q)\ddot{q} + C(q,\dot{q}) + G(q) = U \quad (1)$$

where $q = \begin{bmatrix} q_1 \\ q_2 \end{bmatrix}$ is the coordinates, $A = \begin{bmatrix} I/ml^2 + 1/4 & -\cos(q_1 - q_2)/2 \\ -\cos(q_1 - q_2)/2 & I/(ml^2) + 5/4 + me/m \end{bmatrix}$ the inertia matrix; $C = \begin{bmatrix} -\sin(q_1 - q_2)\dot{q}_2^2/2 \\ \sin(q_1 - q_2)\dot{q}_1^2/2 \end{bmatrix}$ the vector of nonlinear inertial effects (Coriolis, centrifugal), $G = \begin{bmatrix} -g\cos(q_1)/(2l) \\ (me/m + 3/2)g\cos(q_2)/l \end{bmatrix}$ the vector of gravitational forces and $U = \begin{bmatrix} T/(ml^2) \\ -T/(ml^2) + F\sin(q1 - q2)/(ml) \end{bmatrix}$ the vector of external forces. The command is the torque T acting from the stance to the swing leg. F is an impulsive force applied under the post-impact swing leg.

Ground collisions are considered inelastic and without slip. The velocity of each link after a collision is estimated by the law of conservation of momentum stating that the angular momentum of the robot about the impacting foot and the angular momentum of the post-impact swing leg about the hip are conserved. This relation is written in a contracted form in equation 2 where *tf-* and *tf+* are the instants prior and following the impact.

$$\dot{q}(t_f^+) = P\left[\dot{q}(t_f^-) - A^{-1}J^T(JA^{-1}J^T)^{-1}s(t_f^-)\right] \qquad (2)$$

where $P = \begin{bmatrix} 0 & 1 \\ 1 & 0 \end{bmatrix}$ is the permutation matrix, $s = l\begin{bmatrix} -\sin(q_2)\dot{q}_2 + \sin(q_1)\dot{q}_1 \\ \cos(q_2)\dot{q}_2 + \cos(q_1)\dot{q}_1 \end{bmatrix}$ the swing foot speed and J the Jacobian matrix defined as $J_{ij} = \frac{\partial s_i}{\partial q_j}$.

3 The controller

3.1 Simplifying hypotheses

To define the controller of the compass gait, some simplifying hypotheses are made. Equation 1 can be rewritten in the form of equation 3.

The term $A^{-1}(q)U$ quantifies the effect of the input on the acceleration vector. The torque as less effect on the stance leg than on the swing leg (at least 31 times for our robot). This ratio is minimal when θ is null. We will therefore assume that the evolution of the stance leg is uncontrollable and dependent only on initial conditions at each step.

Secondly, the legs evolve within small angles about the vertical. Conditionally to sufficient kinetic energy, the loss of speed of the stance leg during a step is negligible. Its position varies almost linearly through time as shown on **Fig. 8**, where τ is directly proportional to the angle of the stance leg about the vertical.

$$\ddot{q} = -A^{-1}(q)C(q,\dot{q}) - A^{-1}(q)G(q) + A^{-1}(q)U \qquad (3)$$

3.2 Time scaling control

Fig. 2 presents the structure of the control system used to implement the time-scaling control. A supervisor manages the robot with a 6 stages sequential function chart. Stage 1 represents the impulse under the left foot, stage 2, the control of the left leg and stage 3, the ground impact of the left foot. Stages 4 to 6 are the equivalent for the right leg. The applied impulsion is constant; hence, the initial conditions at each step are not commanded. The only command is the torque T controlled by a sate feedback in order to track the angle and angular speed reference of the hip joint generated according to the parameter A delivered by the supervisor.

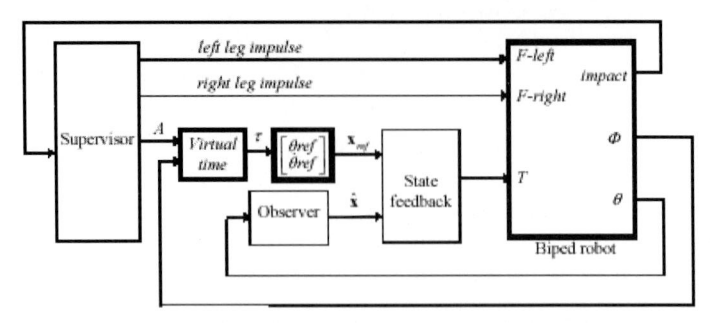

Fig. 2. Structure of the control system

The main objective is to achieve stable and strongly attractive limit cycles while maintaining efficient, smooth and natural-like walking patterns. The control of the robot can be defined as a problem of planning optimal trajectories as a function of time which requires reaching desired initial conditions at each step, or adapting the trajectory to the initial conditions. It is also possible to define the trajectory as a function of the geometric evolution of the robot [6]. This is equivalent to a time scaling control [7].

The main originality of this paper lies in the generation of a reference for the swing leg as a monotone function of the stance leg angle instead of a function of time. For this purpose, the function τ ("virtual time") proportional to Φ and function of A, the imposed amplitude of a step, is created.

The function τ varies linearly and undisturbed by small torques. It is used to build a reference for the swing leg. A sinusoidal trajectory is what is expected from a pendulum with a hinge moving at a near constant speed.

An angular velocity reference, with low authority and dedicated to smoothing the swing leg oscillations is built using the constant ω set to the approximate time derivative of τ. This bypasses the derivation of noisy signals hard to filter without producing destabilizing delays. These relations are presented in equations 4 to 6. A saturation bounds τ between -1 and 1.

$$\tau(t) = sat\left(\frac{2\phi(t)}{A}\right) \quad (4)$$

$$\Theta ref(t) = q_2(t) + \frac{A}{2}\sin\left(\tau(t)\frac{\pi}{2}\right) \quad (5)$$

$$\dot{\Theta} ref(t) = \dot{q}_2(t) + \frac{A\pi}{4}\omega\cos\left(\tau(t)\frac{\pi}{2}\right) \quad (6)$$

4 Prototype and experimental setup

To experiment with the model, a prototype (**Fig. 3**), was realized. The telescopic legs are driven by three-positions pneumatic cylinders. They generate impulsive forces and avoid unwanted ground collisions. A brushless motor generates the torque acting at the hip joint. The system is controlled using *dSpace* prototyping tool and programmed using *Matlab/Simulink*. Two potentiometers and two switches are used so that the angle of each leg and the moments of ground collision might be known.

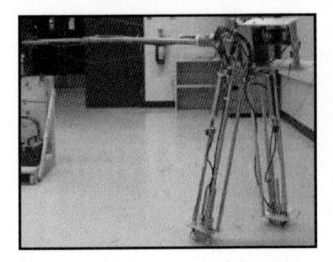

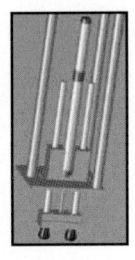

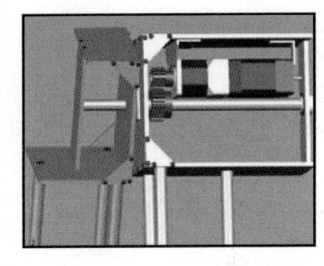

Fig. 3. The prototype

5 Simulation results

To solve the robot dynamics, *Matlab/Simulink ODE5* was used. Simulation and experimental models use identical parameters and strongly correlate.

The first tests proved the existence and attracting nature of a limit cycle. **Figs. 4** and **5** are projections of the state phase for a leg during walks.

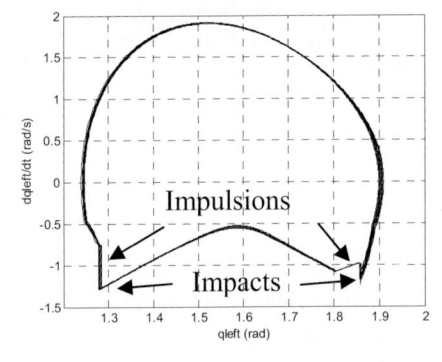

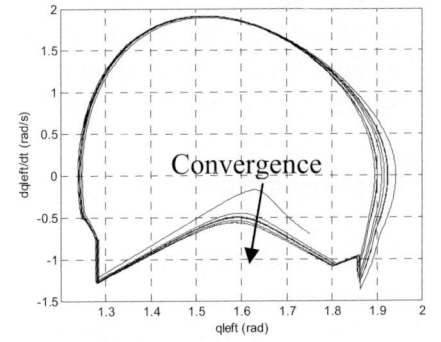

Fig. 4. Projection of a limit cycle

Fig. 5. The attractive nature

A Poincaré map permits a more rigorous visualization. The state space is analyzed every time the stance leg crosses a specific position. The undetermined state variables are thus brought down to three. **Figs. 6** and **7** show the state evolution during first steps with different initial conditions. Data is taken following the impulsions. The robot starts with arbitrary conditions and converges towards a limit cycle defined as the state vector producing an identical state after any number of steps. The data that converges is shown. The axes are set to the span of the testing. The Figs. give a raw idea of the shape of the basin of attraction.

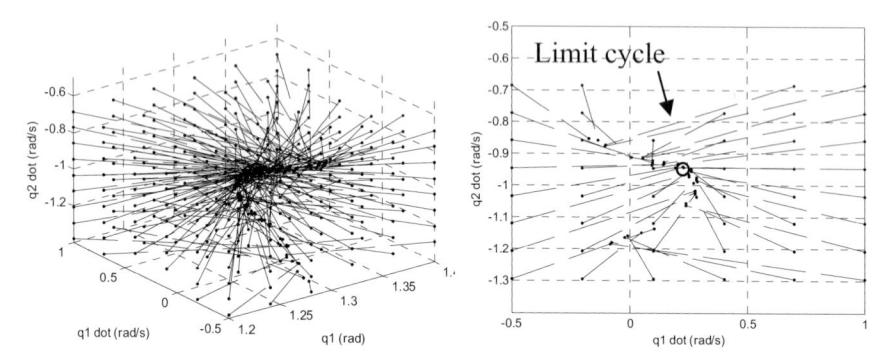

Fig. 6. Poincaré sections

Fig. 7. Projection of Poincaré sections

Interestingly, the convergence is dependant almost only on the speed of the stance leg and very little on the swing leg state. The support leg imposes its dynamics and has a huge influence.

Also, as demonstrated by Goswani et al. [1], the transition rule, applied to a robot with legs of null moment of inertia, reduces the state phase map to a 2D problem. It becomes necessary only to specify the state of the stance leg to fully determine the motion of the subsequent steps. It implies that the transition rule permits convergence even with important variations of the swing leg state for robots approaching this mass distribution.

The natural energy dissipation of the collisions with the ground is used to drive the system towards a stable limit cycle.

6 Experimental results

Fig. 8 shows the evolution of important parameters: τ that is used to build the references, stages of the sequential function chart and the angle from one leg to the other and its speed and their references.

Fig. 9 plots a Poincaré section. The dots are measures taken during a normal walk. The circles pairs are taken during the two first steps of starting sequences where the robot starts itself with a variable impulsion. The Fig. shows the existences of an attractor. Squares do not converge.

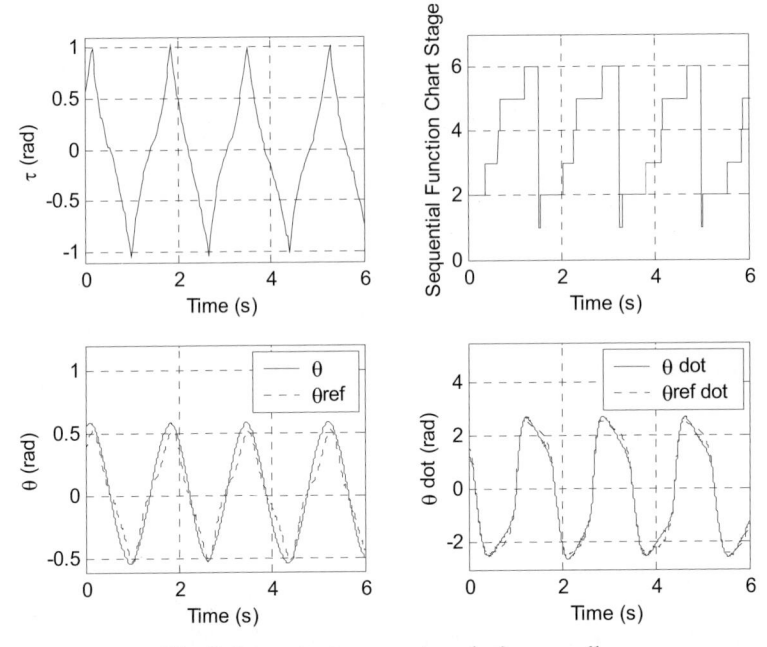

Fig. 8. Important parameters during a walk

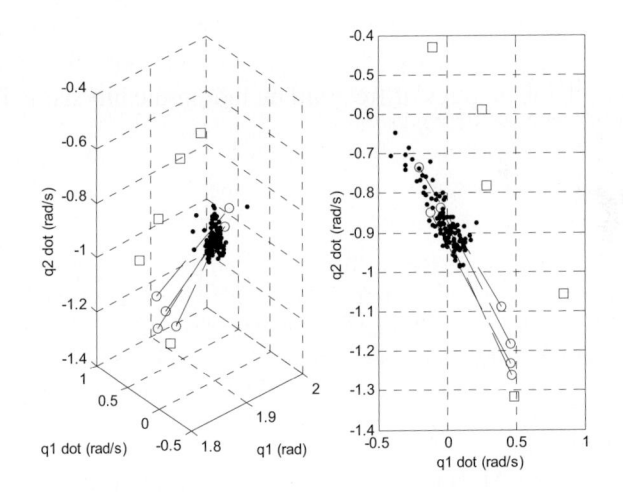

Fig. 9. Poincaré section revealing an attractor

We further tested the stability of the robot by placing free obstacles in its path producing tripping impulses. **Fig. 10** shows a projection of the state phase for one leg during such an event. **Fig. 11** shows the leg angle and reference. The robot reaches its limit cycle during the following step.

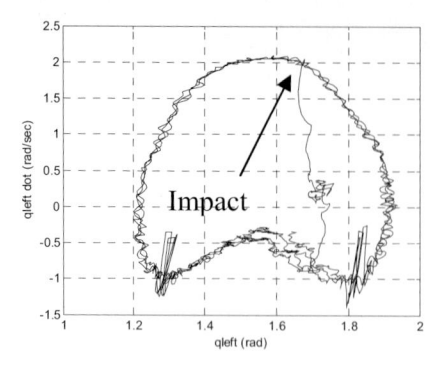

Fig. 10. State phase projection with tripping impulse

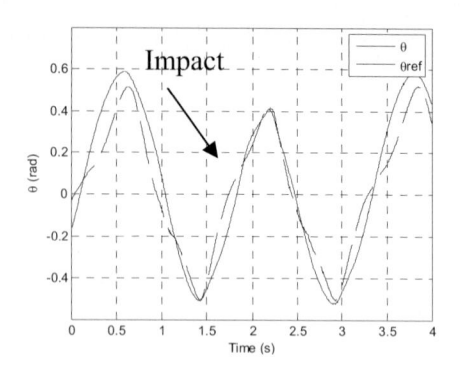

Fig. 11. Leg angle and its reference with tripping impulse

7 Conclusion

The paper presented a possible approach for the control of an actuated walking model. In order to reach a dynamically stable walking mode, a simple control strategy involving references built via a function of the geometric evolution of the robot was used and showed its robustness. The geometry based control proved its simplicity and applicability to systems with uncontrollable degrees of freedom including the bipedal gait.

References

1. Goswami A, Tuillot B, Espiau B (1996) Compass like bipedal robot part 1: Stability and bifurcation of passive gaits. INRIA Research report 2996
2. Garcia M, Chatterjee A, Ruina A, Coleman M (1998) The simplest walking model: stability, complexity, and scaling. J Biomech Eng 120(2): 281-8
3. Micheau P, Buaka P, Bourassa P (2002) Control of the simplest walking model with Lambda model. Automatic Control 15th IFAC World Congress Barcelona, 6 pages
4. Micheau P, Roux M, Bourassa P (2003) Self-tuned trajectory control of a biped walking robot. clawar, 527-534
5. Bourassa P, Micheau P (2002) Tripping impulses – gait limit cycle for biped. clawar. 791-798
6. Chevallerau C, (2003) Time-scaling control for an underactuated biped robot. IEEE trans. robotics and automation. vol.19, no.2, 363-368. April 2003
7. Dahl O, Nielsen L(1990), Torque-limited path following by online trajectory time scaling, IEEE Trans. Robot. Automa., vol. 6, pp. 554-561
8. Web site of P.Micheau, http://mecano.gme.usherbrooke.ca/~pmicheau/

Design Methodology and Gait Generation

Mechanical Design of Step-Climbing Vehicle with Passive Linkages

Daisuke Chugo[1], Kuniaki Kawabata[2], Hayato Kaetsu[3], Hajime Asama[4] and Taketoshi Mishima[5]

[1] The University of Tokyo, 2-11-16, Yayoi, Bunkyo-ku, Tokyo, Japan
chugo@iml.u-tokyo.ac.jp
[2] RIKEN (The Institute of Physical and Chemical Research), 2-1, Hirosawa, Wako-shi, Saitama,Japan kuniakik@riken.jp
[3] RIKEN (The Institute of Physical and Chemical Research), 2-1, Hirosawa, Wako-shi, Saitama,Japan kaetsu@riken.jp
[4] The University of Tokyo, 5-1-5, Kashiwanoha, Kashiwa-shi, Chiba, Japan asama@race.u-tokyo.ac.jp
[5] Saitama University, 255, Shimo-Ookubo, Saimata-shi, Saitama, Japan mishima@me.ics.saitama-u.ac.jp

Abstract

In our current research, we are developing a holonomic mobile vehicle which is capable of running over irregular terrain. Our developed vehicle realizes omni-directional motion on flat floors using special wheels and passes over non-flat ground using the passive suspension mechanism. This paper proposes a mechanical design of passive linkages for increasing the vehicle's mobile performance on rough terrain. The developed vehicle has plural actuated wheels and all wheels are grounded for enough traction force using the passive linkage mechanism. However, according to the mechanical design of the passive linkages, the body configuration cannot fit the terrain surface and wheels cannot transmit its traction force. Therefore, in this paper, we discuss the mechanical design of the passive linkages which enable the body configuration to fit the ground shape and we propose new passive linkage mechanisms for increasing the mobile performance on rough terrain. The performance of our proposed method is verified by the experiments using our prototype vehicle.

Keywords: Omni-Directional Mobile System, Passive Linkage Mechanism, Step-Climbing.

1 Introduction

In recent years, mobile robot technologies are expected to perform various tasks in general environment such as nuclear power plants, large factories, welfare care facilities and hospitals. However, there are narrow spaces with steps and slopes and the vehicle is required to have quick mobility for effective task execution in such environments. Omni-directional mobility is useful for moving in narrow spaces, because there is no holonomic constraint on its motion. [1] Furthermore, the step-overcoming function is necessary when the vehicle runs around in the environment with steps. Therefore, we are developing a holonomic omni-directional vehicle with step-climbing ability. [2]

Our prototype mechanism consists of seven special wheels with free rollers and a passive linkage mechanism. (**Fig.1**(a)) The special wheel consists of twelve cylindrical free rollers [3] and helps to generate omni-directional motion with suitable wheel arrangement and wheel control. The passive linkage mechanism ensures that the vehicle can pass over the step smoothly when the wheel contacts the step, changing the body configuration of the vehicle. No sensors and no additional actuators are required to pass over the nonflat ground.

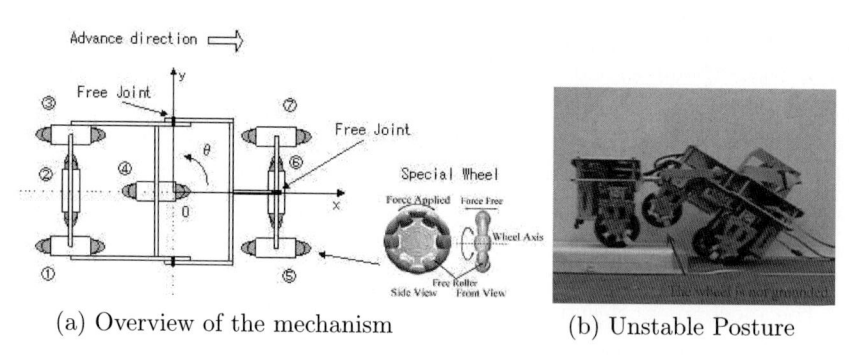

(a) Overview of the mechanism   (b) Unstable Posture

Fig. 1. Our prototype mechanism

Many mobile vehicles which have passive linkages have been developed. Rocker-bogie suspension mechanism is a typical one. [4] For realizing high step-climbing performance, these vehicles have plural drive wheels and all drive wheels are grounded by the change of body configuration. However, according to the mechanical design of passive linkages, the body configuration cannot fit the terrain surface and the drive wheels float from the ground. (**Fig.1**(b)) As a result, the wheels cannot transmit traction force and these actions disturb the mobile performance of the vehicle. Thus, in this paper, we propose a new passive linkage mechanism which ensures that all wheels are grounded thus increasing the step-climbing performance of the vehicle.

2 Mechanical Design of Passive Linkages

2.1 Stability Condition

When the vehicle passes over the step, moment forces are generated as shown in **Fig.2**(a) and (b). **Fig.2**(a) shows the vehicle model as it runs over the step in forward direction and **Fig.2**(b) shows the model in backward direction. Where l_1 and l_2 are distance between wheels, m_l and m_b are mass of the body, r is radius of wheel, μ_1 and μ_2 are friction coefficients, f_i and F_i $(i = 0, \cdots, 3)$ are forces between bodies and ground. α is vertical angle of step and in this case, we set $\alpha = \pi/2$.

M_l and M_b are moment forces of the body when the vehicle contacts the step in forward direction. M_l' and M_b' are moment forces when the vehicle contacts it in backward direction. When the vehicle climbs the step in forward direction, M_l should be positive so that the body configuration can change according to the terrain surface and all wheels can be grounded. Similarly, when the vehicle climbs the step in backward direction, M_b' should be positive. Therefore, for realizing high step-climbing performance, it is required that these moment forces have positive direction as shown in **Fig.2** by mechanical condition.

(a) Forward Direction (b) Backward Direction

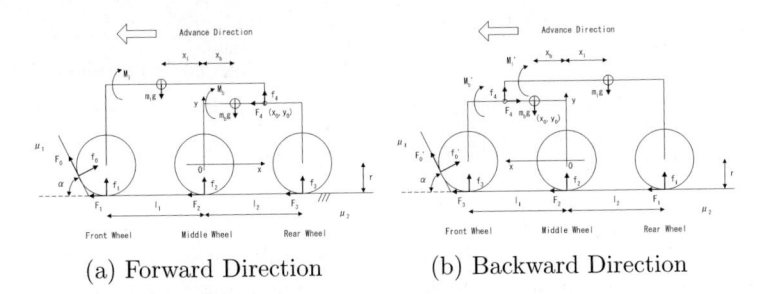

Fig. 2. Moment forces during step-climbing

Now, we derive these moment forces.

We set the position of free joint point as (x_0, y_0) on the coordination in **Fig.2**. When the vehicle contacts the α-degree step from forward direction, these equations are derived from the balance of forces.

$$(f_0 \sin \alpha - F_0 \cos \alpha) - F_1 - F_4 = 0 \tag{1}$$

$$(f_0 \cos \alpha + F_0 \sin \alpha) + f_4 = m_l g \tag{2}$$

$$F_4 - F_2 - F_3 = 0 \tag{3}$$

$$f_2 + f_3 - m_b g - f_4 = 0 \tag{4}$$

From the balance of moment forces on its body, equation (5) is derived.

$$M_b = f_2 l_2 - m_b g (l_2 - x_b) - f_4 (l_2 - x_0) + F_4 (r + y_0) = 0 \tag{5}$$

When the vehicle climbs the step, the vehicle should have enough traction force to lift its own body. Therefore, the reaction force between the front wheel and ground is zero.

$$f_1 = F_1 = 0 \tag{6}$$

We assume that the wheel transmits maximum power to the ground within the range of friction between the wheel and ground.

$$F_0 = \mu_1 f_0, \quad F_2 = \mu_2 f_2, \quad F_3 = \mu_2 f_3 \tag{7}$$

From equation (1) to (7), we can derive the equation (8) and (9).

$$M_l = f_0 (\cos \alpha + \mu_1 \sin \alpha)(r \sin \alpha + l_1 + x_0) - m_l g (x_l + x_0) \\ - f_0 (\sin \alpha - \mu_1 \cos \alpha)(r \cos \alpha + y_0) \tag{8}$$

$$f_0 = \frac{\mu_2 (m_l + m_b) g}{(\sin \alpha - \mu_1 \cos \alpha) + \mu_2 (\cos \alpha + \mu_1 \sin \alpha)} \tag{9}$$

From equation (8) and (9), the moment force when the vehicle passes over the step ($\alpha = \pi/2$) in forward direction are expressed as equation (10). From this equation, moment force M_l is derived from the position of free joint point (x_0, y_0).

$$M_l = \frac{\{\mu_1 \mu_2 (2m_l + m_b) - m_l\} x_0 - \mu_2 (m_l + m_b) y_0 + \{\mu_1 \mu_2 (m_l + m_b)(r + l_1) - (1 - \mu_1 \mu_2) m_l x_l\}}{1 + \mu_1 \mu_2} g \tag{10}$$

Similarly, when the vehicle passes over the step in backward direction, the moment force M_b' is derived as equation (11). The moment force M_b' is derived from the position of free joint point (x_0, y_0).

$$M_b' = \frac{-\{\mu_1\mu_2(m_l+2m_b)-m_b\}x_0 - \mu_2(m_l+m_b)y_0 + \{\mu_1\mu_2(m_l+m_b)(r+l_2)-(1-\mu_1\mu_2)m_bx_b\}}{1+\mu_1\mu_2}g \quad (11)$$

From these equations, it is required to design the position of free joint point (x_0, y_0) for increasing the step-climbing performance of the vehicle.

2.2 Design of Passive Linkage Mechanism

In this section, we design the proposed passive linkage mechanism. For increasing mobile performance on the step, the moment force which applies to the vehicle body should be positive. From equation (10) and (11), the moment force when the vehicle passes over the step is led by only the position of free joint point. Thus, we design the free joint point position so that the moment force is positive when the vehicle climbs the step.

Fig.3 shows the moment force when the vehicle contacts the step. In **Fig.3**, (a) shows the moment force based on x_0 as a variable and (b) is based on y_0. M_l is the moment force which is applied on the vehicle body when the vehicle passes over the step in forward direction. On the other hand, M_b' is the moment force when the vehicle passes over it in backward direction. The parameters of the vehicle are chosen from the prototype vehicle model as shown in **Table1**. The friction coefficients between the wheel and the floor are set as $\mu_1 = \mu_2 = 0.3$ or $\mu_1 = \mu_2 = 0.5$. The former assumes linoleum floor and the latter assumes skidding on the floor.

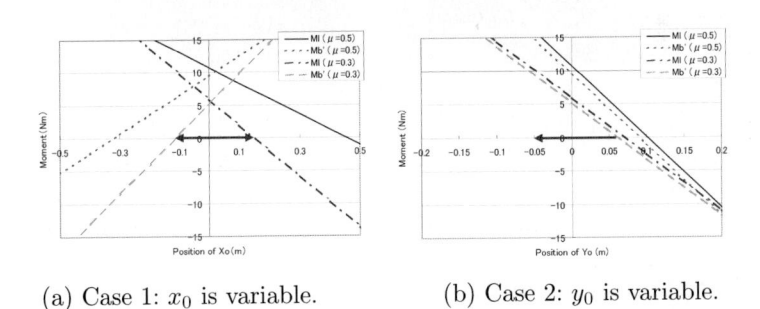

(a) Case 1: x_0 is variable.

(b) Case 2: y_0 is variable.

Fig. 3. Moment forces when the vehicle climbs the step

From **Fig.3**, the position of free joint point (x_0, y_0) should meet the following requirements.

Table 1. Parameters of prototype

	Rocker-Part	Bogie-Part
Friction coefficient (μ_1,μ_2)	Skid: 0.5, Linoleum: 0.3	
Body Weight (m_l,m_b)	13(kg) include Payload	14(kg)
Wheel Diameter (r)	0.132(m)	
Distance between wheels (l_1,l_2)	0.255(m)	0.215(m)
Center-of-gravity position (x_l,x_b)	0.128(m)	0.108(m)

- x_0 should be smaller than 0.15[m] and larger than -0.1[m]. (In the range of the arrow.)
- y_0 should be as smaller than 0.05[m]. (However, there is danger of conflict between the vehicle body and ground if y_0 is too small.)

From two conditions, we set the free joint point as equation (12).

$$x_0 = y_0 = 0 \tag{12}$$

We propose new passive linkage mechanism as shown in **Fig.4**(b). **Fig.4**(a) is an old prototype which has rocker-bogie suspension mechanism.

(a) Rocker-bogie mechanism   (b) Proposed passive linkage mechanism

Fig. 4. Proposed passive linkage mechanism

2.3 Computer Simulations

We verify the effectiveness of our proposed design on passive linkage mechanism. In this simulation, the vehicle model passes over the step in forward direction at 0.25[m/s] and we measure the moment force which is applied on the body when the vehicle contacts the step. We compare the result of proposed mechanism with the result of rocker-bogie mechanism. As

Table 2. Simulation Results

	Rocker-Bogie Mechanism		Our Proposed Mechanism	
Height of Step(mm)	Step Climbing	Moment(Nm)	Step Climbing	Moment(Nm)
40	Success	-0.006	Success	0.037
60	Success(*)	-0.105	Success	0.059
80	Success(*)	-0.111	Success	0.105
100	Failure	-0.129	Success	0.105

(*): Success, however, wheels float from ground.

initial conditions, simulation parameters of test vehicle model are chosen from our prototype model. Parameters are shown in **Table1**.

As the results of the simulation show, the moment forces on the proposed mechanism are positive and all wheels are grounded during step-climbing. On the other hand, the moment forces on the rocker-bogie suspension mechanism are negative and during step-climbing, the middle wheel floats from the terrain surface. Furthermore, the step-climbing performance of the vehicle with our proposed mechanism is improved.

From these results, our design for passive linkage mechanism is useful for increasing the mobile performance.

3 Experiments

In this experiment, the test vehicle passes over the 60[mm]-height step in advance direction at 0.25[m/s] and we verify the tracks of wheels. We compare the result of proposed mechanism with one of the rocker-bogie suspension mechanism. In both cases, PID based controller for traction control is employed. [2]

Fig.5 shows the tracks of the vehicle during step-climbing. Tracks are plotted at every 0.3[sec]. The vehicle with our proposed mechanism can pass over the step more smoothly as shown in **Fig.5**. In this experiments, the vehicle with rocker-bogie suspension mechanism floats the middle wheel for 2.8[sec]. On the other hand, the vehicle with our proposed passive linkage mechanism floats the middle wheel for 0.6[sec].

As the result, the vehicle with proposed mechanism can pass over the 128[mm]-height step maximum. The vehicle with rocker-bogie suspension mechanism can climb up only 60[mm]-height step. From these results, we verify that our mechanism design is effective for increasing the mobile performance of the vehicle.

(a) Rocker-bogie mechanism (b) Proposed passive linkage mechanism

Fig. 5. Tracks of the vehicle during step-climbing

4 Conclusion

In this paper, we proposed the mechanical design for passive linkages. We discuss the moment force which is applied on the vehicle body when the vehicle contacts the step and we derive the moment force using the position of free joint point. From the derivation, we design new passive linkage mechanism and utilize it to our prototype.

We verified the effectiveness of our proposed design on passive linkages by computer simulations and experiments. Utilizing our proposed mechanical design on the prototype, the moment force becomes positive, all wheels are grounded and step-climbing ability increases. From these results, our proposed mechanical design for passive linkages improves the mobile performance.

References

1. Guy C., Georges B. and Brigitte D. A. (1996) Structual Properties and Classification of Kinematic and Dynamic Models of Wheeled Mobile Robots. *IEEE Transaction on Robotics and Automation*, Vol.12, No.1, pp. 47–62.
2. Daisuke C., Kuniaki K., Hayato K., Hajime A. and Taketoshi M. (2005) Development of a Control System for an Omni directional Vehicle with Step-Climbing Ability. *Advanced Robotics*, Vol.19, No.1, pp. 55–71.
3. Hajime A., Masatoshi S., Luca B., Hayato K., Akihiro M. and Isao E. (1995) Development of an Omni-Directional Mobile Robot with 3 DOF Decoupling Drive Mechanism. *Proceedings of the 1995 IEEE International Conference on Robotics and Automation*, pp. 1925–1930.
4. Henry W. S. (1996) Mars Pathfinder Microrover: A Low-Cost, Low-Power Spacecraft. *Proceedings of the 1996 AIAA Forum on Advanced Developments in Space Robotics*, Madison, WI, August.

Integrated Structure-control Design of Dynamically Walking Robots

Petko Kiriazov

Dynamics & Optimisation of Controlled Mechanical Systems Department
Institute of Mechanics, Bulgarian Academy of Sciences, Sofia, Bulgaria
kiriazov@imbm.bas.bg

Abstract

This study is motivated by the need of dynamics-based methodologies for overall design of legged robots (LR). Along with the basic design requirement for strength/load capacity, additional design criteria for LR are needed to meet the continuously increasing demands for faster motion, higher position accuracy and reduced energy consumption. A conceptual framework for their integrated structure-control design is proposed that can be used to create LR with maximum capability to achieve the required dynamic performance. To verify our design optimisation concepts, several interesting examples regarding two- and four-LR are considered.

Keywords: dynamics, controllability, robustness, design optimisation;

1 Introduction

Difficulties to study LR and optimise their dynamic performance are mainly due to complex system dynamics and variable external disturbances. At first, the structure of the input-output relations between their subsystems has to be defined, i.e., numbers and locations of their inputs, outputs and interconnections, [2]. For analysis, design and control purposes, adequate dynamic models are needed and they have to be easily and accurately identifiable.

The dynamic model structure of a robotic system depends on the type of its control subsystem: centralised or decentralised. In practice, the decentralised manner of control has been adopted for its main advantages to the centralised one: simplicity, reliability and faster response. A common feature of the existing control design methods for linear systems with decentralised control structure is that stability in the face of parameter uncertainties can be ensured if the control transfer matrix (TM) is generalised diagonally dominant (GDD), [6]. We extend this property to robotic systems and prove that the GDD-condition is necessary and sufficient for them to be robust against arbitrary, but bounded disturbances, [1]. By using the nonnegative matrix theory [5], relevant design criteria are derived and they are not in conflict with the strength requirement. Moreover, these criteria enable a decomposition of the overall design task into a sequence of design solutions for the LR's components: mechanical structure, actuators, and controls. Several important case studies concerned with the optimisation of two- and four-LR show the efficiency of the proposed design approach.

2 Integrated structure-control design approach

LR are highly non-linear and difficult to model, identify, and control dynamical systems mainly due to gravitation, inertia couplings, friction, elasticity, and actuator limits. In addition, the structure of their dynamics may change during locomotion. The dynamic performance of a LR in a specific locomotion phase can be described, in general, by the following system of differential equations

$$\ddot{q} = M^{-1}(q)\ (Bu - C(q,\dot{q}) + g(q)) \tag{1}$$

where, q is the vector of the links' rotation angles, M is the inertia matrix, is the vector of velocity forces, C stands for friction and gravitation forces, matrix B represents the actuator location, and u is the vector of actuator torques. To identify the coefficients describing these dynamic terms we can apply the approach proposed in [3].

After compensating to some extent for the inertia, velocity, friction and gravitation forces by feedforward control, the following reduced model for the error dynamics can be used for the feedback stabilisation of the programmed motion

$$\ddot{e} = A(q)u + d \tag{2}$$

where $e = q - q^{ref}$, $A = M(q)^{-1}B$ is the control TM and vector d stands for all uncompensated terms, as well as for measurement and environment noises. With A being GDD, the non-negative matrix theory [5] states that there always exists a positive vector $\bar{u}$ of control magnitudes solving the following system of equations

$$A_{ii}\bar{u}_i - \sum_{j \neq i} |A_{ij}| \bar{u}_j = \bar{d}_i \qquad (3)$$

where $\bar{d}_i$ are some upper bounds.

Eqs. (3) present optimal trade-off relations between the bounds of model uncertainties and the control force limits. The greater the determinant Δ of this system of linear equations, the less control forces are required to overcome the disturbances. In other words, Δ quantifies the capability of LR to be robustly controlled in a decentralised manner. For these reasons, Δ can be taken as a relevant integrated design index for the subsystems whose parameters enter the control TM.

3 Design optimisation scheme

The linearity of Eqs. (3) makes it possible a decomposition of the overall design problem into much simpler design problems for LR's components: (1) mechanics, (2) actuators (sensors), and (3) controls. This order will correspond to the hierarchy in a multi-level optimisation procedure in which a series of design problems for these subsystems are to be solved.

A. Mechanics
- design parameters: all inertial/geometrical data of the bodies;
- design constraints: strength and GDD conditions;
- design objective: maximise Δ;

B. Actuators
- design parameters: actuator masses and positions;
- design constraints: strength and GDD conditions;
- design objective: maximise Δ;

C. Controls
- design parameters: control gains;
- design constraints: optimal trade-off relations (3);
- design objective: minimise the control effort;

With a specified structure of the control transfer matrix, the proposed optimisation procedure is feasible and convergent, [1].

4 Case studies

To verify the above concepts for integrated structure-control design, we consider several interesting examples of applying GDD-condition, Δ-criterion, and design relations (3):
- ❖ on shape and mass distribution optimisation of the torso of four-LR;
- ❖ on mass distribution in the trunk of biped robots;
- ❖ control synthesis of a seven-link biped robot;

4.1 On shape and mass distribution of the torso of four-LR

Consider a simplified geometrical model of a four-legged robot in the plane of pitch/heave motion (the plane of symmetry), Fig. 1. The two front legs as well as the two rear legs are considered acting virtually as one leg. The vertical components of the driving forces produced by the rear and front legs are u_r and u_f, respectively. The body has mass m and inertia I and the angle of its rotation θ is assumed small.

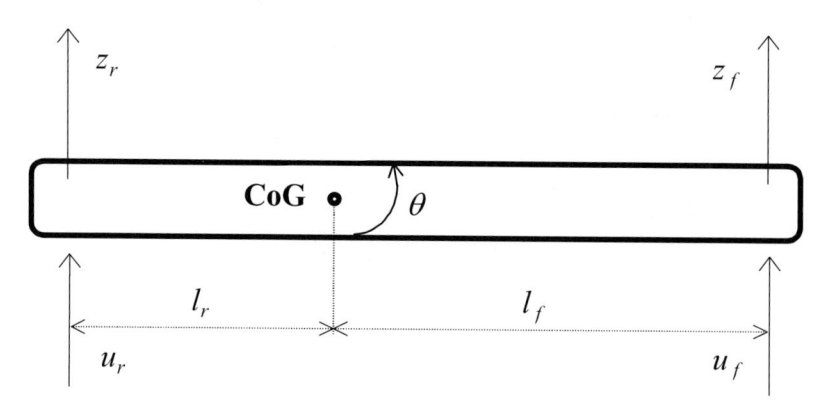

Fig. 1. Robot torso in the plane of pitch/heave motion

As with any two-degree-of-freedom system, the GDD condition is here fulfilled and the index of controllability is $\Delta = l^2/(mI) > 0$. This design optimisation index therefore does not depend on the location of the center of gravity (CoG). To provide the robot with best controllability in the pitch/heave motion, we have to maximise this index. With given l and mass m, the inertia moment I can be reduced if the body has an ellipsoidal shape instead of rectangular one; I can be further reduced if the mass of the body is concentrated around CoG. These design recommendations are again in accordance with the strength/load capacity criterion.

4.2 On mass distribution in the trunk of biped robots

Consider, for definiteness, a planar biped with seven links: a trunk, two thighs, two shanks and two feet, and with six actuators at the legs' joints, Fig. 2. The walking task is analysed and the points in the gait where the LR dynamics changes its structure are identified. Thus the stride is presented as a consequence of several phases, and, in each of them, we consider the LR as a multibody system (MBS) with a relevant control structure. The structure of the biped dynamics changes and the structure of the control system needs to be changed accordingly. From this point of view, four phases in performing steps can be distinguished: double-support, taking-off, single-support (SS) and landing.

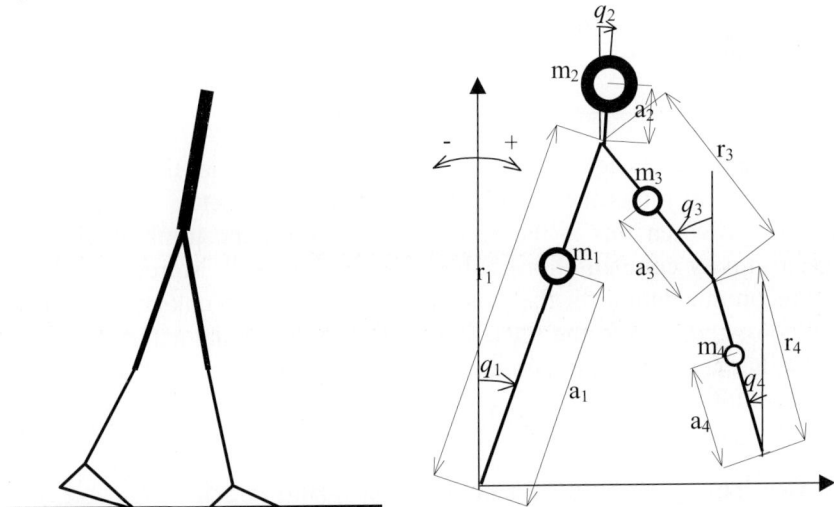

Fig. 2. Biped just before SS-phase   **Fig. 3.** Simplified scheme for SS-phase

In each phase of locomotion, the LR is with specified kinematical, dynamical, and control structures and therefore, the structure of the control transfer matrix is known. We can therefore define the corresponding integrated design criterion and perform design optimisation for any of the controlled MBS representing the LR in the different phases. Design recommendations to the geometry and the mass distribution of the biped as well as to the sizes of its actuators can be given for it to have best controllability and dynamic performance.

The most complicated control problem is posed during the SS-phase when almost all the joints are to be actuated and the joint motions are in general strongly interacting. Moreover, the massive part of the walking mechanism is driven mainly during this time-period and the problems of dynamic stability and time/energy optimization are most important.

In the present case-study, we assume that the design of the legs and the mass of the trunk are already specified and the aim is to optimise the controllability of the biped with respect to the parameter a_2 – the position of the mass center of the trunk, Fig. 3. It is very important to consider such a design optimisation problem because its solution can help designers of autonomous LR when decide where to put some massive parts of the biped like batteries, motors, etc.

We consider mainly the SS-phase, which is the most important from dynamics point of view. Some numerical results showing the dependence of the index of controllability Δ on the parameter a_2 are given in Table 1.

Table 1. Dependence of the index of controllability Δ on parameter a_2

a_2	0.155	0.1	0.07	0.05
Δ	14.073	28.499	30.996	31.141

The main conclusion from this numerical study is that the closer to the axis of the hip joints the trunk's mass center is, the better the biped controllability. When this is the case, less power for the trunk stabilisation (easier gravity compensation) will be required as well. Moreover, our design recommendation for the mass distribution of the trunk can be considered consistent with some related design solutions in the nature like those for the human and some animals (e.g., kangaroo) performing dynamic biped locomotion.

4.3 Control synthesis of a seven-link biped robot

We assume that all geometrical and mass parameters of the biped robot, Fig. 2 and Fig. 3, are already specified and the problem now is to design time/energy efficient control functions for performing dynamic locomotion tasks. To this end, we apply the method of iterative learning control as proposed in [4]. Simple bang-bang or bang-pause-bang control functions can be used and their magnitudes can be determined according to design relations (3). What remains to be done is to find (learn) their switch times.

We decompose the task to perform a step into a sequence of basic point-to-point movements. Especially, the SS-motion can be considered as a composition of two point-to-point movements, Fig.4.

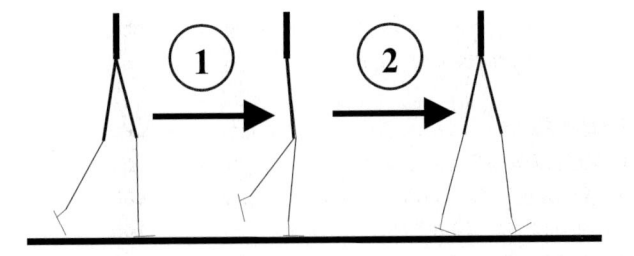

Fig. 4. Motion decomposition in performing a step

The final configuration of the biped for the first point-to-point movement (which is initial for the next one) is characterized with the maximum bending of the trailing leg. The final configuration for the second point-to-point movement is that just before heel striking.

The biped in the SS-phase will be driven by the following three control actions, viz., one at the knee joint of the swinging leg and two at the hip joints. The following final or intermediate values of the state variables are taken as controlled outputs:
- angle of rotation at the ankle of the supporting leg;
- angle between the thighs;
- angle of maximum bending at the knee of trailing leg;
- knee angle of the trailing leg before its landing;

From controllability point of view, we need as many control variables as the controlled outputs are. The control actions at the hip joints will have one switching (from acceleration to deceleration) point and the control actions at the knee of the transferring leg will have two switching points.

Numerical Simulation: Simplified dynamic models were used for all the locomotion phases. With the objective to optimize the locomotion speed, bang-bang test control functions were employed. Applying our control learning approach [4], the required TPBVPs were solved with an acceptable accuracy after 10-20 test movements. The corresponding motion is depicted in Fig.5 (the trunk is not shown – its inclination is kept constant during locomotion).

Fig. 5. Stick diagram of the controlled walking motion

5 Conclusions

We have proposed a generic approach for integrated structure-control design optimisation of LR driven by decentralised joint controllers. The optimisation criteria are defined on the basis of full dynamics, generalised diagonal dominance conditions on the control transfer matrix and optimal trade-off design relations between bounds of model uncertainties and control force limits. The approach makes it possible a decomposition of the overall design problem into design solutions for the robot subsystems: mechanics, actuators, and controls. It can be used in developing new modularity concepts in the design of the mechanical and the drive subsystems of LR. The approach has the necessary mathematical guarantees and its efficiency has been verified considering several important examples. Moreover, its application is found to lead to biologically plausible solutions.

References

1. Kiriazov, P.: Robust decentralized control of mechanical systems, *Solid Mechanics and its Applications*, Vol.52, Ed. D. van Campen, Kluwer Acad. Publ., 1997, 175-182
2. Kiriazov, P., Virk, G.S. On design optimization of legged robots, *Proc. of the 2nd CLAWAR Conf. on Climbing and Walking Robots,* Sept. 1999, Portsmouth, England, Eds. Virk, G.S., M Randall, and D Howard, Professional Engineering Publ. Ltd, 373-381.
3. Kiriazov, P., Efficient approach for dynamic parameter identification and control design of structronic systems, *Solid Mechanics and its Applications*, Vol. 89, Eds. U. Gabbert and H.S. Tzou, Kluwer Acad. Publ, 2001, pp. 323-330.
4. Kiriazov, P. Learning robots to move: biological control concepts, *Proc. 4th CLAWAR Conf. on Climbing and Walking Robots*, Karlsruhe, Germany, Eds. K. Berns and R. Dillman, Prof. Eng. Publ. Ltd, 2001, pp. 419-426.
5. Lunze, J.: *Feedback Control of Large-Scale Systems*, Prentice Hall, UK, 1992
6. Nwokah, O. D. I. and Yau, C.-H.: Quantitative feedback design of decentralized control systems, *ASME Journal of Dynamic Systems, Measurement and Control*, 115 , pp. 452-466, 1993

Intuitive Design and Gait Analysis for a Closed Loop Leg Mechanism of a Quadruped with Single Actuator

Vinayak

Department of Mechanical Engineering, Punjab Engineering College, Sector-12, Chandigarh, India, vinayak.pro@gmail.com.

Abstract

The present work has originated from the idea of developing a single degree of freedom quadruped by using single actuator for all the legs with a novel closed loop leg mechanism. This paper presents the leg mechanism right from its basic form to its kinematical and dimensional details, intuitive in the sense that no specific optimization technique has been adopted to decide the magnitude of various kinematical parameters vis-à-vis link lengths etc. The engineering details of the leg have been given and the gait analysis has been performed for two different sets of link ratios. The author proposes gear trains for achieving the gait of the quadruped. The ultimate aim of this work is to reduce the complex control algorithms for performing dynamically dexterous tasks.

1 Introduction

With time, the development of legged locomotives has matured to more advanced levels, pertaining to stability and at the same time, attaining simplicity of control. Efforts have been made to reduce the control complexity for legged systems by using techniques like reducing the number of actuators [1, 2], and by the use of compliant legs [1, 2]. Recently, an effort was made by Yan and Chen [3] to design a single actuator quadruped by using an optimized version of the Wang type of eight link planar mechanism as leg. The quadruped thus designed was a one degree of freedom vehicle and hence compromised with terrain adaptability.

This paper gives presents a novel leg mechanism, contains the basic formulation pertaining to its kinematics and establishes the basic design of a single actuator quadruped, which in spite of having one actuator, has the potential for achieving the second degree of freedom (steering) by **manual control** and good terrain adaptability due to the compliance in the leg. It uses the advantages of all the previously mentioned designs and it is a contribution towards the design of manually controlled legged vehicles with good terrain adaptability.

2 Basic leg design and kinematical analyses

2.1 Link parameters and length ratios

Figure 1 shows the schematic of the proposed leg mechanism and figure 2 shows the allotment of frames to various links in the leg mechanism in accordance with the Denavit-Hartenberg convention [4]. The vertical line represents a slotted bar fixed to the base frame (x_0, y_0, z_0). The leg of the quadruped is a Revolute-prismatic (R-P) mechanism which is **partially** a closed loop mechanism. There is a compression spring connecting links 1 and two along the direction Y_2 (refer figure 2). The first link of the leg is oscillated to attain the transfer and support phases. During the support phase (clockwise rotation of input link), the mechanism acts as an open kinematical chain and during the transfer phase (counter-clockwise rotation of input link), as a closed kinematical chain as shown in **figure 3**.

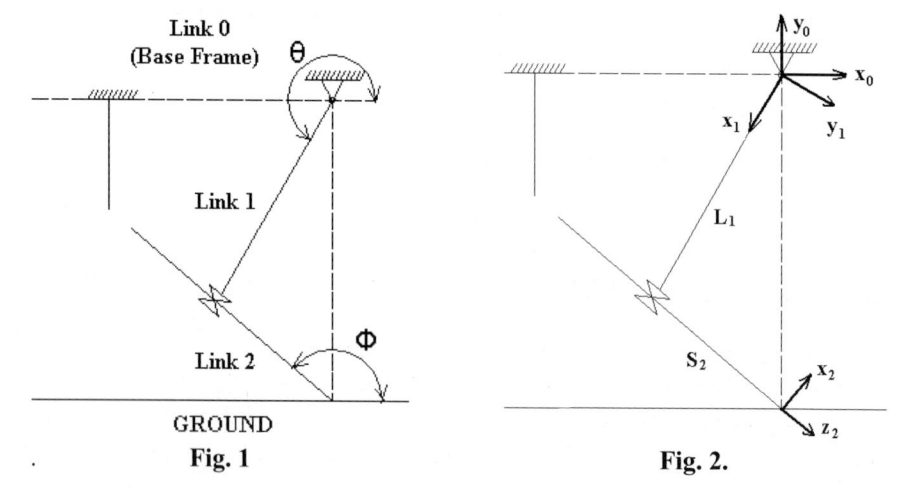

Fig. 1

Fig. 2.

Table 1 contains the link parameters according to the Denavit-Hartenberg convention [4, 5]. There are two distinct sets of link length ratios for which the kinematical analysis has been performed and each one of these has its advantage and disadvantage which will be discussed later. The data pertaining to these has been tabulated below for comparison (refer table 2).

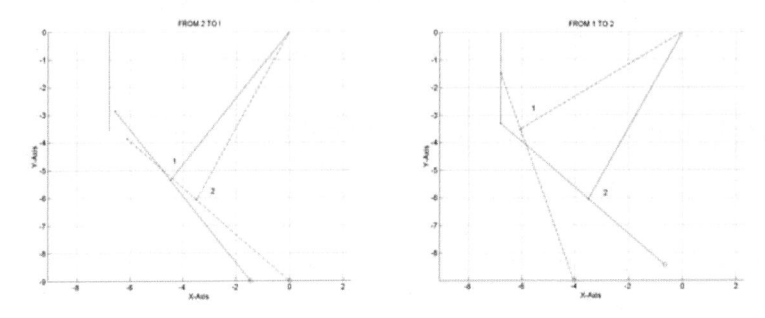

Fig. 3. Open Loop Chan (Left) and Closed Loop Chain (right)

Table 1. Link Parameters for the leg

i	α_{i-1}	a_{i-1}	d_i	θ_i
1	0°	0	0	Θ
2	90°	L_1	S_2	0

Table 2. Geometrical Specifications for leg mechanism

	Configuration A	Configuration B
Length of Link 1 (L1 cm.)	7	7.5
Length of Link 2 (L2 cm.)	7	12
Difference between theta and phi	100	110
Angular range of input link L1	210° – 240°	240° – 260°
Horizontal distance of slotted bar from origin of the base frame in cm.	6.7963	7.7135
Total length of slotted bar in cm.	3.5823	4.0691

The presence of the slotted bar basically facilitates the commencement of transfer phase when link 1, (input link) moves from position 1 to 2 (refer figure 4), i.e. rotates in counter-clockwise sense [5]. Figure 2 shows the allotment of the frames to the links of the leg mechanism. The axes, Z_0 and

Z_1 are coincident, coming out of the plane of paper and the axis Y_2 is also along the same direction, though originating from different point.

It has been assumed that the height of the center of gravity of the robot from the ground will not change and the motion of the leg as an individual system and the quadruped as a whole will take place along a straight horizontal line.

2.2 Position and velocity analyses equations

Equations (1) and (2) represent the position of foot during the support phase and equations (3) and (4) represent the foot position for transfer phase [5].

$$x_Q = L_1 \cos\theta + S_2 \cos\phi \qquad (1)$$

$$y_Q = L_1 \sin\theta + S_2 \sin\phi \qquad (2)$$

$$x_Q = L_2\{\cos(180° + \phi) - \cos 50°\} + x_{Q_0} \qquad (3)$$

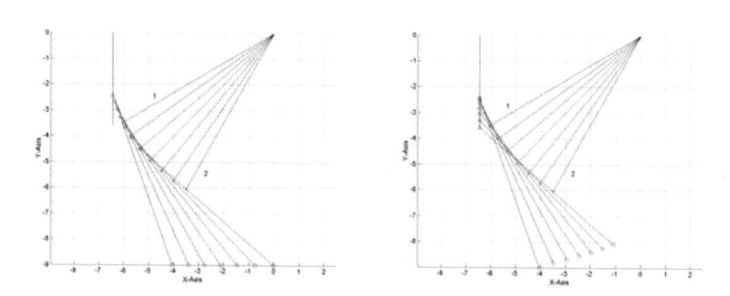

$$y_Q = L_1 \sin\theta + (L_2 + S'_2)\sin(180° + \phi) \qquad (4)$$

Fig. 4. Phase 1 (support) and Phase 2 (transfer) for configuration A

Where:
θ Independent Input Variable,
Φ Dependent Constrained Variable
S_2 Dependent Constrained Variable,
L_1 Length of Link 1
L_2 Length of Link 2,
x_{Q_0} Constant
x_Q Dependent Output Variable,
y_Q Output Variable

3 Engineering details

Figure 11 shows all the engineering details of the leg mechanism in **configuration A**. The same can be applied to configuration B; hence providing the detailed drawing for the second configuration is a redundant step. Instead the data given in table 2 serves the same purpose. All the other dimensions are same (vis-à-vis thickness etc.) for both the configurations.

4 Foot trajectory

Figure 5 shows the foot trajectories for both the configurations (A and B) considering single leg system. Figure 6 shows the motion of the leg with respect to the ground during the gait again considering a single leg system. The assumptions made during the kinematical analysis are applicable.

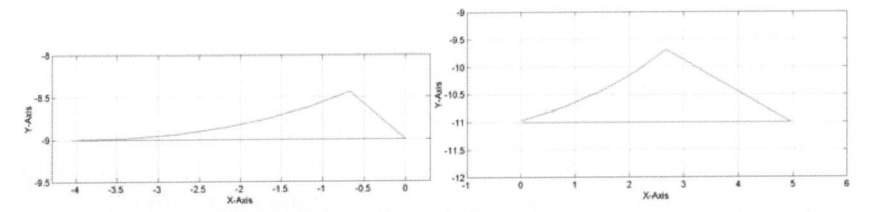

Fig. 5. Foot Trajectories for configuration A and B respectively

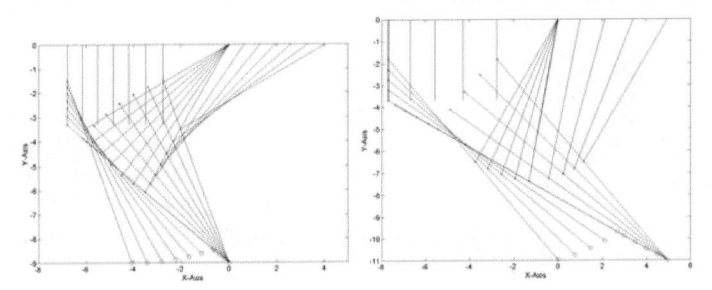

Fig. 6. Positions of legs relative to ground for configurations A and B

5 Gait analyses

The gait of the current quadruped is based on the "trot gait" that uses diagonal pairs of legs in synchrony [6]. But the gait is different from the ideal or square trot in terms of mechanics. In the present case, all the legs come into a momentary **support phase.** Figures 7 and 8, show the gait diagram and the football formula diagram [7, 8] depicting the gait of the quadruped. It can be observed from the figures that during the gait, each leg remains in support for more than half the leg cycle ($\beta > 0.5$).

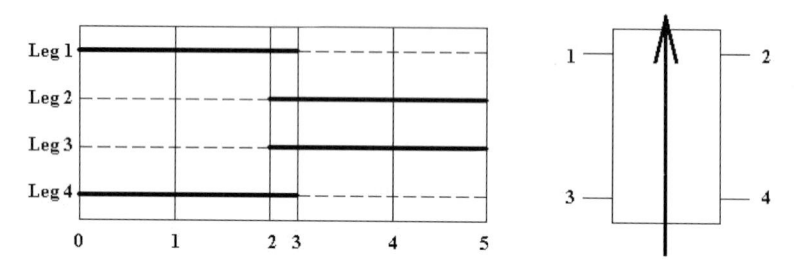

Fig. 7. Gait Diagram and numbering of leg on the quadruped

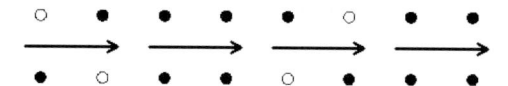

Fig. 8. Football Formula for the Gait of the Quadruped

For achieving the gait discussed above, two distinct gear trains have been designed, one consisting of spur gears and the other consisting of bevel gears. The perspective views of the gear trains can be seen in figure 9. The red spur gear in both the cases represents the input.

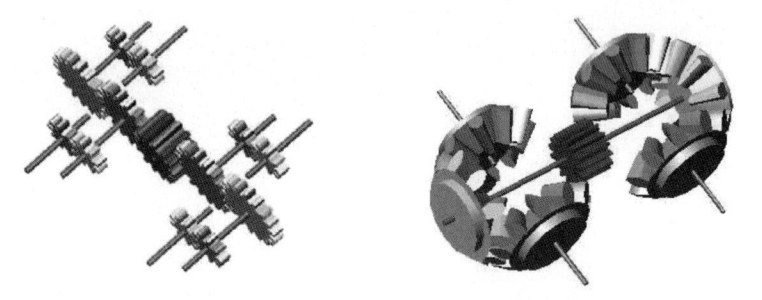

Fig. 9. Gear Trains for achieving the Gait (Bevel on right and Spur on Left)

6 Comparison between configurations A and B

Figure 10 clearly indicates that while in support phase, the actuator will have to work against the normal reaction in configuration B. This data is not sufficient to decide the better configuration. Table 3 shows the comparison of both the configurations in terms of the output & gait parameters.

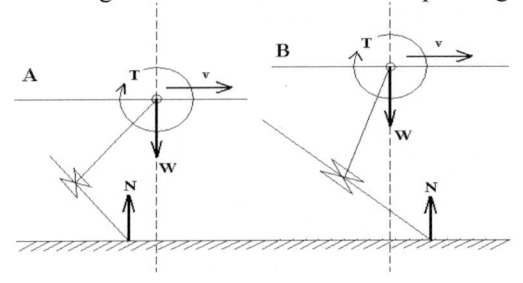

Fig. 10. Diagram showing the comparison between the two leg configurations

Table 3. Comparison b/w output parametyers of the two leg configurations

	Configuration A	Configuration B
Leg Stroke (cm.)	4.0613	4.9572
Max. Height of foot (cm.)	0.5616	1.2797
Effective Length of slot (cm.)	1.8136	1.9129

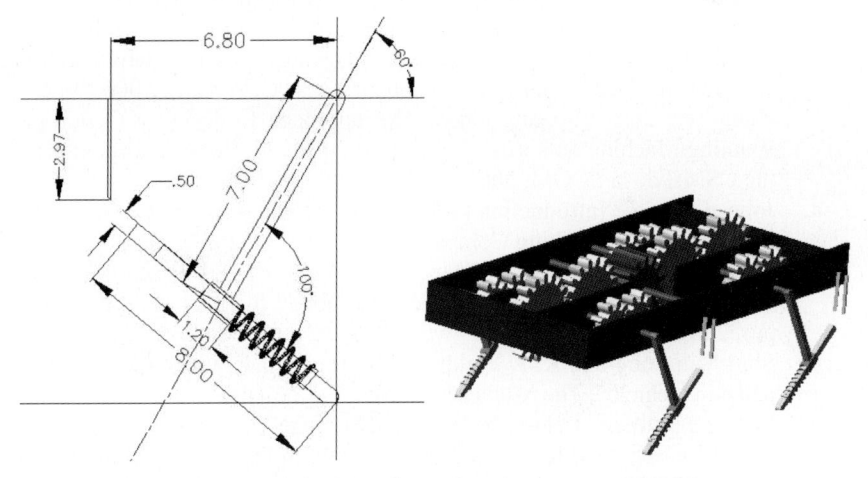

Fig. 11. Engineering Details of Configuration A   **Fig. 12.**

7 Conclusion and scope of work

In this paper the author presented a novel leg mechanism with two different configurations and an overview of the gait, for a quadruped with single actuator. The following issues have the potential to be worked upon vis-à-vis dynamic analysis, modeling of the leg compliance and effect of terrain on the behavior of leg, exploration of the possibility of manual steering and related mechanisms, increase in the number of actuators (one for each leg) to achieve multiple gaits and finally the extension of this work to stable hexapod gaits.

Acknowledgements

This work owes its existence to Dr. Bhaskar Dasgupta, Centre for Robotics and Mechatronics, Indian Institute of Technology, Kanpur. The author shows his gratitude to Dr, Dibakar Sen, Madan Mohan D., Hari Kumar V. and Rajeev Lochana C. G. for their valuable contribution.

References

1. G. Hawker and M. Buehler, "Quadruped Trotting with Passive Knees - Design, Control, and Experiments," IEEE ICRA, California, April 2000.
2. S. Talebi, Buehler, and Papadopoulos, "Towards Dynamic Step Climbing For A Quadruped Robot with Compliant Legs" CLAWAR, 2000.
3. Hong-Sen Yan, Yu-Gang Chen, "An approach to design a Quadruped Walking Machine with a single actuator." Vol. 27, No. 4, Transactions of the CSME/de la SCGM, 2004.
4. John. J. Craig, "Introduction to Robotics: Mechanisms and Control", Addison-Wesley Publishing Company, 1986.
5. Vinayak, Madan Mohan D., Hari Kumar V.," Leg Design and Kinematical Analysis for an under-actuated quadruped" ICTACEM, IIT Kharagpur, 2004.
6. Shin-Min Song and K. J. Waldron, "Machines that Walk: Adaptive Suspension Vehicle", The MIT Press, 1989.
7. Marc H. Raibert, "Legged Robots That Balance", The MIT Press, 1986.
8. Shin-Min Song and K. J. Waldron, "Geometric Design of a Walking Machine for Optimal Stability", Journal of Mechanisms, Transmissions, and Automation in Design, March 1987.

Design of a Cockroach-like Running Robot for the 2004 SAE Walking Machine Challenge

Marc-André Lavoie, Alexis Lussier Desbiens, Marc-André Roux, Philippe Fauteux, Éric Lespérance.

Université de Sherbrooke, Sherbrooke, Qc, Canada, J1K 2R1, marc-andre.lavoie@usherbrooke.ca

Abstract

Captain Basile is a robot inspired from the cockroach and built to participate at the SAE *Walking Machine Challenge* 2004, an undergraduate competition of walking robots.

The robot weighs 35 kg and is 86 cm long, 58 cm wide and 38 cm tall. Initially validated by the use of dynamic simulations, this pneumatic actuated hexapod is characterised by specialized legs, passive visco-elastic elements and a self-stabilizing posture. This mechanical system allows straight line running at 1.11 m/s and turning at 33.2 degrees per seconds with a simple feedforward control system.

This paper presents *Captain Basile*, its design, performances and the results from speed experiment.

Keywords: Cockroach robot, Hexapod, Specialized legs.

1 Introduction

Locomotion over uneven ground has always been a difficult challenge. Many wheel designs exist but have limited performances on uneven ground at high speed. Observations on cockroaches [3, 4, 5 and 6] suggest that legged robots could achieve better performances in this field than they did in the past.

This is why, three years ago, a group of students from the *Université de Sherbrooke* (Canada) was created to take up the challenge of designing and building an insect-like running robot that would perform at the *Society of*

Automotive Engineers (SAE) *2004 Walking Machine Challenge* (WMC 2004). This paper gives an overview of the robot named *Captain Basile*, followed by an analysis of its performances.

2 Problem statement

The *SAE Walking Machine Challenge* is a competition where the speed, the force to weight ratio, the autonomy and the all-terrain performances of walking machines designed by undergraduate students are evaluated.

To complete the competition events, the robot must be able to climb over 20 cm obstacles, to grasp and carry a 10x10x10 cm dense wood block and to carry a charge of 10 kg or more over 18 meters. The rules also limit the overall size to a 1x1x1 meter box.

The fastest robots at the WMC 2002 had a maximum speed of 0.2 m/s on flat ground and 0.02 m/s at the obstacle challenge [1]. As speed is a criterion in all the events of the competition, there is a big opportunity to improve these performances.

3 Design inspiration

Many recent studies on cockroaches [3, 4, 5 and 6] propose a new approach to build walking robots that use insects as a source of inspiration to improve the speed and all-terrain performances. Robots like *Sprawlita* [6], *Whegs I & II* [10] and *RHex* [12] show that rather than just copying the morphology of a cockroach, the solution is to better understand the way it dynamically walks and to implement the fundamental characteristics that explain its performances in a functional and simple robot.

These studies suggest that insect inspired robots should exhibit:

- **Self-stabilizing posture:** A self-stabilizing posture, provided by a low center of mass and a large triangle of support where kinetic energy and leg thrust is used to stabilize the walking pattern [3].
- **Specialized legged function:** Hind legs propel the robot forward, front leg are used to go over obstacles while middle do something in-between [7, 5]. The positioning of the rear foot during climbing prevents the cockroach from falling on its back [9].
- **Compliance:** A well chosen visco-elastic structure provides compliance which is useful to maintain a good contact with the ground [6] and to absorb the foot impact without any active control.

- **Timed, open-loop/feedforward control:** These three characteristics, when embedded in a properly tuned mechanical system, simplify the control system so that only feedforward control is necessary in order to obtain a stable and fast gait.

As a walking machine with less actuators tends to have a better power to mass ratio [10], effort should be made to reduce their number to the strict minimum. This is also useful as this robot has to be power autonomous.

4 Simulations

In order to validate how each of these observations can be applied to reach the targeted goals, *MSC.visualNastran 4D*, a physics-based simulation package and Matlab/Simulink was used.

A simple model was initially used to test the general concept of insect-like locomotion. Then, a second model (**Fig.1.**) was used to select and position the proper components and to tune the feedforward parameters.

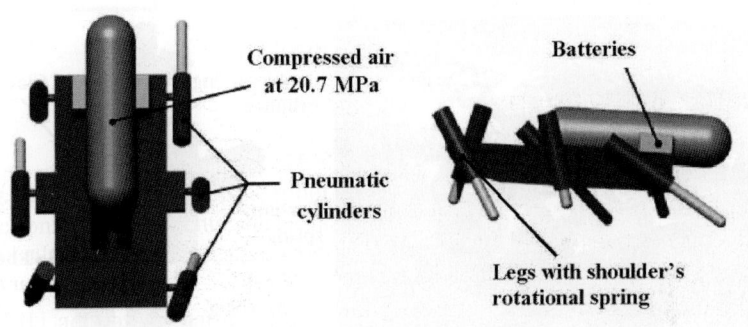

Fig. 1. Model used in the component-sizing simulations

These simulations confirmed that the robot can behave as insects and gave a few more insight on the construction:

- Open-loop control is adequate for straight line running with minimal drifting of about 1% [3].
- Compliance, given by the shoulder spring, is enough to passively produce the proper motion needed by running.
- The position of the center of mass should be between middle and hind legs to allow proper use of those legs for propulsion [3].
- Leg thrusting force should be approximately three times the weight of the robot divided by the supporting legs [8].

7 Performances

After a few simple tuning through simulations and experimental optimisation, *Captain Basile* reaches a top speed of 1.11 m/s (1.25 body length per second). In straight line running, its two air tanks provide about 2.5 minutes of autonomy. Despite its speed, an operator can easily position the robot with a precision of 1 cm. A turning speed of 33.2 degrees per second is achieved by pushing only with the legs on one side.

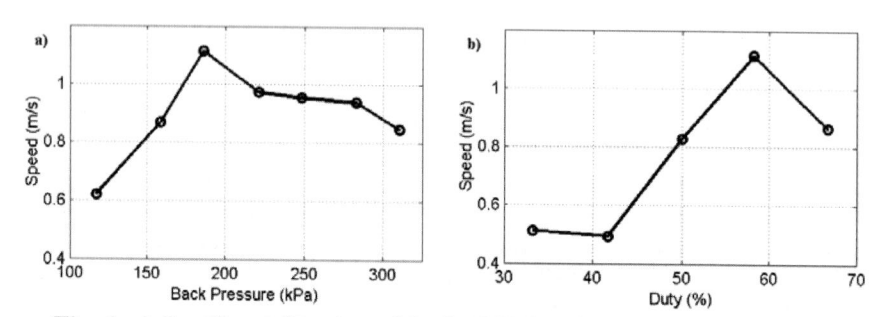

Fig. 4. a) Speed as a function of the back pressure, with an active pressure of 745 kPa. The leg thrusting period is 0.3 s and the duty cycle is 50 %.
b) Speed as a function of the duty cycle, with an active pressure of 745 kPa. The leg thrusting period is 0.3 s and the back pressure is 186 kPa.

Fig.4. a) shows the influence of the pressure that retracts the cylinders on the speed of the robot. To simplify the pneumatic system, the back pressure is always active. The difference between the back and the thrusting pressure should be maximized. However, a too low back pressure can't retract the legs efficiently, slowing the robot. This experiment showed that the optimum pressure for the greatest speed is around 180 kPa, but 220 kPa is more appropriate for better stability with less vertical oscillation of the center of mass.

Fig.4. b) shows the influence of the duty cycle on the speed of the robot. The leg duty cycle represents the percentage of time where each cylinder is commanded to be in extension. A too short duty cycle doesn't give the legs the time to push efficiently. With a too long duty cycle, the legs stay on the ground even after their useful pushing action and slow the robot. The best experimental duty cycle is around 58%.

The robot is also able to grasp a 10x10 cm block or to carry a 10 kg payload, 28% of its own weight, without significant speed loss. Climbing performances aren't as good as simulated; the actual robot is able to climb obstacles 10 cm high while 20 cm obstacles were climbed in simulation.

8 Future work

In addition to various ameliorations, further experiments based on the way insects turn [5] are being done. Climbing and the influence of the claws are analyzed in order to better understand the difference between simulations and experimental results and to improve the all-terrain performances. For now, beacon positioning and inertial guidance are under investigation.

9 Conclusion

There is a new generation of walking robots emerging that, unlike their slow, complex and heavy predecessors, rely on abstracted biological principles to be faster and more agile. There is a feeling that this relatively recent approach could lead to robots with better performances. The current work relies on studies of an insect, the cockroach, to design an agile and strong but yet a simple walking machine.

Initially validated by the use of dynamic simulations, this robot includes specialized-legged functions, passive visco-elastic elements and a self-stabilizing posture. This mechanical system allows straight line running with a simple open-loop feedforward control system. It was the fastest robot at the WMC, and work is underway to further optimize the prototype.

Acknowledgements

We would like to thank our sponsors: AGEG, Faculté de genie de l'Université de Sherbrooke, Numatics and MSC software. Many individuals also provided invaluable support and should be named: François Michaud, Michel Lauria, Richard Cloutier, Frédérick Rivard, Réjean Bernier, Yvon Turcotte, Pierre Savard and Marie-Josée Blackburn.

References

1. http://www.Mines.edu/fs_home/jsteele/wmc, [2002] CSM 2002 SAE Walking Machine Challenge.
2. Éric Lespérance, Alexis Lussier Desbiens, Marc-André Roux, Marc-André Lavoie and Philippe Fauteux, *Design of a Small and Low-Cost Power Management Unit for a Cockroach-Like Running Robot*, IROS 2005.

3. L. Ting, R. Blickhan, and R. Full, "Dynamic and static stability in hexapedal runners," *Journal of Experimental Biology*, no. 197, pp. 251-269, 1994.
4. T. M. Kubow and R. J. Full, "The role of the mechanical system in control: a hypothesis of self-stabilization in hexapedal runners," *Royal Society of London*, no. 354, pp. 849-861, 1999.
5. D. L. Jindrich and R. J. Full, "Many-legged maneuverability: dynamics of turning in hexapods," Journal of experimental biology, no. 202, pp. 1603-1623, 1999.
6. J. E. Clark, J. G. Cham, S. A. Bailey, E. M. Froehlich, P. K. Nahata, R. J. Full, and M. R. Cutkosky, "Biomimetic design and fabrication of a hexapedal running robot," *IEEE International Conference on Robotics and Automation*, May 2001.
7. R. Kram, B. Wong, and R. J. Full, "Three dimensional kinematics and limb kinetic energy of running cockroaches," *Journal of Experimental Biology*, no. 200, pp. 1919-1929, 1997.
8. M. H. Raibert, *Legged Robots that Balance*, M. Press, Ed., Cambridge, MA, 1986.
9. M. B. Binnard, *"Design of a small pneumatic walking robot"*, Master of Science, Massachusetts Institute of Technology, Jan. 1995.
10. Thomas J. Allen, Roger D. Quinn, Richard J. Bachmann, Roy E. Ritzmann, *"Abstracted Biological Principles Applied with Reduced Actuation Improve Mobility of Legged Vehicles"*, Case Western Reserve University
11. James T. Watson, Roy E. Ritzmann, Sasha N. Zill, Alan J. Pollack, "*Control of obstacle climbing in the cockroach, Blaberus discoidalis*". Case Western Reserve University, 2001.
12. U. Saranli, M. Buehler, D.E. Koditschek. *"RHex: A Simple and Highly Mobile Hexapod Robot."* The International Journal of Robotics Research 20 (2001) July 616-631.

Finding Adequate Optimization Criteria to Solve Inverse Kinematics of Redundant Bird Leg Mechanism

L. Mederreg[1], V. Hugel[1], A.Abourachid[2], P. Blazevic[1], R. Hackert[2]

[1]: Laboratoire de Robotique de Versailles (LRV), UVSQ, France
[2]: Muséum National d'Histoire Naturelle USM 302, France
mederreg@lrv.uvsq.fr ; abourach@ mnhn.fr

Abstract

In this paper we shall present the progress of the ROBOCOQ project. This project aims to design a prototype of autonomous biped based on the avian model, capable of exploring cluttered environments. This design relies on experimental kinematics data obtained from the quail. The experimental protocol used to reconstruct a 3D model is presented in this paper as well as the procedure and criteria employed to find an optimal inverse kinematics control.

Keywords: Kinematics, inverse geometric model, bird-like biped walk, trajectories, optimal posture.

1 Introduction

Birds and humans are the only two bipeds that colonized the planet. Birds can be found everywhere on the Earth in every kind of environment. Biologists of the MNHN believe that the locomotion system of birds presents interesting features and is more efficient than the human one in terms of stability, stride length and mobility [1][2][3]. A biped robot presents several advantages in comparison with other kinds of legged locomotion. Having only two legs allows a large polyvalence in terms of clearing obstacles such as stairs and cluttered environments. The concept of designing biped robots encounters a large interest among researchers those last decades. However many projects consider the human model for the robot structure design. Some scientists have tried to explore other types of biped configuration to investigate whether they can be more efficient that the human one. In that spirit the ROBOCOQ project [4] has started and aims to check the efficiency of the bird structure.

In this paper we describe the procedure followed to gather the necessary data, and the tools to make a kinematics analysis. The measurements carried out in the museum with real walking quails are exploited to reconstruct a 3D movement. A geometric model for the bird leg is proposed and calibrated using these measurements. Each leg of the bird features one degree of redundancy. Trajectories issued from the reconstructed 3D data are used to test some optimization criteria that are needed for solving the inverse kinematics model. In this paper we shall present some results of this analysis and give future work prospects.

2 Kinematics and Biological analysis

We conducted several experiments on the birds to understand the way they walk and they react in different environments and their morphology [5][6][7]. We noticed that the quail can adopt different postures depending on the nature of the ground and the environment. When filmed from the lateral side for example, the quail had a relative high-legged posture whereas when being filmed from the top, the posture was slightly low-legged. It was as if the animal was trying to protect itself from the bulky device placed above the walkway to capture video. For kinematics data acquisition small lead bullets were attached to the wanted points in the quail body to observe the motion trajectory of the different segments and joints of the bird during walk. Quails had to move on a walkway. An X-ray device coupled with a video-camera was used to collect the coordinates of the joints. These experiments provided us with two non simultaneous views, a lateral one and a dorsal one.

To overcome this drawback, we set up an experimental protocol to synchronize the two views and reconstruct the 3D cycle. The first step of the work is to filter the collected data. The second step consists of scaling the data so that the number of samples is the same for both signals. The signals must then be interpolated using FFT to get continuous signals or discrete values at the same sampling times [8] [9]. This procedure is described in a previous paper [10]. Once the two views are synchronized, we can easily reconstruct 3D data series of a walking cycle [10].

3 Data exploitation

3.1 Kinematics model

To study the bird locomotion we have focused on the structure of the leg and tried to get a faithful kinematics model. The scheme proposed in **fig.1.** shows the direct geometric model used. It is decomposed into 5 rotations named respectively $(\theta_1, \theta_2, \theta_3, \theta_4, \theta_5)$ starting from the hip. Three are located at the hip, one at the knee and one at the ankle. The three segments of the leg are contained in the same plane, that we call the leg plane. By convention, whatever the direction of motion we set the data so that the animal is walking from the left to the right. We also consider the right leg [11].

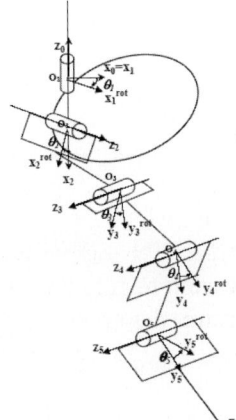

Fig.1. Kinematics scheme and convention of reference frames.

Axis x is along the direction of motion. Axis z is the vertical axis. Axis y is oriented toward the body.

The 3D direct geometric model expresses the 3D coordinates of the foot within the body reference frame as a function of the joint angles.

3.2 Evaluations of the joint angles during a walking cycle

Since our objective is to inverse the kinematics, it is useful to calculate the joint angles related to the kinematics model from the 3D biological data. As a matter of fact, we can get the model joint angle as a function of time and have an estimation of the range and the minimal and the maximum angles for each joint. These values can than be used as non linear constraints for the optimization issue (see § 4). We design a graphical animation interface based on OpenGL™ that enables to play the leg motion cycle and to zoom at it from different viewpoints. Before refreshing the next view, the program calculates all the model joints from the 3D coordinates of the biological joints. The algorithm used can be divided into two main parts. The first one concerns the knee and the ankle joints. The second part deals with the joint angles at the hip. There are n frames beginning with frame 1 taken at time t_0, frame k taken at time t_{k-1}, etc. The biological leg joints in frame k are designated by H_k, K_k, A_k, and F_k.

The 3D joints coordinates in frame k are stored respectively in: $\{(x_H^k, y_H^k, z_H^k),(x_K^k, y_K^k, z_K^k),(x_A^k, y_A^k, z_A^k),(x_F^k, y_F^k, z_F^k)\}$ starting from the hip. The rotation axes are noted z_i, where i designates the number of the rotation, starting from the hip. Five pairs of reference frames (R_i, R_i^{rot}) have been defined, for $i \in \{1,2,3,4,5\}$ such that we have:

$$R_i(x_i; y_i; z_i) \xrightarrow{rot(\theta_i)} R_i^{rot} \qquad (1)$$

An initial reference frame linked to the body is also defined as R_0 having as base: $\{x_0; y_0; z_0\}$ where x_0 and y_0 define the horizontal plane, and z_0 the vertical axis. The expression to find the joint angles of the knee and the ankle can be obtained by using the cross product and the scalar product of the normal to the first two segments, $H^k K^k$ and $K^k A^k$, and the normal to the last two segments, $K^k A^k$ and $A^k F^k$. The normal to the leg plane is the average of these two normals. For the hip we start by calculating θ_2^k. The trick is to express z_3^k inside the reference frame in two different ways. First we express z_3^k in R_3 and apply a rotation of $(-\theta_2^k)$. Second we have z_3^k expressed in R_0 because z_3^k is the normal to the leg plane, the normal being calculated by the average of two cross-products of both pairs of leg segments. We apply a rotation of θ_1^k to $z_3^k(R_0)$ in order to be in the same reference frame for both equations. Manipulating these two equations leads to a sine/cosine equation of the following form:

$$A.\cos(\theta_2^k) + B.\sin(\theta_2^k) = C \qquad (2)$$

Where A, B and C are scalars. θ_1^k doesn't appear here because we focus on the component of the vector z_3^k projected on z_1.

Solving this equation gives two solutions where one of them is physically impossible to reach. To find the remaining two joints angles of the hip namely θ_1^k and θ_3^k, a similar procedure must be adopted for both of them. From one of the previous equations used to find θ_2^k, it is possible to get θ_1^k and θ_3^k. The procedure consists of calculating the angle of rotation, knowing its axis, an initial and a final vector.

If we had a large panel of 3D coordinates data issued from measurements, it would be possible to calculate calibration tables of joint angles that can be used to control the motion of the leg. However it is better to design an algorithm capable of calculating the joint angles only from the foot 3D coordinates by solving the inverse kinematics.

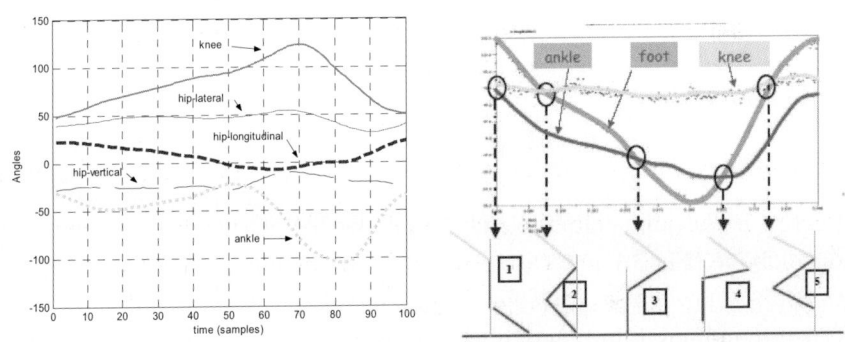

Fig. 2. Calculated joint angles variations

Fig. 3. x-coordinates of knee, ankle and foot as a function of time

Fig.2. presents the model joint angles variation during a walking cycle resulting from the developed algorithm and using a given 3D reconstructed data.

4 Solving the inverse kinematics model

This part deals with the procedure adopted to solve the inverse kinematics model problem. Indeed the leg of the bird features one degree of redundancy. For a given position of the foot we can have many possible postures. In the previous part we have used the 3D data to have an idea of the joint angles variations. We use this result to limit the reachable area of each joint. The program is developed under MATLAB R13 using some functions of the optimization tool-box. The inputs of the algorithms are:

- The 3D coordinates of the wanted foot position,
- Up and down limit angles of each joint,
- The expression of the direct geometric model,
- An initial position for each joint angle,
- An optimization criterion to be followed to reach the convergence.

The outputs are the computed joint angles that can be sent to the actuators that drive the motion.

In this program we exploit five specific configurations of the leg that are extracted from the longitudinal axis coordinates of all biological joint as a function of time. These configurations are: touchdown; middle of the propulsion; beginning of lift-off; end of lift-off; beginning of touchdown (see **fig.3.**). The program allows to perform two operations. The first consists of computing the joint angles of a given single position. It detects the nearest posture stated above and minimizes the following quadratic criterion.

$$C = \sum_{i=1}^{5} (\theta_i^k - \alpha_i^j)^2 \qquad (3)$$

Where i is the joint number, j is the number of the nearest configuration stated above (**Fig.3.**) and k is the number of the current iteration. α_i is the angle related to joint i and configuration j.

The second possibility is to input a trajectory of the foot as a vector. The program performs an optimization between 2 successive positions starting from the nearest posture. The optimization algorithm aims at minimizing the following quadratic criterion:

$$C = \sum_{i=1}^{5} (\theta_i^k - \theta_i^{k-1})^2 \qquad (4)$$

This procedure yields a smooth movement and avoids having a sharp variation of the posture.

5 Simulations and results

Fig.4. presents the results of the simulation that takes two known positions of the foot as inputs. The scheme compares the results with the joint angle trajectories issued from the calculated model (see section 3.2). We note that the hip is fixed because it has been chosen as a reference. We can notice that for each case the calculated postures fit almost the original ones. We have less than 10% errors for the worst cases of the simulated points. These errors may be due to some numerical approximations.

We tried to recover a full trajectory using the developed algorithms. As input we use a measured trajectory of the foot during the walking cycle. **Fig.5.** shows the positions errors for the coordinates of each computed joint. We can notice that even for trajectories the errors are bounded and never exceed 10% of the total length of the segments.

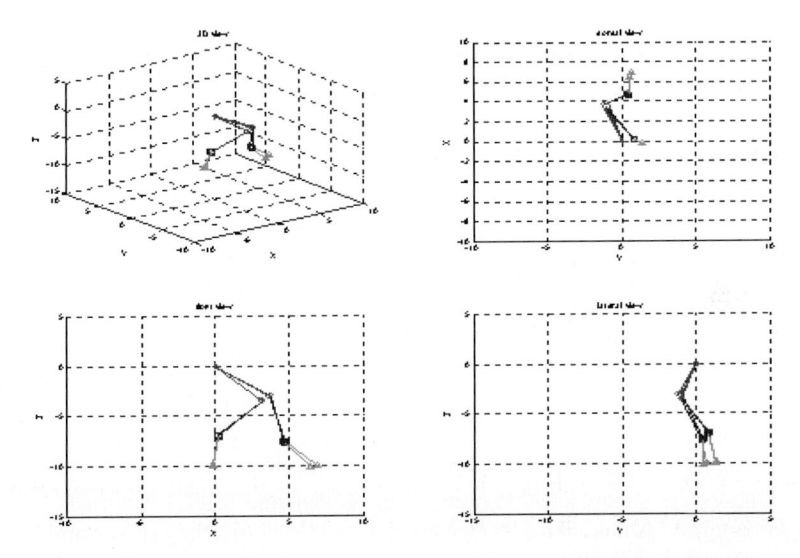

Fig.4. Comparative of the calculated position and the measured ones

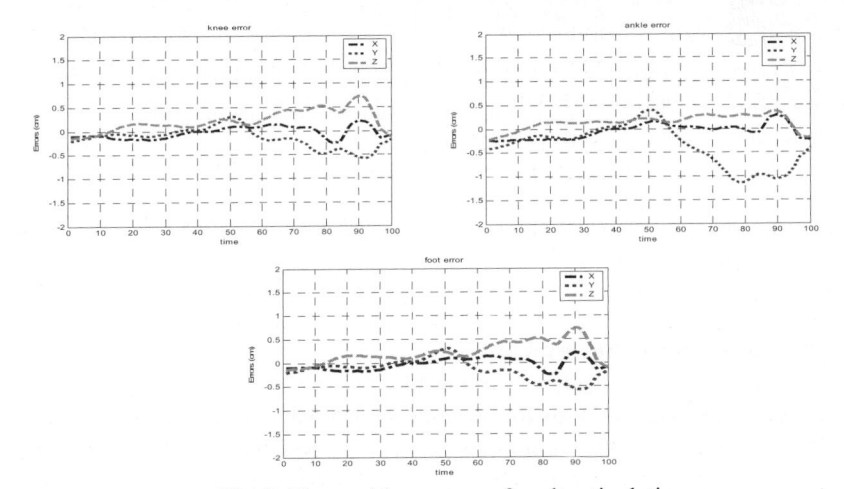

Fig.5. The position errors of each articulation

6 Conclusion

This paper describes the procedure and the tools used to perform a kinematics analysis of a bird leg motion. A solution is proposed to overcome the redundancy of this structure and to get the inverse geometric model.

The perspective of this work is to define a kinematics control in order to reproduce the 3D leg movement which will be decomposed into two parts:
- The movement of the leg plane which will be controlled by the normal to the plane.
- The movement of the leg segments inside the plane.

References

[1] A. Abourachid and S. Renous, 2000. Bipedal locomotion in ratites (Paleognatiform): examples of cursorial birds. *Ibis,* 142: 538-549.
[2] A. Abourachid, F.Lambert, A Msimanga and P.-Y. Gagnier. In press. Adaptations for walking on floating vegetation : the case of the jacana.
[3] A. Abourachid (2001) Comparison of kinematic parameters of terrestrial locomotion in cursorial (Ratites) swimming (ducks) and striding birds (quail and guinea fowl.) *J. Comp. Biol. Physiol.* A 131: 113-119.
[4] The RoboCoq Project (2003) *Modelling and Design of a Bird-like Robot Equipped with Stabilized Vision,* AMAM, Kyoto
[5] Congrès des Neurosciences françaises, Rouen mai (2003) Maurice, M., H. Gioanni, A. Abourachid, Influence du vol et de la marche sur la stabilisation du regard chez le Pigeon
[6] A Abourachid (2000) Bipedal locomotion in birds: importance of functional parameters in terrestrial adaptation in Anatidae. *Canadian Journal of zoology,* 78: 1994-1998
[7] A Abourachid (2001) Comparison of kinematic parameters of terrestrial locomotion in cursorial (Ratites) swimming (ducks) and striding birds (quail and guinea fowl.) J. Comp. Biol. Physiol. A 131: 113-119.
[8] F.Jedrzejewski (2001) *Introduction aux méthodes numériques, Chap 2, 31-61.* Springer, Paris.
[9] W. D. Stanley *Digital signal processing, Chap 10, 295-297.* Prentice-Hall Company.
[10] L. Mederreg, V.Hugel, A.Abourachid, P.Blazevic (CIFA 2004) *Reconstruction par traitement du signal du mouvement 3D d'une patte d'oiseau à partir de plusieurs vues 2D non synchronisées.*
[11] L. Mederreg, V. Hugel, A. Abourachid, O. Stasse, P. Bonnin, P. Blazevic (2003) The RoboCoq Project: Modelling and Design of Bird-like Robot. *6th International Conference on Climbing and Walking Robots, CLAWAR*, Catania – Italy September 17-19.

Integrated System of Assisted Mechatronic Design for Oriented Computer to Automatic Optimising of Structure of Service Robots (SIDEMAR)

C. Castejón[1], A. Gimenez[2], A. Jardón[2], H. Rubio[1], J. C. García-Prada[1], C. Balaguer[2]

[1]Mechanical Department, Universidad Carlos III de Madrid, Spain;
[2]RoboticsLab, Universidad Carlos III de Madrid, Spain {castejon,agimenez,ajardon,hrubio,jcgprada,balaguer}@ing.uc3m.es

Abstract

Service robotics is one of the main priority areas of research in the robotics and automation field. The application of these robots for service tasks (personal assistance, education, social tasks, etc.) has a lot of influence in their design [1], [2]. Several parameters will be crucial like the weight, the kinematics configuration, the layout of masses, etc. These service robots have innovative designs and structures that allow them to move in non structured environments, walking, climbing, etc. One of the main tasks of these robots is to interact and operate with humans. For this reason, their design and development methodology must be different to traditional robots. In this paper we present a new methodology to help the optimization of the process design. This design will allow selecting properly the optimal actuators, materials and the degrees of freedom. This methodology is being developed in a national project between two different areas of this university.

Keywords: Simulation and Design of CLAWAR, Design Modularity and System, Service Robotics.

1 Introduction

The main objective of this work is the development of a methodology that automates the service robot design process, (non available in the market). The design will take place from different points of view: electro-mechanical, control and structural. As a result of the application, this system will optimize several characteristics of the robot: the overall weight, the manoeuvrability, the number of degrees of freedom (DOF), the duty cycle, etc. The *SIDEMAR* system integrates several design tools with others to develop, making an environment easy to use. The used and tried tools are: the design and mechanics CAD synthesis, finite elements design and 3D simulation systems, etc. On the other hand, force calculation at the joints tools, thermal actuator design tools, and refinements mechanics tools will be developed. The integration of these tools will be with a user friendly environment, including the 3D simulation, with the final goal of optimizing the main characteristics of the service robots [3].

1.1 Design coefficient (DC)

The design coefficient allows joining variables related to the mechanism topology in a compact and clear way. The DC formula appears in equation (1).

$$DC = \frac{\text{reach} \cdot \text{DOF}}{\text{weight}} \qquad (1)$$

The coefficient relates the distance that a robot can reach, with the number of actuadores (DOF) and the total mass.

1.2 Iterative principle

The service robot design is an iterative process where the reduction of the weight structure leads the reduction of the torque joints, so actuators and auxiliary equipment (gearboxes, amplifiers, etc) can be lighter [4]. All of this process produces, again, a reduction of the overall robot weight, so we need less torque again at the joints, and so on. Which is the final step of this process? There is an optimal point where there is no possibility to reduce the overall weight. In this case, an optimal DC is said to be reached.

Following the method used till now, figure 1 demonstrates that there is a minimum for a certain number of iterations and from there, the robot weight returns to rise. Figure 1 data is taken of the design of the climbing robot ROMA2 [5]. With the new strategy, proposed in this work, a sequence of possible optimized designs will be provided, up to obtain the requirements established by the service robot.

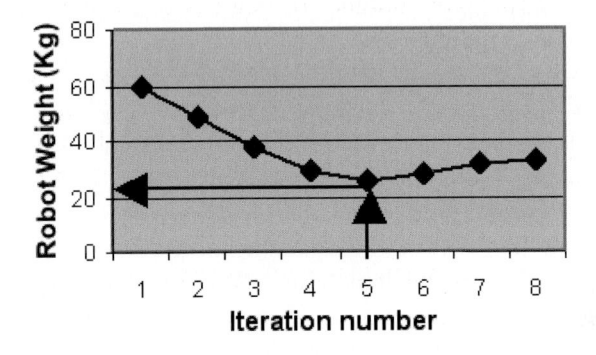

Fig. 1. Number of iterations developed for the ROMA2 design

As an example, in table 1, the DC and the basic characteristics of some industrial robots are presented.

Table 1. Conventional industrial robots characteristics.

Robot	DOF	Reach (mm)	Arm Weight (kg)	On-board control	DC
ABB Irb-2400	6	2000	300	NO	0,04
Kuka-KR 15/L6	6	1863	200	NO	0,05
Comau-S2	6	2208	300	NO	0,04
Mitsubishi PA-10	7	950	35	NO	0,19

All the work developed for the iterative principle is very tedious and takes a long time. The aim of this work is develop an electromechanical design integrated system which allows optimizing in an automatic way the process design of the mechanical structure and the actuator selection for a service robot. All these tasks are being developed with commercial tools of mechanical design (SolidWorks, MATLAB), and structure design (ALGOR), and our own actuator design tool developed using Visual .NET.

2 The methodology

The iterative process design, described above, can be divided in three different steps:

1. Mechanical design: synthesis, modeling, kinematics and dynamics simulation, and design.
2. Actuators selection, torque analysis and working temperature.
3. Structural design.

The design process to develop a prototype of a service robot is presented in this section. First, it is necessary to carry out a study of the environment and the task of the robot. Based on this study, the kinematics chain and the number of degrees of freedom are decided to solve the problem. The third step will be the obtaining of the robot preliminary design, with the different elements that form it. With these basic elements, the cycle continues with the preliminary design of the mechanical structure and the estimation of the robot weight. The following step is the actuators selection that fulfills the torque requirements in the joint, and then the total robot weight including actuators is re-calculated again. In this point, the static reactions can be calculated.

With this result the work factor is selected for every actuator, together with the thermal behavior.

2.1. Design cycle

In the service robots mechanical optimization cycle, the design specifications are developed based on the previous cycle experience, to generate the mechanism kinematics synthesis. In a first approach, kinematics simple chains are modelled depending on the requirements. After the model, a previous static analysis of the mechanism is performed, which consist of the study of links speeds and accelerations. This study provides the information needed for the actuator and reduction system selection, and for the calculation of the static efforts suffered by the structure. In the next step, the simulation of the mechanism is developed, the simulation will be both, kinematics and dynamics, where links will be considered as rigid

Integrated system of assisted mechatronic design for oriented computer to ... 331

kinematics and dynamics, where links will be considered as rigid solid (in a first approach) and with flexibility characteristics.

The figure 2 shows the design process to develop a service robot prototype. First a study of the environment and the application of the robot must be necessary. Based on this study, the kinematics chain, and the number of DOF can be decided to solve the problem. The third step will be to obtain the preliminary robot design, with the different elements that configure the robot. With these basic elements the cycle follows with the mechanical structure preliminary design and the robot weight estimation. The next step is the suitable actuators selection for the torque joints requirements and the re-calculation of overall robot weight including its actuators. From this point static torques can be calculated. With this result the duty cycle for each actuator is selected with the thermal behavior.

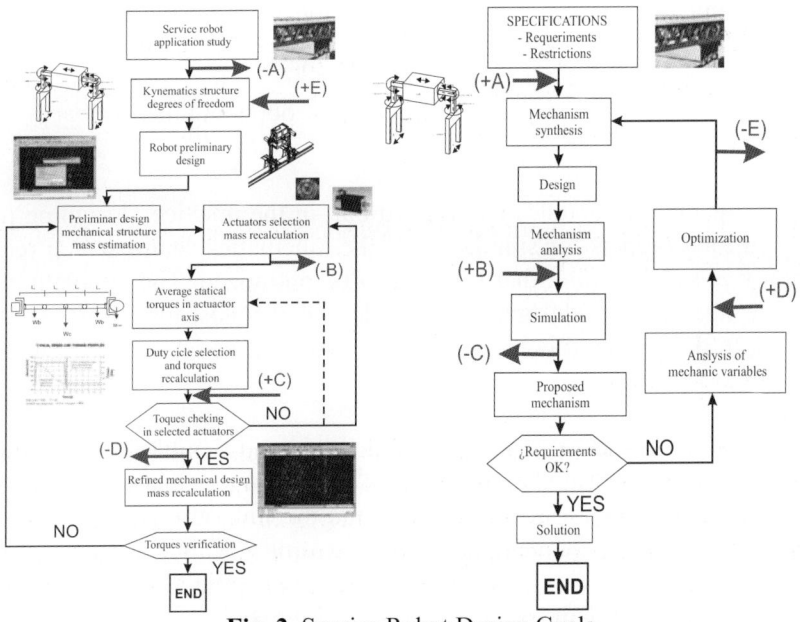

Fig. 2. Service Robot Design Cycle.

With these principal elements, the work of the mechanical structure preliminary design and the estimation of the robot weight are performed. Later, the commercial most adapted actuators to move the joints of the robot are selected. From here, the static torques that are produced in the worst robot positions/configurations can be calculated. These calculations can be done by programs that the actuators manufacturer provides, or by means of the behaviour equations.

In this point, it must be tested if the selected actuators are capable of providing the necessary nominal torque. If this statement is not true, then it is possible to take two alternatives: the shortest, which is to return and re-calculate the new nominal torques and based on then, to choose a new actuators duty cycle and, the second way, longer, returns to the point in which other actuators are selected and returns to re-calculate the whole weight of the robot. In case this statement were positive, it is possible to return and to refine the mechanical design, optimizing the weight in certain parts of the robot mechanical structure. Torques, with the refined design, are again tested and, in affirmative case the robot design process finishes. In negative case, return to the stage of the robot mechanical structure preliminary design.

The simulation result provides information of the static torques in the duty cycle. From here, the process of optimization is started, the specifications are evaluated and parameters corresponding to the electro-mechanical characteristics of the robot are modified up to obtaining the most appropriate robot.

This general methodology is carried out, in the practice, designing partial and total models for simulation of the kinematics chains which represent a real service robot that is going to be optimized. The simulation allows to optimize the designs in a simple and low cost time way, without putting in danger the robot integrity. The work of model and simulation is divided in two stages:

1.- Modelling and simulation of a link: rigid and flexible links are modelled by using MATLAB® and Simulink®, an open tool that allows to have a control of all the variables which define the link, and allows to obtain any system output, defined by a known equation. This part has been already performed, and has been presented in different works in the XVI Spanish National congress of Mechanical Engineering [6], [7], in León. Currently, a more precise model of flexible link, using the mechanics Lagrangian has been developed [8].

2.- Modelling and simulation of a kinematics chain: In this case, programmes of finite elements are used; the tools are SolidWorks® and Algor®. The tools allow to study the kinematics and the dynamics of a kinematics chains formed by the join of a set of links. The results obtained by means of the simulation of a flexible robot with these methods will have to be contrasted, in the first instance, with results measured of the real robot, later; results of the simulation will provide information to modify the geometries of the links and to choose different materials in order to optimize the design.

3 Conclusions

As principal conclusion, it is possible to say that, the kinematics analysis allows a simplified interaction between the mechanical and automation design groups. It will allow a better adequacy of the following goals: weight, torque and actuators selection for every joint. The first experiences performed with service robots, already designed in the university Carlos III of Madrid, state that this methodology reduces the robot design process time. Besides, the use and the interaction among different software tools described in this work make the process more precise and easy to use for the designers of service robots.

Acknowledgements

The authors gratefully acknowledge the funds provided by the Spanish Government through the MCYT Project DPI2003-084790-C02-01 and DPI2003-084790-C02-02.

References

1. Gimenez, A., Abderrahim, M. and Balaguer C., "Lessons from the ROMA1 inspection robot development experience", International Symposium on Climbing and Walking Robots (CLAWAR'01), Karlsruhe (Germany).
2. New trends of walking robotics research and its application possibilities, Tanie, K., 4th International conference on Climbing and Walking Robots, Karlsruhe, Germany, Sept, 2002.
3. Optimal design of manipulator with four-bar mechanism, Memertas, V. Mechanism and Machine theory, N° 39, 2004
4. Service robot applications for elederly people care in home environments, A. Giménez, A. Jardón, R. Correal, R. Cabas, C. Balaguer, 2nd

International workshop on advances in service robotics. Sttugart. Germany. Jun, 2004.
5. Climbing Robots for Inspection of Steel Based Infrastructures. Industrial Robot, C. Balaguer, A. Giménez, M. Abderrahim, Vol. 29. No. 3. pp.246-251. 2002.
6. Castejón, P.J. Lorca, J. Meneses, H. Rubio, L. Rubio, J.C.García-Prada "Modelado y simulación de una cadena cinemática de un grado de libertad, aplicación al diseño de robots". XVI Spanish National congress of Mechanical Engineering, december 2004, vol. 4, pp-2987-2993.
7. J. Lorca, J. Meneses, C. Castejón, H. Rubio, L. Rubio, J.C.García-Prada "Simulación paramétrica de un sistema mecánico flexible de un grado de libertad". XVI Spanish National congress of Mechanical Engineering, december 2004, vol. 4, pp-29995-2300.
8. C. Méndez, C. Castejón, "A one degree of freedom flexible arm modelling". Internal Report., Mechanical Eng. Dep., Universidad Carlos III de Madrid, January 2005.

The Construction of the Four Legged Prototype Robot ARAMIES

Jens Hilljegerdes, Dirk Spenneberg, and Frank Kirchner

Robotics Lab, Faculty of Mathematics and Computer Science,
University of Bremen, Bibliothekstr.1 28329 Bremen, Germany
hillje@informatik.uni-bremen.de

Abstract

This paper describes the mechanical design of the ARAMIES integration study, a four legged robot. It combines newly developed electronic, mechanical, and software components building a fully functional ambulating system. The mechanical concept aims at flexible kinematics and robustness of all components. Therefore, a modular construction kit with specially designed parts is used.

Keywords: Walking Robot, Four Legged, Modular Mechanical Design, Standard Joint, Actuated Foot.

1 Introduction

This paper presents the mechanical details of the integration study of the ARAMIES project[1]. This project is aimed at developing a four-legged robot capable of semi-autonomous operation at difficult sites. The reason to build such a system is that extra-terrestrial sites of high scientific interest are often located in areas of craters or similarly hard terrains. These areas are characterized by cliffs or steep slopes with uneven sandy or rocky soil. Due to the fact that conventional wheeled rovers are not able to provide access to these extremely difficult sites, the ESA and DLR jointly fund the ARAMIES project which started in april, 2004. The aim of the project is to design, build, and programm a system according to the requirements mentioned above. The concept will be tested on terrestrial sites resembling the extra-terrestrial areas of interest.

Fig. 1. The ARAMIES Integration Study.

In various tests, walking robots[2][3][4] have shown that ambulating systems can achieve fast and exceptionally robust mobility in rough outdoor environments.

Because there are no light-weight walking systems with the ability of moving up or down extremely steep surfaces, for example, the volcano explorer DANTE II[4] weighted more than 700 kg, we started to develop a new robot. Its integration study is depicted in **Fig. 1**. The mechanical and electronic design profitted strongly from our experiences with the SCORPION robot[2]. By moving from eight to four legs we intend to master better steep terrain because a four legged system provides more flexibility regarding its possible climbing behaviors. Furthermore, the ARAMIES robot is rather high, enabling it to overcome many of the obstacles typically found at rocky sites.

The integration study combines electronics and mechanics to a fully functional prototype. The robot is driven by 20 DC-motors which are controlled using a MPC565 micro-controller from Motorola and a XCV600 FPGA module from XILINX. For further details concerning the electronics, confer [2].

The rest of this paper is organized as follows. In Section 2 we describe the mechanics in detail. We discuss our design in Section 3 and conclude with an outlook to further developments in Section 4.

2 The Mechanics

The main principles underlying the design of our walking robot are flexibility in the mechanical setup and ease of maintenance. While the latter is desirable for any mechanical system, flexibility is a very important aspect

(a) Most frequently used Standard Joint.

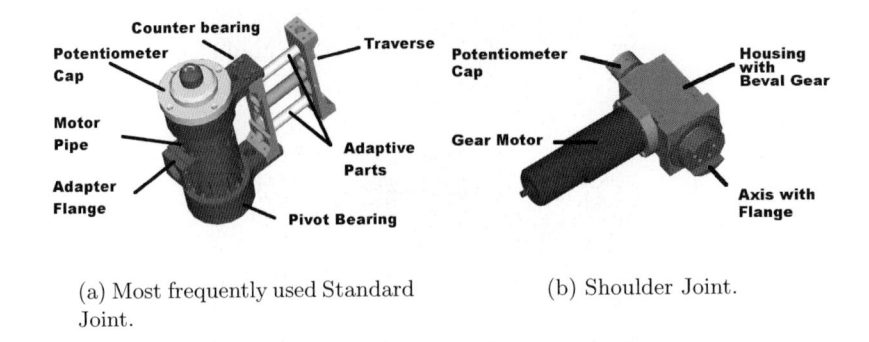

(b) Shoulder Joint.

Fig. 2. Two standard joints of the ARAMIES.

during the design of the extremities of our walking robot. Because of the demanding rocky and steep environment and regarding the fact, that no comparable light-weight system has been developed till now, the optimal setup of the kinematics has still to be found.

In order to achieve the previously mentioned flexibility we developed a construction kit to build different numbers of legs with different numbers and orientations of joints easily. The kit consists of three kinds of joints.

The first and most frequently used is depicted in **Fig.** 2(a). It is built of a pipe containing a Faulhaber DC motor attached to a planetary gear (ratio: 159:1; ∅: 38 mm). This combination drives a pivot bearing with a tappet at one side of the joint. At the other side a counter bearing moves the shaft of a built-in potentiometer. By this construction the absolute angular position of the joint can be measured independently of influences from slackness of the gear or the tappet.

Plastic flange sleeves moving on anodized aluminum are used as bearing.

This system is as light-weighted as the necessary cover against dust and sand allows. The SCORPION System shows that this concept of tribological pairing results in a robust and stable system with abrasion only on the plastic side.

The traverse is mounted to the counter and to the pivot bearing. It shapes a U which operates here as the driven limb. The points of connection are located on the motor pipe and on the traverse. Because of this, the standard joints can be adapted to each other easily. In combination with a limited number of adapters and distance parts, several "'extrem-

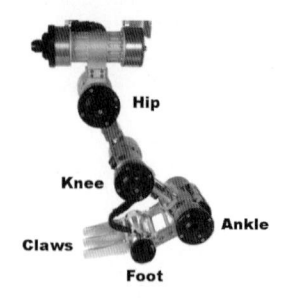

338 Jens Hilljegerdes, Dirk Spenneberg, and Frank Kirchner

Fig. 3. Current Leg Configuration inspired by Four Legged Mammals.

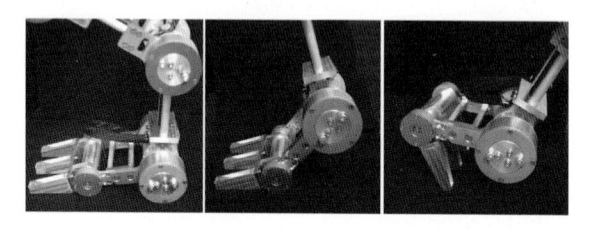

Fig. 4. The Active Foot can be Used for Walking as Well as Clinging to Steep Terrain.

ities"' with different lengths, DOF (degrees of freedom), and orientation to each other can be created as is shown in **Fig. 3**.

The figure also shows that we use a second kind of joint and a static claw to construct an active foot. The new joint represents the second standard joint which is copied from the SCORPION system. It is driven by a 22 mm DC motor combined with a planetary gear (ratio: 333:1). **Fig. 4** demonstrates that this actively controllable foot significantly increases the field of application. It cannot only be used for walking on flat surfaces but allows clinging to potentially slippery surfaces in steep terrain. This ability will be very important in the future, considering the difficult environments for which ARAMIES is developed.

Our third kind of standard component is a shoulder joint as shown in **Fig. 2(b)**. It can be placed between the corpus and a leg to further increase the DOF. Note that the shoulder joint is currently not integrated (cf. **Fig. 1**). The shoulder contains the same kind of DC-motor and a planetary gear (ratio 134:1) as the first standard joint but drives an axis perpendicularly to the motor axis by a bevel gear. The ratio of the latter can be adapted. Like the other joints the shoulder contains a potentiometer for position measurement.

The corpus of the robot is built of system strut aluminum profiles, allowing rapid and cost-effective adaptation of the body. Next to the extremities the corpus carries the electronics of the robot and the accumulators.

Length	700 mm
Width	450 mm
Height	600 mm
Weight	28 kg (including batteries)
38 mm Joint:	
Torque (cont./ peak)	8 (13) / 17 (26) Nm
Rotational Frequency	1 Hz
22 mm Joint:	
Torque (cont. / peak)	2 / 3 Nm
Rotational Frequency	0.5 Hz
Accumulator:	
Capacity / Voltage	9 Ah / 24 V
Power Consumption:	
Standby	$\sim$ 20 W
Average	$\sim$ 100 W
Peak (theoretical)	2800 W

Table 1. Technical Data of ARAMIES. The values in Parentheses Correspond to a New Motor Type we are Currently Testing.

In **Table 1** the mechanical details of the robot are listed for the currently used motor and also for a new motor option featuring twice the continuous torque. The peak torque values can be applied for at least one second, and the values for continuous torque express the maximum continuous load where thermal stability is still provided. The average power consumption corresponds to normal static walking on plain surface, while the peak power consumption is a theoretical value when all motors are blocked and are drawing their maximum current. The standby power consumption occurs when all motors are off but the electronic parts are still in operation.

Summarizing the mechanical design of the ARAMIES integration study, the adaptivity of the corpus and the modular construction kit of

(a) Position

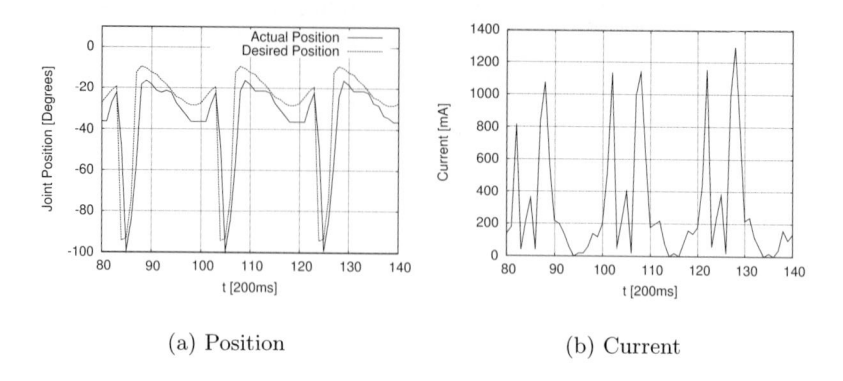

(b) Current

Fig. 5. Positions and drawn currents of the knee joint during walking.

standard joints provide ease of maintenance and flexibility in the design. Especially the latter is extremely important to be able to find an optimal kinematical configuration for walking through extremely hard terrain.

The currently ongoing work on the first integration study is to improve stability and weight. This also includes minor changes in electronics and cabling. The further development will discussed in Section 4.

3 Discussion

The preliminary tests with the integration study showed already in the beginning that upscaling the components used in the SCORPION project to increase the weight to torque ratio is as important as the kinematical setup of the legs and the used walking software approach. First experimental data shows that the current consumption during first walking tests on flat terrain is mostly below the specified continuous allowed current threshold of 1.2 A. However, sometimes the current drawn at the knee joint (see **Fig.** 5(b)), which performs most of the work during walking, has peaks above or near this threshold. All other joints stay below it (more data can be found in [5]). **Fig.** 5(a) shows the corresponding angular positions of the joint.

The peaks can be expected in normal walking and will inevitably become higher in rough terrain. In the worst case the power consumption will go up to its maximum of about 4 A. This cannot be avoided by changing only the mechanical design. Therefore, precautions have to be

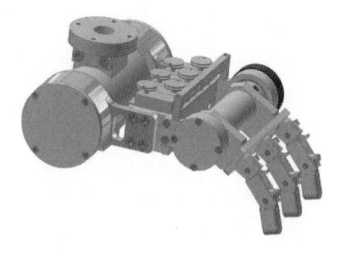

Fig. 6. Foot with Actuated Claws

integrated into the software to realize motion within the mechanical and thermal specifications.

Moreover, for the heavily used joints, especially the knee, we plan to replace the current motor with another motor which has almost the double rate of nominal torque and twice as much allowed continuous current (2,2 A). We will only use these motors in the most stressed joints because the increase in torque comes with the cost of increasing the weight by 60 g. The new motor is slightly longer than the old one but is easy to integrate by using a spacer attached to the potentiometer cap.

Another issue that needs to be discussed is the cabling. Because of our electronical architecture the cable harness is routed through all joints. Currently, we have a direct cabling of all actuators to the motor sensor board. Thus, the cable harness starts with 30 wires at the shoulder unit and ends up with eight wires in the ankle. This means that a reconfiguration of the leg can result in a revision of the cabling. The usage of custom and advanced plugs to avoid this problem is not practical because of extra weight, extra installation space, and the cable routing. In addition, problems with contact resistance and strain relief are arguments against a different cable routing. Consequently, we will keep the current cabling as a good compromise between easy maintenance and robustness of the system.

4 Outlook

The first prototype robot based on the experiences gained with the integration study will be assembled at the end of 2005. The First tests with the integration study showed that parts of the legs can be optimized by weight and function. A second major goal is development of a closed and lighter body for the prototype in contrast to the strut profile frame of the integration study.

Subsequently, we plan to redesign is the foot device to make it even more adaptive to the expected undergrounds. Already in progress is a design which includes actuated claws. With this feature the ground contact and the clinging in steep slopes can be enhanced. A preliminary design study of this concept shown in **Fig. 6**. In this study the claws have two joints driven by cable. Here, the challenge is to design a robust system and still keep the weight of the feet low, because they are the last element in our leg.

Acknowledgments

The presented work is sponsored by the Deutsches Zentrum fuer Luft und Raumfahrt (DLR grant no. 50JR0561) and European Space Agency (ESA contract. 18116/04/NL/PA). Furthermore we like to thank all members of the Robotics Lab for their contributing work.

References

1. Spenneberg, Dirk, Kirchner, Frank, and de Gea, Jose (2004) Ambulating robots for exploration in rough terrain on future extrateresstial missions. In *Proceedings of the 8th ESA Workshop on Advanced Space Technologies for Robotics and Automation (ASTRA 2004)*. European Space Agency.
2. Spenneberg, Dirk, McCullough, Kevin, and Kirchner, Frank (2004) Stability of walking in a multilegged robot suffering leg loss. In *Proceeding of ICRA 04*, vol. 3.
3. Buehler, M., Saranli, U., Papadopoulos, and Koditschek, D.E. (2000) Dynamic locomotion with four and six-legged robots. In *Int. Symp. Adaptive Motion of Animals and Machines*.
4. Bares, J. E. and Wettergreen, D. (1999) Dante ii: Technical description, results, and lessons learned. *Internat. Jornal of Robotics Research*, vol. 18, no. 7.
5. Spenneberg, Dirk, Albrecht, Martin, Backhaus, Till, Hilljegerdes, Jens, Kirchner, Frank, Strack, Andreas, and Zschenker, Heiko (2005) Aramies: A four-legged climbing and walking robot. In *to appear in Proceedings of 8th International Symposium iSAIRAS*, Munich.

Application of Waves Displacement Algorithms for the Generation of Gaits in an All Terrain Hexapod

Alejandro Alonso-Puig

Quark Robotics S.R.L. - Spain.
www.QuarkRobotics.com
alejandro.alonso@quarkrobotics.com

Abstract

Gait generation for legged robots has been studied since decades. Different ways of generating them for rough terrain have been proposed with more or less success. Here we propose a method for gait generation based on triple periodic functions, which parameters are automatically altered to ensure a smooth walking over rough terrain.

Keywords: gaits, hexapod, rough terrain.

1 Introduction

Since decades legged vehicles have been identified as the way of overcoming obstacles [1], [2], [3], [4] instead of other ways like using wheels or treads. Although many studies have been published about the gait generation for legged robots in rough terrain, the problem of making robots walk smoothly over any kind of terrain is still under consideration. In this work we will explain a gait generation method applied to an hexapod that will give to it the capability of walking over rough terrain, adapting the gait to the surface while the body of the robot continue moving with no alteration on its linear speed, direction and elevation.

This method has been applied to the robot Melanie-III [5] developed by the author, demonstrating its practical results.

The gait explained in this document is in base a periodic gait, that is altered in order to adapt it to the obstacles [6]. Therefore we get a sort of free gait [7] while walking over rough terrain, that becomes periodic when the robot walks on flat terrain.

2 The periodic gait generation algorithm

The algorithm is based on the determination of the point *(x,y,z)* in the space of the extreme of each leg (see **Fig. 1.**).

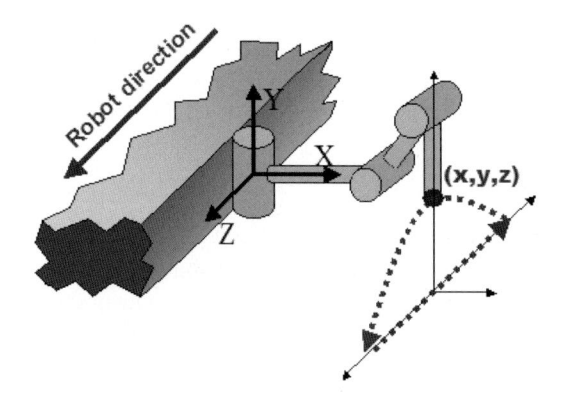

Fig. 1. Leg movement

The generation of such points is based on the displacement of three functions in time. Each function is defined as *f1(t)*, *f2(t)* and *f3(t)* that determines the value of each of the coordinates *x*, *y* and *z*.

The functions are all of periodic sort and may be altered in phase, amplitude and wave length. Therefore the equation (1) of movement could be defined as the matrix:

$$f(t) = (f1(t), f2(t), f3(t)) \qquad (1)$$

Such function gives the position *(x,y,z)* in the space of the extreme of each leg.

As it was our intention to implement a system as simple as possible, we decided to use a function that simply moves each leg backwards in parallel to the body when the leg is touching the floor and then trace an arc with the leg to the front (see **Fig. 1.**).

One of the function that makes this movement possible is one were:

- The x component is always constant in order to make the extreme of the leg move always parallel to the body (see **Fig. 2.**),
- The y component is a half sinusoidal wave, so it traces an arc when going to the front, while maintaining the position when moving backwards (see **Fig. 2.**) and
- The z component is a sinusoidal wave that makes the leg move from back to front and front to back (see **Fig. 2.**).

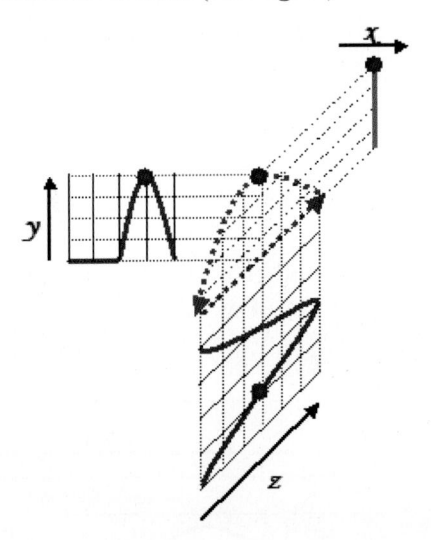

Fig. 2. Function of movement

At each moment t in time it is calculated the result of the three functions, obtaining the coordinates of the extreme of the leg. Then inverse kinematics algorithms are applied [8] in order to move the three motors of each leg to the correct angle.

Modifying the phase out between legs, and therefore between their functions, and changing the shape of the wave it is possible to get different gaits for the robot to walk. Again, as we are looking for the simplest way of implementing and demonstrating the method, we have used the tripod gait, which is obtained just ensuring a phase out of 180° in three legs in respect to the other three (see **Fig. 3.**).

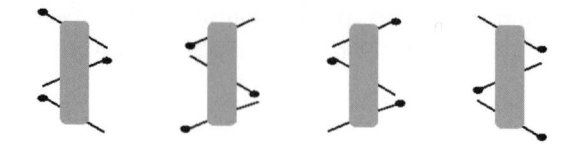

Fig. 3. Steps using tripod Gait

Up to now this method allow us to make the robot walk in flat terrain with a tripod gait, although we have the possibility to change the parameter to walk with different gaits. In next section we will explain how to use this method also to walk on rough terrain.

3 Rough terrain gait algorithm

As explained in the previous section, the coordinates of the legs are obtained from three functions (see **Fig. 4.**).

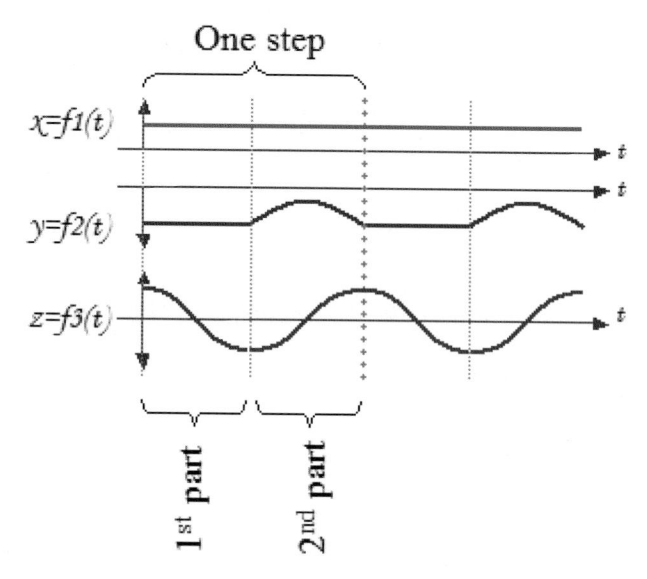

Fig. 4. Three coordinate generator functions

Varying the phase out between the waves of the different legs and the relation between the length from the first part and second part of the cycle of the wave different gaits are obtained.

Application of WDA for the generation of gaits in an all terrain hexapod 347

The first part of the cycle corresponds to the moment at which the end of the leg is touching the ground and moving backwards (body advancing), whereas the second part of the cycle corresponds to the elevation and displacement ahead in arc of the leg (positioning for a new step). Thus for example in the previous figure, the first and second part of the cycle is of the same length (in time), so it will take the same time in moving backwards the leg than in positioning it again ahead.

The robot is prepared for the detection of obstacles on the ground and to adapt the gait to walk over these obstacles. The way it is done is by measuring the pressure sensors values at the moment each leg is going down to put its end on the ground. If the pressure measurement is over a given threshold, the leg stops a fraction of second, its position is obtained and the parameters of the functions are suitably altered to overcome the obstacle (see **Fig. 5.**).

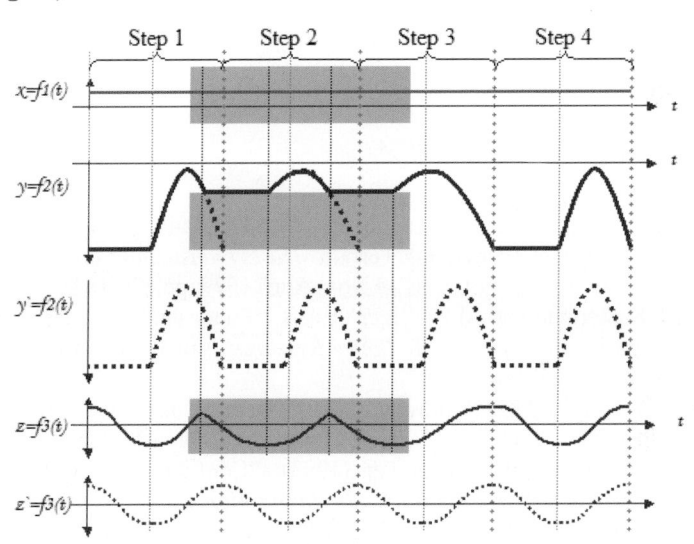

Fig. 5. Functions altered by an obstacle (grey area)

f1(t) is not altered and remain constant as in any case the leg will continue moving parallel to the body.

f2(t) and *f3(t)* are affected by the obstacle (grey area). The obstacle of the figure is long enough to affect to consecutive steps. As it could be seen, the obstacle is reached in the second part of the first step. This means, when the leg was moving forward in arc and going down. Therefore the evolution of the second part of the cycle could not continue and hence the leg start moving backwards over the obstacle (first part of the next cycle hence next step). The first part of a cycle is always of the same length because this ensure that all legs move at the same speed while they are on the ground, so the robot could continue moving smoothly with no slip of the legs. After the first part of the cycle is complete, the second part of the cycle starts, but this time the wave length is increased to make possible the leg to go to it's original position if no obstacle is detected. In the case shown in the figure, an obstacle is again detected by the leg, so the process is repeated. After the obstacle is overcome, the leg goes to the original position and the parameters of the functions are normalized, with no disruption in the continuity of the waves during the transition.

x, y, z show the functions affected by the obstacles while x, y' and z' show how would evolve the functions in time if no obstacle would have been detected, so it is possible to compare both cases in one graph. The result is that applying this algorithm, the robot could walk over many kind of complex surface.

References

1. R.B. McGhee and G.I.Iswandhi (1979): "Adaptative Locomotion of Multilegged Robot over Rough Terrain" *IEEE Transactions on Systems, Man and Cybernetics*, Vol. SMC-9 No4, April 1979, pp.176-182
2. I.E.Sutherland and M.K.Ullner (1984): "Footprints in the Asphalt", *The International Journal of Robotics Research*, Vol.3, No.2, Summer 1984, pp. 29-36
3. S-M. Song and K.J.Waldron (1989): "Machines that walk: The adaptive suspension vehicle". MIT Press, Cambridge, Massachusetts.
4. J.E.Bares and W.L.Whittaker (1993): "Configuration of Autonomous walkers for extreme terrain". *The International Journal of Robotics Research*, vol. 12, No. 6, December 1993, pp. 535-559.
5. URL http://www.mundobot.com/projects/melanie/v3/enmelanie3.htm
6. D.Wettergreen, C.E. Thorpe, (1992) "Gait Generation for Legged Robots". *Proceedings of the IEEE International Conference on Intelligent Robots and Systems, 1992, pp 1413-1420.*
7. M. Dohi, T. Fujiura, A. Yano. (2002) "Hexapod Walking Robot for Agricultural field (I) - Mechanism of leg and gait control". Pp. 196-203 in *Automation Technology for Off-Road Equipment, Proceedings of the July 26-27, 2002 Conference* (Chicago, Illinois, USA)
8. A.Ollero Baturone. (2001) "Manipuladores y robots móviles". Ed. Marcombo. Pp 74-85.

Extensive Modeling of a 3 DOF Passive Dynamic Walker

M. A. Roa, C. A. Villegas and R. E. Ramírez

Department of Mechatronics and Mechanical Engineering, National University of Colombia   maroag@unal.edu.co

Abstract

Passive dynamic walkers can achieve a cyclic gait behavior basically under the influence of gravity. Different researchers have studied a number of passive walkers, but we still lack an extensive numerical modeling for the kneed passive walker, a walker with three degrees of freedom. This article shows preliminary results on the modeling of the kneed passive dynamic walker and compares the model results with known experimental facts from the human gait analysis.

Keywords: passive dynamic walk, kneed passive walker.

1 Introduction

Human walk can be understood as the influence of a neuromotor control system acting on a mechanism moved basically under the influence of gravity (mainly in the swing phase), as stated by Mochon and McMahon [1]. McGeer [2] demonstrated by numerical and physical simulations that some mechanisms could achieve bipedal walking on small slopes without actuators or control systems. These passive systems can generate stable walking patterns just by gravitational effects; besides, these motions are natural and energy-optimal. Research on such motions may lead us to produce better mechanical designs and control strategies for active walking machines.

Since McGeer [2] first studied the passive walk through simple models, different researchers have achieved dynamical insights on these models

[3], [4]. García [4] studied the simplest walking model, a double pendulum with a punctual mass at the hip and two smaller point masses at the feet. Goswami [3] presented the extensive modeling for the passive walk with two degrees of freedom, the straight legged walker (exhibiting the so-called compass gait) by choosing some adimensional numbers. However, to the best of our knowledge, we lack the same approach to model the passive dynamical walker with three degrees of freedom, the kneed passive walker. We present the simulation process and resume the preliminary results and first insights into the limit cycles for the kneed-walker. Also, we compare the model results with known experimental facts for human gait.

The rest of the paper is organized as follows. Section 2 introduces the dynamic model for the kneed passive walker. Section 3 describes the modeling process, i.e. the normalized equations and the simulation sequence. Finally, Section 4 shows the main results of the modeling process.

2 Dynamic model

We study the passive kneed-walker, original of McGeer [2] and also used by other authors, such as Yamakita [4]. The model has three links (stance leg, thigh and shank), and four punctual masses (each link has a concentrated mass, and an additional mass at the hip); we use punctual feet (round and without mass). Fig. 1 shows the notation used in the model: link 1 is the stance leg, link 2 is the thigh and link 3 is the shank; m_1, m_2, m_3 and m_c are the stance leg, thigh, shank and hip masses, respectively; a and b denote the distal and proximal distances to the punctual mass in each of the links; θ denotes the link angles with respect to the vertical, and γ is the slope angle.

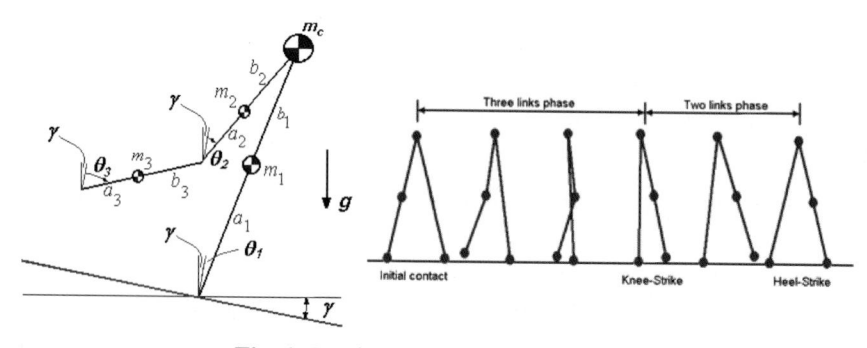

Fig. 1. Passive walking model with knees.

Fig.1 also shows the diagram of a gait cycle. The gait cycle is started when both feet are on the ground. The swing leg (thigh and shank) begins to swing until the knee strike, when thigh and shank are aligned and begin to move as a single link (preventing hyperextension in the knee). This is the beginning of the two-link phase; the walker behaves as a compass-gait walker until the swing leg touches the ground in the heel strike. At this point, swing leg and stance leg are interchanged and a new cycle begins.

Dynamic equations and transition conditions for the model can be found in [5] and [3]. Their matrix form is

$$D(\theta)\ddot{\theta} + H(\theta,\dot{\theta})\dot{\theta} + G(\theta) = 0 \quad (1)$$

with $D(\theta)$ the matrix of inertial terms, $H(\theta,\dot{\theta})$ the matrix with coriolis and centripetal terms and $G(\theta)$ the vector of gravitational effects. There are two transition events: the knee strike and the heel strike. Knee strike can be included in the above differential equation by using an artificial restriction λ_r.

$$D(\theta)\ddot{\theta} + H(\theta,\dot{\theta})\dot{\theta} + G(\theta) = -J_r^T \lambda_r \quad (2)$$

$$\lambda_r = -\chi_r^{-1} J_r D(\theta)^{-1}(H(\theta,\dot{\theta})\dot{\theta} + G(\theta)), \quad \chi_r = J_r D(\theta)^{-1} J_r^T \quad (3)$$

Thus, the same dynamic model can be used for the three links phase (before the knee strike) and the two links phase (after the knee strike). The transition equations for the heel strike are

$$\begin{pmatrix} m_c l_1^2 + m_1 a_1^2 + m_1 l_1 (l_1 - b_1 \cos 2\alpha) & m_1 b_1 (b_1 - l_1 \cos 2\alpha) \\ -m_1 b_1 l_1 \cos 2\alpha & m_1 b_1^2 \end{pmatrix} \begin{pmatrix} \dot{\theta}_1^+ \\ \dot{\theta}_3^+ \end{pmatrix} = \begin{pmatrix} (m_c l_1^2 + 2m_1 a_1 l_1)\cos 2\alpha - m_1 a_1 b_1 & -m_1 a_1 b_1 \\ -m_1 a_1 b_1 & 0 \end{pmatrix} \begin{pmatrix} \dot{\theta}_1^- \\ \dot{\theta}_3^- \end{pmatrix} \quad (4)$$

and this event happens when $\theta_1^- + \theta_3^- = 0$.

3 Extensive modeling

In order to make an extensive simulation of the model, we normalize the dynamic and transition equations using the following adimensional parameters:

Mass numbers:

$$\mu = \frac{m_c}{m_1}, \quad \mu_1 = \frac{m_2}{m_1}, \quad \mu_2 = \frac{m_3}{m_1} \tag{5}$$

Length numbers:

$$\beta = \frac{b_1}{a_1}, \quad \beta_1 = \frac{a_2}{a_1}, \quad \beta_2 = \frac{b_2}{a_1}, \quad \beta_3 = \frac{b_3}{a_1} \tag{6}$$

However, not all of these numbers are independent; because of the symmetry of the walker there are some restrictions:

$$m_1 = m_2 + m_3, \quad l_1 = l_2 + l_3, \quad b_1 = \frac{m_2 b_2 + m_3 (l_2 + b_3)}{m_2 + m_3} \tag{7}$$

Normalized equations for the dynamical model are

$$m_1 a_1^2 \left[D_n(\theta)\ddot{\theta} + H_n(\theta,\dot{\theta})\dot{\theta} + \frac{1}{a_1} G_n(\theta) \right] = -J_r^T \lambda_r \tag{8}$$

with

$$D_n(\theta) = \begin{pmatrix} 1+(\mu+\mu_1+\mu_2)(1+\beta)^2 & -(1+\beta)[\mu_1\beta_2+\mu_2(\beta_1+\beta_2)\cos(\theta_1-\theta_2)] & -\mu_2\beta_3(\beta+1)\cos(\theta_1-\theta_3) \\ -(1+\beta)[\mu_1\beta_2+\mu_2(\beta_1+\beta_2)\cos(\theta_1-\theta_2)] & \mu_1\beta_2^2+\mu_2(1+\beta)^2 & \mu_2\beta_3(\beta_1+\beta_2)\cos(\theta_2-\theta_3) \\ -\mu_2\beta_3(\beta+1)\cos(\theta_1-\theta_3) & \mu_2\beta_3(\beta_1+\beta_2)\cos(\theta_2-\theta_3) & \mu_2\beta_3^2 \end{pmatrix} \tag{9}$$

$$H_n(\theta) = \begin{pmatrix} 0 & -(1+\beta)(\beta_2+\mu_2\beta_1)\sin(\theta_1-\theta_2)\dot{\theta}_2 & -\mu_2\beta_3(\beta+1)\sin(\theta_1-\theta_3)\dot{\theta}_3 \\ -(1+\beta)\beta_2+\mu_2\beta_1\sin(\theta_1-\theta_2)\dot{\theta}_1 & 0 & \mu_2\beta_3(\beta_1+\beta_2)\sin(\theta_2-\theta_3)\dot{\theta}_3 \\ \mu_2\beta_3(\beta+1)\sin(\theta_1-\theta_3)\dot{\theta}_1 & -\mu_2\beta_3(\beta_1+\beta_2)\sin(\theta_2-\theta_3)\dot{\theta}_2 & 0 \end{pmatrix} \tag{10}$$

$$G_n(\theta) = \begin{pmatrix} -[1+(\mu+\mu_1+\mu_2)(1+\beta)]g\sin(\theta_1+\gamma) \\ [\mu_1\beta_2+\mu_2(\beta_1+\beta_2)]g\sin(\theta_2+\gamma) \\ \mu_2\beta_3 g\sin(\theta_3+\gamma) \end{pmatrix} \tag{11}$$

The model is finally described by the set of differential nonlinear equations (8), along with the normalized heel-strike transition equations (4).

The previous model is solved in Matlab using a 4th order Runge Kutta routine with numerical error of 10^{-6}. The code implemented solves the equivalent system of first order equations for the balancing dynamics. It checks for the knee strike event; from this moment on, λ_r is calculated (λ_r is zero before knee strike), thus including the blocking effect of the knee in the dynamic model. The system is solved until heel strike; at this point, heel strike transition equations are applied and a new gait cycle begins.

In order to get a stable motion, the joint variables (displacement and velocity) must follow a cyclic trajectory. Such trajectory can be found with appropriate initial conditions at the start of one step. The characteristics of the gait (velocity, time period, step length) depend on the geometry and inertial properties of the robot and the slope of the plane. We choose arbitrarily ranges for the adimensional numbers and solve repeatedly the equations to find the stable limit cycle for each particular walker. Results for each particular model include the angle and velocity progression for each link vs. time (Fig.2) and the phase planes for each of the links (Fig.3); the example uses the same numerical values as in [5]. Stability is verified through the jacobian eigenvalues, as described in [8].

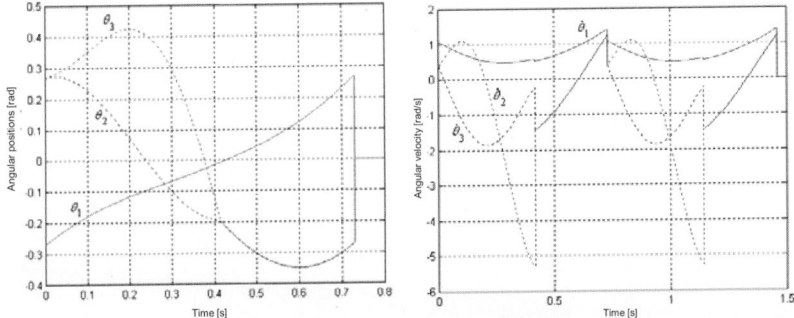

Fig. 2. Angles and angular velocity vs. time.

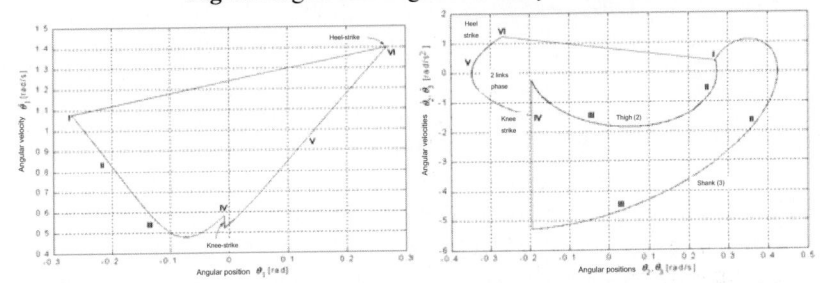

Fig. 3. Phase planes for the three links. Left: link 1. Right: links 2 and 3.

4 Simulation results

The extensive simulation of the dynamical model gives some insight of the dynamic behavior of the passive walker, and establishes some interesting limits to get a periodic limit cycle.

First of all, the dynamical model shows a cascade of periodic doubling bifurcations, as previously verified for the compass gait [3]. Figure 4 shows this behavior for the same numerical values used in the previous example; graphs show 1, 2, 4 and 8 period cycle limits in the periodic gaits for the step length and velocity of progression as a function of the slope angle. As the angle increases, the gait turns out to be chaotic (not shown in the figure). It is interesting to note that although the walker is totally symmetric, we can get asymmetric gaits, similar to limping in humans, perhaps suggesting that limping is a natural behavior in humans due to natural dynamics implicit in the walking mechanism.

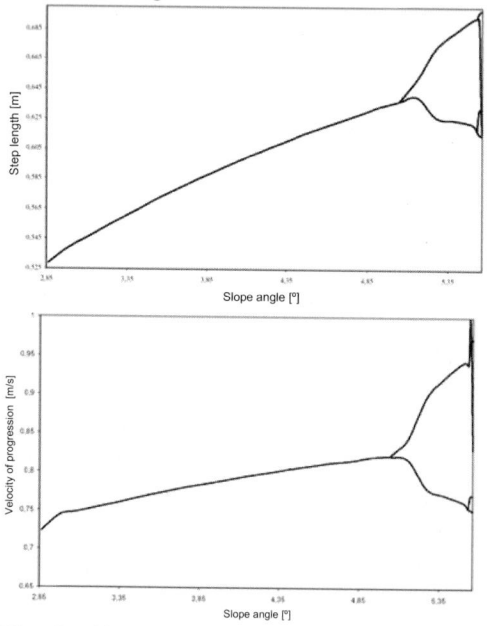

Fig. 4. Bifurcation diagrams for the kneed walker.

Figure 5 shows the influence of the adimensional parameter μ in the time and length of a step, also as a function of the slope angle. Perhaps more interesting are the limits in the mass relations to get a stable limit cycle:

$$m_c > m_2 + m_3 \,,\, \mu > 1 \,,\, m_2 \geq m_3 \,,\, \mu_1 \geq \mu_2 \quad (12)$$

Additional restrictions can be imposed on the model lengths when considering adimensional length numbers in order to get a stable limit cycle:

$$a_2 \geq b_2 \,,\, \beta_1 \geq \beta_2 \,,\, a_3 \geq b_3 \quad (12)$$

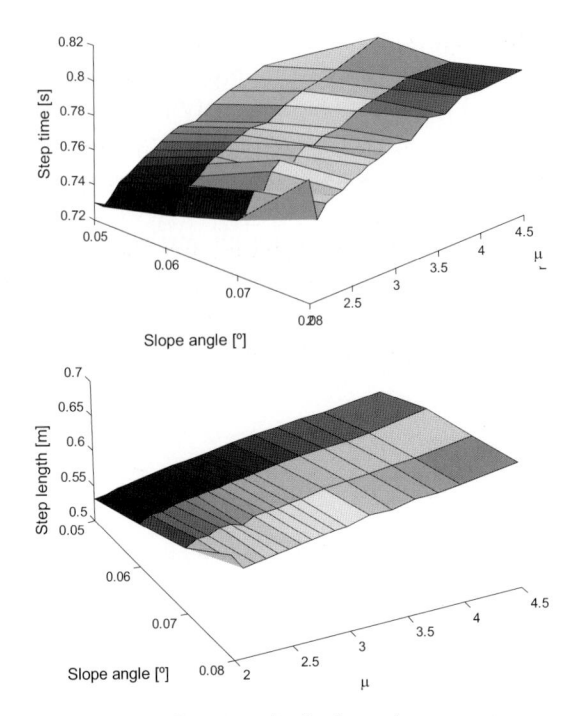

Fig. 5. Influence of μ in the gait parameters.

Goswami [2] probed some relations between the gait parameters between two models with similar mass and length proportions in the compass gait. For example, the walking features are identical between two models with the same mass proportion μ. Moreover, the features for a walker with length proportion $\beta = b_1'/a_1'$ can be compared with the features for other walker with the same ratio following the relations in Table 1, based on the scalar $k_a = a_1/a_1'$.

Table 1. Parameter relations for models with the same ratio β.

Model with lengths a_1' and b_1'	Model with lengths a_1 and b_1
θ	θ
$\dot{\theta}$	$\dfrac{1}{\sqrt{k_a}}\dot{\theta}$
L	$k_a L$
T	$\sqrt{k_a}\, T$
v	$\sqrt{k_a}\, v$

As an interesting result, the same relations hold for the kneed walker. If two kneed passive robots have the same proportions (i.e. the same adimensional mass numbers) the robots have exactly the same walking behavior. Also, relations resumed in Table 1 hold for the kneed walker; modifications in the length of link 1 keeping the same proportion with a reference model affect all the gait parameters by a scale factor, except for the angle progression vs. time. This fact gives an interesting insight into one of the supporting hypotheses in human gait analysis: that every person, no matter his height, and in consequence, his limb length (considering a normal biotype), describes the same angle trajectory for each link in his gait cycle.

Acknowledgements

This work was partially supported by the DIB-UN project 308005.

References

1. S. Mochon and T. McMahon (1980). Ballistic Walking: An Improved Model. *Mathematical Biosciences*, Vol. 52, pp 241 – 260.
2. T. McGeer (1990). Passive Dynamic Walking. *Int. J. of Robotic Research*, Vol. 9, No.2, pp 62 – 82.
3. A. Goswami, B. Thuilot and B. Espiau (1996). *Compass – Like Biped Robot. Part I: Stability and Bifurcation of Passive Gaits*, Research Report, INRIA (Institut Nacional de Recherche en Informatique et en Automatique).
4. M. García, A. Chatterjee, A. Ruina and M. Coleman (1998). The Simplest Walking Model: Stability, Complexity and Scaling. *ASME J. Biomechanical Engineering.*Vol.120, No.2, pp 281-288.
5. M. Yamakita y F. Asano (2001). Extended Passive Velocity Field Control with Variable Velocity Fields for a Kneed Biped. *Advanced Robotics*, Vol. 115, No.2, pp 139 – 168.
6. H.K. Lum, M. Zribi y Y.C. Soh (1999). Planning and Control of a Biped Robot. *J. Eng. Science*, Vol.37, pp 1319 – 1349.
7. A. Goswami, B. Espiau and A. Keramane (1996). Limit Cycles and their Stability in a Passive Bipedal Gait. *Proc. Int. Conf. Robotics and Automation – ICRA 1996*.
8. M. García (1999). *Stability, Scaling and Chaos in Passive – Dynamic Gait Models*. Ph.D. Thesis, Cornell University.

Development of Biped Robots at the National University of Colombia

M. A. Roa, R. E. Ramírez and D.A. Garzón

Department of Mechatronics and Mechanical Engineering, National University of Colombia   maroag@unal.edu.co

Abstract

This article presents the design process used in the NUC to build and control biped robots. There are two main highlights in our methodology. First, the design process allows us to get an appropriate mechanical structure (and appropriate actuators) to get the desired walk because it relies on the modeling of bipedal gait. Second, the trajectory control in the robots follows a gait pattern obtained trough the simulation of a 3 dof passive dynamic walker, joining two different lines of research in bipedal gait in a common ground. We also present the three biped robots built until now at NUC using the presented methodology.

Keywords: biped robots, gait pattern, passive walk.

1 Introduction

The National University of Colombia (NUC) has been working on the design and control of biped robots within the Mechatronics and Mechanical Engineering Department, supported by two research groups, Biomechanics and Mobile Robots. The joint effort of the groups has produced three biped robots with successful walks, based on a single idea: if an appropriate design methodology exists, the resulting hardware must have appropriate dynamical characteristics, making easier the control of the walking movements. The design process successfully merges two lines of research in bipedal walk, passive an active walks, by using gait patterns obtained trough simulation of a kneed passive walker to create the trajectory followed by the control of an active biped robot.

The rest of the paper is organized as follows. Section 2 introduces the design methodology used to create the biped robots. Section 3 presents the models used to simulate passive and active walks, and Section 4 shows three biped robots built at the National University of Colombia.

2 Design methodology

Biped robot design should be based on a design methodology which produce an appropriate mechanical structure to get the desired walk. We use a design methodology that groups passive and active walk relying on dynamic models for bipedal gait [1]. The methodology is an iterative process, as shown in Figure 1. The knowledge of biped robot dynamics allow us to develop simple and efficient control systems, based on the system dynamics and not on assumptions of a simplified model, such as a simple inverted pendulum.

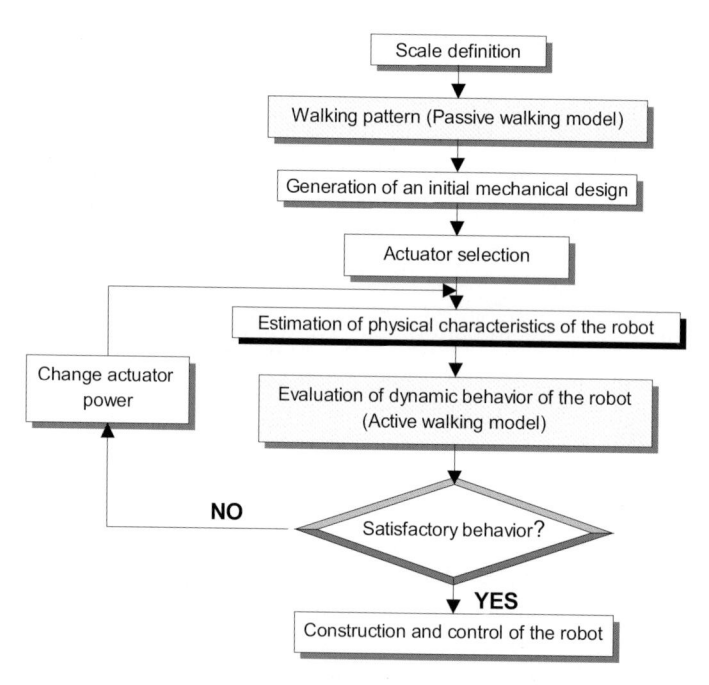

Fig. 1. Design methodology for biped robot design.

The dynamical model for actuated walk is the base in the design methodology used here; it is presented in Section 3. Geometrical and kinematical data are used to solve the model. Geometric variables of the robot can be defined with different criteria, e.g. if the biped robot is intended to be a model of human gait, it is useful to scale anthropometrical proportions with an adequate scale factor. Geometrical data required in the dynamical model include mass, inertia moment, length, and position of the centre of mass for each segment. These data can be easily acquired trough a CAD solid modeler software for a preliminary design. Kinematical data conform the gait pattern for the robot. This pattern can be acquired from two approximations. First, data can be obtained trough gait analysis of normal people in a gait laboratory. Second, a gait pattern can be generated trough the simulation of passive walking models, as shown in the next section. Actually, this approach has probed useful to obtain gait patterns at different speeds. The methodology outlined here assures that the controlled system is mechanically appropriate to get the desired walking patterns.

3 Dynamic models for biped walkers

The main step in the development of a biped robot is the study and modeling of biped walk. The dynamic study of bipedal walk can be accomplished from two points of view: passive and active walk. In passive walk the main issue is gravitational influence over artificial mechanisms, getting a device to walk in a slope without actuators or control. In active walk there are different actuators which give energy to the mechanism so it can walk as desired. The models, passive and active, begin with a symmetry assumption: the spatial parameters of movement of the two legs are identical. Besides, the two legs are composed of rigid links connected trough pin joints, so each joint has just one degree of freedom. Although real walk is a three dimensional process, the models will consider a planar walk, describing the movements in the sagital plane (progression plane) of motion.

3.1 Passive walk

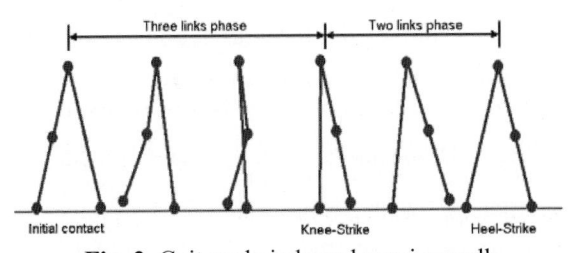

Fig. 2. Gait cycle in kneed-passive walk.

McGeer [2] presented the passive walk concept based on the Mochon and McMahon hypotheses [3] of understanding human gait as the influence of a neuromotor control mechanism acting on a device moved only by the gravity influence. McGeer first studied the passive walk through simple models, developed subsequently by different researchers [4], [5]. The model used in this work is the passive dynamic walker with knees, original of McGeer [2] and also used by Yamakita [6]. The model has three links (stance leg, thigh and shank), and four punctual masses (each link has a concentrated mass, and one additional mass is at the hip). Figure 2 shows the diagram of a gait cycle. The three links swing freely (under gravity action) until the knee – strike, when the thigh and shank are aligned and become one single link (knee-lock), preventing a new knee flexion. There is a second phase (two – links phase) when the robot can be considered a walker with compass gait. The gait cycle ends when the swinging leg hits the ground (heel-strike).

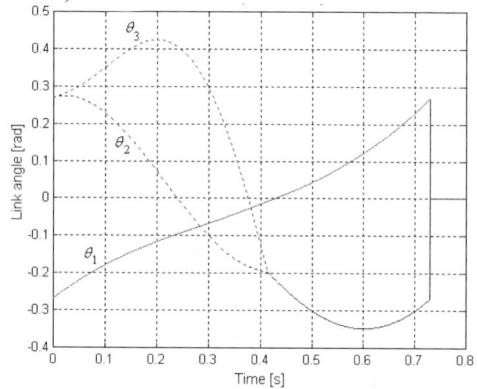

Fig. 3. A gait pattern for the kneed-passive walker.

Dynamic equations for the kneed-passive walker have the general form

$$D(\theta)\ddot{\theta} + H(\theta,\dot{\theta})\dot{\theta} + G(\theta) = 0 \qquad (1)$$

with $D(\theta)$ the matrix of inertial terms, $H(\theta,\dot{\theta})$ the matrix with coriolis and centripetal terms and $G(\theta)$ the gravitational terms. Transition equations and conditions for knee-lock and heel strike are modeled as algebraic equations; further modeling details can be found in [6] - [8]. Using this passive dynamic model, we can get different trajectories of link angles, and we use them to get hip and knee angles in time by using different configuration parameters for the robot. The model analysis and numerical simulation give as result gait patterns for different slope angles in the movement of the robot (used to get different walking speeds in the real robots); one example of such gait pattern is shown in Figure 3, with θ_1, θ_2 and θ_3 the stance leg, thigh and shank angle, respectively.

3.2 Actuated walk

In dynamic walk, we consider robots moving under the effect of an actuator (servomotor, pneumatic muscle, elastic actuator, etc). The model used in the design process for biped robots is a seven segments model, shown in Figure 4. The robot has two feet, two thighs, two shanks and a HAT (Head, Arms & Trunk). The model of the robot has a planar walk (in the sagital plane). Physical parameters of the model include the link mass (m), inertia moment (I), length (l), distance between the center of mass and distal point of the link (a) and the segment angle with respect to the vertical (φ).

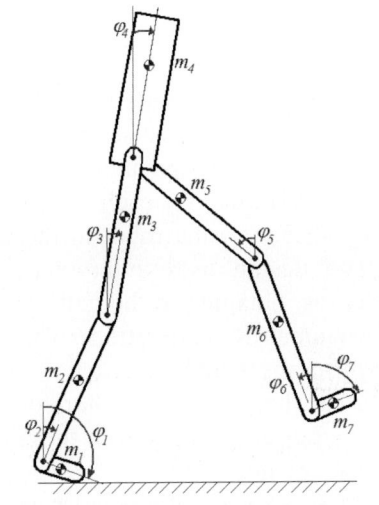

Fig. 4. Seven segments model for biped walk.

The gait equations are obtained with lagrangian dynamics; they can be expressed in the form

$$D(\varphi)\ddot{\varphi} + H(\varphi,\dot{\varphi}) + G(\varphi) = T \qquad (2)$$

with D the inertial matrix, H the coriolis and centripetal forces vector, G the gravitational effects vector and T the vector for moments (generalized external torques) over each segment. This model contains the effect of external torques; however, torques exerted on the segments of the biped walker (human or robot) are internal moments caused by muscular forces or joint actuators; it is necessary to do an angle and torque transformation to get the internal torques in each joint [1].

The model is solved using as inputs 1) segments dimensions and 2) kinematic data. The kinematic data are angular position, velocity and acceleration for each joint; they can be either data from a passive walk model (as the kneed passive walker) or data from a human gait analysis. Model can be simulated to get torque charts for each of the actuators composing the robot, and act as a guideline for the appropriate actuator selection in the design process. Movement in the frontal plane can be modeled with the behavior of a simple inverted pendulum coupled to the movement in the sagittal plane.

4 Three biped robots

Using the proposed design methodology, we have currently developed three biped robots, called UNROCA (*Universidad Nacional – RObot CAminador*, National University – Walking Robot), shown in Figure 5. The first prototype was conceived as the simplest actuated robot that could achieve two dimensional walk; the dynamic model for he robot is a five links model: a HAT and two links for each leg (the minimum required to change the supporting surface of the robot in the walking movement); it has punctual feet. The design uses anthropometrical proportions and physiological gait patterns, obtained in the gait lab at CIREC (Colombian Research Institute for Human Rehabilitation). The robot achieves two dimensional walk through a mechanical restriction, a walking guide. Robot moves with a guiding car that moves on a horizontal supporting bar. The walking guide prevents the hip oscillation in the frontal plane and the hip rotation in the transversal plane.

In the second prototype we also used a planar gait, but now describing a circle around a central support, getting more freedom in the robot movements. Robot proportions are still anthropometrical, but we used normal human gait as well as passive walking gait patterns to produce the trajectory for each of the joints, so we could change the velocity of progression for the movement. The robot has two ankles, two knees and two joints at the hips; the dynamic model for the robot is the seven links model presented before. As in the previous prototype, the control system follows a trajectory based on a predefined speed, but this robot modifies the gait pattern according to the disturbances in the environment (e.g. a slippy surface). The third prototype has four actuated degrees of freedom (two knees and hips), plus passive ankles. This version does not need any external help to walk; achieves stability through an oscillating mass in top of the hip, so it actually attains a 3D walk. The oscillating mass couples and controls the gait in the sagital and frontal plane. We used big feet to keep static stability while not moving, but they are not used to get stability in the walking process, as can be seen in Figure 5; stability is assured through the appropriate movement of the counterweight in the hip according to a valid ZMP (*Zero Moment Point*) trajectory.

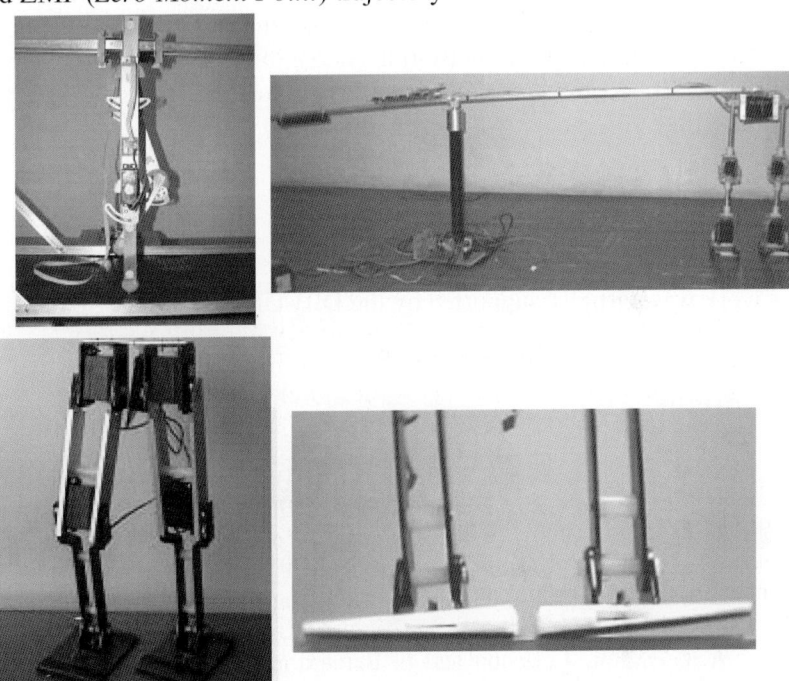

Fig. 5. Biped walking robots; above: UNROCA-I (left) and II (right), below: UNROCA-III, with a detail of its feet.

All of the robots use servomotors as joint actuators. We use proportional-derivative control to cause the prescribed gait: the position for each joint at every instant is given by the gait pattern, and the control system enforces this position. The internal potentiometer in the servos was used to get the reference signal for the closed loop position control. The control was implemented with microcontrollers, one for each couple of servos (for example, one micro controls the hip movements), plus a central controller to synchronize joint controllers.

The fundamental achievement of these prototypes is the use of a simple control system that allows them to get a stable walk, and makes them able to overcome small perturbations. This achievement was possible thanks to the previous dynamical study, using an appropriate scaling to implement the human gait pattern or a passive walker gait, and using a design methodology that takes into account the physical and dynamical features of the implemented system. Robots become an example of control systems adapting to the system dynamics, thus optimizing the power requirements on the actuators, the sensors and electronics used on the robot. The implemented gait does not use a complex sensorial system, but the robot can extend its capabilities to become more robust to environment changes (for example, to walk stairs).

Acknowledgements

This work was partially supported by the DIB-UN project No. 308005.

References

1. M. Roa, C. Cortés and R. Ramírez (2004). Design Methodology for Biped Robots with an Application. *Proc. CARS & FOF 2004*, pp 438 – 446.
2. T. McGeer (1990). Passive Dynamic Walking. *Int. J. of Robotic Research*, Vol. 9, No.2, pp 62 – 82.
3. S. Mochon and T. McMahon (1980). Ballistic Walking: An Improved Model. *Mathematical Biosciences*, Vol.52, pp 241 – 260.
4. A. Goswami, T. Benoit and E. Bernard (1996). *Compass – Like Biped Robot. Part I: Stability and Bifurcation of Passive Gaits*, Research Report, INRIA (Institut Nacional de Recherche en Informatique et en Automatique).
5. M. García, A. Chatterjee, A. Ruina and M. Coleman (1998). The Simplest Walking Model: Stability, Complexity and Scaling. *ASME J. Biomechanical Engineering*.Vol.120, No.2, pp 281-288.
6. M. Yamakita y F. Asano (2001). Extended Passive Velocity Field Control with Variable Velocity Fields for a Kneed Biped. *Advanced Robotics*, Vol. 115, No.2, pp 139 – 168.
7. H.K. Lum, M. Zribi y Y.C. Soh (1999). Planning and Control of a Biped Robot. *J. Eng. Science*, Vol.37, pp 1319 – 1349.
8. M.A. Roa, C.A. Villegas and R.E. Ramírez (2005). Extensive Modeling of a 3 DOF Passive Dynamic Walker. *Proc. 8th Int. Conf. CLAWAR* 2005.

Design of a Low Cost Force and Power Sensing Platform for Unmanned Aerial Vehicles

David M. Alba, Hector Montes, Gabriel Bacallado, Roberto Ponticelli, and Manuel Armada

Instituto de Automática Industrial (CSIC), Carretera de Campo Real km 0.200, La Poveda, Arganda del Rey, 28500 MADRID
dmalba@iai.csic.es

Abstract

This paper describes the design of a platform for measuring forces in a model helicopter using a calibration method, and experimental verification of the strategy. It is found that the platform is able to measure the forces in x, y and z direction. Another characteristic of the platform is that, it is constructed using small-size pieces, thus allowing easy and economic manufacturing of the platform.

1 Introduction

A thorough mechanical design can significantly improve the sensibility of a sensor. With more sensibility less amplification is needed, and as such less filtering is required for the acquisition of useful measurements. All such considerations result in a faster and accurate sensor. Although the mechanical resistance of the overall sensor is decreased; the resistance is chosen considering the amplitude of the measurement. The design is based on thin plates in cantilever. In this way the stress is maximized where the strain gauges are attached. The basic design is inspired by [1], although there are other designs with advantages over this idea [2], [3], [4], but the simplicity of this design is in a sensor that is made with separated pieces.

The most common practice in manufacturing of sensors with strain gauges is to make the sensor from one billet of material, in order to avoid errors introduced by assembling different pieces. But with a careful construction, errors due to coupling between the different axes of the sensor can be controlled. In this paper the experience with a sensor made of several pieces is shown and discussed.

The use of unmanned aerial vehicles (UAVs) for different applications is increasing. However, there are problems related to the mathematical modelling of such vehicles, and also to their practical use, where training of operators is seen crucial and requires high dexterity. This platform is a tool for helping in the training and acquisition of data about the forces produced in UAVs.

2 Design of the sensor

The sensor uses a full bridge of strain gauges, the strain gauges are attached in a zone with a reduction of the transversal section, with which the sensibility of the bridge is increased. It is composed by three different pieces, the play between the pieces have a maximum value of 0.2mm. The sensor is composed of three bridges [1], each bridge is located perpendicularly from each other. The complete sensor is shown in **Fig. 1**.

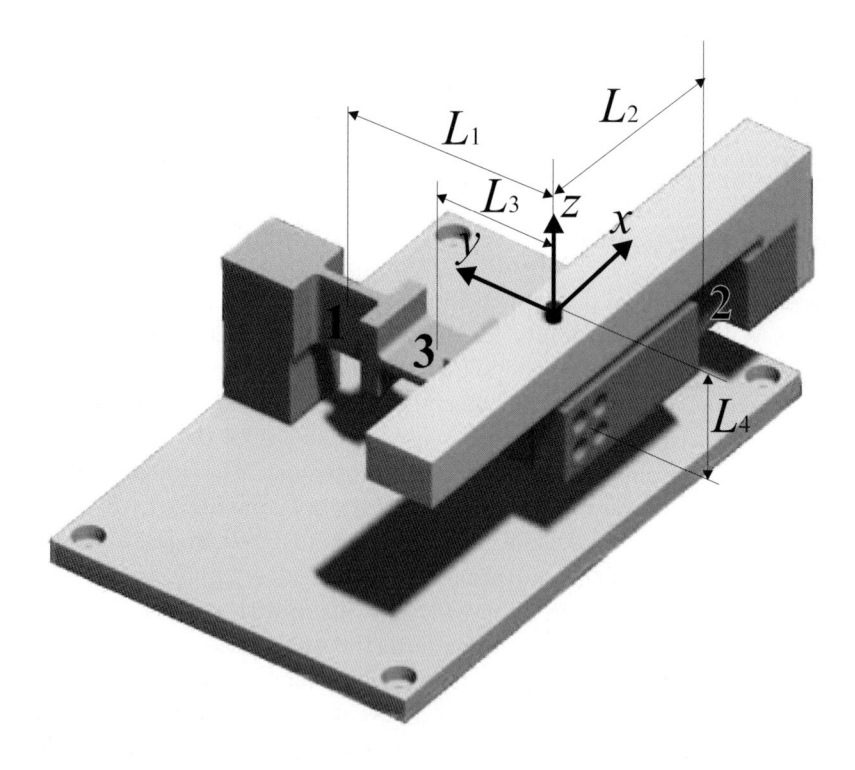

Fig. 1. General view of the sensor

The following system of equations is ontaine from **Fig. 1.**:

$$\begin{bmatrix} L_1 & 0 & 0 \\ 0 & L_2 & 0 \\ 0 & L_4 & L_3 \end{bmatrix} \begin{bmatrix} F_x \\ F_y \\ F_z \end{bmatrix} = \begin{bmatrix} M_1 \\ M_2 \\ M_3 \end{bmatrix} \quad (1)$$

The matrix of constants L_i in Equation (1) provide the basic form of the matrix calibration. Since the amplifying card has a maximum gain of 1000, an initial analysis of the sensitivity was required. The difference in voltage can be obtained by combining the strain gauge formula [1], the wheatstone bridge relationship and the beam bending formula [5], giving

$$\Delta V = V_{EX} F \varepsilon \quad (2)$$

where V_{EX} is strain gauge's supply voltage (2.5V), F is the applied force, and ε is the strain (1.76×10^{-4} obtained with $L_3 = 32 \times 10^{-3}$m, the dimensions of the plate are thickness=2mm and width=12mm. These values are taken from the bridge 3, that is the less sensitive one). Considering that the minimum value that can be distinguished is 4mV it is found that the sensor can detect a minimum value of 0.05N

Keeping in mind that the measured values are voltages and the measured voltages are proportional to the applied moments, in can be inferred from Equation (1) and the values from **Fig. 2.** that

$$\begin{bmatrix} 13.8122 & 0 & 0 \\ 0 & 9.0498 & 0 \\ 0 & 6.5041 & 14.7275 \end{bmatrix} \begin{bmatrix} V_1 \\ V_2 \\ V_3 \end{bmatrix} = \begin{bmatrix} F_x \\ F_y \\ F_z \end{bmatrix} \quad (3)$$

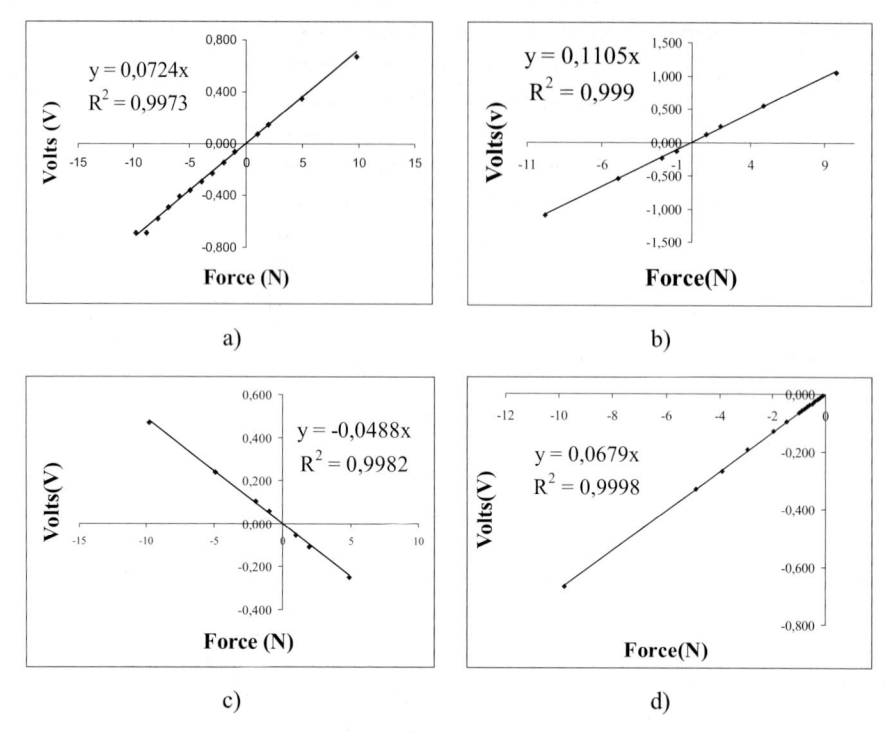

Fig. 2. Calibration curve for a) bridge 1 with load in x; b) bridge 2 with load in y; c) bridge 3 with load in y and d) bridge 4 with load in z

A significant amount of data was found experimentally using calibrated weights by applying loads in each axis separately, then with linear regression the values of the constants were found. The values of the coefficient of determination for all constants were over 0.99.

3 Experiments

Three experiments were conducted. The first one was an experiment over the z axis. The objective of this experiment was to determine the standard deviation of a series of measurements with different values. The second experiment determined the coupling between the axes, in other words the errors introduced by the loads in the other bridges. Finally, the sensor was tried under real conditions, and was used with a UAV.

3.1 Determination of the standard deviation

Since observations during the calibration of the sensor showed that the results from one axis can be extended to other axess, this experiment was performed only over the z axis. The results are presented in **Table 1**. It is important to note that the CV was very low.

Table 1. Standard deviation

F(mN)	σ (mN)	CV(%)
98	0,00741	0,0076
196	0,00439	0,0022
294	0,01041	0,0035
392	0,01196	0,0030
490	0,01484	0,0030
588	0,01022	0,0017
686	0,01586	0,0023
784	0,00745	0,0010
882	0,01361	0,0015
980	0,00828	0,0008
1470	0,02054	0,0014
1960	0,02016	0,0010
2940	0,03576	0,0012
3920	0,02134	0,0005
4900	0,04282	0,0009
9800	0,28011	0,0029

3.2 Coupling between axes

Several experiments with different sets of forces were performed. The results are shown in **Table 2**. The results show that the error is small due to the coupling between bridges; the largest error registered was 6%.

Table 2. Coupling between axes

	Experimental Measure (mN)	Real Measure (mN)	Error (%)
	Test 1		
F_x	980	980	0
F_y	-2136	-2156	2
F_z	-4851	-4900	1
	Test 2		
F_x	980	980	0
F_y	1881	1960	4
F_z	-6007	-5880	2
	Test 3		
F_x	1999	1960	2
F_y	-1038	-980	6
F_z	-1479	-1470	<1

3.3 Measurements with a UAV

The UAV is a model helicopter, the sensor was protected against electromagnetic noise produced by the RC controller. The sensor was able to detect forces in the helicopter in every direction in agreement with the supposed direction of the real force. The amplitudes of the forces were between 0.5N to 5N.

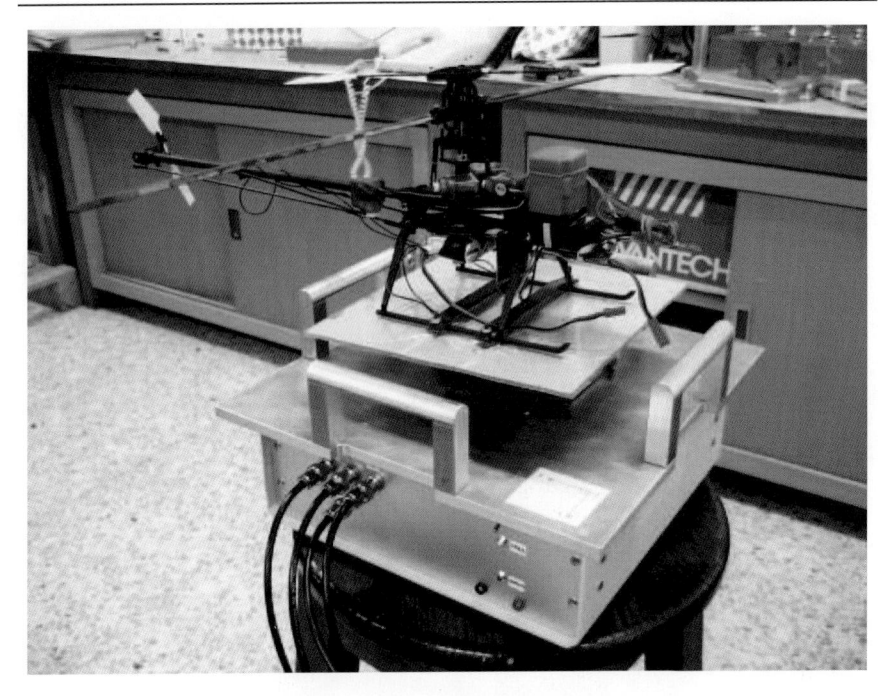

Fig. 3. Sensor with the UAV

4 Conclusions and future work

Strain gauges provide an excellent solution for Design and manufacture of force sensors, thanks to their linear Characteristics over the full amplitude measurement range. The play of 0.2mm between the pieces of the sensor had little impact on the precision and the coupling between axes. The measurements taken with the UAV were considered satisfactory because the sensor was able to detect the power variations of the UAV. In the future it is expected to carry out measurements of the forces and power of the helicopter in real time so as to establish a mathematical model of the system.

Acknowledgements

The authors would like to acknowledge Prof. J. M. de la Cruz and Prof. J. A. Lpez Orozco and their group from the University Complutense of Madrid for suggesting this project.

References

1. Gorinevsky, D., Formalsky, A., and Shneider, A. (1997) *Force Control of Robotics Systems.* 0-8493-2671-0. CRC Press, 1 edition.
2. IBM (1983) An inteligent, self-adaptive robot. *Epsilonics*, pp. 3–4.
3. Folchi, G. A., Shelton, G. L., and Wong, S. S. (1976) Six-degree-of-freedom force transducer for manipulator system. In *Us. Patent 3948093*.
4. Bentley, John (1995) *Measurement Systems.* ISBN 0-582-23773-3. Prentice Hall, 3 edition.
5. Beer, F.P. and Johnston, R. (1981) *Mechanics of Materials.* ISBN 968-451-414-X. McGraw-Hill, Inc., 1 edition.
6. Dally, James W., Riley, William F., and G., McConnell. Kenneth (1993) *Intrumentation for Engineering Measurements.* John Wiley and Sons, 2 edition.
7. Stokic, D., Vukobratovic, M., and Hristic, D. (1999) Implementation of force feedback in manipulator robots. In *IJRR*. vol. 5, pp. 2055–2062, Oulu University Press.
8. Wang, S. S. and Will, P. H. (1978) Sensors for computer controlled mechanical assembly. *Industrial Robot*, vol. 5, pp. 9–18.

Hopping and Legged Robots

Observer Backstepping for Height Control of a Resonance Hopping Robot

Roemi Fernández, Teodor Akinfiev, and Manuel Armada

Industrial Automation Institute, Spanish Council for Scientific Research. Automatic Control Department. La Poveda 28500 Arganda del Rey, Madrid, Spain. {roemi,teodor,armada}@iai.csic.es

Abstract

This paper addresses the height control problem of a one-legged resonance hopping robot. The particular construction of the robot with compensation of energy losses during robot's flight, the use of a special dual drive with changeable transmission ratio for allowing an additional decrease of energy spent, and the unavailability of part of the state for measurement, impose some specific requirements on the design of height controller. The proposed solution utilizes an observer based backstepping algorithm to produce a nonlinear controller that ensures asymptotic tracking of the desired law of motion. Simulations and experimental tests are carried out to validate the proposed control approach and to demonstrate the effectiveness and feasibility of the resonance hopping robot design.

1 Introduction

The hop height of a hopping robot with electric motor and elastic devices is, in general, controlled by measuring and manipulating the system's energy. In this way, most of the hopping height controllers deliver a specified thrust value during the stance to compensate for the energy lost due to friction and air resistance [1], [2]. However, for such kind of control, it is necessary to use an electric motor powerful enough to compensate for the losses by the end of the stance. Bearing in mind that the stance phase is between five and ten times shorter than a complete cycle time, it seems to be advantageous to make the compensation of losses during the flight phase [3], [4], [5]. The hopping robot described in this paper and its height controller have been designed by taking such consideration into account. Moreover, the robot has a special drive with dual properties that enables it to increase the motor efficiency by adjusting the transmission ratio for two different movements, namely, during the process of stretching of the spring in the presence of external load, and during the process of releasing the spring, several times faster and without an external load (see **Fig. 1**). A detailed description of the resonance hopping robot and a first height control approach were presented in [6] and [7]. The height controller presented in [7] was obtained

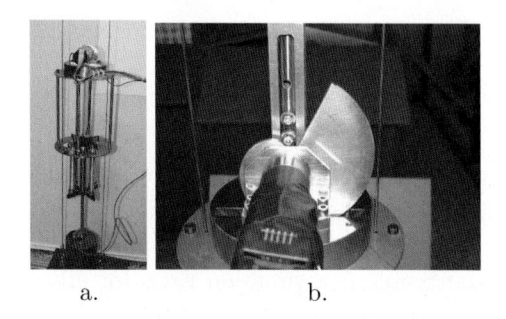

Fig. 1. a. Prototype of the resonance hopping robot. b. Dual Drive.

under the assumption that the full state of the system was measured. Nevertheless, the prototype of the hopping robot only has an optical encoder for the tracking application and it was necessary to numerically reconstruct the velocity from position measurements. This was achieved by simply taking backwards differences of the position measurements to approximate the differentiation of position. However, the noise due to the encoder quantization of the position measurement resulted in significant noise in the differentiated signal. To limit the amount of this noise, the differentiated signal was filtered, in which case one has to seek a trade-off between minimizing the filter's delay (high cutoff frequency) and filtering out unwanted noise (low cutoff frequency). In order to improve the previous tracking results, and to overcome the problem of unavailability of part of the state for measurement, a combination of optimal control and observer backstepping is proposed, where, the optimal state trajectories are used as reference values in the controller, which is designed usigng the observer based backstepping scheme [8], [9], [10]. It is natural to consider the use of this systematic design procedure, allowing to incorporate a nonlinear observer, which provides exponentially convergent estimate of the unmeasured state in feedback control. In this way, the observer backstepping procedure is applied to a new system in which the equation of the unmeasured state is replaced by the corresponding equation of its estimate from the observer. Although the nonlinear observer provides exponential convergence, it is necessary to redesign the control law to make it robust against observer error. Then, at each step of the procedure, observation errors are treated as disturbances and accounted for using nonlinear damping. The resulting Lyapunov-based nonlinear controller achieves global asymptotic tracking of the reference trajectories.

2 Dynamic Equations

The robot's equations of motion are based on a progressive use of the corresponding mechanical energy conservation laws for each stage of the movement, including the impact between the leg and the robot's body. The equations describing the robot's motion in the ith cycle are presented in detail in [6] and [7]. In this section a state-space model, required for the control system design, is derived. In this way, the dynamic equation of the cam is given by:

$$J\ddot{\theta} = K_G M_o - mgL\sin(\theta) - M_{ext}, \qquad (1)$$

where J is the inertia of the cam, L is the distance to the center of mass, $\ddot{\theta}$ is the angular acceleration of the cam, K_G is the (constant) transmission ratio, M_o is the moment that actuates over the cam, and M_{ext} is the external moment. The rotor equation is given by:

$$(J_M + J_G)\ddot{\theta}_M = \frac{K_m}{R_M}\left[u - K_E\dot{\theta}_M\right] - M_o - (b_M + b_G)\dot{\theta}_M, \qquad (2)$$

where J_M is the rotor inertia, J_G is the gearhead inertia, $\ddot{\theta}_M$ is the angular acceleration of the rotor, $\dot{\theta}_M$ is the angular velocity of the rotor, M_o is the moment that actuates over the rotor, b_M is the viscosity friction coefficient of the motor shaft, b_G is the viscosity friction coefficient of the gearhead, K_m is the torque constant, u is the input control motor terminal voltage, R_M is the motor resistance, and K_E is the back - EMF constant.

The angular position of the rotor as a function of the angular position of the cam is given by:

$$\theta_M = K_G \theta. \tag{3}$$

Combining equations (1) to (3), gives in the dynamic model of the system as:

$$\left(\frac{J}{K_G^2} + J_M + J_G\right)\ddot{\theta}_M + \left[\frac{K_E K_m}{R_M} + b_M + b_G\right]\dot{\theta}_M$$
$$+ \frac{mgL}{K_G}\sin\left(\frac{\theta_M}{K_G}\right) + \frac{M_{ext}}{K_G} - \frac{K_m}{R_M}u = 0. \tag{4}$$

If the state variables are denoted by: x_1 - angular position of the rotor θ and x_2 - angular velocity of the rotor $\dot{\theta}$, then the state-space model is given by:

$$\dot{x}_1 = x_2$$
$$\dot{x}_2 = -K_1 x_2 - K_4 \sin\left(\frac{x_1}{K_G}\right) - K_2 + K_3 u \tag{5}$$

where

$$K_1 = \frac{1}{J_{eq}}\left(\frac{K_E K_m}{R_M} + b_M + b_G\right), \quad K_2 = \frac{M_{ext}}{J_{eq} K_G}, \quad K_3 = \frac{K_m}{J_{eq} R_M} \tag{6}$$

$$K_4 = \frac{mgL}{J_{eq} K_G}, \quad J_{eq} = \frac{J}{K_G^2} + J_M + J_G. \tag{7}$$

3 Observer Backstepping

In order to solve the tracking problem, and to overcome the difficulties that arise due to the unavailability of one of the state variables for measurement, a nonlinear tracking controller is proposed using the observer based backstepping algorithm [8], [9], [10]. Firstly, a coordinate transformation is introduced for the system (5):

$$\dot{e}_1 = e_2 + x_{d2} - \dot{x}_{d1}$$
$$\dot{e}_2 = -K_1(e_2 + x_{d2}) - K_4 \sin\left(\frac{e_1 + x_{d1}}{K_G}\right) - K_2 + K_3 u - \dot{x}_{d2} \tag{8}$$

where $e_1 = x_1 - x_{d1}$, $e_2 = x_2 - x_{d2}$ denote the position and velocity tracking errors, and x_{d1}, x_{d2} the optimal reference trajectories. Since the angular velocity of the motor x_2 is not measured, e_2 cannot be chosen as the virtual control. Nevertheless, the state e_2 can be estimated by $\hat{e}_2$, where

$$\dot{\hat{e}}_2 = -K_1\left(\hat{e}_2 + x_{d2}\right) - K_4 \sin\left(\frac{e_1 + x_{d1}}{K_G}\right) - K_2 + K_3 u - \dot{x}_{d2} \quad (9)$$

To utilize this estimate, e_2 is replaced by $\hat{e}_2 + \tilde{e}_2$ in (8):

$$\dot{e}_1 = \hat{e}_2 + \tilde{e}_2 + x_{d2} - \dot{x}_{d1}. \quad (10)$$

It is now possible to take $\hat{e}_2$ as a virtual control, which means that in the backstepping procedure $\dot{e}_2$ must be replaced with (9):

$$\begin{aligned}\dot{e}_1 &= e_2 + x_{d2} - \dot{x}_{d1} \\ \dot{\hat{e}}_2 &= -K_1\left(\hat{e}_2 + x_{d2}\right) - K_4 \sin\left(\frac{e_1 + x_{d1}}{K_G}\right) - K_2 + K_3 u - \dot{x}_{d2} \\ \dot{\tilde{e}}_2 &= -K_1 \tilde{e}_2\end{aligned} \quad (11)$$

The next step is to design the control law for (11). Using $\hat{e}_2$ as the virtual control law, the stabilization function α_1 is modified as:

$$\alpha_1 = -x_{d2} + \dot{x}_{d1} - (1 + d_1) e_1. \quad (12)$$

This leads to the error variable

$$z = \hat{e}_2 + x_{d2} - \dot{x}_{d1} + (1 + d_1) e_1, \quad (13)$$

and the closed-loop expression

$$\dot{e}_1 = z - (1 + d_1) e_1 + \tilde{e}_2. \quad (14)$$

Then, the derivative of $V_1 = \frac{1}{2} e_1^2$ gives,

$$\begin{aligned}\dot{V}_1 &= -e_1^2 + e_1 z - d_1 \left(e_1 - \frac{1}{2d_1}\tilde{e}_2\right)^2 + \frac{1}{4d_1}\tilde{e}_2^2 \\ &\leq -e_1^2 + e_1 z + \frac{1}{4d_1}\tilde{e}_2^2.\end{aligned} \quad (15)$$

The last inequality implies that if $z_i \equiv 0$, e_1 will remain bounded if $\tilde{e}_2$ is bounded. Exploiting the fact that $\tilde{e}_2(t)$ is the error of an exponentially converging observer, the function $V_1(e_1)$ is augmented with a quadratic term in $\tilde{e}_2$:

$$V_2(e_1, \tilde{e}_2) = \frac{1}{2} e_1^2 + \frac{1}{2d_1}\tilde{e}_2^2. \quad (16)$$

Using (15), it is seen that the derivative of V_2 satisfies

$$\dot{V}_2 = -e_1^2 - \frac{3}{4d_1}\tilde{e}_2^2 + e_1 z - d_1\left(e_1 - \frac{1}{2d_1}\tilde{e}_2\right)^2$$
$$\leq -e_1^2 + e_1 z - \frac{3}{4d_1}\tilde{e}_2^2. \tag{17}$$

The derivative of z is now expressed as:

$$\dot{z} = -K_1(\hat{e}_2 + x_{d2}) - K_4 \sin\left(\frac{e_1 + x_{d1}}{K_G}\right) - K_2 + K_3 u - \dot{x}_{d2}$$
$$-\frac{\partial \alpha_1}{\partial e_1}(\hat{e}_2 + x_{d2} - \dot{x}_{d1}) - \frac{\partial \alpha_1}{\partial e_1}\tilde{e}_2. \tag{18}$$

In the above equation, the state estimation error appears again, so its effect will have to be compensated for by a nonlinear damping term. This is reflected in the corresponding Lyapunov function, which is augmented not only by a z^2-term, but also by an additional $\tilde{e}_2^2$-term:

$$V_3(e_1, z, \tilde{e}_2) = \frac{1}{2}[e_1^2 + z^2 + \left(\frac{1}{d_1} + \frac{1}{d_2}\right)\tilde{e}_2^2]. \tag{19}$$

the derivative of which is:

$$\dot{V}_3 = -e_1^2 - \frac{3}{4d_1}\tilde{e}_2^2 - d_1\left(e_1 - \frac{1}{2d_1}\tilde{e}_2\right)^2 + z\left[e_1 - K_4\sin\left(\frac{e_1 + x_{d1}}{K_G}\right)\right.$$
$$\left.-K_1(\hat{e}_2 + x_{d2}) - K_2 + K_3 u - \dot{x}_{d2} - \frac{\partial \alpha_1}{\partial e_1}(\hat{e}_2 + x_{d2} - \dot{x}_{d1})\right]$$
$$-z\frac{\partial \alpha_1}{\partial e_1}\tilde{e}_2 - \frac{1}{d_2}\tilde{e}_2^2. \tag{20}$$

It suffices now to take:

$$u = \frac{1}{K_3}\left[-d_3 z - e_1 + K_1(\hat{e}_2 + x_{d2}) + K_2 + K_4 \sin\left(\frac{e_1 + x_{d1}}{K_G}\right)\right]$$
$$+\frac{1}{K_3}\left[\frac{\partial \alpha_1}{\partial e_1}(\hat{e}_2 + x_{d2} - \dot{x}_{d1}) + \dot{x}_{d2} - d_2 z\left(\frac{\partial \alpha_1}{\partial e_1}\right)^2\right] \tag{21}$$

to ensure the negative of (20):

$$\dot{V}_3 \leq -e_1^2 - \frac{3}{4}\left(\frac{1}{d_1} + \frac{1}{d_2}\right)\tilde{e}_2^2 - d_3 z^2. \tag{22}$$

This implies asymptotic stability according to Lyapunov stability theorem.

4 Simulations and Experimental Results

To evaluate the performance of the proposed nonlinear control algorithm, simulations and various experiments were carried out using a prototype of the resonance hopping robot (see **Fig. 1**a). **Fig.(2)** and **Fig.(3)** show the tracking performance using control law (21) during the stance phase and the flight phase respectively. **Fig.2**a and **Fig.3**a show the time evolution of the angular position of the rotor during the stance phase and the flight phase respectively. The dotted lines represent the desired values and the solid lines represent the actual values. **Fig.2**b and **Fig.3**b show the time evolution of the angular velocity of the rotor, where the dotted line represents the desired values and the solid line represents the estimated values. In all the experiments a satisfactory tracking performance was achieved with a reasonable control effort.

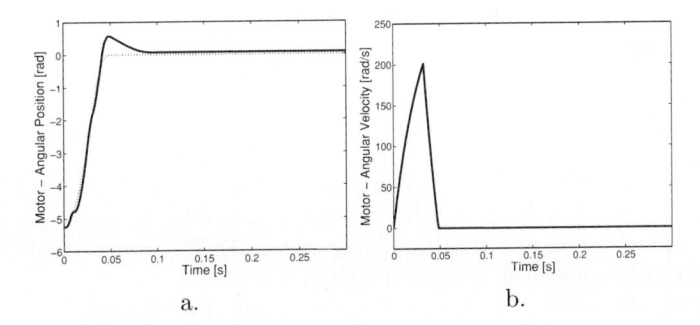

Fig. 2. Experimental results for the stance phase.

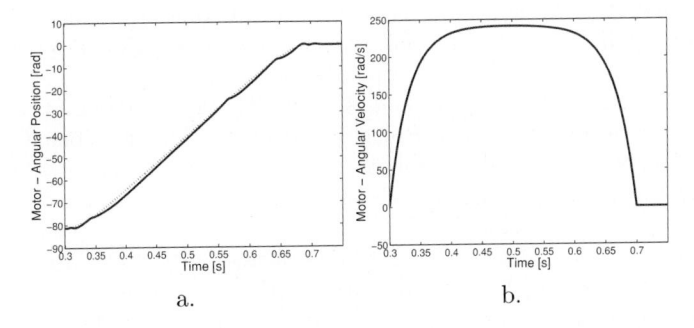

Fig. 3. Experimental results for the flight phase.

5 Conclusions and Future Developments

In this paper, a method for controlling the apex height of a special resonance hopping robot has been presented. The solution includes the design of a tracking controller using an observer based backstepping procedure. The resulting scheme achieves asymptotic tracking of the reference trajectories in spite of availability of only a part of the states of the system for measurement. Simulations and experimental results show the effectiveness of the control strategy and demonstrate that the control objectives were accomplished.

6 Acknowledgment

The authors would like to acknowledge the financial support of the Spanish Ministry of Education and Science: Fellowship F.P.U. and Project "Theory of optimal dual drives for automation and robotics".

References

1. Raibert M. (1986) Legged Robots that Balance. MIT Press, Cambridge.
2. Harbick K. and Sukhatme G. (2002) Controlling Hopping Height of a Pneumatic Monopod. In: Proceedings of the IEEE International Conference on Robotics and Automation, 4:3998–4003.
3. Zeglin G. (1999) The bowleg-hopping robot. PhD Thesis, The Robotics Institute, Carnegie Mellon University, Pittsburgh, Pennsylvania, USA.
4. Ringrose R. (1997) Self-stabilizing running. In: Proceedings of the IEEE International Conference on Robotics and Automation, IEEE Press.
5. Akinfiev T., Armada M., Fernández R. and Montes H. (2002) Hopping robot and its control algorithm. Patent Application, P200201196, Spain.
6. Akinfiev T., Armada M., Fernández R., Gubarev V. and Montes H. (2004) Dual Drive for Vertical Movement of Resonance Hopping Robot. International Journal of Humanoid Robotics (In press).
7. Fernández R., Akinfiev T. and Armada M. (2004) Height Control of a Resonance Hopping Robot. Proceedings of the 7th International Conference on Climbing and Walking Robots. Springer. ISBN: 3–540–22992–2.
8. Kokotović P. (1992) The Joy of Feedback: Nonlinear and Adaptive. In: IEEE Contr. Sys. Mag., 12:7–17.
9. Krstić M., Kanellakopoulos I. and Kokotović P. (1995) Nonlinear and Adaptive Control Design. Wiley, New York.
10. Khalil H. (2002) Nonlinear Systems. Prentice Hall, New York.

Standing up with Motor Primitives

V. Hamburger[1], K. Berns[1], F. Iida[2] and R. Pfeifer[2]

[1]AG Robotersysteme, TU Kaiserslautern, Germany; [2]Artificial Intelligence Laboratory, University of Zürich, Switzerland

Abstract

As observed in nature, complex locomotion can be generated based on an adequate combination of motor primitives. In this context, the paper focused on experiments which result in the development of a quality criterion for the design and analysis of motor primitives. First, the impact of different vocabularies on behavioural diversity, robustness of pre-learned behaviours and learning process is elaborated. The experiments are performed with the quadruped robot MiniDog6M for which a running and standing up behaviour is implemented. Further, a reinforcement learning approach based on Q-learning is introduced which is used to select an adequate sequence of motor primitives.

Keywords: Motor primitive, morphology, behavioural diversity, reinforcement learning, quadruped locomotion

1 Introduction

The theory of having a basic set of motor primitives which can be composed into a broad and general movement repertoire is an appealing organizational principle to avoid online trajectory planning. After all the representation of a function as a linear combination of less complex functions is a well established theory in mathematics and physics. Evidence in nature can be found in the human spinal cord as well as in various animals [3][4] e.g. about a dozen of non-adaptive primitives are enough to produce a frog's entire movement repertoire by means of sequence and superposition [5][6]. Moreover, it is widely believed that all sorts of movements seem to be designed as cyclic motions. Analysis of animal locomotion suggests that these motions are generated by neural networks which are capable of generating basic rhythmic motor activity [8]. These concepts motivated us to create a methodology that helps to derive several design principles for cyclic motor primitives. Questions which have to be answered are how can a basic set of fixed primitives, which is called vocabulary, be derived from morphological properties and how it affects behavioural diversity, robustness and learning process. Robustness in this context means that behaviours are tolerant against changes in morphology, environment and posture. Further, abstract, task and platform independent measures are provided to categorise and evaluate single motor primitives, entire vocabularies and behavioural diversity. This methodology is applied to the quadruped machine MiniDog6M (see fig. 1), for running and standing up behaviour. Fast running is obtained by a minimalist biologically inspired design which means that most of the control is compensated by exploiting some simple physics, such as the resilient properties of a spring yielding a passive degree of freedom in each leg. The six motors (servos in shoulders, hips and spine) are controlled by a simple sinusoidal function without having to differentiate stance and flight phases.

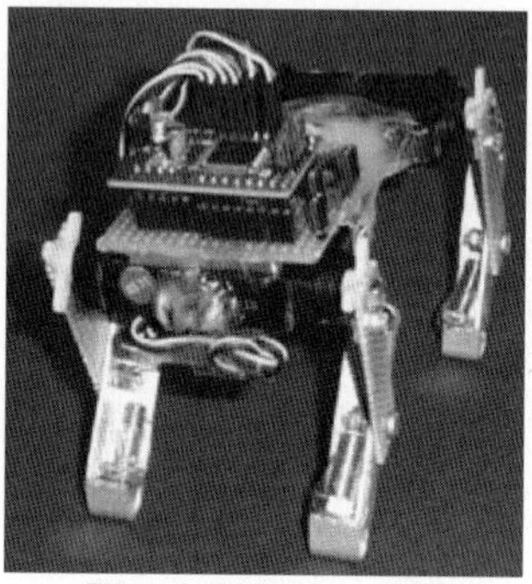

Fig. 1. MiniDog6M

2 Classification of vocabularies

The measures Flexibility-Index Flx and Coherence-Index Co are used to classify distinct vocabularies. In case a motor primitive does not consider all motors, a Don't-Care term is introduced, which causes a motor to remain in its current position. This means that their position depends on the last motor primitive which explicitly specified and set these motors in the past. Thus the recent history of selected components must be taken into account in order to get the current angles of all six motors.

To honour this flexibility, a special Flexibility-Index Flx is introduced in Equation (1) that represents the ratio between the actual amount and the theoretically maximum of Don't-Care terms 'x'.

$$Flx = \frac{actual\ amount\ of\ x}{max\ amount\ of\ x} \qquad (1)$$

A second criterion for motor primitives is the inner correlation respectively homogeneity of a posture. Therefore, a root posture needs to be defined, for instance home posture (here: all motors in mid position). The similarity to this root posture and the symmetry of motor positions is expressed by the Coherence-Index Co. In doing so, symmetry is regarded as primary criterion and mid position as secondary one. Symmetry in this connotation denotes parallel movement of front, hind, left and/or right hand legs.

First, the motors are grouped into "clusters of interest" attributing the fact that certain motors serve different purposes regarding their position and effective direction within the robot. For MiniDog6M we identify three groups: leg motors, bend motor, twist motor. In the special case of one motor per group (here: bend and twist motor), the goal position needs to be compared to the root posture only.

Tab. 1. Coherence distribution for each cluster (Bend, Twist, Legs)

Bend/Twist	0	x	+/-
Co(Bend/Twist)	1.0	0.33	0.0

legs	(0 0 / 0 0)	(* * / * *)	(0 * / 0 *)	(+ - / + -)	(0 * / 0 *)	(* x / * x)	(x x / x x)
Co(legs)	1.0	0.833	0.667	0.5	0.333	0.167	0

The tables in Table 1 list the coherence distribution for each cluster. '0' stands for mid position (high correlation), 'x' Don't-Care (correlation estimated with the probability of mid position) and '*' stands for minimum or maximum (low correlation). The circle denotes parallel movement. Note that the orientation of the robot is not given, so the head may be set on either side. Hence, an encircled pair of legs can be deemed front, hind, left or right hand, without loss of generality. Merging the Coherence-Index of spine and legs, we weight the corresponding values in ratio of the number of motors as shown in Equation (2).

$$Co(motor\ primitive) = \frac{1}{n} \cdot \sum_{all\ groups} Co(group) \cdot n_{grp} \quad (2)$$

The Coherence-Index of an entire vocabulary as set of motor primitives is defined in Equation (3) as ratio between the summed up Coherence-Index and the theoretical maximum.

$$Co(vocabulary) = \frac{\sum_{all\ motor\ primitives} Co(motor\ primitive)}{max} \quad (3)$$

3 A measure for behavioural diversity

Despite the great variety of definitions of the term intelligence, behavioural diversity is a central aspect in most of them. In order to provide a good basis of behaviours, it would be desirable to maintain a balance between diversity and heterogeneity. In order to evaluate behavioural diversity of a given task, the Behavioural-Diversity-Index BDI can be calculated as follows:

1. Delete all sequences that consist of another (shorter) sequence plus some prefix. (It is always possible to lengthen a sequence by performing senseless actions beforehand.)
2. Extract groups of sequences which have the same beginning or ending. Each sequence may appear in more than one group.
3. Divide every group into subgroups of identical length.
4. Calculate the diversity factor D for every group as in Eq. (4).

$$D = W_{nonmbr}(n) \cdot (1 - \frac{n_{grpmbr}}{n}) + W_{mbr}(n) \cdot \sum_{all\ groups} \frac{n_{submbr}}{n} \cdot (1 - \frac{n_{equal}}{l}) \quad (4)$$

D is a measure for the heterogeneity of the corresponding group. n_{grpmbr}, n_{submbr}: number of group resp. subgroup members, n: total amount of solutions, n_{equal}: number of equal steps, l: length of sequence (= number of steps)

5. Calculate BDI in Equation (5) as the product of the total amount of legal sequences with the mean diversity factor:

$$BDI = \frac{n_{total}}{n_{group}} \cdot \sum_{all\ groups} D \quad (5)$$

In our case study, we set $W_{nonmbr}(n) = W_{mbr}(n) = 1$. One might argue that, under certain circumstances, equal weights are inappropriate. For these cases it is recommended to set $W_{mbr}(n)$ and $W_{nonmbr}(n)$ as in Equations (6) and (7).

$$W_{mbr} = \begin{cases} 1 & n < 1 \\ 1 + \sin(\frac{4n}{\pi}) & 1 \leq n < 2N \\ 2 & 2N \leq n \end{cases} \quad (6)$$

$$W_{nonmbr} = \begin{cases} 2 & n < 1 \\ 1 + \cos(\frac{4n}{\pi}) & 1 \leq n < 2N \\ 1 & 2N \leq n \end{cases} \quad (7)$$

The threshold N can be chosen intuitively or as average amount of sequences over all vocabularies to be compared.

4 Experiments

In our experiments, the robot dog will be toppled by a random force applied to its head while running. Then MiniDog6M shall get up and carry on its way. Six different vocabularies have been examined with three different shapes of the head (see fig. 2), since changes in the shape of the head have biggest impact on the agent's behaviour. In order to overview the behavioural diversity, a full search algorithm was performed. These solutions have been tested on nine different ground configurations with slopes of 0°, 22.5° and 30°. Generally the support of the learning progress is an important feature of a good basic vocabulary. Here, a Reinforcement learning approach which linearly approximated a Q-Function with a Radial Basis Function network (RBF) was selected. This approach merges the advantages of both concepts. RBF has great generalisation abilities combined with optimal computational efficiency, but are highly non-linear and very difficult to analyse. Q-Learning is well formulated in mathematical terms and easy to understand for the external observer. The learning process was performed by the Reinforcement Learning Toolbox [8] with the following parameters: Epsilon-Greedy-Policy with $\varepsilon = 0.3$, learning rate $\alpha = 0.4$, discount factor $\gamma = 0.95$ and replacing eligibility traces with $\lambda = 0.9$. The reward function assigned 150 for running, 100 for standing, -1 for failure and 0 otherwise. The initial position (standing on its

Head, lying on the *Left*, *Right* or *Back*) for each learning episode is determined by applying a random force to knock over the running dog.

Fig. 2. Simulated versions of MiniDog6M

5 Results

The vertical head cannot roll over on the back, whereas the round head cannot stand on the head or lying on the back, but rolls over on the back most easily. Generally, Flx>0 results in a higher BDI, a primitive employing Don't-Care can occur in many different variants. Depending on the chosen vocabulary, there are many examples where the solutions for one initial position are equal or a subgroup of the solutions of another initial position within the same or different head forms. The solutions for the round head intersect to a large extent with the solutions of the original head. The solutions for *Right* of the original head are strongly related to the ones for *Left* of the vertical head.

Sometimes, even symmetrical tendencies are observed. This means for the original head that all solutions for the *Left* are also valid for *Back* and about half of the solutions for *Right* are valid for *Head*. Being an implication of the form, this relation is inverted for the vertical head which means that the solutions for *Back* are identical to those for *Right* instead of *Left*. Moreover, single solutions are valid for all initial positions of one head ('position stable'), some for the same initial position of different heads ('form stable'). Most surprisingly, some solutions are even valid for all heads and all initial positions (form and position stable). Depending heavily on the shape of the head, only the vertical head supports position stability for all vocabularies. An interesting fact is that sequences, that are stable in only one way, involve motor primitives with Flx$\geq$0, but that primitives belonging to form and position stable sequences all have Flx=0.

On inclines, it was found, that the original and the round head perform equal or better in most cases, irrespective of the underlying vocabulary. The robustness for the vertical head can be divided into different categories. For some vocabulary the solutions for *Left* and *Right* are switched. For the second category, this is only true for either *Left* or *Right*. The other side behaves as if with the original head. After all, the original and the round heads' solutions are really robust for applications in environment with slopes of various degrees and directions. Those for the vertical head underlie a strange morphological effect that it is somehow grounded in the shape.

Comparing the results gained from learning progress and behavioural diversity, a BDI>10 guarantees success in not more than 100 episodes, whereas in many cases vocabularies with a lower BDI do not even reach 100% success at all. Finding a proper sequence for these configurations equals the notorious search for the proverbial needle in a haystack. Interestingly, Flx>0 has different consequences for different morphologies. It was expected that this type of flexibility hinders the learning process for all heads because the goal posture is no longer unambiguous. Considering the restrictions of the distinct heads, it becomes obvious that the learning progress is the faster the less initial positions are supported, but only for vocabularies with Flx=0. Thus the round head performed best and original head poorest. For Flx>>0, it is just the other way around. This insight is especially astonishing, since the solutions for round are for the most part subsets of the solutions for the original head. This flexibility outweighs the additional effort resulting from more initial positions. Anyway, this effect comes effortlessly if it is considered in the design phase already.

6 Conclusion

While it is obvious that the introduction of discrete actions alone reduces the complexity of a learning task by avoiding online trajectory planning, it was also testified that learning and control processes are closely related to the morphological properties of the executing agent. As an effect of proper or improper shape, the given task can be simplified, complicated or even be ruled out by creating situations from which it is impossible to solve the task at all. This effect can easily be exploited if considered early in the design phase. Looking at toys a robot's infrastructure can be hidden under bizarre shaped plastic covers. If designed properly the decorative shell can serve as effortless enhancement of performance. Introducing inclines, the feasibility of the gained knowledge in changing environment without adaptation was demonstrated. This means no longer learning and still being able to succeed in new situations by making use of behavioural diversity (rating of vocabularies is listed in tab. 2). The numbers determine the rank (1 best, 6 worst) of the respective vocabulary depending on the shape of the head and the property. Co and Flx are ordered numerically (1 least, 6 highest value). O, V and R stand for original, vertical and round head. The resultant ranking of the vocabulary-head-combinations leads to the assumption of a hidden system behind the qualitative results which encourages to follow up matter.

Table 2. Final ranking of vocabularies

Vocabulary	Co	Flx	Learning Progress			BDI			Robustness		
			O	V	R	O	V	R	O	V	R
1	1	4	6	3	1	5	4	4	4	5	5
2	2	4	2	2	2	2	2	2	2	2	2
3	5	4	5	4	4	6	6	6	6	4	4
4	3	1	4	5	5	3	5	3	6	6	6
5	4	2	1	1	3	1	1	1	1	1	1
6	6	3	3	6	6	4	3	5	3	3	3

References

1. Verena Hamburger. Locomotion of a quadruped robot based on motor primitives. Diploma thesis, Robotic System Group, TU Kaiserslautern, April 2005
2. Rolf Pfeifer, Christian Scheier. *Understanding Intelligence*. MIT Press, 1999.
3. E. Bizzi, N. Acornero, W. Chapple, N. Hogan. Posture control and trajectory formation during arm movement. 4:2738–2744, 1984.
4. Feldman. Superposition of motor programs, I. Rhythmic forearm movement in man. 5:81–90, 1980.
5. Simon F. Giszter, Fernando A.Muss-Ivaldi, Emilio Bizzi. Convergent Force Fields Organized in the Frog's Spinal Cord. *Journal of Neuroscience*, Februar 1993.
6. F.A.Mussa-Ivaldi, S.F. Giszter, E. Bizzi. Linear combinations of primitives in vertebrate motor control. 1994.
7. Ansgar Büschges. Sensory Control and Organization of Neural Networks Mediating Coordination of Multisegmental Organs Organs for Locomotion. *Journal of Neurophysiology*, Februar 2005.
8. Gerhard Neumann. The Reinforcement Learning Toolbox, Reinforcement Learning for Optimal Control Tasks. Diploma thesis, Institute of Computer Science, TU Graz, May 2005.

Multiple Terrain Adaptation Approach Using Ultrasonic Sensors for Legged Robots

S Nabulsi, M Armada, and H Montes

Automatic Control Department of the Industrial Automation Institute (IAI-CSIC) Madrid, Spain. snabulsa@iai.csic.es

Abstract

In this article, different methods are discussed in order to define a system for a legged robot to adopt an attitude when, while moving, finds an obstacle, irregularities and/or different kinds of soil. On-board ultrasonic sensors are used to detect the surface conditions and to determine the behaviour of the robot while walking, and therefore, with the help of the information obtained from other sensors, the system can choose between different types of gaits in order to optimise the walking conditions in each particular situation. Some experiments are presented to show the performance of the proposed approach of multiple terrain adaptation for legged robots.

Keywords: Walking robot, terrain adaptation, ultrasonic sensors.

1 Introduction

Even though the adaptation over uneven terrain is one of the main advantages of legged robots, still there is no autonomous machine suitable for walking on different types of terrain or soil with variable mechanical properties [1] in a completely autonomous way. In most of the cases the control unit is predefined to behave according to a specific application as seen widely in literature [2, 3, 4, 5], but this problem increases when the robot have to move through more than one type of terrain.

Roboclimber [6, 7] is a 3500kg, four legged climbing and walking robot designed to work remotely on rocky sloped mountains where the possibility of landslides requires the tasks of firming up the slopes by drilling 20m. depth holes.

In this application, sometimes the conditions are favourable enough to get near the mountain and manipulate the robot with a truck crane and hang up the robot from the ropes placed on the top of the mountain. But sometimes there is no possibility for a truck to approach close enough to the mountain to make this operation, so the robot should make the necessary walking positioning tasks to be able to get near and hang itself up from the mountain; in this case the robot can get through many types of terrain. This situation may arise also in many other applications for walking machines. For a remote operator it is not always possible to make the best decision about what type of control is optimal to operate in a certain type of terrain knowing that, when it is a natural one, the conditions can vary unexpectedly.

Fig. 1. Roboclimber in walking process: a) irregular soft terrain; b) obstacle step; c) plain hard terrain

Previous works show control strategies where the adaptation on natural terrain can be done in real time without external sensors but, for specific types of terrain; at the moment the terrain condition changes, the control is not often optimal according to the relationship of the position between the body of the robot and the surface, the working space of the legs, velocity, and even more important, stability.

For example, as shown on the sequence above (see **Fig. 1**), the robot must change from a regular soft terrain (a), to overcome an obstacle (b), and finally to start walking again on a regular terrain but with more stiffness (c).

This situation can lead to many problems of stability and efficiency, and the operator must choose between the different walking configurations to reach the desired position in each case as shown on **Fig. 2**.

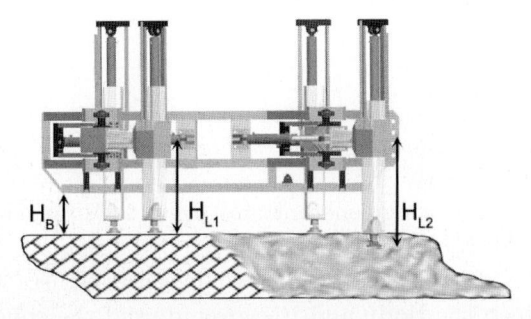

Fig. 2. Example of a robot over a multiple soil surface. H_B distance between the body and the surface; H_{L1} extension of Leg 1; H_{L2} extension of Leg 2

2 Body/Ground Distance Control

Like the sensorial systems used by animals or humans, in robots external sensors are used to identify the environment and to adopt the attitude needed to overcome an obstacle or to avoid complicated situations. Many kinds of sensor are used to determine external conditions in robotics (artificial vision systems, distance detection systems, ground detection, etc.). The idea of this development is to generate a not very complex sensor control to determine the properties of a natural landscape where a robot can work.

Ultrasonic distance sensors were chosen, to acquire information to detect and measure possible obstacles on the ground. This kind of sensors can give great accuracy and a very long distance detection (up to 3000mm). Some experiments were designed to show the performance of these kind of sensors for a real-time body position control. The control strategy has an external force control loop that is converted, through the desired distance q_m(e.g. between the floor and the robot body), into velocity V_m (see **Fig. 3**).

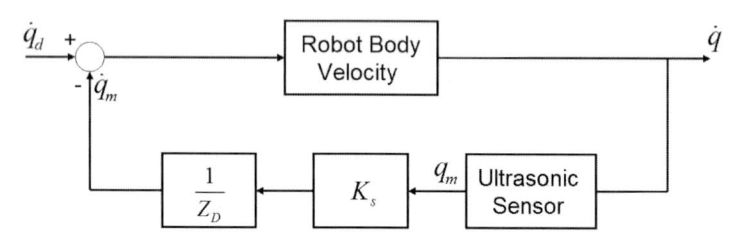

Fig. 3. Block diagram for robot velocity control.

Where K_s is the spring constant and Z_D is the impedance constant of the system which converts the position error to a velocity signal as shown in Equation (1). In this case an ultrasonic sensor is used to measure the distance between the body and the ground, and the robot will adopt a springy behaviour to overcome or to increase the distance between the body and a possible irregularity on the ground. The position signal is used as feedback for the velocity control, which is defined by the control algorithm,

$$\dot{q}_m = K_s[\frac{1}{Z_D}(q_d - q_m)] \quad (1)$$

The next image (see **Fig. 4a**), shows an experiment made to keep the distance between the uneven floor and the robot while the body is moving horizontally, avoiding possible ground interferences. The behaviour is described by the phase diagram shown in the figure (see **Fig. 4b**) where E_z is the error of the distance and E_{dz} is the velocity of body of the robot, and the velocity control (see **Fig. 3**) is design to keep always the same distance between the robot and the floor; this, of coarse, depends on the position of the sensor in the robot. For practical uses the best position is in the front part of the robot where it is possible to detect an irregularity or an obstacle before reaching it.

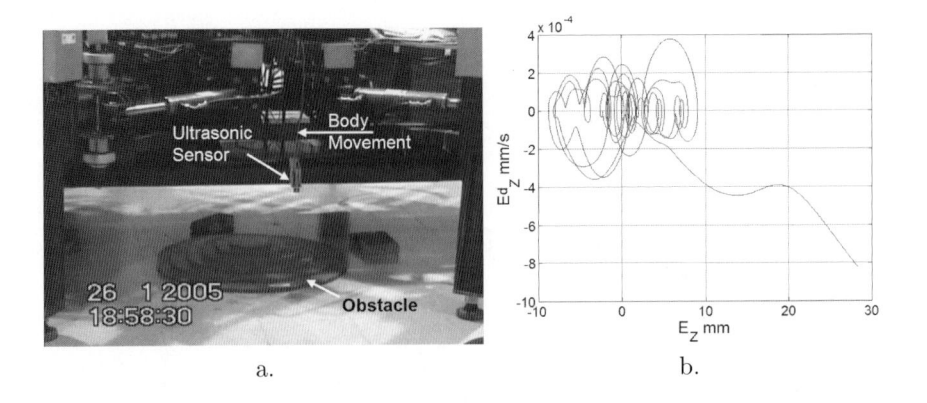

Fig. 4. Behaviour of the system controlled by the ultrasonic sensor.

3 Soil Properties Analysis

Some important problems emerge while walking over natural terrain, and even more when it is over soft soil. The most common solution is to level the inclination of the robot's body after a leg transfer [2]. This leveling is used to correct the joint position misreading after a ground contact, caused by the mechanical deformations of the robot's body, but on soft terrain different approaches must be considered; the walking gait is designed to optimize the stability of the quadruped robot when a leg is lifted and the robot is supported by the other three legs, meaning that the forces supported by the ground can be bigger and burying of the legs can occur. This situation may be given again while the body is moving because the center of gravity of the robot is constantly changing, and consequently the weight distribution between the legs.

Coordinated leg movements are used to optimize the time for the leg transfer. If any of the legs is buried in the ground some coordinated leg movements can be blocked and can produce malfunctioning in the robot systems or instability (see **Fig. 5**). In the first case (1.), the trajectory of the coordinated movement can be blocked by the surface where the leg is buried. In the second case (2.) a trajectory is selected in a way obstacles can be avoided, and in the third one (3.) is a highly secure trajectory but the time consumption is very high.

In hard soil type terrain (see **Fig. 5b**), like in a rocky surface, the trajectories can be planned as it follows and in general there is no difficulty in generating coordinated movements. So, leg trajectories may be selected by the system in order to generate optimum velocity.

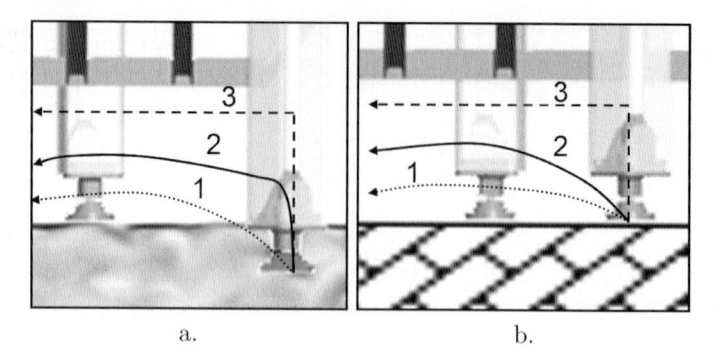

Fig. 5. Possible leg trajectories over a) soft soil and b) hard soil.

Our approach consist in using a ultrasonic sensor capable of measuring the difference between the body and the surface and the desired position of the joints of the leg (see **Fig. 2**), where it is possible to compute the stiffness of the soil knowing the depth after touching the ground. In this way a specific coordinate movement can be selected in each situation in order to optimize not only stability but the robot velocity avoiding further complications.

4 Multiple Terrain Adaptation and Obstacle Avoidance

In order to synthesize the combination between the different data generated by the procedures explained before and the relationship with the basic gait elements, preferably, the minimum amount of sensors should be used. So the walking strategy through natural terrain, with different kinds of soil and obstacle avoidance must start with an ultrasonic array of sensors placed in the front, or walking direction, of the robot in order to scan the presence of obstacles adjusting the distance between the body and the floor. With this procedure decisions about attainable obstacle avoidance must be done, generating an attitude according to the kind of difficulty left to overpass by the legs and the basic gait position generation.

The control system must be prepared in order to realize a real time soil adaptation while the body is in movement or while a leg transfer, is necessary for the obstacle avoidance assistance. The control system of the robot must proceed in order to keep the body level and select the optimal leg trajectory, and the predefined distance respect the surface decided

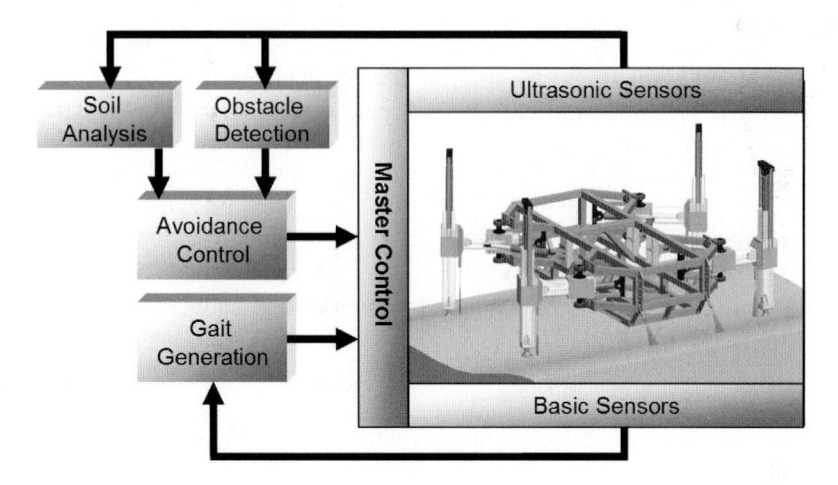

Fig. 6. Block diagram for robot velocity control.

by the control system according to the soil properties combined with the level inclination methods, the position, ground contact and force control of the legs used by the basic gait generation (see **Fig. 6**).

5 Conclusions

The solutions given are appropriate for a robot to work over natural surface conditions. Instead of a manual selection made by the user (as shown in the first example (see **Fig. 1**), a complete autonomous operation could be made in order for the robot to adapt to multiple terrain and soil circumstances by this new approach.

These approaches are based on the experiences gained with Roboclimber, which is a heavyweight robot but can be considered also for any kind of medium weight robot operation. Predefined gaits are used to generate the robot movements, so the system must be capable of running the exposed approaches from an external control loop. Some developments are in progress to create in real time, a map surface to optimize free gaits generation.

Acknowledgments

ROBOCLIMBER project is funded by the EC under Contract N: G1ST-CT-2002-50160. The project partnership is as follows: ICOP S.p.a.,

Space Applications Services (SAS), Otto Natter Prazisionenmechanik GmbH, Comacchio SRL, Te.Ve. Sas di Zannini Roberto & Co. (TEVE), MACLYSA, D'Appolonia S.p.a., University of Genova-PMAR Laboratory, and CSIC-IAI.

References

1. Maza M, Fontaine J G, Armada M, Gonzales de Santos P, Papantoniou V, Mas M (1997) Wheel+Legs: A New Solution For Traction Enhancement Without Addaptive Soil Compaction. IEEE: Robotics and Automation Magazine, Vol. 4, No. 4.
2. Jimenez M A, Gozalez de Santos P (1997) Terrain-Adaptive Gait for Walking Machines. The International Journal of Robotics Research, Vol. 16, No. 3, pp 320-339.
3. Bares J E, Whittaker W L (1993) Configuration of Autonomous Walker for Extreme Terrain. The International Journal of Robotics Research, Vol. 12, No. 63, pp 535-559.
4. Song S, Waldron J K (1989) Machines That Walk: The Adaptive Suspention Vehicle. The MIT Press.
5. Hirose S (1984) A Study of Design and control of a Quadruped Walking Vehicle. The International Journal of Robotics Research, Vol. 3, No. 2.
6. Anthoine P, Armada M, Carosio S, Comacchio P, Cepolina F, Gonzlez P, Klopf T, Martin F, Michelini RC, Molfino R M, Nabulsi S, Razzoli R P, Rizzi E, Steinicke L, Zannini R, Zoppi M (2003) ROBOCLIMBER. In: ASER03, 1st International Work-shop on Advances in Service Robotics, March 13-15, Bardolino, Italy.
7. Nabulsi S, Armada M (2004) Climbing Strategies for Remote maneuverability of ROBOCLIMBER. In: 35th International Syposium on Robotics, 23-26 March, Paris-Nord Villepine, France.

Sliding Mode Observer with No Orientation Measurement for a Walking Biped

V. Lebastard, Y. Aoustin, and F. Plestan

IRCCyN, Ecole Centrale de Nantes, CNRS, Université de Nantes - BP 92101, 1, rue de la Noë, 44321 Nantes cedex 03, France
{Vincent.Lebastard, Yannick.Aoustin, Franck.Plestan}@irccyn.ec-nantes.fr

1 Abstract

An observer-based controller is proposed to estimate the absolute orientation of a three-link biped without feet, during a cyclic walking gait composed of single supports and impacts. The angular moment is used for the design of the observer, based on second-order sliding mode approach, and for analysis of the stability of the walking.

Keywords: Second order sliding mode observer, Poincaré's sections, angular momentum, finite time convergence, biped.

2 Introduction

This work proposes an observer-based control of a biped robot, supposing that only joint variables are available. Usually, a precise measurement of the absolute orientation of a walking biped is, by a technical point-of-view, quite difficult to get. However, the generalized coordinates are often necessary for the control of the walking robot [1, 2, 3, 4]. Then, there is a real interest to develop observers in order to estimate the absolute orientation from only the knowledge of joint variables. To our best knowledge, very few works have been done for the design of such observers, observers being traditionally design for the estimations of velocities (for noiseless differentiation) by supposing that all the angular variables are measured [5]. In the present paper, an original observer, based on second-order sliding mode control ([6]) because a property of this class of observers is the finite-time convergence of the estimation error, witch simplifies the stability proof of the walking gait of the biped under the observer based-control [7]. The article is organized as follows: the dynamical model of the under interest robot is presented in Section 3. The control law of the biped is presented in Section 4. A second order sliding mode observer, based on the *twisting algorithm*, with its associated canonical form in order to estimate the absolute orientation of the biped, is detailed in Section 5. Conclusion and perspectives finish this paper.

3 Model of the biped

The complete model of the biped robot consists of two parts: the differential equations describing the dynamics of the robot during the swing phase and an impulse model of the contact event.

3.1 Swing motion equations

A planar three-link biped is considered (see **Fig.1**) and is composed by a torso and two identical legs without knee and foot. The joint between the torso and each leg is actuated by an actuator located in the hip. Then, the biped is under-actuated in single support phase. Assume that during the swing phase, the stance leg is acting as a pivot; the contact of the swing leg with the ground results in no rebound and no slipping of the swing leg. Then, the dynamic model, based on Lagrange's equations reads as

$$D\ddot{q} + H\dot{q} + G = B\Gamma \qquad (1)$$

where the vector q is composed by the joint variables and the absolute orientation of the trunk, $q := [\delta_1 \ \delta_2 \ \psi]^T$ (see **Fig.1**). $D(\delta_1, \delta_2)(3 \times 3)$ is the symmetric positive inertia matrix, $H(q, \dot{q})(3 \times 3)$ the Coriolis and centrifugal effects matrix, and $G(q)(3 \times 1)$ the gravity effects vector. $B(3 \times 2)$ is a constant matrix composed of 1 and 0. System (1) can be written as a nonlinear dynamical system

$$\dot{x} := \begin{bmatrix} D^{-1}(-H\dot{q} - G + B\Gamma) \\ \dot{q} \end{bmatrix} =: f(x) + g(q_{rel}) \cdot \Gamma \qquad (2)$$

with $x := [\dot{q}^T, q^T]^T = [\dot{\delta}_1 \ \dot{\delta}_2 \ \dot{\psi} \ \delta_1 \ \delta_2 \ \psi]^T$ and $q_{rel} := [\delta_1 \ \delta_2]^T$ the relative angles. The state space is taken such that $x \in \mathcal{X} := \{x := (\dot{q}^T, q^T)^T \mid \dot{q} \in \mathcal{N}, \ q \in \mathcal{M}\}$, where $\mathcal{N} = \{\dot{q} \in {I\!\!R}^3 \mid |\dot{q}| < \dot{q}_M < \infty\}$ and $\mathcal{M} = (-\pi, \pi)^3$.

3.2 Passive impact model

The impact occurs at the end of a single support phase, when the swing leg tip touches the ground. State the subscript 2 for the swing leg and 1 for the stance leg during the single support phase. An impact occurs when δ_2 equals a desired value δ_{2f}, i.e. $x \in \mathcal{S} = \{x \in \mathcal{X} \mid \delta_2 = \delta_{2f}\}$ [8]. The choice of δ_{2f} directly influences the length of the step. Assume that the impact is passive and absolutely inelastic [9], the swing leg touching the ground does not slip and the previous stance leg takes off the ground. Then the angular positions are continuous, the angular velocities discontinuous. The ground reactions at the impact can be considered as impulsive forces and defined by Dirac delta-functions. Considering $x^+ := [\dot{q}^{+T}, q^T]^T$ (state just before the impact) in terms of $x^- := [\dot{q}^{-T}, q^T]^T$ (state just after the impact), the impact can be described through the algebraic relation $x^+ = \Delta(x^-)$ [8]. The overall biped model can be expressed as a nonlinear system with impulse effects as $\begin{cases} \dot{x} = f(x) + g(q_{rel})\Gamma & x^-(t) \notin \mathcal{S} \\ x^+ = \Delta(x^-) & x^- \in \mathcal{S}. \end{cases}$ where $\mathcal{S} = \{x \in \mathcal{X} \mid \delta_2 = \delta_{2f}\}$.

4 Control law

The control for the walking gait [8] consists in maintaining the angle of the torso at some constant value ψ_d and in controlling the swing leg such that it behaves as a mirror image of the stance leg, $\theta_2 = -\theta_1$ (see **Fig.1**). During the single support phase, the degree of the underactuation equals one: only two outputs can be driven. Then, the robot gets a walking motion if the controller drives to zero the following outputs $v := [v_1 \ v_2]^T = [\psi - \psi_d \ \theta_2 + \theta_1]^T =: h(x)$. The control consists

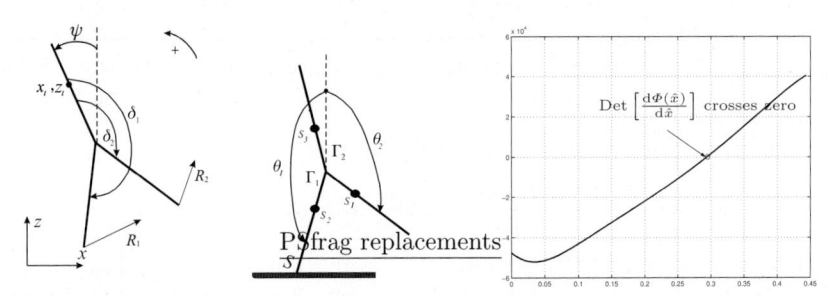

Fig. 1. Three-link biped's diagram: generalized coordinates, torques, forces applied to the leg tips.

Fig. 2. $\det(\frac{d\Phi(\tilde{x})}{d\tilde{x}})$ versus time (sec.) along one step

in decoupling the system and in imposing a desired dynamic response. Note that, in $\mathcal{X}$, the decoupling matrix $L_g L_f h$ is always invertible. The control law Γ is then $\Gamma := [L_g L_f h]^{-1}[-L_f^2 h + w]$, in v such that $\ddot{v} = w$. In the present work, the control law w is chosen to be *finite time convergent*, which could be done with for example sliding mode approach. The feedback function used in the present work comes from [10]: $w = \Upsilon(v, \dot{v}) := \frac{1}{\epsilon} \cdot \begin{bmatrix} \Upsilon_1(v_1, \epsilon \cdot \dot{v}_1) \\ \Upsilon_2(v_2, \epsilon \cdot \dot{v}_2) \end{bmatrix}$. Each function $\Upsilon_i(v_i, \epsilon \cdot \dot{v}_i)$, is defined as $\Upsilon_i := -\text{sign}(\vartheta_i(v_i, \epsilon \cdot \dot{v}_i)) \cdot |\vartheta_i(v_i, \epsilon \cdot \dot{v}_i)|^{\frac{\alpha}{2-\alpha}} - \text{sign}(\epsilon \cdot \dot{v}_i) \cdot |\epsilon \cdot \dot{v}_i|^\alpha$, with $\vartheta_i(\cdot) = v_i + \frac{1}{2-\alpha}\text{sign}(\epsilon \cdot \dot{v}_i) \cdot |\epsilon \cdot \dot{v}_i|^{2-\alpha}$ and $0 < \alpha < 1$. The real parameter $\epsilon > 0$ allows the settling time of the controllers to be adjusted.

5 Second-order sliding mode observer

In this section is presented a second order sliding mode observer [6]. The use of this class of observers is motivated by the following reasons. First, as dual properties of sliding mode control, these observers have robustness and finite time convergence properties. Furthermore, the "second order" interest consists in the reduction of chattering phenomenon. The angular momentum evaluated at the leg tip of the biped is $\sigma = \sum_{i=1}^{2} f_i(\delta_1, \delta_2)\dot{\delta}_i + f_3(\delta_1, \delta_2)\dot{\psi}$. Consider system (2), with the outputs $[y_1 \ y_2 \ y_3 \ y_4]^T = [s_1 \ s_2 \ \delta_1 \ \delta_2]^T$ (with $s_1 = f_1(\delta_1, \delta_2)\dot{\delta}_1$ and $s_2 = f_2(\delta_1, \delta_2)\dot{\delta}_2$). The observability indices [11] are respectively $[k_1 \ k_2 \ k_3 \ k_4]^T = [2 \ 2 \ 1 \ 1]^T$, and system (2) is generically observable [7]. There exists a local state coordinates transformations $z = \Phi(x)$ defined as $z = [z_1 \ z_2 \ z_3 \ z_4 \ z_5 \ z_6]^T := [s_1 \ \dot{s}_1 \ s_2 \ \dot{s}_2 \ \delta_1 \ \delta_2]^T$, such that system (2) can be rewritten as

$$\dot{z} = \underbrace{[z_2 \ 0 \ z_4 \ 0 \ 0 \ 0]^T}_{Az} + \underbrace{[0 \ \varphi_2(z) \ 0 \ \varphi_4(z) \ \varphi_5(z) \ \varphi_6(z)]^T}_{\varphi(z)} \quad (3)$$

Observer design :
Suppose that there exists a system defined as

$$\dot{\hat{z}} = A\hat{z} + \varphi(\hat{z}) + \chi(\hat{z}, y) \quad (4)$$

which is an observer of (3), $\hat{z}$ being the estimation of z. Let $\chi(\hat{z}, y)$ define as $\chi = [0 \ \chi_2 \ 0 \ \chi_4 \ \chi_5 \ \chi_6]^T$. The goal is to define functions χ_i such that the estimation error $e = \hat{z} - z$ reachs zero in a finite time. One uses the standard sliding mode approach for χ_5 and χ_6 by $\chi_5 = -\lambda_5 \text{sign}(\hat{z}_5 - z_5)$, $\chi_6 = -\lambda_6 \text{sign}(\hat{z}_6 - z_6)$. Functions χ_2 and χ_4 read as $\chi_2 = -\Lambda_2 \text{sign}(e_1)$,

$\chi_4 = -\Lambda_4 \, \text{sign}(e_3)$. The choice for Λ_2 and Λ_4, based on the *twisting algorithm* [12], allows to ensure that the previous system converges to zero in finite-time. Then, from (4) and the inverse state transformation $\hat{x} = \Phi^{-1}(\hat{z})$, a second-order sliding mode observer for (2) reads as $\dot{\hat{x}} = f(\hat{x}) + g(y_3, y_4)\Gamma + \left[\frac{d\Phi(\hat{x})}{d\hat{x}}\right]^{-1} \chi(\hat{x}, y)$, with $\chi(\hat{x}, y) = [0 \; \chi_2 \; 0 \; \chi_4 \; \chi_5 \; \chi_6]^T = [0, \Lambda_2 \text{sign}(\hat{s}_1 - y_1), 0, \Lambda_4 \text{sign}(\hat{s}_2 - y_2), \lambda_5 \text{sign}(\hat{x}_4 - y_3), \lambda_6 \text{sign}(\hat{x}_5 - y_4)]^T$

Loss of observability : During the swing phase, and along the desired trajectories, the determinant of $\frac{d\Phi(x)}{dx}$ crosses zero (**Fig.2**), which implies a loss of observability. Around this singular point, it is necessary to make adaptations/corrections of the observer, which is not valid at exactly the singularity. An intuitive and quite natural solution is proposed in [7]. A smooth corrective term[1] $\Theta \in \mathbb{R}$ is added at the observer such that $\dot{\hat{x}} = f(\hat{x}) + g(y_3, y_4)\Gamma + \Theta\chi(\hat{x}, y)$.

5.1 Simulations

The control law described in Section 4 is applied with parameters $\alpha = 0.9$ and $\epsilon = 20$. The initial real and estimated values have been respectively stated as : $\begin{bmatrix}\dot{\delta}_1 & \dot{\delta}_2 & \dot{\psi} & \delta_1 & \delta_2 & \psi\end{bmatrix} = [-1.53, 1.53, -0.05, -2.9374, -2.4033, -0.6283]$, $\begin{bmatrix}\dot{\hat{\delta}}_1 & \dot{\hat{\delta}}_2 & \dot{\hat{\psi}} & \hat{\delta}_1 & \hat{\delta}_2 & \hat{\psi}\end{bmatrix} = [-1.5, 1.5, 0, -2.8798, -2.3562, -0.5236]$. Parameter D_{min} has been stated to $0.5 \cdot 10^4$, and the sliding mode observer parameters are $\lambda_m^2 = 500$, $\lambda_M^2 = 2500$, $\lambda_m^4 = 50$, $\lambda_M^4 = 250$, $\lambda_5 = 1$, $\lambda_6 = 0.01$. The choice of observer and control law parameters has been made with respect to closed-loop dynamics and admissible maximum value for input (saturation). **Fig.3** displays the absolute position

---
[1] This term depend of a parameter D_{min} (see [7])

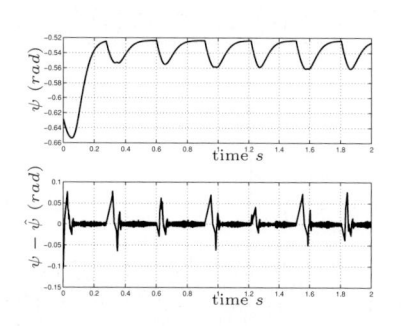

Fig. 3. Absolute orientation ψ (top), estimation error $\psi - \hat{\psi}$ (bottom)

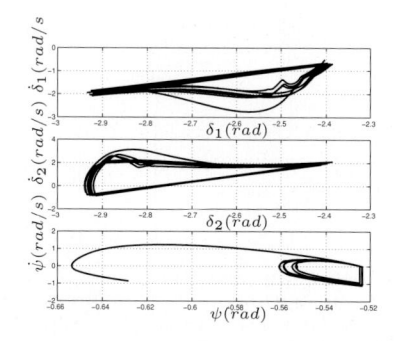

Fig. 4. Phase plan $(\dot{\delta}_1, \delta_1)$, $(\dot{\delta}_2, \delta_2)$ and $(\dot{\psi}, \psi)$.

ψ and the estimation error $\psi - \hat{\psi}$, the absolute position converge to the desired value ψ_d and the estimation error has converge towards a neighbourhood of zeros before each impact. **Fig.4** displays the convergence, to a stable cycle, of walking gait over several steps. During first step, a large transient appear and it was induce by the estimated state variables have converging to real values. This transient appear in the first step only. Once the observer has converged, the control law ensures that the biped robot reaches a stable walking cycle, as the initial conditions have been adequately chosen.

6 Stability and angular moment

Fig.4 shows the effect of the observer and the initial conditions (error between the real and the estimate state). The proof of the stability was developed in the paper [13]. In the present paper, only a sketch of this proof is displayed. The approach uses the angular momentum. In Section 5, the angular momentum and the output are defined. Let $\bar{\sigma}$ define as $\bar{\sigma} = y_1 + y_2 = f_1(\delta_1, \delta_2)\dot{\delta}_1 + f_2(\delta_1, \delta_2)\dot{\delta}_2$. and its estimated value $\hat{\bar{\sigma}} = f_1(\hat{\delta}_1, \hat{\delta}_2)\dot{\hat{\delta}}_1 + f_2(\hat{\delta}_1, \hat{\delta}_2)\dot{\hat{\delta}}_2$. Then, the angular momentum can be rewritten as $\sigma = \bar{\sigma} + f_3(\delta_1, \delta_2)\dot{\psi}$. The stability proof is made over two steps (see **Fig.5**) and is based on the existence of $T_O^1, T_C^1, T_I^1, T_C^2$ and T_I^2, such that $0 \leq T_O^1 \leq T_C^1 \leq T_I^1 \leq T_C^2 \leq T_I^2$ and during the first step, one has

- For $0 \leq t < T_O^1$ (observer finite time), the estimated state converges to the real one.
- For $t \geq T_O^1$, then $\hat{\bar{\sigma}} = \bar{\sigma}$: the estimated state equals the real one.
- At $t = T_C^1$, the controller has converged, which induces that the outputs equal zero: $\hat{\bar{\sigma}} = \bar{\sigma} = \sigma$.

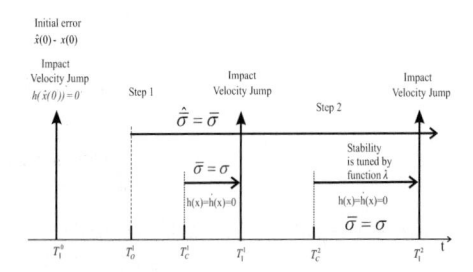

Fig. 5. Scheme defining the observer convergence time T_O^i, the controller convergence time T_C^i, the impact time T_I^i and the behavior of the observer (through the estimation error) and the controller (through the output $h(x)$ and its time derivative), over several steps.

- At $t = T_I^1$, an impact occurs.

The existence of T_O^1 and T_C^1 depends of the initial estimation errors $(e_\psi, e_{\dot\psi})$ and value of $\hat{\bar{\sigma}}$ (see **Fig.6**). If the system is initialized such that the estimation errors are in the 3D area displayed in **Fig.6**, the system converges to the limit cycle. During the second step, one has

- For $t > T_C^2$: $\hat{\bar{\sigma}} = \bar{\sigma} = \sigma$: the controller has converged and ensures that the outputs go to zero before the next impact, $T_C^2 \leq T_I^2$.

The existence of a stable limit cycle is shown through the existence of a fixed point of the function λ defined as $\lambda\left[\hat{\bar{\sigma}}_1^-(T_I^1)\right] := \hat{\bar{\sigma}}_1^-(T_I^2)$. This fixed point is highlighted in **Fig.7** (for details see [7])

7 Conclusion

The measurement of the absolute orientation of a biped is a difficult problem from a technical point of view. In order to overcome this drawback, a nonlinear observer based on sliding mode method is designed for the estimation, in finite time of the absolute orientation of the biped. In previous works, its has been observed that the walking gait for a biped without feet is essentially due to the gravity effects. Then, in this paper the design of the surface of the sliding mode observer is defined as a function of the angular momentum evaluated at the stance leg tip, which is explicitly function of the gravity. The simulation results lead to think that an experimentation of this observer is possible.

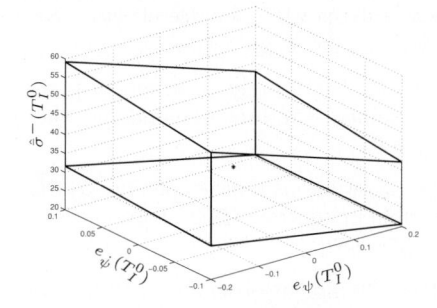

Fig. 6. $\hat{\bar{\sigma}}^-(T_I^1)$ in terms of $e_\psi(T_I^0)$ and $e_{\dot\psi}(T_I^0)$. Each point of this 3D-area allows a convergence to the stable limit walking cycle. The star-point corresponds to the conditions for which the biped evolves on the stable limit cycle.

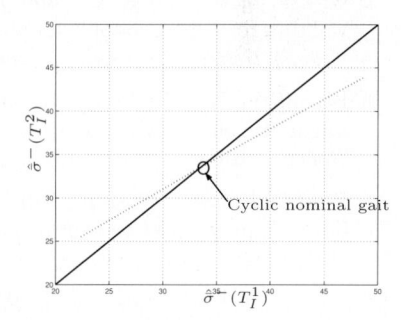

Fig. 7. Function λ (dotted line) and identity function (bold line) versus $\hat{\bar{\sigma}}^-(T_I^0)$. This graph describes the existence of an asymptotically stable walking motion.

References

1. Aoustin, Y. and Formal`sky, A.M. (2003) Control design for a biped: Reference trajectory based on driven angles as functions of the undriven angle. *Journal of Computer and Systems Sciences International*, vol. 42, no. 4, pp. 159–176.

2. Chevallereau, C., Abba, G., Aoustin, Y., Plestan, F., Westervelt, E.R., de Wit, C. Canudas, and Grizzle, J.W. (2003) Rabbit: a testbed for advanced control theory. *IEEE Control Systems Magazine*, vol. 23, no. 5, pp. 57–79.

3. Mu, X. and Wu, Q. (2003) Synthesis of a complete sagittal gait cycle for a five-link biped robot. *Robotica*, vol. 21, pp. 581–587.

4. Plestan, F., Grizzle, J.W., Westervelt, E.R., and Abba, G. (2003) Stable walking of a 7-dof biped robot. *IEEE Transactions on Robotics and Automation*, vol. 19, no. 4, pp. 653–668.

5. Micheau, P., Roux, M.A., and Bourassa, P. (2003) Self-tuned trajectory control of a biped walking robot. In *Proc. International Conference on Climbing and Walking Robot*, Catania, Italy, pp. 527–534.

6. Floquet, T., Barbot, J.P., and Perruquetti, W. (2002) A finite time observer for flux estimation in the induction machine. In *Proc. IEEE Conference on Control Applications*, Glasgow, Scotland.

7. Vincent, V., Plestan, Y., and Plestan, F. (2005) Second order sliding mode observer for stable control of a walking biped robot. In *Proc. IFAC World Congress 2005*, Praha, Czech Republic.

8. Grizzle, J.W., Abba, G., and Plestan, F. (2001) Asymptotically stable walking for biped robots : analysis via systems with impulse effects. *IEEE Transactions on Automatic Control*, vol. 46, no. 1, pp. 51–64.

9. Hurmuzlu, Y. and Marghitu, D.B. (1994) Rigid body collisions of planar kinematic chains with multiple contact points. *The International Journal of Robotics Research*, vol. 13, no. 1, pp. 82–92.

10. Bhat, S.P. and Bernstein, D.S. (1998) Continuous finite-time stabilization of the translational and rotationnal double integrator. *IEEE Transaction on Automatic Control*, vol. 43, no. 5, pp. 678–682.

11. Krener, A.J. and Respondek, W. (1985) Nonlinear observers with linearizable error dynamics. *SIAM J. Contr. Optim.*, vol. 2, pp. 197–216.

12. Levant, A. (1993) Sliding order and sliding accuracy in sliding mode control. *International Journal of Control*, vol. 58, no. 6, pp. 1247–1263.

13. Lebastard, V., Aoustin, Y., and Plestan, F. (2004) Observer-based control of a biped robot. In *Proc. Fourth International Workshop on Robot Motion and Control*, Puszczykowo, Poland, pp. 67–72.

Humanoid Robots

Detection and Classification of Posture Instabilities of Bipedal Robots

O. Höhn[1], J. Gačnik, and W. Gerth

Institute of Automatic Control, University of Hannover
hoehn@irt.uni-hannover.de

Abstract

This paper deals with a pattern recognition approach to detect and classify falls of bipedal robots according to intensity and direction. Reflex motions, that are initiated by the classified state, are intended to prevent the robot from falling.

Keywords: Fall detection, pattern recognition, reflex motions

1 Introduction

A legged service robot that is working autonomously in the human environment is at a permanent risk of loosing its balance. To give an example the ground can be uneven or slippery. Also, there can be impacts against the robot's torso that can cause a downfall.
The reactions that have to be executed to keep the robot from falling depend on the degree of instability, the situation and the robot's current state. So the very first thing before initiating a stabilizing reaction is to detect and classify the fall according to its direction and intensity.
Stability predications with the 'Zero Moment Point' which is usually considered to evaluate the postural stability of humanoid robots have turned out to be little significant. Especially with stronger disturbances the ZMP permits no statement about the degree of instability. More suitable for the recognition and the classification of a forthcoming fall are procedures which are based on the foot tilting angle and its angular velocity or the combination of different types of sensor data. Appropriate for this task are for instance algorithms that originate from the field of pattern recognition.

2 Simulation Environment

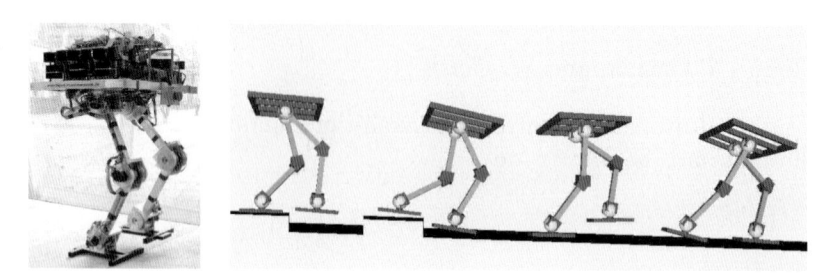

Fig. 1. Left: Biped BARt-UH. **Right:** Simulated robot model with four additional active joints. After toppling over a step, a reflex motion is performed.

The research of falling down motions entails a high risk of seriously damaging the robot. For this reason a simulation tool was developed to simulate the robot's behaviour if falling down. The simulation model is related to our 6 DOF bipedal robot BARt-UH [1] with exception of two additional active joints in the hip and the ankle of each leg (**Fig. 1**). So this model provides the possibility to examine more general (especially sideward) motions of the robot and with some restrictions it is still comparable with the real biped.

Including the six virtual joints the robot is fixed to the world coordinate system, the equation of motion has 16 degrees of freedom. It can be written as

$$\mathbf{M(q)\ddot{q}} + \mathbf{c(q,\dot{q})} + \mathbf{g(q)} + \mathbf{Q}_{friction}(\mathbf{q,\dot{q}}) = \mathbf{Q}_{motor} + \mathbf{Q}_{ext} \quad (1)$$

The generalized joint torques are termed as $\mathbf{Q}$ where $\mathbf{Q}_{ext}$ are the torques that are caused by external forces acting on the robot due to ground contact or external disturbances, e.g. punches. In the simulation model N explicit points exist where external forces $\mathbf{F}_{ext}$ are allowed to act. The resulting joint torques caused by these forces can be calculated with the jacobian matrices as $\mathbf{Q}_{ext} = \sum_{p=1}^{N} \mathbf{J}_p^T \mathbf{F}_{ext}$.

The ground reaction forces are calculated using a virtual spring damper model. This requires only little computation time and allows to change the ground characteristics quite easily. The simulator allows to apply a variety of disturbances to the simulated robot. These are for instance an inclined, uneven or slippery ground or several kinds of punches.

3 Reflex motions

The avoidability of falls highly depends on the capabilities of the robot. In former experiments [2] it could be shown that it is possible to prevent our robot BARt from falling over by executing a reflex step. It turned out that the typical step execution time is about 400-500 ms and that the time needed to detect and classify a fall should be shorter than 100 ms. A reflex motion consists of two 3rd order splines describing the trajectory of the torso and the swing foot (**swf**) with respect to the stance foot (**stf**). The splines start at the actual position of the torso resp. foot and end with a specified velocity in predefined positions where the robot should remain statically stable. Furthermore, the reaction force caused by the acceleration of the swing foot towards the direction of fall has a stabilizing influence on the robot's dynamics.

For these examinations 5 reflex motions (step backward, step forward, step to swf site, step to stf site, crouching) where taught to the biped, whereas only the simulated biped is able to execute the sidewards motions.

4 Classification

Pattern classification is the task of giving names to objects based on observations [3]. The names correlate to predefined classes $C_{1..N} \in \mathbb{S}^1$ the classification algorithm is supposed to distinguish between. In the context of fall detection it is now possible to classify the robot's state or the optimal reaction that should be executed to prevent a fall resp. to minimize damages when falling down. In the first case, termed as state classification, the classes would look like this:

$$\mathbb{S}_N^1 = [\{\text{fall ahead}\}_1, \{\text{heavy fall ahead}\}_2, \{\text{fall to stf site}\}_3, \ldots, \{\text{stable state}\}_N].$$

The latter case is referred to as response classification. Here the following classes are thinkable:

$$\mathbb{S}_N^1 = [\{\text{reflex step ahead}\}_1, \{\text{reflex crouch}\}_2, \{\text{reflex step to stf site}\}_3,$$
$$\{\text{reflex step to swf site}\}_4, \ldots, \{\text{no reflex reaction}\}_N].$$

The underlying observations for the classification task are the measurements $m_{i...M}$ available from the robot as will be discussed in Sect. 4.1. These measurements are combined to the measurement vector $\mathbf{v} = [m_1, m_2, \ldots, m_M]^T$. The mapping of $\mathbf{v}$ to C can be described as

$$C = f(\mathbf{v}) \text{ with } \mathbf{f}(\cdot) : \mathbb{R}^M \mapsto \mathbb{S}^1. \tag{2}$$

A quite simple function $\mathbf{f}(\cdot)$ can be realized when a prototype vector $\mathbf{p}_k$ for each class C_k is defined. With the distance d_k of $\mathbf{v}$ to $\mathbf{p}_k$

$$d_k(\mathbf{v}, \mathbf{p}_k) = \left[\sum_{i=1}^{M} |v_i - p_{ki}|^\gamma \right]^{\frac{1}{\gamma}} \tag{3}$$

$\mathbf{f}(\mathbf{v}) = C_i$ if $d_i < d_j$ for all $j \neq i$ [4]. Additional information is obtainable by regarding the distance d_n of $\mathbf{v}$ to the nearest prototype vector $\mathbf{p}_n$. This provides an indication of the classification confidence.

4.1 Feature Selection

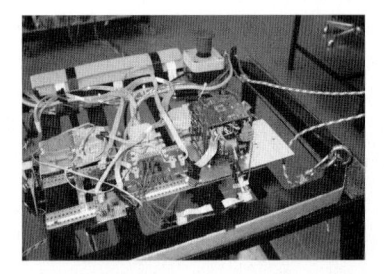

Fig. 2. **Left:** Foot of BARt-UH with force sensors and μIMU. **Right:** Torso of the robot. The electronic parts were replaced to prevent damages during the experiments. The IMU mounted on the torso is used to measure the robot's attitude and velocities.

The right choice of features is very significant for the classification performance. The optimal subset of features should increase the interset distance of the classes and decrease their intraset distances.

Possible features m_i contained in $\mathbf{v}$ should provide significant information of the robot's state and should be measurable in practice. In [5] was shown that for instance the torso velocity can be used to detect and distinguish falls. This information was gathered with an inertial measurement unit (IMU) [6] placed on the robot's torso (**Fig. 2**). Apart from the velocity this sensor allows to measure the attitude and acceleration of the torso. In former experiments the foot-ground angle and its rotating speed turned out to be good fall indicators. Therefore a smaller version of the IMU was developed. This so-called μIMU was mounted on the robot's foot (**Fig. 2**). Force sensors below the feet were used to measure the CoP. To include the actual gait phase, which has a relevant influence to the fall sensitivity, a signal T_{step} was generated running from -1 to 1 during one step.

In the simulation other features were examined but had no significant influence to the classification results. These were e.g. the overall ground reaction force and the stance leg's control deviation. As another distinguishing mark the conversion of potential to kinetic energy of the robot was investigated but turned out to respond too late to disturbances and was only considered in the simulation. After all, the following features were chosen:

1	torso velocity	4	CoP	7	step time
2	torso rotating speed	5	foot tilt angle	8	potential energy
3	torso tilt angle	6	foot rotating speed	9	kinetic energy

The features 1-6 were recorded as well for the sagittal as for the transversal plane. With it the measurement vector $\mathbf{v}$ had the dimension $M = 15$. In order to distinguish between stance / swing foot side and not between right / left side, the signs of the features sensitive to sideward falls were adapted. The foot related data was taken from the sensors attached to the actual stance foot. Since the features have different units a standardization is required. So each feature was normalized with its standard deviation (**Fig. 3**).

4.2 State Classification

The goal of our first experiments with BARt was to distinguish falls to different directions from the normal, undisturbed walk. Therefore several punches were applied to the robot. **Fig. 3** shows the distribution of the features recorded approx. between 0 ms and 80 ms after the acceleration peak of the punch. The data recorded for the same type of disturbance form up well separable clusters in the feature space. Comparing **Fig. 3d:** and e: it can be seen that features sensitive in one direction are hardly influenced by orthogonal disturbances. **Fig. 3a:** and b: show the influence of the swing foot's position. Obviously two clusters are formed for the sideward falls depending on the step phase. So two prototype vectors, one for $-1 \leq T_{step} < 0$ and one for $0 \leq T_{step} \leq 1$ were set up for each disturbance.

Furthermore it is possible to estimate the degree of instability. As visible in **Fig. 3b:** to e: the distance of the measurement vectors to the center of the normal classes vary. With stronger disturbances the distance is increasing. So a separation of the classes that represents unstable states into sub-classes depending on the degree of instability is quite easy.

The classification results achievable with this set of prototypes can be seen in **Fig. 4**. The left diagram shows the distances $d_{1..10}$ of the measurement vector $\mathbf{v}$ to the prototype vectors $\mathbf{p}_{1..10}$. A small uncertainty during the

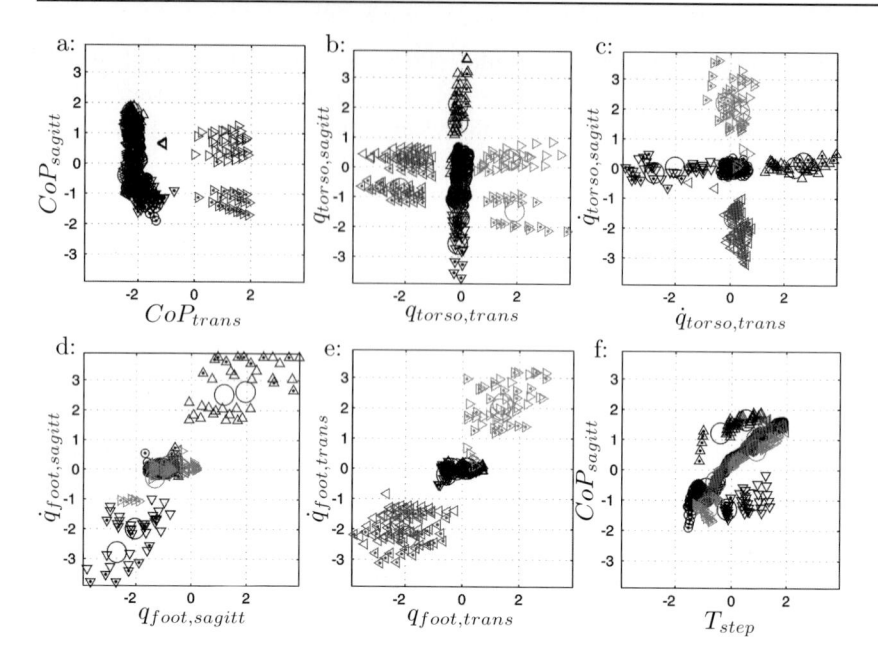

Fig. 3. The figures show the distribution of the normalized measurements recorded during punch experiments with BARt-UH. The symbols denote the direction of the induced disturbances ($\triangle$: ahead, $\triangledown$: back, $\triangleleft$: stf, $\triangleright$: swf, $\circ$: no disturbance). Samples that were recorded in the first step phase are marked with dots. The circles illustrate the prototypes.

double support phase ($|T_{step}| \approx 1$) can be noticed. As expected the transition between the two 'normal'-classes occurred for $T_{step} = 0$.
In the right diagram the distances $d_{1..10}$ for a forward fall are plotted. At $T_{step} = 0.1$ the applied punch can be detected. A confident classification of the state takes place approx. 60 ms after the impact.

4.3 Response Classification

The classification approach discussed in Sect. 4.2 paid no attention to the reactions necessary to stabilize the robot. To execute a stabilizing motion a downstream path planning algorithm is needed.
In the next approach it is intended to classify the optimal reaction to execute. So the robot's states that need to be distinguished, are directly linked to an optimal reflex reaction. Due to its extended abilities to move, our simulated robot model was used for the following reflex experiments. The simulation tool and the behaviour of the robot was controlled by a perl based script. Various impulses could be applied at several step phases

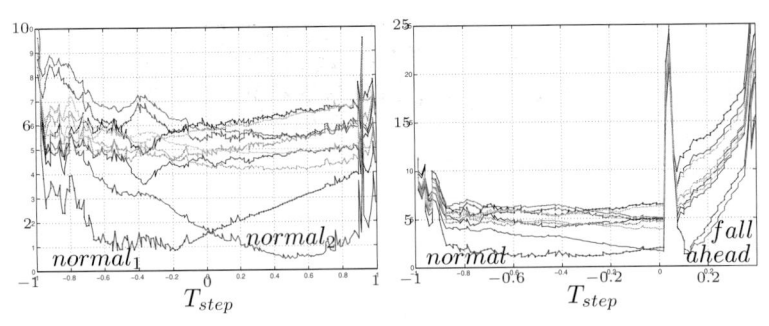

Fig. 4. Distances $d_k(\mathbf{v}, \mathbf{p}_k)$ of the measurement vector $\mathbf{v}$ to the prototype vectors $\mathbf{p}_k$. **Left:** The data was recorded during an undisturbed step. **Right:** At $T_{step} = 0.1$ the robot was pushed from behind. A reliable classification of the fall occurred approx. 60 ms after the impact.

to the robot. The impulse duration was about 60 ms, what we found out to be typical in former experiments. The robot was forced to execute a reflex motion approx. 100 ms after the beginning of the impulse. In the period between impulse and reflex multiple measurement vectors $\mathbf{v}$ were recorded. These vectors were assigned to classes depending on the success of the reflex motion. For each direction of fall two classes were formed. The reflex motion was assigned to the class where the robot managed to stabilize. A protection reflex (crouching) was linked to the other class. To enhance the algorithm, the classes were divided into sub-classes for which the prototype vectors were calculated (cp. Sect. 4.2)[1]. Additionally multiple measurement vectors $\mathbf{v}$ were recorded for the undisturbed walk and assigned to the 'normal' classes. The behaviour of the robot with the so trained prototypes was tested in several scenarios in the simulation.

5 Results

The proposed fall detection turned out to be quite robust. The induced disturbances could be detected and classified early enough for initiating a reflex reaction. Even though the algorithm was trained with data gathered from simulated impact-experiments, instabilities due to other disturbances, e.g. stumbling, were recognized as well (**Fig. 1**).
In table **1** the stabilizing success is presented. The data was recorded with a benchmark consisting of several pushes to the back and front of the biped. It can be seen that the stabilizing rate is significantly higher comparing to the robot's behaviour without reflex reaction. With stronger

---
[1] Among other things the pattern recognition toolbox 'tooldiag' was used for the clustering. URL: www.inf.ufes.br/ thomas

Impulse [Ns]	4	8	12	16	20	24	28	36
without reflex	100	78	56	56	33	33	11	0
forced reflex 100 ms after punch	100	100	100	94	83	72	56	22
reflex triggered by classification	100	100	100	94	78	72	56	22
motions initiated by class.								
no reaction	100	56	17	11	6	6	0	0
reflex step	0	44	83	89	78	67	56	39
crouching	0	0	0	0	16	27	44	61

Table 1. Stabilizing success in percent. The robot was pushed from behind and front with various impulses ranging from 4 Ns to 36 Ns at 9 different step phases.

disturbances affecting the robot, a fall is not always avoidable. But considering the stabilizing success, when a reaction is forced 100 ms after an impact, it can be seen that the stabilizing rates nearly comply. From this follows that a fall caused by stronger impacts is rather a lack of sufficient reflex motions than a fault of the classification algorithm.

So it is planned in further experiments to improve the algorithm by providing a larger amount of possible reflex motions.

Acknowledgements

This work emerged in the framework of the project GE 451 founded by the DFG. The authors would like to thank the DFG for the support.

References

1. Albert, A. (2002) *Intelligente Bahnplanung und Regelung für einen autonomen, zweibeinigen Roboter.* VDI Verlag.
2. Höhn, O., Schollmeyer, M., and Gerth, W. (2004) Sturzvermeidung von zweibeinigen Robotern durch reflexartige Reaktionen. *Eingebettete Systeme*, pp. 60–69.
3. Schürmann, J. (1996) *Pattern Classification.* John Wiley & Sons, Inc.
4. Liedtke, Claus E. and Erder, Manfred (1989) *Wissensbasierte Bildverarbeitung.* Springer-Verlag.
5. Wu, G. (2000) Distinguishing fall activities from normal activities by velocity characteristics. *Journal of Biomechanics*, vol. 33, pp. 1497–1500.
6. Strasser, R., Seebode, M., and Gerth, W. (2003) A very small low cost Inertial Measurement Unit (IMU) for robotic applications. *Symposium Gyro Technology*, pp. 18.0–18.9.

Development of a Low-Cost Humanoid Robot: Components and Technological Solutions

Vítor M. F. Santos, Filipe M. T. Silva

[1]University of Aveiro, DEM, TEMA, Portugal, vsantos@mec.ua.pt
[2]University of Aveiro, DET, IEETA, Portugal, fsilva@det.ua.pt

Abstract

The paper presents a set of solutions to build a humanoid robot at reduced costs using off-the-shelf technology, but still aiming at a fully autonomous platform for research. The main scope of this project is to have a working prototype capable of participating in the ROBOCUP humanoid league, and to offer opportunities for under and pos-graduate students to apply engineering methods and techniques in such an ambitious and overwhelming endeavor. The most relevant achievements on this implementation include the distributed control architecture, based on a CAN network, and the modularity at the system level. These features allow for localized control capabilities, based both on global and local feedback from several sensors, ranging from joint position monitoring to force sensors. Force sensors on the feet were designed and integrated using strain gauges properly calibrated and electrically conditioned. Although some issues are yet to be completed, the stage of development is already enough for practical experiments and to obtain positive conclusions about the solutions proposed.

Keywords: Humanoid robot; Biped locomotion, Distributed control; Force sensors

1 Introduction

In recent years, there has been a large effort in the development of humanoid robot prototypes and in the control and analysis of biped gaits. Research in bipedal walking can be divided into two categories: passive mechanisms and active walkers. The passive mechanisms are interesting because of their simplicity, energetic efficiency and consistency of the resulting gaits, but only in a limited range of operational conditions [1][2].

In the other extreme of the spectrum are the active walkers, falling largely into two groups: time-dependent and time-invariant. By far, the most popular are time-dependent that involve the tracking of pre-computed trajectories [3][4][5]. One of the most prominent schemes used to enhance trajectory tracking controllers or to analyse their stability is the so-called Zero Moment Point criterion [6]. In addition to the various time-dependent algorithms, there have been several other time-invariant control schemes proposed [7][8]. The results obtained with time-invariant schemes are impressive by inducing dynamic walking, but it is unclear how stability is achieved and how robustness or efficiency can be improved.

The paper presents the design considerations of a small-size humanoid robot under development. The main scope of the project beneath this paper has been the development of a humanoid platform to carry out research on control, navigation and perception, and to offer opportunities for under and pos-graduate students to apply engineering methods and techniques. Purchasing a commercial platform carries prohibitive costs and it would reduce the involvement at the lowest levels of machine design, which was posed as a relevant pursuit for the desired engineering approach. Moreover, recent advances in computing hardware have promoted research of low-cost and easy-to-design humanoids [9][10].

In this line of thought, the project aimed at building a prototype capable of participating in the ROBOCUP humanoid league. A wide range of technologies need to be integrated and evaluated, giving added value for project-oriented education. The design and development of the autonomous humanoid platform has considered three phases:

- Definition of functional and physical requirements, *i.e.*, mechanical structure, dimensions and degrees of freedom (DOFs);
- Selection and integration of hardware and software to achieve these requirements;
- Development of low and intermediate level tasks (*i.e.*, hardware and sensor oriented).

In what concerns the physical and functional requirements, the initial considerations were largely imposed by the rules of the ROBOCUP, namely, the robot dimensions, the mobility skills, the high level of autonomous operation and the selected tasks (e.g., walk, turn, kick a ball). As consequence, many technologies need to be integrated and a number of technical breakthroughs must be accomplished. The demands for limited costs gave rise to the selection of off-the-shelf materials and components.

One major concern of the project is to provide modularity at the system level. The main advantage is the possibility of reusing specific modules, in terms of both hardware and software, with no major efforts. A key concept for the control architecture is the distributed approach, in which independent and self-contained tasks may allow a standalone operation. At the same time, the increase of computational power may allow the development of more sophisticated sensor fusion schemes.

The design process has revealed much about the several problems, challenges and tradeoffs imposed by biped locomotion. As in most systems, the design options that are taken deeply influence the used technology, and vice-versa. Here, the emphasis is made on the mechanical design, the selection of actuators and batteries, the sensorial integration and the control system architecture. Though much work remains to be done in exploiting the potential of the proposed tools, the first results achieved at the present stage of the project are also presented.

2 Mechanical Structure for the Robot

When conceiving a robotic platform, namely a humanoid, countless decisions have to be made. Specifications and target applications must be defined and applied to impose limits both on skills and overall objectives.

After the structure height (ca. 60 cm) and remainder body proportions, the very first issue has been the number of degrees of freedom, namely to ensure proper and versatile locomotion. Walking concerns can range from simply ensuring robust equilibrium for static walking up to, hopefully, dynamic walking which will be a must for energetic efficiency.

The most versatile humanoids presented in ROBOCUP, and elsewhere, show up six DOFs per leg, namely one universal joint at the foot, a simple joint on the knee and a spherical joint on the hip, where, nonetheless, a simpler universal joint can still deal with many of the walking demands. Connecting the legs to the upper structure of the abdomen was decided to be done with two DOFs mainly aiming at greater flexibility in control to balance and account for the perturbations of the center of mass (CoM). So far, arms have been poorly defined and the head accounts for two DOFs for the future vision based perception. A complete humanoid model and a view of the current stage of implementation are illustrated in Fig. 1.

Fig. 1. Model of the humanoid robot and the current stage of implementation

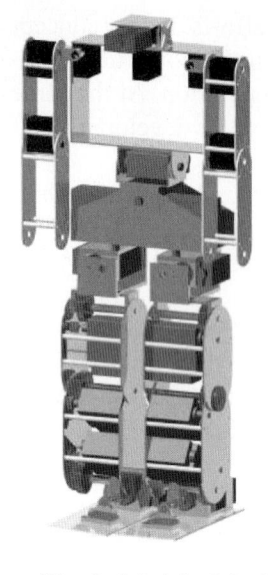

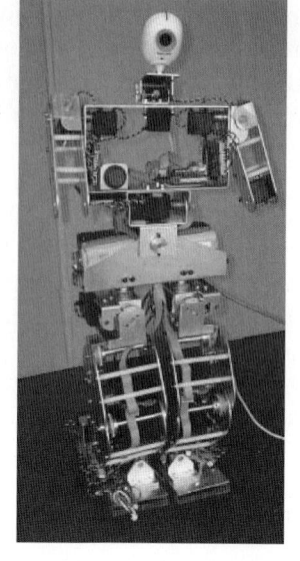

Degrees of freedom	
Head	2
Arms	3 (2×)
Trunk	2
Hip	3 (2×)
Knee	1 (2×)
Foot	2 (2×)
Total	**22**

3 Motors and Batteries

For the dimensions involved, and for good autonomy, low cost off-the-shelf technologies for actuation do not offer significant alternatives other than the small servomotors, such as those from FUTABA, HITEC, and similar, used worldwide. To pick the adequate motors, several static (and some dynamic) simulations were carried out to estimate motor torques. The 3D structural model developed in CATIA furnished the CoM of the several links of this multi-body system, and were then fed into a pseudo-static model developed in Matlab, and calculated all motor torques along a sequence necessary to produce a locomotion step. Torques were obtained using $T_k = \left\| \sum_{i=k}^{N} m_i \mathbf{r}_i \times \mathbf{g} \right\|$, where $\mathbf{r}_i$ is the relative vector position of the CoM of link i, m_i is its mass and $\mathbf{g}$ the acceleration of gravity vector. The Matlab model was based on the superimposition of several open kinematics chains built using the Denavit-Hartenberg methodology. The simulation results shown in Table 1 indicate that the most demanding situations occur at the hip joint responsible for lateral opening of the legs, showing up torques on some joints greater than 2.5 Nm.

Table 1. Extreme angles and motor torques during one step in locomotion

Motor/Joint	Θ_1 [°]	T_1 [N.m]	Θ_2 [°]	T_2 [N.m]	Θ_3 [°]	T_3 [N.m]
Foot 1 roll	0.0	2.37	7.1	0.98	7.1	0.95
Foot 1 tilt	4.7	0.30	10.1	0.20	10.1	0.17
Knee 1	10.1	0.76	21.8	1.17	21.8	0.80
Hip 1 tilt	5.4	0.35	11.7	0.30	11.7	0.07
Hip 1 roll	0.0	**2.26**	7.1	**2.57**	7.1	**2.54**
Foot 2 roll	0.0	0.00	7.1	0.00	7.1	0.00
Foot 2 tilt	4.7	0.12	10.1	0.12	38.0	0.12
Knee 2	10.1	0.17	21.8	0.23	52.9	0.29
Hip 2 tilt	5.4	0.07	11.7	0.02	15.0	0.35
Hip 2 roll	0.0	0.01	7.1	0.30	7.1	0.27

The HITEC servos (as FUTABA and their "clones") occur in several variants of torques, dimensions and power. The models chosen for this system are described in Table 2.

Table 2. Selected HITEC Motor models for the robot

Application	Model	Mass (g)	Torque (Nm)
Arms and small torque joints	HS85BB	19.8	0.35
Legs and high torque joints	HS805BB	119	2.26

It is obvious from Table 1 and Table 2 that several joints require gear ratios greater than 1:1. That was done mainly for the leg joints, and gear ratios of 2.5 (and more) were used. This would give a theoretical maximal torque of 5.6 Nm, which accounts for overall efficiencies as low as 46%, thus giving some room for less efficient implementations. Early simple gear couplings were later replaced by toothed belt systems for improved transmission and tuning facility. Ball bearings and copper sleeves contribute to improve mechanical efficiency. Actuating with this servomotor has the disadvantage that velocity can not be automatically controlled. That is being overcome by an algorithm based on dynamic PWM tracking using the servo own potentiometer for feedback information. Velocity is now going to be controlled in slots of 10 ms, or less.

Power to drive the motors is a crucial issue since servos require a relatively high current, namely at startup and when producing motion in some configurations. Two ion-lithium batteries were installed and the system counts with a 7.2 V/9600 mAh pack, with maximal sustained current specified by the vendor at more than 19A. Each one of the two battery sets weights circa 176 g and confines to a box of 37x37x65 mm³. Proper fusing, polarity protection and charge monitoring were also implemented.

4 Sensors

As all systems intended to be autonomous, this robot has both proprioceptive and exteroceptive sensors. For now, only the former exist and the following data is monitored: joint position, motor electric current, force sensors on both feet to measure ground reaction forces, accelerometers, used mainly as inclinometers (ADXL202E from Analog Devices), and a gyroscope for instant angular velocity measuring (GYROSTAR ENJ03JA from MURATA). Accelerometers and gyro are of the integrated type based on MEMS technology. Force sensing was custom made using strain gauges properly calibrated and electrically conditioned; a device with four strain gauges was arranged near the four corners of the foot base. The model and a prototype of the sensitive foot base are shown in Fig. 2.

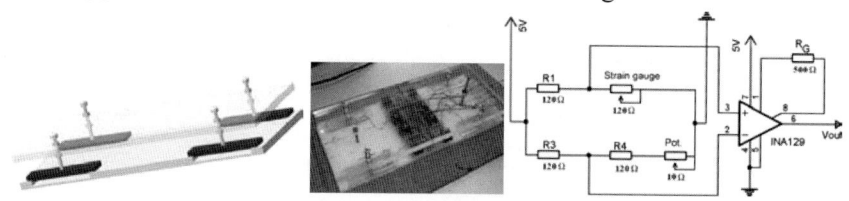

Fig. 2. Sensitive foot: model, device prototype, and electrical interface.

Current monitoring in the motors is made through a 0.47Ω power resistor in series with the motor power supply line, and an analog value is acquired by the local microcontroller. The hardware implementation can have other sensors, and a piggy-back electronic board has been left prepared for that.

5 Control system architecture

One of the major challenges of the project was to conceive and implement a distributed control system. To allow for short and possibly longer term developments, the platform was given a network of controllers connected by a CAN bus in a master/multi-slave arrangement. The master unit performs no device low-level control, but dispatches orders and collects information to be exchanged with a central system that currently is still located on an external computer, but is expected to be implemented in a PC104-based board or similar.

Each slave controller can control up to three joints and consists of a PIC18F258 device (from Microchip) with its own program made up of local low-level actuator control. This possibility of local units with their own control ability allows for more elaborate strategies, since they can simply accept directives from upstream controllers or implement their own control decisions (or a combination of both). This ability releases the higher level control units from the burden of being aware of all details of control and perception (control laws, PWM generation, sensor processing, etc.). A simplified version of the control system is shown in Fig. 3.

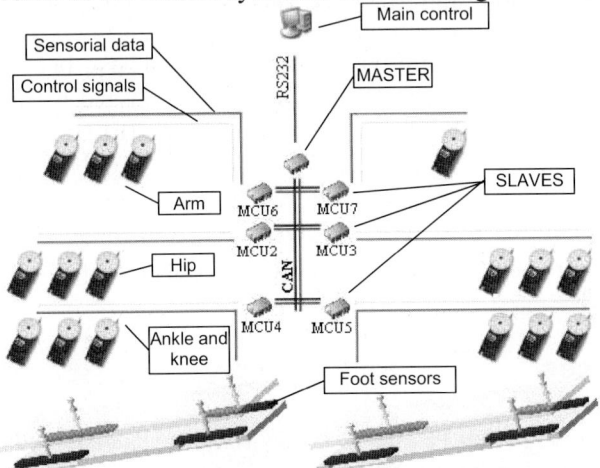

Fig. 3. Simplified Diagram of Robot Control Architecture

All communications are asynchronous and occur at two levels: among master and slaves on the CAN bus (1 Mbit/s) and between master and high level controller (currently serial RS232 at 38400 baud). Data exchanged between the master and each of the (currently) eight slaves refers to position and velocity set-points for motors and sensorial feedback from sensors associated to each slave controller. Each CAN message has 8 bytes, which allows control and perception refresh periods of circa 150 µs per slave, or a grand total 1.2 ms for 8 slaves, that is, a global network control cycle over 800 Hz. Of course, the serial link to upper controller does not need to be that fast since not all data and commands are expected to be exchanged throughout the entire control architecture. The upper control level will deal with vision and global motion directives, as well as any kind of planning and navigation to develop in the future. All this, however, is much slower than the microcontrollers which operate at 40 MHz and ensure a PWM resolution of about 1 µs, and sensorial acquisition rates at tenths or hundreds of kHz.

6 Conclusions and perspectives

This paper presented specific technological solutions and approaches to build a relatively low cost humanoid robot based on off-the-shelf components. The main features of this 22-DOF system include a distributed control architecture with local control possibility, based on a CAN network, and is prepared to use several types of sensors, ranging from joint position monitoring to force sensors on the feet made with standard low cost strain gauges. Most of the final platform hardware has been built and results are promising since the system now is able to stand, lean on the sides and forward-backward, and primitive locomotion steps have been achieved. Ongoing developments cover the inclusion of vision and its processing, possibly with a board based on PC104 (USB or IEEE1394 camera to be selected). Currently, what has been developed is only a platform for research; for the next few years, the research will cover distributed control, alternative control laws, like neural computation, and also deal with issues related to navigation of humanoids and, hopefully, cooperation.

References

1. McGeer, T. (1990) Passive Dynamic Walking, *International Journal of Robotics Research*, Vol. 9, No. 2, pp. 62-82.
2. Garcia, M., et al. (1998) The Simplest Walking Model: Stability, Complexity, and Scaling", *ASME J. Biomech. Eng.*, Vol. 120, No. 2, pp. 281-288.
3. Hirai, K. et al. (1998) The Development of Honda Humanoid Robot, *Proc. IEEE Int. Conf. on R&A*, pp. 1321-1326.
4. Yamaguchi, J-I., et al. (1999) Development of a Bipedal Humanoid Robot – Control Method of Whole Body Cooperative Dynamic Biped Walking, *Proc. IEEE Int. Conf. Robotics & Automation*, pp. 368-374.
5. Kuffner, J., et al. (2002) Dynamically-Stable Motion Planning for Humanoid Robots, *Autonomous Robots*, Vol. 12, pp. 105-118.
6. Vukobratovic, M., Borovac, B., Surla, D., Stokik, D. (1990) Biped Locomotion – Dynamics, Stability, Control and Application, *Springer-Verlag*.
7. Pratt, J., Pratt, G. (1998) Intuitive Control of a Planar Bipedal Walking Robot, *Proc. IEEE Int. Conf. on R&A*, pp. 2014-2021.
8. Kajita, S., Tani, K. (1996) Experimental Study of Biped Dynamic Walking, *IEEE Control Systems*, vol. 16, n. 1, pp. 13-19.
9. Yamasaki, F., Miyashita, T., Matsui, T., Kitano, H. (2000) PINO the Humanoid: A Basic Architecture, *Proc. Int. Workshop on RoboCup*, Australia.
10. Furuta, T., et al. (2001) Design and Construction of a Series of Compact Humanoid Robots and Development of Biped Walk Control Strategies, *Robotics and Automation Systems*, Vol. 37, pp. 81-100.

Analysis of Humanoid Robot Lower Extremities Force Distribution in Standing Position

H Montes, P Alarcon, R Ponticelli and M Armada

Automatic Control Department, Industrial Automation Institute - CSIC, Madrid, Spain   hmontes@iai.csic.es

Abstract

This paper introduces an approach oriented to obtain an analogous value to the ZMP (measured with force sensors on the feet soles) by means of the measurement of the robot lower extremities link forces. A preliminary study of force distribution in non-linearly actuated humanoid robot lower extremities is presented. The goal of this study is to relate the link forces with the ZMP in the sagittal plane. A ZMP model based on the link forces and the robot leg posture in the sagittal plane is obtained. The robot link forces are indirectly estimated through the measurement of the forces in the non-linear actuator, that are then directly related to the values of the ZMP in the sagittal plane. Experimental evaluations are presented.

Keywords: Force distribution, ZMP, humanoid robot, stability.

1 Introduction

Force distribution in robotic systems has been investigated by many researchers in the last years. These investigations have been focused, mainly, in legged robots [1-4], and in another kind of robotic systems [5-6]. The main goal is to know the contact forces acting on the robot extremities in order to carry out force control algorithms to distribute the forces in the robot structure, and, therefore, to stabilise the robot.

The force distribution problem in walking machines is statically undetermined (robot with four o more legs); in other words, there is an infinite number of possible force distributions that maintains the robot in balance

[4]. On the other hand, in biped robot dynamically balanced locomotion, the most important criterion utilized is the ZMP [12].

In this paper, it is presented an analysis of humanoid robot lower extremities force distribution using as experimental platform SILO2 robot [7]. The lower extremities of this robot have 12 DOF and six of them are driven by non-linear actuators (SMART) [8]. Fig. 1 shows SILO2 lower extremities in two complementary views. The upper part of the robot (not shown in figure 1) will be employed also in the experiments for inputting various load conditions.

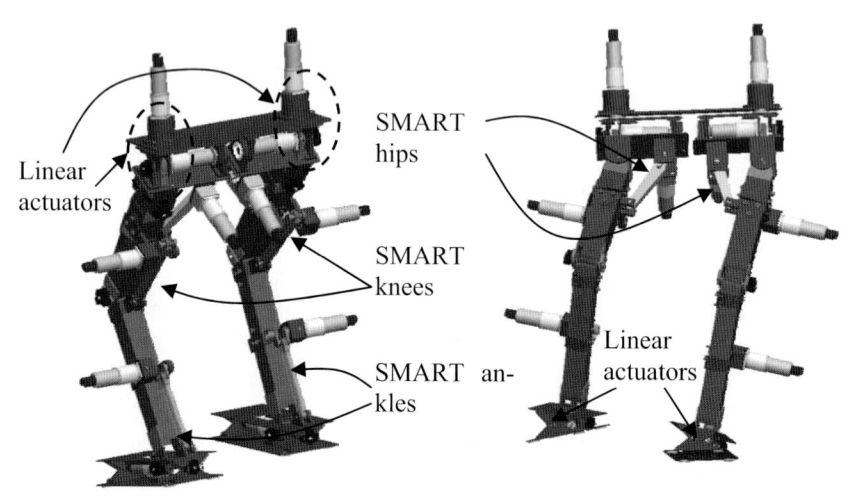

Fig. 1. View of SILO2 lower extremities, employing linear and non-linear actuators

This robot has been provided with built-in force sensors on each SMART rod [9]. The output of these sensors is amplified ten thousand times by means of a high precision modular instrumentation amplifier [10]. With this force sensor, the extension and compression forces in each SMART rod are measured. The SMART rod forces in the ankles and knees are influenced by external forces parallel to the sagittal plane. On the other hand, SMART rod forces in the hip are influenced by external forces parallel to the lateral plane. Moreover, force measurements depend on the robot accommodative stance. Also, there are four load cells on each robot foot sole in order to carry out the measurement of the ZMP.

2 Kinematics and physical model of the robot leg

One SILO2 robot leg in the sagittal plane is formed of three links joined by two monoaxial rotating joints (see **Fig. 2**). Each link has its own mass (m_1, m_2, and m_3) and each joint has a restricted angular displacement (q_1 and q_2). The rotary displacement of the knee joint is $-63.95° \leq q_1 \leq -3.37°$ and the ankle joint excursion is $-14.9° \leq q_2 \leq 20.2°$ [11].

Knowing the mass of each link and the leg posture it is possible to obtain the reaction force ($\vec{F}_R$), the ZMP, the force in each link ($\vec{F}_1$ and $\vec{F}_2$), and the tangential force ($\vec{F}_X$). With these values it is possible to relate the ZMP against the link forces. For the SILO2 robot, the ZMP value can be obtained with the load cell sensor placed on each foot sole, and the forces in each link can be measured by mean of a force sensor implemented in each SMART rod [11].

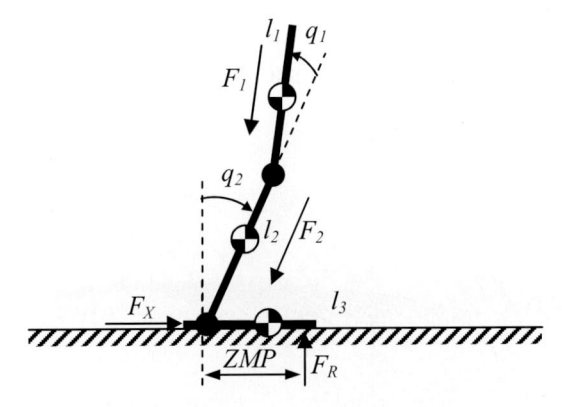

Fig. 2. Scheme of the robot leg

According to **Fig. 2**, the acting forces in the biped robot leg in the sagittal plane are,

$$\vec{F}_R = \sum_{i=1}^{n} m_i g \tag{1}$$

$$\vec{F}_1 = m_1 g \cos(q_1 + q_2) \tag{2}$$

$$\vec{F}_2 = m_1 g \cos(q_1 + q_2)\cos(q_1) + m_2 g \cos(q_2) \tag{3}$$

$$\vec{F}_x = m_1 g \cos(q_1+q_2)\sin(q_1+q_2) + m_2 g \cos(q_2)\sin(q_2) \qquad (4)$$

And the ZMP where the reaction force actuates is calculated with,

$$ZMP = \frac{m_1 g\left(l_2 \sin(q_2) + l_{1(c)} \sin(q_1+q_2)\right) + m_2 g l_{2(c)} \sin(q_2) + m_3 g l_{3(c)}}{\sum_{i=1}^{n} m_i g} \qquad (5)$$

where, $l_{i(c)}$ is the distance of the link origin to its centre of gravity.

3 Force distribution analysis

Several simulations were carried out with some postures of the robot leg in the sagittal plane. For each ankle joint displacement, the knee joint was swept within the boundaries $-63.95° \le q_1 \le -3.37°$. The relationship between the link forces and the ZMP is shown in **Fig. 3**.

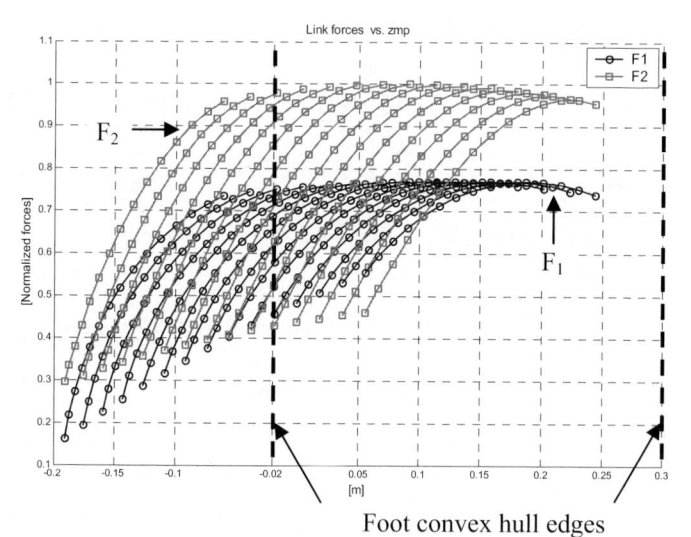

Fig. 3. Relationship between the link forces and the ZMP

In some leg postures the ZMP does not exist. This happens because the point where $M_x = 0$ and $M_y = 0$ is outside the foot convex hull [12]. In these postures, the leg rotates about the foot edge and will fall down. This overturns should be compensated with the posture of the robot upper part, so that ZMP exists, in other words, that the reaction force is in the foot convex hull. **Fig. 3** shows the limits of the foot convex hull of the SILO2 leg. It is possible to relate the links forces and the leg posture with the ZMP value, and, for that reason, with the robot stability.

As the SILO2 leg in the sagittal plane has force sensors in each SMART rod located in ankle and knee joints, then, it is not difficult to relate it to the ZMP measurement. Moreover, when in some times the force sensors of the foot sole does not have a real (good) contact with the ground, it is possible to know the ZMP by the force measurements at the SMART rod and the joints positions.

Combining the Equations (1), (2), (3), and (5) the ZMP function based on the links forces is obtained,

$$ZMP = \frac{F_1\left(l_2 s_2 c_2 + l_{1(c)} s_{12} c_2 - l_{2(c)} s_2 c_1 c_{12}\right) + F_2\left(l_{2(c)} s_2 c_{12}\right) + W_3 l_{3(c)} c_2 c_{12}}{F_1\left(c_2 - c_1 c_{12}\right) + F_2 c_{12} + W_3 l_{3(c)} c_2 c_{12}} \quad (6)$$

where, as usual, s_2 is $sin(q_2)$, s_{12} is $sin(q_1+q_2)$, c_i is $cos(q_i)$, c_{12} is $cos(q_1+q_2)$, and $W_3 = m_3 g$.

4 Experimental results

One experiment performed to analyse force distribution in SILO2 humanoid robot was to place the robot in several postures in standing position, with different masses over the hips. Afterwards, the robot carries out movements up and down.

During these movements forces in the ankles, knees and hips of the SMART rods are measured. In this example the hips' forces are very small, because the robot movement is parallel to the sagittal plane (see **Fig. 4**), but in other cases (when the robot movement is in the lateral plane) it may happen the opposite.

In addition to the measured forces in SMART rods, the ZMP measurements were registered. The SMART rod forces were compared with the corresponding ZMP values, with the purpose of obtaining the relation between them. The relationships between the SMART rod forces and the ZMP in the sagittal plane are shown in the **Fig. 5**.

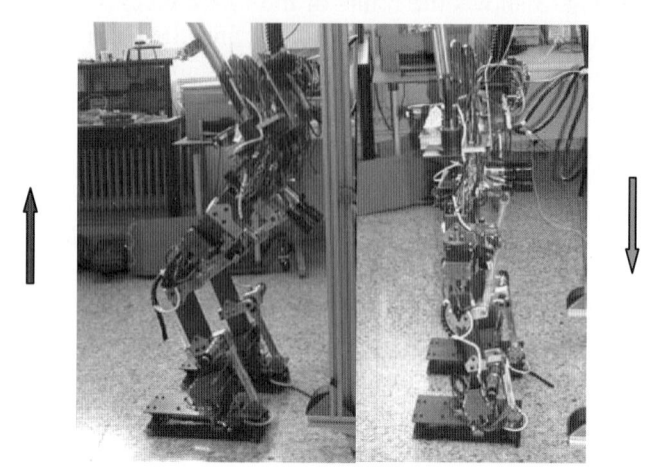

Fig. 4. Two different standings (end and start) of SILO2 during the experimental evaluation

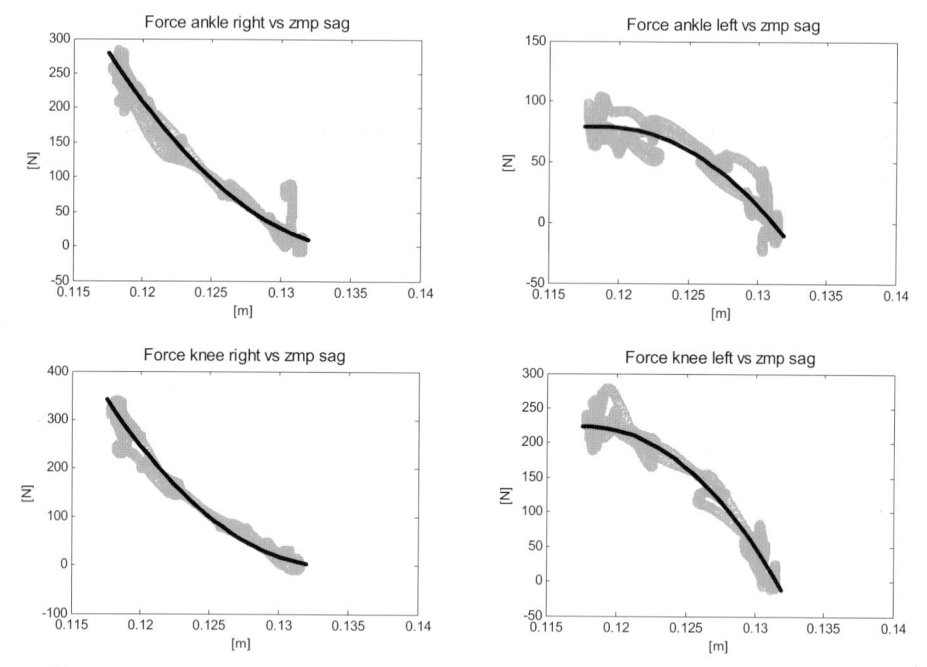

Fig. 5. Experimental relations between SMART rod forces and the sagittal ZMP during up and down movement of SILO2 legs

As a consequence of these experimental results, a matrix function can be obtained $\mathbf{F} = [\mathbf{A}][\mathbf{ZMP}]$,

$$\begin{bmatrix} F_1 \\ F_2 \\ F_3 \\ F_4 \end{bmatrix} = \begin{bmatrix} 8.29 & -2.26 & 0.15 \\ 1.29 & -0.34 & 0.02 \\ -1.19 & 0.28 & -0.02 \\ -5.11 & 1.21 & -0.07 \end{bmatrix} \begin{bmatrix} ZMP^2 \\ ZMP \\ 1 \end{bmatrix} \times 10^5 \qquad (7)$$

where, F_1, F_2, F_3, and F_4 are right ankle, right knee, left knee, left ankle SMART rod forces, respectively.

So, from this preliminary study it is possible to relate directly measured forces in the non-linear actuators to the ZMP. Further investigations are oriented to obtain a family of curves which represents each ZMP position knowing only the SMART rod forces. With this approach it could be possible to use only the SMART rod forces sensors to calculate the ZMP of the SILO2.

5 Conclusions

In this article a preliminary study of the force distribution of the SILO2 lower extremities is presented. The links forces can be measured, indirectly, through the SMART rod forces [11].

For different robot postures, the SMART rod forces can be related with the ZMP values, obtained from the load cell measurement placed on the robot feet sole. The idea is to obtain mathematical functions where it is possible to calculate the ZMP through the robot postures and the forces measured in the SMART rods. Additionally, it is possible to make a sensorial fusion with the SMART rod forces and the load cells (placed on the feet soles) in order to know the SILO2 stability in unusual postures (i.e. not perfect sole-ground contact).

Acknowledgements

Authors would like to acknowledge the help provided by the mechanical and electronic shops of the IAI-CSIC.

References

1. Gorinevsky, D.M. and Schneider, A.Y. (1990) Force Control in Locomotion of Legged Vehicles over Rigid and Soft Surfaces. *The International Journal of Robotics Research*, vol. 9, no. 2, MIT, 4-23.
2. Kumar, V. and Waldron, K.J. (1990) Force Distribution in Walking Vehicles Trans. of the ASME, *Journal of Mechanical Design*, vol. 112, pp. 90-99.
3. Jiang, W, Liu, A, and Howard, D. (2001) Foot-force distribution in legged robots. *In Proceedings of International Conference on Climbing and Walking Robots,* Germany, pp. 331-338
4. Gálvez, J. A. (2002) Percepción, control y distribución de fuerzas en robots móviles con patas. *Ph.D. Thesis*, Universidad Politécnica de Madrid, Spain.
5. Cheng, F. and Orin, D. (1990) Efficient algorithm for optimal force distribution-the compact dual LP method. *IEEE Transaction on robotics and automation*, vol. 6, no. 2, pp. 178-187.
6. Hung, M., Orin, D., and Waldron, K. (2000) Efficient formulation of the force distribution equations for general tree-structured robotic mechanism with a mobile base. *IEEE transactions on systems. Man, and cybernetics*, vol. 30, no. 4, pp.529-538.
7. Armada, M., Caballero, R., Akinfiev, T., Montes, H., Manzano, C., Pedraza, L., Ros, R., and González de Santos, P. (2002a) Design of SILO2 Humanoid Robot. *In Proc. of IARP Workshop on Humanoid and Friendly Robotics*, December 11-12, Tsukuba, Japan, pp. 37-42.
8. Caballero, R. (2002) Control de robots bípedos con accionamientos no lineales. *Ph.D. Thesis*, Universidad Politécnica de Madrid, Spain.
9. Montes, H., Pedraza, L., Armada, M., Caballero, R., and Akinfiev, T. (2003) Force sensor implementation for enhanced responsiveness of SMART non-linear actuators. *In Proceedings of International Conference on Climbing and Walking Robots*, September 17-19, Catania, Italy, pp. 887-894.
10. Montes, H., Pedraza, L., Armada. M., Akinfiev, T., and Caballero, R. (2004a) Adding extra sensitivity to the SMART non-linear actuator using sensor fusion. *Industrial Robot: An Intl. Journal*, vol. 31, no. 2, pp. 179-188.
11. Montes, H. (2005) Analysis, design, and evaluation of force control strategies in walking robots. *Ph.D. Thesis*, Universidad Complutense de Madrid, Spain.
12. Vukobratović, M. and Borovac, B. (2004) Zero-Moment Point – Thirty five years of its life. *Intl. Journal of Humanoid Robotics*, vol. 1, no. 1, pp. 157-173.

ZMP Human Measure System

M. Arbulú, F. Prieto, L. Cabas, P. Staroverov, D. Kaynov, C. Balaguer

Robotics Lab, Department of Systems Engineering and Automation University Carlos III of Madrid, Spain   marbulu@ing.uc3m.es

Abstract

This paper describes a measuring system to calculate the ZMP in humans. The main goal is to do several tests with a hardware measure system in order to analyze whether it will be possible to implement it in the Rh-0 humanoid robot. The humanoid robot Rh-0 under development at the Robotics Lab of the University Carlos III of Madrid, will have a movement control system which will measure its ZMP on-line. First of all, several tests of these possible control systems on people have been developed, measuring their ZMP while they are walking on a plane floor.

Keywords: Humanoid gait, ZMP, COM

1 Introduction

The humanoid robot Rh-0 has 21 DOFs, all of them active and electrically driven. The robot's height is 1.4 m and its weight is about 40 kg. **Fig. 1** shows the general view of the robot. Rh-0 is expected to realize many different functions, but, the fact that the humanoid could fall down, must be avoided. For stable humanoid robot gait is neccessary need to control the position of the Zero Moment Point (ZMP) [1]. The robot will have a movement control system for its ZMP. The main objective of this paper is to describe a method to measure the ZMP of a person while walking with a hardware system in order to use this information in the future Rh-0's stability control. The dynamic model of the humanoid robots with high number of DOFs is a complex and generally unsolved problem. Specially developed algorithms and strategies had been used. Some of them are based on the direct modeling the robot using complex dynamic formulation taking in account the center of masses, angular moments, speed an acceleration, etc [2,3]. Other approach is the simplification of the robot dynamics by using the 3D linear inverse pendulum model (3D-LIMP) [4].

Fig. 1. The Rh-0 robot

2 Dynamic human walking model

In this experiment we are going to use the 3D-LIMP in order to model the human body walking. It will suppose that the center of mass will move along a plane with constant height ($z = z_{cte}$). Moreover, all the mass of the body is in the pendulum's ball.

Therefore, the dynamics will be given by [5]:

$$zmp_x = x_p - \frac{z_{cte}}{g}\ddot{x}_p \qquad (1)$$

$$zmp_y = y_p - \frac{z_{cte}}{g}\ddot{y}_p \qquad (2)$$

where g is the gravity, (x_p, y_p) the coordinates of the center of mass projection on the floor (Pseudo-ZMP) [6], (zmp_x, zmp_y) the ZMP position.

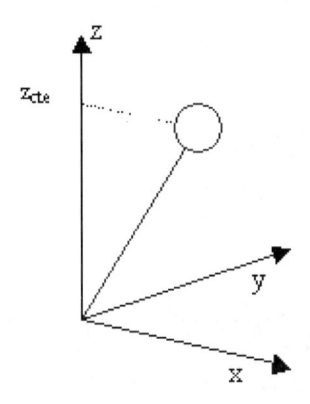

Fig. 2. Inverted Pendulum

As the main goal is to calculate the ZMP position, first the Pseudo's position and its acceleration is calculated.

3 Hardware system

The experiment will be simple. The Pseudo-ZMP of a person while walking is measured. Next, the accelerations of the COM is calculated, finding the ZMP's position. With the purpose of finding the Pseudo-ZMP a hardware system is used, formed by:

- 2 metal platforms
- 8 load cells
- 2 data acquisition cards

3.1 Platforms

A human will put the metal platforms on his feet, and then he/she will walk with them. One platform and four load cells for one foot is shown in the **Fig. 3**. Rectangular platforms are used in order to minimize errors [7], and they were designed to have the minimum friction with the floor.

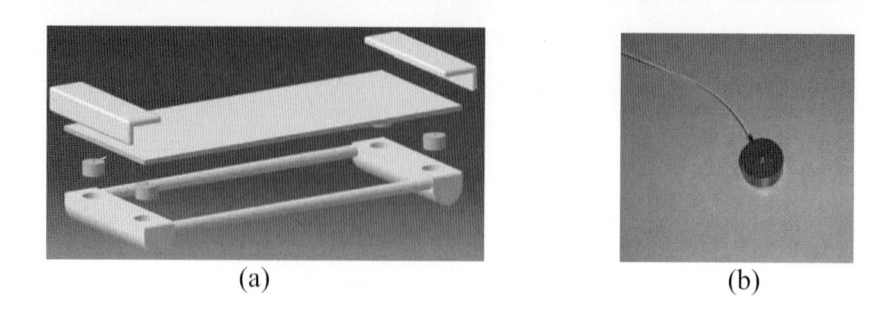

Fig. 3. (a) Platforms; (b) Load Cell

3.2 Load cells

There are four load cells for every platform (eight in total), located at the platform's corners. This lay out with four contact point is enough to cover the stability area.

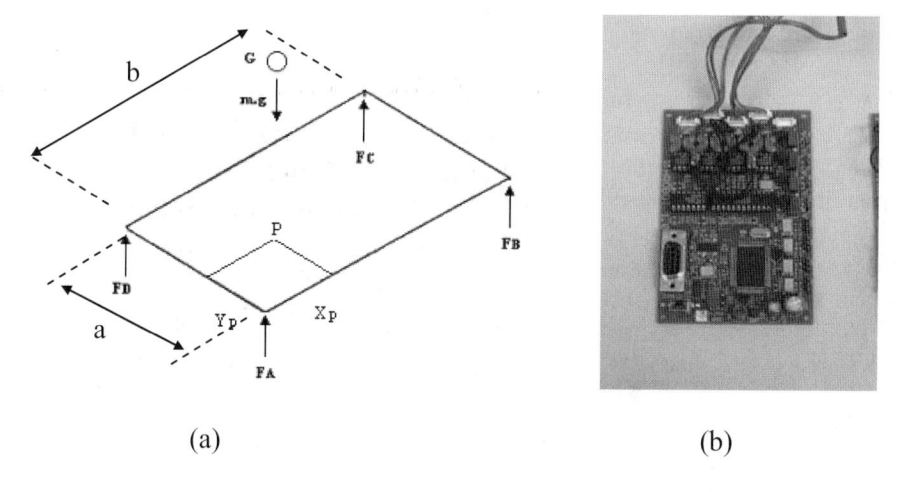

Fig. 4. (a) Force balance; (b) Acquisition card

Kyowa LM-50KA load cells are designed to resist until 51 kgf (see **Fig. 3b**). Load cells work like a full Wheatstone brigde. This one axis sensor is quite good to achieve our first objective: To watch the On-Line trajectory of the Center of Mass projection (Pseudo-ZMP). A forces-diagram is visualized in **Fig. 4a**. Where G is the Center of Mass, and P is the Pseudo-ZMP. The system's origin is the point A [7], where *a* is the width of the platform, *b* is the length, and $W = m.g$ is the total body weight. The next equations (3), (4) and (5) calculate the COM (Pseudo-ZMP):

$$F_A + F_B + F_C + F_D = W \tag{3}$$

$$\sum M_Y = 0; \quad x_P = \frac{F_B + F_C}{W}.a \; ; \tag{4}$$

$$\sum M_X = 0; \quad y_P = \frac{F_D + F_C}{W}.b \; ; \tag{5}$$

3.3 Data acquisition cards

The output of the load cells will go to a data acquisition cards to convert the signal in a digital one. These cards can get until 100 samples per second coming from the load cells (see **Fig. 4b**).

3.4 Overall system

The data path will be the next: While a person is walking on the platforms, he is pressing the load cells which provokes a voltage in the Wheatstone bridge's output. This will go to the acquisition cards, amplifying the signal and converting it into a digital. Data will arrive at the computer through the serial port and they will be processed by an Interface (see **Fig. 6**).

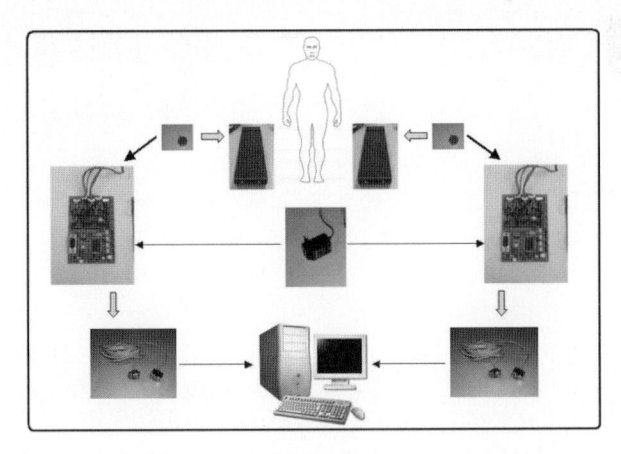

Fig. 6. Overall Hardware System

4 Interface

One of the most important advantages of this experiment is the possibility of watching on-line the walking movement. An interface has been programmed to show in the PC screen the platforms movement and the Pseudo-ZMP and ZMP's trajectories (see **Fig. 7**). MatLab's GUIDE (Graphical User Interface Development Environment [9]) has been chosen for programming.

5 Experimental results

Several tests have been developed. An example and a brief explanation is as follows: At first, the hardware elements have been connected, and push the offset button. Then, the load cells will take their origin. Next, the person who is going to do the test has to put the platforms on his/her feet and the interface operator will introduce his/her height and the weight. At this time the test is ready to start. The person will start to walk straight on the floor [9]. Now, it is possible to see in the screen on-line the movement of the platforms and the discrete trajectory of the Pseudo (COM projection). Pseudo coordinates (x_p, y_p) have been calculated with eq. (3), (4) y (5). Finally, the result after having walked will be show.

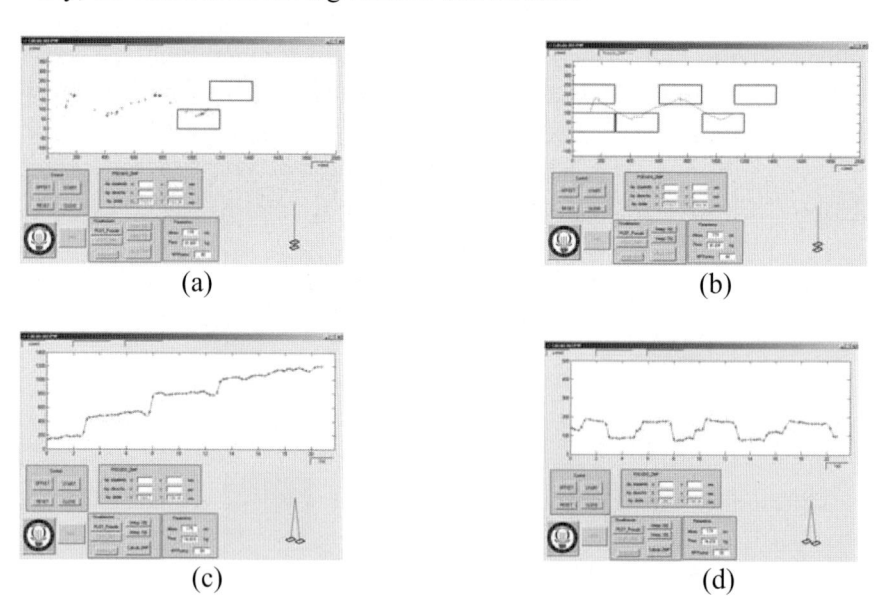

(a)

(b)

(c)

(d)

Fig. 7. Interface Window: (a) Walk Finished, (b) Steps, (c) $x_p(t)$ continuous, (d) $y_p(t)$ continuous

The screen will show at every time the discrete position of the Pseudo in the transversal plane (see **Fig. 7a**). Moreover, at the down-right corner there is a model as a inverted pendulum which it is going to move on-line too. Once walk time is finished, off-line all the steps and the continuous Pseudo trajectory are shown (see **Fig. 7b**). Now, the transversal acceleration of the Pseudo is calculated to find the ZMP position (zmp$_x$, zmp$_y$), using eq. (1) and (2). The coordinates x$_p$(t), y$_p$(t) are shown **Fig. 7c** and **7d**. In this case to get the accelerations have used the following relationships:

$$\dot{x}_p = \frac{\Delta x_p}{\Delta t}; \quad \ddot{x}_p = \frac{\Delta \dot{x}_p}{\Delta t} \qquad (6)$$

$$\dot{y}_p = \frac{\Delta y_p}{\Delta t}; \quad \ddot{y}_p = \frac{\Delta \dot{y}_p}{\Delta t} \qquad (7)$$

In **Fig. 8a** the ZMP continuous trajectory (black line) and the continuous trajectory of the Pseudo (red line) are drawn. It is possible to be appreciated the stability area for single and double support too. Those trajectories are very similar because the transversal accelerations were very small. It is right because in all tests the speed of the walking was not very high.

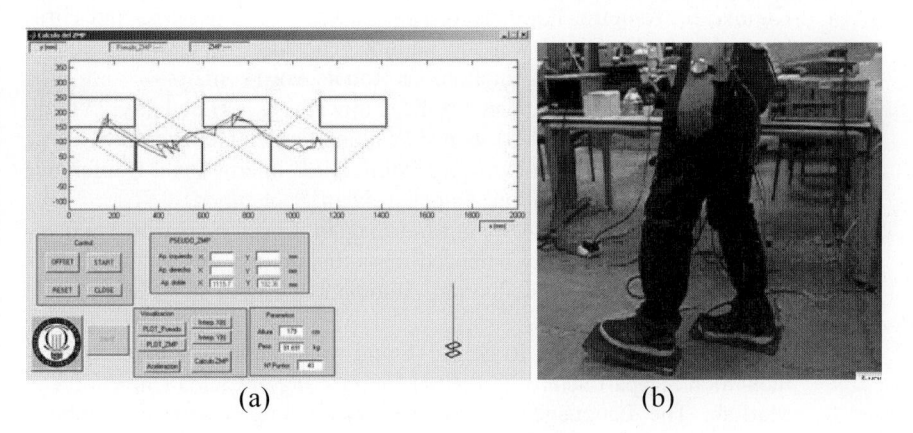

(a)   (b)

Fig. 8. (a) ZMP; (b) Walking Test

For that, it`s known to ensure the walking stability, if ZMP is always inside the stability area. The graph shows a stable movement. **Fig. 8b** gives us an image of a test being performed in the laboratory

6 Conclusions

This paper describes a method to measure a ZMP person while walking. Overall, this has been an experimental project trying to known more about human movement to use it in humanoids, in this case in the robot Rh-0. Future work is to measure the ZMP online, in order to get a dynamic control of humanoid gait

References

1. Vukobratovic, M., Borovac, B., Surla, D and Stokic D. (1990) Biped Locomotion – Dynamics, Stability, Control and Application, Springer-Verlag, Berlin.
2. Hirai, K., Hirose, M., Hikawa, Y. and Takanaka, T. (1998) The development of Honda humanoid robot, IEEE International Conference on Robotics and Automation (ICRA 1998) Leuven (Belgium).
3. Yamaguchi, J., Soga, E., Inoues, S. and Takanishi, A. (1999) Development of a bipedal humanoid robot control method of whole body cooperative dynamic bipedal walking, IEEE International Conference on Robotics and Automation (ICRA' 1999), Detroit, (USA).
4. Kajima, S., Kaneiro, F., Kaneko, K., Fujiwara, K., Yokoi, K. and Hirukawa, H. (2003) Biped walking pattern generation by a simple 3D inverted pendulum model, Autonomous Robots, vol 17, n^a2.
5. Kajita, S., Kanehiro, F., Kaneko, K., Fujiwara, K., Harada, K., Yokoi, K., and hirukawa, H. (2003) Biped Walking Pattern Generation by using Prewiew Control of Zero-Moment Point, IEEE International Conference on Robotics and Automation, Taipei, Taiwan, September.
6. De Torre, S., Cabas, L., Arbulú, M., Balaguer, C. (2003) Inverse dynamics of humanoid robot by balanced mass distribution method". Robotics Lab. Department of Systems and Automation. University Carlos III of Madrid, Spain.
7. Palacín, J., Donaire, O., Roca, J., Static walker foot design and implementation . Department of Industrial Engineering of Lleida University.
8. MatLab. The Language of Technical Computing. Creating Graphical User Interfaces. Version 6.
9. http://roboticslab.uc3m.es/roboticslab/gallery.php?albumname=gaits

Mechanical Design and Dynamic Analysis of the Humanoid Robot RH-0

L. Cabas, R. Cabas, D. Kaynov, M. Arbulu, P. Staroverov, and C. Balaguer

Robotics Lab., Department of Systems Engineering and Automation University Carlos III of Madrid c. Butarque 15, Leganés, 28911, Madrid - Spain.
`lcabas@ing.uc3m.es`

Abstract

This paper presents the design process and the dynamic analysis of the 21 DOF humanoid robot RH-0, developed in the Robotics Lab. at the University Carlos III of Madrid.

The main goal of the RH-0 prototype is to obtain a robot able to walk and to manipulate light objects and also to have recognition capability by means of a series of sensors located in its head.

The present article tries to briefly describe the design and analysis process of the humanoid robot RH-0, and in detail the dynamic analysis, showing the experimental results which were carried out with the humanoid robot RH-0.

Keywords: humanoid robot, mechanics analysis, inverse dynamic analysis

Introduction

Humanoid Robots have been investigated for decades. At the moment they present a surprising degree of development and progress in different fields and activities. Following this line, the RH-0 has been designed to assist disabled people and as a first step in its development it has been designed with measurements adapted to the ergonomics that a person in a wheelchair presents. This condition is specific and fundamental, but the applications for which a humanoid robot can be developed are very varied since it was also thought to extend its application field.

Prototype RH-0 (**Fig. 1**) is able to walk and to manipulate light objects and to recognize objects by means of a series of sensors located in its head.

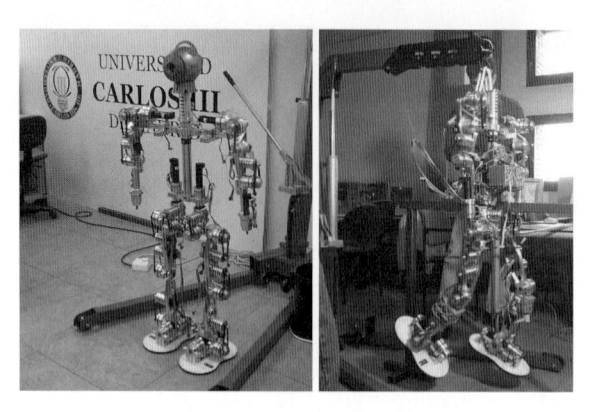

Fig. 1. Humanoid Robot Rh-0

Within the design of the structure, certain structure characteristics were assumed. Later, a series of considerations and results of previous studies were taken in order to choose the optimal trajectory the robot should follow, without loosing its balance and without falling down.

Afterwards, a dynamic study was carried out. The goal of this study was the calculation of the torsion efforts that are going to take place in the joints, in order to select the motors that provide the robot with movement, avoiding both unnecessary over-dimensioning o the motor and its failure due to excessive efforts the motors are unable to support. The dynamic analysis carried out with the chosen structure for the robot left the final design as far as the physical composition of the humanoid, (system performance, motors and outer appearance) with optimal results.

1 Model Selection

The main goal of the humanoid robot is to obtain a robot able to Walk, manipulate light objects and recognize objects by means of a series of sensors located in its head. With these goals and trying not to unnecessarily complicate the robot analysis and control, the robot was designed with 21 degrees of freedom (DOF) with a height close to 1.20m, able to walk in any direction, overcoming heights, and manipulating light objects of up to 500 grams at the end of the arm and recognizing by means of a series of sensors located in its head both with motion and voice commands.

At the present time consideration is given to the possibility of including an additional degree of freedom located in the neck to allow the nodding movement of the head and another one in the in the frontal plane of the batteries. Within the chosen scheme, it is possible to highlight the arrangement of the joints of the shoulder and of the ankle, in which the axes of the frontal and sagital planes are crossed to simplify the kinematics, optimize the movement and allow the most anthropomorphic gait possible, the pelvis configuration, in cantilever or beam in projection, providing the robot with considerable rigidity in a very complicated zone.

2 Kinematic analysis

Although in the human beings walking is something mechanical and practically intuitive, a detailed analysis of the way and why human beings walk, is difficult to obtain. For this reason, prior to the design a kinematic analysis was made, and this is continuously being updated with further developments.A series of considerations and results of previous studies were taken into account, when choosing the optimal trajectory to be followed by each one of the robot joints, trying to avoid the robot loosing its balance and falling down. Therefore, the ZMP (Zero Moment Point) concept arises for consideration.

It is noted that the stability of the robot will be set by the location of the ZMP. In the simple support stage, this one could not be outside the support area (area covered by a foot), whereas in the double support stage it could be wherever it wants within the area defined by both feet.

Therefore, the knowledge of this point is essential when designing the robot movement, but it cannot be known a priori because of the influence of thee robot's dynamic parameters.In order to correct this deficiency an approach can be carried out considering a quasi-static system, that is, a system which would only be influenced by the robot's kinematics and its mass. Pseudo-ZMP thus obtained and used in the definition of the trajectory, is the projection in the ground of the center of masses of the robot. As an approach, the position of the biped will be stable if the ZMP stays within both feet during the double support stage and if it stays within the footplant support during the simple support stage.

With these considerations it was decided to design a trajectory that ZMP assured the quasi-static stability of the robot throughout the route and giving a threshold so that if the accelerations and inertia effects were taken into account, the robot will not fall (**Fig. 2**), (**Fig. 3**), (**Fig. 4**).

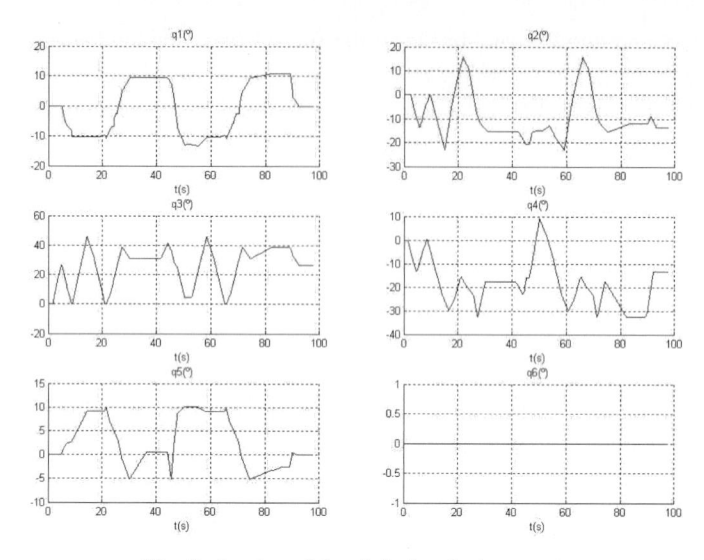

Fig. 2. Angles of the right leg during a gait

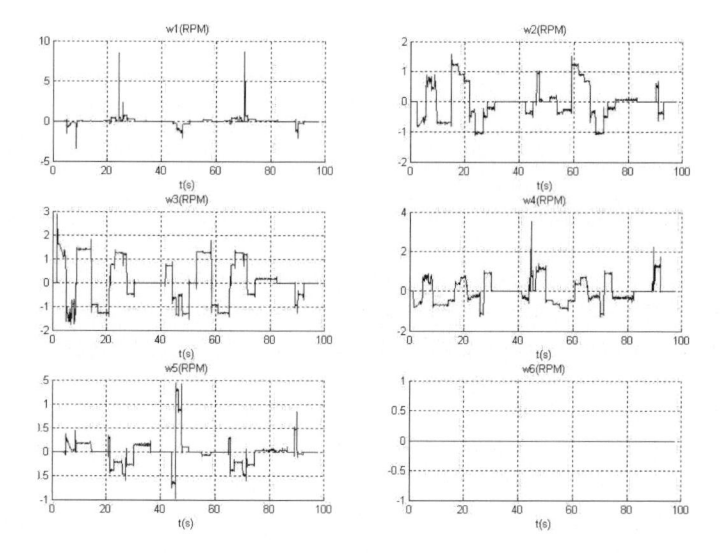

Fig. 3. Speeds of the right leg during a gait

3 Dynamic Analysis

First, the dynamic analysis is approximated, but the obtained results serve as an idea of the torques that are going to take place in the critical joints such as: the ankles and the knees of the biped.

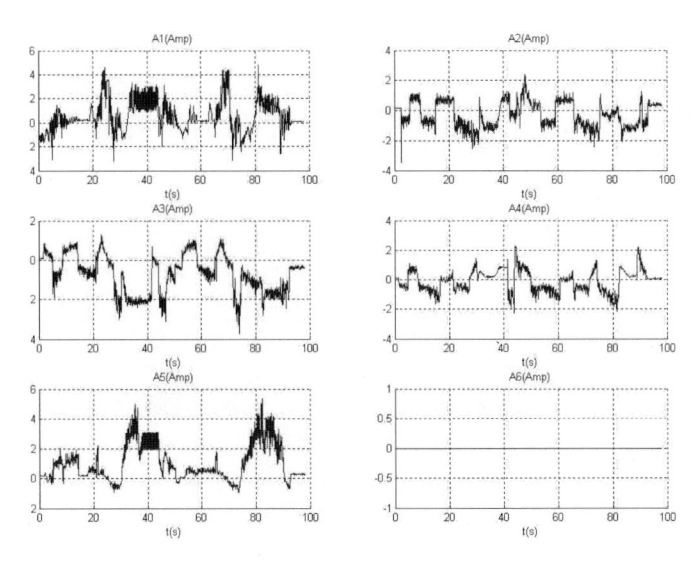

Fig. 4. Current of the right leg during a gait

As a first step in the design of the humanoid robot, an exhaustive analysis of the long walk was carried out. This was widely analyzed since long ago but nowadays, thanks to the powerful computers and the sophisticated electronics that can be accomplished.

During this analysis a disadvantage was found. Human walking is made up of two differentiated stages which alternate in a repetitive and periodic way. These are: the simple support stage, when a foot is in the air moving towards its position, while the other one is in contact with the ground acting as the supported base and the double support stage, when both feet are supported in the ground and the body is balanced laterally to relocate its center of gravity in a stable position. Therefore, the dynamic model was processed for both stages separately.

In order to obtain the dynamic model of the biped a series of functions available in the MatLab robotics toolbox, were used. It is totally integrated for a mechanical design and it performs programs of CAD/CAM, which assures a total optimization of the functions and therefore, totally reliable and precise results (**Fig. 5**).

The need of carrying out a dynamic study arises. The goal of this study is the calculation of the torsion efforts that are going to take place in the joints, in order to choose the most optimal motors that provide the robot with movement, avoiding both unnecessary over-dimensioning of the motor and its failure due to excessive efforts that the motors are unable to support. The study was divided into two sections corresponding to the upper and lower parts of the humanoid robot. At the moment, the lower part is the only one that is moving, but the integration with the upper part is immediate.

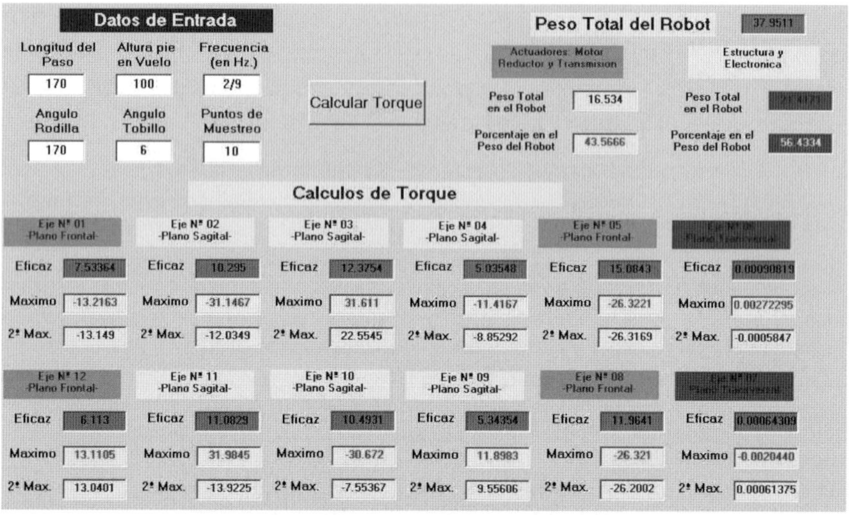

Fig. 5. General Results

Thanks to the kinematics the position, speed and acceleration matrices are obtained. where the angular columns correspond to each one of the twelve joints and where the rows contain the evolution of the angles, angular velocities and accelerations during gait.

The calculation of the torques generated in the joints is defined in a precise and optimal way by means of the dynamic model of Newton-Euler type. The difficulty of the calculation during the double support is increased due to the appearance of a degree of freedom not controlled by any actuator. A closed chain the solution of which is not unique is obtained; the number of unknown parameters is greater than the number of equations. Therefore, two ways of estimating the evolution of the torques and trying to calculate the effective values for each motor, during the gait were studied. The mass, the pair and the inertia are intrinsic parts of the structure of the robot. In this sense the correct choice of each part of the robot is as important as the chosen electronics or the implemented software.

Finally, the structural calculation is carried out separately for each leg, obtaining the torques that the motors of each joint are going to support. With the obtained results the effective torque that the motors should give during the evolution of the gait for each method, is calculated. The choice of the motors needed for the biped was made by considering the maximum effective torque between both methods as critical value, to assure its resistance. The dynamic analysis of the upper structure of the humanoid robot does not require estimation methods for calculation of the torques since it does not present a closed chain. The only exception is the neck joint, that simultaneously generates the movement to turn both arms around the body.

4 Results and conclusions

From the analysis carried out with the chosen structure for the robot, experimental results of the final design have been obtained so far as the humanoid robot structure, system performance and the outer appearance, although the latter is subject to manufacturing modifications.

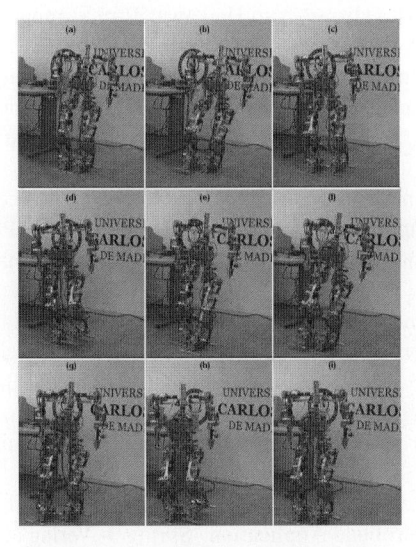

Fig. 6. Rh-0 making his first passings

5 Future Improvement

Currently, the arms movement is not considered during the step evolution since the kinematics for the arms have not yet been developed. This aspect will be considered in future studies.

References

1. L. Cabas, S. de Torre, I. Prieto, M. Arbulu, C. Balaguer, "Development of the light-weight human size humanoid robot RH-0". CLAWAR 2004, Madrid - September 2004.
2. S. de Torre, L. Cabas, M. Arbulú, C. Balaguer "Inverse dynamics of humanoid robot by balanced mass distribution method". Robotics Lab. Department of Systems and Automation. University Carlos III of Madrid, Spain.
3. M. Hirose, Y. Haikawa, T. Takenaka, and K, Hirai, "Development of Humanoid Robot ASIMO", Proc. Int. Conference on Intelligent Robots and Systems, Workshop2 (Oct. 29, 2001), 2001.
4. Kazuo Hirai, Masato Hirose, Yuji Haikawa, Toru Takenaka. "The development of Honda Humanoid Robot". Proceedings of the 1998 IEEE. International Conference on Robotics and Automation, Leuven, Belgium - May 1998.
5. Ambarish Goswami. "Postural stability of biped robots and the foot-rotation indicator (FRI) point". Department of Computer and information science. University of Pennsylvania
6. Giusseppe Carbone, Hun-ok Lim, Atsuo Takanishi, Marco Ceccarelli. "Stiffness analysis of the humanoid robot Wabian RIV: Modelling. Proceedings of the 2003 IEEE. International Conference on Robotics and Automation, Taipei, Taiwan - September 2003.
7. K. Hirai, "Current and Future Perspective of Honda Humanoid Robot". Proc. IEEE/RSJ Int. Conference on Intelligent Robots and Systems, pp. 500-508, 1997.
8. Gordon Wyeth, Damien Kee, Mark Wagstaff, Nathaniel Brewer, Jared Stir-zaker, Timothy Cartwright, Bartek Bebel. "Design of an autonomous Humanoid Robot". School Computer science and electrical engenieering, University of Queensland, Australia
9. M. Vukobratovic, B. Borovac, D. Surla, D. Stokic, "Biped Locomotion – Dynamics, Stability, Control and Application" Springer- Verlag (1990), Berlin.

Advanced Motion Control System for the Humanoid Robot Rh-0

D. Kaynov, M. A. Rodríguez, M. Arbulú, P. Staroverov, L. Cabas and C. Balaguer

Robotics Lab, Department of Systems Engineering and Automation, University Carlos III of Madrid, Spain   dkaynov@ing.uc3m.es

Abstract

This paper presents an advanced motion control system for the Rh-0 humanoid robot. Rh-0 is a humanoid platform which we developed on the first phase of the Rh project, which was launched by the Robotics Lab at the University Carlos III of Madrid in 2002. The motion control system of Rh-0 is based on standard commercial available hardware components. It provides scalability, modularity and application of standardized interfaces. As a result, easy upgrade and further development of reduced weight humanoid robot is possible. In this paper the hardware system, control level software and experimental results are discussed.

Keywords: Humanoid Robots, Motion Control, Rh-0.

1 Introduction

Recently the development of sophisticated humanoid robots has increased and it has become very active area. There is growing interest not only in academic area but in some industrial areas too. Several humanoid robots have been developed in these years. One of them is WABIAN (WAseda Bipedal humANoid) constructed by Waseda University [1]. This robot has 43 D.O.F, is 1890 mm. height and 131.4 kg weight. JOHNNIE is another well known humanoid robot project realized by Technical University of Munich [2]. It has 17 D.O.F., 1800 mm height and 37 kg weight. One of the best humanoid robots is HRP-2 designed by Kawada Industries [3]. HRP-2 has 30 D.O.F., 1540 mm height and 58 kg weight. And the most impressive humanoid robot should be ASIMO constructed by HONDA [4]. The new generation of ASIMO robots presented in 2004 has 34 G.D.L. It is 1200 mm height and 43 kg weight. It is necessary to mention that the great success of ASIMO makes the current research of humanoid robots to become very promising working field for scientists and engineers. One of the recently presented successful projects is KHR-2 humanoid robot constructed by KAIST (Korea Advanced Institute of Science and Technology) [5]. It has 41 D.O.F., 1200 mm height and 56 kg weight.

However the area of humanoid robots is still limited by very high cost of maintenance and development. In this context, the Robotics Lab of the university Carlos III de Madrid has launched the first phase of Rh project. The main goal of this project is development of a reduced weight human size robot which can be a reliable humanoid platform for implementing different control algorithms, human interaction, etc. The control system should be designed using the conventional electronic components of automation industry in order to reduce development time and cost and to have flexible and easy upgradeable hardware system.

We have constructed a humanoid robot Rh-0 – the first version of Rh series humanoid platform. It can walk stably on the flat surfaces in the open body control loop mode. This paper presents the Rh-0 robot, introduces its control system which includes hardware and software architectures and shows some experimental results.

2 Humanoid robot Rh-0

Rh-0 is a new humanoid robot in the first phase of the Rh project. We develop this robot for several principal tasks such as human care, maintenance of dangerous for human health plants, entertainment etc. We have designed and constructed Rh-0 in order to fulfill several conditions. The robot has to be able to:
- walk straight and turn its body on the plane surface
- go upstairs and downstairs
- simulate human hands movements
- gesticulate with its hands (give signals, salute)
- take and manipulate with objects up to 0.5 kg

The main design concept of Rh-0 robot is a light weight and compact (comparable with a human) size. **Fig. 1** shows an overview of Rh-0 and its mechanical configuration.

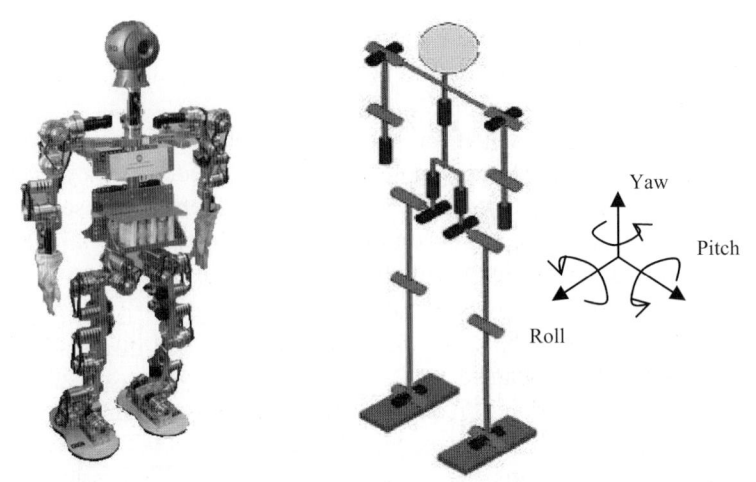

Fig. 1 (a) Developed Rh-0 robot **(b)** Mechanical configuration

Fig. 1 (a) confirms that the mechanical and dynamical design of Rh-0 is completed. It has 1200 mm height (without head), 50 kg weight and 21 D.O.F. (23 with camera). The detail hardware and control software design will be presented in followings sections.

3 Control system of Rh-0 robot

The main goal of the humanoid robot control system is provide it with stable walking and avoid fallings down. To do this we generate motion patterns for each articulation according to the ZMP (Zero Moment Point) theory [6]. The humanoid robot do not falls down when the target ZMP is inside of the support polygon made by the supporting leg(s). We have developed an advanced motion control system to control humanoid movements and provide it with human appearance in the locomotion capacity aspect. It includes both, hardware and bottom level control and communication software.

3.1 Hardware architecture

The hardware architecture for the humanoid robot has some important restrictions posed by the limited availability of space. In general, basic requirements for hardware architecture of humanoid robot are: scalability, modularity and standardized interfaces [7]. In the case of Rh-0 robot with 21 D.O.F., which supposes the use of 21 DC motors first of all it is necessary to choose the appropriate control approach. An electrical design of robot is based on distributed motion control philosophy there each node is an intelligent device.

Fig. 2 shows an overview of the hardware structure. Presented architecture is provided with large level of scalability and modularity by dividing the hardware system into three basic layers. Each layer is represented as a controller centered on its own task such as external communications, motion controller's network supervision, and general control.

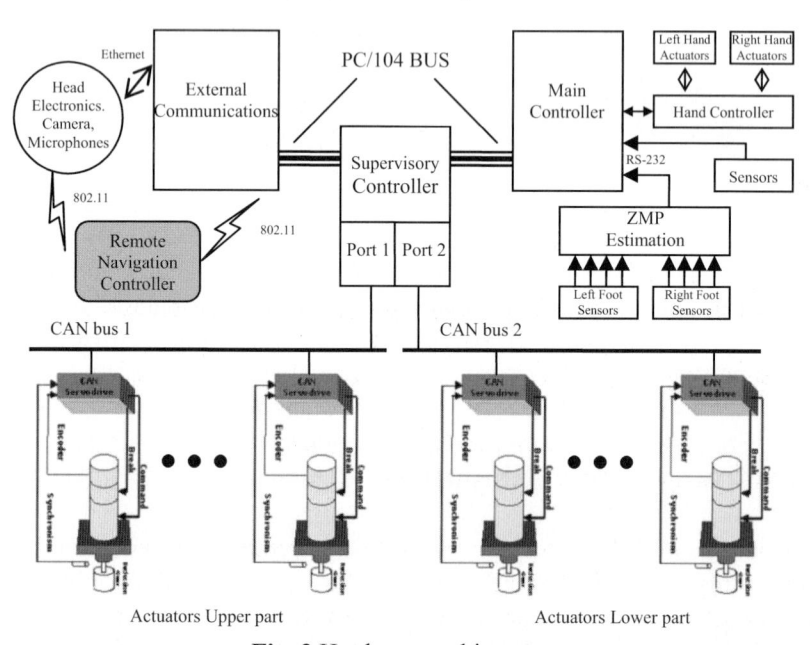

Fig. 2 Hardware architecture

Main Controller is a commercial PC/104 single board computer because of it small size and low energy consumption. We use it instead of a DSP controller because it has different peripheral interface such as Ethernet and RS-232, and easy programming environment. As well there is a great variety of additional extension modules for PC/104 bus like CAN-bus, digital and analog input-output, PCMCIA cards. Selecting criterions were fast CPU speed, low consumption and availability of expansion interfaces. Main Controller provides general synchronization, updates sensor data, calculates trajectory patterns and sends it to the servo controllers of each joint. It also supervises data transmission for extension boards like Supervisory Controller and External Communications via PC/104 bus.

Table 1 shows specifications of the main controller of Rh-0 robot.

Table 1. Specifications of Main Controller

CPU	VIA EDEN ESP 4000 400Mhz
System Memory	512 MB
Chipset	VIA ProSavage TwisterT PN133T + VT82C686B
Expansion	PC/104 16 bit standard 64 + 40 pins connector
Power Consumption	5V, typical 1,5 A
Size/Weight	PC/104 form factor, 90x96 mm, 0.150 kg
I/O	1 × RS232C and 1 RS232/422/485
2 USB 1.1 ports
Fast Ethernet 10/100baseT
EIDE interface with UltraDMA 66 mode

On the motion controller's level each servo drive not only closes the servo loop but calculates and performs trajectory online, synchronizes with other devices and can execute different movement programs located in its flash memory card. This kind of devices could be located near the motors gaining the benefit of less wiring that is one of the requirements for energy efficiency, lightweight and small effort in cabling. We implemented advanced commercial available motion controllers in order to reduce development time and cost. Continuous evolution and improvements in electronics and computing have already made it possible to reduce the controller's size for using it in the humanoid development project. Furthermore, it gives an advantage of using well supported and widely used devices from the industrial automation field.

Communication Supervisory Controller uses a network bus to reliably connect distributed intelligent motion controllers with the Main Controller. Control system adopted in the Rh-0 robot is a distributed architecture based on CAN bus. CAN bus was chosen because of its characteristics such as bandwidth up to 1 MBit/s that is sufficient speed to control axes of humanoid robot, large number of nodes (Rh-0 has 21 controllable D.O.F.), differential data transmission that is important to reduce EMI effects caused by electric motors and the possibility for another devices like sensors to reside on the same control network.

Controllers' network of the Rh-0 is divided into 2 independent CAN buses in order to reduce the load of the bus. The Lower part bus controls 12 nodes of two legs and the Upper part bus controls 10 nodes of two arms and the trunk. By this, the communication speed of CAN bus used in Rh-0 is 1MBit/s. Synchronization of both parts is realized by Supervisory Controller.

The External Communications board provides Ethernet communication with the head electronics which comprises independent vision and sound processing system. It also provides wireless communicates with Remote Navigation Controller which planes motion and sends operating commands for the humanoid robot.

3.2 Software architecture

We developed the bottom level software for the advanced motion control system. It configures intelligent motion controllers, establishes CAN communication, controls trajectory execution and collects motion data which is used in humanoid robot control process. **Fig. 3** shows the bottom level software architecture.

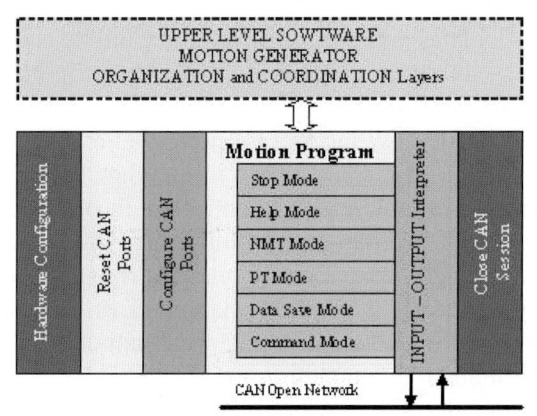

Fig. 3 Software architecture

Motion patterns generated on the upper level by the Motion Generator using Lie logics or conventional Denavit-Hartenberg inverse kinematics method are processed in the Motion Program as a reference input data. Different modes of interaction with the robot on the control level were developed.

It is possible to control and manipulate as a single joint in the Command Mode interaction as well as the whole robot in the PT (Position - Time) Mode. The latter is the fundamental one for the synchronized multi-axis walking applications.

In the PT mode, motion patterns as a sequence of the absolute encoder's positions is interpolated by the third-order polynomial (1) with one continuous derivative:

$$P(t) = a(t-t_0)^3 + b(t-t_0)^2 + c(t-t_0) + d \qquad (1)$$

The time space of interpolation is an integer multiple of the servo drive sampling time T which for the implemented controller equals 0.026 s. The speed at the point k inside the trajectory path is calculated as:

$$V(k) = \frac{P(k+1) - P(k-1)}{2 \cdot T} \qquad (2)$$

Main Controller divides the joints' trajectory into tables of PT sets. The tables are either pre-loaded to the servo drives memory or are sent on-the-fly with the joints position corrections at a rate of at less the sampling period of these. To start the motion the Main Controller sends a "broadcast" start message. As the motion controllers' internal clocks are not identical, a synchronization mechanism is required to synchronize these every 0.5 second with a clock of a network Supervisory Controller.

The communication is realized using Process Data Objects (PDO) of CANopen protocol [8]. Developed Input-Output Interpreter provides the encoding of the motion data to the CANopen communication objects. The CAN 2.0 A with 11 bit identifier addressing method was implemented.

4 Experiments

The basic experiments were carried out to confirm whether Rh-0, provided with a developed mechanical structure, hardware and software architectures, can make motions like steps. **Fig.** 4 shows the experimental results of motor current variations of three leg joints at the forward walking mode.

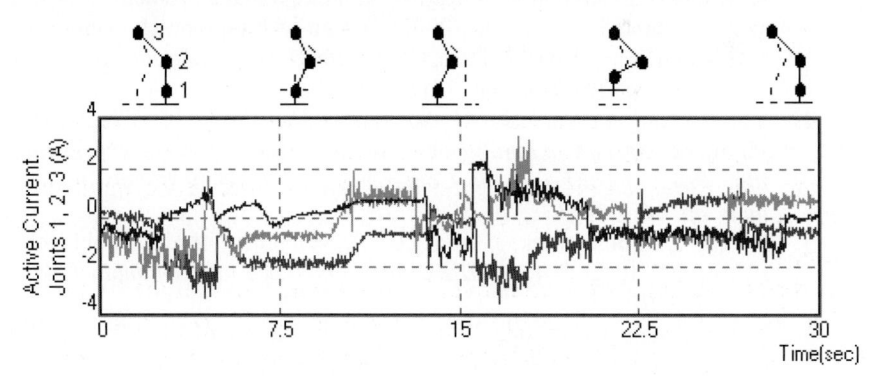

Fig. 4 Working current at forward walking mode

Current's variation of the most loaded joints stays at a low level indicating appropriateness of the mechanical design and the control architecture.

5 Conclusions

This paper presented the development of the control system of the humanoid robot Rh-0. It was proved that Rh-0 robot provided with the current mechanical, hardware and software control system can walk.

In the future, developed software motion control system will be integrated into the global software system of RH-0 robot which comprises wireless communication module, HMI, sound and image processing systems, the motion planning module. Further work also includes integrating a new sensory system into the robot. For example force torque sensors and accelerometers will be incorporated into the body of the RH-0 humanoid robot in order to close its body inclination control loop. Furthermore, the analysis and improvement of the hardware and software architecture will also be continued.

References

1. J. Yamaguchi, E. Soga, S. Inoue and A. Takanishi (1999) Development of a Bipedal Humanoid Robot – Control Method of Whole Body Cooperative Dynamic Biped Walking. *Proc. of IEEE Int. Conference on Robotics and Automation, pp. 368-374.*
2. M. Giender, K. Löffler, and F. Pfeifer (2001) Towards the Design of Biped Jogging Robot, *Proc. of IEEE Int. Conference on Robotics and Automation, pp. 4140-4145.*
3. K. Kaneko, F. Kanehiro, S. Kajita, K. Yokoyama, K. Akachi, T. Kawasaki, S. Ota and T. Isozumi (2002) Design of Prototype Humanoid Robotics Platform for HRP, *Proc. of IEEE/RSJ Int. Conference on Intelligent Robots and Systems, pp. 2431-2436.*
4. Y. Sakagami, R. Watanabe, C. Aoyama, S. Matsunaga, N. Hikagi and K. Fujimura (2002) The intelligent ASIMO: System overview and integration, *Proc. of IEEE/RSJ Int. Conference on Intelligent Robots and Systems, pp. 2478-2483.*
5. I. Park, J. Kim, S. Park, J. Oh (2004) Development of Humanoid Robot Platform KHR-2, *Proc. of IEEE Int. Conference on Humanoid Robots, pp. 292-310.*
6. K. Regenstein and Rudiger Dillmann (2003) Design of an open hardware architecture for the humanoid robot ARMAR, *Proc. of IEEE Int. Conference on Humanoid Robots*
7. M. Vukobratovic, D. Juricic (1969) Contribution to the Synthesis of Biped Gait, *IEEE Tran. On Bio-Medical Engineering, Vol. 16, No. 1, pp. 1-6.*
8. Elmo Motion Control GmbH. (2004) *Harmonica Software Manual.* Villingen-Schwenningen, Germany.

Humanoid Vertical Jump with Compliant Contact

Victor Nunez[1,2] and Nelly Nadjar-Gauthier[1,2]

[1] AIST/IS - CNRS/STIC Joint French-Japanese Robotics Laboratory
[2] Laboratoire de Robotique de Versailles (LRV), Université de Versailles.
10-12 avenue de l'Europe 78140 Vélizy, France.
nunez, nadjar@lrv.uvsq.fr

Abstract

The objective of the authors is to achieve a vertical jump in experimentation using the humanoid robot HRP-2. After satisfactory results considering rigid foot ground contact on a Matlab simulator, this paper studies the vertical jump considering a compliant contact and discusses the differences between both models, using Matlab and compared with openHRP (realistic humanoid simulation software)

Keywords: humanoid jump, ground contact, compliant contact, control

1 Introduction

Nowadays, with the mature stage of the biped walking topic, more interest is given to advanced locomotion tasks and in many examples humanoid robots are considered as testbeds. In this paper the jumping motion for humanoid robots is considered. In [5] the first jumps of the robot HRP-2L are presented. Kajita *et all* used the Resolved Momentum Control in order to specify the jumping motion but the impact at land-in is not specifically treated.

However, our approach is closer to that presented in [9]. In this work the authors present the Variable Impedant Inverted Pendulum (VIIP) model, which allows to express the desired vertical ground force and vertical velocity of the CoM as a reference movement of the CoM. Again the impact at land-in is not treated on detail. A recent work [8] treats the vertical jump on humanoids robots. The main drawback of the method is that the impact at landing is not explicitly reduced, and there is not a closed loop control which takes into account the dynamics of the robot. On the other hand the authors considers a whole body movement, finding an optimal movements of the arms.

The work presented here was done within the frame of the French-Japanese Joint Robotics Laboratory (JRL). This laboratory is devoted to autonomy of humanoid robots and uses as testbed the robot HRP-2 [1]. The objective of the authors is to achieve a vertical jump in experimentation using this humanoid robot.

Within this objective, and after satisfactory results considering rigid foot-ground contact (f-gC) obtained using Matlab, this paper studies the vertical jump considering a compliant contact and discusses the differences between both models, using Matlab and compared with openHRP [2] (realistic humanoid simulation software).

The influence of the contact on the velocity of the center of mass (CoM) at lift-off and -as a consequence- on the flight time, is shown. One major consequence of neglecting this influence may be a large impact at landing.

In section 2, a brief presentation of the HRP-2 humanoid robot is done. The considered model and the control strategy for vertical jump are exposed.

In section 3, our results are illustrated via MATLAB simulations - considering a rigid contact model- and compared with openHRP which considers a compliant contact between the foot and the ground. The differences between the two simulations are explained based on a simple dynamic analysis. Then in section 4, the MATLAB model is modified to take into account a compliant contact and we finally show how the simulation on openHRP is improved. Section 5 gives conclusions and develops our future works.

2 Previous works

2.1 Main assumptions

The humanoid robot HRP-2 [1] has 30 articulations. Considering only the left half (using the symmetry of the vertical motion), 17 articulations (6 articulations on the leg, 4 on the body and 7 for the arm) still remains and 18 masses have to be considered (see **Fig. 1**). We used a simplified structure which considers only the movement of the legs (it is intuitively clear that the main effort is concentrated there), and furthermore we will only consider the sagittal movement of the legs, i.e. only the pitch of the hip, knee and ankle will be allowed to move.

All the other articulations will be fixed (at zero for the most of them). In order to have an accurate model of the robot in this conditions, all the masses and inertias have been considered.

2.2 Control strategy for a vertical jump

This section summarizes the work presented in [4]. With the vertical jump of humanoid robots as objective, the authors proposed a control method (applied to a simplified biped structure) based on sliding mode [3]. The key points of the approach are: a) to keep the CoM of the robot exactly above the ankle, b) to require the trajectory tracking of the vertical component of the CoM, and c) to minimize the impact at landing by inducing almost zero absolute velocity of the foot at landing. The last point is a key objective, because the large impacts at landing are the major obstacles to perform aerial phases in human size humanoid robots. In [4] the whole humanoid structure is considered and the foot orientation is controlled during flight to have landing with flat foot. In both works MATLAB simulations are presented and a rigid *f-gC* is considered.

Before going to experimental phase using the HRP-2 humanoid robot, a previous simulation using openHRP is necessary. One of the most important aspects in this software is that the dynamics of the HRP-2 robot are simulated in an accurate manner, and the experiments after simulations are done in an easy way.

As a first instance, we passed the articular positions obtained on MATLAB simulations as references to the openHRP simulator. This references are followed precisely and as a consequence, the desired trajectory of the CoM *with respect to the ankle* is obtained. However the flight time on openHRP simulation is bigger than expected and therefore the landing happens in an unexpected instant, producing large impacts at landing.

In experimentation it will be crucial to guarantee a predicted flight time, and to obtain reduced impacts at landing. Actually in [5] (where experiment for jumping of biped robot HRP-2L are presented) the flight time is different from expected. Furthermore in [6] (section III B) the problem is solved by explicitly considering a compliant contact. Nevertheless in this work the impact at landing is not explicitly reduced, and the solution is proposed with no further study.

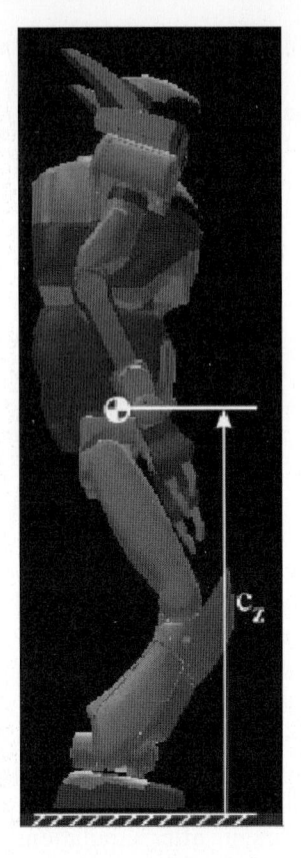

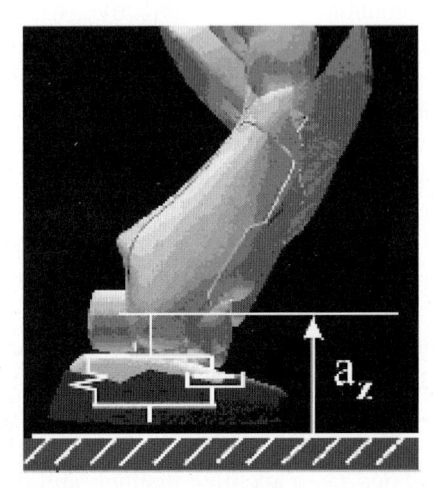

Fig. 1. Total CoM height

Fig. 2. Ankle height

The compliant f-gC considered in OpenHRP simulator, realistic because based on experimentation, is considered as concentrated at the foot (see **Fig. 2**). However it is rather the result of the compliance of the ground, the sole and the non negligible compliance of the robot servos working at high speeds [6]

In **Figs. 1** and **2**, the height of the CoM and the ankle position for the HRP-2 robot in simulation are shown. The distance from the ankle to the CoM is denoted by l, ($l = c_z - a_z$). In [4], we proposed a trajectory for l to achieve the vertical jump. The time derivative of l at the instant of lift-off is specified as a desired value: $\dot{l}(t_{LO}) = \dot{l}^{des}(t_{LO})$ with t_{LO} being the instant when the robot leaves the ground.

3 Rigid contact versus a compliant one

In this section, the MATLAB simulations are obtained using a rigid ground –robot contact. The articular positions are passed to the openHRP simulator as desired positions. Since openHRP is realistic, a compliant contact model is implemented. The openHRP results are compared to MATLAB results and analyzed (see **Fig. 3**).

Notice that if we suppose a rigid contact (in MATLAB), then the ankle height is fixed $\dot{a}_z = 0$ and actually $\dot{l} = \dot{c}_z$. If it is so, we control the velocity of the CoM at lift-off. During flight the CoM will follow a ballistic trajectory and, therefore it is possible to control the absolute foot position by changing l. With the objective of reducing the impact force at landing, the trajectory of l is chosen such that the absolute foot velocity is almost zero at this instant.

However, with compliant contact the velocity of the ankle during lift-off is not zero, so the velocity of the CoM at lift-off is:

$$\dot{c}_z(t_{LO}) = \dot{l}(t_{LO}) + \dot{a}_z(t_{LO}) \quad (1)$$

with $\dot{a}_z(t_{LO}) \neq 0$

Consequently, the CoM goes higher than expected and the flight time grows in consequence. With the landing occurring at a time after the predicted one, foot velocity is not controlled any more and the impact forces are too important.

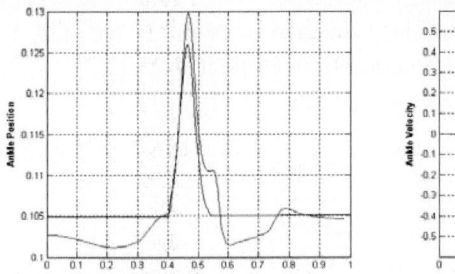

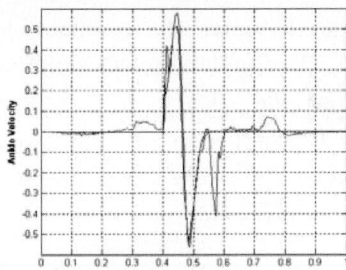

Fig. 3. Ankle position (left) and velocity (right) during jump
Comparison between MATLAB (-) and OpenHRP (-)

Fig. 3 shows the ankle position and velocity during jump, from both MATLAB simulator and openHRP simulator. During the take-off phase (from t=0 to 0.4 s), the ankle positions are slightly different due to the depression in the ground caused by the compliant model of the openHRP simulator. At time of liff-off, $t_{LO} = 0.4$ s, the ankle velocity in OpenHRP is greater than zero as planned by MATLAB. Therefore, the ankle maximal height in OpenHRP is greater than the one planned by MATLAB (0.13 m instead of 0.126 m), and the flight duration is longer. Then, the landing time arrives after and the robot in openHRP touches the ground too late: the ankle touch-down velocity is not zero and thus it appears an impulsive force around 1000 N (see **Fig. 4**) damageable for the robot.

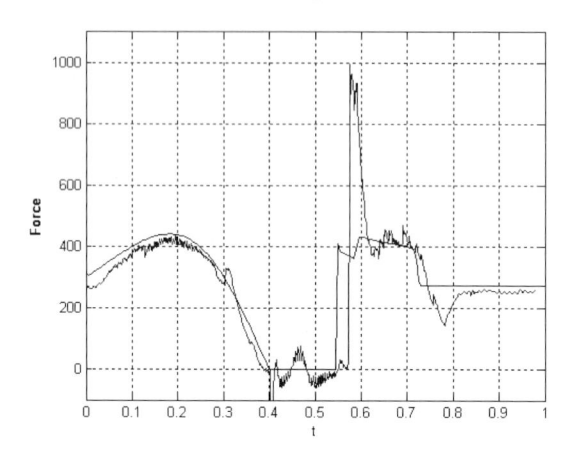

Fig. 4. Vertical force during jump
Comparison between MATLAB (-) and OpenHRP (-)

4 Comparison of compliant contacts

In order to solve the problem described in section 3, we add to the MATLAB simulator a very simple compliant contact model :

$$F_z = -ke - b\dot{e} \qquad (2)$$

where k and b are two constant parameters and e the depression in the ground.

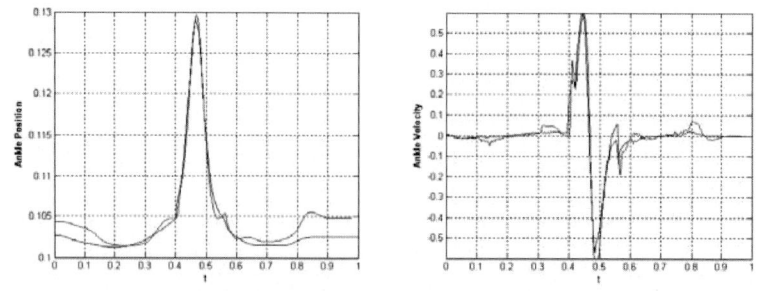

Fig. 5. Ankle position (left) and velocity (right) during jump
Comparison between MATLAB (-) and OpenHRP (-)

In **Fig. 5**, we can see that the tracking of the ankle position and velocity is better than the one obtained in **Fig. 3**. In particular, the ankle openHRP maximal height is equal to the one planned by MATLAB, and at t = 0.55s, corresponding to the landing, the ankle velocity is zero. Thus the contact force shown in **Fig. 6** is minimized compared to **Fig. 4** and it is never greater than 550 N.

These simulations proves that it is absolutely necessary to take into account the compliant contact if a real experiment is envisaged.

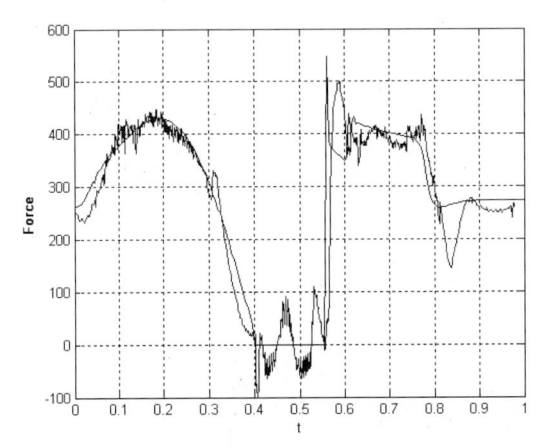

Fig. 6. Vertical force during jump
Comparison between MATLAB (-) and OpenHRP (-)

5 Conclusions and future works

In this paper, we have compared two simulators, one in MATLAB, used to understand main features of the vertical jump of a complex humanoid robot, and the other one called openHRP simulator taken as our reference to envisage an experimentation on the real robot.

In our first studies, we programmed in MATLAB a rigid contact. However, when we tried to transpose the obtained results on the openHRP simulator, big differences appeared. We have supposed that the main reason was the contact model and proved that, with a very simple compliant model in MATLAB, these differences vanished.

Now, our next stage is to make an experiment. We work on the robustness of our approach to be sufficiently secure and safe for the real robot. In the same time, we are interested in adding arms in the vertical or forward motion of the jump.

References

1. Kaneko K., Kanehiro F., Kajita S., Yokoyama K., Akachi K., Kawasaki T., Ota S., Isozumi T.(2002) Design of prototype humanoid robotics platform for HRP, in Proc. of IEEE/RSJ Int. Conf. on Intelligent Robots and Systems.
2. Kanehiro F., Fujiwara K., Kajita S., Yokoi K., Kaneko K., Hirukawa H., Nakamura Y., Yamane K. (2002) Open architecture humanoid robotics platform, in Proc. of IEEE Int. Conf. on on Robotics and Automation.
3. Nunez V., Drakunov S., Nadjar-Gauthier N., Cadiou J. C. (2005) Control strategy for planar vertical jump, in Proc. of the IEEE/RSJ International, Conference on Advanced Robotics. (Accepted paper).
4. Nunez V., Nadjar-Gauthier N. (2005) Control Strategy for Vertical Jump of Humanoid Robots, in Proc. of the IEEE Int. Conf. on Intelligent Robots and Systems. (Accepted paper).
5. Kajita S., Nagasaki T., Kaneko K., Yokoi K., Tanie K. (2004) A hop towards running humanoid biped, in Proc. of the IEEE Int. Conf. on Robotics and Automation.
6. Nagasaki T., Kajita S., Kaneko K., Yokoi K., Tanie K. (2004) A running experiment of humanoid biped, in Proc. of the IEEE/RSJ Int. Conf. On Intelligent Robots and Systems.
7. Kaneko, K.; Kajita, S.; Kanehiro, F.; Yokoi, K.; Fujiwara, K.; Hirukawa, H.; Kawasaki, T.; Hirata, M.; Isozumi, T. (2002) Design of advanced leg module for humanoid robotics project of METI, in Proc. of the IEEE Int. Conf. on Robotics and Automation.
8. Sakka S.; Yokoi K (2005) Humanoid vertical jump based on force feedback and inertial forces optimization, in Proc. of the IEEE Int. Conf. On Robotics and Automation.
9. Sugihara, T.; Nakamura, Y (2003) Contact phase invariant control for humanoid robot based on variable impedant inverted pendulum model, in Proc. of the IEEE International Conference on Robotics and Automation, Vol: 1.

Locomotion

A 3D Galloping Quadruped Robot

Darren P. Krasny and David E. Orin

Department of Electrical and Computer Engineering
The Ohio State University
Columbus, OH, 43210, USA
krasny.1@osu.edu, orin.1@osu.edu

Abstract

In this work, we present a practical approach for producing a stable 3D gallop in a simulated quadrupedal robot which includes the prominent characteristics of the biological gait. The dynamic model utilizes biologically-based assumptions, and the resulting 3D gallop contains the prominent biological features of early leg retraction, phase-locked leg motion, a significant gathered flight phase, unconstrained spatial dynamics, and a smooth gait. A multiobjective genetic algorithm is used to find control parameters in a partitioned search space. During stance, a simple energy control law ensures a fixed amount of energy in the knee springs during each stride, which is a key factor for stabilization.

Keywords: gallop, quadruped, legged robot.

1 Introduction

The allure of high-speed dynamic locomotion has motivated the design of many legged quadrupedal robots. Although some were capable of dynamic running ([1]-[3]), the achievement of a well-controlled, biological gallop has remained elusive. The gallop is the preferred high-speed gait of most cursorial[1] mammals [4]. It consists of a rotary (e.g., RR-LR-LF-RF [2]) or transverse (e.g., RR-LR-RF-LF) asymmetric footfall pattern,

---
[1] Cursorial animals stand and run with humerus and femur nearly vertical.
[2] "RR" means right-rear, "RF" means right-front, etc.

single-leg duty factors less than 50%, and, almost always, at least one significant flight phase, called gathered flight, occuring after the two front footfalls [5]. Furthermore, galloping involves early leg retraction (rearward leg rotation just prior to touchdown) and phase-locked leg motion, both of which are integrated into a smooth gait executed with heading and velocity control.

The lack of good analytical models and proven control strategies are two major reasons for the scarcity of examples that truly epitomize the biological gallop. Recently, Smith and Poulakakis [6] demonstrated what appears to be the first rotary gallop in the quadruped robot Scout II, although their gait lacked several of the defining features above and had no heading control, a fundamental feature of biological locomotion, as the robot moved in a tight circular trajectory. In another example, a transverse footfall sequence was employed in a walking quadruped robot [7]. Because the gait lacked a flight phase, however, it is not generally considered a gallop. In simulation, there are several notable examples of gallops [8] - [12], although each one made one or more significant simplifying assumptions.

In the sections that follow, we will describe our approach for generating a 3D gallop that demonstrates its important biological characteristics. We first describe the dynamic model of the quadruped, followed by a description of the state-based 3D gallop controller. Next, the evolutionary optimization method is explained, followed by the results, conclusions, and a summary.

2 Dynamic Model

The dynamic model (**Fig. 1**) is based on the physical characteristics of a small dog. The model has nonzero leg mass, an asymmetric body mass distribution, and articulated knee joints with passive torsional springs to model biological legs. A compliant contact model is used to compute contact forces using linear springs and dampers in the vertical and horizontal directions, and slipping is computed by static and kinetic friction coefficients which model rubber on concrete. Ideal actuators are modeled at the abductor/adductor (hereafter, "ab/ad"), hip fore-aft (hereafter, "hip"), and knee joints, and dynamic simulation is implemented using software for tree-structured robots [13] with Runge-Kutta fourth-order numerical integration. Yaw, pitch, and roll are defined using the ZYX Euler angle convention described in [14].

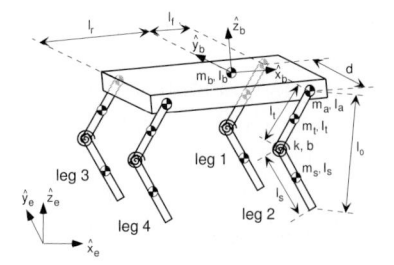

Fig. 1. Dynamic 3D quadruped model with compliant knees.

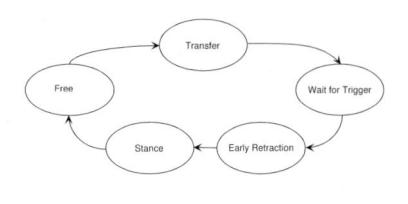

Fig. 2. State diagram for each leg for the 3D gallop controller.

3 The 3D Gallop Controller

3.1 Transfer

The state diagram for the controller is shown in **Fig. 2**. The Transfer state occurs while each leg is in flight, during which the hip and knee angles are rotated to fixed touchdown angles found from preliminary experimentation. The ab/ad touchdown angles are computed using an "outward legrotation" scheme, where the ab/ad joints are rotated outward from the body to correct for roll and yaw errors. For example, if the body is rolling over onto its right side, ab/ad joints 2 and 4 (**Fig. 1**) are rotated outward using PD gains k_{p_γ} and k_{d_γ} applied to the roll and roll rate errors. Ab/ad joints 1 and 3 are rotated outward by a scaled amount f_γ. Yaw control works similarly, although the ab/ad joints are rotated outward in diagonal pairs. For example, if the body is rotating CCW about $\hat{z}_e$ (**Fig. 1**), ab/ad joint 1 is rotated outward by applying PD gains k_{p_α} and k_{d_α} to the yaw and yaw rate errors; ab/ad joint 4 is rotated outward by a scaled amount f_α. Here, ab/ad joints 2 and 3 are not changed, as an outward rotation of these joints would not counter the body's yaw motion.

3.2 Wait for Trigger

Following the Transfer state, the leg enters the Wait for Trigger state, where all joints are held at nominal touchdown angles. When the trigger is detected, the leg enters the Early Retraction state. The triggering scheme is based roughly on [11], where specified delay times (found experimentally) enforce the desired 4-3-2-1 footfall sequence. To make the triggering less immune to variations in body state, our system uses a spatial cue to trigger leg 4 when its hip height is below a certain threshold.

3.3 Early Retraction

Once a leg has triggered, it enters the Early Retraction state, where it is retracted (rotated rearward about the hip) to match the tangential foot velocity with the desired running speed. While biological early retraction is not completely understood, one advantage appears to be a lower foot velocity relative to the ground, which reduces impact losses. In our case, early retraction also enforces the transverse leg phasing for the 3D gallop. Without it, front and rear leg pairs would touch down nearly simultaneously, as in a bound.

3.4 Stance

At touchdown the leg enters the Stance state, where several critical control functions are performed. First, a hip servo controls the body's forward velocity by commanding a tangential foot velocity equal to the desired running speed of 4.15 m/s (the preferred galloping speed for an animal with similar mass [15]) plus a bias. A negative front bias (v_{b_f}) and positive rear bias (v_{b_r}) generates a "shoulder braking" and "hip thrusting" effect, which also stabilizes the body's pitch [11].

The second control function is the injection of energy into the knee springs at maximum compression to account for impact and other losses throughout the stride. The total energy E_0 is distributed among the four knee springs by applying fore-aft and lateral distribution factors d_f and d_l. d_f is comprised of a bias factor d_{f_0} plus PD gains k_{p_β}, k_{d_β}, k'_{p_β}, and k'_{d_β} applied to pitch and pitch rate errors for the current and last stride. d_l is comprised of a bias factor of 0.5 plus PD gains k_{p_l} and k_{d_l} applied to roll and roll rate errors for the current stride. The total amount of energy in the springs remains constant at E_0, although the fore-aft and lateral distributions are modulated to correct for pitch and roll errors. The knee spring rest position is adjusted instantaneously at maximum compression to produce the desired amount of energy. PD position control is used to maintain the ab/ad touchdown angles throughout Stance.

3.5 Free

At liftoff, the leg enters the Free state, where the hip joint is allowed to rotate freely to reduce jerk in the system after the leg breaks contact. The spring rest position, which was moved during Stance, is now returned to its nominal value over $0.025\,s$, the duration of the Free state.

4 The Evolutionary Optimization Problem

The evolvable control parameters are given as follows:

$$\phi = \begin{bmatrix} d_{f_0}, v_{b_f}, v_{b_r}, \dot{\beta}_d, k_{p_\gamma}, k_{d_\gamma}, f_\gamma, k_{p_l}, k_{d_l}, k_{p_\alpha}, k_{d_\alpha}, f_\alpha, \\ k_{p_\beta}, k_{d_\beta}, k'_{p_\beta}, k'_{d_\beta} \end{bmatrix}^T, \quad (1)$$

where v_{b_f} and v_{b_r} are the velocity biases (Sect. 3.4), k_{p_γ}, k_{d_γ}, f_γ, k_{p_α}, k_{d_α}, and f_α are the ab/ad touchdown angle gains for roll and yaw (Sect. 3.1), k_{p_l} and k_{d_l} are the lateral energy distribution gains for roll (Sect. 3.4), d_{f_0}, k_{p_β}, k_{d_β}, k'_{p_β}, and k'_{d_β} are the bias fore-aft energy distribution factor and pitch error gains (Sect. 3.4), and $\dot{\beta}_d$ is the desired pitch rate used to compute the pitch rate error. A staged approach is used to find the parameters, where d_{f_0}, v_{b_f}, v_{b_r}, and $\dot{\beta}_d$, which affect the sagittal plane dynamics, are found first. Second, the roll control parameters k_{p_γ}, k_{d_γ}, f_γ, k_{p_l}, and k_{d_l} are evolved, followed by the yaw control parameters k_{p_α}, k_{d_α}, and f_α. Finally, the pitch control parameters k_{p_β}, k_{d_β}, k'_{p_β}, and k'_{d_β} are evolved to further stabilize the sagittal plane dynamics.

Evolutionary multiobjective optimization (EMO) was chosen because there were multiple evaluation criteria in the fitness function: accuracy, stability, and correctness. Accuracy is measured by the Euclidean distance between the average body state at the top of gathered flight (TOF) and the desired state. Stability is based on the dispersion of TOF state variables from their linear regression lines across multiple strides, and correctness is the average amount of time per stride before an error (e.g., untriggered touchdown) occurs.

5 Results

In this section we present the results of evolving the control parameters listed in Eq. (1). Thirty-two individuals and 250 generations were selected for the evolution, and up to twenty trials were run for each stage. **Figure 3** shows several TOF state variables for the first 60 seconds of 3D galloping. After an initial error in heading, the quadruped gallops in a straight line (**Fig. 3 (a)**). The quadruped runs at a height of 0.235 m (**Fig. 3 (b)**), which provides adequate ground clearance for leg transfer but minimal height excursion for smoothness. The average TOF pitch rate is about -0.15 rad/s (**Fig. 3 (c)**), which is considerably smaller than in other studies [16]-[17], further illustrating the smoothness of the solution. **Figure 3 (d)** shows an average forward running velocity close to the desired value of 4.15 m/s, a result of using velocity servos at the hips. Finally, **Fig. 4** shows the results after each stage of evolution. While roll and yaw control dramatically improve stability, the final stage of pitch control is required for a completely stable solution. A series of screen-captures for one stride of the gait is given in **Fig. 5**, showing biological features like asymmetric footfalls, a significant gathered flight phase, and a smooth gait with minimal vertical excursion and pitch motion.

6 Summary

This paper has demonstrated stable 3D galloping in a simulated quadrupedal robot with biological characteristics such as early leg retraction, a significant gathered flight phase, phase-locked leg motion, smoothness defined by minimal height excursion and pitch motion, and unconstrained spatial dynamics. Furthermore, the solution was accomplished without the simplifying assumptions made in prior work, an important advantage of using an evolutionary method. While the study of the 3D gallop has just begun, the approach described here promises to provide an effective means of further investigation in both simulation and hardware.

Acknowledgements

This work was supported by the National Defense Science and Engineering Graduate fellowship, The Ohio State University Dean's Distinguished University fellowship, and the National Science Foundation under Grant IIS-0208664.

References

1. Raibert M. H. (1990) Trotting, Pacing, and Bounding By a Quadruped Robot. *Journal of Biomechanics*, vol. 23, pp. 79-98.
2. Furushu J., Akihito S., Masamichi S. and Eichi K. (1995) Realization of a Bounce Gait in a Quadruped Robot with Articular-Joint-Type Legs. *Proceedings of the 1995 IEEE International Conference on Robotics and Automation (ICRA)*, Nagoya, Japan, pp. 697-702.
3. Kimura H., Akiyama S. and Sakurama K. (1999) Realization of Dynamic Walking and Running of the Quadruped Using Neural Oscillator. *Autonomous Robots*, vol. 7, pp. 247-258.

A 3D Galloping Quadruped Robot   473

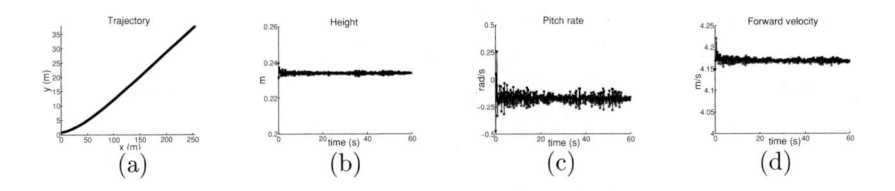

Fig. 3. Selected TOF state variables for the 3D gallop for the first 60 sec.

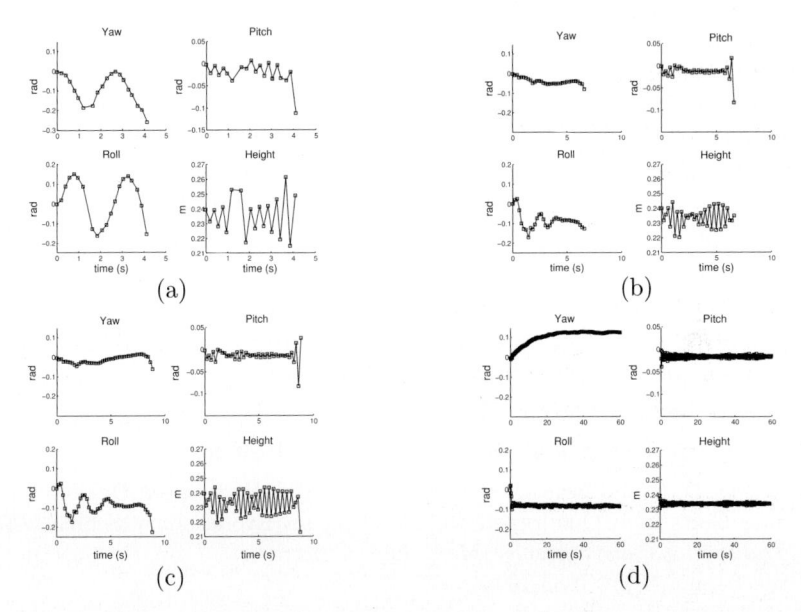

Fig. 4. Stability results after evolving (a) sagittal plane parameters (d_{f_0}, v_{b_f}, v_{b_r}, and $\dot{\beta}_d$), (b) roll parameters (k_{p_γ}, k_{d_γ}, f_γ, k_{p_l}, and k_{d_l}), (c) yaw parameters (k_{p_α}, k_{d_α}, and f_α), and (d) pitch parameters (k_{p_β}, k_{d_β}, k'_{p_β}, and k'_{d_β}).

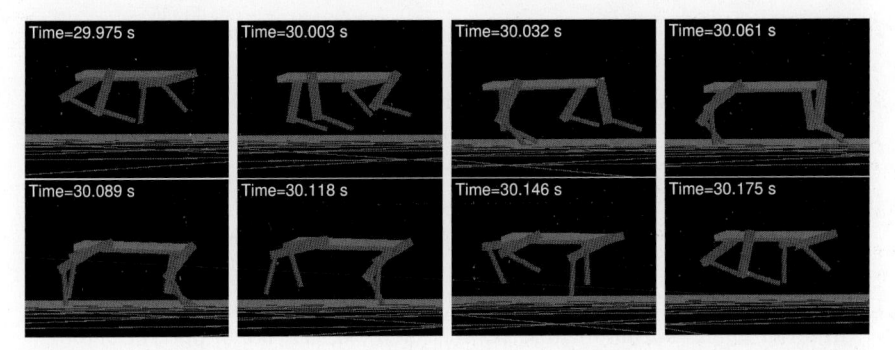

Fig. 5. Screen captures of the 3D gallop over one stride, starting from TOF.

4. Hoyt D. F. and Taylor C. R. (1981) Gait and the Energetics of Locomotion in Horses. *Nature*, vol. 292, pp. 239-240.

5. Gambaryan P. P. (1974) *How Mammals Run: Anatomical Adaptations*. John Wiley & Sons, New York, NY, pp. 23-37.

6. Smith J. A. and Poulakakis I. (2004) Rotary Gallop in the Untethered Quadrupedal Robot Scout II. *Proceedings of the International Conference on Intelligent Robots and Systems (IROS)*, Sendai, Japan, pp. 2556-2561.

7. Morita K. and Ishihara H. (2005) A Study of Mechanisms and Control of Underactuated 4-Legged Locomotion Robot: Proposal of Walking Model by Underactuated System. *Video Proceedings of the 2005 IEEE International Conference on Robotics and Automation (ICRA)*, Barcelona, Spain.

8. Nanua P. and Waldron K. J. (1994) Instability and chaos in quadruped gallop. *Journal of Mechanical Design*, vol. 116, pp. 1096-1101.

9. Ringrose R. (1997) Self-Stabilizing Running. *Proceedings of the 1997 IEEE International Conference on Robotics and Automation (ICRA)*, Albuquerque, NM, pp. 487-493.

10. Marhefka D. W., Orin D. E., Schmiedeler J. P. and Waldron K. J. (2003) Intelligent Control of Quadruped Gallops. *IEEE/ASME Transactions on Mechatronics*, vol. 8, no. 4, pp. 446-456.

11. Herr H. M. and McMahon T. A. (2001) A Galloping Horse Model. *The International Journal of Robotics Research*, vol. 20, no. 1, pp. 26-37.

12. Krasny D. P. and Orin D. E. (2004) Generating High-Speed Dynamic Running Gaits in a Quadruped Robot Using an Evolutionary Search. *IEEE Transactions on Systems, Man, and Cybernetics, Part B: Cybernetics*, vol. 34, no. 4, pp. 1685-1696.

13. McMillan S., Orin D. E. and McGhee R. B. (1995) DynaMechs: An Object Oriented Software Package for Efficient Dynamic Simulation of Underwater Robotic Vehicles. *Underwater Robotic Vehicles: Design and Control*, (Yuh J., ed.), Albuquerque NM, TSI Press, pp. 73-98.

14. Craig J. J. (1989) *Introduction to Robotics: Mechanics and Control*, Addison-Wesley Publishing Co., Reading MA, pp. 48-49.

15. Heglund N. C. and Taylor C. R. (1988) Speed, Stride Frequency and Energy Cost per Stride: How Do They Change with Body Size and Gait? *Journal of Experimental Biology*, vol. 138, pp. 301-318.

16. Marhefka D. W. (2000) *Fuzzy Control and Dynamic Simulation of a Quadruped Galloping Machine*. Ph.D. thesis, The Ohio State University, Columbus, Ohio.

17. Berkemeier M. (1998) Modeling the Dynamics of Quadrupedal Running. *International Journal of Robotics Research*, vol. 17, no. 9, pp. 971-985.

Kineto-static Analysis of an Articulated Six-wheel Rover

Philippe Bidaud, Faiz Benamar, and Tarik Poulain

Laboratoire de Robotique de Paris - Université Paris 6 / CNRS
18, route du Panorama - 92265 Fontenay Aux Roses - France
bidaud@robot.jussieu.fr

Abstract

In this paper, a kineto-static analysis for an articulated six-wheeled rover called RobuRoc is investigated. A methodology based on reciprocal screw systems is developed for the kinematic modeling and analysis of such multi-monocycle like kinematic structure. A six dimensional force ellipsoid is introduced for the evaluation of traction performances.

1 Introduction

RobuRoc (shown in **Fig.1**) has been designed and built by RoboSoft [1] in response to a Research Program launched in 2004 by the French Defense Agency (DGA - Délégation Générale l'Armement) called MiniRoc whose aim is to develop and evaluate experimentally several semi-autonomous system serving as an extensions of the human soldier. RobuRoc belongs to this class of robot vehicle named Tactical Mobile Robot (TMR). TMRs are basically high mobility small vehicles supposed to operate in highly uncertain urban environments including outdoors and indoors as well. TMR development did not truly begin until the early 1990s. Until then, the military's primary focus for ground robotics was in developing Unmanned Ground Vehicles (UGVs). Netherless, various families of TMRs have been developed during the last decade [2]. Their design is more compact and robust than exploration robotics vehicles and have to satisfy specific operational requirements (see [3]).

One of the features of RobuRoc is its ability to operate in extremely rough terrain and negotiating stairs or clear obstacles with height greater than its wheel radius. Moreover, the vehicle concept was designed to offer reconfiguration capabilities for providing either a maximum of ground adaptation for traction optimization or a high manoeuvrabilty. RobuRoc kinematics can be seen as a series of 3 unicycle modules linked together by two orthogonal rotoid passive joints. It has been optimized for stair-climbing as well as several typical bumps and jumps clearance.

The paper introduces a methodology for the kinematic modeling and analysis of such locomotion system. An equivalent kinematic model which encapsulate monocycle sub-system kinematics is proposed. The mobility of the whole mechanism is then analyzed using the theory of reciprocal screws. From the kinematic model the concept of traction ellipsoid is introduced for evaluating quantitatively the obstacle clearance capabilities when the configuration of the system and the contact conditions are changing.

2 Kinematic modeling and analysis

2.1 Mobility requirements

The RobuRoc is an articulated wheeled robot vehicle designed for use in urban environments. Urban environments typically include open spaces such as city streets and building interiors. Common obstacles that robots encounter in urban environments include : curbs, stairs, small rubble piles, pipes, railroad tracks, furniture, and wires. The ability to surmount these obstacles is essential for the success of their missions.

2.2 Kinematic description

RobuRoc has a multi-monocycle like kinematic structure as depicted in **Fig.1**. It consists of three pods steered and driven by 2 actuated conventional wheels on which a lateral slippage may occur. The rear and the front pods are symmetrically arranged about the central pod. They are attached to this later one by 2 orthogonal passive rotoid joints providing a roll/pitch relative motion for keeping the wheels on the ground to maintain traction of the pod when traversing irregular surfaces.

2.3 Kinematic modeling

Kinematics plays a fundamental role in design, dynamic modeling, and control. In this section, we illustrate a methodology for modeling and

Kineto-static analysis of an articulated six-wheel rover   477

Fig. 1. MiniRoc in a 4 wheels (left) and 6 wheels (right) configurations

analysis of articulated multi-monocycle mobile robots. The obtention of the relationship between the central-pod velocity in a reference frame and wheel velocity vector can be greatly facilitated by extending the methodology used for parallel mechanisms [4] to articulated multi-monocycle asymmetric mechanisms. A parallel manipulator typically consists of several limbs, made up of an open loop mechanism, connecting a moving platform to the ground. Here, the body S_0 of central module can be seen as the moving platform. It is connected to the ground by a differential steering system as well as by the rear and the front modules which can be seen as 2 others "limbs" connecting the S_0 to the ground.

Jacobian of a module: Each individual module j ($j = 0$ for the central, $j = 1, 2$ for the others) can be modeled as an equivalent serial open-chain mechanism (see **Fig.2**).

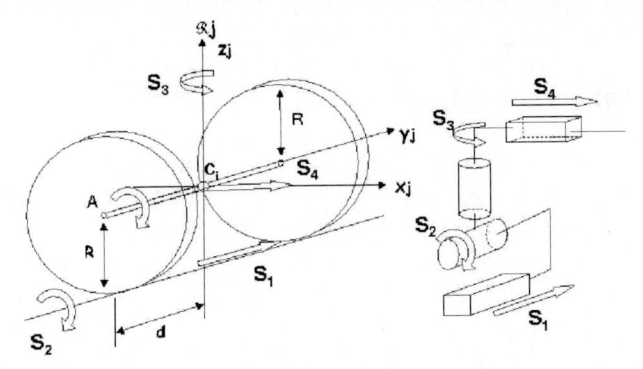

Fig. 2. The differential driving wheels mechanism (left) and its equivalent open-chain mechanism (right)

The differential driving wheels mechanism kinematics can be represented in the $\mathcal{R}_i$ local frame by a set of four unit instantaneous twists

$\hat{\$}_j = (s_i, s_{0i}^* = p_0 \wedge s_i + \mu_i s_i)^t$ $i = 1, 4$ where s_i is a unit vector along the direction of the screw axis, p_0 is the position vector of any point of the screw axis with respect to the origin O of a reference frame and μ_i is the pitch of the screw. The scalar coordinates of these screws written at the axle middle point C_j are:

$$\hat{\$}_1 = [0,0,0,0,1,0]^t \quad \hat{\$}_2 = [0,1,0,R,0,0]^t \quad \hat{\$}_3 = [0,0,1,0,0,0]^t \quad \hat{\$}_4 = [0,0,0,1,0,0]^t$$

The orthogonal complement in the screw space of the system $\{\hat{\$}_1, \hat{\$}_2, \hat{\$}_3, \hat{\$}_4\}$ has a dimension 2. It represents the pod motions which are not feasible due to the wheel/soil constraints. For the module j, the amplitude ω_{z_j} and v_{x_j} of the screws $\hat{\$}_3^j$ and $\hat{\$}_4^j$ are related to left and the right differentially driven wheels velocities $(\dot{\theta}_{j1}, \dot{\theta}_{j2})$ by the relationship:

$$\begin{pmatrix} \omega_{z_j} \\ v_{x_j} \end{pmatrix} = 1/2 \begin{pmatrix} R/d & -R/d \\ R & R \end{pmatrix} \begin{pmatrix} \dot{\theta}_{1_j} \\ \dot{\theta}_{2_j} \end{pmatrix}$$

or in a compact matrix form:

$$\dot{q}_j = J_{a_j} \dot{\theta}_j \qquad (1)$$

where R is the radius of the wheels and d is half length of the width of unicycle module. J_{a_j} is Jacobian matrix of the active part of the pod mechanism.

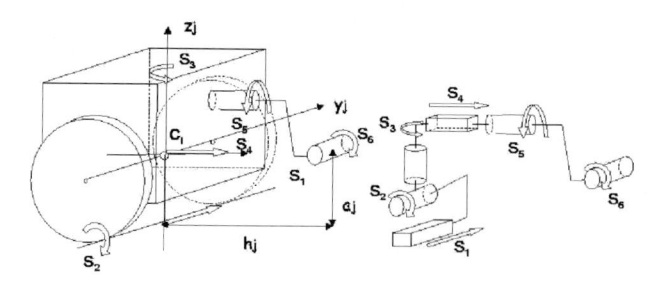

Fig. 3. The differential driving wheels mechanism (left) and its equivalent open-chain mechanism (right) (front/rear module)

Reciprocal Screws of the 3 limbs By referring to the notation introduced in **Fig.3**, the instantaneous twist $\$_P$ of the central body can be expressed

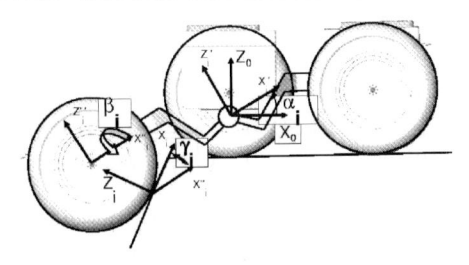

Kineto-static analysis of an articulated six-wheel rover   479

Fig. 4. Definition of α_j, β_j and γ_j angles.

as a linear combination of the n actuated and non-actuated joints screws of each j sub-chain:

$$\$_P = \sum_{i=1}^{n} \dot{q}_i^j \hat{\$}_i^j$$

where $\dot{q}_i^j$ and $\hat{\$}_i^j$ denote the intensity and the unit screw associated with the ith joint of the jth limb. By considering alternatively the $j = 0, 1, 2$ (which denote respectively the labels for the central, front and rear pods), we obtain:

$$\$_P = \dot{q}_1^0 \hat{\$}_1^0 + \dot{q}_2^0 \hat{\$}_2^0 + \omega_{z_0}^0 \hat{\$}_3^0 + v_{x_0}^0 \hat{\$}_4^0 \tag{2}$$

$$\$_P = \dot{q}_1^1 \hat{\$}_1^1 + \dot{q}_2^1 \hat{\$}_2^1 + \omega_{z_1}^1 \hat{\$}_3^1 + v_{x_1}^1 \hat{\$}_4^1 + \dot{q}_5^1 \hat{\$}_5^1 + \dot{q}_6^1 \hat{\$}_6^1 \tag{3}$$

$$\$_P = \dot{q}_1^2 \hat{\$}_1^2 + \dot{q}_2^2 \hat{\$}_2^2 + \omega_{z_2}^2 \hat{\$}_3^2 + v_{x_2}^2 \hat{\$}_4^2 + \dot{q}_5^2 \hat{\$}_5^1 + \dot{q}_6^2 \hat{\$}_6^1 \tag{4}$$

Equations (2) (3) (4) contain many unactuated joint rates that must be eliminated to reach a relationship between the instantaneous screw which defines the absolute motion $\$_P$ of the central body and the wheel's velocities vector $\dot{\theta}$. This can be done very efficiently by using a set of reciprocal screws $\$_i^{rj}$ associated with all the actuated joints (of order i in the equivalent open chain) of the $j_t h$ limb. A detailed description of screw systems can be found in [5]. We will only recall that two screws $\hat{\$}_1$ and $\hat{\$}_2$ are said reciprocal if they satisfy the condition:

$$s_1.s_{02}^* + s_2.s_{01}^* = 0$$

To simplify the expression, we will consider the particular situation where $\beta_j = 0$ ($j = 1, 2$). The coordinates of screw $\hat{\$}_3^{rj}$ and $\hat{\$}_4^{rj}$ that are reciprocal to all screws except for respectively $\hat{\$}_3^j$ and $\hat{\$}_4^j$ for $j = 1, 2$ expressed at the point 0_0(origin of $\mathcal{R}_I$ frame) and in the $\mathcal{R}_j$ frame are (see figure 4)([1]):

---
[1] $C \equiv cos$ and $S \equiv sin$

3 Obstacle clearance capacities and force transmission

A key factor involved in the design of TMR is generating sufficient amounts of force and moment from the actuation system for climbing critical obstacles such as stairs.

The position of the pitch passive joint has been determined to assist the step climbing of the front wheel. From static equilibrium conditions

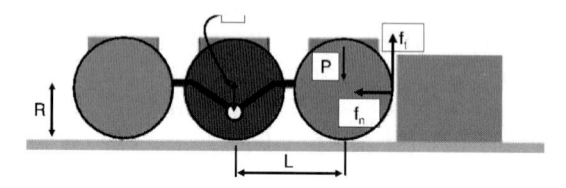

Fig. 5. Pitch passive axis location in the central pod

of the front pod wheels in contact with the front of the step when starting to climb:

$$(L+R)f_t + hf_n = PL \qquad (11)$$

When $h > 0$ the normal force component, f_n helps the passive motion of the front wheel about the axis passing through C_0. **Fig.6** shows sequences of a step clearing using dynamic simulation with Adams software. In this simulation the height of the obstacle is greater than the wheel radius (40cm vs 24cm) and the friction coefficient in wheel-ground contact is 0.6.

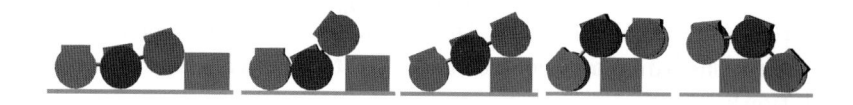

Fig. 6. Dynamic simulation with Adams software of a high obstacle clearing.

In the other part, force-moment transmission on the central body is reflecting by the D matrix. Its lines represent the elementary actions (here in the reduced wrench space $(f_{x_0}, f_{z_0}, m_{y_0}, m_{z_0})$) developed by the driving torques $(\tau_{11}, \tau_{12}, \tau_{01}, \tau_{02}, \tau_{21}, \tau_{22})^t$ on the central body. Similarly as when multiple robots exert forces or carry an object in cooperative way, D is equivalent to G^t, G representing the grasp matrix that contains the contact distribution and the way the active forces are transmitted on the central body. G can be partitioned into 2 blocks $G = (G_f \quad G_m)^t$, G_f and G_m representing respectively the force and the moment transmission. It is interesting to be able to compare the traction capabilities for different contact conditions. Hence, the set of forces and moments realizable by τ such $\|\tau\| \leq 1$ form an ellipsoid. A representative measure σ of the traction derived from the image of this unit ball of active joint torques:

$$\sigma = \left[\det(G_f^t G_f)\right]^{1/2} \quad (12)$$

When $\beta_j = 0$ $(j = 1, 2)$, the traction index σ is:

$$\sigma = \left[(S\gamma_1 C\gamma_2 - C\gamma_1 S\gamma_2)^2 + S\gamma_1^2 + S\gamma_2^2\right]^{1/2}$$

This index is equal to zero when G_f is singular. Then the force transmission in the vertical direction becomes null, this happens in configurations where $\gamma_j = 0$ $(j = 1, 2)$.

4 Conclusion

This paper presents a general framework for study wheeled modular vehicle composed by monocycles. We show that by using an equivalent kinematic model of a monocycle and the reciprocal screws theory, we derive easily the inverse velocity model that could be used for control and trajectory tracking. The force transmission in these systems is also investigated by making analogy to parallel manipulators and the concept of the manipulability ellipsoid. This theoretical study should be generalized to other vehicle kinematics including those with wheels, legs or with both. Future works should be focused on minimization of torques and energy consumption during manoeuvring or steering along curved trajectory.

5 Acknowledgments

The authors would like to acknowledge DGA/SPART and RoboSoft for offering them the opportunity to develop a a scientific activity in the Miniroc framework.

References

1. Robosoft (2005) Robotsoft website. http:robosoft.fr.

2. Department of Defense (1989) Dod robotics master plan for unmanned ground vehicles. Conference Report on the Department of Defense Appropriations Act.
3. Pierce, G.M. (2005) Robotics: Military applications for special operations forces. .
4. Mohamed, M.G. and Duffy, J. (1985) A direct determination of the instantaneous kinematics of fully parallel robot manipulators. *ASME Journal*, vol. 107, no. 2.
5. Hunt, K.H. (1978) Kinemaric geometry of mechanisms. Clarendon Press.

Momentum Compensation for the Dynamic Walk of Humanoids Based on the Optimal Pelvic Rotation

H.Takemura[1], A. Matsuyama[2], J. Ueda[3], Y. Matsumoto[3], H. Mizoguchi[1] and T. Ogasawarahi[3]

[1]Tokyo University of Science, JAPAN; [2]DENSO Co. Ltd, JAPAN; [3]Nara Institute of Science, JAPAN   takemura@rs.noda.tus.ac.jp

Abstract

In this paper, a method of determining the optimal rotation of the humanoid's waist for momentum compensation around the perpendicular axis of the stance foot during dynamic walk is proposed. In order to perform a task using the arms during a walk, it is desirable that the upper body part, i.e., the arms and the trunk, should not be used for the momentum compensation and should be dedicated to achieving a task. The proposed walk achieves a whole walking motion including momentum compensation only by the lower body. The characteristics of the trunk-twistless walk are analyzed by using the mathematical model. The optimal relative phase of the swing leg and the pelvic rotation appears to be in an angle around. And we also confirm that the torque around the perpendicular axis is reduced in the proposed trunk-twistless walk of humanoid when compared to a standard humanoid walk without the twisting of the trunk or swinging of the arms.

Keywords: Humanoid robot, Biped walking, Momentum compensation, Trunk-twisting, Optimal pelvic rotation

1 Introduction

Biped walking for humanoid robot has almost been achieved through ZMP theory [1, 2 and 3]. The research on humanoids has begun to focus on achieving tasks using the arms during walking [4]. In order to achieve a stable biped-walking, the momentum around the perpendicular axis generated by the swing leg must be counterbalanced. In a normal human walk, the upper body compensates this momentum, i.e., by rotating the thorax (or shoulders) and swinging the arms in an anti-phase of the swing leg [5, 6 and 7]. For humanoid control, some researches have been presented for momentum compensation using the motion of the entire body including the arms [8, 9, 10 and 11]. However, momentum compensation by the upper body is undesirable for a humanoid that uses its arms to achieve a task since this type of compensation limits the degree of freedom (DOF) for the task. In addition, the fluctuation of the upper body has a bad effect not only on the task accomplishment, but also on visual processing since most vision systems are attached to the head part. As a result, it is desirable to preserve as many degrees of freedom of the upper body as possible, and to suppress the fluctuation of the body at the same time. The walking action including momentum compensation should be completed only by the lower body, which leads to a simplification of motion planning.

Recently, however, in the field of exercise and sports science, a clarification of efficient motion in the human has begun, and this clarification has been accompanied by improvements in the measuring equipments used for this endeavor. Many common features can be observed in the motion of contact sport athletes, i.e., they move so as not to twist their trunks as much as possible. The particular pelvic rotation walk called a trunk-twistless walk has been empirically investigated from the observation of contact sport athletes [12]. The walking action including the momentum compensation is completed only by the lower body. The upper body DOF can be used for accomplishing a task. It is said that this trunk-twistless walk tend to have an advantage in the energy efficiency in humanoids and human athletes. Furthermore, a relative phase of the swing leg and the pelvic rotation tends to be in an anti-phase when compared with the normal walk of humans. However, there seems to be no analysis result to explain these tendencies.

In this paper, the characteristics of the trunk-twistless walk are analyzed by using the mathematical model. The optimal relative phase of the swing leg and the pelvic rotation appears to be in an angle around π. A method of determining the optimal rotation of the humanoid's waist is proposed based on a minimization of the momentum around the perpendicular axis. In this paper we confirm that the torque around the perpendicular axis is reduced in the proposed trunk-twistless walk of humanoid when compared to a standard humanoid walk without the twisting of the trunk or swinging of the arms.

Momentum Compensation for the Dynamic Walk of Humanoids ... 487

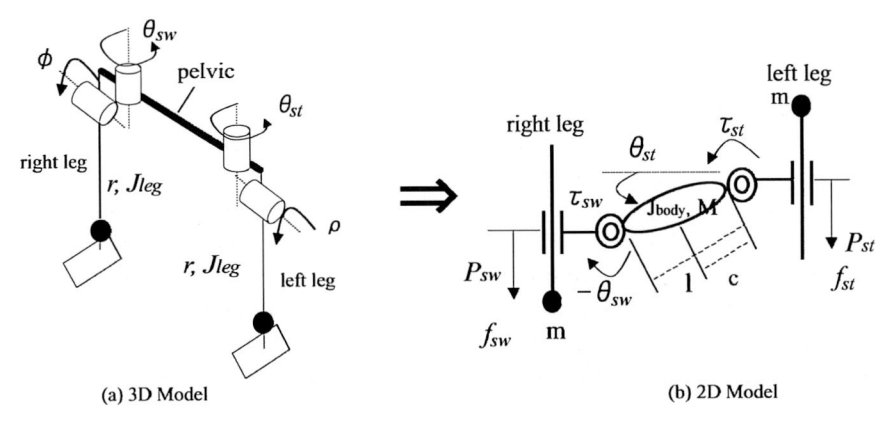

Fig. 1. Lower body analysis model

2 Lower Body Analysis Model

In order to analyze of the trunk-twistless walk, a lower body analysis model is proposed. We especially focus on hip joints of human. Fig. 1(a) shows the four joints link model which has only two joints as each hip joint. To simplify the analysis, the 2D model as viewed from top of the 3D model is also proposed (Fig. 1(b)). In the Fig.1, J_{body} and J_{leg} are the moment of inertia of waist and leg. θ_{st} and θ_{sw} are the pelvic rotation angle of support leg and swing leg, respectively. ρ and ϕ are the pitch angle of left and right legs, respectively. τ_{st} and τ_{sw} are the torque of support left and swing pelvic joint, respectively. f_{st} and f_{sw} are the load force of left and right leg, respectively. A condition of constraint ($\theta_{st} = -\theta_{sw}$) is added for keeping the direction of foots. p_{st} and p_{sw} are the position of the support leg and swing leg, respectively. p_{st} and p_{sw} are given by

$$p_{st} = r\sin\rho, \quad p_{sw} = r\sin\phi \tag{1}$$

2.1 Equation of motion of the model

An equation of motion of the proposed 2D model is derived from the Lagrange's equation as follow;

$$L = \frac{1}{2}J_{body}\dot{\theta}_{st}^2 + \frac{1}{2}J_{leg}\dot{\theta}_{sw}^2 + \frac{1}{2}MV_G^2 + \frac{1}{2}m(l\dot{\theta}_{st} + \dot{p}_{st} + \dot{p}_{sw})^2 + \frac{1}{2}m\dot{p}_{st}^2 \tag{2}$$

where $\mathbf{V}_G$ is the velocity of center of gravity. The forces and the torques are given by

$$\begin{bmatrix} J_{body} + Mc^2 + ml^2 & 0 & ml - Mc\sin\theta_{st} & ml \\ 0 & J_{leg} & 0 & 0 \\ ml & 0 & m & m \\ ml - Mc\sin\theta_{st} & 0 & M + 2m & m \end{bmatrix} \begin{bmatrix} \ddot{\theta}_{st} \\ \ddot{\theta}_{sw} \\ p_{st} \\ p_{sw} \end{bmatrix} + \begin{bmatrix} 0 \\ 0 \\ 0 \\ -Mc\dot{\theta}_{st}^2\cos\theta_{st} \end{bmatrix} = \begin{bmatrix} \tau_{st} \\ \tau_{sw} \\ f_{st} \\ f_{sw} \end{bmatrix} \quad (3)$$

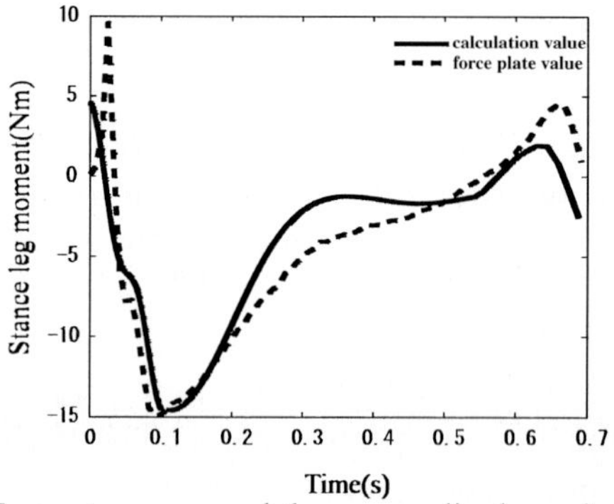

Fig. 2. Momentums around the perpendicular axis of one step

where M and m are the mass of body and leg, respectively. c is the length from joint to waist position. Therefore, the momentum around the perpendicular axis of the stance foot is given by

$$\tau_{stance} = \tau_{st} + f_{st}l\cos\theta_{st} \quad (4)$$

2.2 Verification of the proposed model

In order to evaluate the proposed model, a verification experiment is confirmed comparing the momentums around the perpendicular axis of the stance foot, which measured by force plate and calculated by using the proposed model. Three healthy male subjects served as subjects. The normal walk was measured for ten steps.

A motion capture system with twelve cameras (Vicon Motion Systems Ltd.) was used to measure three dimension kinematics data (sampling frequency 120Hz) for calculating the momentum around the perpendicular axis by using the proposed model. The floor reactive force was measured concurrently by using the force plate (Kistler Co Ltd.). The momentums around the perpendicular axis of one step are shown in Fig. 2. The both moment peaks are around 0.1 second. The sharp of the lines is almost same. The mean square error is less than 5%. Consequently, the proposed model can be use for analyzing the trunk-twistless walk.

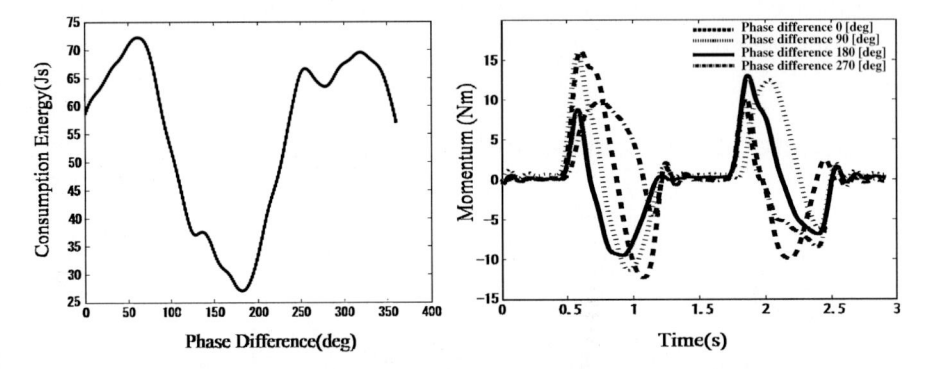

Fig. 3. Energy consumption

Fig. 4. Momentums in each phase difference

3 Analysis of Trunk-twistless Walk for Momentum Compensation

In order to analysis of trunk-twistless walk, a pseudo-trunk-twistless walk is generated by using 3D motion capture data as following steps. First, the motion captures system was used to measure three dimensional kinematics data. All subjects were given several minutes to set used. The treadmill velocity was set to 1.5 km/h, 3.0 km/h and 4.0 km/h. The normal walk was measured for 30 seconds. Second, the pelvic rotation trajectory θ_{st} and the pitch angle of the support leg ρ are calculated from the measured 3D motion capture data. Last, to generate the pseudo-trunk-twistless walk, the ρ is fixed, and the θ_{st} is forcibly changed to produce the phase difference. Then, the generated pseudo-trunk-twistless walk is applied to the proposed model to calculate the momentums around the perpendicular axis of the support foot and energy consumption. The energy consumption $E(t)$ is given by

$$E(t) = \int_0^t U dt \tag{5}$$

where U is the kinetic energy. Results of one subject are shown in Fig.3 and Fig.4. Fig.3 shows the energy consumption and Fig.4 shows the momentums around the perpendicular axis when the waking velocity is 4.0 km/h. The phase difference of normal walking is in an angle around $\pi/6$. However, the phase difference of minimum energy consumption is in an angle around π (see in Fig.3). The momentum around the perpendicular axis is reduced when compared to the normal walking (see in Fig.4).

These results are suggested that the possibility of improving efficiency of momentum compensation and energy consumption to change the phase difference between the pelvic rotation and the leg's trajectory at the normal walking pattern. This result strongly supports Ueda et al.'s reports [12].

4 Generation of Optimal Pelvic Rotation

In section 3, the generated pseudo-trunk-twistless walk is generated by forcibly producing the phase difference. In this section, the optimal pelvic rotation is determined based on a minimization of the momentum around the perpendicular axis as following equation:

$$J = \int_0^T \tau_{stance}^2 dt \rightarrow \min \tag{6}$$

τ_{stance} is the momentum around the perpendicular axis of support leg. A perturbed trajectory $\theta_{st}(t)+\delta\theta_{st}(t)$ is considered. A functional of the perturbed trajectory is given by

$$\tilde{J} = \int_0^T f(\theta+\delta\theta, \dot{\theta}+\delta\dot{\theta}, \ddot{\theta}+\delta\ddot{\theta}, t)dt$$

$$= \int_0^T f(\theta,\dot{\theta},\ddot{\theta},t)dt + \int_0^T \left(\frac{\partial f}{\partial \theta}\delta\theta + \frac{\partial f}{\partial \dot{\theta}}\delta\dot{\theta} + \frac{\partial f}{\partial \ddot{\theta}}\delta\ddot{\theta}\right)dt + o(\delta\theta^2, \delta\dot{\theta}^2, \delta\ddot{\theta}^2)$$

$$= J + \delta J$$

A necessary condition is Equation (7) calculated based on Euler-Poisson equation.

$$\frac{\partial f_{st}}{\partial \theta_{st}} - \frac{d}{dt}\left(\frac{\partial f_{st}}{\partial \dot{\theta}_{st}}\right) + \frac{d^2}{d^2 t}\left(\frac{\partial f_{st}}{\partial \ddot{\theta}_{st}}\right) = 0 \tag{7}$$

The optimal pelvic rotation is calculated by using the motion capture data and Equation (7). We confirmed that the peak momentum around the perpendicular axis is reduced by 42% on an average in the optimal pelvic rotation walk when compared to the normal walk.

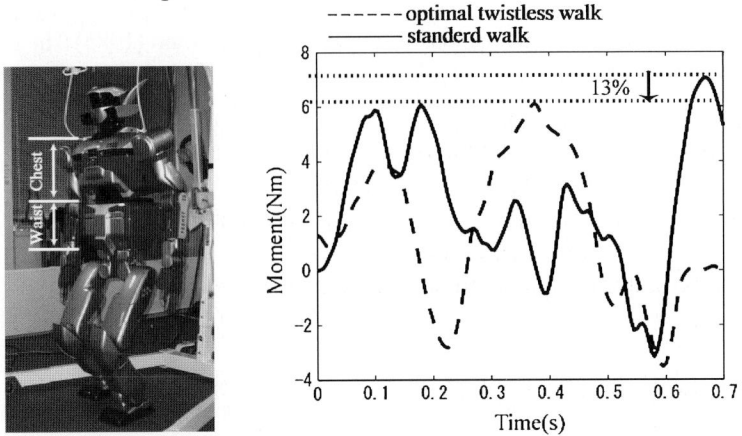

Fig. 5. Humanoid robot HRP-2

Fig. 6. Perpendicular axis momentum

5 Evaluation of Humanoid Walk

We apply the optimal pelvic rotation to humanoid robot HRP-2 (Fig.5) [13 and 14]. The standard walk of a humanoid is defined that a single supporting time is 0.7[s], a double supporting time is 0.1[s], a step width is 0.25[m], the phase difference is 0, and without arm swing. The momentums around the perpendicular axis are shown in Fig. 6. The peak momentum around the perpendicular axis of the proposed walk decrease in 13% is observed, and the amount of integration of the momentum is also reduced by 18% when compared to the standard walk.

6 Conclusion

In this paper, the trunk-twistless walk was analyzed by using the 2D mathematical model. The proposed optimal relative phase of the swing leg and the pelvic rotation was applied to the walk of humanoid. The walking action including the momentum compensation was completed only by the lower body, so that the upper body DOF can be used for accomplishing a task.

The future work includes an evaluation of the energy efficiency of the trunk-twistless walk, both in humanoids and human. An optimization program for an efficient walking pattern should be investigated.

References

1. A. Takanishi, M. Ishida, Y. Yamazaki, and I. Kato (1985) The realization of dynamic walking by biped walking robot WL-10RD. *Proc. Int. Conf. Advanced Robotics*, pp. 459-466.
2. A. Goswami (1999) Postural Stability of Biped Robots and the Foot Rotation Indicator (FRI) Point. *Int. J. Robotics Research*, vol.18, no.6, pp. 523-533.
3. S. Kajita, et al. (2002) A Realtime Pattern Generator for Biped Walking. *Proc. Int. Conf Robotics and Automation*.
4. K. Harada, S. Kajita, K.Kaneko, and H.Hirukawa (2003) ZMP Analysis for Arm/Leg Coordination. *Proc. IEEE/RSJ Int. Conf. Intelligent Robots and Systems*.
5. R. E. A. van Emmerik, and R. C. Wagenaar (1996) Effects of walking velocity on relative phase dynamics in the trunk in human walking. *J. Biomech*, vol. 29, no. 9, pp. 1175-1184.
6. C.J.C. Lamoth, P.J. Beek, and O.G. Meijer (2002) Pelvis-thorax coordination in the transverse plane during gait. *Gait & Posture*, vol. 16, pp. 101-114.
7. M. LaFiandra, R.C. Wagenaar, K.G. Holt, J.P. Obusek (2003) How do load carriage and walking speed influence trunk coordination and stride parameters? *Journal of Biomechanics*, vol. 36, no. 1, pp. 87-95.
8. J. Yamaguchi, A. Takanishi, I. Kato (1993) Development of a biped walking robot compensating for three-axis moment by trunk motion. *Proc. Int. Workshop Intelligent Robotics and Systems,* pp. 561-566.
9. S. Kagami, F. Kanehiro, Y. Tamiya, M. Inaba, H. Inoue (2000) AutoBalancer: An Online Dynamic Balance Compensation Scheme for Humanoid Robots. *Proc. 4th Int. Workshopon Algorithmic Foundation on Robotics*, pp. 329-340.
10. K. Yamane and Y. Nakamura (2003) Dynamics Filter-Concept and Implementation of online Motion Generator for Human Figures. *IEEE Trans. Robotics and Automation*, vol.19, no.3, pp.421-432.
11. S. Kajita, F. Kanehiro, K. Kaneko, K. Fujiwara, K. Harada, K. Yokoi, and H. Hirukawa (2003) Resolved Momentum Control: Humanoid Motion Planning based on the Linear and Angular Momentum. *Proc. IEEE/RSJ Int. Conf. Intelli. Robots and Systems* pp. 1644-1650.
12. J. Ueda, K. Shirae, Y. Matsumoto, S. Oda, T. Ogasawara (2004) Momentum Compensation for the Fast Dynamic Walk of Humanoids based on the Pelvic Rotation of Contact Sport Athletes. *Proc. of IEEE/RSJ Int. Conf. on Humanoid Robots*.
13. H. Inoue, et al. (2000) HRP, Humanoid Robotics Project of MITI. *Proc. IEEE-RAS Int. Conf. Humanoid Robots*.
14. H. Hirukawa, F. Kanehiro and S. Kajita (2001) OpenHRP, Open Architecture Humanoid Robotics Platform. *Proc. Int. Symp. Robotics Research*.

Walk Calibration in a Four-legged Robot

Boyan Bonev[1], Miguel Cazorla[1], and Humberto Martínez[2]

[1] Robot Vision Group. University of Alicante. P.O. Box 99. 03080 Alicante, Spain. `bib@alu.ua.es, miguel@dccia.ua.es`
[2] Departamento de Ingeniería de la Información y las Comunicaciones. Universidad de Murcia. P.C. 30100 - Murcia, Spain. `humberto@um.es`

1 Introduction

The main problem of legged robots is their locomotion and odometry. Locomotion has many parameters subjected to calibration, and depending on the surface, the speed response is different. On the other hand odometry also varies depending both on the characteristics of the surface, as well as on the locomotion parameters. We have developed a calibration process for both walk parameters and odometry. We use a simulated annealing learning technique and the application developed allows us to make such a calibration without human aid. As the input data for the learning algorithm is the robot's instant speed, the walk calibration is relatively fast, taking about 40 minutes, and the odometry calibration 60 minutes.

The scope of our problem is the RoboCup domain. RoboCup is an international competition whose final goal is to develop a team of autonomous robots able to compete with the human world champion by the year 2050. The current soccer field in the Four-Legged Robot League has a size of approximately $4,5m \cdot 3m$, and the only allowed robot is the SONY AIBO [1] . The exteroceptive sensor of the robot is a camera, which can detect objects on the field. Objects are color coded: there are six uniquely colored beacons, two goal nets of different color, the ball is orange, and the robots wear colored uniforms.

The calibration tool developed will be used in RoboCup 2005, Osaka (Japan), by the "TeamChaos" [2]. This team is a cooperative effort which involves the Örebro University from Sweden and the spanish universities Rey Juan Carlos University (Madrid), Murcia University and Alicante

---

* This research is funded by the project DPI2004-07993-C03-02 of the Spanish Government.

University. The robot we use for the RoboCup and four the experiments realized is Sony's AIBO ERS-7 four-legged robot [1]. The walk implementation used is an omnidirectional locomotion walking style [3] and the parameters we consider are represented on the **Fig.1**.

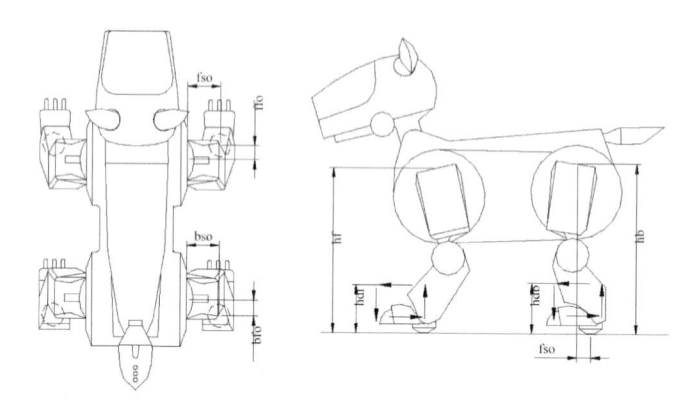

Fig. 1. Scheme of Sony's walking robot AIBO with the 8 PWalk parameters subjected to calibration: { HF, HB, HDF, HDB, FSO, BSO, FFO, BFO } [4].

Below we first describe the experimental procedure (Section 2). Then we describe the learning process (Section 3) for both speed maximization (Section 3.1) and response optimization (Section 3.2). Finally we describe the error measuring and results (Section 4), concluding in section 5.

2 Experiment setup

The experiment is realized measuring the robot's speed while walking. To measure speed and guide the robot we use the camera which the robot has on its head. We work with a 208x160 resolution to segment the white colour [5] and make a blobs analysis. A white line on the ground determines a straight trajectory for the robot (**Fig.2**). When it reaches the end of the line it pauses the measuring and turns back to follow the line again. If the robot gets deviated it finds the line again, allowing the process to be unsupervised by the user. There are also distance marks along the line, separated 20cm from each other. The robot detects them and calculates its instant speed based on the milliseconds it takes to reach each consecutive mark. We also apply a filter for erroneous measures, in case the robot misses a mark. Finally, the learning algorithm's input consists of the mean value of each three instant speed measures, to improve the stability of the received data.

There are three different walking modes (**Fig.2**): Forward walking, lateral walking, and turning. The most important for us is forward walking, which is the one taken into account for the learning process. During the execution of the on-line learning algorithm the walk parameters are continuously being changed to obtain different speed responses. For measuring the error of forward, lateral and rotational walking the parameters remain the same, as the robot takes several speed measures of each different speed requested.

Fig. 2. Robot and distance marks for forward, lateral, and rotational walking.

3 Learning procedure

As mentioned before, the objective is to find sets of parameters which both maximize the maximum speed and improve the response for the entire range of possible speeds. An appropiate speed response allows a more precise control and odometry.

Different techniques could be applied to find a suitable set of walk parameters. This is an optimization problem in a continuous parametric space with 8 dimensions. Setting parameters by hand takes many hours and the results are not satisfactory. On the other hand, obtaining a general analytical solution is not feasible. Varying each parameter independently and selecting the best value of each parameter is possible, but not optimal because the parameters are not independent, due to kinematics constraints (for example, rear knee's angles are determined by the HB, HDB, and FSO parameters together, see **Fig.1**). This method was used to establish the limits of the parametric space and to determine an initial set of parameters for a hill-climbing algorithm.

The algorithm we use is a simulated annealing[6]. In the following subsections we explain our solutions to the two already mentioned objectives.

between intervals. To make such a calibration we modify the learning algorithm in order to achieve simultaneously the real speeds (v'_l, v'_u) for the upper and lower limit of the definite speed interval $[v_l, v_u]$. When the goal for the simulated annealing is a concrete value of speed, the difference between goal and speed has to be minimized. We have two goals so we alternate measures for the speeds v_l and v_u, being the function to minimize:

$$f_{min}(s_1, s_2) = ((s_1 - v_l) - (s_2 - v_u))^2 \qquad (4)$$

where s_1 and s_2 are the speed measures, and v_l and v_u are the goals, as already explained.

Then, for $n > 1$ sets of parameters the calibration process would be as follows:

1. Find a parameters set by means of speed maximization. Thus the achieved maximum speed is v_{max}. Determine the values $v_{n-1}, v_{n-2} \ldots v_1$ so that $v_{max} > v_{n-1} > v_{n-2} > \ldots > v_1 > 0$.
2. Optimize a parameters set S_n for the speed interval $[v_{n-1}, v_{max}]$
3. Similarly, for each i, $n > i \geq 0$ optimize a parameters set S_n for the speed interval $[v_{i-1}, v_i]$

Thus we try to approximate the speed responses of the lower limit of an interval S_i and the upper of the next interval S_{i-1}. The resulting speed responses for 1, 2, 3, and 6 sets are graphically represented on the **Fig.5**. It may result difficult to avoid discontinuities between the different speed intervals. However, with two parameters sets it resulted feasible to avoid the discontinuity at 15 cm/s, or in other words, to minimize difference between requested and obtained speed.

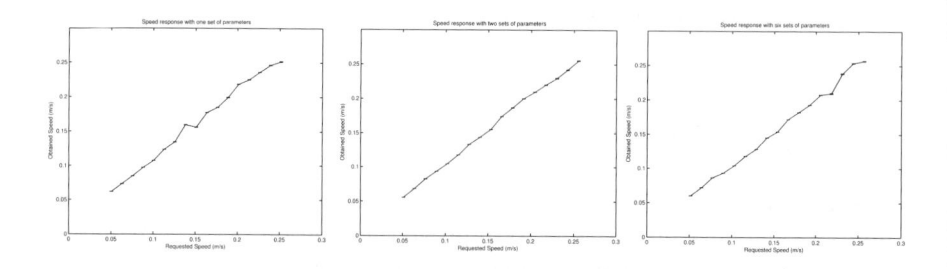

Fig. 5. Relation between requested and obtained speed for: a) 1 set of parameters; b) 2 sets of parameters; c) 6 sets of parameters.

4 Error Measure

Although we have improved significantly the speed response, still the odometry is not perfect. Measuring error is necessary to make a precise estimation of the uncertainty accumulated each time the robot walks. The error depends both on the surface and the parameters set, so these measures have to be taken after each new calibration of the walk. To model the error we discretize the speed range into 20 values. We take 10 speed measures for each one of them, to obtain their associated error and variance. On the **Fig.6** we represent the speed responses of forward, lateral and rotational walking and their associated error models.

5 Conclusions and future work

The use of walk calibration using machine learning proved to be feasible, as well as necessary, specially when changing the walking surface. The TeamChaos improved its maximum speed a 30% using simulated annealing. On the other hand, using multiple parameters sets allowed us to improve the speed response by fixing the discontinuities in the speed space. We obtained our best result using 2 sets of parameters. Moreover, the precise error measures enable the odometry system with more precise information about the motion uncertainity.

Calibrating the walk parameters for forward speed only is a good solution for the RoboCup domain. Using the infrastructure of our experiments, a study on a more complete calibration could be done, by calibrating simultaneously forward, lateral and rotational walking. This is not necessary in our domain and it would take much more time; still it is possible because of our instant speed measuring procedure. Finally, a calibration on curve lines could be considered, instead of separating calibration into three different walking types.

References

1. SONY (2005) Sony AIBO ERS7. http://www.aibo.com/.
2. Team Chaos (2005) Team Chaos website. http://www.aass.oru.se/Agora/RoboCup/.
3. B. Hengst, B. Ibbotson, Pham, P., and Sammut, C. (2002) Omnidirectional locomotion for quadruped robots. In *RoboCup 2001*.

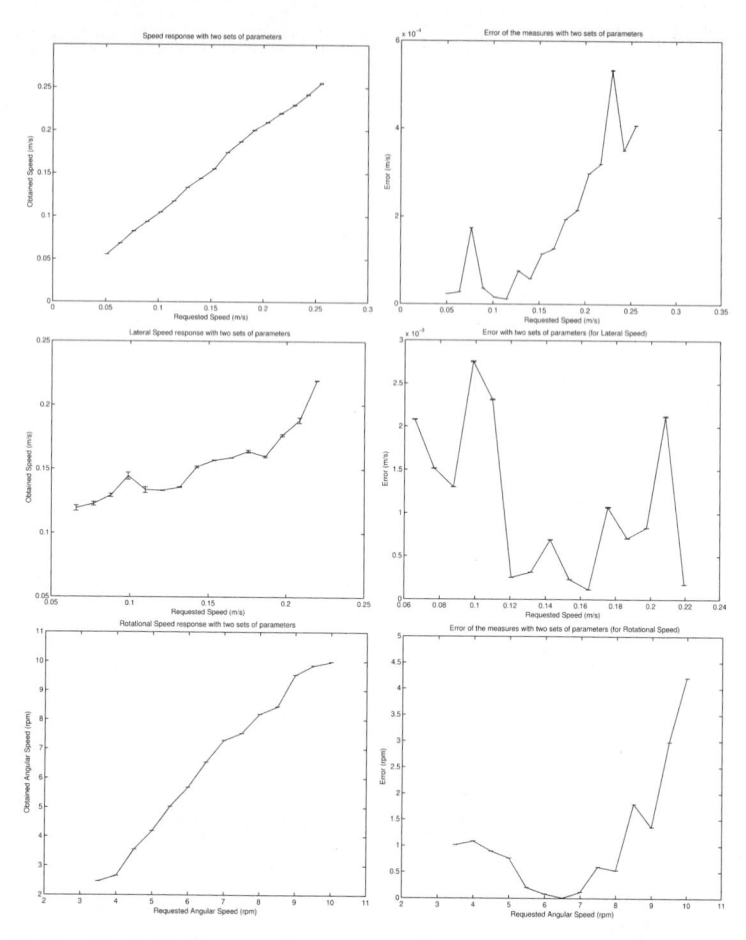

Fig. 6. On the left, the speed responses for forward, lateral and rotational walking; on the right, their associated error models (requested speed against the obtained error).

4. Hafmar, H. and Lundin, J. (2002) Quadruped locomotion for football robots.

5. Wasik, Z. and Saffiotti, A. (2002) Robust color segmentation for the robocup domain. In *IEEE Int. Conf. on Pattern Recognition (ICPR)*.

6. Kirkpatrick, S., Gelatt, C. D., and Vecchi, M. P. (1983) Optimization by simulated annealing. *Science*, , no. 4598, pp. 671–680.

Peristaltic Locomotion: Application to a Worm-like Robot

Federico Cotta, Flavio Icardi, Giorgio T. Zurlo, Rezia M. Molfino,

PMAR laboratory of Design and Measurement for Automation and Robotics, Dept. of mMechanics and Machine Design, via All'Opera pia 15A - 16145 Genova, Italy.   {molfino; zurlo}@dimec.unige.it

Abstract

The paper investigates the use of peristaltic locomotion mainly adapted to inspection and rescue applications. A state of the art on peristaltic locomotion is introduced and a short overview about suitable smart materials is given. An effective solution using Shape Memory Alloys (SMA) springs is discussed. The working principle is described and a preliminary prototype is presented.

Keywords: Peristaltic locomotion, SMA actuation, modularity.

1 Introduction

Terrestrial locomotion strategies can be divided into three main groups:
- Locomotion with wheels or tracks
- Leg locomotion
- Crawling\waving (peristaltic, snake-like or inchworm-like)

The last two can be included in the biologically inspired ones, mimicking natural beings. What attracts and holds attention on animal world is the efficiency of bodies structures of many insects and other lower animals, how they accomplish tasks with minimum energy consumption, obtaining results impossible to traditional robot configurations. Many efforts have been done in science history to mimic animal behaviour and locomotion, but first tries were simply puppet-like devices. Today, researchers work to replicate both animal gaits and structures in mechanisms, by using new technologies available for actuators and materials [1].

Several crawling animals have been analysed: snakes, slugs, snails, earthworms and inchworms. After a detailed motion analysis we focused on the earthworm, thus the peristaltic locomotion. In fact, this kind of locomotion satisfies the needs of several mini and micro robotics application fields. Wheels offer smooth and efficient locomotion on almost plain terrains; even all-wheel-drive mechanisms are limited in the type and scale of obstacles that they can overcome. Walking mechanisms can overcome bigger obstacles but they provide discrete, rather than continuous, contact surfaces. This highly affects static and dynamic stability, so that control is critical.

2 Fundamental of peristaltic locomotion

Earthworms' body can be considered as sequence of segments, surrounded by a hydrostatic exoskeleton. This constrains segments to have constant volume, so that they can move by changing segments shape. Looking at a cross section of an earthworm's body, we can recognize two sets of muscles: an inner, longitudinally oriented set, and an outer, circularly oriented set. As earthworms crawl forward, waves of circumferential and longitudinal contractions of muscles pass posteriorly along their constant-volume body segments, forming retrograde waves. When the longitudinal muscles of a segment contract, the segment becomes shorter (along the anterior–posterior axis) and wider. On the contrary, when circumferential muscles contract, the segment becomes long and thin. Peristaltic locomotion is not only characterized by the typical behaviour of single segments, but especially by the sequence in which they work, generating waves that run in the inverse direction than movement [2].

For all these reasons to have peristaltic locomotion, body must have at least three segments because while a segment elongates, the other two have to be anchored to the ground. If segments are more than three, the wavelength of peristaltic waves is greater. The basic motion sequence, however, is the one for a three-segmented body (see **Fig. 1 c)**.

Peristaltic locomotion shows its best when working in somehow severe environment where traditional machines are precluded due to size or shape and where appendages such as wheels or legs may cause entrapment or failure. For example in tight spaces, long narrow interior traverses, and travelling on loose materials or dense vegetation.

3 Previous work

New classes of robots with non conventional locomotion, such as peristaltics, hyper-redundant, worm-like, bio-inspired robots are becoming increasingly popular. *PADeMIS* [3] has been designed to operate in spinal canal on column problems. Spinal canal is filled with soft tissue, so the robot structure has to be flexible, and works with a great contact surface. For these reasons it crawls by filling with fluid its body segments in a defined sequence, generating peristaltic waves along its body. A peristaltic colonoscope with inchworm-like locomotion and with clamping devices used to stick on intestinal walls has been proposed [4]. Norihiko Saga and Taro Nakamura used magnetic fluid cells to simulate body segments of a earthworm like robot. Elastic rod-like bodies connect segments. The locomotion is obtained by shifting a magnetic field along the robot body [5]. A further step towards the realization of an artificial moving plartform designed to replicate the peristaltic locomotion mechanism is given in [6]. The platform is composed of four modules, each module is actuated by one SMA spring. The robot is covered by a shaped silicone material that can be endowed with tiny legs in order to obtain differential friction conditions.

Gan's work provides the conceptual design of a mobile platform composed of a series of modules that look like two cones placed base to base. Each couple of these cones is connected by a piston-like assembly actuated by parallel sets of spring-opposed SMA wires. When the piston is actuated the outer cone is expanded providing the anchor mode. Otherwise the module extension provides the forward mobility [7]. *Moccasin* has been designed to rescue survivors under collapsed buildings after an earthquake or a bombing; it runs with compressed air. This robot is designed especially to crawl into pipes, because pipes are often left intact by building collapse; its locomotion is inchworm-like, as it crawls by putting its soft padded feet on pipe's walls. When a module (segment) of the robot shortens, bent rubber "foot" bulges out and goes in contact with substratum: a sequence of these movements generates the desired peristaltic crawling [8]. A different robotic modular platform has been realized: modules are contained in a silicone "skin" which makes them volume-constant, simulating real earthworm's ones. SMA coil actuator is placed along the axis of every section, providing actuating force. Extension and shortening of segments is generated by cooling and heating up SMA coils [9]. An inchworm-like biomimetic robot designed for axial locomotion that provides the propulsive force by changing the shape and its interaction with the environment has been recently proposed [10]. Some worm like robotic solutions developed at the PMAR laboratory of the University of Genova are presented in [11].

4 Overview of the proposed worm robot

The paper presents the design of an earthworm-like robot using SMA actuation. The locomotion is achieved by connecting in series trust-orientation modules mimicking the peristaltic waving, and inspection modules carrying instrumentation and sensors. With reference to the rescue application, the sensorial system is composed by one mini-camera, a microphone, a loudspeaker, CO_2 and dangerous gas sensors, chemical analysers. Mainly, the head module is an inspection module. In the case presented the mini-camera is mounted on a 2dof active support. Modularity has been used for mechanical, electric and electronic interfaces in order to reduce the cost and improve the maintenance and re-configurability of the platform.

The robot is teleoperated and monitoring data are collected by an umbilical cable that guarantees the communication robustness.

5 Locomotion module

5.1 Actuation

The mechatronic architecture of this worm-like robotic system has been defined paying attention to the possible use and advantages of smart materials, such as magneto-rheologic fluids (MR), electro active polymers (EAP) and SMA (Shape Memory Alloys). Compared to traditional actuation systems, smart materials join actuation and structural functionalities, with advantages in terms of mass, dimensions and simplicity. This choice allows to build a light and high force-density device. The working principle of SMAs is well known: they exploit a solid-state transition between two crystal lattices forms: austenite, stable at high temperatures, and martensite, stable at low temperatures. Martensitic variant of the crystal shows a plastic-like behaviour due to twinning of the lattice, while austenite is characterized by a linear elastic stress-strain slope [12]. During the transition between these states, caused by a change in temperature, the material can recover a previously memorized shape thanks to de-twinning typical of martensite-austenite transformation. Memorization of the shape occurs upon a thermal process similar to tempering, characterized by a long period at high temperature and sudden cooling. In proposed solution Nitinol traction springs are used to increase actuation displacement, that for a straight wire is of the magnitude order of 8% of its own length.

5.2 Architecture

To obtain the required behaviour for the single locomotion module, three SMA springs and a steel bias spring are used. The actuators are placed with a 120° symmetry. When Nitinol springs are cool, thus deformable, the bias spring lengthen them and the module elongates. When traction springs become austenitic (heated), they regain their memorized shortened shape and contract segment. Heating of one spring causes a 30° bending of ther module in that direction. A Pro/E virtual mock-up of three segments is shown in **Fig. 1** together with the peristaltic motion sequence.

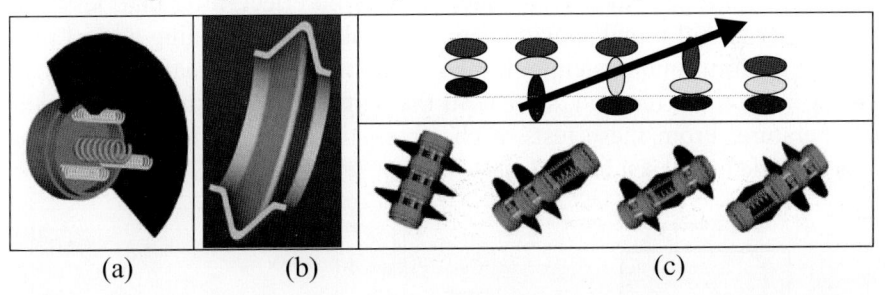

(a) (b) (c)

Fig. 1. Pro/E model of one module of the robotic worm actuation module **(a)**, bellow **(b)** and actuation sequence for a three segments robotic worm **(c)**; actuation sequence for a three segmented body **(d)**

The external covering has to accomplish two different tasks: providing insulation from external environment and working as thickening/clamping element when segment shortens. Different solutions were considered. The best solution from all points of view is the use of a bellow, providing grip and granting sealing, **Fig. 1 (b)**. The presence of a covering brings problems too, dealing with the increased reaction force and dealing with heat exchange.

5.3 Tests and fabrication

A session of tests has been done on the SMA springs to obtain a more precise knowledge of their behaviour, their characteristics have been introduced in the dynamic model to dimension the locomotion module. Following, we did another set of tests, to evaluate the right duty cycle for actuators, in order to avoid superheating and find possible speed of the SMA actuated device.

The characteristics of the SMA springs are the following: length: from 16mm to 140mm; diameter: 6mm; activation temperature: 45-50°C; reaction speed: some 0,1 s in heating, cooling time is much more complicated, depending from ambient condition, reached temperature during heating phase.

We built a test bench equipped with load cells Burster type 8435 to measure force. LabView has been used to acquire data and for a first rough filtering of noise.

As first tests actuation forces versus displacement were measured. For every length the registered trend was the same as in the square in **Fig. 2 (a)**. This comply with the well known non-linear behaviour of SMA. The reason is this: percentage of austenite rapidly increases with the first part of thermal energy provided to material by Joule effect; after that, less and less material still remains martensitic. To make this remaining part of material transform into austenite, a thermal diffusion in the material for conduction has to be realised, so that all the parts of alloy to reach activation temperature. From these tests, a characteristic force versus displacement has been deduced, see **Fig. 2 (a)**.

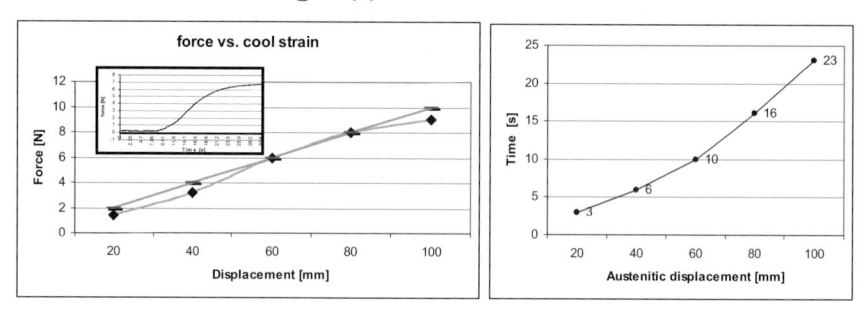

Fig. 2. **(a)** force vs. displacement: experimental result and linear characteristic; **(b)** time to reach change of slope in the heating curve

This characteristics (a) is nearly linear; this confirms that in austenitic phase material behaves as perfectly elastic, at least for little displacements, while at larger strains a shift from linearity can be noticed. This shows that from the point of view of actuation forces, there are no preferential displacements to work with, since slope is constant as for traditional springs.

More critical for our goal is the time SMA takes to reach flex point, The results are shown in the graph (b). From this graph, it can be noticed that time do not grows linearly with starting displacement, but it has an increasing slope. Therefore, from the point of view of speed of actuation, it is favourable to set low displacements for the springs, in order to work with shorter times of heating for following shortening.

Using this data, it is possible to determine the needed bias spring.

The main problem using SMA for repeated actuation is long time for SME to regain martensitic, thus plastic, state. So, generally speaking, SMAs are unsuitable for actuation of fast devices, but speed is not a peculiarity of peristaltic actuation; however, an optimal duty cycle has to be found in order to determine how fast the robot will crawl, that is the highest frequency of peristaltic waves that can be applied.

From the point of view of readiness of actuation, thus in growing slope, we noticed that response times have a magnitude of some 0,1s. Then we evaluated the time needed by coils to relax to their residual strains, by heating them for a time long enough to reach complete austenitic phase, then measuring time taken from springs to regain martensitic state.

Finally, from the results obtained within the duty cycle tests campaign, two non-saturating cycles were found for a 60 mm elongation, which is suitable for our robot. These were 4-20 and 3-15 cycles, with 0,166 duty factor. These sequences proved good for their stability, in fact, force even after various cycles stops to increase its medium value, keeping a good amplitude that is needed for actuation.

On the base of these evaluations, the physical prototype of the locomotionn modules has been successfully built and a control system with non-linearities compensation is defined and tested.

Conclusions and future work

The design of a peristaltic locomotion module for worm like robot has been presented. After suitable parametric 3D modeling, computational analyses and laboratory experiences, actuation by SMA springs disposed in a symmetrical configuration has been chosen. The physical prototype of a module has been built using commercial materials, this solution proved good during tests and it has been also found how, using tailored materials instead of commercial ones, its behaviour could still improve.

Acknowledgement

The civil protection units and fire department of Liguria region are kindly acknowledged for the definition of specifications and requirements of our worm like robots. We thank CRF and GE for the critical discussion on the materials choice and tests results.

References

1. Kevin J. Dowling (1997) Legless Locomotion: Learning to Crawl with a Snake Robot *Doctoral dissertation, tech. report CMU-RI-TR-97-48*, Robotics Institute, Carnegie Mellon University
2. Quillin K. J. (1999) Kinematic scaling of locomotion by hydrostatic animals: ontogeny of peristaltic crawling by the earthworm lumbricus terrestris. *The Journal of Experimental Biology* 202, 661–674
3. Dietrich, J.; Meier, P.; Oberthür, S., Preuß R.; Voges, D. Zimmermann, K. (2004) Development of a peristaltically actuated device for the minimal invasive surgery with a haptic sensor array. *Micro- and Nanostructures of Biological Systems*, Halle, Shaker-Verlag,
4. Phee L., Arena A., Gorini S., Menciassi A., Dario P., Jeong Y.K., Park J.O., (2001) Development of Microrobotic Devices for Locomotion in the Human Gastrointestinal Tract, *International Conference on Computational Intelligence, Robotics and Autonomous Systems*
5. Saga N., Nakamura T., (2004) Development of a peristaltic crawling robot using magnetic fluid on the basis of the locomotion mechanism of the earthworm, *Smart Materials and Structures*, Issue 3, pp566-569
6. Menciassi A., Gorini S., Pernorio G., Dario P., (2004) A SMA Actuated Artificial Earthworm *IEEE International Conference on Robotics and Automation*, April 26-May 1; Vol 4, pages 3282-3287
7. Thakoor S., Kennedy B., Thakoor A. P., (1999) Insectile and vermiform exploratory robots, *J.P.L., California Institute of Technology, Pasadena, California*, Vol. 23, N° 11, November 99, pp.1-22
8. Xiang LIU, Jianhua MA, Mingdong LI, Peisun MA, (2000) Microbionic and peristaltic robots in a pipe" *Chinese Science Bulletin* Vol. 45 No. 11 June
9. Fang-Hu Liu, Pei-Sun Ma, Jian-Ping Chen, Jie Zhu, Qin Yao Jianix, (2002)Locomotion characteristics of an SMA-actuated micro robot simulating a medicinal leech in a pipeline. *Journal of Robotic Systems*, Volume 19, Issue 6, pages 245 - 253
10. D. M. Rincon, J. Sotelo, "Dynamic and Experimental Analysis for Inchwormlike Biomimetic Robots", IEEE Robotics and Automation Magazine, December 2002, pp. 53-57
11. Cepolina E., Molfino R.M., Zoppi M.. (2004) The PMARlab in humanitarian effort. *Int. Workshop on Robotics and mechanical assistance in humanitarian demining and in similar risky interventions*, Brussels-Leuven, Belgium, June 16-18.
12. Peirs J., Reynaerts D., Van Brussel H., (2001) The true power of SMA microactuator. *MME 2001*, Cork, Ireland, 2001, 217-220

Impact Shaping for Double Support Walk: From the Rocking Block to the Biped Robot

J.-M. Bourgeot[1], C. Canudas-de-Wit[1], and B. Brogliato[2]

[1] Laboratoire d'Automatique de Grenoble, INPG-ENSIEG, UMR CNRS 5528, BP 46, 38402 St Martin d'Hères - France
   `jean-matthieu.bourgeot@inrialpes.fr`
[2] INRIA Rhône-Alpes, 655 av de l'Europe, 38330 Montbonnot - France

Abstract

This paper presents a study on how a biped needs to be configured before an impact to achieve double support. To model the biped behavior during the double impact, a novel approach based on the rocking block is proposed. Moreover, an application to the double support walk impact shaping for planar under-actuated biped robot is proposed.

1 Introduction

The focus of this paper is to study how to configure a biped robot just before an impact to obtain a phase of double support just after the impact.

The biped robot studied in this article is under-actuated. The biped does not have an actuator in the ankle. This means that during a cycle of walk, it is falling by rotating around its support foot (foot A in **Fig. 1**(a)). This type of robot can be studied by using the equivalent representation of an inverted pendulum [5] (as seen in **Fig. 1**(b)).

1.1 Double support walk

In [3] and [7] the authors used this approach successfully to make the Rabbit biped robot walk. In these works the walking gaits were single support walks. This means that when the free flight foot of the biped hits the ground, the support foot changes immediately (from A to B) and the rotation continues (see **Fig. 2**). In this case the duration of the double support phase is zero. Human biomechanical observations show

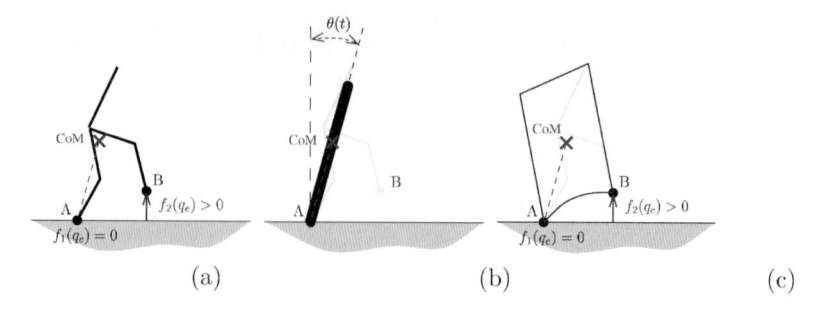

Fig. 1. Inverted pendulum / Rocking block equivalence

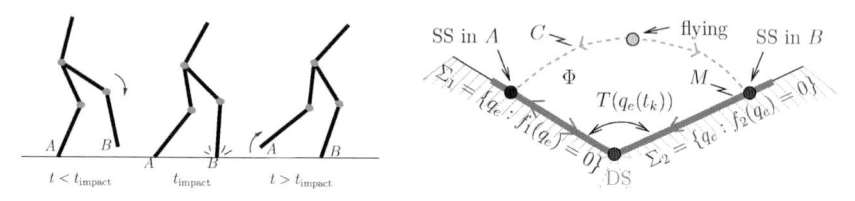

Fig. 2. Instantaneous transfer of support foot

Fig. 3. Two constraints in the space coordinates

that double support phases constitute 20% of the total cycle of walk. Thus to design human-like walking gait, it is required to take the double support phase in the bipedal locomotion into account. In any case, double support walk is needed for transitions between standing phases (biped with no velocity) and walking (or running) phases. The central purpose of this article is to determine conditions for having a phase of double support of non-zero time.

The double support walking problem and the biped robot Rabbit, used as an application example in this paper, are presented in this section. Section 2 deals with the deformable rocking block, conditions for non-topple. Section 3 applies the proposed approach to the biped robot Rabbit.

The aim of this work is to achieve a double support phase. Clearly the so used inverted pendulum representation cannot be used to model the biped during double impact. This pendulum equivalence can be used during the single support phase to describe the evolution of the biped but an alternative approach for the double impact is required. In this work a novel approach based on the rocking block is proposed. The biped is symbolized by a rocking block at impact time (see **Fig. 1**(c)). The width of the block corresponds to the position of the feet (points A and B), the center of mass (CoM) of the block is the same as that of the biped, and the block has the same moment of inertia. The two unilateral constraints associated with each foot are shown in **Fig. 1**: $f_1(q_e) \geqslant 0$ and $f_2(q_e) \geqslant 0$ characterize the fact that the solids are rigid and the robot's feet cannot penetrate into the ground, (q_e is the state vector, $f_1(\cdot)$ and $f_2(\cdot)$ are $\mathcal{C}^1$ functions). These two unilateral constraints are displayed in the space coordinates in **Fig. 3**, which represents a section of the space, and in which the two constraints define two half spaces that are not accessible by the system. Thus, the two constraints restrict the evolution of the biped in the sub-space Φ of $\mathbb{R}^{\dim q_e}$. For each position $q_e(t)$, the tangent cone $T(q_e(t))$ is defined by the cone of the admissible velocity. The orbit C (dashed grey line) in **Fig. 3** presents the orbit of a running biped (an orbit with flying phase).

On the other hand the orbit M (large grey line) presents the orbit of a walking biped (at least one unilateral constraint is active at any moment). A single support phase occurs if the transit by the intersection $\Sigma_1 \cap \Sigma_2$ is instantaneous, and a double support phase occurs if not. The purpose of this work is to give conditions for achieving a post impact velocity so that the system remains in the intersection DS^3 after a double contact.

Before studying the condition for a double support phase with the rocking block, the biped robot Rabbit will be briefly described.

1.2 Rabbit Robot

The robot Rabbit (see **Fig. 4**) is a biped composed of 5 links and 4 actuators (2 on knees and 2 on hips). As there is no actuator in ankle, this robot is underactuated when it is in single support. The dynamics of the Rabbit robot can be describe as:

$$M(q_e)\ddot{q}_e + C(q_e, \dot{q}_e)\dot{q}_e + G(q_e) = U + \nabla f_1(q_e)\lambda_1 \\ + \nabla f_2(q_e)\lambda_2 \quad (1)$$
$$f_1(q_e) \geqslant 0 , \ f_1(q_e)\lambda_1 = 0 , \ \lambda_1 \geqslant 0$$
$$f_2(q_e) \geqslant 0 , \ f_2(q_e)\lambda_2 = 0 , \ \lambda_2 \geqslant 0$$

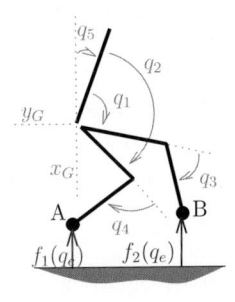

Fig. 4. Rabbit biped

where, the first line describes the Lagrangian dynamic where $M(q_e)$, $C(q_e, \dot{q}_e)$, $G(q_e)$, U are the inertia matrix, Coriolis and centripetal effect

---

[3] **Notation:** *DS*- Double Support, *SS*- Simple Support and SS_X- Simple Support on foot (or corner) X.

matrix, gravity and input torque vectors. $\nabla f_1(q_e)\lambda_1$ and $\nabla f_2(q_e)\lambda_2$ are the ground reaction of each foot. The last two lines correspond to the two unilateral constraints and complementarity relations associated with each foot.

2 Deformable Rocking Block

The dynamics of a mechanical system subject to unilateral constraints as in (1) are complete when an impact law is added (a relation which gives the post-impact velocities as functions of the pre-impact ones). In this work, the frictionless multiple impacts law of Moreau is used [6]:

$$\dot{q}_e(t_k^+) = -e_n \dot{q}_e(t_k^-) + (1+e_n) \operatorname*{argmin}_{Z \in T(q_e(t_k^-))} \frac{1}{2}(Z - \dot{q}_e(t_k^-))^T M(q)(Z - \dot{q}_e(t_k^-)) \quad (2)$$

where e_n is the coefficient of restitution $e_n \in [0,1]$, t_k is the impact time, and $T(q_e(t_k))$ is the tangent cone at $q_e(t_k)$ (see **Fig. 3**).

In the case of plastic impact ($e_n = 0$), the post impact velocity corresponds to the nearest velocity in the set of admissible velocity $T(q_e(t))$ from the pre-impact velocity. In the elastic case ($e_n > 0$), a supplementary term is added to symbolize the elasticity.

With this law, the post impact evolution of the rocking block will be studied. Rocking blocks have been often studied in the literature: In [2, section 6.1.1] the evolution of homogenous rocking block has been studied, comparable results are shown in [4] with another impact law.

In the next section the case of a non-homogeneous block (the CoM is not in the middle of the diagonal of the block) is studied. It will be noted that four different evolutions are possible.

2.1 Four possible post-impact evolutions

If a system subject to two unilateral constraints is considered with the impact law (2), then the evolution of the system after a double impact (an impact at the intersection of the two constraints in the space coordinate in **Fig. 3**) is one of the following four possible motions:

- If the shock is plastic ($e_n = 0$) and,

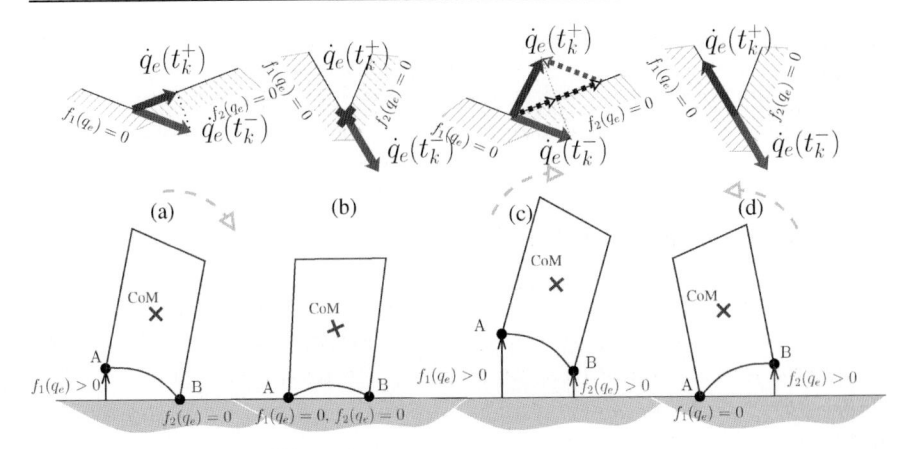

Fig. 5. Four post-impact evolutions

- If the kinetic angle between the constraint (θ_{kin}^{12}) is obtuse, the projection[4] of $q_e(t_k^-)$ on $T(q_e(t_k))$ is in Σ_2 and not in Σ_1, see **Fig. 5**(a) then after the impact the block continues to rotate around point B.
- If the kinetic angle is acute, the projection of $q_e(t_k^-)$ on $T(q_e(t_k))$ is in Σ_2 and is in Σ_1, then the two points A and B remain on the ground, see **Fig. 5**(b). In this case there is a double support phase.

• If the shock is elastic ($e_n > 0$) an elastic term is required, and

- If θ_{kin}^{12} is obtuse, the post impact velocity goes away from the constraint surface, see **Fig. 5**(c), then the block takes off.
- If θ_{kin}^{12} is acute, then the block bounces back around the point A, see **Fig. 5**(d).

From these four evolutions, it can be concluded that the conditions for having double support phase are:

• having a plastic impact,
• having a kinetic acute angle at the time of impact.

For this study on the biped it is supposed that impacts are always plastic (this assumption was verified through different experiments conducted with the rabbit biped before this work). The conditions to have kinetic acute angle at the time of impact are given below.

---

[4] The projection is defined in the sense of the metric defined by the matrix $M(q)$.

2.2 Conditions for non-topple

Consider the rocking block described earlier, and define the state vector by $X = (x_G, y_G, \theta)^T$ where x_G and y_G are the coordinates of the center of mass, and θ is the orientation of the block.

Propose the following:

Proposition 1. *Assume that the block is defined as above, then a non-instantaneous double support phase in A and B exists if and only if:*

$$(x_G - x_A)(x_B - x_G) \geqslant \frac{I_{block}}{m_{block}} \qquad (3)$$

where I_{block} is the moment of inertia of the block, m_{block} is the mass of the block, G is its center of mass, and x_i is the x-coordinate of point i. ∎

Proof.

For sake of brevity, the proof, which are reported in [1], is not presented here.

Remark 1. $(x_G - x_A)(x_B - x_G)$ is maximal when the ground projection of G is exactly between the contact points A and B. If the biped is compared with a block, the most advantageous configuration is when the center of gravity is between the feet.

3 Application to the double support walk

In this section, the proposition above is applied to the biped locomotion. Conditions are found on the impact configuration with the goal of obtaining non-instantaneous support phases.

3.1 Conditions for double support walk

Assumption 1 *At the impact time t_k, the articulations of the system are considered completely rigid. As such the biped is considered as a block with the same mass, the same moment of inertia, and with the same ground contact points (corners A and B).*

Let assumption 1 hold, the condition (3) of the proposition 1 becomes:

$$(x_G - x_A)(x_B - x_G) \geqslant \frac{I(q)}{M_{biped}} \qquad (4)$$

From section 2.1, the two sufficient conditions to obtain a double support phase are: have a plastic impact, and equation (4) to hold.

Remark 2. Assumption 1 is not senseless if the local feedback loops which drive each actuator have sufficient high gains, and if the actuators are chosen sufficiently powerful to be able to block the associated link.

In the inequality (4), the values of x_A, x_B, x_G and $I(q)$ are function of the configuration $q(t_k)$ of the biped at the impact time. To resolve the double support walk problem, it is required to find the configuration q_{impact} which maximizes $(x_G - x_A)(x_B - x_G)$ and minimizes $I(q)$. In the next section, this result will be applied to the biped robot Rabbit.

3.2 Application to robot Rabbit

First the moment of inertia $I(q(t))$ needs to be computed. Considering the biped on a plane, the Huygens-Steiner theorem can be used to compute the moment of inertia: $I(q) = \sum_{i=1}^{5} I_i + \sum_{i=1}^{5} m_i GG_i^2$. Now, a valid pre-impact configuration can be shaped by choosing a configuration which assures condition (4).

3.2.1 Computation of valid configurations

Valid configurations $q(t)$ are such that condition (4) holds. It is required to find $q(t)$ which maximizes $(x_G - x_A)(x_B - x_G)$ and minimizes $\frac{I(q)}{M_{biped}}$. The best configuration for a given length of the footstep is given by solving the following optimization problem:

$$\min_{q \in [q_{min}, q_{max}]} \frac{I(q)}{(x_G(q) - x_A(q))(x_B(q) - x_G(q))}$$
$$\text{subject to } x_A(q) - x_B(q) = 0$$
$$y_A(q) - y_B(q) + l_{footstep} = 0$$
(5)

where $l_{footstep}$ is the desired length of the footstep.

A valid configuration can be seen in **Fig. 6**. As explained in remark 1, it is shown that the center of mass is in the middle between the two feet. This method gives results for shaping valid configuration. But workable configurations are found only for footsteps longer than 79cm. These configurations are slightly at extremes for design of natural walking gaits. The minimum footstep length found by the minimization algorithm is 79cm, whcih gives postures that are not anthropomorphic and non-useable for design of walking gaits (in **Fig. 6** the chest of the biped is too much backward).

4 Conclusion

This paper has delt with double support walking problem for a planar Under-actuated biped robot. The problem is to find a pre-impact Configuration, which gives non-instantaneous double support phases in view of design of double support walking gaits. In this work an original approach has been presented. The biped has been considered as a rocking block at the time of impact. After a study of the non-homogeneous rocking block, results have been applied to the biped Rabbit. The proposed method has allowed to shape workable configurations. But these configurations are so extreme to be useful in human-like walking gaits.

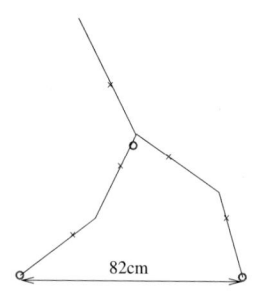

Future work will be directed to improve the workable configurations. First the rigidity of the articulations may be relaxed. Indeed a lax articulation may improve the dissipation of energy at the time of impact. Secondly, the impact law could be modified to integrate more dissipativity, actually accurate impact laws for multiple impacts remain an open problem.

Fig. 6. Workable configuration

References

1. J.-M. Bourgeot. *Contribution à la commande de systèmes mécaniques non-réguliers*. PhD thesis, INP de Grenoble, October 2004.
2. B. Brogliato. *Nonsmooth Mechanics*. Springer, London, 2nd edition, 1999.
3. C. Chevallereau, G. Abba, Y. Aoustin, F. Plestan, E.R. Westervelt, C. Canudas-de-Wit, and J.W. Grizzle. Rabbit: A testbed for advanced control theory. *IEEE Control Systems Magazine*, 23(5):57–79, October 2003.
4. M. Fremond. *Non-Smooth Thermomechanics*. Berlin, London: Springer, 2002.
5. S. Kajita and K. Tani. Experimental study of biped dynamic walking. *IEEE Control Systems Magazine*, 16(1):13–19, 1996.
6. J.-J. Moreau. Unilateral contact and dry friction in finite freedom dynamics. In *Nonsmooth Mechanics and Applications*, CISM Courses and Lectures no 302. Springer-Verlag, 1988.
7. E.R. Westervelt. *Toward a Coherent Framework for the Control of Planar Biped Locomotion*. PhD thesis, University of Michigan, 2003.

Proposal of 4-leg Locomotion by Phase Change

Kazuo MORITA[*1] and Hidenori ISHIHARA[*2]

[*1]Kagawa University, Japan   s05d506@stmail.eng.kagawa-u.ac.jp
[*2]Kagawa University, Japan   ishihara@eng.kagawa-u.ac.jp

Abstract

This paper discusses the possibility to realize both of walking and running at a single robot, which is made by the combination mechanism of linkages and actuators. This linkage mechanism have actuators below the number of joint, therefore it enables to reduce the weight of the robot. This study aims at realizing the 4-legged locomotion by controlling the phase to manipulate the legs made by linkage mechanism. This paper firstly shows the similarity of walking and running of 4-legged animal, and proposes the way of realizing the both actions by the change of phase of motion. In order to design the leg mechanism with effective linkage systems, the kinematics simulator was developed and the performance of the prototype was tested.

Keywords: 4-legged Locomotion, Control by Phase Change

1 Introduction

Several concrete applications of robots have been introduced such as Mine Cleaning and Disaster Relief. Therefore, The important factor is locomotion on any natural environment. Above all, the legged robot has some suitable features. These are as follows.
1. It is hardly influenced by the condition of ground;
2. It is able to walk stably;
3. It can dynamically run with high-speed.

Legged pattern have different feature by the number of leg. 2-leg have dynamic motion of high speed, but stability is low. Multi-leg have high stability, but legged system have to need complex mechanism therefore robot's weight is very heavy. 4-leg can have both dynamic motion of high speed and static motion of high stability. So, 4-leg is an advantageous mechanism of high versatility movement.

Some interesting robots have been reported around the related fields recently. Hirose et al. developed the quadruped robot, TITAN [1], which has the 4 spider-like legs. Kimura et al. is investigating the 4-legged robot like a dog, TEKKEN [2], of which shape is similar to a dog and which has feet of spring plate. Iwamoto et al. is aiming at the running performance by their robot, RYUMA [3].

Each robot has the distinct character, but they have the common features and problems as follows.

(a) Weight is heavy because 3-6 actuators are needed per one leg.
(b) These have much redundancy of links so control system is complication.
(c) Enlargement of a battery is not avoided because many actuators are driven.
(d) Realization of kicking is difficult because of shortage of torque and lack of momentary power.

This research purposes are design of light robot by linkage mechanism and optimizes foot locomotion for walking and running. Therefore this robot solves the above-problems. As a result, a high-speed movement in irregular terrain and jumps over the ditch and the puddle are achieved where it cannot run with usual wheel mechanism.

2 Walking and Running Locomotion.

Generally, walking and running are defined as "A walking is the legged locomotion which one or more legs is in contact with the ground all the time, and a running have a moment when all legs separate from the ground." Although the different kinds of animals have the different structure and operating pattern of legs, we focused on 4-legged locomotion of the small mammals such as a dog. The legs of such mammals are attached to the underside of their body vertically. Considering these structure by the theory of structure, it is easier to support the body with the leg structure of mammal than reptile. That is, it is possible to minify the stiffness of legs to support its own body.

As for the action pattern, **Fig. 1** shows the typical patterns of 4-legged mammals. In **Fig. 1**, the numbers typed around the tips of legs indicate the phase of leg motion against the leg of 0. 1 is one cycle. Leg of 0.5 moves delaying half wavelength from the standard leg. With many dogs, the Trot (**Fig1- a**) is used for walking locomotion, and Bound (**Fig. 1- c**) or the Gal-

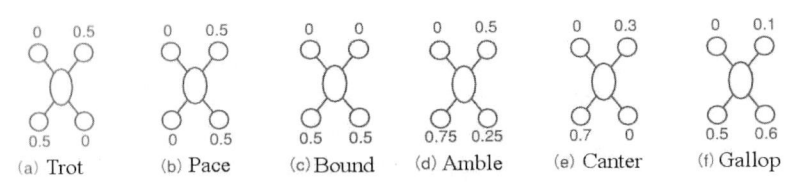

(a) Trot	(b) Pace	(c) Bound	(d) Amble	(e) Canter	(f) Gallop
0 0.5 / 0.5 0	0 0.5 / 0 0.5	0 0 / 0.5 0.5	0 0.5 / 0.75 0.25	0 0.3 / 0.7 0	0 0.1 / 0.5 0.6

Fig. 1: Pattern of 4-legged locomotion.

lop (**Fig1- f**) is used for running locomotion. **Fig. 1** indicates that the motions of respective leg are similar and that the differences of pattern are generated by the change of phase. The interesting feature is that these patterns change from one to other dynamically without the intermediate motion from walking to running. As the results of above considerations, shifting the phase to operate the legs enables to realize the entire operating pattern, though the amplitude of legs are different.

Finally, a possibility to realize both of walking of running performance is derived, if the basic motion can be generated by the linkage mechanism without mounting the actuator at all the joints. Although kinds of motions in such system are limited in compared with the mechanism with actuators at all the joints, it is sufficient and effective for the walking and running behavior.

3 Leg Mechanisms

The design conditions are derived through some basic requirements:
1. The robot consists of as same number of joints and link ratio of legs as an animal.
2. Stride length of forelegs and back legs are equal.
3. The number of inputs should be minimized.

For designing the link system fulfilling these requirements, the numerical simulator based on kinematics was developed.

Fig. 2 shows the basic link architectures, which were determined by parameters of mammal model and preparative experiments. In both legs are imaged by actual wolf's frame. Each fixed point has position of A (front) = the root of scapula, E(front) = the joint of scapula and humeral, A(back) = the root of pelvis and E(back) = the joint of pelvis and femoral. Both E are the origin, and both E are the driven points.

In this mechanism, position of both A and each length of link is calculated as follows in order to equalize the stride lengths of both legs.

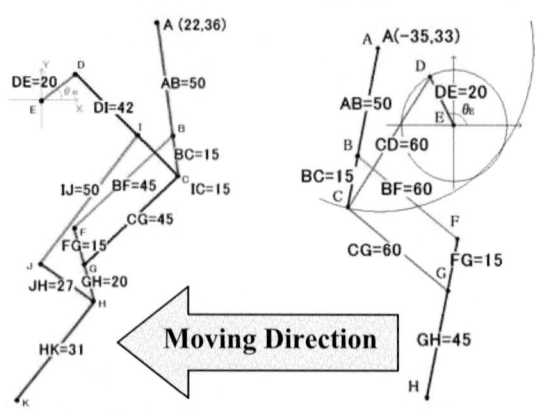

Fig. 2 The link architectures of both legs.

In this mechanism, position of both A and each length of link is calculated as follows in order to equalize the stride lengths of both legs.

$$e = \left|\left(K_{fx}^{max} - K_{fx}^{min}\right) - \left(H_{bx}^{max} - H_{bx}^{min}\right)\right|$$

K_{fx}^{max} : The maximum value of K_{fxL} in $K_{fyL} < K_{fyR}$ (K_{fxL}, K_{fyL}) :Coordinates of left foreleg.

K_{fx}^{min} : The minimum value of K_{fxL} in $K_{fyL} < K_{fyR}$ (K_{fxR}, K_{fyR}) :Coordinates of right foreleg.

H_{bx}^{max} : The maximum value of H_{bxL} in $H_{byL} < H_{byR}$ (H_{bxL}, H_{byL}) :Coordinates of left back leg.

H_{bx}^{min} : The minimum value of H_{bxL} in $H_{byL} < H_{byR}$ (H_{bxR}, H_{byR}) :Coordinates of right back leg.

Fig. 3 shows the screen of link simulator for calculating the lengths of links as above-mentioned. **Fig. 4** shows the trajectories of joints at the back leg respectively.

As the results of basic consideration through the simulation, the link architecture with the following features was designed.
(1) The tips of legs can be moved in parallel with the ground when they attach the ground.
(2) The idling legs are forwarded quickly.

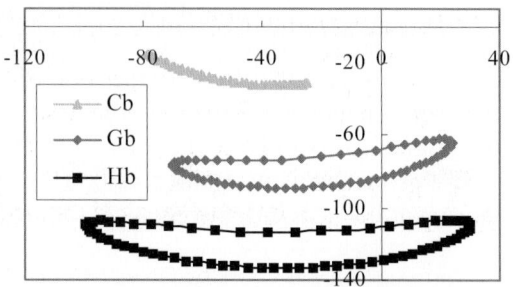

Fig. 3 4-legged link simulator.   Fig. 4 Trajectories of joints of back leg.

Additionally, the difference of strides of foreleg and back leg was successfully designed in minimum (0.28% of length of step). As shown above, the designed simulator successfully derived the link parameters for the legs satisfying the requirements.

4 Change of Locomotion Pattern by Phase Change

In order to confirm the effectiveness of generation of locomotion pattern by change of phase, the locomotion simulator was developed. Basic model of link system and parameter are as same as **Fig. 1**. This simulator targeted the visualization of performance of the virtual robot designed by the kinematics model derived by the link simulator. The motions of forelegs and back legs are same, and the phase to manipulate each leg was controlled as shown in **Fig. 1**. The number in **Fig. 1** shows the phase of other legs, when one cycle of based on the front left leg. This simulator is calculated by these parameters and move amount "M_Values" of following equation. In this connection, "Before_**" in this equation shows the former position of each leg, and "+20" is that because the point of back leg origin is 20 millimeter higher than foreleg.

$$M\_Values = \begin{cases} K_{fxR} - Before\_K_{fxR} \\ \quad \text{if } K_{fyR} \leq \min(K_{fyL}, H_{byR} + 20, H_{byL} + 20) \\ K_{fxL} - Before\_K_{fxL} \\ \quad \text{if } K_{fyL} \leq \min(K_{fyR}, H_{byR} + 20, H_{byL} + 20) \\ H_{bxR} - Before\_H_{bxR} \\ \quad \text{if } H_{byR} + 20 \leq \min(K_{fyR}, K_{fyL}, H_{byL} + 20) \\ H_{bxL} - Before\_H_{bxL} \\ \quad \text{if } H_{byL} + 20 \leq \min(K_{fyR}, K_{fyL}, H_{byR} + 20) \end{cases}$$

As a result of this simulation, mammal's six locomotion could be confirmed by phase change. **Figs. 5** and **6** show the trot, bound locomotion of simulation result, as space is limited.

Fig.5 Trot locomotion of movement simulator.

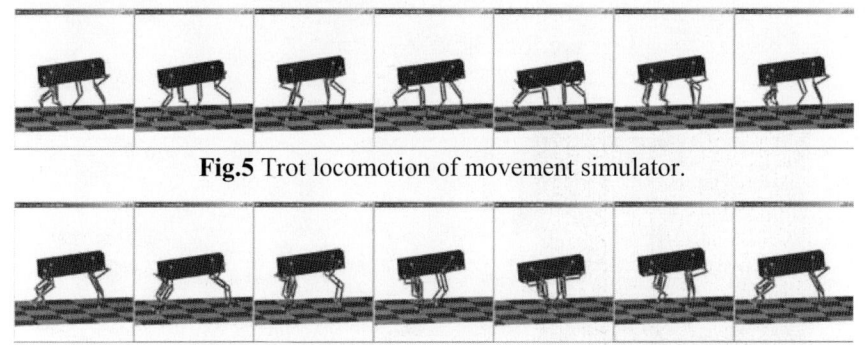

Fig.6 Bound locomotion of movement simulator.

5 Basic Experiments by Prototype

Fig. 7 shows the prototype for testing the concept that locomotion patterns are generated by the change of phase. **Table 1** indicates the details of developed prototype. In this prototype, the material of each mechanism is the foam vinyl chloride that is high processing to carry out the design trying and erring. Just a single motor drives all legs, but this prototype does not have the mechanism that changes the phase of motions of legs. Therefore, the phase of leg's motion is adjusted by changing the engagements of gear manually.

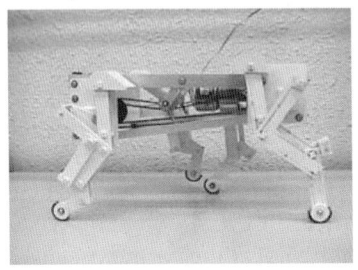

Fig. 7 Prototype for basic locomotion.

Table. 1 The details of prototype.

Motors	SAYAMA Geared Motor RB-35CM-GL1-B
Power source	18[V] (External)
Material	Foam vinyl chloride (Legs) Aluminum (Body)
Size	255[mm] X 190[mm] X 180[mm]
Weight	0.6[kg]

Figs. 8 and **9** show the pictures of Trot, Bound locomotion of developed prototype among 6 patterns of locomotion, as space is limited. As the results of these basic experiments, the motions of Trot and Bound patterns are confirmed. However, slips of tips of legs and shortage of stiffness of legs disturb the smooth motion.

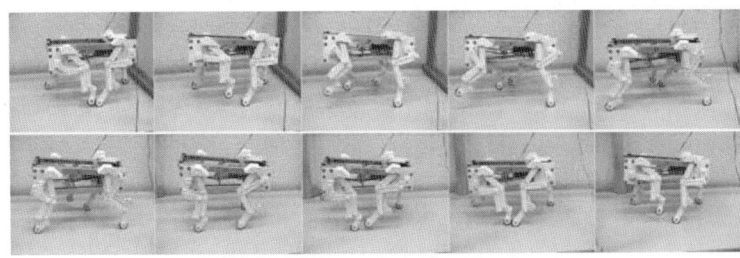

Fig. 8 Trot locomotion of prototype.

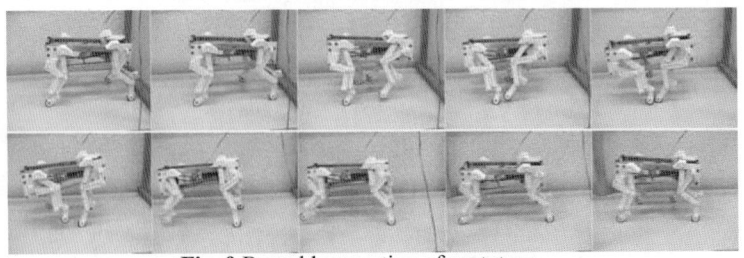

Fig. 9 Bound locomotion of prototype.

Fig. 10 shows the graph of three kinds of prototype locomotion speed. An enough result is not obtained in gallop locomotion cause the above issue, though it shows reference. In the result, the speed of Trot is faster than Bound in low voltage (6-20V). As this cause, it thought not to be able to have enough idling leg time in Bound because the locomotion speed of leg is slow. Moreover, the result of step that Trot was wider than Bound was confirmed too. However, the speed of Trot have peak in high voltage (20-32V), afterward this speed was decreased. In addition, it slipped on ground in actual Trot locomotion in high voltage, and the kicking motion was not done at all. It is thought that a load do not enough working in the kick back leg because Trot support deadweight by two legs in the diagonal, and in high voltage the locomotion speed is too fast so that it begins to kick before the center of gravity moves to support leg. On the other hand, Bound locomotion is landing the foreleg and kicking back leg, therefore all deadweight can be almost used as kicking motion. Moreover in high voltage, movement speed was confirmed rising in proportion to the voltage, not have peak at the speed like Trot, because the high speed locomotion can use enough idling leg time. As a result, the locomotion pattern are changing Trot that is static locomotion in low speed into Bound that is dynamic locomotion in high speed like an actual animal, consequently it is thought that the possibility of the optimization of locomotion.

On another front, when paying attention the stride length graph (**Fig. 11**), the length of Trot is longer than Bound in low voltage. And when high voltage, the length of Bound is longer than Trot. However, the length of Bound have peak, too, afterward this stride length was decreased. As this cause, we thought there did not have enough ground contact area and kicking power. Moreover, the rotational speed of the leg could not be controlled with this prototype. Therefore, the action speed of the leg was con-

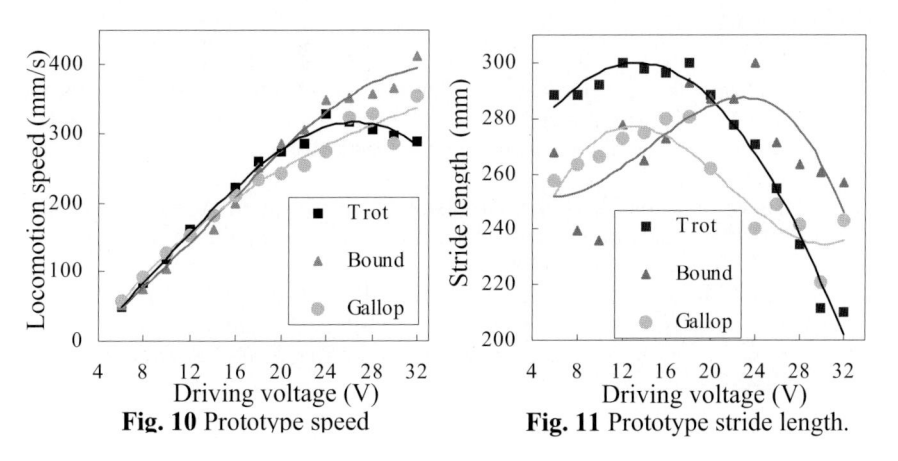

Fig. 10 Prototype speed

Fig. 11 Prototype stride length.

stant. For that purpose, we thought prototype couldn't have action of corresponding to animal's "stretch of body". The improvement of the speed can be expected by improving these.

6 Conclusions

This paper proposed the 4-legged locomotion robot with under-actuated linkage mechanism for realizing the real walking and running like as real dogs and cats. As the first approach, the link simulator was designed due to the determination of the effective link structure therefore the difference of strides of foreleg and back leg was very minimum. The motions of respective patterns were also confirmed by the locomotion simulator. In parallel to the numerical simulation, the basic performance were tested by the prototype developed based on the results of simulations. Through the simulation and basic experiments, the effectiveness of proposed mechanism was confirmed. However, the structures have not been optimized yet. Therefore it was impossible for dynamic running locomotion of Amble and Gallop to smooth and continuation movement both simulation and basic experiment by prototype. However, it was confirmed that the movement speed was increase in current Bound, in addition kicking motion and operation of "Stretch up" by speed control are bigger movement speed improvement is expected. As the problem in the future, the system that can correspond to dynamic locomotion is designed by considering dynamics and the weight shift etc. In addition, the mechanism of kick motion using the ankle, and it aims at a dynamic running locomotion.

References

1. K. Kato and S. Hirose, "Development of Quadruped Walking Robot, TITAN-IX -Mechanical Design Concept and Application for the Humanitarian Demining Robot-" Advanced Robotics, 15, 2, pp.191-204 (2001)
2. H, Kimura and Y, Fukuoka, "Biologically Inspired Adaptive Dynamic Walking in Outdoor Environment Using a Self-contained Quadruped Robot: 'Tekken2'", IROS, 2004.
3. K.Kadomae, Y.Tokuyama, K.Shibuya, K.Tsutsumi, T.Iwamoto "Realization of Trot Gait by Bio-Mimetic 4-legged Running Robot Ryuma equipped with Torsion Coil Springs in the Leg Joints" ROBOMEC, June 2005

Introducing the Hex-a-ball, a Hybrid Locomotion Terrain Adaptive Walking and Rolling Robot

Phipps, C.C. and Minor, M.A.

University of Utah Department of Mechanical Eng. 50 S Central Campus Dr. Rm. 2110, cristianphipps@gmail.com, minor@mech.utah.edu

Abstract

The purpose of this paper is to introduce the Hex-A-Ball (HAB), a robot capable of hybridizing walking and rolling locomotion for improved adaptability, velocity, efficiency, and range of operation. The HAB is a hexapod robot with a spherical exoskeleton which uses its legs and a minimal set of actuators for both walking and rolling locomotion. This paper discusses the mechanical design of the HAB, and walking and rolling locomotion strategies.

Keywords: hybrid walking rolling spherical robot

1 Introduction

The purpose of this paper is to introduce the Hex-A-Ball (HAB), **Fig. 1**, which is a mobile robot motivated towards improving adaptability, efficiency and speed in exploration and reconnaissance applications. This is accomplished by an innovative design that allows the robot to provide both legged and rolling locomotion using a common set of actuators. The HAB consists of a spherical exoskeleton that provides rolling locomotion and six segments of the sphere that unfold to form legs for walking and climbing on rugged terrain. The six legs provide a stable configuration for locomotion and allow for rolling control authority in all three spatial dimensions.

This research is derived from past experience with miniature climbing robots [1, 2] that has shown that even though they are suited to accessing confined locations and climbing over obstacles, they are frequently too slow and inefficient. Our recent research dealing with the Rolling Disk Bi-

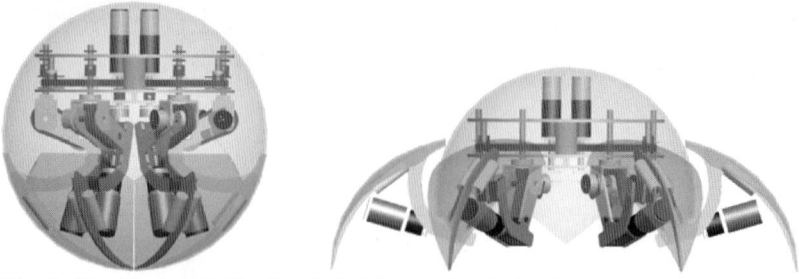

Fig. 1. The Hex-A-Ball robot hybridizes spherical rolling (left), and hexapod walking and climbing locomotion (right).

ped [3], which is the HAB predecessor, has shown that hybridized rolling and climbing locomotion has the potential to more than double the velocity of travel and increase range by at least an order of magnitude given a limited power source.

The HAB platform provides numerous advantages for adaptability, durability, and efficiency. Since the HAB can provide multiple modes of locomotion it can adapt its behavior to the particular terrain that it is traversing. Rolling is adapted to efficient high-speed locomotion, and in fact the rate of travel is limited by rolling resistance and the magnitude and frequency of force applied by the legs. The legs can apply force via direct ground interaction, they can shift the Center of Gravity (CG) to apply a rolling torque, they can create dynamic forces, and they can be used to generate a running gait followed by a dynamic transition to the spherical shape. This form of locomotion is brethren to the escape mechanism employed by Namibia Desert Wheeling spider, which curls its legs to form a spherical shape after running or while descending a dune [4]. The HAB spherical shape also allows the robot to be completely enclosed to protect itself, and the hexapod configuration provides a low profile for crawling in recesses or climbing inclined surfaces. This paper introduces the HAB mechanical design and describes basic walking and rolling locomotion.

2 Design of the Hex-A-Ball

Fig. 2a contains an exploded view of the HAB showing the major subassemblies. These consist of six leg subassemblies, the Base-Ring subassembly, and a spherical shell. The six legs are arranged radially about the z'-axis (attached to the geometric center of the Top Shell) and are attached to the Base-Ring. Integral to the leg sub-assemblies, **Fig. 2b**, is a curling axis of rotation, $y_n"$ (attached to the Shoulder), that allows the leg to fold into the spherical shape as well as control leg height in the hexapod configuration. Actuation for each curling joint is provided by a geared DC motor that drives a self-locking single-lead worm-gear. This allows the motor to be mounted perpendicular to the curling axis and fit within the leg package space such that it can be attached to the leg for increased leg mass. Further, the worm gearing is coupled by chain drive to the curling axis, y_n'', which provides additional reduction and allows the motor to be mounted near the tip of the leg while not interfering with other components. This motor position, in conjunction with two batteries mounted to each leg, provides appreciable mass moment to the legs for controlling rolling locomotion. Synchronized curling thus allows for excellent rolling control authority by shifting the net robot CG in three spatial dimensions.

The Base-Ring supports the legs, internal components, and the Top Shell. Integral to the Base-Ring are timing belts, pulleys, and motors, **Fig 2a** and **c**, that actuate leg gaiting (z_n''-axis of rotation, **Fig. 2b**). Gaiting articulation is under-actuated by mechanical coupling of Legs 1, 2 and 3 to Gaiting Motor 1 and Legs 4, 5 and 6 to Gaiting Motor 2, thus requiring only two gaiting actuators, which significantly reduces weight and improves weight distribution. The legs are coupled such that the front and

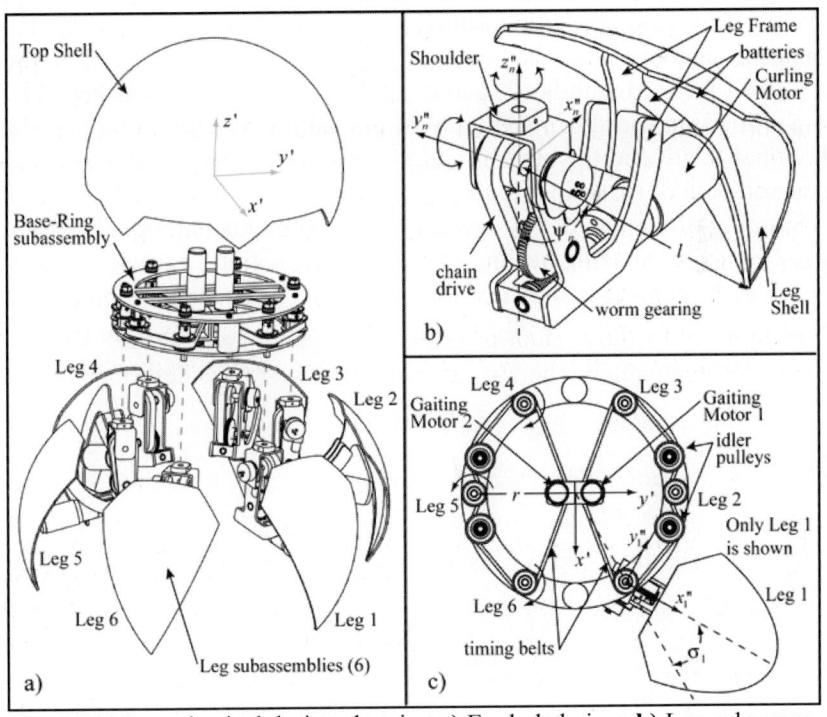

Fig. 2. HAB mechanical design showing **a)** Exploded view, **b)** Leg subassembly, and **c)** Base-Ring subassembly (top plate omitted for clarity).

rear legs (Legs 1 and 3, or 4 and 6) rotate in the same direction, while the middle legs (2 or 5 respectively) rotate in the opposite. Specifics of leg gaiting and locomotion are described in Sec. 3.

The spherical exoskeleton consists of a top section mounted to the Base-Ring and six identical leg sections. The exoskeleton is sectioned in such a way as to eliminate interference between sections as the legs move. This is achieved in part by the curling joint placement and Leg Frame, which allow the leg to rotate out of the spherical shape and provide sufficient clearance near the top portion of the leg during hexapod gaiting. The complete assembly is designed such that the reachable space of the CG (via curling actuation) surrounds the geometric center of the spherical shell.

3 Walking Strategies

The HAB walks by employing a tripod gaiting strategy for maximum stability; three legs (either Legs 1, 3 and 5 (Set 1), or Legs 2, 4 and 6 (Set 2)) are always in contact with the ground. To walk forward along the x'-axis, the robot: (1) lifts Set 1 off the ground via curling actuation, (2) propels itself forward via gaiting actuation of Set 2 while Set 1 simultaneously rotates in the opposite direction, (3) lowers Set 1 and raises Set 2, (4) propels itself forward with Set 1, and (5) the process repeats from step (1). To turn, either the speed or magnitude of the gaiting actuation on one side of the robot relative to the other is varied such that the robot can travel along a curved path or rotate in place.

Slipping of the tips of the legs (feet) caused by leg gaiting may be mitigated by synchronizing curling and gaiting actuation to cause the feet to sweep paths parallel to the direction of travel. The equation relating $\hat{y}_n$, the y' component of the nth foot position, to its gaiting and curling angles, σ_n and ψ_n, respectively, in the x'-y'-z' coordinate system is described by,

$$\hat{y}_n = \sin(60n-30)r + l\sin(\psi_n)\sin(60n-30+\sigma_n), \quad (1)$$

where r is the radial distance of the gaiting axis, z_n'', from the z'-axis and l is the orthogonal distance from the curling axis, y_n'', to the leg tip. For a path parallel to the y'-axis, $\hat{y}_n$ is held constant. Solving (1) for ψ_n yields

$$\psi_n = \sin^{-1}\left(\frac{\hat{y}_n - \sin(60n-30)r}{l\sin(60n-30+\sigma_n)}\right), \quad (2)$$

which provides curling angle trajectories for a given gaiting angle trajectory and y' position component within the reachable space of the leg tip.

4 Quasi-Static Rolling Strategies

Our initial studies of the HAB's rolling locomotion are from a quasi-static perspective. The rolling HAB is at equilibrium when its CG is positioned directly above the exoskeleton's contact point with the ground, and below the geometric center of the exoskeleton. A change in the position of the CG causes the robot to become unstable and consequently roll to restore equilibrium. The HAB rolls quasi-statically by reconfiguring its legs in order to reposition its CG. Assuming a slow rolling rate, dynamic effects may be ignored. This method of quasi-static spherical rolling is similar to that discussed by [5, 6], who have proven its viability.

For the quasi-static analysis, the robot is considered as consisting of seven components: the six identical legs and the body which connects them. Each of these components is regarded as having a mass concentrated at the individual components' CG. Only the curling actuators are used to actuate quasi-static rolling. **Fig. 3** shows the kinematic model of the HAB used for quasi-static rolling calculations. All position references are in the x'''-y'''-z''' coordinate frame attached to the robot, at the geometric center of the spherical exoskeleton.

The equation relating a new set of curling angles, ψ_{k+1}, to a change in CG position, $\Delta \mathbf{C}$, and an original set of curling angles, ψ_k is

$$\frac{\sum_{n=1}^{6} \mathbf{C}_{leg,n,k+1} m_{leg} + \mathbf{C}_{body} m_{body}}{m_{tot}} = \frac{\sum_{n=1}^{6} \mathbf{C}_{leg,n,k} m_{leg} + \mathbf{C}_{body} m_{body}}{m_{tot}} + \Delta \mathbf{C}, \qquad (3)$$

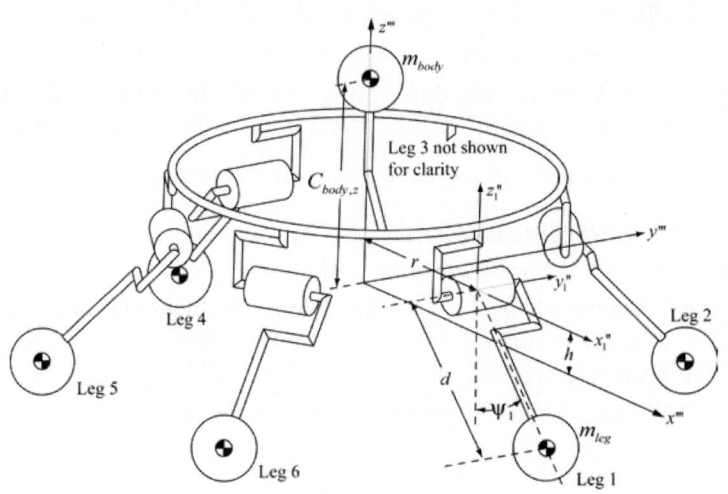

Fig. 3. Kinematic model used for quasi-static rolling calculations.

where the LHS is the new position of the CG and the RHS is the initial position of the CG plus $\Delta \mathbf{C}$. Eq. (3) can be simplified to

$$\sum_{n=1}^{6} \mathbf{C}_{leg,n,k+1} = \frac{m_{tot}}{m_{leg}} \Delta \mathbf{C} + \sum_{n=1}^{6} \mathbf{C}_{leg,n,k}, \qquad (4)$$

where m_{leg} is the mass of a single leg, m_{body} is the mass of the body and m_{tot} is the entire mass of the robot ($m_{tot} = 6m_{leg} + m_{body}$). $\mathbf{C}_{body}$ is the position of the CG of the body and $\mathbf{C}_{leg,n}$ is the position of the CG of leg n,

$$\mathbf{C}_{leg,n} = \begin{bmatrix} \cos(60(n-1))(r + d\sin(\psi_n)) \\ \sin(60(n-1))(r + d\cos(\psi_n)) \\ h - d\cos(\psi_n) \end{bmatrix}, \qquad (5)$$

where r is the distance of the curling joints from the z'-axis, d is the distance from the curling joints to the CG's of the legs, h is the distance of the curling joints from the x''-y'' plane, and ψ_n is the curling joint angle of leg n. The terms $\cos(60(n-1))$ and $\sin(60(n-1))$ account for the orientations of the legs in the circular arrangement around the base ring.

When decomposed into scalar form, (4) becomes a system of three non-linear equations with six unknowns, the elements of ψ_{k+1}. The other three equations necessary to make the system solvable may be chosen as necessary to prevent interference between the legs and the ground, and to ensure continuity in the spherical shape as the robot rolls between exoskeleton segments.

The resulting set of non-linear equations may be solved for ψ_{k+1} using a non-linear solver; we use a multivariate form of the Newton-Raphson method, providing ψ_k as the initial guess for the calculation of ψ_{k+1}. If the $\Delta \mathbf{C}$'s are small, smooth leg trajectories may be generated for any CG trajectory within the reachable space (**Fig. 4a**).

5 Results and Discussion

Based upon the design and algorithms described above, we present fundamental results on simulated rolling and walking performance. Gaiting motors have been selected that allow the HAB to scale 20° slopes by employing the walking strategies detailed in Sec. 3. On a flat surface, the HAB is capable of a walking speed of about 7 cm/s.

Our rolling simulation is limited to simple one-dimensional locomotion about the y''-axis. Constraining ψ_1 and ψ_4, to zero ensures a continuous circular shape of the exoskeleton for the robot to roll along, and for the sixth necessary equation, we simply set ψ_3 and ψ_5 to be equal. Assuming no deviation from the straight line path, ψ_2 and ψ_6 are equal as well, but they could be used to compensate for deviations. We then define a circular trajectory of discrete points for the CG about the geometric center of the spherical exoskeleton (**Fig. 4a**), and use Matlab^{TM} to solve for curling angles for each CG trajectory point.

The resulting curling angle trajectories are shown in **Fig. 4b**. **Fig. 4c** shows chronological snapshots of the solid-model prototype quasi-static rolling simulation. Curling motors have been selected with the capability to move the legs through the aforementioned trajectories at a rate of 1.2 Hz. The HAB's diameter is 0.3 m which results in its ability to roll quasi-statically at a rate of 1.2 m/s, at which the quasi-static assumption degrades. As can be seen from **Fig. 4a** the longest moment arm which can be achieved for any HAB orientation is 0.6 cm. Combined with the HAB's 5.4 kg mass, a maximum continuous rolling torque of 0.32 Nm is obtainable, allowing for controlled ascent and descent of slopes up to 2.3°, assuming no rolling friction. Additionally, the curling motors are capable of speeds as high as 2.5 rad/s, facilitating more dynamic forms of rolling locomotion such as applying impulses against the ground with the legs.

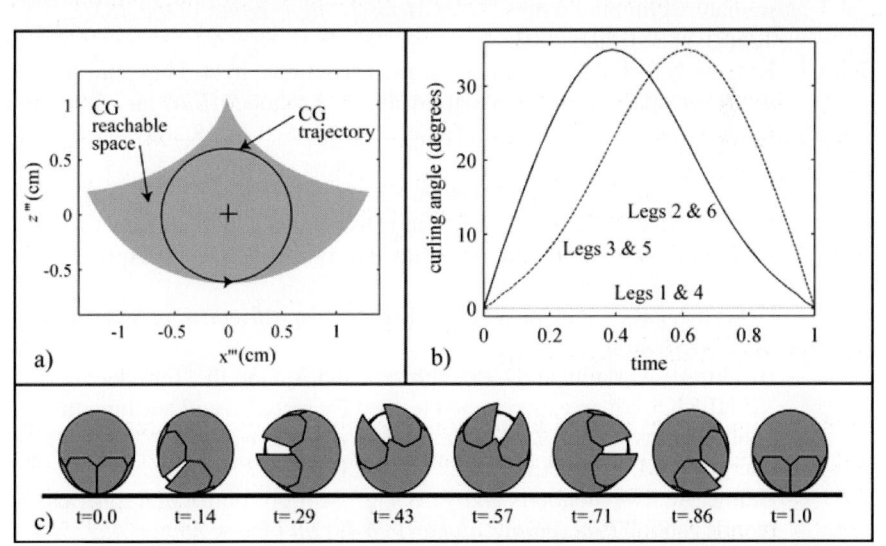

Fig. 4. One dimensional quasi-static rolling: **a)** net CG reachable space and trajectory used in simulation, **b)** simulation curling joint trajectories, **c)** simulation chronological progression.

6 Conclusions

Mechanical design of the HAB is complete and simulations of its locomotion strategies prove its viability. Future work is to include: optimization of the design for minimum weight and maximum rolling control authority, completion of the physical prototype, and testing of developed walking and quasi-static rolling strategies on the physical prototype. Further development of locomotion strategies is to include: quasi-static rolling in three dimensions, dynamic rolling actuated by shifting the CG or via interaction between the legs and ground, and inclusion of feed-back control.

Acknowledgements

The authors gratefully acknowledge the contributions of Ryan Ferrin, Jon Meikle, Ben Ruttinger, and Brigham Timpson, who helped to realize the mechanical design and walking strategies of this robot.

References

1. Minor, M.A. and R. Mukherjee, "Under-actuated kinematic structures for miniature climbing robots," *J. of Mech. Design, Trans. of the ASME*, 125(2): p. 281-291, 2003.
2. Krosuri, S.P. and M.A. Minor. "A multifunctional hybrid hip joint for improved adaptability in miniature climbing robots," *IEEE Int'l Conf. on Rob. and Autom. (ICRA '03)*. Taipei, Taiwan. Sep 14-19 2003. pp. 312-317. 2003.
3. Shores, B. and M. Minor. "Design, Kinematic Analysis, and Quasi-Steady Control of a Morphic Rolling Disk Biped Climbing Robot," *IEEE Int'l Conf. on Rob. and Autom (ICRA 05)*. Barcelona, Spain. April 18-22. pp. 2732-37. 2005.
4. Henschel, J.R., "Spiders Wheel to Escape," *So. African J. of Science*, 86: p. 151-153, 1990.
5. Bicchi, A., A. Balluchi, D. Prattichizzo, and A. Gorelli. "Introducing the "SPHERICLE": an experimental testbed for research and teaching in nonholonomy," *IEEE Int'l Conf. on Rob. and Autom. (ICRA '97)*. Albuquerque, NM, USA. April 20-25. pp. 2620-5. 1997.
6. Halme, A., T. Schonberg, and Y. Wang. "Motion control of a spherical mobile robot," *Proceedings of the 1996 4th Int'l Workshop on Adv. Motion Control, AMC'96. Part 1 (of 2)*, Tsu, Jpn. Mar 18-21 1996. pp. 259-264. 1996.

Stability Control of an Hybrid Wheel-Legged Robot

G. Besseron, Ch. Grand, F. Ben Amar, F. Plumet, and Ph. Bidaud

Laboratoire de Robotique de Paris (LRP)
CNRS FRE 2507 - Université Pierre et Marie Curie, Paris 6
18 route du Panorama - BP61 - 92265 Fontenay-aux-Roses, FRANCE
{besseron,grand,amar,plumet,bidaud}@robot.jussieu.fr

Abstract

For exploration missions with autonomous robotic systems, one of the most important goal of any control system is to ensure the vehicle integrity. In this paper, the control scheme of an hybrid wheel-legged robot is presented. This scheme uses artificial potential field for the on-line stability control of the vehicle. First, the principle of this method is outlined and, then, simulation results in 2-D are presented.

Keywords: Control, stability, potential field, hybrid wheel-legged robot.

1 Introduction

The general field of this research is the mobility of autonomous robots navigating over an unknown natural environment like those come accross during exploration missions.

Autonomous exploration missions, like planetary and volcanic exploration or various missions in hazardous areas or sites under construction, require mobile robots that can move on a wide variety of terrains while insuring the system integrity.

One of the main difficulties in this kind of environment is due to the geometrical and physical soil properties (large slopes, roughness, rocks distribution, soil compaction, friction characteristics, etc). Hybrid robots (like Gofor [1], SRR [2], Workpartner [3] robots), which combine both the advantages of wheeled and legged robotic systems, are new locomotion systems specially designed to overcome these difficulties. They are articulated vehicles with active internal mobilities which can be used to improve the stability.

HyLoS I [4] and HyLoS II, which have been designed and built at the LRP, are other examples of high mobility redundantly actuated (16 degrees-of-freedom) hybrid robots.

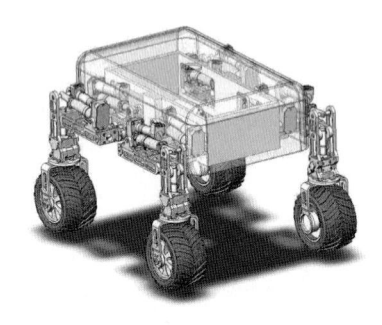

(a) HyLoS I   (b) HyLoS II (CAD View)

Fig. 1. Hybrid Robots of LRP

Both systems have the ability to adapt their configuration and locomotion modes to the local difficulties of the crossed terrain. Previous published works have focused on the kinematic-based decoupling control of HyLoS [5] as well as on the comparison of locomotion performance [6].

However, uses of the internal mobilities to guarantee the robot integrity (i.e. stability) at all time is clearly inefficient from the energy consumption point of view.

In this paper, a new control scheme derived from the potential field approach is proposed. The scheme takes advantage of the capability of this method to merge different operational constraints and of the zero-band of the used potential fonctions.

In the next three sections, the kinematic model of the robot and the control scheme will be outlined. In the last section, simulation results of this control law will be shown and discussed.

2 Kinematic Model of HyLoS II

The kinematic of Hylos II is similar to HyLoS I one [4]. The main difference in HyLoS II (**Fig. 1(b)**) is the internal passive revolute joint linking the left and right part of the robot. Each part is composed of two articulated legs, which are made up of two degrees-of-freedom suspension mechanism and a steering and driven wheel. The control strategy is based on the velocity model of the vehicle. On the assumption that the rolling is ideal, Grand [5] proposes a velocity model of the system HyLoS I. The same formalism is used and adapted to the specific kinematic of HyLoS II, leading to $\mathbf{L}\dot{\mathbf{x}} + \mathbf{J}\dot{\boldsymbol{\theta}} = \mathbf{0}$ which can also be written as:

$$[\mathbf{L}\ \mathbf{J}] \begin{bmatrix} \dot{\mathbf{x}} \\ \dot{\boldsymbol{\theta}} \end{bmatrix} = 0 \quad \text{or} \quad \mathbf{H}\dot{\mathbf{q}} = \mathbf{0} \tag{1}$$

where $\mathbf{L}$ is the Locomotion matrix which gives wheel contribution to plateform movement, $\mathbf{J}$ corresponds to the Jacobian matrix of wheel-legged kinematic chain, and $\mathbf{q}^T = [\mathbf{x}^T, \boldsymbol{\theta}^T]$ is the vector of robot parameters where $\mathbf{x}$ and $\boldsymbol{\theta}$ are respectively vectors of plateform parameters and articular-joint parameters of each wheel-legged subsystem.

Then virtual potential forces $\mathbf{F_i}$, described in the next section, are used to control the $\dot{\mathbf{q}}$ in the vehicle space configuration by considering the kinematic constraints of the system.

3 Proposed Approach

The main idea of this approach consists in using a driving virtual force $\mathbf{F_i}$ parallel to the opposite gradient of some potential field U_i.

$$\mathbf{F_i} = -\boldsymbol{\nabla} U_i \tag{2}$$

Khatib [7] was the first to describe this approach, which has been extensively used over the last two decades. This method is quite flexible since it can include other operational and functional constraints. The total potential field U can result from numerous functions U_i defined by:

$$U = \sum \alpha_i U_i(\mathbf{q}) \tag{3}$$

where U_i express potential function for obstacle avoidance, path tracking [8], joint limit avoidance or, in the present case, to guarantee the stability of the system. An influence coefficient α_i is allocated to each potential function to give more or less importance to the considered potential function.

3.1 Stability Constraint

Navigation on rough terrain requires to define a stability margin index. The "tipover stability margin" proposed by Papadopoulos in [9] is used.

This tipover stability margin takes into account both the distance of the projected center-of-gravity (CoG) to the support polygon and its vertical position relatively to the average plane defined by contact points P_i. Moreover, all the external forces including gravity are considered to work on the CoG of the vehicle. The formalism can be described briefly as follows (**Fig. 2(a)**) : the line joining two consecutive terrain-contact points P_i defines a tipover axis. The unit vector $\mathbf{l_i}$ of the axis joining the vehicle CoG, G, to the center of each tipover axis is computed. Then, the angle v_i between each $\mathbf{l_i}$ and the total external force τ_t applied to the vehicle gives the stability angle over the corresponding tipover axis.

Considering only quasi-static evolution of the vehicle here, the total external force τ_t is reduced to its own weight. The stability margin m_{s_i} is therefore defined as the angle between $\mathbf{l_i}$ and the gravity resultant $\mathbf{g}$.

However, contrary to the original tipover stability margin, which uses the overall vehicle stability margin (defined as the minimum of all stability angles), one stability margin m_{s_i} per tumbling axis is defined in order to get a differential form for the potential field.

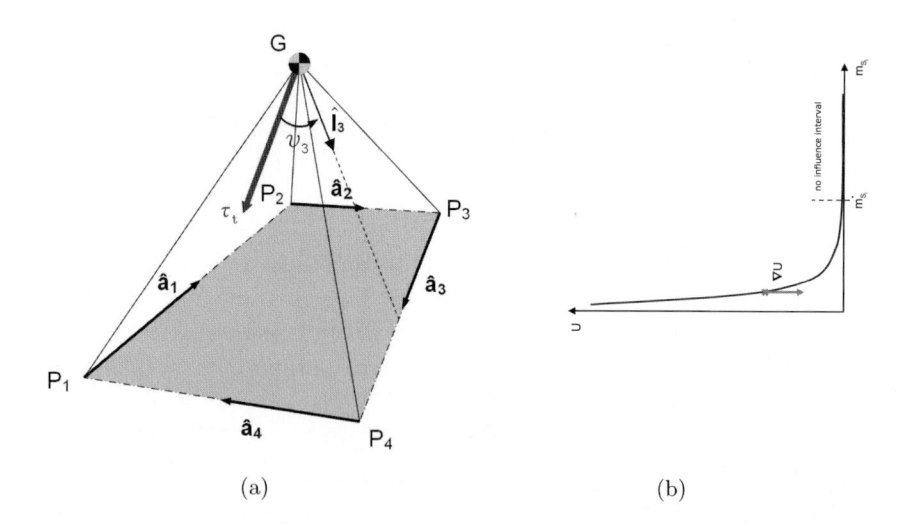

Fig. 2. Stability Margin and associated Potential Function

So as to control the stability of the vehicle, a potential function is associated with each tipover axis and is defined by :

$$U_i = \begin{cases} \frac{1}{2} \xi \left(\frac{1}{m_{s_i}} - \frac{1}{m_{s_i}^*} \right)^2 & \text{if } m_{s_i} \leq m_{s_i}^* \\ 0 & \text{if } m_{s_i} > m_{s_i}^* \end{cases} \quad (4)$$

where m_{s_i} is the computed stability margin corresponding to each tumbling axis, $m_{s_i}^*$ is a stability margin limit, and ξ is a constant gain. This function has a zero-band (**Fig. 2(b)**), that results in the robot control of having a correction only is necessary. That is more efficient to the control from the energy consumption point of view.

3.2 Overall Control Scheme

The overall control scheme is shown in **Fig. 3**.

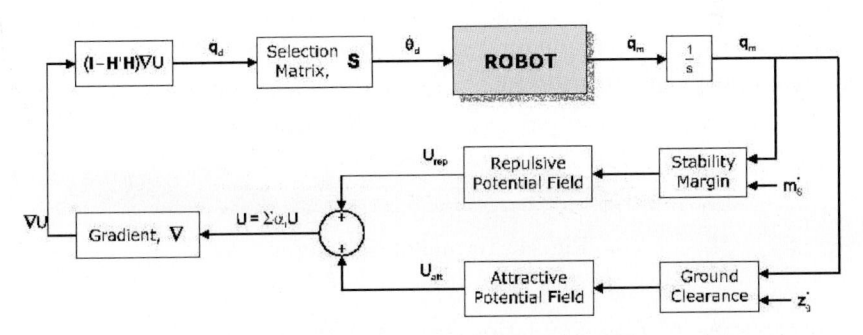

Fig. 3. Synoptic Control Scheme of robot

The control loop applied to the robot consists in moving away from a stability margin limit $m_{s_i}^*$ (defined in section 3.1) and, in the same time, in keeping the ground clearance – height of plateform – near the nominal value z_g^*. As defined in section 3.1, this control used potential functions associated to each criterion. The movement of the robot is not given here by any potential function but the angular speed of the wheels is forced. Therefore, a repulsive potential function relative to the stability margin $U_{rep_{sm}}$ and an attractive potential function to the ground clearance $U_{att_{gc}}$ are computed in this control loop.

The command set point **u** is then computed to follow the opposite gradian function of the required potential field projected on a constraints surface:

$$\mathbf{u} = -\mathbf{S}(\mathbf{I} - \mathbf{H}^+\mathbf{H})\boldsymbol{\nabla} U \tag{5}$$

where $\mathbf{H}^+\mathbf{H}$ refers to term of projection on surface of kinematic contraints in contact, U is the total potential field developed previously and $\mathbf{S}$ is a selection matrix.

4 Results

The proposed strategy is validated in 2-D simulation. This assumption could be explained by the kinematic of the robot, briefly presented in section 2. This 2-D consideration leads to the calculation of two stability margins (one per leg) m_{s_1} and m_{s_2} – in 2-D, there are only two tipover axis.

The validation of the stability control has been made on a terrain, which has a lot of slopes higher and higher (**Fig. 4**). A simulation in open loop (without reconfiguration) and another in feedback loop have been computed on this terrain.

Fig. 4. Terrain of Simulation

The results of this simulation show the overall stability margin M_s comparison between open loop and feedback loop behaviour – this overall stability margin is defined as the minimum of all stability angles: $M_s = min(m_{s_i})$. Unlike open loop behaviour of robot, in feedback loop, the stability margin is constrained to move away from a stability margin limit that is ensured by associated potential function (Eq. 4). Each time the stability margin m_{s_i} gets closer to the stability margin limit $m_{s_i}^*$, the robot configuration is modified to move away from it. Thus, at the start of the simulation (**Fig. 5**(a): Time < 6 sec) when the slopes are not too high (**Fig. 4**), the potential field has no influence on the robot configuration.

Compared with open loop behaviour, the overrun of stability margin limit is very small in feedback loop (**Fig. 5**(a)): the vehicle integrity in closed loop control is ensured.

Concerning the results of ground clearance measurement (**Fig. 5**(b)), as for stability, there is no correction by the associated potential function

(a) Stability Margin (b) Ground Clearance

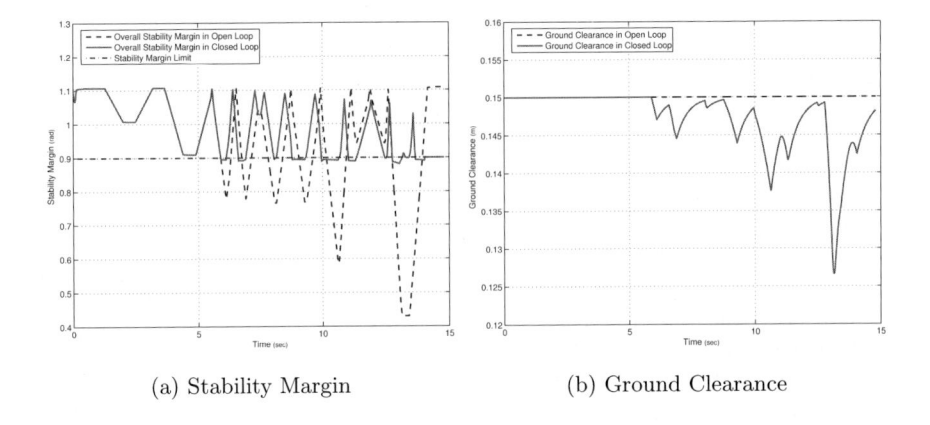

Fig. 5. Simulation Results

at the beginning of the simulation. The ground clearance is evaluated as the mean height of the robot legs, that is the reason why it is constant in open loop control. In feedback loop, the robot control attempts to bring ground clearance back to z_g^*. The results show than more sloping the terrain is, higher is the overrun of initial ground clearance z_g^*. This can be explain by the more important reconfiguration of the robot on high slopes, what involves one higher disturbance on ground clearance and a greater difficulty not to go over the stability margin limit $m_{s_i}^*$. The used control is a combination between two opposite actions, a compromise between the two defined potential functions – one which attempts to lift down the robot plateform and the other which lifts it up.

5 Conclusion

An original method to control the stability of a robotic system was proposed in this paper. This method gives interesting first results. Nevertheless some parameters of control must be sharpened. An extended simulation to 3D is planned in order to take into account the overall robot behaviour. Another interesting side, which could be evaluated, is the expected gain of energy consumption with this control strategy. Lastly, an experiment campaign will be hold with HyLoS II plateform.

References

1. Sreenivasan, S.V. and Wilcox, B.H. (1994) Stability and traction control of an actively actued micro-rover. *Journal of Robotics Systems*, vol. 11, no. 6, pp. 487–502.
2. Iagnemma, K., Rzepniewski, A., Dubowsky, S., and Schenker, P. (2003) Control of robotic vehicles with actively articulated suspensions in rough terrain. *Autonomous Robots*, vol. 14, no. 1, pp. 5–16.
3. Halme, A., Leppänen, I., Salmi, S., and Ylönen, S. (2000) Hybrid locomotion of a wheel-legged machine. In *3rd Int. Conference on Climbing and Walking Robots (CLAWAR'00)*.
4. BenAmar, F., Budanov, V., Bidaud, P., Plumet, F., and Andrade, G. (2000) A high mobility redundantly actuated mini-rover for self adaptation to terrain characteristics. In *3rd Int. Conference on Climbing and Walking Robots (CLAWAR'00)*, pp. 105–112.
5. Grand, C., BenAmar, F., Plumet, F., and Bidaud, P. (Oct. 2004) Stability and traction optimisation of a reconfigurable wheel-legged robot. *Int. Journal of Robotics Research*, vol. 23, no. 10-11, pp. 1041–1058.
6. Besseron, G., Grand, Ch., BenAmar, F., Plumet, F., and Bidaud, Ph. (2004) Locomotion modes of an hybrid wheel-legged robot. In *7th Int. Conf. on Climbing and Walking Robots*, Madrid, Spain.
7. Khatib, O. (1986) Real-time obstacle avoidance for manipulators and mobile robots. *Int. Journal of Robotics Research*, vol. 5, no. 1.
8. rensen, M.J. Sø Artificial potential field approach to path tracking for a non-holonomic mobile robot. .
9. Papadopoulos, E.G. and Rey, D.A. (1996) A new mesure of tipover stability for mobile manipulators. In *IEEE Int. Conf. on Robotics and Automation, ICRA'96*, pp. 3111–3116.

Manipulation and Flexible Manipulators

Hybrid Control Scheme for Tracking Performance of a Flexible system

F. M. Aldebrez, M. S. Alam and M. O. Tokhi

Department of Automatic Control and Systems Engineering, The University of Sheffield, UK   f.aldebrez@sheffield.ac.uk

Abstract

This paper introduces a hybrid control scheme comprising a proportional and derivative (PD)-like fuzzy controller cascaded with a proportional, integral and derivative (PID) compensator. The hybrid control scheme is developed and implemented for tracking control of the vertical movement of a twin rotor multi-input multi-output system in the hovering mode. The proposed control scheme is designed in a way that the output of the PD-type fuzzy controller is fed as a proportional gain of the PID compensator. Genetic algorithm (GA) is used to tune simultaneously the other two parameters of the PID compensator.

The performance of the proposed hybrid control strategy is compared with a PD-type fuzzy controller and conventional PID compensator in terms of setpoint tracking. It is found that the proposed control strategy copes well over the complexities of the plant and has done better than the other two controllers. The GA optimization technique is also found to be effective and efficient in tuning the PID parameters.

Keywords: Flexible systems, fuzzy control, genetic algorithms, hybrid control.

1 Introduction

Since 1974, when the first fuzzy logic (FL) controller was proposed by Mandani [7], fuzzy logic control (FLC) has emerged as one of the most active research areas in control theory [5, 6, 8, 11]. Moreover, FLC has been successfully applied in the control of various physical processes [1, 9, 10]. On the other hand, their similarity with conventional control schemes is still under investigation. FL controllers use rules in the form the "IF [condition] THEN [action]" to linguistically describe the input/output relationship. The membership functions convert linguistic terms into precise numeric values. The control method of modelling human language has many advantages, such as simple calculation, as well as high robustness, lack of a need to find the transfer function of the system and suitability for nonlinear systems [8, 11].

There are three most popular types of FLC structure that have been studied and investigated: PD-type, PI-type and PID-type. The PD-type FLC generates a control action (u) from system error (e) and change of system error *(Δe)*. The PI-type FLC generates an incremental control action *(Δu)* from error *(e)* and change of error *(Δe)*. The PID-type FLC generates control action (u) from error (e), change of error *(Δe)* and sum of the accumulative errors *(δe)*. The difficulty with this type of fuzzy controller is that it needs three inputs, which will greatly increase the rule-base and make the design of the controller more complicated. Therefore, such types of PID fuzzy controllers are rarely used. PD-type fuzzy control is known to be more practical than PI-type, which is slow and gives poor performance in the system transient state for higher order processes due to the internal integration operator. On the other hand PD-type fuzzy control cannot eliminate the system steady-state error.

This paper investigates the development of hybrid control scheme that comprises a PD-type FLC cascaded with conventional PID compensator for setpoint tracking control. The motivation of cascading the PD-type FLC with PID is to overcome the drawbacks of pure PD-type FLC as mentioned above. The hybrid control scheme is designed and tested for tracking control of the vertical movement of a twin rotor multi-input multi-output system (TRMS). The parameters of the PID compensator are tuned using genetic algorithm (GA) optimisation. GA is a robust search technique proven to be an effective optimisation mechanism in complex search spaces [2, 4]. In control engineering, GA is usually utilized to overcome the difficulty in obtaining an optimal solution when using a manual tuning of the control parameters based on a heuristic (trail-and-error) approach.

2 Experimental setup

The TRMS is a laboratory scale set-up designed for control experiments by Feedback Instruments Ltd. [3]. In certain aspects it behaves like a helicopter. The TRMS rig consists of a beam pivoted on its base in such a way that it can rotate freely both in the horizontal and vertical directions producing yaw and pitch movements, respectively (**Fig. 1**). At both ends of the beam there are two rotors driven by two d.c. motors. The main rotor produces a lifting force allowing the beam to rise vertically making a rotation around pitch axis (*vertical angle*). While, the tail rotor is used to make the beam turn left or right around the yaw axis (*horizontal angle*).

The laboratory set-up is constructed so that the angle of attack of the blades is fixed and the aerodynamic force is controlled by varying the speed of the motors. Therefore, the control inputs are supply voltages of the d.c. motors. A change in the voltage value results in a change in the rotational speed of the propeller, which results in a change in the corresponding position of the beam [3].

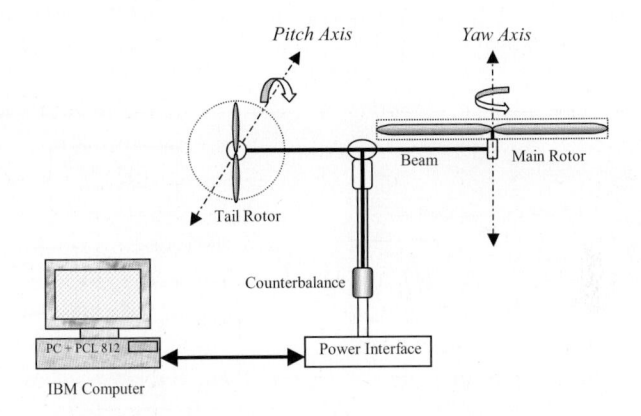

Fig. 1. The schematic diagram of the TRMS.

Although the TRMS permits multi-input multi-output (MIMO) experiments, this paper addresses a tracking control problem of a single-input single-output (SISO) mode in the longitudinal axis (i.e. vertical movement). A modelling exercise to characterise the system was carried out in a previous investigation assuming no prior knowledge of the model structure or parameters relating to the physical system, i.e. black-box modelling [1]. This was realised by minimising the prediction error of the actual plant output and the model output. The extracted continuous transfer function of

4th order parametric model that represents the system's vertical movement is given as:

$$\frac{y(s)}{u(s)} = \frac{-0.08927s^3 + 2.249s^2 - 45.57s + 595.1}{s^4 + 3.469s^3 + 519.6s^2 + 35.95s + 2189} \quad (1)$$

where $u(s)$ represents the main rotor input (volt) and $y(s)$ represents pitch angle (radians). This transfer function is utilized to represent the actual system and simulate the controllability and performance of the proposed control methods in this work.

3 Hybrid Control Scheme

The current study is confined to the development of a hybrid control scheme for tracking performance of the vertical movement in the TRMS. The proposed control strategy depicted in **Fig. 2** comprises PD-type fuzzy control and PID compensator in a cascade form. The output of PD-type fuzzy controller is fed as a proportional gain K_p of the PID compensator. Thus, K_p is continuously updated by the PD-type fuzzy controller.

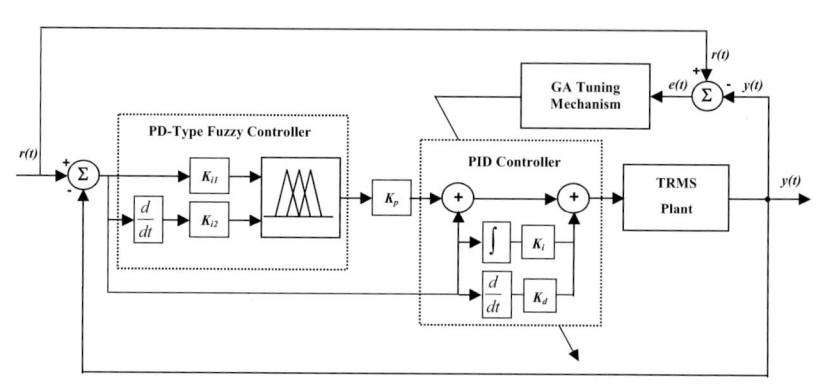

Fig. 2. Hybrid control scheme

The potential of GA optimization is demonstrated in further tuning of the conventional compensator to improve the system response. An objective function is created to tune the other two parameters; integral gain (K_i) and derivative gain (K_d) of the PID compensator in order to obtain a satisfactory performance in terms of overshoot, rise-time, settling-time and steady-state error. In order to combine all objectives (minimum overshoot, small rise and settling times and minimum steady-state error), multiple objective functions are used to minimise the output error of the controlled system. The GA optimisation process is initialised with a random population consisting of 40 individuals. The population is represented by real value numbers or binary strings each of 16 bits called chromosomes. The integral of time-weighted absolute error (ITAE) and integral of absolute error (IAE) are used as the performance criteria:

$$IAE = \int_0^T |r(t) - y(t)| dt \quad , \quad ITAE = \int_0^T t|r(t) - y(t)| dt \quad (2)$$

where $r(t)$ represents the reference input and $y(t)$ represents the system response. The chromosome is assigned an overall fitness value according to the magnitude of the error; the smaller the error the larger the fitness value.

4 Simulation and results

In this investigation three control strategies namely; PD-type fuzzy logic controller (Fpd), hybrid Fpd PID controller (HFpd_PID) and conventional PID compensator (PID) were developed and implemented within the simulation environment of the TRMS using MATLAB. An appropriate control rules between the two inputs and output of the fuzzy inference system were defined based on the rule-base array shown in **table 1**.

Table 1. Rule-base of the fuzzy inference

Δe \ e	PB	PS	Z	NS	NB
PB	PVB	PB	PM	PS	Z
PS	PB	PM	PS	Z	NS
Z	PM	PS	Z	NS	NM
NS	PS	Z	NS	NM	NB
NB	Z	NS	NM	NB	NVB

The membership functions for the two fuzzy input (control error and its derivative) sets and output (angle), used for designing the PD-type FLC are depicted in **Figs. 3a, b & c**, respectively. The input and output universes of discourse of the PD-type fuzzy controller were normalized on the range [-1,1]. The input gains K_{i1} and K_{i2} were used to map the actual inputs (control error and its derivative) of the fuzzy system to the normalized universe of discourse [-1,1] and are called normalizing gains. Similarly, K_p is the output gain that scales the output of the controller. Singleton fuzzification and centre of gravity (COG) difuzzification were used in the design of fuzzy control system throughout this investigation.

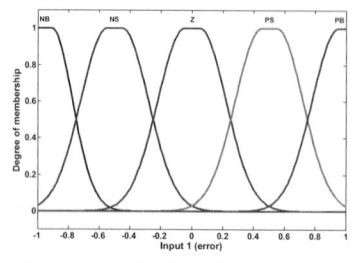

a) Membership function of Input 1

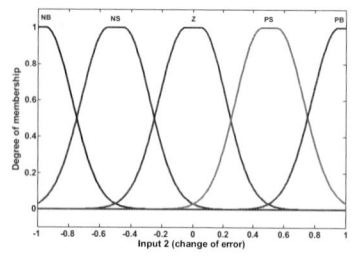

b) Membership function of Input 2

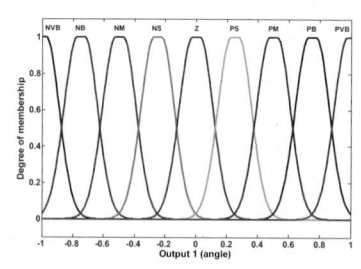

c) Membership function of output

Fig. 3. The membership functions of the two inputs and output of PD-like fuzzy control

The performances of the three employed control methods were assessed in terms of time domain specifications. **Table 2** summaries the performance of the controllers and **Fig. 4** shows unit step system response for the three control approaches. Among the three controllers, the hybrid controller recorded the best performance, followed by PID compensator and then PD-type fuzzy controller. It successfully eliminated the steady-state error and provided a satisfactory performance in both the steady-state and transient responses of the system.

Table 2. Performance of the three employed control schems in time domain specifications

Control Strategy	Time domain specifications			
	O.S. (%)	RT (sec)	ST (sec)	SSE (%)
Conventional control (PID)	0.97	1.47	2.37	0
PD-type Fuzzy Control (Fpd)	-15.97	-	-	16.13
Hybrid control (Fpd+PID)	0.34	0.57	1.12	0

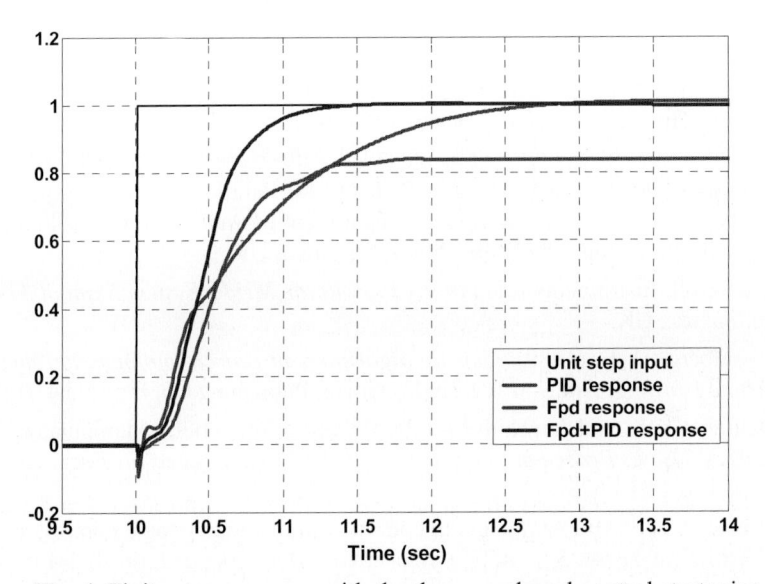

Fig. 4. Finite-step response with the three employed control strategies

5 Conclusions

A hybrid control strategy comprising PD-type fuzzy control and GA-tuned PID compensator has been proposed for set-point tracking control of the vertical movement of the TRMS. The performance of the hybrid control scheme has been compared with two feedback control strategies namely, PD-like fuzzy control and conventional PID compensator. It was observed that the hybrid control scheme is superior over the other two employed control strategies in terms of time-domain specifications namely, response over-shoot, rise-time, settling-time and accumulated steady-state error.

The investigation has also witnessed the ability of GA optimisation in tuning the PID parameters within the hybrid scheme. The obtained results revealed that the GA optimization technique is effective and efficient in tuning the PID parameters. The GA optimization process can save time as compared with conventional trial-and-error tuning procedures.

As an extension to this investigation, the proposed control scheme will be implemented on the actual TRMS in a real-time environment. Furthermore, GA-based optimization process will also be extended to tune the membership functions of fuzzy inference system within the hybrid scheme.

References

1. Aldebrez F. M., Alam M. S., Tokhi M. O. and Shaheed M. H. (2004) Genetic modelling and vibration control of a twin rotor system. *Proceedings of UKACC International Conference on Control-2004, Bath,* 6-9 September.
2. Chipperfield A. J. and Fleming P. J. (1994) Parallel Genetic Algorithms: a Survey, *Research report no. 518*, Department of Automatic Control and Systems Engineering, The University of Sheffield, UK.
3. Feedback Instruments Ltd. (1996) *Twin Rotor MIMO System Manual 33-007-0.* Sussex, UK.
4. Goldberg, D. E. (1989). *Genetic algorithms in search, optimisation and machine learning*, Addison Wesley Longman, Publishing Co. Inc., New York.
5. Kumbla K.K. and Jamshidi M. (1994) Control of Robotic Manipulator Using Fuzzy Logic. *Proceedings of the Third IEEE Conference on Fuzzy Systems*, vol. 1, pp. 518-523.
6. Lee C. C. (1990) Fuzzy logic in control systems: Fuzzy logic controller-- Part I&II. *IEEE Trans. Sys., Man, Cybern.*, vol. SMC-20, no. 2, pp. 404-435.
7. Mamdani E. H. (1974). Application of Fuzzy Algorithms for Control of Simple Dynamic Plant. *Proceedings of IEEE*, vol. 121, no. 12, pp. 1585-1588.
8. Mudi R. K. and Pal N. R. (1999) A robust self-tuning scheme for PI and PD-type fuzzy controllers. *IEEE Trans. On Fuzzy Systems*, vol. 7, no. 1, pp. 2-16.
9. Pedrycz W. (1991) Fuzzy Modelling: Fundamentals, Construction, and Evaluation. *Fuzzy Sets and Systems*, vol. 41, no. 1, pp. 1-15.
10. Sugeno M. (Ed.) (1985). *Industrial applications of fuzzy control*. Elsevier Science, North-Holland, Amsterdam.
11. Verbruggen H. B. and Bruijin P. M. (1997) Fuzzy control and conventional control: what is (and can be) the real contribution of fuzzy systems? *Fuzzy Sets and Systems*, vol. 90, pp. 151-160.

Predesign of an Anthropomorphic Lightweight Manipulator

C. Castejón[1] D. Blanco[2] S.H. Kadhim[2] L. Moreno[2]

1 Mechanical Dept.; 2 System Engineering and Automation Dept., Carlos III University. Madrid, Spain.
{caston, dblanco, salah, moreno} @ing.uc3m.es

Abstract

One of the main goals in the service robotics field is the design of mobile robots which are able to operate in human environments and interact with people in a safety way. The basic functionalities that an assistance robot must fulfil, from a mechanical point of view, are the mobility and the manipulation. Currently, assistance robotic, is based on the development of mobile manipulator robots with the basic characteristics of reliability, safety and easiness of the use. The mobile base and the manipulators optimum mechanical design is the key to obtain the aims. To go more deeply into the manipulator, the optimum design of the elements and joints is based on the previous and detailed kinematical and dynamical studies. The goal of this work consists of designing the preliminary mechanical analysis of the light robotic arm LWR-UC3M-1 placed in the mobile manipulator MANFRED, in order to observe the arm dynamical behaviour with and without load applied in the end of the kinematical chain. The model and simulations presented in this manuscript have been developed based on the COSMOS MOTION® of SolidWorks® tool. A static analysis for the links has been also developed, and in this work we present the results for the links with more fault probability to be used as experience for a future redesign.

Keywords: Design Modularity and System, Simulation and Design, Service Robotics, Mechanism Predesign.

4. Simulation results

Simulation results based on the model developed are presented. Among the simulations running, only the most interesting are presented.

4.1. Statical analysis. The statical analysis results are presented, performing over all the links that suffer more stress. First and second joints suffer the most reactions with load.

First joint: *Application of a vertical load of 100N at the end of the chain:* As can be seen in figure 5, the maximum stress reached is of 0.732 MPa, this value correspond with a strain of 8.8 μm. The piece which suffers in this joint is the aluminium extension that is situated between the reduction gear and the fixed mount. The motor would be placed inside the mount with a shaft that transmits the movement to the reduction gear.

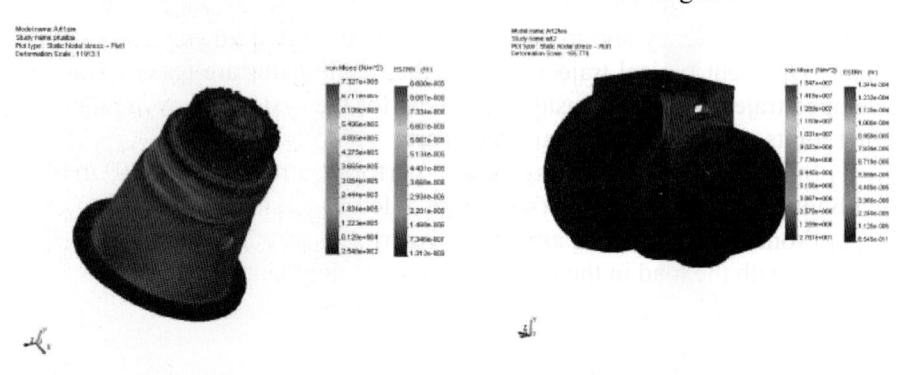

Fig. 5. Stress-strain for the first joint, with vertical load of 100N at the end of the chain.

Fig. 6. Stress-strain in the second joint, for a 75N force at the end.

Second joint: *Application of a 75N force at the end of the chain:* Figure 6 shows how the maximum stress value, which is 16 MPa., is greater than the first joint value. This value is far from the strain limit for aluminium. In the same figure, a zone with a concentration of elevated stress can be seen.

4.2. Dynamical analysis.

Joints reaction strains, due to both trajectories selected, are presented and analyzed. Several experiments has performed (simulation of the movement for the model for different joints rotational speed and for different trajectories, measure of the reaction toques and forces in joints, study with and without load, etc.).

It must point out that joints near to the load application point hold up less reactions, because the motor must perform less torque. For this reason, in this paper only joint 1 and 2 are studied, because they support the majority of the efforts during the movement. In figure 7 maximum values that suffer both trajectories in force and torque, versus the applied load are presented.

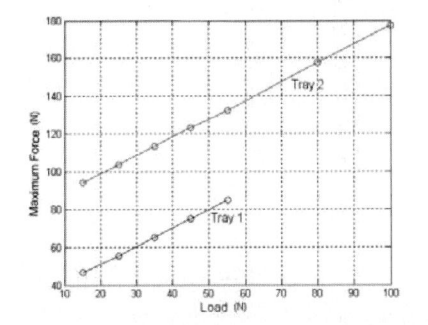

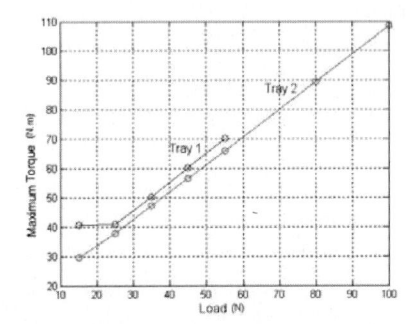

Fig. 7. Reaction force (left) and Torque (right) maximum value versus the applied load.

5. Conclusions

This paper describes some results obtained from a mechanical pre-design of an anthropomorphic mobile manipulator. The robot configuration is made up of a mobile base and the LWR-UC3M-1 lightweight manipulator. Hardware architecture has been designed to improve safety properties and performance. For physical safety purpose, two main objectives in the hardware design are considered: to improve the system stability and to develop lightweight manipulator reducing the possible injury due to the impact.

The first result obtained is the LWR-UC3M-1 manipulator, a new robotics arm which features are similar to human arm characteristics and which has a smaller weight than commercial industrial manipulators. On the other hand, a completely open hardware architecture with open controller system based directly on motor control boards has been implemented to increase robot's functionality. The system is completed with an advanced sensorial system including vision, 3D laser telemetry and a force-torque sensor.

Acknowledgements

The authors gratefully acknowledge the funds provided by the Spanish Government through the CICYT Project TAP 1997-0296 and the DPYT project DPI 2000-0425. Further, author would thank to I. Casillas for his experimental work with the robot.

REFERENCES

[1] N. Roy, G. Baltus, D. Fox, F. Gemperle, J. Goetz, T. Hirsch, D. Margaritis, M. Montemerlo, J. Pineau, J. Schulte, and S.Thrun, "Towards personal service robots for the elderly," in *Workshop on Interactive Robots and Entertainment (WIRE 2000)*, 2000.
[2] A. Gimenez, A. Jardon, and C. Balaguer, "Light weight autonomous service robot for disable and elderly people help in their living environment," in *The 11th International Conference on Advanced Robotics*, 2003.
[3] T. Rofer and A. Lankenau, "Architecture and applications of the Bremen autonomous wheelchair", in *Proceedings of the Fourth Joint Conference on Information Systems* , pp. 365–368, 1998.
[4] L.M. Bergasa, M. Mazo, A. Gardel, J.C. Garcia, A. Ortuno, and A.E. Mendez, "Guidance of a wheelchair for handicapped people by face tracking", *IEEE International Conference on Emerging Technologies and Factory Automation,* 1999.
[5] S. Fioretti, T. Leo, and S. Longhi, "Navigation systems for increasing the autonomy and security of mobile bases for disabled people" *Proceedings of the1998IEEE International Conference on Robotics& Automation*, Leuven, Belgium. May 1998.
[6] O. Khatib, K. Yokoi, O. Brock, K. Chang, and A. Casal, "Robots in human environments: basic autonomous capabilities," *International Journal of Robotics Research*, vol. 18, no. 7, pp. 684–696, july 1999.
[7] O. Khatib, O. Brock, K. Chang, F. Conti, D. Ruspini, and L. Sentis, "Robotics and interactive simulation," *Comunications of the ACM*, vol. 45, no. 3, pp. 46–51, march 2002
[8] V.Fernández; C.Balaguer; D.Blanco; M.A.Salichs. Active Human-Mobile Manipulator Cooperation Through Intention Recognition. "*IEEE International Conference on Robotics and Automation*". Seoul. Korea. May, 2001. pp.2668-2673.

Design of a "Soft" 2-DOF Planar Pneumatic Manipulator

M. Van Damme, R. Van Ham, B. Vanderborght, F. Daerden, and D. Lefeber

Robotics and Multibody Mechanics Research Group, Department of Mechanical Engineering, Vrije Universiteit Brussel, Belgium
michael.vandamme@vub.ac.be

Abstract

This paper presents the concept of a lightweight manipulator that can interact directly with an operator in order to assist him in handling heavy loads. The advantages of the system, ergonomics, low weight, low cost, ease of operation and operator safety are a consequence of the use of Pleated Pneumatic Artificial Muscles as actuators. The design of a small-scale model of such a manipulator using these actuators is presented, as well as a sliding mode controller for the system.

Keywords: Pneumatic artificial muscles, sliding mode control.

1 Introduction

Manual material handling tasks such as lifting and carrying heavy loads, or maintaining static postures while supporting loads are a common cause of lower back disorders and other health problems. In fact, manual material handling has been associated with the majority of lower back injuries, which account for 16-19% of all workers compensation claims, while being responsible for 33-41% of all work-related compensations [1]. The problem has an important impact on the quality of life of affected workers, and it presents an important economic cost.

The traditional solution is using a commercially available manipulator system. Most of these systems use a counterweight, which limits their use to handling loads of a specific mass.

In order to increase safety and productivity of human workers, several other approaches to robot-assisted manipulation have been studied in the robotics community [2, 3, 4]. The devices developed in the course of these studies belong to a class of materials handling equipment called Intelligent Assist Devices (IADs). Most of these systems, however, are heavy, complex and expensive.

In this paper we present the initial design and control of a manipulator that will eventually combine ergonomics, operator safety, low cost, low weight and ease of operation. All of this can be achieved through the use of an actuator, developed at the Department of Mechanical Engineering at the Vrije Universiteit Brussel: the Pleated Pneumatic Artificial Muscle (PPAM) [5], a contractile device operated by pressurized air.

We are working towards a system that behaves as follows: when the operator wants to move a load attached to the manipulator, he/she starts moving it as if there were no manipulator. By measuring the muscle gauge pressures, the system continuously estimates the forces applied by the operator and assist him/her in accomplishing the desired load movement. The direct interaction between operator and load (without intermediary control tools) allows for very precise positioning.

The main requirement for any mechanical device that is used in the immediate environment of people is safety. The PPAM actuators greatly contribute to the overall safety of the manipulator system: they allow for a lightweight construction, there is no danger of electrocution and, most important of all, the muscles are inherently compliant.

In this paper, the design of a small-scale proof-of-concept model of such a manipulator, consisting of two PPAM actuated links in inverse elbow configuration, is presented and a sliding-mode controller for the system is developed and tested. The system is shown in **Fig. 1**.

Fig. 1. The manipulator scale model.

2 Manipulator design

2.1 Introduction

The goal is to design a machine that will provide assistance in the vertical plane. This means that two actuated degrees of freedom are sufficient. Three possible link configurations were considered: elbow, inverse-elbow and rhombic. Since the design should be as lightweight and simple as possible, the rhombic configuration isn't suitable. As operator and manipulator will be interacting directly, it's important that the manipulator doesn't obstruct the operator's movements. For this reason, the elbow-up configuration was chosen.

For easier development and testing, and to gain experience with this type of system, we decided to develop a small-scale manipulator first. The length of both links was chosen to be 30 cm.

2.2 Design

Fig. 2 shows a schematic representation of the two links in inverse elbow configuration. The conventions used in the rest of this document regarding

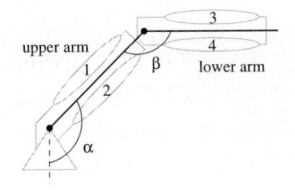

Fig. 2. The inverse elbow configuration.

to how both joint angles are defined and how the different pneumatic muscles are numbered are also included in the figure.

Since we have four PPAMs, there are eight attachment points. The location of each of these points can be described by two coordinates. Each muscle has two parameters (slenderness and maximum length). This means there are a total of 24 parameters to be determined. Determining the best design means finding a global optimum in a 24-dimensional parameter space, subject to conditions such as producibility, absence of space conflicts, avoiding excessive muscle loading, ensuring a large enough working area,.... This has proven to be computationally intractable. Therefore, the different parameters were chosen manually, mainly with ease of production in mind, after extensive computer experiments.

2.2.1 Torque characteristics

Once all attachment point locations and PPAM parameters are known, we can determine the torque characteristics of both joints. Using the nonlinear force-pressure-contraction relation of the PPAM muscle (see [5, 6]), torque generated by a muscle can be written as

$$\tau = p \cdot m(\gamma) \qquad (1)$$

with $\gamma = \alpha$ for muscles 1 and 2 and $\gamma = \beta$ for muscles 3 and 4. Equation (1) provides a clear separation between the two factors that determine torque: gauge pressure and a torque function m, that depends on the design parameters and the position. The torque functions are shown in **Fig. 3**. More details can be found in [7].

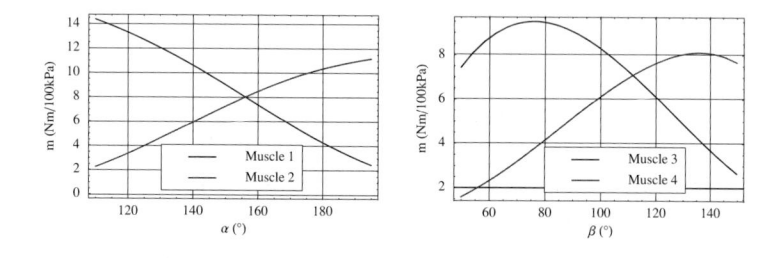

Fig. 3. Torque functions.

3 Control

3.1 Introduction

When using pleated pneumatic artificial muscles, controller design is not straightforward. Difficulties encountered when designing a controller include the following:

- Both the manipulator and its actuators are strongly nonlinear systems. Measurements on PPAMs also show a slight hysteresis in the

force-pressure characteristic. This makes it hard to estimate actuator force when only pressure measurements are available.
- Actuator gauge pressures can take a relatively long time to settle (around 100 ms for large pressure steps).
- Actuator paramters (slenderness) are not very well known.

In this paper, we describe a sliding mode approach to control the system.

3.2 Δp - approach

To reduce the number of actuator outputs that have to be calculated, the Δp-approach was used [6, 8]. This involves choosing an average pressure p_m for both muscles of an antagonistic pair, and having the controller calculate a pressure difference Δp that is added in one muscle ($p+\Delta p$) and subtracted in the other ($p - \Delta p$). The choice of p_m influences compliance while Δp determines joint position.

The control of the actuator pressures themselves is handled by off-the-shelf proportional pressure regulating valves with internal PID controllers.

3.3 Controller

The dynamical model of a 2-DOF planar arm is well known, and can be written as

$$H(q)\ddot{q} + C(q,\dot{q})\dot{q} + G(q) = \tau \quad (2)$$

where $q = [q_1 \, q_2]^T$ is the vector of joint angles, H is the inertia matrix, C is the centrifugal matrix (centrifugal and coriolis forces) and G is the gravitational force vector. τ is a vector representing the actuator torques, and can be written as

$$\tau = \begin{bmatrix} p_1 m_1(q_1) + p_2 m_2(q_1) \\ p_3 m_3(q_2) + p_4 m_4(q_2) \end{bmatrix} \quad (3)$$

with p_i ($i = 1\ldots 4$) the gauge pressure in muscle i, and m_i the torque function associated with that muscle (see section 2.2.1).

The gauge pressures are determined by the pressure regulating valves. For simplicity, we model the valves as first order systems. Combined with the Δp-approach, this gives us the following valve model:

$$\begin{aligned} T_1 \dot{p}_1 &= -p_1 + p_{m1} + \Delta p_1 \\ T_2 \dot{p}_2 &= -p_2 + p_{m1} - \Delta p_1 \\ T_3 \dot{p}_3 &= -p_3 + p_{m2} + \Delta p_2 \\ T_4 \dot{p}_4 &= -p_4 + p_{m2} - \Delta p_2 \end{aligned} \quad (4)$$

Δp_1 and Δp_2 are the inputs for upper and lower arm joints, respectively. Combining equations (2), (3) and (4) gives us the complete model of the system to be controlled.

This system is not in a form that allows direct application of sliding mode control techniques[1], such as described in for instance [9]. To solve this, we treat the system as consisting of two (coupled) SISO systems, writing them as

$$\dot{x}_i = f_i(x_i) + g_i(x_i) u_i \quad (5)$$
$$y_i = h_i(x_i) \quad (6)$$

with $i = 1, 2$ (1 for upper arm, 2 for lower arm), $x_i = [q_i \; \dot{q}_i \; p_{2i-1} \; p_{2i}]$ the state vector, $u_i = \Delta p_i$ the scalar input and $y_i = h_i(x_i) = q_i$ the system output. We can now transform these systems to the normal form using a procedure described in [10]. Written in the coordinates $\xi_{i1} = h_i(x_i)$, $\xi_{i2} = L_{f_i} h_i(x_i)$, $\xi_{i3} = L_{f_i}(L_{f_i} h_i)(x_i) = L_{f_i}^2 h_i(x_i)$, $\eta_i(x_i)$ with $\eta_i(x_i)$ satisfying $L_{g_i} \eta_i(x_i) \equiv 0$ [2], we get the following two systems (again, $i = 1, 2$):

$$\dot{\xi}_{i1} = \xi_{i2}$$
$$\dot{\xi}_{i2} = \xi_{i3}$$
$$\dot{\xi}_{i3} = b_i(\xi_i, \eta_i) + a_i(\xi_i, \eta_i) u_i$$
$$\dot{\eta}_i = r_i(\xi_i, \eta_i)$$

with $b_i(\xi_i, \eta_i) = L_{f_i}^3 h_i(x_i)$, $a_i(\xi_i, \eta_i) = L_{g_i} L_{f_i}^2 h_i(x_i)$ and $r_i(\xi_i, \eta_i) = L_{f_i} \eta_i(x_i)$.

To design a sliding mode controller that makes these systems track their respective desired output trajectories $y_{im}(t) = q_{im}(t)$, we use $e_{i0}(t) = y_{im}(t) - y_i(t)$ to define the sliding surfaces $s_i(x_i, t)$:

$$s_i(x_i, t) = \ddot{e}_{i0} + \alpha_{i1} \dot{e}_{i0} + \alpha_{i0} e_{i0} \quad (7)$$
$$= \ddot{y}_{im} - L_{f_i}^2 q_i + \alpha_{i1}(\dot{y}_{im} - L_{f_i} q_i) + \alpha_{i0}(y_{im} - q_i) \quad (8)$$

In going from (7) to (8), we have used the fact that both systems have strict relative degree 3 (see [10]), which implies $L_{g_i} q_i = 0$ and $L_{g_i} L_{f_i} q_i =$

---

[1] The technique outlined in [9] (for a SISO system) supposes the system is in the form $x^{(n)} = f(\mathbf{x}) + b(\mathbf{x})u$, with state vector $\mathbf{x} = \begin{bmatrix} x \; \dot{x} \cdots x^{(n-1)} \end{bmatrix}^T$, x being the scalar output and u the scalar input. Thus, the state vector only contains the output and its first $n-1$ derivatives, and one has to differentiate the output n times for the input to appear (which means the system has strict relative degree n [10]). Our system's state vector also contains the gauge pressures, which are of course no derivatives of the joint angles (outputs) of the system.

[2] $L_f h(x) = \frac{\partial h}{\partial x} f$ stands for the Lie derivative of h with respect to f.

0. The coefficients α_{i0} and α_{i1} are chosen so that the polynomials $p^2 + \alpha_{i1}p + \alpha_{i0}$ are Hurwitz[3]. If the trajectory is on the sliding surface (if $s_i = 0$), the error will tend to zero exponentially. By selecting a control law that makes the sliding surface attractive to the initial conditions in finite time, we can achieve our control objective. One possibility is (see [10]):

$$u_i = \frac{1}{a_i}\left(\dddot{y}_{im} - b_i + \alpha_1 \ddot{e}_{i0} + \alpha_0 \dot{e}_{i0} + K \operatorname{sgn}\left(s_i\left(x_i, t\right)\right)\right)$$

If K is large enough to overcome system uncertainty and perturbations, s_i will tend to zero in finite time. To reduce chattering, a boundary layer (see [9]) is introduced by replacing $\operatorname{sgn}(s_i)$ with $\operatorname{sat}(s_i/\Gamma_i)$, where

$$\operatorname{sat}(z) = \begin{cases} z, & |z| \leq 1 \\ \operatorname{sgn}(z), & \text{otherwise} \end{cases}$$

and Γ_i are constants determining the width of the boundary layers.

3.4 Results

To evaluate the tracking performance of the proposed sliding mode controller, it was used to track a circle in $x - y$ space. The desired trajectory was tracked in a period of 5 seconds. In order to deal with chattering, significant boundary layers were necessary ($\Gamma_1 = 4$, $\Gamma_2 = 3$), which of course increases tracking error. The resulting path is shown in **Fig. 4**.

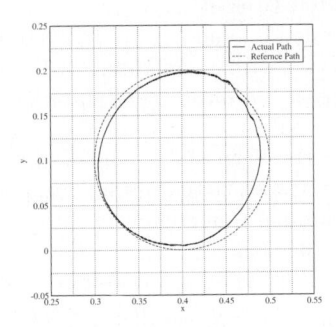

Fig. 4. Spatial tracking behaviour.

---

[3] A polynomial with real positive coefficients and roots which are either negative or pairwise conjugate with negative real parts.

4 Conclusion

The design of a small-scale, lightweight manipulator actuated by Pleated Pneumatic Artificial Muscles was presented. A sliding mode tracking controller for the system was also proposed, and initial tracking results were presented. The problem of chattering limits the tracking precision that can be achieved.

References

1. W.S. Marras, K.P. Granata, K.G. Davis, W.G. Allread, and M.J. Jorgensen (1999) Effects of box features on spine loading during warehouse order selecting. *Ergonomics*, vol. 42, no. 7, pp. 980–996.
2. Kevin M. Lynch and Caizhen Liu (2000) Designing Motion Guides for Ergonomic Collaborative Manipulation. *IEEE International Conference on Robotics and Automation*.
3. H. Kazerooni (1996) The human power amplifier technology at the University of California, Berkeley. *Journal of Robotics and Autonomous Systems*, vol. 19, pp. 179–187.
4. Jae H. Chung (2002) Control of an operator-assisted mobile robotic system. *Robotica*, vol. 20, no. 4, pp. 439–446.
5. Daerden F. and Lefeber D. (2001) The concept and design of pleated pneumatic artificial muscles. *International Journal of Fluid Power*, vol. 2, no. 3, pp. 41–50.
6. Frank Daerden (1999) *Conception and Realization of Pleated Pneumatic Artificial Muscles and their Use as Compliant Actuation Elements*. Ph.D. thesis, Vrije Universiteit Brussel.
7. Van Damme M., Daerden F., and Lefeber D. (2005) A pneumatic manipulator used in direct contact with an operator. In *Proceedings of the 2005 IEEE International Conference on Robotics and Automation*, Barcelona, Spain, pp. 4505–4510.
8. Daerden F., Lefeber D., Verrelst B., and Van Ham R. (2001) Pleated pneumatic artificial muscles: actuators for automation and robotics". In *IEEE/ASME International Conference on Advanced Intelligent Mechatronics*, Como, Italy, pp. 738–743.
9. J.-J. Slotine and W. Li (1991) *Applied Nonlinear Control*. Prentice Hall.
10. Sastry, S. (1999) *Nonlinear Systems Analysis,Stability and Control*. Springer.

Simulation and Experimental Studies of Hybrid Learning Control with Acceleration Feedback for Flexible Manipulators

M. Z. Md Zain[1], M. S. Alam[1], M. O. Tokhi[1] and Z. Mohamed[2]

[1] Department of Automatic Control and Systems Engineering, The University of Sheffield, UK; [2] Faculty of Electrical Engineering, 81310 UTM Skudai, Universiti Teknologi Malaysia, Malaysia
cop02mzm@sheffield.ac.uk,

Abstract

This paper presents investigations at developing a hybrid iterative learning control scheme with acceleration feedback (PDILCAF) for flexible robot manipulators. An experimental flexible manipulator rig and corresponding simulation environment are used to demonstrate the effectiveness of the proposed control strategy. In this work the dynamic model of the flexible manipulator is derived using the finite element (FE) method. A collocated proportional-derivative (PD) controller utilizing hub-angle and hub-velocity feedback is developed for control of rigid-body motion of the system. This is then extended to incorporate iterative learning control with acceleration feedback and genetic algorithms (GAs) for optimization of the learning parameters for control of vibration (flexible motion) of the system. The system performance with the controllers is presented and analysed in the time and frequency domains. The performance of the hybrid learning control scheme without and with acceleration feedback is assessed in terms of input tracking, level of vibration reduction at resonance modes and robustness with various.

Keywords: Acceleration feedback, flexible manipulator, genetic algorithms, iterative learning control.

1 Introduction

The control of flexible manipulators to maintain accurate positioning is an extremely challenging problem. Due to the flexible nature and distributed characteristics of the system, the dynamics are highly non-linear and complex [1].

Many industrial applications of robot manipulators involve iterative repeated cycles of events. Thus, it is important to minimize errors in trajectory tracking of such manipulators, and this can be achieved with suitable learning strategies. The basic idea behind iterative learning control (ILC) is that the controller should learn from previous cycles and perform better every cycle. Such ideas were first presented by Arimoto et al [2] in 1984 who proposed a learning control scheme called the improvement process, and since then several researchers have addressed robot control in combination with ILC, [3]. The convergence properties when using ILC control form another very important aspect, addressed in [2], and further covered in [4]. In this paper ILC is studied to complement conventional feedforward and feedback control and the effectiveness of the resulting scheme is assessed in input tracking and vibration reduction in a flexible robot manipulator.

The paper presents investigations into the development of hybrid learning acceleration feedback control (PDILCAF) using genetic algorithms (GAs) for optimization of the learning parameters for input tracking and end-point vibration suppression of a flexible manipulator system. A comparative assessment of the hybrid learning control scheme in input tracking and vibration suppression of the manipulator is presented.

2 The Flexible manipulator system

Figure 1 shows a laboratory-scale single-link experimental rig used in this work. The flexible arm is constructed using a piece of thin aluminium alloy with length $L = 0.9$ m, width = 19.008 mm, thickness = 3.2004 mm, Young's modulus $E = 71 \times 10^9$ N/m², area moment of inertia $I = 5.1924$ m⁴, mass density per unit volume $\rho = 2710$ kg/m³ and hub inertia $I_h = 5.8598 \times 10^{-4}$ kgm². The manipulator can be considered as a pinned-free flexible arm, which can bend freely in the horizontal plane but is relatively stiff in vertical bending and torsion [5].

The digital processor used is an IBM compatible PC based on an Intel(r) celeron ™ processor. Data acquisition and control are accomplished through the utilization of PCL-812PG board. This board can provide a direct interface between the processor, actuator and sensors. A simulation algorithm characteristising the dynamic behaviour of the manipulator has previously been developed using the finite element (FE) method [6]. This is used in this work as a platform for theoretical test and evaluation of the proposed control approaches.

duction at the end-point acceleration was achieved as compared with the other two methods. Moreover, vibration of the system settled within 4 s, which is better as compared to those with PD and PDILC.

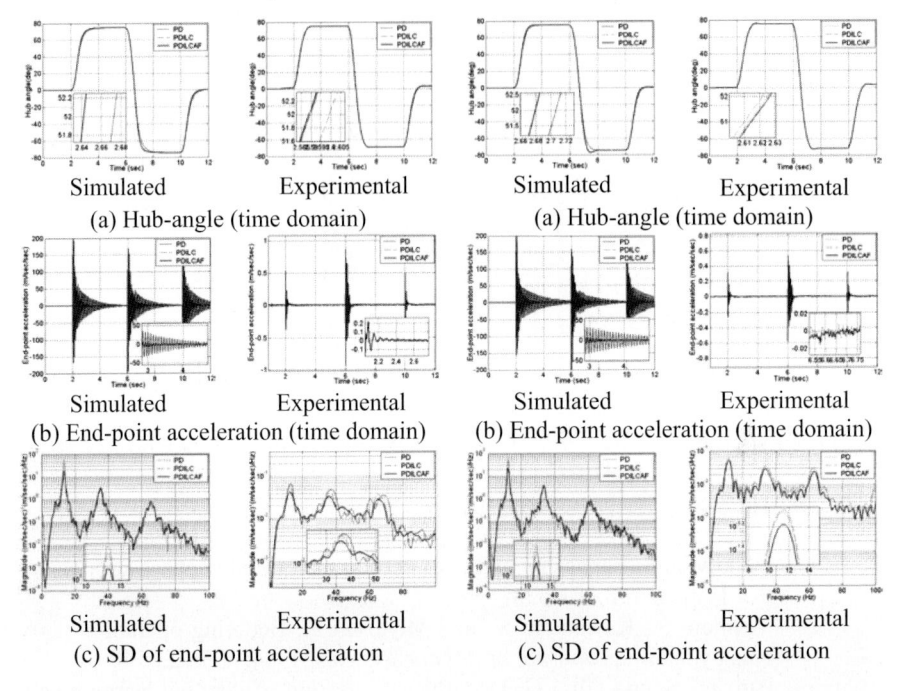

Simulated — Experimental
(a) Hub-angle (time domain)

Simulated — Experimental
(a) Hub-angle (time domain)

Simulated — Experimental
(b) End-point acceleration (time domain)

Simulated — Experimental
(b) End-point acceleration (time domain)

Simulated — Experimental
(c) SD of end-point acceleration

Simulated — Experimental
(c) SD of end-point acceleration

Fig. 6. Response of the simulated and experimental manipulator system without payload

Fig. 7. Response of the simulated and experimental manipulator system with 10 g payload

6 Conclusion

The development of PDILC with acceleration for input tracking and vibration suppression of a flexible manipulator has been presented. Control schemes have been developed on the basis of collocated PD with ILCAF based on GA optimisation. The control schemes have been implemented and tested within the simulation and experimental environments of a single-link flexible manipulator with various payloads. The performances of the control schemes have been evaluated in terms of input tracking capability and vibration suppression at the resonance modes of the manipulator. A comparative assessment of the control techniques has shown that the PDILCAF scheme results in better performance than the PDILC control in respect of input tracking and speed of response and vibration suppression of the manipulator.

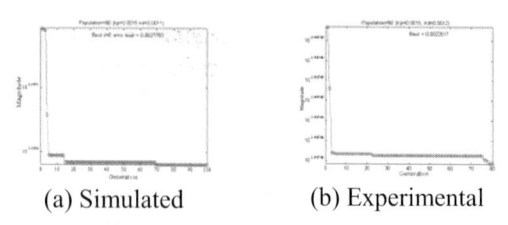

(a) Simulated   (b) Experimental
Fig. 8. Objective value vs number of generation

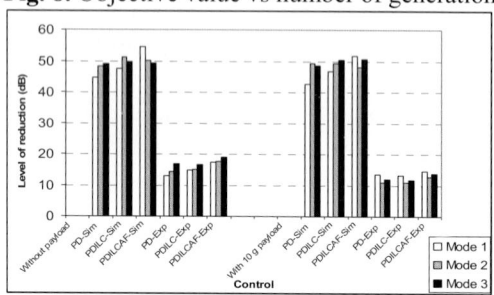

Fig. 9. Vibration reduction with the control techniques

References

1. Yurkovich S. Flexibility effects on performance and control. *Robot Control* 1992; Part 8:321-323.
2. Arimoto S., Kawamura S., and Miyazaki F.. Bettering operation of robots by learning. *Journal of Robotic Systems,* 1984;1(2):123-140
3. Panzieri S. and Ulivi G. Disturbance rejection of iterative learning control applied to trajectory tracking for a flexible manipulator. In *Proceedings of 3rd European Control Conference,* ECC, September 1995, pages 2374-2379.
4. Amann N., Owens D. H., and Rogers E. *Iterative learning control for discrete time systems with exponential rate of convergence.* Technical Report 95/14, Centre for Systems and Control Engineering, University of Exeter, 1995.
5. Tokhi M. O., Mohamed Z. and Shaheed M. H. Dynamic characterisation of a flexible manipulator system. *Robotica,* 2001; 19(5): 571-580.
6. Azad K. M. *Analysis and design of control mechanisms for flexible manipulator systems.* PhD thesis, Department of Automatic Control and Systems Engineering, The University of Sheffield, 1994.
7. Chipperfield, A.J., Flemming P.J., & Fonscea, C.M. 'Genetic algorithms for control system engineering', *Proceeding Adaptive Computer in Engineering Design and Control,* September 1994: pp.128-133.
8. Linkens, D.A., & Nyongesa, H.O., 'Genetic algorithms for fuzzy control', *IEE Proceeding Control Theory Application,* Vol. 142(3): pp. 161-185.

BNN-based Fuzzy Logic Controller for Flexible-link Manipulator

[1]M.N.H. Siddique, [2]M.A. Hossain and [3]M.O. Tokhi

[1]School of Computing and Intelligent Systems, University of Ulster;
[2]Department of Computing, University of Bradford, [3]Department of Automatic Control and Systems Engineering, University of Sheffield.
nh.siddique@ulster.ac.uk, m.a.hossain1@bradford.ac.uk, o.tokhi@sheffield.ac.uk.

Abstract

As is well recognized, rule acquisition has been regarded as a bottleneck for implementation of fuzzy logic controller. Moreover, defuzzification is a time consuming procedure. Though Roger Jang's adaptive neuro-fuzzy and Sugeno's fuzzy systems eliminated those shortcomings but both require a set of input-output data, which may not be available always. This paper reports on a backpropagation neural network based fuzzy logic controller, where neural network is trained using linguistic description.

Keywords: Fuzzy logic controller, backpropagation neural network, linguistic variable, fuzzy number, flexible-link manipulator.

1 Introduction

A significant amount of interest has been shown by researchers in the control of flexible-link manipulators over the past decade. The dynamic model of the system is highly non-linear which greatly complicates controller design and development [1]. To address these problems, research efforts have evolved adopting fuzzy logic and neural networks as alternative methods to conventional mathematical model based control [1-4]. Most of the fuzzy logic controllers (FLC) reported for flexible link manipulator are Mamdani-type rule-based controllers [1, 2, 4, 8]. These controllers require processing of the rule-base consisting of $n \times m$ rules, where n and m are the number of primary fuzzy sets of the inputs. In most existing applications, the fuzzy rules are generated by an expert in the area, especially for the control problems with only a few inputs. With an increasing number of inputs and linguistic variables, the possible number of rules for the system increases exponentially, which makes it difficult for experts to define a complete set of rules [8]. Another well-known problem with fuzzy rule-based systems is that the processing of such a rule-base is time consuming. Consequently, most of the time is spent in calculating the control output using center of gravity method of defuzzification, which in some applications can degrade system response, especially in the case of a flexible-link manipulator. The problem of such defuzzification methods has been eliminated by the use of Sugeno-type fuzzy systems, where each consequent fuzzy set is replaced by a linear function [5]. This imposes a further set of consequence parameters to be estimated. Current neuron-fuzzy systems are mainly Sugeno-type fuzzy systems with the rule-base replaced by a neural network. The most widely used neuro-fuzzy system is the adaptive neuro-fuzzy inference system (ANFIS) proposed by Roger Jang [6]. The learning of the parameters of the neuro-fuzzy system requires a set of input-output data and uses several passes of backpropagation and/or LMS algorithm to estimate the antecedent parameters of MFs and consequent parameters of the system [5-6]. The problem is now how to cope with developing an FLC where *a priori* information such as a set of input-output data or expert knowledge for constructing rule-base is not available, partially available or very poorly structured.

The objective of this research lies in constructing a rule-base represented by a backpropagation neural network (BNN) and training of the BNN from information granules such as linguistic data and/or description. The developed controller is then applied to a flexible-link manipulator to verify the performance of the methodology.

2 Experimental Rig

The experimental rig constituting the flexible manipulator system consists of two main parts: a flexible arm and measuring devices. The flexible arm contains a flexible link driven by a printed armature motor at the hub. The measuring devices are shaft encoder, tachometer, accelerometer and strain gauges along the length of the arm. The shaft encoder, tachometer and accelerometer are essentially utilised in this work. The experimental flexible-link manipulator is shown in Figure 1. The flexible arm consists of an aluminium-type beam. The outputs of the sensors as well as a voltage proportional to the current applied to the motor are fed to a computer through a signal conditioning circuit and an anti-aliasing filter for analysis and calculation of the control signal. Physical parameters of the flexible arm are given in Table 1.

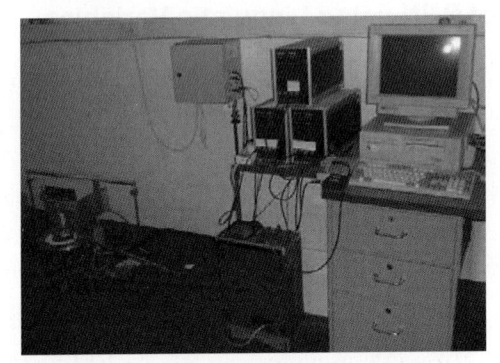

Fig. 1. Flexible manipulator system.

Table 1. Physical parameters of flexible manipulator system.

Parameter	Value
Length	960.0 mm
Width	19.008 mm
Thickness	3.2004 mm
Mass density/ unit volume	2710 kgm^{-3}

3 BNN-based Fuzzy Controller

The basic principle of the BNN-based fuzzy logic controller is that for a given rule-base a functional mapping from the fuzzy logic-based controller to the network-based approach can be established [7]. In other words, the inference mechanism of a rule-based FLC is implemented by a BNN. A block diagram of the proposed BNN-based fuzzy control system is shown in Figure 2. All linguistic variables of both antecedent and consequent part are translated into fuzzy sets described by membership functions and the fuzzy sets are then converted into fuzzy numbers or numeric values. The BNN network is trained off-line by presenting all rules sequentially to the network. Once the BNN is trained, the network is then inserted in the control loop for on-line operation. The inputs come directly from measured inputs, fuzzified into fuzzy numbers, presented to the network and the control input is calculated using defuzification.

3.1 Training of BNN using Linguistic Variables

The rule-base of a Mamdani-type fuzzy logic controller with two inputs and a single output consisting of the following rules is used to train the BNN.

$$\text{IF } X_1 \text{ is } A_i \text{ AND } X_2 \text{ is } B_j \text{ THEN } U \text{ is } C_k \tag{1}$$

where $X_1 \triangleq$ error, $X_2 \triangleq$ change of error, $U \triangleq$ control input, A_i, $i = 1, 2, \cdots, N$, B_j, $j = 1, 2, \cdots, M$ are the input linguistic variables and C_k, $k = 1, 2, \cdots, L$ are the output linguistic variables. N, M and L are the maximum number of linguistic variables (primary fuzzy sets).

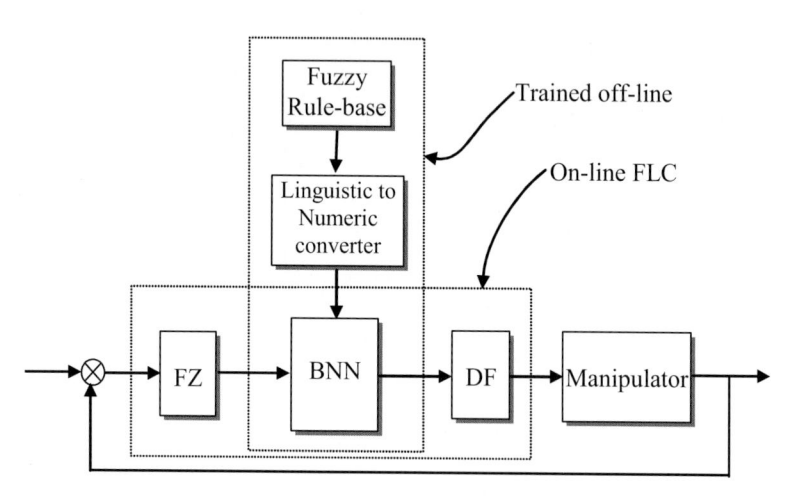

Fig. 2. Block diagram of the BNN-based FLC.

The training procedure of the BNN, using these fuzzy numbers, is shown in Figure 3. Training the BNN is equivalent to rule-base construction in a Mamdani-type fuzzy controller, where the user has to decide the number of fuzzy sets (linguistic variables) for each input and output. The rule-base of such Mamdani-type fuzzy controller, which is to be learnt, is shown in Table1, where $C_r \in \{C_1, C_2, \cdots, C_L\}$. The neural network will, eventually, learn the rule-base with fixed predefined linguistic variables defined by membership function.

Table 2. Rule-base of FLC.

	U	B_1	B_2	...	B_M
Input X_1	A_1	C_r	C_r	...	C_r
	A_2	C_r	C_r	...	C_r
	⋮	⋮	⋮	⋱	⋮
	A_N	C_r	C_r	...	C_r

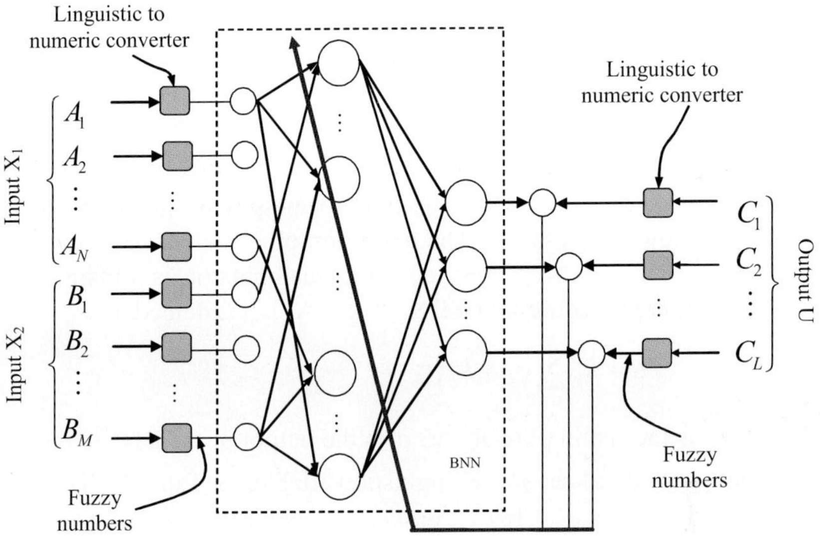

Fig. 3. Training of BNN using linguistic description.

3.2 Linguistic to Numeric Converter

A linguistic variable, such as large, medium or small, is typically a fuzzy set described by a membership function. The linguistis variables in (1) is defined by the following triangular MF

$$\mu_i(A_i) = \max\left(\min\left(\frac{x_1 - a_i}{m_i - a_i}, \frac{b_i - x_1}{b_i - m_i} \right), 0 \right) \tag{2}$$

Where m denotes the central value, $\{a,b\}$ denotes left and right values of a triangular linguistic variable. The fuzzy numbers are typically characterised by a central value with an interval around the center [9]. The width of the associated interval determines the degree of fuzziness [10]. The mapping of linguistic variable to triangular fuzzy number is shown in Figure 4. $\Gamma\{a_i, m_i, b_i\}$ is the converter function, which convert linguistic variables into fuzzy numbers.

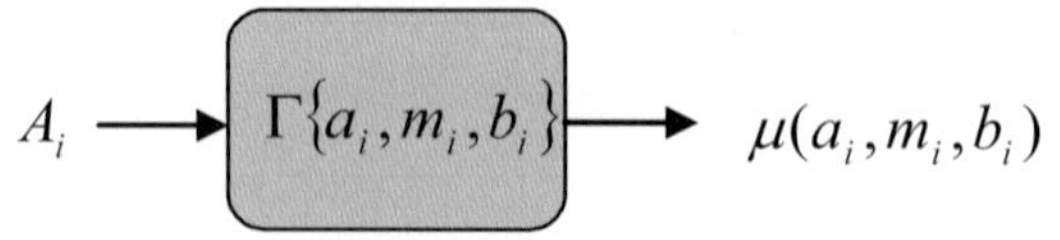

$$A_i \longrightarrow \boxed{\Gamma\{a_i, m_i, b_i\}} \longrightarrow \mu(a_i, m_i, b_i)$$

Fig. 4. Linguistic to numeric converter

4 Experimental Results

The BNN-based fuzzy controller architecture with two inputs, hub angle error (e) and change of error (Δe), and one output, torque input (u), thus utilized is shown in Figure 5. The definition of e and Δe is obvious from the Figure. The control input u to the manipulator is obtained using defuzzification procedure in the BNN-FLC, which is deined as

$$u = k_c \sum_{i=1}^{L} c_i = k_c (c_1 + c_2 + \cdots + c_L) \quad (3)$$

Where k_c is the scaling factor and c_i is the output of the BNN comparable to predefuzzified values of the linguistic variables in Table 2. The scaling factor k_c can be broken down into components $k_c = [k_{c1} \ k_{c2} \ \cdots \ k_{cL}]$ and equation (3) is rewritten as

$$u = [k_{c1} \ k_{c2} \ \cdots \ k_{cL}] \begin{bmatrix} c_1 \\ c_2 \\ \vdots \\ c_L \end{bmatrix} = \sum_{i=1}^{L} k_{ci} \cdot c_i \quad (4)$$

From equations (3)-(4), it is established that the tuning of the scaling factor k_c is equivalent to tuning the vector of scaling factors $[k_{c1} \ k_{c2} \ \cdots \ k_{cL}]$.

Five linguistic variables were chosen for each inputs and output defined within the universe of discourse [-36, +36] degree, [-25, +25] and [-3, +3] volts for the hub angle error, change in hub angle error and torque input respectively. Using the central values of the linguistic variables, a BNN with

10 inputs and 5 outputs was trained. The BNN with 11 hidden neurons was able to successfully learn the rule-base. The response of the manipulator for a demanded hub-angle of $36°$ with $k_c = 76$ is shown in Figure 6.

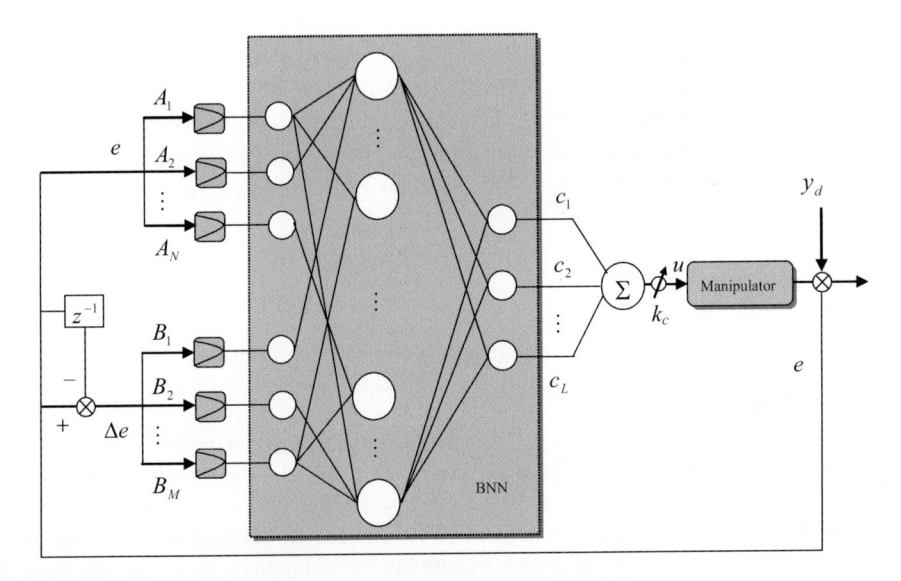

Fig. 5. BNN-based FLC in operation.

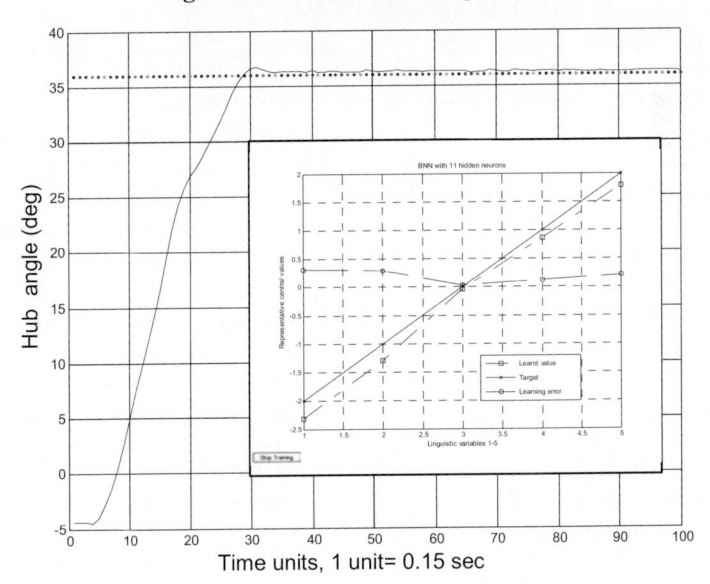

Fig. 6. Response of the manipulator by BNN-based FLC.

The training of the rule-base using representative central values of the linguistic variables is shown in the inset of Figure 6.

5 Conclusion

A procedure is developed to learn the rule-base of an FLC using a BNN, which is trained by linguistic description. Though the experimentation showed very promising results but the system is still lacking some formal methods of pruning the BNN, which will be addressed in a future research.

References

1. V.G. Moudgal, W.A. Kwong, K.M. Passino, and S. Yurkovich (1995). "Fuzzy learning control of a flexible-link robot"; *IEEE Transaction on Fuzzy Systems;* vol.3, No.2, 1995, pp. 199-210.
2. J.X. Lee, G. Vukovich, and J.Z. Sasaidek, (1994). "Fuzzy control of a flexible link manipulator"; *Proceeding of American Control Conference*; Baltimore, Maryland, June 1994, pp.568-574; USA.
3. H.A. Talebi, K. Khorasani, and R.V. Patel, (1998). "Neural network based control schemes for flexible-link manipulators: simulations and experiments", *Neural Networks*, vol.11, 1998, pp. 1357-1377.
4. M.O Tokhi, M.N.H. Siddique and M.S. Alam (2004). "Neuro-Fuzzy Control of Flexible-Link Manipulators", *Proceedings of IEEE SMC UK-RI Chapter Conference on Intelligent Cybernetic Systems* (ICS'04), September 7-8, University of Ulster, Londonderry, UK, ISSN 1744-9189, pp.108-113.
5. T. Takagi and M. Sugeno (1985). "Fuzzy identification of systems and its applications to modeling and control", *IEEE Transaction on System, Man and Cybernetics*, Vol. 15, 1985, pp. 116 -132.
6. J.-S. Roger Jang, (1993). "ANFIS: Adaptive-network-based fuzzy inference system", *IEEE Transaction on Systems, Man and Cybernetics*, Vol. 23, No. 3, 1993, pp. 665-685.
7. D.A. Linkens and J Nie (1994). "Backpropagation Neural Network based Fuzzy Controller with a Self-learning Teacher", *International Journal of Control*, vol. 60, No. 1, 1994, pp. 17-39.
8. M.N.H. Siddique, (2002). "Intelligent Control of Flexible-link Manipulator Systems", PhD Thesis, Department of Automatic Control and Systems Engineering, The University of Sheffield, England, UK.
9. D. Dubois, and H. Prade, (1987). "Fuzzy Numbers: An Overview", in Analysis of Fuzzy Information 1, pp.3-39, *CRC Press*, Boca Raton.
10. D. Dubois, and H. Prade, (1987). "Mean Value of a Fuzzy Number", *Fuzzy Sets and Systems* 24(3), pp. 279-300.

Design Constraints in Implementing Real-time Algorithms for a Flexible Manipulator System

M.A. Hossain[1], M.N.H. Siddique[2], M.O. Tokhi[3] and M.S. Alam[4]

[1]Department of Computing, University of Bradford, [2]School of Computing and Intelligent Systems, University of Ulster, [3,4]Department of Automatic Control and Systems Engineering, University of Sheffield,
m.a.hossain1@bradford.ac.uk, nh.siddique@ulster.ac.uk, o.tokhi@sheffield.ac.uk, cop03msa@sheffield.ac.uk

Abstract

This paper presents an investigation into the design constraints of the algorithm of a flexible manipulator system for real-time implementation. A dynamic simulation algorithm of a single link manipulator system using finite difference (FD) method is considered to demonstrate the critical real-time design and implementation issues. The simulation algorithm is analyzed, designed in various forms and implemented to explore the impact. Finally, a comparative real-time computing performance of various forms of the algorithms is presented and discussed to demonstrate the merits of different design mechanisms through a set of experiments.

Keywords: Algorithm design, real-time computing, flexible manipulator, finite difference simulation.

1 Introduction

It is reported earlier [1], [2] that the ideal performance of a computer system demands a perfect match between machine capability and program behaviour. Program performance is the turnaround time, which includes disk and memory accesses, input and output activities, compilation time, operating system overhead, and central processing unit (CPU) time. In order to shorten the turnaround time, one can reduce all these time factors. Minimising the run-time memory management, efficient partitioning and mapping of the program, and selecting an efficient compiler for specific computational demands, could enhance the performance. Compilers have a significant impact on the performance of the system [3]. This means that some high-level languages have advantages in certain computational domains, and some have advantages in other domains. The compiler itself is critical to the performance of the system as the mechanism and efficiency of taking a high-level description of the application and transforming it into a hardware dependent implementation differs from compiler to compiler [4].

The performance demand in modern real-time signal processing and control applications has motivated the development of advanced special-purpose and general-purpose hardware architectures. However, the developments within the software domain have not been at the same pace and/or level as within the hardware domain. Thus, although advanced computing hardware with significant levels of capability is available in the market, these capabilities are not fully utilised and exploited at the software level. Efficient software coding is essential in order to exploit the special hardware features and avoid associated shortcomings of the architecture. There has been a substantial amount of effort devoted to this area of research over the last decade [5], [6].

This paper presents an investigation into the analysis and design mechanisms that will lead to reduction of the execution time in implementing real-time algorithms. The proposed mechanisms are exemplified by means of one algorithm, which demonstrates the applicability of these mechanisms to real-time applications. A simulation algorithm characterising the dynamic behaviour of a flexible manipulator system, developed using the finite difference (FD) method, is considered to demonstrate the effectiveness of the proposed methods. A comparative performance evaluation of the proposed design mechanisms is presented and discussed through a set of experiments.

2 The Flexible Manipulator System

A schematic representation of a single-link flexible manipulator is shown in Figure 1. A control torque τ is applied at the pinned end (hub) of the arm by an actuator motor. θ represents the hub angle, POQ is the original co-ordinate system (stationary coordinate) while $P'OQ'$ is the co-ordinate system after an angular rotation θ (moving coordinate). I_h is the inertia at the hub, I_p is the inertia associated with a payload M_p at the end-point and u is the flexible displacement (deflection) of a point at a distance x from the hub. The dynamic equation of the flexible manipulator, considered as an Euler-Bernoulli beam equation, can be expressed as [7]:

$$\rho \frac{\partial^2 y(x,t)}{\partial t^2} + EI \frac{\partial^4 y(x,t)}{\partial x^4} = \tau(x,t) \quad (1)$$

where, $y(x,t)$ is the displacement (deflection) of the manipulator at a distance x from the hub at time t, ρ is the density per unit length of the manipulator material, E is Young modulus, I is the second moment of inertia, $\tau(x,t)$ is the applied torque. The product EI represents the flexural rigidity of the manipulator.

Discretising the manipulator in time and length using central finite difference methods, a discrete approximation to equation (1) can be obtained that can be stated in matrix notation as [8]:

$$\mathbf{Y}_{i,j+1} = \mathbf{A}\mathbf{Y}_{i,j} + \mathbf{B}\mathbf{Y}_{i,j-1} + \mathbf{CF} \quad (2)$$

where $\mathbf{Y}_{i,j+1}$ is the displacement of grid points $i = 1,2,\cdots,n$ of the manipulator at time step $j+1$, $\mathbf{Y}_{i,j}$ and $\mathbf{Y}_{i,j-1}$ are the corresponding displacements at time steps j and $j-1$ respectively. $\mathbf{A}$ and $\mathbf{B}$ are constant $n \times n$ matrices whose entries depend on the flexible manipulator specification and the number of sections the manipulator is divided into, $\mathbf{C}$ is a constant matrix related to the given input torque and $\mathbf{F}$ is an $n \times 1$ matrix related to the time step Δt and mass per unit length of the flexible manipulator.

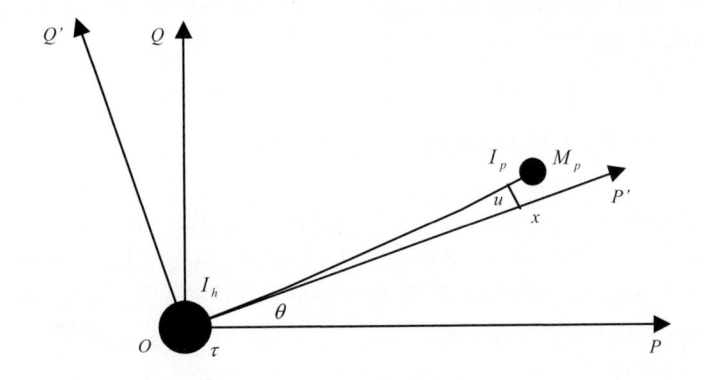

Fig. 1. Schematic representation of the flexible manipulator system

3 Algorithm Design

The FD simulation algorithm of the manipulator system as shown in equation 2 is considered to demonstrate the design impact in implementing the algorithm in real-time. The algorithm is designed in three different methods. These are briefly described below.

Algorithm–1: The 'Algorithm–1' is listed in Figure 2. It is noted that complex matrix calculations are performed within an array of three elements each representing information about the beam position at different instants of time. Subsequent to calculations, the memory pointer is shifted to the previous pointer in respect of time before the next iteration. This technique of shifting the pointer does not contribute to the calculation efforts and is thus a program overhead. Other algorithms were deployed to address this issue at further levels of investigation.

Fig. 2. Algorithm-1

```
// Calculate the common parameters
loop {
//Step 1 :
y0[2]= (A[1][1]*y0[1]+ A[1][2]*y1[1] + A[1][3]*y2[1])- (B[1][1]*y0[0]+
B[1][2]*y1[0]) + tau[j]*C;
y1[2]= (A[2][1]*y0[1]+ A[2][2]*y1[1] + A[2][3]*y2[1] + A[2][4]*y3[1])-
(B[2][1]*y0[0]+B[2][2]*y1[0]+ B[2][3]*y2[0]);
// --------------------------- continue
y18[2]=(A[19][17]*y16[1]+A[19][18]*y17[1]+A[19][19]*y18[1]+A[19][20]*y19
[1])-(B[19][18]*y17[0]+B[19][19]*y18[0]+ B[19][20]*y19[0]);
y19[2]= (A[20][18]*y17[1]+ A[20][19]*y18[1] + A[20][20]*y19[1])-
B[20][20]*y19[0];
//Step 2 : Shifting memory locations
y0[0]=y0[1]; y0[1]=y0[2]; y1[0]=y1[1]; y1[1]=y1[2];
// -------- continue
y18[0]=y18[1]; y18[1]=y18[2]; y19[0]=y19[1]; y19[1]=y19[2];
}
```

Algorithm–2: The 'Algorithm–2' is listed in Figure 3. The 'Algorithm–1' is rewritten here with reduced instruction set form. Instead of writing an equation for each segment, this algorithm contains a loop for the middle sixteen segments. Moreover, the memory shifting operation also written in reduced form to demonstrate the merit of the reduced instruction set algorithm. However, it is worth noting that this design method requires more memory allocation as compared to the 'Algorithm-1'.

Fig. 3. Algorithm-2

//Calculate common parameters
//Step 1
Loop {
y[0][2]= (A[1][1]*y[0][1]+ A[1][2]*y[1][1] + A[1][3]*y[2][1])- (B[1][1]*y[0][0]+ B[1][2]*y[1][0]) + tau[j]*C;
y[1][2]= (A[2][1]*y[0][1]+ A[2][2]*y[1][1]+A[2][3]*y[2][1]+A[2][4]*y[3][1])- (B[2][1]*y[0][0]+B[2][2]*y[1][0] + B[2][3]*y[2][0]);
 Loop{
y[i][2]=(A[i+1][i-1]*y[i-2][1]+A[i+1][3]*y[i][1]+A[i+1][i+2]*y[i+1][i]
+A[i+1][i+3]*y[i+2][i])-(B[i+1][i]*y[i-1][0]+B[i+1][i+1]*y[i][0]
+B[i+1][i+2]*y[i+1][0]);
}
y[18][2]=(A[19][17]*y[16][1]+A[19][18]*y[17][1]+A[19][19]*y[18][1]+A[19][20] *y[19][1])- (B[19][18]*y[17][0] + B[19][19]*y[18][0]+ B[19][20]*y[19][0]);
y[19][2]= (A[20][18]*y[17][1]+ A[20][19]*y[18][1] + A[20][20]*y[19][1])- B[20][20]*y[19][0];
// Step 2 : Shifting memory locations
Loop{ y[i][0]=y[i][1]; y[i][1]=y[i][2]; y[i+1][0]=y[i+1][1]; y[i+1][1]=y[i+1][2];}}

Algorithm–3: A listing of the 'Algorithm-3' is given in Figure 4. In this case, each loop calculates three sets of data. Instead of shifting the data of the memory pointer (that contains results) at the end of each loop, the most current data is directly recalculated and written into the memory pointer that contains the older set of data. Therefore, re-ordering of array in the 'Simulation Algorithm–1' is replaced by recalculation. The main objective of the design effort is to achieve better performance by reducing the dynamic memory allocation and, in turn, memory pointer shift operation. Thus, instead of using a single code block and data-shifting portion, as in 'Simulation Algorithm–1', to calculate the deflection, three code blocks, are used with the modified approach in 'Algorithm–3'. It is worth noting that in 'Algorithm–3', the overhead of 'Algorithm-1' due to memory pointer shift operation is eliminated and every line of code is directed towards the simulation effort.

Fig. 4. Algorithm-3

//Calculate common parameters
//Step 1 : Loop {
y0[2]= (A[1][1]*y0[1]+ A[1][2]*y1[1] + A[1][3]*y2[1])- (B[1][1]*y0[0]+ B[1][2]*y1[0]) + tau[j]*C;
// --------------------------- continue as step one of Fig.2

y19[2]= (A[20][18]*y17[1]+ A[20][19]*y18[1] + A[20][20]*y19[1])-
B[20][20]*y19[0];
//Step 2 : Calculate y0[0]
y0[0]= (A[1][1]*y0[2]+ A[1][2]*y1[2] + A[1][3]*y2[2])- (B[1][1]*y0[1]+
B[1][2]*y1[1]) + tau[j]*C;
y1[0]= (A[2][1]*y0[2]+ A[2][2]*y1[2] + A[2][3]*y2[2] + A[2][4]*y3[1])-
(B[2][1]*y0[1]+B[2][2]*y1[1]+ B[2][3]*y2[1]);
//---------------------------- Continue
y18[0]= (A[19][17]*y16[2]+ A[19][18]*y17[2] + A[19][19]*y18[2] +
A[19][20]*y19[1])- (B[19][18]*y17[0]+ B[19][19]*y18[1]+ B[19][20]*y19[1]);
y19[0]= (A[20][18]*y17[2]+ A[20][19]*y18[2] + A[20][20]*y19[2])-
B[20][20]*y19[1];
// Step 3 : Calculate y0[1]
y0[1]=(A[1][1]*y0[0]+ A[1][2]*y1[0] + A[1][3]*y2[0])- (B[1][1]*y0[2]+
B[1][2]*y1[2]) + tau[j]*C;
y1[1]=(A[2][1]*y0[0]+A[2][2]*y1[0]+A[2][3]*y2[0]+A[2][4]*y3[0])-
(B[2][1]*y0[2]+B[2][2]*y1[2]+ B[2][3]*y2[2]);
// -------------------------------- continue
y18[1]= (A[19][17]*y16[0]+ A[19][18]*y17[0] + A[19][19]*y18[0] +
A[19][20]*y19[0])-(B[19][18]*y17[2]+ B[19][19]*y18[2]+ B[19][20]*y19[2]);
 y19[1]= (A[20][18]*y17[0]+ A[20][19]*y18[0] + A[20][20]*y19[0])-
B[20][20]*y19[2]; }

---

4 Experiments and Results

The algorithms for the manipulator system based on three different design methods were implemented for similar specification. An AMD-K6(TM) 3D processor based PC is used as a computing domain. Turbo C++ version 3.0 compiler is used to implement the simulation environment. It is worth noting that a fixed number of segments for various iterations were considered in implementing all the algorithms for the sake of consistence. Moreover, the sampling time *0.000217904 sec* is considered for each iteration, therefore, required real-time performance of *5000* iterations, as an example, should be *5000X0.000217904 = 1.0895 sec*.

Figure 5 depicts the comparative performance of the three algorithms for 20 segments. It is noted that the execution time for the algorithms increases almost linearly with the increment of iterations. It is also noted that the Algorithm-2 performs best among the algorithms. In contrast, Algorithm-3 performs waste among the three algorithms. It is also observed that Algorithm-1 and Algorithm-3 have not achieved required performance to implement in real-time.

Table 1 demonstrates the relative performance of the three algorithms. It is worth mentioning that the performance in Table 1 is presented as ratio of the Algorithm-1 and Algorithm-3 relative to the Algorithm-2. It is observed that the Algorithm-3 is about 13 to 56 times slower depending on number of iterations, as compared to the Algorithm-2 for various iterations. On the other hand, the Algorithm-1 is about 5 to 22.5 times slower as compared to the Algorithm-2. In general, execution time of the Algorithm-3 is more then double as compared to the Algorithm-1. Thus, the design mechanism employed in Algorithm–2 can offer potential advantages for real-time implementation.

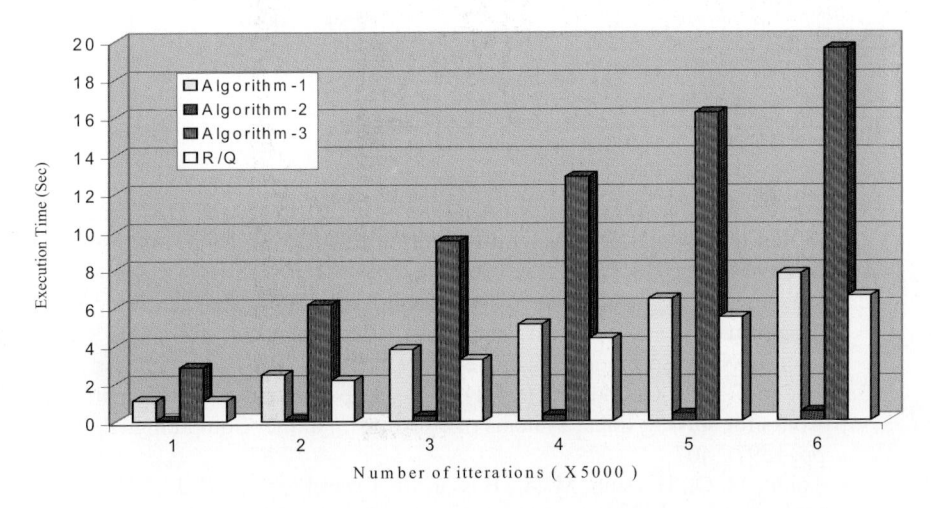

Fig. 5. Performance comparison of the various forms of the Algorithms and real-time requirement (R/Q)

5 Conclusion

This paper has presented an investigation into the algorithms analysis, design, software coding and implementation so as to reduce the execution time and, in turn, enhance the real-time performance. Three different design approaches for real-time implementation have been proposed and demonstrated experimentally. It has been observed that the execution time and in turn, performance of the algorithms varies with different approaches in a real-time implementation context. It is also noted that only Algorithm-2 with reduced instruction sets has achieved real-time performance for various number of iterations. Although, the other two algorithms can also be implemented in real-time by using high performance computer domain However, identification of the suitability of Algorithm design and implementation mechanism for best performance is a challenge, in particular for a specific architecture.

Table. 1 Performance of the Algorithm-1 (Alg-1) and Algorith-3 (Alg-3) relative to the Algorithm-2(Alg-2)

Iterations	20000	30000	50000	10000	15000	20000	25000	30000
Alg-1/Alg-2	5.0	11.0	20	22.5	13.8	15.5	16.71	15.67
Alg-3/Alg-2	12.9	26	52	55.99	33.59	38.99	42.42	39.56

References

1. Tokhi, M. O. and Hossain, M. A. (1995), "CISC, RISC and DSP processors in real-time signal processing and control", Journal of Microprocessors and Microsystems, **19**(5), UK. pp. 291-300.
2. Hossain, M. A., Tokhi, M. O. and Dahal, K. P, (2004), Impact of algorithm design in implementing real-time active control systems, Computer Science Lecture Note, Springer Verlag, Germany, pp: 247-253
3. Bader, G. and Gehrke, E. (1991), "On the performance of transputer networks for solving linear systems of equation", Parallel Computing, 1991, **17**, pp. 1397-1407.
4. Tokhi, M. O., Hossain, M. A., Baxter, M. J. and Fleming, P. J. (1995), "Heterogeneous and homogeneous parallel architectures for real-time active vibration control", IEE Proceedings-D: Control Theory and Applications, **142**, (6), pp. 1-8.
5. Clader, B., Krintz, C., John, S. and Austin, T. (1998), "Cache -concious data placement", Proceedings of Eighth International Conference on Architectural Support for Programming Languages and Operating Systems (ASPLOS'98), San Jose, Canada, pp. 139-149.
6. Kabir, U., Hossain, M. A. and Tokhi, M. O. (2000). "Reducing memory access time in real-time implementation of signal processing and control algorithms", Proceedings of AARTC00: IFAC Workshop on Algorithms and Architectures for Real-time Control, Palma de Mallorca (Spain), 15-17 May 2000, pp. 15-18.
7. Azad AKM (1994) Analysis and design of control mechanisms for flexible manipulator systems, PhD Thesis. Department of Automatic Control and Systems Engineering, The University of Sheffield, UK.
8. Porwanto, H. (1998), Dynamic simulation and control of flexible manipulator systems, PhD Thesis. Department of Automatic Control and Systems Engineering, The University of Sheffield, UK

Pay-Load Estimation of a 2 DOF Flexible Link Robot

N. K. Poulsen[1] and O. Ravn[2]

[1]Informatics and Mathmatical Modelling, Technical University of Denmark, nkp@imm.dtu.dk; [2]Ørsted•DTU, Technical University of Denmark., or@oersted.dtu.dk.

Abstract

The paper presents a new method for online identification of pay-loads for a two-link flexible robot. The method benefits from the close correspondence between parameters of a discrete-time model represented by means of the Delta-Operator, and those of the underlying continuous-time model. Although the applied principle might be general in nature, the paper is applied to the well-known problem of identifying a pay-load of a moving flexible robot. The presented method benefits from the close correspondence with the continuous-time representation to allow a scalar and implicit adaptive technique which based on flexibility measurements leads to the online estimation of the pay-load.

Keywords: Flexible Link Robot; Delta-Operator; System Identification; Parameter Estimation; Adaptive Control.

1 Introduction

Flexible robot systems are motivated by a desire for better arm-weight to pay-load ratios, shorter travel-times and lower energy consumption.
The desire for high-performance manipulators and the benefits offered by a light-weight flexible link capable of maneuvering large pay-loads have lead to analysis of the behavior of the dynamics in which flexibility is the essential issue. The high-performance requirements will inevitably pro-

duce designs that during operation will excite vibrations in the manipulator structure.

The aim of the controller is to suppress the structural vibration while in addition to minimize the cycle time of the manipulator system. In this work we will use a model-based controller in order to mitigate the first harmonics. However, changes in pay-load degrade the model and consequently the performance of the control system, unless some sort of adaptation or gain-scheduling is taken into account to estimate these effects.

In order to investigate different aspects of control of flexible links robot configurations an experimental setup has been made and a simulation model has been developed. The setup consists of two very flexible links with two actuators located in the joints. In this work the links are moving in the horizontal plane making gravity ignorable. The actuators are DC-motors with a sufficient gear ratio and tachometers making an analog velocity feedback feasible. Apart from the tachometers there are also two sensors on the setup, a potentiometer in each joint enabling a measurement of the position of the joint and a number of strain gauges located on each link enabling the measurement of the deflection of the link.

The literature contains experimental results featuring gravity compensation is presented for a double link robot with a flexible forearm. But like other references, practical algorithms seem to ignore the fact that one could construct an adaptation technique that directly gains insight to the pay-load parameter. This is topic for the present work. By deducting a linear state-space model describing the pay-load parameter's influence in the continuous-time model, it is possible to apply a Delta-Operator technique for estimation of this parameter in discrete-time.

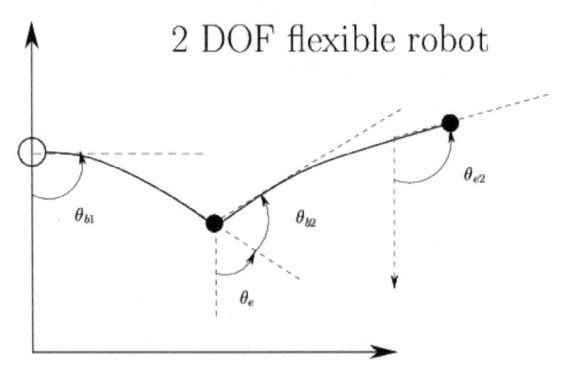

Fig. 1. The flexible robot system consists of two flexible links that are coupled together through actuator 2.

2 Design model of the system

The flexible manipulator system studied here, see Fig 1, carries a pay-load, m_p, at its tip and moves in the horizontal plane. The active degrees of freedom are the two rotational angles θ_1 and θ_2.

In literature the equations of motions are commonly modeled by either a Finite Element Method (see. e.g. [1]) or the Eigenvalue Method (see e.g. [2]). As the latter method is normally considered more accurate when only a limited number of modes are included, cf. [3], the following description is be based on this approach, cf. [4].

The model of the flexible link robot consists of four parts; namely the models for the two actuators and the two arms. The dynamics of the flexible arms can be described by a PDE which can be transferred into a ODE by using the method of separation of variable. In that case the deflection, $\omega_j(x,t)$, j=1,2, of the arms is approximated by a finite sum of contributions

$$\omega_j(x,t) = \sum_{i=1}^{n} \phi_{ji}(x) q_{ij}(t) \quad (1)$$

where $\phi_{ji}(x)$ and $\phi_{ji} q(t)$ are the normal and harmonic function of mode i and arm j, respectively.

2.2 Equations of Motion in Compact Form

If the actuator equations are used for obtaining the angular accelerations the four main equations can be written in a more compact form
Introducing the notation:

$$\underline{q} = [q_{11},...q_{1n}, q_{21},...q_{2n}]^T, u = \begin{bmatrix} u_1 \\ u_2 \end{bmatrix}, \underline{\theta} = \begin{bmatrix} \theta_1 \\ \theta_2 \end{bmatrix} \quad (2)$$

The the descripton of the flexibility can be linearized and be brought into the following compact form:

$$\underline{\ddot{q}} = M_1 \underline{q} + M_2 \underline{\dot{q}} + M_3 \underline{q} + M_4 \underline{u} + M_5 \underline{\dot{u}} \quad (3)$$

Where the matrices M_2 M_3 and M_5 are linearly dependant on m_p. Notice the matrices depend on the linearization point. In this case the matrices depend only on q_2. Also notice the angular acceleration q occurs on both sides of the i equation.

Also the actuator dynamics can be written in a compact form.

$$\dot{\theta} = M_6 \underline{q} + M_7 u \tag{4}$$

Now the compact description in (3) and (4) is to be transformed into a state space description. It is possible to establish the following linear dependencies of m_p

$$M_2 = M_2^0 + m_p M_2^m \tag{5}$$
$$M_3 = M_3^0 + m_p M_3^m$$
$$M_5 = M_5^0 + m_p M_5^m$$

while M_1, M_4 M_6 and M_7 are independent of m_p. If we define:

$$\underline{A}^0 = \begin{bmatrix} 0 & M_6 & 0 \\ 0 & 0 & I \\ 0 & \Lambda M_1 & \Lambda M_2^0 \end{bmatrix} \quad \underline{A}^m = \begin{bmatrix} 0 & 0 & 0 \\ 0 & 0 & 0 \\ 0 & 0 & \Lambda M_2^m \end{bmatrix} \tag{6}$$

$$\underline{B}^0 = \begin{bmatrix} M_7 \\ \Lambda M_5 \\ \underline{B}_3^0 \end{bmatrix} \quad \underline{B}^m = \begin{bmatrix} 0 \\ \Lambda M_5^m \\ \Lambda M_2^m \Lambda M_5^0 + \Lambda M_2^0 \Lambda M_5^m \end{bmatrix}$$

where

$$\underline{B}_3^0 = \Lambda M_2^0 \Lambda M_5^0 + m_p^2 \Lambda M_2^m \Lambda M_5^m + \Lambda M_4 \tag{7}$$

then the system is described on a state space form.

The measurement system consists of two potentiometers and four strain gauges. The potentiometers give measurements of the link angles θ_{b1} and θ_{b2}, whereas the strain gauges are located tactically on the links in order to give measurements of the deflections i.e. q.

3 Simulation experiment

In order to investigate the properties of the algorithm there have been performed a series of simulations. One simulation is described in the following.

The simulation (and the design) model of the deflection is a two mode approximation. The reference signals are square waves for the link angles and the signals are shown in Figure 2. The sampling period was chosen to T=0.01 sec.

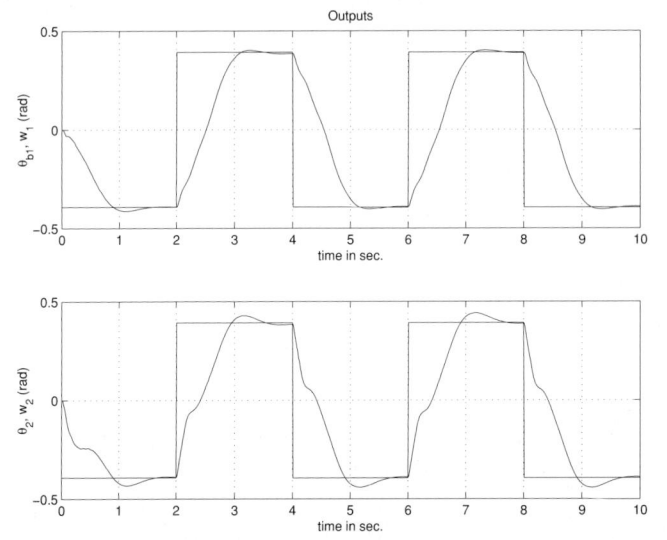

Fig. 2. Angular positions θ_{b1} and θ_{b2} presented as measured outputs (solid line) and as reference trajectories (dashed line).

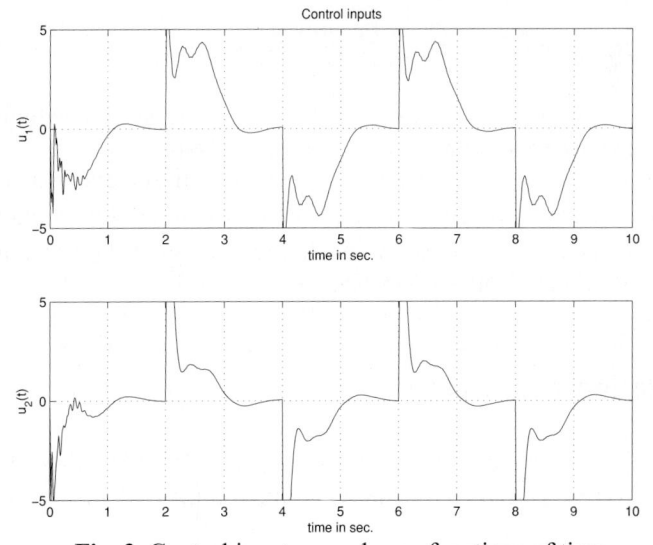

Fig. 3. Control inputs u_1 and u_2 as functions of time.

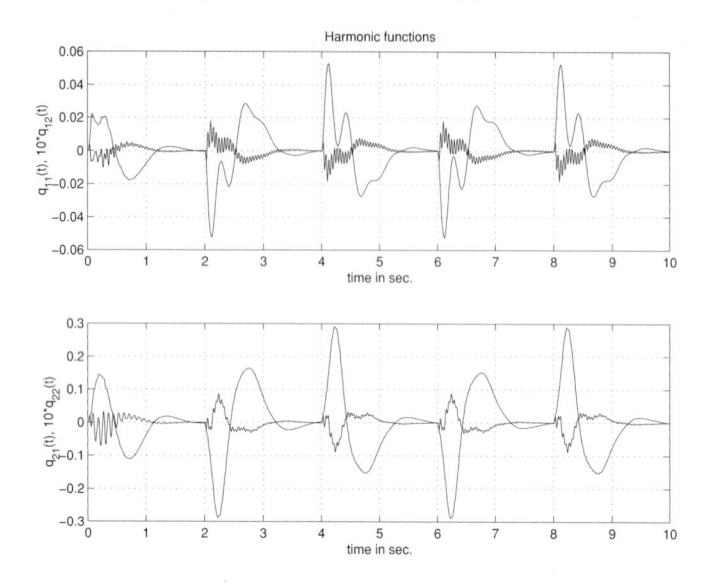

Fig. 4. Harmonic time-functions q_{i1} and $10q_{i2}$ for each of the beams, i=1,2 presented as functions of time

The controller is a state space LQG controller which parameters are iterated one step per sample in order to follow the adaptation on m_p. This would asymptotically producing the optimal feed-back gain based on the loss weights

$$Q_1 = C^T C \quad Q_2 = 0.005I \tag{8}$$

The control is based on the estimated state of the system which are obtained by means of the predictive Kalman filter with

$$R_1 = 10^{-4}I \quad R_2 = 10^{-4}I \tag{9}$$

where R_1, R_2 denote respectively the process and the measurement covariance matrices. In spite of the deterministic simulation, the process noise covariance is needed in order to compensate for the unknown pay load mass in the model.

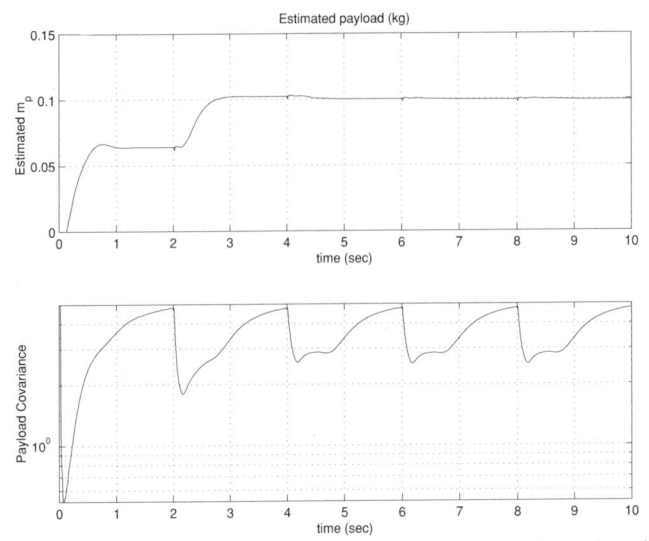

Fig. 5. Estimated pay-load $m_p(t)$ and variance of estimate in a closed-loop experiment.

The payload mass was estimated using the following parameters

$$\alpha_0=0.1 \quad \alpha_1=10 \quad P_o=10 \tag{10}$$

The value of m_p=0.1 kg can be compared to e.g. the mass of the lower arm, m_{l2} =0.133 kg, saying that the manipulator is heavily loaded. A better performance can be seen after the second step reflecting that m_p is almost estimated after 3 seconds, see Figure 5. From here it also appears that only when the set-points are varied new information is obtained.

4 Conclusions

This paper has presented a method for online identification of pay-loads for a two-link flexible robot. The method is based on a state-space model of the flexible link which has been transformed into the discrete time domain using the Delta-Operator. This enables a close correspondence between the parameters in the discrete-time model and the underlying continuous-time model. Due to the close correspondence, it is shown that both domain models can produce almost the same linearity with respect to a pay-load. This fact is used in a pay-load estimation technique. By simulation it is demonstrated that it is possible to identify a time-varying pay-load of a two link flexible robot during closed-loop control.

References

1. Sakawa, Y., Matsuno F. and Fukushima S. (1985): Modelling and control of a Flexible arm. Journal of Robotic Systems, 2, 453-472.
2. Kruise L. (1990): Modelling and control of a flexible Beam and Robot Arm. Ph.D. Thesis, University of Twente
3. Baungaard, J.R. (1996): Modelling and Control of Flexible Robot Links. Ph.D. thesis. Ph.D. thesis. Department of Automation, The Technical University of Denmark.
4. Rostgaard M (1995): Modelling, Estimation and Control of Fast Sampled Dynamic Systems. Ph.D. theses, Department of Mathematical Modelling, The Technical University of Denmark
5. Caspersen, Morten Keller (2000): Control of a Flexible Link Robot. Master Thesis, Department of Automation, The Technical University of Denmark.
6. Luca, A.D. and Panzieri, S. (1994): An iterative scheme for learning gravity compensations in flexible robot arms. *Automatica*, **30**(6), 993-1002.
7. M'Saad, M, Dugard, L. and Hammand, S. (1993): A suitable generalized predictive adaptive controller case study of a flexible arm. *Automatica*, **29**(3), 589-608.
8. Ravn, O and Poulsen, N.K. (2004): Analysis and design environment for flexible manipulators. Chapter 19 in Tokhi, M.O. and Azad, A.K.M.: Flexible robot manipulators – modeling, simulation and control.
9. Rostgaard ,M, Poulsen, N.K. and Ravn, Ole (2001): Pay-load estimation of a 2DOF flexible link robot using a delta operator technique. IEEE Conference on Control Application, Mexico, (248-253).
10. Timoshenko, S., Young, D.H. and Weaver, J.W. (1974). Vibration problems in Engineering. John Wiley and Sons.

Design of Hybrid Learning Control for Flexible Manipulators: a Multi-objective Optimisation Approach

M. S. Alam, M. Z. Md Zain, M. O. Tokhi and F. Aldebrez

Department of Automatic Control and Systems Engineering, The University of Sheffield, UK.   cop03msa@sheffield.ac.uk

Abstract

This paper presents investigations at development of a design approach of a hybrid iterative learning control scheme for flexible robot manipulators using the multi-objective genetic algorithm (MOGA) approach. A single-link flexible manipulator system is considered in this work. This is a high order, nonlinear and single-input multi-output system with infinite number of modes each with associated damping ratios. Moreover, rise time, overshoot, settling time and end-point vibration are always in conflict in the flexible manipulator since the faster the motion, the larger the level of vibration. A collocated proportional-derivative (PD) controller utilising hub-angle and hub-velocity feedback is developed to control rigid-body motion of the system. This is then extended to incorporate iterative learning control with acceleration feedback to reduce the end-point acceleration of the system. The system performance largely depends on suitable selection of controller parameters. Single objective optimisation techniques can hardly provide good solution in such cases. Multi-objective GAs with fitness sharing technique is used to find optimal set of solutions for iterative learning control parameters, which trade off between these conflicting objectives. The performance of the hybrid learning control scheme is assessed in terms of time-domain specifications and level of vibration reduction at resonance modes.

Keywords: Flexible manipulator, Iterative learning control, Multi-objective GA, Pareto optimal set.

1 Introduction

Flexible robotic manipulators pose various challenges in research as compared to rigid robotic manipulators, ranging from system design, structural optimization, and control. In order to achieve high-speed and accurate positioning, it is necessary to control the manipulator's vibratory response in a cost effective manner. A number of methods have been attempted to improve system response in terms of overshoot, rise time, settling time and vibration at the end-point which are often in conflict. [1-3, 11, 12]. The objective of this work is to devise optimum solutions for a hybrid learning control which trades-off among these conflicting features. Multi-objective genetic algorithm (MOGA) with fitness sharing technique was employed for this purpose.

2 The flexible manipulator system

The single-link flexible manipulator system considered in this work is shown in **Fig. 1**, where X_oOY_o and XOY represent the stationary and moving co-ordinates respectively, τ represents the applied torque at the hub. E, I, ρ, V, I_H and M_P represent the Young modulus, area moment of inertia, mass density per unit volume, cross-sectional area, hub inertia and payload of the manipulator respectively. In this work, the motion of the manipulator is confined to the X_oOY_o plane. In this study, an aluminium type flexible manipulator of dimensions $900 \times 19.008 \times 3.2004$ mm^3, $E = 71 \times 10^9$ N/m^2, $I = 5.253 \times 10^{-11}$ m^4, $\rho = 2710$ kg/m^3 and $I_H = 5.8598 \times 10^{-4}$ kgm^2 is considered [1].

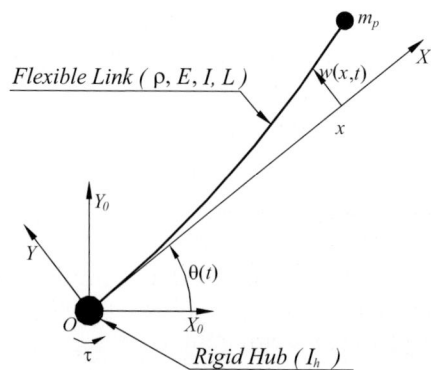

Fig. 1. Schematic representation of the single-link flexible manipulator.

3 Control schemes

A collocated proportional-derivative (PD) controller utilising hub-angle and hub-velocity feedback is developed to control the rigid-body motion of the system. To reduce the end point acceleration of the system, a PD type iterative learning control (ILC) with acceleration feedback is incorporated. Details of this control strategy can be found in [11]. The schematic diagram of this hybrid learning control with MOGA is shown in **Fig. 2**.

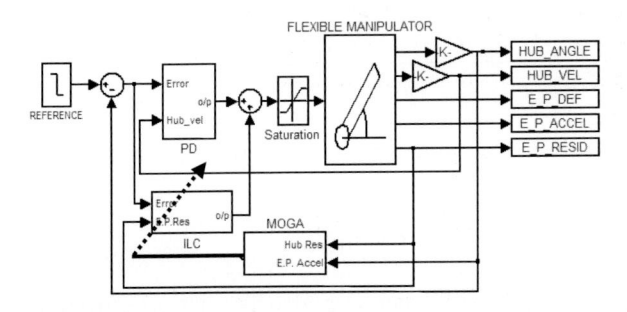

Fig. 2. Schematic representation of the hybrid learning control using MOGA

4 The multi-objective genetic algorithm

GAs are suitable for solving multi-objective problems, where objectives are often in conflict. GAs can search for multiple solutions in parallel, producing a family of non-dominated solutions to a problem known as the Pareto-optimal set [4, 5, 8, 9]. The Pareto front yields a set of candidate solutions, from which the desired one can be picked up under different trade-off conditions. A MOGA algorithm employing rank-based fitness sharing technique [5-7] is discussed in this paper.

4.1 Initialisation and evaluation

A randomly selected population is generated within a specific range. Each individual of the population is evaluated by the objective functions. Each individual is then ranked according to their degree of dominance [5, 6]. Individuals on the Pareto front have a ranking of one, as they are non-dominated. An individual's ranking equals the number of individuals that it dominates by plus one (see **Fig. 3**).

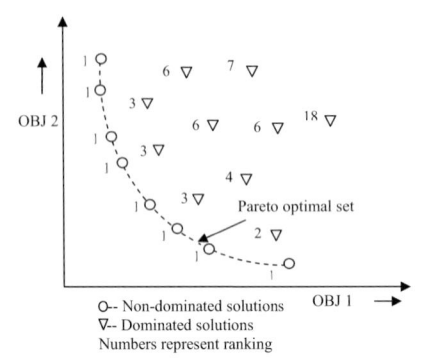

Fig. 3. Dominated and non-dominated solutions with rank values

4.2 Fitness assignment and sharing

Fitness is understood here as the number of offspring an individual is expected to produce through selection. The procedure is as follows [5, 6]:
a. Sort population according to ranking
b. Assign fitness by interpolating from the best individual to the worst according to some function, in the form of fitness function, such as linear or exponential, possibly other types. The lower rank of the individual is the smaller the fitness of the individual.
c. Average the fitness assigned to individuals with the same rank, so that all of them are sampled at the same rate while keeping the global population fitness constant.

Genetic diversity of a population can be lost due to the stochastic selection pressure. Fitness sharing based on niching method can overcome it. The basic idea of fitness sharing is that all the individuals within the same region (called a niche) share their fitness. Therefore, individuals in over-populated regions will experience a greater fitness decrease than isolated individuals. Fitness sharing may be genotypic or phenotypic. Genotypic sharing uses the distance between the chromosomes (e.g. the Hamming distance). Phenotypic sharing uses the distance between the phenotypes. Goldberg suggested a new fitness function based on a ranking process [8, 9]. A non-dominated sorting based fitness sharing technique has been used in this work. Here share counts are computed based on individual distance in the objective domain, but only between individuals with the same rank. Detail of this method can be found in [5, 6].

The stochastic universal sampling method is used to select the best individuals [9]. However, mating restrictions are employed in order to protect lethals [4, 9]. GA operators, namely crossover and mutation are employed on the selected individuals to form the next generation [9].

5 Implementations

The initial population consists of 50 individuals. Each individual consists of two randomly generated binary strings each of 20 bits, called chromosome which are converted into real values within a defined range that ensures stability of the closed loop system. The two values of each row, termed as Kp and Kd, are used to form the PD type ILC as shown in **Fig. 2**. The output of the ILC is added with acceleration feedback taken from the end-point. This is then added to the output of a collocated PD controller to form the final actuating signal which is applied to the flexible manipulator. A dynamic model of the flexible manipulator is derived using the finite element (FE) method. The whole experiment is carried out in the Simulink environment because it allows for simple construction of the hybrid control system with discrete digital filters and saturation components. Filters are used with all the three outputs with the aim to filter out flexible motion, keeping the rigid-body motion intact. The filtered output of the end-point acceleration is considered as residue error. The system response parameters, namely overshoot, rise time, settling time and steady-state error are calculated from the hub-angle response. Rise time and end-point acceleration are considered as two conflicting objectives to find the Pareto optimal set for the hybrid controller. The overshoot and settling time are later incorporated to form a four objective optimisation problem. Here the goals of all objectives were set to zero and assigned the same priority [6, 7]. The crossover rate and mutation rate for this optimization process were set at 80% and 0.01% respectively.

6 Results and discussion

The MOGA optimisation process was run for a maximum generation of 200 and corresponding results are shown in **Figs. 4** and **5**. A bang-bang signal at a sampling rate of 5000 Hz was chosen as reference input of the hybrid learning control system. A Pareto optimal set for two objectives is shown in **Fig. 4**. The x-axis shows objective 1, which is end-point acceleration, while the y-axis indicates rise-time as objective 2.

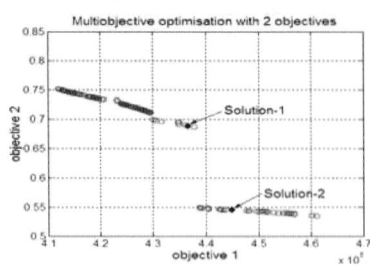

Fig. 4. Pareto optimal solutions for 2 objectives

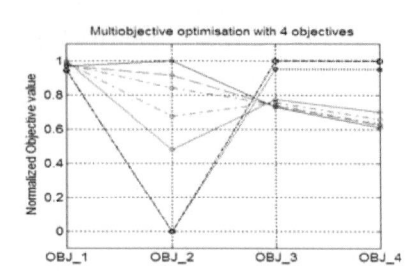

Fig. 5. Pareto optimal solutions for 4 objectives

A 4 objectives Pareto set is shown in **Fig. 5**. The x-axis shows the design objectives, the y-axis shows the performance of controllers in normalised form for each objective domain. Crossing lines between adjacent objectives indicates that there is a trade-off between those two objectives while parallel lines show that there is no conflict in the current population of solution estimates. The explicit objective values are shown in **Table 1**.

Table 1. Controller parameters and objective values for 4-objective MOGA

ILC parameters		Time domain specifications			
Kp	Kd	Overshoot (%) (OBJ_2)	Rise-time (sec) (OBJ_3)	Settling time (sec) (OBJ_4)	End-point acceleration (OBJ_1)
0.001	0.0031	-0.0018	0.7574	1.5758	499.0143
0.0036	0.0040	11.0555	0.5742	1.0380	524.1652
0.0038	0.0028	14.9605	0.5572	0.9816	524.3208
0.0034	0.0011	15.2157	0.5556	0.9868	521.7773
0.0037	0.0031	13.7435	0.5622	0.9968	517.6247
0.0040	0.0034	15.5657	0.5552	0.9712	517.0444
0.0012	0.0025	-0.0008	0.7240	1.4980	501.2038
0.0039	0.0029	16.3275	0.5520	0.9648	511.8355
0.0011	0.0025	-0.0010	0.7356	1.5194	497.7129
0.0027	0.0010	7.8697	0.5884	1.1072	528.8507

To assess the controller performance in terms of input tracking capability and vibration suppression at the end-point, two solutions were tested as indicated in **Fig. 4**. The controller parameters and objective values of solutions 1 and 2 are shown in **Table 2**.

Table 2. Controller parameters and objective values for solution 1 and 2

Solution	ILC parameters		Time-domain specifications	
	Kp	Kd	Rise time (sec)	EP_Acceleration
1	0.001236	0.001001	0.6886	4.3657×10^5
2	0.003941	0.002033	0.5452	4.4486×10^5

It was observed that rise-times for solutions 1 and 2 were 0.6886 and 0.5452 sec while the end-point accelerations were 4.3657×10^5 and 4.4486×10^5 respectively. Solution 2 was faster than solution 1 but the former had higher error in the end-point acceleration. The hub-angle response for solutions 1 and 2 are shown in **Fig. 6**. To investigate attenuation of vibration at the end-point, solutions 1 and 2 are compared with a conventional PD controller. The time-domain responses (for convenience, from 2-6 sec.) are shown in **Fig. 7**. The power spectral density (PSD) plots at the end-point for PD controller, with solutions 1 and 2 are shown in **Fig. 8** for three main resonance modes, which are at 13, 35 and 65 Hz respectively. It was observed that substantial attenuation was achieved with hybrid learning controller; both for solution 1 and 2, at mode1 compared to PD controller whereas attenuation patterns are quite arbitrary at modes 2 and 3.

Fig. 6. Hub-angle response (time domain)

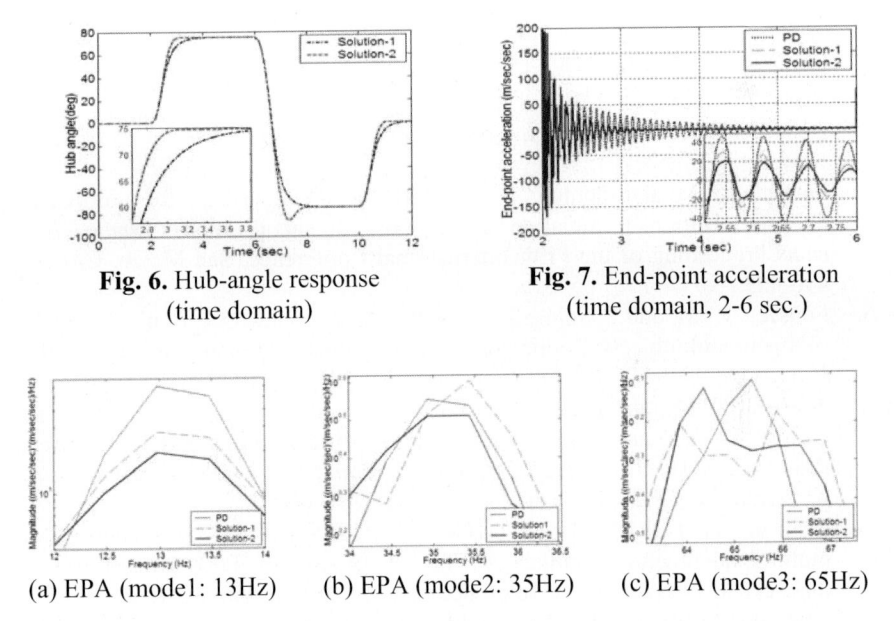

Fig. 7. End-point acceleration (time domain, 2-6 sec.)

(a) EPA (mode1: 13Hz)   (b) EPA (mode2: 35Hz)   (c) EPA (mode3: 65Hz)

Fig. 8. PSD plots of end point acceleration at three resonance modes

7 Conclusion

This paper has investigated the use of MOGA to design a hybrid ILC scheme for the control of a flexible manipulator. It has been shown that MOGA can produce a good set of results that form a Pareto solution set, allowing the system designer the flexibility of trading one solution against others to achieve a desired performance. The main advantage of MOGA is its versatility for including a variety of objectives and constraints while designing the controller. The performances of the control schemes have been evaluated in terms of input tracking capability and vibration suppression at the resonance modes of the manipulator. Acceptable input tracking control and vibration suppression have been achieved with the control strategies.

References

1. Azad, A. K. M. (1994). *Analysis and design of control mechanism for flexible manipulator systems*, PhD thesis, Department of Automatic Control and Systems Engineering, University of Sheffield, UK.
2. Benosman, M. and Vey, L. (2004). Control of flexible manipulators: A survey, *Robotica*, Vol. 22, pp. 535-545.
3. Cannon, R. H. and Schmitz, E. (1984). Initial Experiments on the End-Point Control of a Flexible One-Link Robot, *The International Journal of Robotics Research*, Vol. 3, No. 3, pp. 62-75.
4. Deb K. (2001) *Multi-objective optimization using evolutionary algorithms*. New York; Chichester: Wiley.
5. Fonseca, C. M. and Fleming, P. J. (1993). Genetic algorithms for multiobjective optimization: formulation, discussion and generalization, *Genetic Algoritms:*Proceeding of the Fifth International Conference, San Mateo, CA, pp. 416-423.
6. Fonseca, C. M. and Fleming, P. J. (1995). An overview of evolutionary algorithms in multiobjective optimization, Evolutionary Computation, Vol. 3, No. 1, pp.1-16.
7. Fonseca, C. M., and Fleming, P. J. (1998). Multiobjective optimization and multiple constraint handling with evolutionary algorithms-part I: A unified formulation, *The IEEE Transaction on Systems, Man and Cybernetics- part A: Systems and Humans*, Vol. 28, No. 1, pp. 26-37.
8. Goldberg D. E, Richardson J. (1987). Genetic algorithms with sharing for multimodal function optimization, In J. Grefenstette, (Ed.), Proceedings of the Second International Conference on Genetic Algorithms, Hillsdale, NJ: Lawrence Erlbaum Associates, 41-49.
9. Goldberg, D. E. (1989). *Genetic algorithms in search, optimisation and machine learning*, Addison Wesley Longman, Publishing Co. Inc., New York.
10. Tokhi, M. O., Alam, M. S., Zain M. Z. Md. and Aldebrez, F. M. (2005). Adaptive command shaping using genetic algorithms for vibration control of a single link flexible manipulator, *Proceedings of 12th International Congress on Sound and Vibration*, Lisbon, Portugal, 11-14 July.
11. Zain M. Z. Md,, Tokhi M. O. and Alam M. S. (2005). Robustness of hybrid learning acceleration feedback control scheme in flexible manipulators, The Fourth World Enformatica Conference, Istanbul (Turkey) 24-26 June.

Intelligent Modelling of Flexible Manipulator Systems

M. H. Shaheed[1], Abul K. M. Azad[2], M. O. Tokhi[3]

[1]DDepartment of Engineering, Queen Mary, University of London, UK;
[2]Department of Technology, Northern Illinois University, USA;
[3]Department of Automatic Control and Systems Engineering, The University of Sheffield, UK;   m.h.shaheed@qmul.ac.uk

Abstract

Intelligent techniques, such as genetic algorithms (GAs) and neural networks (NNs) have attracted the attention of the wider control community due to their various advantageous features in relation to system identification and control. The major advantage of utilising GAs for system identification is that they simultaneously evaluate many points in the parameter space and converge towards the global solution. On the other hand, the use of NNs is inspired by their ability to mimic the capabilities of the brain such as learning, adaptation, association and generalisation. More importantly, NNs can address the nonlinearity of a system. This paper presents an investigation into the use of GAs and NNs to model a single-link flexible manipulator. The GA and NN based identification of the system are realised in this investigation by minimising the prediction error of the actual plant output and the model output. However, to allow interactive and user friendly features, that are desired especially in computer aided teaching and research, be incorporated a modelling, simulation and control environment is developed in this work for flexible manipulators using Matlab and Simulink. To this end the authors have developed an interactive and user-friendly environment referred to as SCEFMAS (**S**imulation and **Co**ntrol **E**nvironment of **F**lexible **Ma**nipulator **S**ystems) [1]. As an on-going development process, the SCEFMAS environment is enhanced by the addition of intelligent modelling using NNs and GAs.

Keywords: intelligent modeling, neural networks, genetic algorithms, flexible manipulators.

1 Introduction

Climbing and walking robots (CLAWAR) are slowly changing their showbiz image to specific task performer image. To increase the deployment time researchers are constantly striving to reduce the weight and hence minimize the energy consumption in CLAWAR. These have been done by using energy efficient electronics, smaller and lighter sensors and actuators, lighter construction material, and efficient physical configurations. In addition to these there are other areas one can look to reduce the weight of these machines.

In most of the cases a mobile robot needs to have a manipulator to perform their intended tasks when arrived at a target location. Employing a traditional rigid robot with larger weight to payload ratio can be a burden in terms of energy consumption and manageability. Considering the lightweight properties and capabilities, flexible manipulators stand for a clear challenge in opening new robotic applications. A flexible manipulator offers faster system response, lower energy consumption, requires relatively smaller actuators, amounts to reduced nonlinearity due to elimination of gearing, less overall mass and, in general, less overall cost [2]. However, their flexible nature and associated difficulties in positioning control restricts their wide spread use.

Developing an effective controller for vibration and position control is a major task ahead for the research community. Reasonably accurate modelling is one of the prerequisites to development of an effective control strategy for efficient vibration suppression and positioning [3]. A considerable amount of research has been carried out on the modelling of flexible manipulators over the last two decades [4]. These include the Lagrange's equation and modal expansion (Ritz-Kantrovitch) or assumed modes method, the Lagrange's equation and finite element method, the Euler-Newton equation and modal expansion, the Euler-Newton equation and finite element, the singular perturbation and frequency domain techniques. Considering the complexity of the system, and although most of the models are truncated, they are still considered to be computing intensive.

To address the model accuracy and computing time, intelligent techniques such as genetic algorithms (GAs) and neural networks (NNs) can be employed. These intelligent modelling techniques have proved to be effective in modelling non-linear systems or if the system possesses nonlinearities to any degree. This paper addresses the design and development of intelligent models of a flexible manipulator system and verifies their performances. Moreover, the model development and verification system is incorporated within a computerized simulation environment.

2 The flexible manipulator system

A schematic representation of the manipulator is shown in **Fig. 1**, where X_oOY_o and XOY represent the stationary and moving co-ordinate frames respectively.

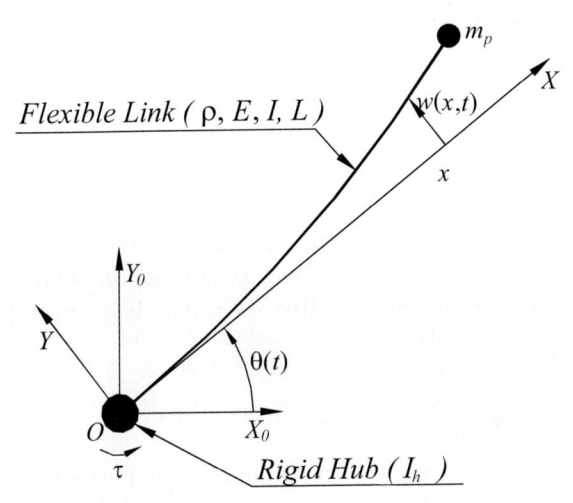

Fig. 1. Schematic representation of the flexible manipulator system

The axis OX coincides with the neutral line of the link in its undeformed configuration, and is tangent to it at the clamped end in a deformed configuration. The τ represents the applied torque at the hub. E, I, ρ, S, I_h and m_p represent the Young modulus, area moment of inertia, mass density per unit volume, cross sectional area, hub inertia and payload of the manipulator respectively. $\theta(t)$ denotes an angular displacement (hub-angle) of the manipulator and $w(x,t)$ denotes an elastic deflection (deformation) of a point along the manipulator at a distance x from the hub of the manipulator.

3 Intelligent modelling

In many cases, when it is difficult to obtain a model structure for a system with traditional system identification techniques, intelligent techniques are desired that can describe the system in the best possible way. GAs and NNs are two intelligent techniques commonly used for system identification and modelling. The major advantage of utilising GAs for system identification is that GAs simultaneously evaluate many points in the parameter space and converge towards the global solution [5]. In contrast, NN approaches for system identification offer many advantages over traditional ones especially in terms of flexibility and hardware realization [6]. This technique is quite efficient in modelling non-linear systems or if the system possesses nonlinearities to any degree. As mentioned before NN and GA approaches are used in this study to model the flexible manipulator. These modelling techniques are described below.

3.1 NN modelling

Non-linear autoregressive process with exogeneous modelling technique is used with multilayer perceptron or radial basis function neural networks to model the manipulator. Mathematically, the model is given as [7]:

$$\hat{y}(t) = f[(y(t-1), y(t-2), \cdots, y(t-n_y),\\ u(t-1), u(t-2), \cdots, u(t-n_u)] + e(t) \quad (1)$$

where, $\hat{y}(t)$ is the output vector determined by the past values of the system input vector, output vector and noise with maximum lags n_y and n_u respectively, and $f(\cdot)$ is the system mapping.

3.2 GA modelling

For parametric identification of the manipulator with GA, randomly selected parameters are optimised for different, arbitrarily chosen order to fit to the system by applying the working mechanism of GA. The fitness function utilised is the sum-squared error between the actual output, $y(n)$, of the system and the predicted output, $\hat{y}(n)$, produced from the input to the system and the optimised parameters:

$$f(e) = \sum_{i=1}^{n} \left(|y(n) - \hat{y}(n)|\right)^2 \qquad (2)$$

where, n represents the number of input/output samples. With the fitness function given above, the global search technique of the GA is utilised to obtain the best set of parameters among all the attempted orders for the system. The output of the system is thus simulated using the best sets of parameters and the system input.

4 The simulation environment

This is an interactive and user friendly environment referred to as SCEFMAS (**S**imulation and **C**ontrol **E**nvironment of **F**lexible **Ma**nipulator **S**ystems).

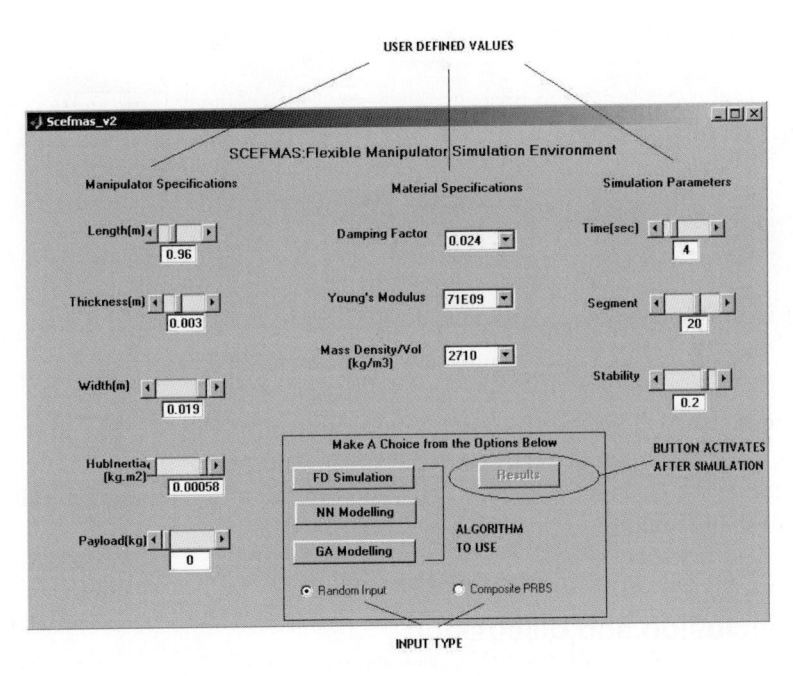

Fig. 2. Startup GUI for the SCEFMAS environment

The main graphical user interface (GUI) for the environment is shown in **Fig. 2**. This environment is developed using Matlab and its associated toolboxes [8], [9]. The user can provide details of a flexible manipulator and perform a simulation using finite difference method. The simulation data is then used as the training data for NN and GA models.

The GUI used for NN modelling is shown in **Fig. 3**. There is a provision for the user to choose the number of layers of neurons within the NN structure along with the properties of neurons in each layer. This is then followed by modelling and model validation.

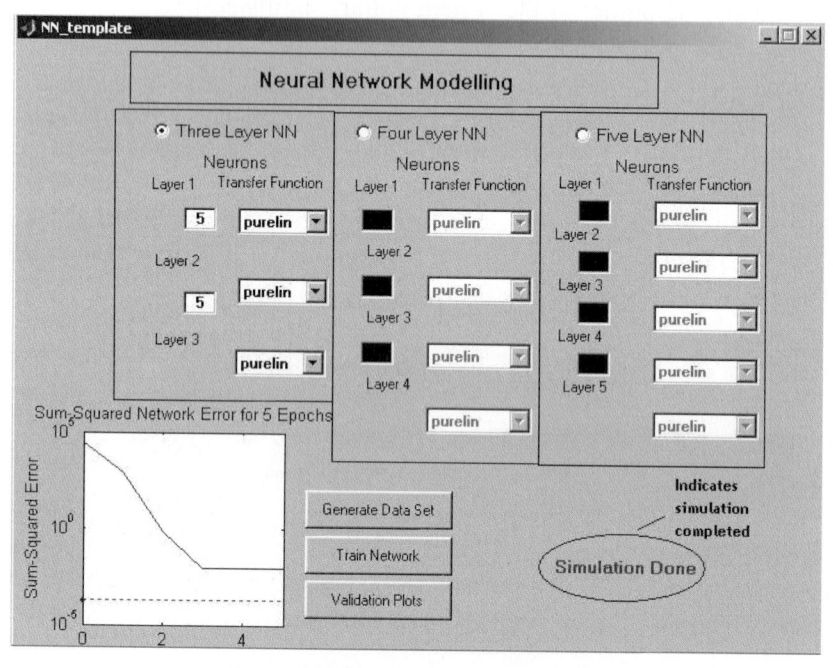

Fig. 3. Neural network modeling GUI

The GUI used for the GA modelling is shown in **Fig. 4**. The user can enter the model parameters and model type and perform the simulation. Similar to the NN model there is provision in the environment for validation of the model.

5 Conclusion and discussion

The paper presents intelligent modelling techniques for flexible manipulator systems that can be used for mobile robotic applications. Within the intelligent techniques only the GA and NN modelling approaches have been used. The developed algorithm has also been incorporated within a user friendly simulation environment using Matlab Simulink and Guide toolboxes. The environment allows the user to choose a flexible manipulator of choice and at the same time to adjust the modelling parameters through an interactive interface. This will enable the user to choose a model structure and monitor the developed model performance without going into the programming details. Moreover, a data analysis provision has been made within the package, to enable users to analyse data obtained from a test run. These make the environment more user-friendly, and saves time and effort to transfer the data to another environment for analysis.

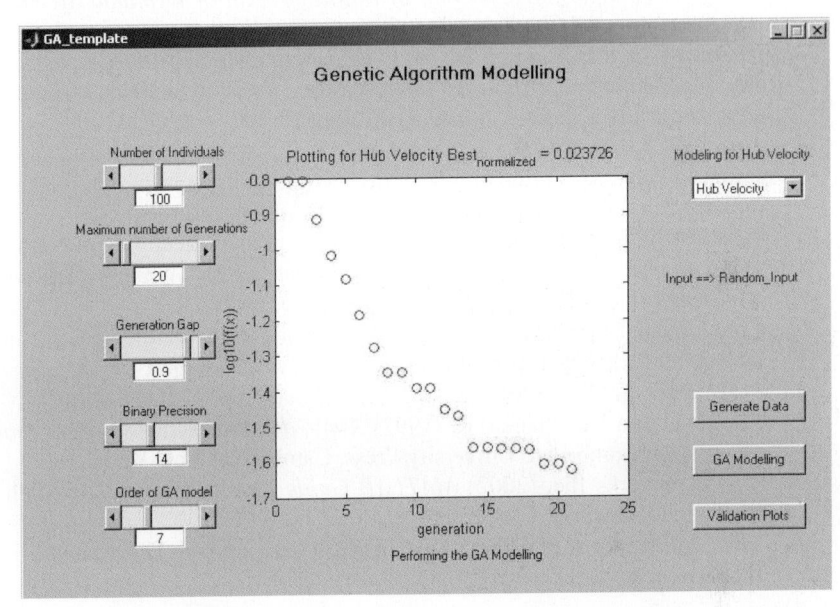

Fig. 4. The GUI used for GA modelling process

The environment has already proven to be a valuable education tool for understanding the behaviour of flexible manipulator systems and development of various controller designs. The GA and NN modelling features of this environment can easily be extended to the development of intelligent controllers within the environment. With the advent of Internet technology, the package can be further used as a distance teaching learning facility. Moreover, the built in Simulink block can be utilized to investigate various other aspects of active vibration control in flexible manipulator systems. Furthermore, users can design their own Simulink blocks and couple them to specific requirements.

References

1. Tokhi, M. O., Azad, A. K. M., and Poerwanto, H. (1999) SCEFMAS: A Simulink environment for dynamic characterisation and control of flexible manipulators. *International Journal of Engineering Education,* vol. 15, no. 3, pp. 213-226.
2. Book, W. J. and Majette, M. (1983) Controller design for flexible distributed parameter mechanical arms via combined state-space and frequency domain techniques, *Transaction of ASME Journal of Dynamic Systems, Measurement and Control*, vol. 105, pp. 245-254.
3. Tokhi, M. O. and Azad, A. K. M. (1995) Active vibration suppression of flexible manipulator systems - Closed-loop control methods. *International Journal of Active Control*, vol. 1, pp. 79-107.
4. Tokhi, M. O., Poerwanto, H., and Azad, A. K. M. (1995) Dynamic simulation of flexible manipulator systems incorporating hub inertia, payload and structural damping, *Machine Vibration*, vol. 4, pp. 106-124.
5. Kristinsson, K. and Dumont, G. (1992) System identification and control using genetic algorithms. The *IEEE Transactions on Systems, Man and Cybernetics*, vol. 22, no. 5, pp. 1033-1046.
6. Ljung, L. and Sjöberg, J. (1992) A system identification perspective on neural networks. *Neural Networks for Signal Processing II.-Proceedings. of the IEEE-SP Workshop,* Helsingoer, Denmark, pp. 423-435.
7. Luo, F-L. and Unbehauen, R. (1997) *Applied neural networks for signal processing*. Cambridge University Press, Cambridge, New York.
8. The Mathworks Inc. (2002) *MATLAB Guide Users Manual.* The Mathworks Inc. Natwick.
9. The Mathworks Inc. (2001) *SIMULINK Users Guide.* The Mathworks Inc. Natwick.

Wafer Handling Demo by SERPC

N. Abbate, A. Basile, S. Ciardo, A. Faulisi, C. Guastella, M. Lo Presti, G. Macina, N. Testa

MLD Central Lab, STMicroelectronics, Stradale Primosole 50, 95121 Catania – ITALY, Tel. +39 095 740 4286, Fax. +39 095 740 4031, E-Mail: adriano.basile@st.com

Abstract

The wafer handling is a complex operation that requires wafer positioning, accuracy and repeatability needed for reliable results. Every action taken with a wafer must be exactly what the operator wants, no more and no less. In semiconductor environment the motor control is fundamental, because the trajectory on the work-piece must be compensated in real-time. The authors would present a demo capable to demonstrate that all command tasks for the movement of a manipulator robot could be generalized to any other robot.

Keywords: wafer handling, motor control, sensor fusion, embedded platform, motion analysis.

1 Introduction

Real-time robot control requires efficient inverse kinematics transformations to compute the temporal evolution of the joint coordinates from the motion of the end-effector. The development of a coherent, efficient and general-purpose hardware framework for a generic robot facilitates its prototyping. Moreover, a good framework could be re-usable for many applications. The proposed hardware system is a Single Board Computer named Super Embedded Robotic PC (SERPC) [1], based on two microcontrollers which are STPC [2] and ST10 [3]. The first one is a powerful x86 class processor able to host the Microsoft Windows CE Operating System [4], meanwhile the second one is a 16 bit MCU capable to manage up to three motors by three dedicated connectors, see **Fig. 1**.

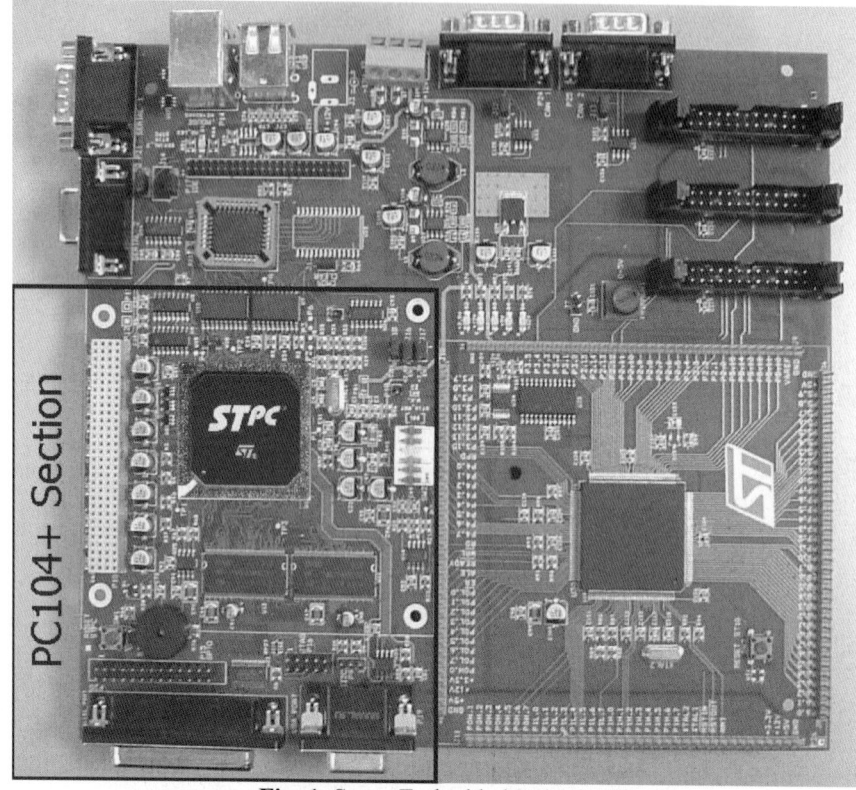

Fig. 1. Super Embedded Robotic PC.

SERPC performs all command tasks for the movement of the wafer handler manipulator, its pre-aligner stage, and tilt sensors. The workload above mentioned is structured in this way: a Graphical User Interface (GUI) runs on the STPC; the kinematics chain, the instantaneous motion analysis and the motion control is managed by ST10. The GUI permits to choose the PID parameters, to drive the end-effectors movements, to observe the end-effectors trajectory and many other internal variables as the encoder signals, the speeds and so on.

The present paper will be organized as follows. The next Section summarizes the modularity of SERPC. Section III shows the wafer handling demo, particular emphasis to its performance will be given. Finally the Section IV concludes the paper highlighting the re-usability of the concepts introduced.

2 SERPC Modularity

It is clear that a modular hardware framework pass through the capability to plug or remove components easily. Moreover, a system in conformity with the modularity concepts identifies the specific requirements, the design parameters and the design specifications of a large amount of applications.

Fig. 2. ST10 Start Development Kit

The Printed Circuit Board of SERPC is mainly composed by two parts: a PC104+ (see the boxed area in Fig. 1) hosting the STPC Atlas [2]; a low-level microcontroller area, where the ST10 DSP [3] takes place. This last one is identical to the stand-alone ST10 Start Development Kit board (see Fig. 2), it is devoted to perform all the real-time tasks directly related to the motors, in fact up to three motor boards could be connected to SERPC. These motor boards are essentially based on an Control core, that includes all logic and analog circuitry, that actually manage the specified physical variable (i.e. current, speed) in a fashion that best suits the specific needs of very different motor types - Stepper, Brushed DC and Three Phase Brushless. The control core includes also all the protection sensing and processing blocks (i.e. Under Voltage, Over Temperature, Over Current) and the optimized buffer stages devoted to properly drive the 60V rated integrated Power MOS. These latter are arranged into arrays of either eight – four high side and four low side – or six – 3 high side and three low side.

The approach used to design these motor boards is simple: to offer a common *motion connector* to the MCU for a wide spectrum of motors in terms of size, current and typology. Two of these boards are shown in Fig. 3: L6208BD [5] board is a fully integrated two phase stepper motor driver, meanwhile L6235BD [6] is a three phase brushless dc motor driver. These boards transmit on the *motion connector* all the feedback from the motors: incremental encoders, tachymeter dynamos, and so on.

Fig. 3. Two motor boards: L6208BD and L6235BD.

As above discussed the real-time motor control is implemented on the ST10, this leaves the STPC to manage only the Operating System and the Graphical User Interfaces. Practically, it is free to interface with the PC104+, the I2C and the UARTs where several daughter boards could be connected.

3 Wafer Handling Demonstrator

As discussed in the previous Section SERPC could be easily used in many robotic application: roving robots, factory automation robots, ... Here a hard trajectory control has been considered as border line real-time task, in other word by satisfying the restrictive constrains of the wafer manipulation (error less than micron) the hardware demonstrates a high degree of competence in all the application inherent the movement control.

The considered manipulator, depicted in Fig. 4, is a dual arm robot with two edge grip end effectors. Its physical characteristics are the following: a weight of 49.9Kg distributed on a structure with a diameter of 247 mm and a height of 768 mm. It is capable to cover an area of 355 x 362 mm (lateral x Z travel). It is designed to move and position stages with sub-micron repeatability over large travels. The motion devices incorporate DC motors with optical encoder position feedback for smooth and accurate motion; it has two motors for each arm, a lift up motor and a shoulder motor. Totally six motor are present, the third joint of each arm is not actuated and its positioning depend on the movement of the second joint.

Fig. 4. The manipulator adopted in the Cleaning Room for 6" Wafer Manipulation.

All the typical tasks performed in a cleaning room have been replicated in our lab, particular attention has been paid to the accuracy of the movements, this precision should be retained at any speed of the end-effectors. An example of the obstacle avoiding is shown in Fig. 5, four frames of a video sequence are there reported. The experiment is a cyclic surrounding of an obstacle, this in order to observe the trajectory deviation from the planned one. This test works in similar way of the UMBmark benchmark [7], a further executed test has been a composition of eight consecutive s-quare trajectories, they are depicted in Fig. 6, this composition is executed several time in sequence, no deviation has been registered.

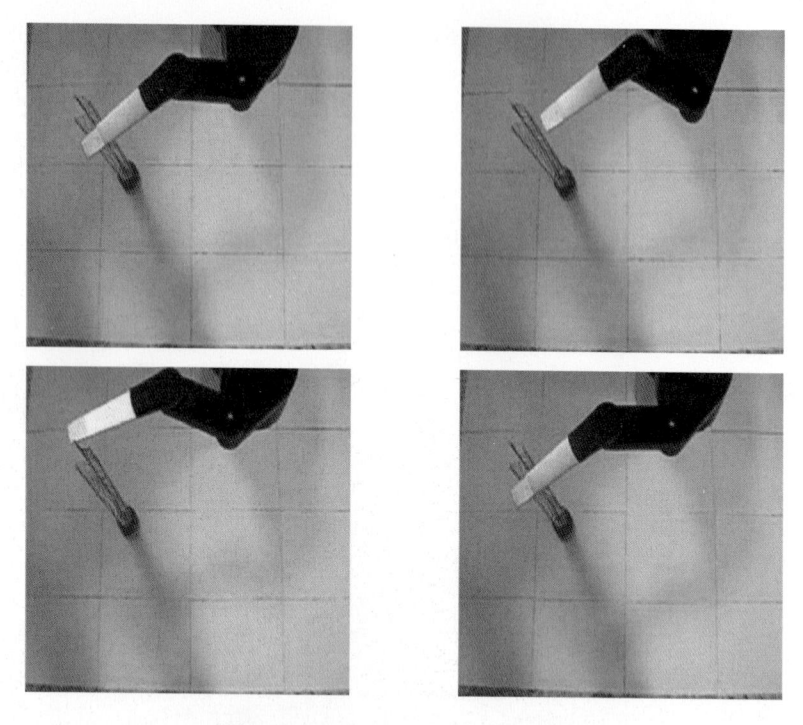

Fig. 5. Cyclic obstacle surrounding.

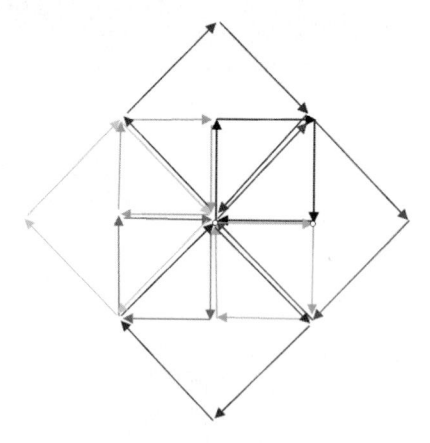

Fig. 6. This trajectory is composed by eight squares, each one begins at the end of the previous one. It is repeated several times.

All the information from the manipulator are collected and managed by some simple GUIs running on STPC, these have been designed with Microsoft Embedded Visual C++, in Fig. 7 is reported an example of these GUIs.

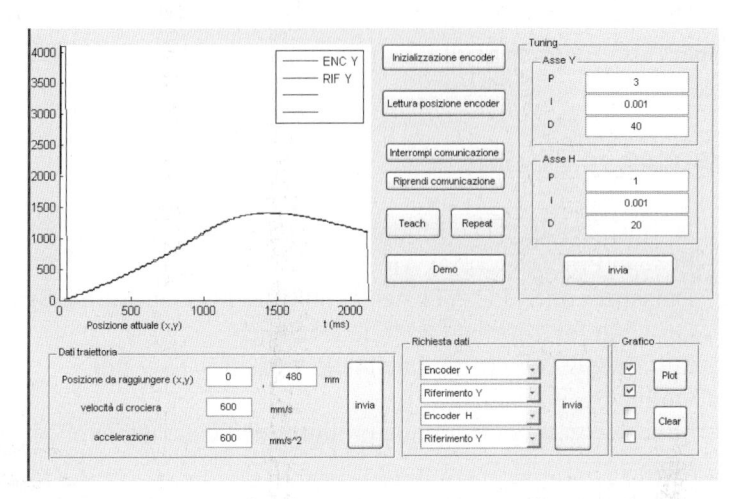

Fig. 7. Graphical User Interface of the Wafer Handling Manipulator.

4 Conclusions

The paper would demonstrate the potentialities of SERPC in a real robotic task, this board is able to work as a Soft PLC and manage directly in a real-time fashion all the quantity of the wafer handler manipulator. This demonstrator has been selected as the hardest task for SERPC, the good results reveal the capability to be applied in any robotic job. Its modularity permits to manage more than three axis simply by adding an other ST10 to the I2C buses.

Finally, some preliminary tests of SERPC platform on a roving robot are ongoing, this will navigate in a structured environment. The environment includes some fixed objects that are loaded as map inside the memory and a no-finite series of possible moving object: peoples, chairs, baskets, small cables, etc. The results are satisfactory and clarify that this task is easy to be managed by the SERPC platform.

Acknowledgements

The authors gratefully acknowledge Eng. Dino Costanzo and Eng. Giuseppe Vasta for their helpful contributions to this work.

References

1. A. Basile et al., "The Modularity of Super Embedded Robotic PC", in Proc. of CLAWAR 2004, Madrid, Spain, Sept. 2004.
2. Datasheet of STPC ATLAS – x86 Core PC Compatible System-On-Chip for Terminal, Online Document: http://www.st.com/stonline/books/pdf/docs/7341.pdf
3. Online document page: http://www.st.com/stonline/prodpres/dedicate/auto/embedded/st10.htm
4. Online document page: http://msdn.microsoft.com/embedded/windowsce/default.aspx
5. Datasheet of L6208 DMOS driver for bipolar stepper motor, online document: http://ccdeu01n.sgp.st.com/stonline/products/literature/ds/7514/16208.pdf
6. Datasheet of L6236 DMOS driver for three phase brushless dc motor, online document: http://ccdeu01n.sgp.st.com/stonline/books/pdf/docs/7618.pdf
7. J. Borenstein and L. Feng, "UMBmark: A Benchmark Test for Measuring Odometry Errors in Mobile Robots", in Proc. of SPIE Conference on Mobile Robots, Philadelphia, October 22-26, 1995.

Vision Control for an Artificial Hand

M. Kaczmarski[1] and D. Zarychta

[1]Institute of Automatic Control, Technical University of Łódź, POLAND,
mkaczmar@wpk.p.lodz.pl

Abstract

In the paper some method of a vision control is considered. Although used algorithms were tested for the control of an artificial hand they could be applied also in the case of navigating a walking robot. The mechanisms of selection a visual information to process and their implementation on the laboratory stand were presented together with the experimental data.

Keywords: Vision control feed-back, artificial hand, pneumatic muscles.

1 Introduction

Biological inspirations are often used in the robotics research - from mobile robots based on primitive animal locomotion to constructions reflecting the structure and method of movement of higher complex organisms such as humans.

Many institutions and laboratories all over the world are currently working on construction of dexterous bioprosthesis for people after an amputation of the hand. One of the more popular ones – a three-finger bioprosthesis of the palm - is constructed in the Oxford Orthopedic Engineering Center. Similar structures include: „Sensor Hand TM" from Otto Bock company, the artificial hand DLR from Germany, Japanese structure from Complex Systems Engineering of Hokkaido University and the Mitech Lab bioprosthesis from Italy. Anthropomorphic five-fingers grippers for special tasks are also being constructed. The NASA robot "RoboNaut"[7] is an example of such a project or a construction of dexterous, five-finger driven with air muscles hand of the Shadow Company [8], having the prehensile capabilities very similar to the human hand.

The artificial hand, constructed in the Institute of Automatics Control in Technical University of Lodz (see **Fig. 2.**), has a very simplified structure. The chosen actuators - pneumatic muscles, are also a simplified substitute for the human muscles. Modelling the work of the human hand can in this case be conducted only in a limited range.

2 General remarks on the applied vision algorithms

In the case of vision analysis in robotic context very often it is known exactly what kind of robots actions can be performed by the robot in the nearest future and these actions can directly determine the way a vision analysis can be applied. Such situation is relevant in the case of manipulating robots as in the case of walking ones. The common denominator of both cases is that the machine parts are move along trajectories that can be predicted to some degree. These predicted trajectories can be transform into curves visible in the image. Using this knowledge the whole image can be decomposed into a limited set of curves instead of its direct matrix representation. Such decomposition is usually a more time-effective approach to the problem.

The method requires the visibility of the moving parts of the robot (e.g. fingers or legs) and potential contact object which should be discerned from the background (objects to be manipulated, obstacles on a route of a walking machine) at all times. In the case of a walking robot a ground surface can be additionally used as a reference plane which makes the calculations more unique than these made in 3 dimensional space. In this case the lines of the trajectories of the legs are scanned. Possible collision with obstacles can be detected so the robot can modify its strategy of gait (see **Fig. 1.**)

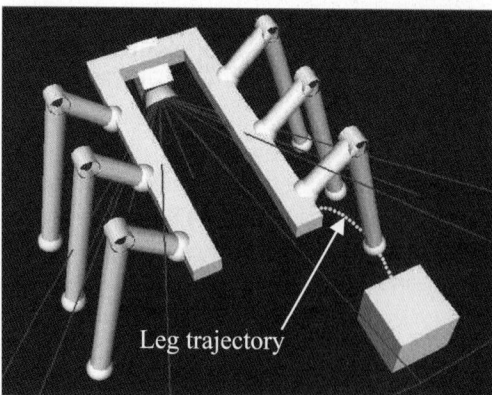

Fig. 1. Hypothetical use of a vision system in a walking machine.

In paper however a more general case is considered – a robot manipulating objects.

3 The artificial hand's construction basics

Every finger (except the thumb) contains three simple rotary joints with 1 DOF, while the thumb contains four. The wrist is made of three rotary joints, placed close to each other and rotated in such way, that they can be treated as one joint with 3 DOF. The presented artificial hand model allows fingers to bend completely, which is obtained by simultaneous rotation of all three rotary joints to the same angle. Thanks to this construction it was possible to use only one pair of actuators to drive the first active joint in the finger. This decreased the number of used pneumatic muscles [8] and simplified the controlling methods.

Direct drive for the simple rotary joint in each finger is received by a pair of artificial, pneumatic muscles, which work antagonistically. Thanks to the damping and inertia it was possible to obtain a smooth movement of the finger tips. The use of Mckibben's muscles [2] as actuators for the grasping manipulator is to reflect the methods of finger movement in a biological hand.

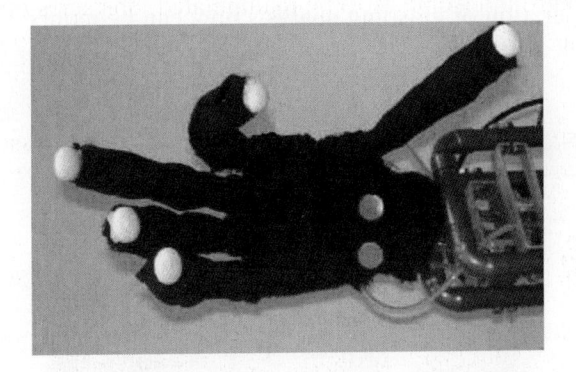

Fig. 2. Real view of artificial hand

4 Mathematical model of the artificial hand

Using Denavit-Hartenberg notation the direct kinematics task for the artificial hand can be easily solved [4].

The transformation matrix A determines a relationship between the general position description of a selected point in the coordinate frame {i-1} and {i} which can be written in the short form as

$$^{i-1}\hat{r} = A_i{}^i\hat{r} \qquad (1)$$

where A_i takes the form (2)

$$A_i = \begin{bmatrix} \cos\theta_i & -\sin\theta_i \cos\alpha_i & \sin\theta_i \sin\alpha_i & a_i \cos\theta_i \\ \sin\theta_i & \cos\theta_i \cos\alpha_i & -\cos\theta_i \sin\alpha_i & a_i \sin\theta_i \\ 0 & \sin\alpha_i & \cos\alpha_i & d_i \\ 0 & 0 & 0 & 1 \end{bmatrix} \qquad (2)$$

The transformation from the end coordinate frame of reference to the base coordinate frame is described by the equation:

$$^0\hat{r} = A_1(q_1)A_2(q_2)...A_n(q_n){}^n\hat{r} \qquad (3)$$

This model allows to directly obtain the trajectories along the moving finger tips which is of great importance in the process of image analysis. The whole operation of grasping analysis now can be reduced to searching only along selected lines.

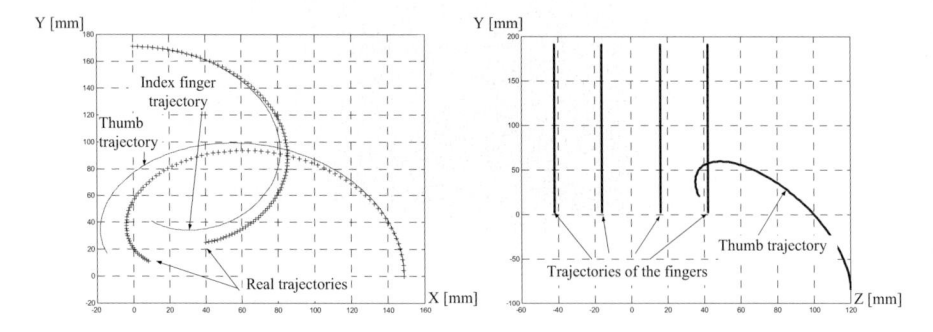

Fig. 3. Trajectories of the index finger and the thumb in the XY plane (side view - Fig.1b)

Fig. 4. Trajectories of all fingers in the ZY plane (front view - Fig. 1a)

The theoretical and the real trajectories of index finger and thumb are presented in Fig 3. The real curves are only slightly different from the theoretical ones, which are obtained by assuming equal revolutions in all the joints. This difference however cannot be entirely neglected so appropriate corrections have to be made.

Trajectories of the finger tips are presented in Fig 4. – angle of the camera is a frontal one. The main disadvantage of this view is the lack of possibility to predict perspective distortions. Without these corrections modelling of the finger tips is a simplification and resulted in obtaining a very innacurate approximation of the position of the finger tips markers.

5 Algorithms of vision analysis

A vision system used for control of the artificial hand is performing several basic tasks that allows to identify each finger. Firstly an acquired image is transformed with early- processing vision functions. At this stage all the RGB colours are extracted and the minimal value of brightness level is calculated for the extraction markers [5] from the background. Secondly a selection of the RGB component best suited for further processing is made.

At the next stage the positions of the bright areas are calculated. The segmentation is made using region merging method. This approach is good for a small number of bright fields and the image frame is search only once. The procedure is known as the Praire Fire [6]. Using this method positions of base markers are obtained. The base markers are needed as a base for the calculation of trajectories (see **Fig. 5.**).

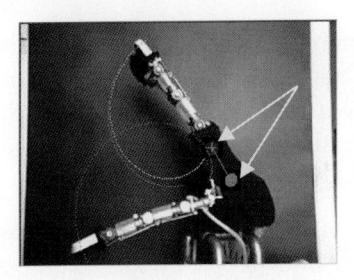

Fig. 5. Base markers

In the next stage the rotation of the trajectory is calculated and translation to the, before mentioned, base marker.

$$x' = x \cdot \cos\theta + y \cdot \sin\theta \quad (4.a)$$
$$y' = -x \cdot \sin\theta + y \cdot \cos\theta \quad (4.b)$$
$$x'' = x_b + k \cdot x' \quad (4.c)$$
$$y'' = y_b + k \cdot y' \quad (4.d)$$

The coefficient k is calculated using a similarity scale, where d stands for the real distance between the markers and (x_{b1},y_{b1}), (x_{b2},y_{b2}) are the coordinates of the base points in the Fig. 5.

$$k = \frac{\sqrt{(x_{b2}-x_{b1})(x_{b2}-x_{b1})+(y_{b2}-y_{b1})(y_{b2}-y_{b1})}}{d} \qquad (5)$$

By this way trajectories independent from the wrist bend are obtained. In this case it is possible to scale the framework since the assumption that the index finger and thumb are moving along the same plane is fulfilled. Additionally it is assumed that the perspective distortions can be neglected. Then the calculated trajectory lines are searched through for the markers [1] like in the case of searching for base points. In the case when the line along which the marker is moving intersects with the object the distance between the object and the finger is measured. When the distance is less than a previosuly defined value it is assumed that the object was grabbed.

6 Experiments conducted with the use of the vision system

The main goal of the constructed vision system is the tracking of the movement of the hand and on-line verification of the assumed strategy of grasping [3]. Two cases are considered – observing the hand from the front and side position. Analysis of the side view allows to confirm a grasp of the object with use of the generated trajectory lines.

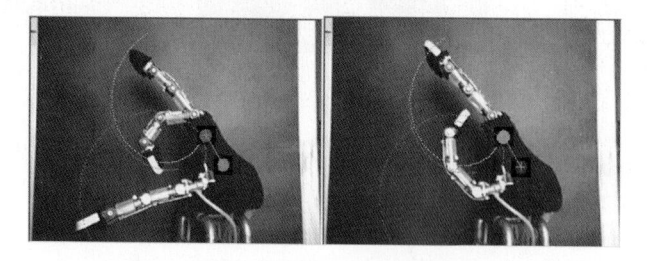

Fig. 6. Side view of real and theoretical trajectories.

Photographies in Fig. 6 present a range of index finger movement and thumb movement. The visible trajectories are generated from the simplified model and the experimentally corrected model. Every point of the curve maps the change of the angle by about 1 degree. Due to the knowledge of the position of the marker on the trajectory we are capable to calculate an angle of bend with the accuracy 1 degree.

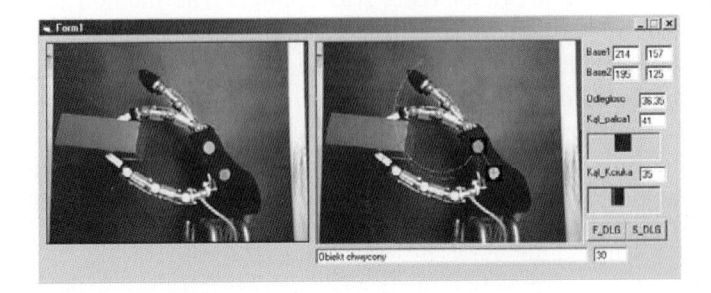

Fig. 7. View of the confirmed grasp.

The screenshot in Fig. 7 presents the window of the control program for the artificial hand. The moment of grasping an object is shown. At each image refresh the distance between the egde of a marker and the nearest object point is measured. In the case when the distance is less than 4 angle degrees the signal for the confirmation of the grasp is sent. Such situation however is definitive only in a limited number of cases i.e. in the case of cylinders or cubes. The grasp of e.g. a cone or a pyramid when the base is directed towards the camera can cause a misinterpretation. In side view only the two-point grasp can be applied. The main cause of this situation is the occluding of the fingers by the nearest one.

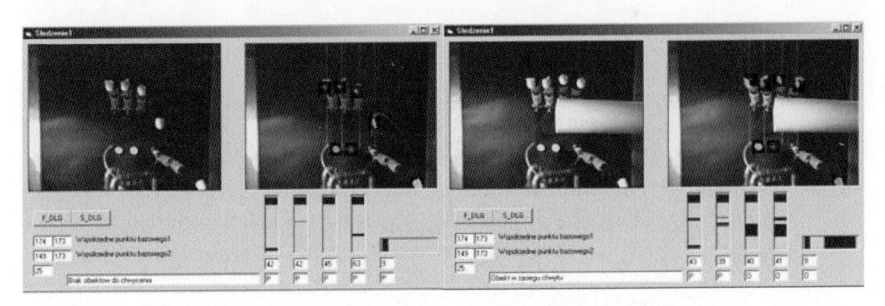

Fig. 8a   **Fig. 8b.**

Fig. 8. Front view of artificial hand without an object a), and with an object b).

Different possibilities for the use of the vision system can be obtained when using the front view of the hand as it is shown in Fig. 8. From this picture we can determine the shape of the object along lines of grasping and to set the most suitable configuration of the fingers.

However this view has many disadvantages. The fact of grasp cannot be confirmed as definitely in this case as it was in the previous view. In this case a method of indirect inferring was applied. The method is based on possibility of movement of the finger. If the finger stays in the same position despite the activation for further movement it means that an obstacle is located on its way. After a few seconds of ineffective attempts to continue the movement the situation is interpreted as a firm contact. This procedure is applied to all the fingers. The calculated trajectories are used for tracking the finger tip markers and checking which trajectories are intersected by the object.

7 Summary

The examples of controlling of the artificial hand presented in the paper are based upon quite significant simplifications in the vision control system. For example perspective distortions are not taken into account and the possibility of the movement of the thumb in the additional DOF is neglected. These problems have not been solved yet but are to be investigated in the future. A particularly promising concept is the possibility of applying a stereovision system. Another interesting area of the research will be the application of the presented methods to walking machines.

References

1. Brzeziński M., Zarychta D. (2001) Vision feedback for didactic robot with using Webcamera [in Polish], *Proceedings of VII National Robotics Conference*, Vol.2 pp. 175-180, Wrocław.
2. Chou C.P., Hannaford B. (1996) Measurement and Modeling of McKibben Pneumatic Artificial Muscles. *IEEE Transactions on Robotics and Automation*, Vol. 12, No. 1, pp. 90-102, Feb
3. Horaud R., Dornaika F., Espiau B. (1998) Visually guided object grasping, *IEEE Transactions on Robotics and Automation*, Vol.14, No.4, pp.525-532
4. Jezierski E. (2002) *Basic course on Robotics* [in Polish], Publisher PŁ
5. Jezierski E., Zarychta D. (1995) Tracking of moving robot arm using vision system. *Proceedings SAMS*, Vol. 18-19, Berlin, pp. 534-546
6. Tadeusiewicz R. (1992) *"Vision systems of industrial robots"* [in Polish], WNT Warszawa
7. http:\\robonaut.jsc.nasa.gov\hand.htm [2005-06-28]
8. http:\\www.shadow.org.uk\products\newhand.shtml [2005-06-28]

Robotic Finger that Imitates the Human Index Finger in the Number and Distribution of its Tendons

David M. Alba, Gabriel Bacallado, Hector Montes, Roberto Ponticelli, Theodore Akinfiev, and Manuel Armada

Instituto de Automática Industrial (CSIC), Carretera de Campo Real km 0.200, La Poveda, Arganda del Rey, 28500 MADRID
dmalba@iai.csic.es

Abstract

This paper presents the design of a robot finger driven by tendons, inspired in the human index finger tendon distribution. The design tries to minimize the number of tendons required, without sacrificing the number of actuated degrees of freedom. In this paper we will take the finger only in the planar position, we will not consider the adduction and abduction movements.

1 Introduction

As was demonstrated by [1] the minimum number of tendons needed for a n degrees of freedom manipulator is $n+1$. In this paper we analyze a design with a distribution of tendons similar to the human hand. Generally in other robot fingers every joint is actuated by two tendons [2]. In order to reduce the number of tendons required, other designs use links for coupling some DOFs [3]. There is other kind of designs like the one shown in [4] with innovative ideas, but this design use a different approach of the one used in this paper. Although in [1] several designs of four tendons manipulators are analyzed, one of them is used in [5], we decide to use a different approach because we want to find a better similarity with the human hand.

2 Kinematics and mechanical design of the robot finger

In biological articulations the behavior is non linear, and constructing such structures is very difficult, this is the reason because other solution have to be adopted. The human hand have three extrinsic muscles, this muscles provide the energy for powerful movements. This muscles are flexor digitorum profundus (FDP), flexor digitorum superficiales (FDS) and extensor digitorum (ED). The attachments of this muscles are shown in **Fig. 1**. The design will try to copy the attachments, distribution, and the number of tendons.

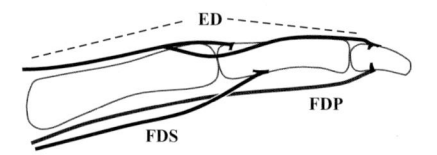

Fig. 1. Tendon and pulley

2.1 Attachments and articulations

An important characteristic of the human hand is that articulations are conjugated surfaces, this differs from robotic systems because in this systems the most common solution is to use hinges or similar designs. An scheme of the solution used for our design can be seen in **Fig. 2**.

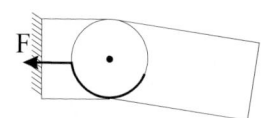

Fig. 2. Tendon and pulley

In this case the articulation have a linear behavior, because the torque always is $T = FR$, where R is the radius of the pulley. As we said before, the design tries to imitate the number and distribution of tendons in the human hand, we decided to distribute the tendons as it is shown in **Fig. 3**, in this figure we can see the "equivalence" between the robot tendons and the human index tendons.

There are differences between the robot finger design and the human index, first the path of the tendon 3 is not the same path of FDS, other

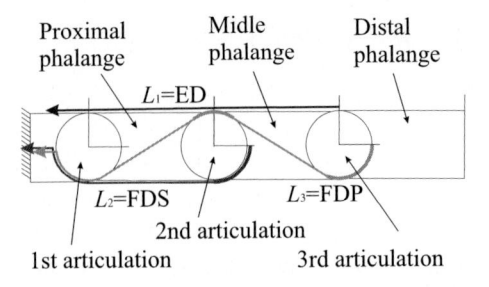

Fig. 3. Tendons in the robot finger

important difference is that the tendon 1 only have one attachment however the ED tendon have two attachments. If we take the finger in the position shown in **Fig. 3** the length of the tendons could be written as follows,

$$\begin{aligned} L_1 &= L_{1,\min} \\ L_2 &= L_{2,\max} \\ L_3 &= L_{3,\text{ini}} \end{aligned} \qquad (1)$$

In this position the robot finger is totally extended. Now if we take the robot finger in a random position, as it can see in **Fig. 4**, then the equations of the length of tendons can be written as,

$$\begin{aligned} L_1 &= L_{1,\min} + R\theta_1 + R\theta_2 + R\theta_3 \\ L_2 &= L_{2,\max} - R\theta_1 - R\theta_2 \\ L_3 &= L_{3,\text{ini}} - R\theta_1 + R\theta_2 - R\theta_3 \end{aligned} \qquad (2)$$

Where R is the radius of pulleys in each articulation. If we reorder the last expression and write it in matrix notation,

$$\begin{bmatrix} R & R & R \\ -R & -R & 0 \\ -R & R & -R \end{bmatrix} \begin{bmatrix} \theta_1 \\ \theta_2 \\ \theta_3 \end{bmatrix} = \begin{bmatrix} L_1 \\ L_2 \\ L_3 \end{bmatrix} - \begin{bmatrix} L_{1,\min} \\ L_{2,\max} \\ L_{3,\text{ini}} \end{bmatrix} \qquad (3)$$

The determinant of the square matrix is $-2R^3$, this means that the robot finger have only one position for a fixed tendon lengths. If we want to be able to move every joint in both directions [1], a fourth tendon is required, this fourth tendon is shown in **Fig. 5**. This path is chosen in order to imitate the second attachment of the ED in a human hand. The equation of tendon 4 is the same one of tendon 2 but with opposite signs.

The final system of equations is,

$$\begin{bmatrix} R & R & R \\ -R & -R & 0 \\ -R & R & -R \\ R & R & 0 \end{bmatrix} \begin{bmatrix} \theta_1 \\ \theta_2 \\ \theta_3 \\ \theta_2 \end{bmatrix} = \begin{bmatrix} L_1 \\ L_2 \\ L_3 \\ L_4 \end{bmatrix} - \begin{bmatrix} L_{1,\min} \\ L_{2,\max} \\ L_{3,\text{ini}} \\ L_{4,\min} \end{bmatrix} \quad (4)$$

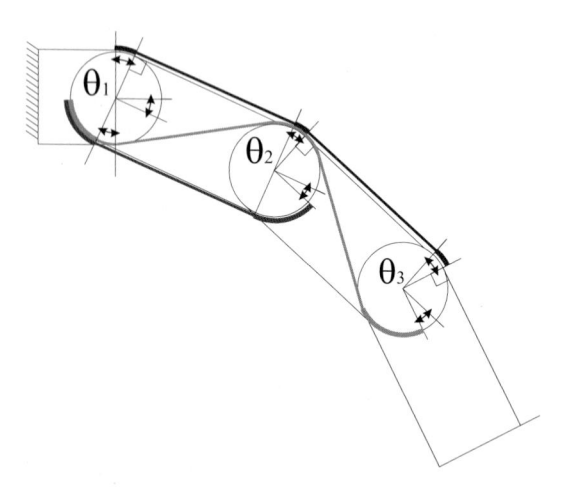

Fig. 4. Finger in a random position

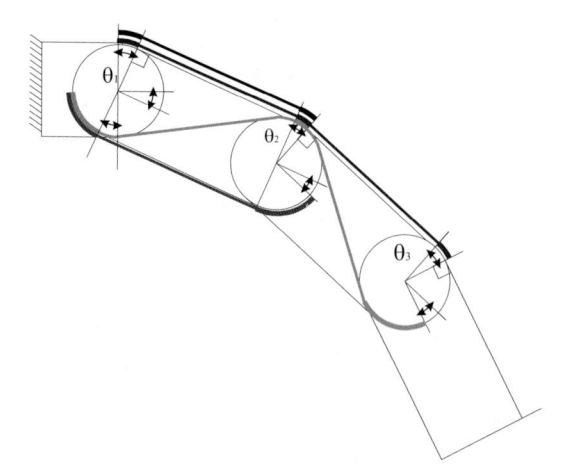

Fig. 5. Finger with all of its tendons

2.2 Jacobian and static analysis

Now we are going to find the relationship between the tension in tendons and the torques in joints. First we will dismiss the equation of tendon 4, because we want a square matrix. This is possible as long as we consider the sign of the answer in the tension of tendon 2, if the sign of this tension is positive then the load is applied to tendon 2, and if the sign of the tension in tendon 2 is negative then the load is applied to tendon 4. Using the principle of virtual work, we can write in matrix notation the virtual work of tension vectors inside the tendons, and the moment reactions inside the joints,

$$\mathbf{T}^T \delta \mathbf{L} - \tau^T \delta \theta = 0 \qquad (5)$$

$\mathbf{T}$ is the tension vector of tendons, $\delta \mathbf{L}$ is the infinitesimal tendons displacement vector, τ is the moment vector at the joints, and $\delta \theta$ is the infinitesimal rotation vector. Now if we differentiate the Eequation (3), we can write the following expression

$$\mathbf{R} \delta \theta = \delta \mathbf{L} \qquad (6)$$

Replacing expression 6 in 5, then ordering the terms, and finally using $\tau = \mathbf{J}^T \mathbf{F}$, the relation that we were looking for is

$$\mathbf{R}^T \mathbf{T} = \mathbf{J}^T \mathbf{F} \qquad (7)$$

Here $\mathbf{F}$ is the force vector at the fingertip, and $\mathbf{J}$ is the jacobian of the finger. The relation of the velocity of the fingertip and the velocity of the tendons is given by

$$\mathbf{v} = \mathbf{J} \mathbf{R}^{-1} \dot{\mathbf{L}} \qquad (8)$$

From Equation (7) we can conclude that there is only one set of tendon tensions for a given force at the finger tip, this is interesting because no optimization methods are required as for example in [6] or [5]. Other aspect to be take in count is the null space of the matrix $\mathbf{R}^T$ in Equation (4) [1], the base vector of the null space is,

$$\mathbf{N} = [0\ 1\ 0\ 1] \qquad (9)$$

This ensures that all the joints in the finger can be manipulated in both directions, because all the terms in the base vector are non negative [1].

3 Conclusions

The robotic finger have only one set of tension for every force vector at the fingertip, a more detailed model based in this analysis could explain some observations of different subjects have similar patterns of muscle excitation with large fingertip forces [7].

Although the robotic finger can be manipulated in both directions, there is two zeros in the null space vector in Equation (9), due to this zeros a complete dynamic model is required to control the finger correctly, other solution could be extend this model by including the intrinsic muscles effect, in order to improve the controllability of the robot finger.

4 Future work

Several experiments have been started, in order to try the design; the experiments will let us know the real behavior of the robot finger. The next step is to introduce a model that includes a simulation of the intrinsic muscles, then with a fully functional design of one finger a complete hand can be made. The only new in the future hand is that the design of the thumb has the joint axes rotated for simulating the workspace of the human thumb.

Fig. 6. Complete robot hand

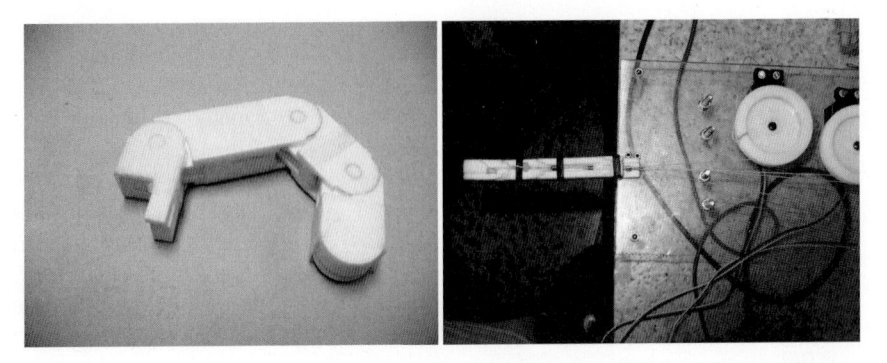

Fig. 7. Prototype of the finger

Acknowledgements

Authors would like to acknowledge the help provided by the mechanical and electric shops of the IAI-CSIC for the manufacture of the prototype.

References

1. Lee, J. and Tsai, L. (1991) The Structural Synthesis of Tendon-Driven Manipulators having Pseudotriangular Structure Matrix. *The International Journal of Robotic Research*, vol. 10, no. 3, pp. 255–262.
2. Jacobsen, S.C (1986) Design of the Utah/MIT dextrous hand. In *Proc. IEEE Int. Conf. Robotics and Automation*, pp. 1520–1528.
3. Lovchik, C., Aldridge, H., and Diftler, M. (1999) Design of the nasa robonaut hand. In *Proceedings of the ASME Dynamics and Control Division*, pp. 813–830.
4. Biagiotti, L., Lotti, F., Melchiorri, G., and Vassura, G. (2003) Mechatronic design of inovative fingers for anthrompomorphic robot hands. In *Proc. IEEE Int. Conf. Robotics and Automation*, vol. 4, pp. 3755–3762.
5. Caffaz, A. and Cannata, G. (1998) The design and development of the DIST-Hand dextrous gripper. In *International Conference on Ro-botics an Automation*, pp. 2075–2080.
6. Pollard, Nancy S. and Gilbert, Richards C. (2002) Tendon arrangement and muscle force requirements for human-like force capabilities in a robotic finger. In *Proc. IEEE Int. Conf. Robotics and Automation*, vol. 4, pp. 3755–3762.
7. Valero-Cuevas, Francisco J., Zajac, Felix E., and Burgar, Charles G. (1998) Large index-finngertip forces are produced by subject-independent patterns of muscle excitation. *Journal of Biomechanics*, , no. 31, pp. 693–703.

8. Salisbury, J.K. and Craig, J.J. (1982) Sensors for computer controlled mechanical assembly. *The International Journal of Robotic Research*, pp. 4–20.
9. Ishikawa, Y., Yu, W., Yokoi, H., and Kakazu, Y. (2000) Development of robot hands with an adjustable power transmitting mechanism. In *Intelligent Engineering Systems Trough Neural Networks*. pp. 631–636, Asme Press.
10. Namiki, Y., Imai, M., Ishikawa, M., and Kaneko (2003) Development of a high-speed multifingered hand system and its application to catching. In *International Conference on Intelligent Robots and Systems*, pp. 2666–2671.
11. Fukaya, N., Shigeki, T.and Afour, T., and Dillmann, R. (2000) Design of the tuat/karlsruhe humanoid hand. In *International Conference on Intelligent Robots and Systems*.
12. Mouri, T., Kawasaki, H., Yoshikawa, K., Takai, J., and Ito, S. (2002) Anthropomorphic robot hand: Gifu hand III. In *International Conference on Computer Applications in Shipbuilding*, pp. 1288–1293.

Modular, Reconfigurable Robots

Methods for Collective Displacement of Modular Self-reconfigurable Robots

Elian Carrillo[1,2], Dominique Duhaut[1]

[1]Université de Bretagne-Sud, Campus de Saint-Maudé B.P.92116, 56325 Lorient Cedex, France; [2]Universidad Romulo Gallegos, San-Juan de Los Morros, Ciudad Universitaria, Venezuela.
{elian.carrillo , dominique.duhaut} @univ-ubs.fr

Abstract

Two methods for collective displacement of modular self-reconfigurable robots have been developed. They are based on different approaches: a reactive behavior model and a hybrid learning control system, by using the flexibility of neural networks architectures and the capacity of genetic algorithms for numerical search.

This paper present two methods and their implementation. We show the results obtain in simulation reactive multiagent displacement control and

Keywords: self-reconfigurable, multiagent systems, reactive, neural network, and genetic algorithms.

1 Introduction

Modular self-reconfigurable robots and multiagents systems [1], [2], [7] are currently being subjects of investigation in several research institutes in the world. The aim of this research is the self-reorganization of modular robots to achieve a task under environmental conditions. One of the difficulties that can be found with this type of robots is the setting up of the control architecture for the distributed systems.

This paper considers modular self-reconfigurable robots developed in MAAM project [3], [4], as homogeneous and self-reconfigurable multi-agent systems. All the agents have the same competencies, same perceptions and same capacities.

The primary target is to study the collective displacement of the multi-agent systems according to a process of reorganization. The collective displacement is considered a very complex problem [5]. The number of possible solutions gives a combinative explosion in the graph of possible displacements. This problem has been addressed by using several control architectures inspired by artificial intelligence techniques.

Two different approaches are proposed: reactive algorithms and learning algorithms based on neural networks.

2 Simulated robot models

The simulation of the multiagent systems is carried out in a discreet 2D environment. Each agent makes the representation of the environment by its 8 neighbour cells. The input informations are memorized in 2 arrays of 8 cells. It is assumed that an agent can perceive the following:
- The direction of the attractor since it can detect the gradient's augmentation of the attractor appeal in its neighbour.
- Presence or not of another agent in its neighborhood (boolean).

An example of inputs of the world for a current agent is presented:

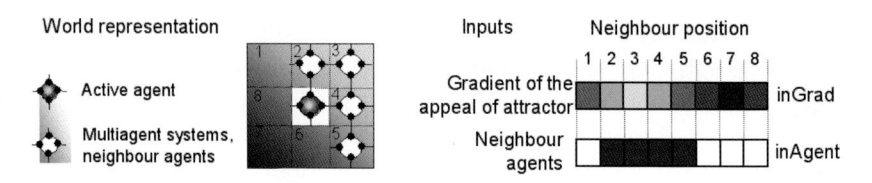

Fig. 1. Inputs representation of the world for an agent

The displacements of the agents are always through a face of another agent, performing connection in its four principal directions (four connections are 2,4,6,8 in **Fig. 1**). It is necessary to introduce geometrical restrictions for the displacement, authorize or not to go on the appeal's attractor direction and insure the group's cohesion in the process of reorganization, as shown in **Fig. 2**.

In the simulation, one module moves at a time so the problem of two modules moving into the same position at the same time is not addressed.

Methods for collective displacement of modular self-reconfigurable robots   643

Authorized displacements →

1. The active agent can move in its two possible free spaces

2. The agent can move in its two possible free spaces

Attractor

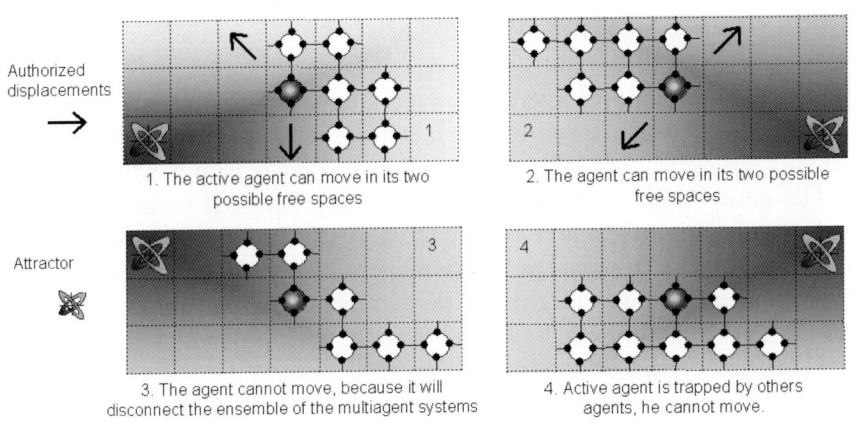

3. The agent cannot move, because it will disconnect the ensemble of the multiagent systems

4. Active agent is trapped by others agents, he cannot move.

Fig. 2. Authorized and unauthorized displacements

3 Reactive algorithms

Reactive algorithms are defined by a set of laws; it is a powerful technique when used with multiagent systems [6], [7]. The reactive control of agents is designed to act in its environment accordingly, using local's objectives defined by a set of actions. Emergent displacement behaviour is obtained through the process of the multiagent's reorganization.

The main principle of this model is that the agents will move by themselves in the attractant direction while geometrical constraints are respected. The representation of the world and competencies of the reactive agents is given in **Fig. 3**.

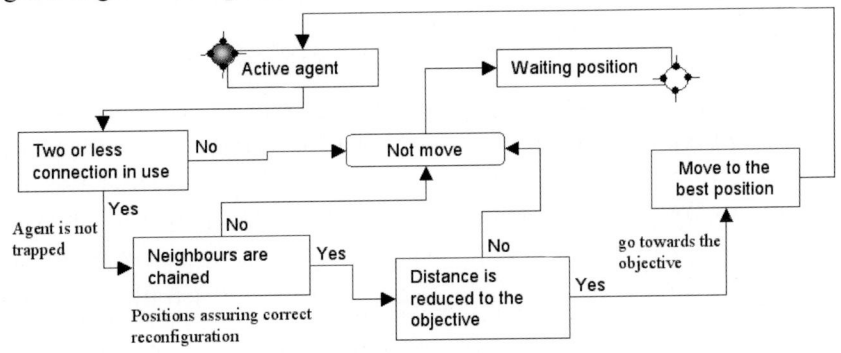

Fig. 3. Reactive algorithms

The multiagent systems framework is assumed to be without explicit communication i.e. the agents cannot interchange messages with each others.

4 The neural network

The brain is a body characterized by the interconnection of a high number of specialized cells: the neuron [8]. It is proposed to use an artificial neural network for each agent to learn about global displacement behavior.

The network architecture is inspired from biological neural functions, using the stretch reflex as reference to define the network architecture [9]; When muscle are stretches by external factor, this stretch is detected by the sensory neurons and transferred to the interneuron in the spinal chord. The stretch sensor could be defined by the input layers, which takes the input perceptions of the world. The interneuron is charged to analyze and command to the motor neuron, this interneuron is represented by the hidden layers. The command to contract or relax the muscle is send out by the motor neurons, represented by the eight neurons output layer.

Inputs to the neural network can be easily seen as circular sequence of connected pixels, each input give a row of pixels. The introduction of quality inputs for the correct learning of the neural network is addressed as a problem of image processing [13].

The input from the gradient is raised extending the levels of the values, to return to a dynamics ranging between 0-1 as the following mask:

$$maskGrad(i) = \frac{inGrad(i) - inGrad_{min}}{inGrad_{max} - inGrad_{min}} \quad (1)$$

The vector from neighbours agents is derived with equations (2) and (3). Derivation peaks represent the number of variations (free cells, agents), and the information of the neighbour chaining. In a second time we obtain the possible displacements with the positives results of the second derivation (4), using the following equations:

$$inAgent'(i - 1/2) = inAgent(i) - inAgent(i-1) \quad (2)$$

and

$$inAgent'(i + 1/2) = inAgent(i+1) - inAgent(i) \quad (3)$$

$$inAgent''(i) = inAgent'(i + 1/2) - inAgent'(i - 1/2) \quad (4)$$

In the case of **Fig. 1**, vector of neighbours agents is [0 1 1 1 1 0 0 0], and agent's mask of second derivation result in: [1 0 0 0 0 1 0 0], witch indicate the possible displacements.

The neural network layer's transfer function are tangential sigmoid, except for the neurons in the output layer, that use linear transfer function. This makes possible to learn non-linear and linear relationships between the input and the output of the neural network. A visualization of the number of the layers in the neural network, the number of neurons per layer, and the connections between layers are shown in **Fig. 4**. Tangential sigmoid (tansig) calculate their outputs according to:

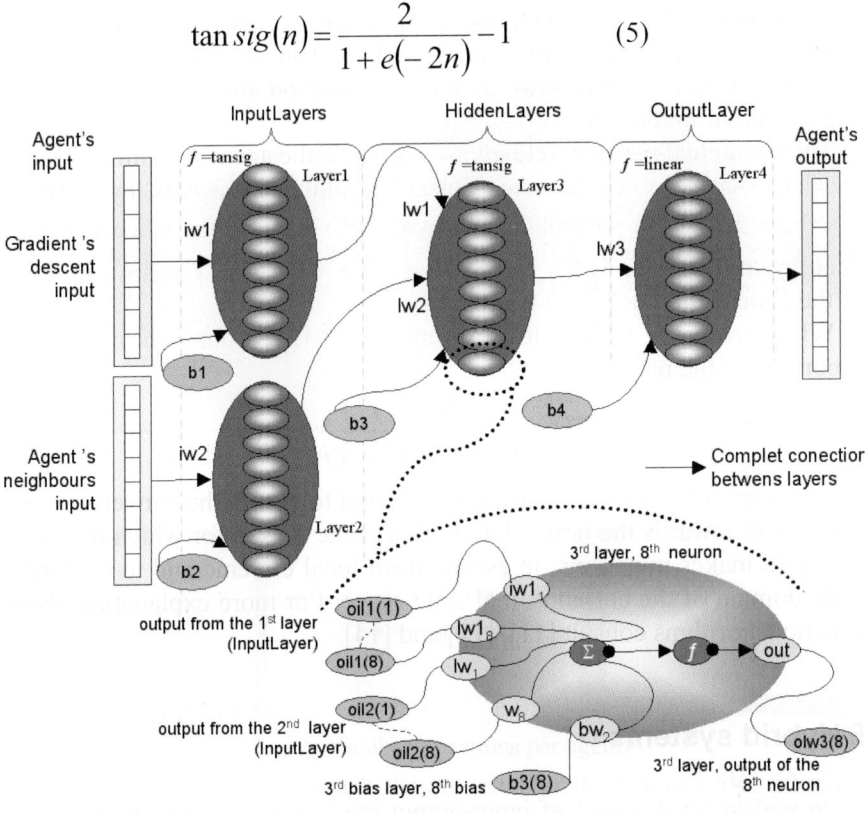

$$\tan sig(n) = \frac{2}{1+e(-2n)} - 1 \qquad (5)$$

Fig. 4. Neural network

Layers consist of several neurons, each neuron is constituted of inputs weight or layers weight neurons receiving information from external input or neuron's layer output, and carry out a weighting of the entries. The bias is a neuron parameter that is added with the neuron's weighted inputs and passed through the neuron's transfer function to generate the neuron's output. The number of outputs from a layer is equivalent to that of layer's neurons. The network output is a vector that represents the 8 directions of possible movements.

The learning method used for the neural network is the Levenberg-Marquardt backpropagation and genetics rules. The backpropagation algorithms updates weight and bias values according to Levenberg-Marquardt optimization. The application of Levenberg-Marquardt to neural network training is described in [10].

References

1. Brener N., Ben Amar F., Bidaud P. (2004) Analysis Of Self Reconfigurable Modular Systems A Design Proposal For Multi Modes Locomotion, , *Proceedings of the IEEE International Conference Robotics & Automation (ICRA'04)*, New Orleans, LA.
2. Bojinov H., Casal A., Hogg T. (2001) *Multiagent Control of Self-reconfigurable Robots*, Xerox Palo Alto Research Center
3. Gueganno C., Duhaut D. (2004) A hardware/software architecture for self reconfigurable robots, *Proceedings of the 7th international Symposium on Distributed Autonomous Robotics Systems (DARS 04)*.
4. www-valoria.univ-ubs.fr/Dominique.Duhaut/maam/publications.htm, MAAM project web site **[17/06/2005]**
5. Yoshida E., Murata S., Kamimura A., Tomita K., Kurokawa H. and Kokaji S. (2001) A Motion Planning Method for a Self-Reconfigurable Modular Robot, *Proceedings of the 2001 IEEE/RSJ International Conference on Intelligent Robots and Systems*.
6. Montreuil V., Duhaut D., Drogoul A. (2005) A collective moving algorithm in modular robotics: contribution of communication capacities, *Proceedings of the 6th IEEE International Symposium on Computational Intelligence in Robotics and Automation*, Espoo, Finland, June 27-30.
7. Stoy K., Nagpal R. (2004) "Self-Reconfiguration Using Directed Growth," *7th International Symposium on Distributed Autonomous Robotic Systems (DARS)*, France, June23-25.
8. Cornuéjols A., Miclet L. (2003*) Apprentissage artificiel, concepts et algorithmes*. EYROLLES.
9. Purves, Augustine, Fitzpatrick, Katz, Lamantia, McNamara, Williams (2003) *Neurosciences & cognition*, De Boeck Diffusion S.A.
10. Hagan, M. T., Menhaj M. (1994) "Training feedforward networks with the Marquardt algorithm," *IEEE Transactions on Neural Networks, vol. 5, no. 6, pp. 989-993*.
11. D. Montana (1995) Neural Network Weight Selection Using Genetic Algorithms, in *Intelligent Hybrid Systems, S. Goonatilake and S. Khebbal (eds.)*.
12. Lucidarme P. (2004) An evolutionary algorithm for multi-robot unsupervised learning, *Proceedings of the CEC, Portland, Oregon*.er.
13. Milgram M. (1993) *Reconnaissance des formes : méthodes numériques et connexionnistes*. A. Colin, Paris.
14. Goldberg D. E. (1998) *From Genetic and Evolutionary Optimization to the Design of Conceptual Machines*, Department of General Engineering, University of Illinois at Urbana-Champaign.

Suboptimal System Recovery from Communication Loss in a Multi-robot Localization Scenario using EKF Algorithms

P. Kondaxakis, V. F. Ruiz, W. S. Harwin

Department of Cybernetics, The University of Reading, UK
p.kondaxakis@reading.ac.uk

Abstract

This paper describes a multi-robot localization scenario where, for a period of time, the robot team loses communication with one of the robots due to system error. In this novel approach, extended Kalman filter (EKF) algorithms utilize relative measurements to localize the robots in space. These measurements are used to reliably compensate "dead-com" periods were no information can be exchanged between the members of the robot group.

Keywords: Multi-robots, Localization, EKF, Communication.

1 Introduction

The last decade has witnessed an increase in research on multi-robot localization. Often, team localization approaches rely on probabilistic Markov architectures and particle filtering [1], [2] or other stochastic methods, like distributed EKF techniques [3], [4].

In [1] relative measurements between N robots are fed to particle filters for pose estimation. A major drawback of this method is that each robot must maintain N-1 separate filters at all times, resulting in heavy computational requirements for large teams of robots. Another example can be found in [2] where the authors propose a fully distributed particle filter algorithm, which can cope with non-Gaussian distributions and dynamic environments. In this work only nearby platforms exchange information through an interactive communication protocol aimed at maximizing information flow by sending the most informative piece of evidence in their local memory.

EKF techniques for multi robot localization can be found in [3], [4] where the authors present a distributed Kalman filter localization approach assuming that robots can sense one another. Thus, whenever robots meet they measure relative distance and orientation difference.

Although the above methods rely heavily on communication between the robots in a group to exchange measurement and error belief information, they fail to provide an efficient solution in case of communication loss. This research provides a robust, novel approach to this problem occurring between two robot groups. Each group maintains a fully centralized state vector and utilizes some of the relative exteroceptive measurements to propagate the states of the robots unable to communicate and transmit their own odometry information. When the communication link is re-established, a simple merging Gaussian distribution algorithm [5] is employed to reintegrate the estimated states and error covariances into a single system.

2 Multi-robot localization before any communication loss

This work continues previous research conducted by the authors in the area of modular multi-robot localization [6]. The centralized EKF algorithm developed earlier takes into account all the critical cross-covariance terms and information interdependencies between the robots in the group and produces the most optimal position and orientation estimates. However, a real-time implementation of this algorithm for systems with large numbers of robots becomes prohibited due to the extensive communication and computational requirements. It is needed to reformulate this centralized estimator in order to be distributed amongst the robots. Due to limited space, the formulation of the distributed EKF for periods when communication between robots is available will not be presented here. Further details can be found in [3].

The motion of each robot used in these scenarios, is described by a "differential drive" configuration employing two passive wheels [6]. Each robot's state vector is composed by position (x, y) and orientation θ components, $\mathbf{X}_i(k) = [x_i(k), y_i(k), \theta_i(k)]^T$ with $k \in \mathbb{N}$. The recursive state space model is governed by non-linear stochastic difference equations, which describe the trajectory $\mathbf{X}_i(k+1)$ of each robot $(i, j = 1...I)$, $i \neq j$ in the group.

The measurement components of the non-linear function $h_n(\mathbf{X}_{i,j}(k+1), \mathbf{v}_n(k+1))$ express, in the local reference frame of robot i, the relative Cartesian distance and the orientation difference between robots i and j, as follows:

$$\mathbf{z}_n(k+1) = h_n(\mathbf{X}_{i,j}(k+1), \mathbf{v}_n(k+1)) = \begin{bmatrix} z_x(k+1) \\ z_y(k+1) \\ z_\gamma(k+1) \end{bmatrix} = \begin{bmatrix} (x_j - x_i)\cos\theta_i + (y_j - y_i)\sin\theta_i \\ -(x_j - x_i)\sin\theta_i + (y_j - y_i)\cos\theta_i \\ \arctan\left(\dfrac{y_j - y_i}{x_j - x_i}\right) - \theta_j \end{bmatrix} \quad (1)$$

where $\mathbf{z}_n(k+1)$, $n = 1, 2 \ldots N$, is the nth measurement vector.

The above models are subjected to standard linearization procedure using first order Taylor expansion (detailed in [6]) and fed in the distributed version of the centralized EKF algorithm to collectively localize a group of robots into an unknown environment. Until now it is assumed that all robots can broadcast their estimated state and covariance elements to the other members of the team.

3 Multi-robot localization with communication failures

While the distributed EKF approach provides an ideal solution for real-time implementation of large robot groups and assumes that groups can obtain relative measurements at all times. Thus, it is not sufficient to deal with situations of limited communication.

To better understand the problem, the first step is to analyze the communicational requirements of the distributed EKF algorithm. Firstly, every robot keeps its own state and error covariance matrix estimations. At every update, the error covariance interdependencies of the initial centralized algorithm are calculated locally to the robots associated with them. Every time a new measurement is introduced into the centralized system, the distributed algorithm needs to exchange the robots' predicted error covariances and state vectors in order to further update them. Thus at every update, each robot must possess the covariance and cross-covariance terms of the centralized matrix related with it, as well as the robots' state values linked to the available relative or absolute measurement. If a single robot loses communication with the rest of the team, it cannot exchange the necessary information to conduct an update.

Our egocentric approach reformulates the state space model equations in order to algebraically decouple the centralized EKF algorithm. With this technique each robot can individually propagate and update the centralized state vector and covariance matrix even if it cannot obtain the necessary odometry and exteroceptive measurement information from the other robots. To facilitate this, each robot can use the relative exteroceptive measurements combined with its own odometry data to indirectly propagate the poses of the robots with limited communication capabilities.

If in a group of three mobile robots (R1, R2, R3) R1 suffers from communication failures, the group is viewed as two separate teams. The first team (A) contains R1 while the second team (B) R2 and R3. Each of the teams propagates and updates its own version of the previously common centralized state vector and error covariance matrix by using some of the relative measurement vectors associated with the robots in the opposite team.

However, exteroceptive relative measurements may not be obtained in the same frequency as odometric data. Therefore, the position and orientation of the robots in an opposite team is propagated and updated only when relative measurements associated with them become available. At all other times they are assumed to remain stationary. An ideal algorithm that can reduce the processing requirements for stationary objects in a centralized EKF formulation can be found in [7] and was initially used to keep track of stationary landmarks in the SLAM framework. The authors describe an alternative formulation of the standard EKF equations that "compress" the elements of covariance, cross-covariance, and state components associated with stationary landmarks as long as no measurements are obtained between a moving robot and the landmarks. Employing such an algorithm in our approach reduces the processing and memory requirements by "storing" the above elements related to a particular robot from which no relative measurements and communication can be obtained at that time.

For the group of three mobile robots described above, we illustrate how R1 state and covariance elements are propagated from team's (B) point of view using relative measurements collected from R2. Utilizing relative measurements obtained from R2 the non-linear state transition function $f_1(\mathbf{X}_2(k), \mathbf{w}_2(k), \mathbf{w}_2(k+1))$ is given by:

$$f_1(\mathbf{X}_2(k), \mathbf{w}_2(k), \mathbf{w}_2(k+1)) = \begin{bmatrix} z_{x21}(k+1)\cos(\theta_2(k)+\dot{\theta}_2(k)\delta T) - z_{y21}(k+1)\sin(\theta_2(k)+\dot{\theta}_2(k)\delta T) + x_2(k) + SP_2(k)\delta T \cos\theta_2(k) \\ z_{x21}(k+1)\sin(\theta_2(k)+\dot{\theta}_2(k)\delta T) + z_{y21}(k+1)\cos(\theta_2(k)+\dot{\theta}_2(k)\delta T) + y_2(k) + SP_2(k)\delta T \sin\theta_2(k) \\ \arctan\left(\frac{z_{x21}(k+1)\sin(\theta_2(k)+\dot{\theta}_2(k)\delta T) + z_{y21}(k+1)\cos(\theta_2(k)+\dot{\theta}_2(k)\delta T)}{z_{x21}(k+1)\cos(\theta_2(k)+\dot{\theta}_2(k)\delta T) - z_{y21}(k+1)\sin(\theta_2(k)+\dot{\theta}_2(k)\delta T)}\right) - z_{y21}(k+1) \end{bmatrix} \quad (2)$$

where $SP_2(k)$ and $\dot{\theta}_2(k)$ is the R2's linear speed and rotational velocity respectively. To linearize equation (2), the first order Taylor expansion is applied about the R2's current estimated state vector $\hat{\mathbf{X}}_2^+(k)$. Furthermore, the *state noise transition matrix* $\mathbf{G}_1(k)$ is obtained by linearizing (2) about the rotational velocity of the R2's wheels $\omega_{mes\,L,R}(k)$, distance $z_{\rho 21}(k+1)$, bearing $z_{\varphi 21}(k+1)$, and orientation $z_{\gamma 21}(k+1)$ measurements contaminated by white noise (e.g. $\omega_{mes\,L,R}(k) = \omega_{L,R}(k) - w_{\omega L,R}$) resulting in:

$$\begin{bmatrix} \Delta \hat{x}_1^-(k+1) \\ \Delta \hat{y}_1^-(k+1) \\ \Delta \hat{\theta}_1^-(k+1) \end{bmatrix} = \begin{bmatrix} 1 & 0 & -z_{x21}(k+1)\sin\hat{\theta}_2^-(k+1) + z_{y21}(k+1)\cos\hat{\theta}_2^-(k+1) + SP_2(k)\delta T \sin\hat{\theta}_2^+(k) \\ 0 & 1 & z_{x21}(k+1)\cos\hat{\theta}_2^-(k+1) - z_{y21}(k+1)\sin\hat{\theta}_2^-(k+1) + SP_2(k)\delta T \cos\hat{\theta}_2^+(k) \\ 0 & 0 & 1 \end{bmatrix} \begin{bmatrix} \Delta \hat{x}_2^+(k) \\ \Delta \hat{y}_2^+(k) \\ \Delta \hat{\theta}_2^+(k) \end{bmatrix}$$

$$+ \begin{bmatrix} -r/d\,\delta T(z_{x21}(k+1)\sin\hat{\theta}_2^-(k+1) & r/d\,\delta T(z_{x21}(k+1)\sin\hat{\theta}_2^-(k+1) & \frac{1}{z_{\rho 21}(k+1)}(z_{x21}(k+1)\cos\hat{\theta}_2^-(k+1) & -z_{x21}(k+1)\sin\hat{\theta}_2^-(k+1) & 0 \\ +z_{y21}(k+1)\cos\hat{\theta}_2^-(k+1)) & +z_{y21}(k+1)\cos\hat{\theta}_2^-(k+1)) & -z_{y21}(k+1)\sin\hat{\theta}_2^-(k+1)) & -z_{y21}(k+1)\cos\hat{\theta}_2^-(k+1) & \\ +r/2\,\delta T\cos\hat{\theta}_2^+(k) & -r/2\,\delta T\cos\hat{\theta}_2^+(k) & & & \\ r/d\,\delta T(z_{x21}(k+1)\cos\hat{\theta}_2^-(k+1) & r/d\,\delta T(-z_{x21}(k+1)\cos\hat{\theta}_2^-(k+1) & \frac{1}{z_{\rho 21}(k+1)}(z_{x21}(k+1)\sin\hat{\theta}_2^-(k+1) & -z_{x21}(k+1)\cos\hat{\theta}_2^-(k+1) & 0 \\ -z_{y21}(k+1)\sin\hat{\theta}_2^-(k+1)) & +z_{y21}(k+1)\sin\hat{\theta}_2^-(k+1)) & +z_{y21}(k+1)\cos\hat{\theta}_2^-(k+1)) & +z_{y21}(k+1)\sin\hat{\theta}_2^-(k+1) & \\ +r/2\,\delta T\cos\hat{\theta}_2^+(k) & -r/2\,\delta T\cos\hat{\theta}_2^+(k) & & & \\ r/d\,\delta T & -r/d\,\delta T & 0 & 1 & 1 \end{bmatrix} \begin{bmatrix} w_{\omega L 2} \\ w_{\omega R 2} \\ w_{x21} \\ w_{y21} \\ w_{\gamma 21} \end{bmatrix} \quad (3)$$

or

$$\Delta \hat{\mathbf{X}}_1^-(k+1) = \mathbf{\Phi}_1(\hat{\mathbf{X}}_2^+(k)) \times \Delta \hat{\mathbf{X}}_2^+(k) + \mathbf{G}_1(\hat{\mathbf{X}}_2^+(k), \hat{\mathbf{X}}_2^-(k+1)) \times \mathbf{w}_2(k)$$

where $z_{\rho 21}^2 = z_{x21}^2 + z_{y21}^2$ and $\mathbf{\Phi}_i(\hat{\mathbf{X}}_i^+(k))$, is the *state transition matrix* of robot i. The radius of left and right wheels is denoted by r and d is the distance from the centre of the robot to the wheels.

From (2), (3) it is clear that the proposed algorithm requires an alternative operating sequence than the usual *prediction-update* EKF cycle. Some of the relative measurements, otherwise obtained before any update of the Kalman filter, are used to propagate the state vectors of any robots unable to communicate odometry information. Thus, it is essential to first acquire the relative pose measurements before performing the *prediction-update* EKF cycle.

3.1 Collective localization with limited communication

Having obtained the linearized recursive state transition model for both robot teams (A) and (B), when each team can collect at least one relative measurement vector for every robot in the opposite party, the version of the centralized system for (B) can be written in a block matrix form resulting in:

$$\begin{bmatrix} \Delta \hat{\mathbf{X}}_1^-(k+1) \\ \Delta \hat{\mathbf{X}}_2^-(k+1) \\ \Delta \hat{\mathbf{X}}_3^-(k+1) \end{bmatrix}_{23c} = \begin{bmatrix} 0 & \mathbf{\Phi}_1(\hat{\mathbf{X}}_2^+(k)) & 0 \\ 0 & \mathbf{\Phi}_2(\hat{\mathbf{X}}_2^+(k)) & 0 \\ 0 & 0 & \mathbf{\Phi}_3(\hat{\mathbf{X}}_3^+(k)) \end{bmatrix}_{23c} \begin{bmatrix} \Delta \hat{\mathbf{X}}_1^+(k) \\ \Delta \hat{\mathbf{X}}_2^+(k) \\ \Delta \hat{\mathbf{X}}_3^+(k) \end{bmatrix}_{23c} + \begin{bmatrix} 0 & \mathbf{G}_1(k) & 0 \\ 0 & \mathbf{G}_2(k) & 0 \\ 0 & 0 & \mathbf{G}_3(k) \end{bmatrix}_{23c} \begin{bmatrix} \mathbf{w}_1(k) \\ \mathbf{w}_2(k) \\ \mathbf{w}_3(k) \end{bmatrix}_{23c} \quad (4)$$

For (A) the centralized approach estimates the motion of the robot group in a $2 \times I$-dimensional space as follows:

$$\begin{bmatrix} \Delta \hat{X}_1^-(k+1) \\ \Delta \hat{X}_2^-(k+1) \\ \Delta \hat{X}_3^-(k+1) \end{bmatrix}_{1c} = \begin{bmatrix} \Phi_1(\hat{X}_1^+(k)) & 0 & 0 \\ \Phi_2(\hat{X}_1^+(k)) & 0 & 0 \\ \Phi_3(\hat{X}_1^+(k)) & 0 & 0 \end{bmatrix}_{1c} \begin{bmatrix} \Delta \hat{X}_1^+(k) \\ \Delta \hat{X}_2^+(k) \\ \Delta \hat{X}_3^+(k) \end{bmatrix}_{1c} + \begin{bmatrix} G_1(k) & 0 & 0 \\ G_2(k) & 0 & 0 \\ G_3(k) & 0 & 0 \end{bmatrix}_{1c} \begin{bmatrix} w_1(k) \\ w_2(k) \\ w_3(k) \end{bmatrix}_{1c} \quad (5)$$

From (4) and (5), both robot teams (A) and (B) estimate the centralized robots' trajectory from their own point of view with no need of communication between them. Equation (4) shows that although R2 and R3, which can broadcast information with each other, are estimated independently, the trajectory of R1 depends only on the motion of R2. The same holds in (5) were R2's and R3's trajectories depend on the motion of R1.

When no relative measurements are available between the groups, the *state transition matrices* $\Phi_i(\hat{X}_i^+(k)), i = 1...I$, related to the robots in the opposite party, become unity matrices and an average time invariant value for $G_i(k)$ is calculated according to [4]. The centralized system formulation for (A) and (B) assumes its previous state [6].

Upon restoration of communication between (A) and (B), the individual centralized state vectors and error covariance matrices of the two teams are merged into one global state vector and error covariance matrix as proposed in [5]:

$$P_c^+(k) = Pl_c^+(k) - Pl_c^+(k)[Pl_c^+(k) + P23_c^+(k)]^{-1} Pl_c^+(k) \quad (6)$$

$$\hat{X}_c^+(k) = \hat{X}l_c^+(k) + Pl_c^+(k)[Pl_c^+(k) + P23_c^+(k)]^{-1}(\hat{X}23_c(k) - \hat{X}l_c(k)) \quad (7)$$

If communication is reestablished at a point where no relative measurements can be obtained between teams (A) and (B), they have to propagate the last known pose of the formerly no-communicating robots, into the new robot location. It is assumed that each robot in the group is capable of sensing when its location is relatively measured by another robot and saves its sequential odometric data until a next relative measurement is detected. This technique allows every robot to transmit the stored odometric information whenever an opposite team requires to post-process this data and propagate its position for centralized state and covariance reintegration.

4 Simulated results

This section illustrates the simulated results of a 2D localization system, featuring three mobile robots, degraded by communication failures. A comparison between an optimal distributed EKF with the proposed suboptimal approach suggests that this method can effectively manage "deadcom" periods in a multi-robot scenario with acceptable localization results.

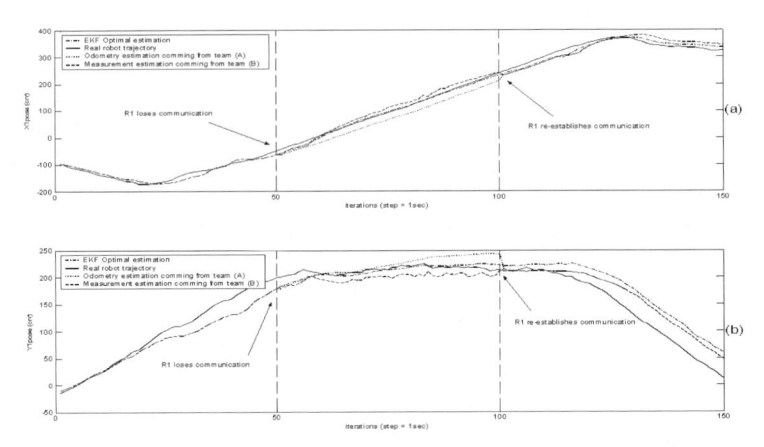

Fig. 1. Graphs (a) and (b) illustrate the x and y- actual and estimated trajectories of R1 for the optimal and suboptimal scenarios.

This experiment runs for 150 iterations and the filter initial conditions assume that the same level of uncertainty corrupts the robots' starting positions $x_i(0|0)$ and $y_i(0|0)$. The robots can move freely, with a constant linear speed of $6 cm/\sec$, in an $10m \times 10m$ obstacle-free rectangular environment. Each of the robots' wheel encoders is subject to noise errors characterized by a zero-mean normal distribution with a standard deviation of $\sigma_{\omega L,R} = 1 rad/\sec$. The exteroceptive relative measurements are corrupted by Gaussian noise with $\sigma_\rho = 3cm$ for the distance, $\sigma_\varphi = 5°$ for bearing and $\sigma_\gamma = 10°$ for relative orientation measurements. An assumption of this simulation is that relative measurements are available at every filter's *prediction-update* cycle.

For the first 50 EKF iterations, the robots maintain their regular communication capabilities producing the most optimal pose estimation. At the 50th iteration and for the next 50 seconds, R1 is incapable to communicate information with the rest of the robots and thus the group is separated into two different teams (A) and (B). For that period of time the proposed egocentric approach produces near optimal pose estimates that are re-integrated into one global state and error covariance estimate at the 100th iteration when team (A) re-establishes communication with team (B). After the 100th iteration the system assumes its previous optimal collective localization approach.

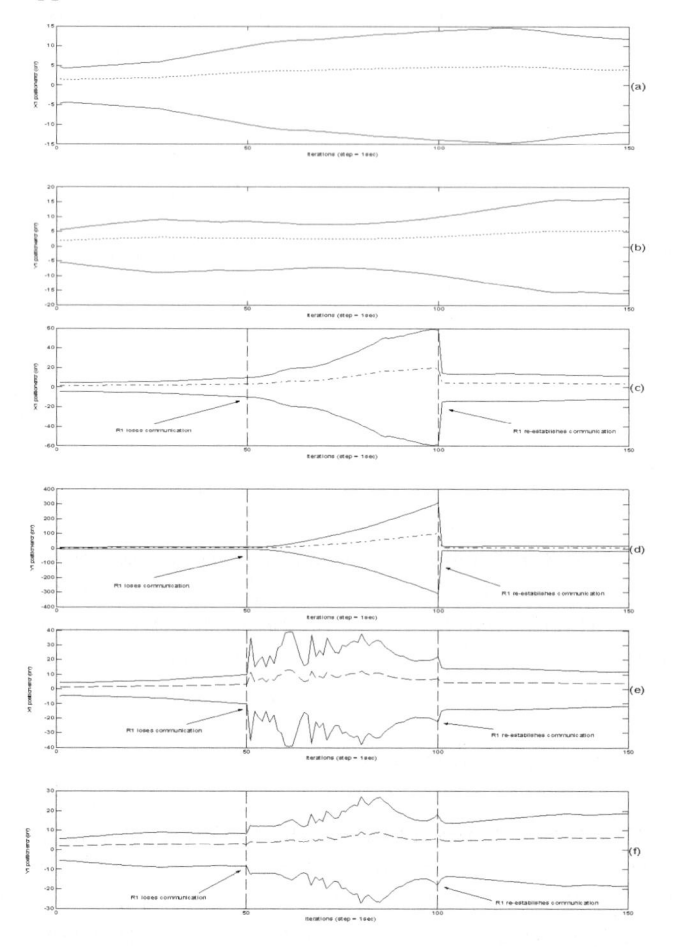

Fig. 2. The positioning errors of x and y- components are displayed in graphs (a) and (b) for optimal scenario and (c) to (f) for the suboptimal case. The solid bounding lines in (a)-(f) determine the 3σ regions of confidence.

In **Fig. 1.**, graphs (a) and (b) illustrate the x and y- actual and estimated trajectories of R1 obtained for the cases when: i) communication is constantly available for all 150 iterations and ii) R1 cannot communicate between iterations 50 and 100. **Fig. 2.**, graphs (a) through (f) display the positional uncertainty associated with the optimal and suboptimal cases. In the same figure and for the suboptimal scenario, graphs (c) and (d) illustrate the positional uncertainty of R1 (team (A)). The recorded error depends exclusively on R1's odometry and it grows rapidly without bounds due to lack of any other useful exteroceptive information. Once communication is re-established the error drops instantaneously obtaining its near optimal levels. Graphs (e) and (f) indicate the positional uncertainty of R1 (team (A)) again, but this time from team's (B) point of view using R2's relative measurement vector.

During this experiment, the maximum recorded estimated error for the x and y- components of the optimal case (**Fig. 2.**, graphs (a) and (b)) is 4.9cm and 5.36cm respectively with mean values of 3.59cm and 3.33cm. Although in the suboptimal scenario the error obtains larger maximum and mean values due to communication-limited periods (e.g. 19.71cm max and 5.9cm mean values for **Fig. 2.**, graph (c) and 9.08cm max and 4.65cm mean for graph (e)), at the point where communication is re-established between all the robots in the group the positional uncertainty drops to its near optimal values. For example around the 100th iteration the positional error acquired from graphs (a) and (b) in **Fig. 2.**, is 4.64cm and 3.36cm for the x and y-directions respectively. On the other hand, around the same iteration, the recorded uncertainty of the suboptimal approach is 4.95cm and 4.99cm for the x and y-directions respectively, as indicated in graphs (c) through (f). All these results suggest that the proposed framework provides a near optimal solution to the communication loss in a multi-robot localization scenario.

5 Conclusion and future work

This publication describes a novel approach on the problem of recovering from temporary communication failures in a multi-robot localization scenario using EKF algorithms. The proposed method divides a group of robots, initially viewed as a unified entity, into teams unable to communicate with each other for periods of time. Each team maintains a fully centralized state vector and utilizes some of the relative exteroceptive measurements to indirectly propagate the states of the robots in the other available teams that are unable to transmit their own odometry information. Upon communication re-establishment, the centralized state vector and error covariance matrix of each of team is merged into a single unified system using a simple merging Gaussian distribution algorithm. The simulated results indicate that this approach achieves near optimal pose estimation values when communication is re-established amongst the robots.

It is planned to test the efficiency of the proposed framework under SLAM scenarios where each robot team will have access to relative and absolute landmark measurements. Furthermore, a complete study will evaluate the framework's suboptimality levels for cases where each of the robots acquires relative measurements with different frequencies or where a mobile robot group is divided into more that two teams and each team is occupied by more than one robot.

Finally, it is in the authors' intentions to implement and evaluate this framework in a real robot scenario.

Acknowledgements

This work is supported by The Leverhulme Trust under a Research Project Grant (Ref F/00239/O).

References

1. Howard, A., Mataric, M. J. and Sukhatme, G. S. 2003. "Cooperative Relative Localization for Mobile Robot Teams: An Ego-centric Approach". In *Multi-Robot Systems: From Swarms to Intelligent Automata*, Kluwer, pp. 65-76.
2. Rosencrantz, M., Gordon, G. and Thrun, S. 2003. "Decentralized sensor fusion with distributed particle filters" In *Proceedings of the Conference on Uncertainty in AI (UAI)*, Acapulco, Mexico.
3. Roumeliotis, S. I. and Bekey, G. A., 2002. "Distributed Multi-Robot Localization", *IEEE Transactions on Robotics and Automation*, Vol 18, No 5, Oct, pp. 781-795.
4. Roumeliotis, S. I. and Rekleitis, I. M., 2003. "Analysis of Multirobot Localization Uncertainty Propagation", *Proceedings of the 2003 IEEE/RSJ International Conference on Intelligent Robots and Systems(IROS'03)*, 27-31 Oct, Las Vegas, Nevada, pp. 1763-1770.
5. Marcelino, P., Nunes, P., Lima, P. and Ribeiro, M. I. 2003. "Improving Object Localization Through Sensor Fusion Applied to Soccer Robots". In *Actas do Encontro Científico 3º Festival Nacional de Robótica - ROBOTICA2003*, Lisboa, 9 de Maio.
6. Kondaxakis, P., Ruiz, V. F. and Harwin, W. S. 2004. "Compensation of Observability Problem in a Multi-Robot Localization Scenario using CEKF". In *Proceedings of 2004 IEEE/RSJ International Conference on Intelligent Robots and Systems(IROS'04)*, Vol 2, Sendai, Japan, 28 Sept- 2 Oct, pp. 1762-1767.
7. Guivant, J. and Nebot, E. M. 2002. "Improved Computational and Memory Requirements of Simultaneous Localization and Map Building". In *Proceedings 2002 IEEE International Conference on Robotics and Automation*, Zurich, Switzerland, May, pp. 2731-2736.

ORTHO-BOT: A Modular Reconfigurable Space Robot Concept

V. Ramchurn, R. C. Richardson and P. Nutter

School of Computer Science, The University of Manchester, Kilburn Bldg., Oxford Road, Manchester, M13 9PL - UK.
vramchurn@cs.man.ac.uk

Abstract

A new set of challenging tasks are envisaged for future robotic planetary space missions. In contrast to conventional exploration rovers, industrial robotic roles are required for object manipulation and transportation in e.g. habitat construction. This prompts research into more robust failsafe robot designs, having greater mission redundancy for cost-effectiveness, with adjustable structures for multi-tasking. A Modular Reconfigurable design is investigated to meet these requirements using linear actuation over revolute since this alternative approach to modular robotics can form truss type structures providing inherently stable structures appropriate to the given task type. For ease of reconfiguration a connectivity solution is sought that may be simple enough to allow self-reconfiguration thus enabling extremely remote autonomous operation. In effort to meet this challenge the *ORTHO-BOT* developmental concept is introduced in this paper. Based on the core module developed thus far, a walking design has been successfully demonstrated in simulation to fulfil the key requirement of locomotion. Though the focus for this research is aimed at space-based roles conceptual solutions developed should also find useful application in terrestrial remote or hazardous environments.

Keywords: Modular robot, Prismatic actuation, Connectivity, Industrial, Simulation.

1 Introduction

Future planetary exploration plans envisage increased robotic involvement in Martian as well as intermediate Lunar missions [1]. Precursor robotic missions ahead of manned missions to Mars are one example where autonomy is required versus teleoperation due to the time lags in signals to and from any Earth-based mission control. The typical autonomous tasks envisaged are more akin to industrial activity [2] as compared with previous exploration missions (e.g. use of rovers). Task types include load transportation and handling, site preparation and power system/resource utilisation deployment, upon the basic requirement for locomotion.

Previous work has been undertaken to investigate solution concepts including Modular Reconfigurable (MR) options [3]. However a level of future technology maturity has been assumed in what are often concepts that exhibit many complex moving parts, whereas the approach outlined in this paper attempts a solution based on currently available technologies with novel geometrical considerations to produce a physically feasible concept.

With industrial type tasks being required a traditional approach based upon the use of trussed structures for strength and stability is taken as a starting point, where prismatic joints, also known as linear actuators (or "linacts" for the purposes of this report), are considered best to form the basis of the robot modules and primary robot dynamic.

Robotic solutions capable of performing the above types of functions can also be considered applicable to terrestrial remote or hazardous tasks, such as deep underwater or mining operations and disaster relief in earthquake or landslide zones.

Within this project two main areas are considered for development: configuration planning and physical connectivity solutions. Configuration planning investigates how to restructure the MR robot to perform a given task, such as number of modules required and potential operational optimisation given the inherently adaptable architecture in order to carry out its job better. Physical connectivity, representing the primary area of this research to date, investigates docking through a simple mechanism, simple enough to potentially allow for self-reconfiguration, this being balanced against the need for dynamic flexibility to carry out a variety of tasks. Degrees Of Freedom (DOF) issues in conceptualising novel joint characteristics are considered as an overriding factor here.

2 MR approaches

The basic design principle behind MR robotics replaces a single very capable robotic unit with many individual, simpler modules that, though not necessarily being very capable alone, can be connected together to cooperate in a more complex, capable machine with benefits including reconfigurability, ease of defective component replacement and economies of scale.

A variety of MR robot types have been developed. A recent review of the state of development in the field of MR robotics was carried out [4]. There exist two options for modular design approach: homogeneous – all modules identical, and heterogeneous – a variety of module types constituting the robot system. The vast majority of MR robots are of the homogeneous type, such as *Polybot* [5]. Here each and every module comprises numerous integrated functions.

For practicality in space-based roles, a heterogeneous MR approach is favoured where a core module design could constitute the main robot body parts with other specialised types being available for attachment e.g. power supply and end-effector modules e.g. cameras or tools. For ease of connectivity, genderless modular interfaces are desirable as are failsafe mechanisms for autonomous rejection/replacement of faulty modules.

Generally, designs tend to be small-scale, lacking the desired robustness to load-bearing tasks, since being driven by locomotive rather than construction type task considerations. Centralised control also poses a redundancy problem as does a lack of genderless connections for easy docking.

Of previously developed MR concepts, a relevant robust and scale appropriate design is *Tetrobot* [6]. Employing prismatic joints (linacts), this design is based upon a parallel robotic architecture providing extensive reconfiguration. However, awkward modular connectivity results from implementing the maximum three Degrees of Freedom (DOF) of rotation in their spherical joints that also allow for large multi-link potential at each of these vertices. The use of electric linacts would be suited to the extreme conditions of space and they can accurately hold fixed lengths. They are also very scalable designs [7].

Docking is the key issue in producing self-reconfigurable robots [8]. This has to be a global operation (rather than local to a couple of modules), a distributed action, and efficiency and reliability are required since docking will need to be performed frequently in a self-reconfigurable robot.

3 Connectivity towards reconfiguration

It is considered that there is a trade-off between ease of connectivity and the number of DOFs possible at attachment sites between modules (joint space DOF). It is proposed that multi-functional MR systems can still be realised with reduced joint DOF in order for simpler connectivity towards enabling self-reconfiguration. This is considered possible through novel geometrical arrangements though capability is also recognised as being dependent upon parallel control scheme development.

Firstly, a linact based parallel modular structure is selected for investigation for the aforementioned reasons, providing 1-DOF translation. Multi-link capability is also reduced to consideration of singular connectivity to simplify the problem for the time being. Given this modular simplification, designs are investigated that explore the extent to which truss type structures can be realised in useful configurations.

With only a 1-DOF joint space rotation capability, only two-dimensional dynamic structures are possible – any three-dimensional arrangement conceivable is over-constrained and incapable of motion. Moving up to 2-DOF rotational designs allows viable three-dimensional structures. Fig. 1. depicts the level of freedom capable in this regime.

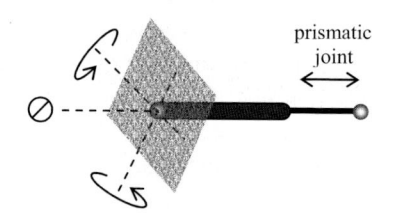

Fig. 1. Prismatic joint (linact) with 2-DOF rotation capability

One such 2-DOF design based on an angular bevel gear chain has been developed that balances strength with high dexterity [9]. However this, as in other examples, suffers from mechanical complexity in aligning the joints and would therefore be difficult to operate in a self-reconfiguring scheme.

While the above focuses on physical design, it is recognised that control scheme development is a critical aspect in developing self-reconfigurable designs.

4 Reduced DOF concept: *ORTHO-BOT*

A new modular concept is designed based on the use of linacts and a novel split-toroidal joint mechanism as depicted in Fig. 2. The module body comprises an in-line linact [7] with split-toroids capable of opening at each end but being spring-loaded to remain closed in the absence of any force.

The toroidal geometry allows one end to connect to another module end producing a genderless joint capable of 2-DOF rotation. By moulding the axis of the linact into a series of grooves with diameters equal to that of the toroids, additional connection sites are enabled along the module's length, orthogonally constrained with 1-DOF rotation. These axial sites could also include positions on the extended part through the use of a pop-up constraining wire (as pictured) or possibly a telescopic mechanism to the same effect. A passive (but lockable) revolute joint is included at one end for additional flexibility.

Three modes are conceived for connection: (1) free to *Latch*/unlatch from a connection site by pushing on/pulling off; (2) *Docked* – free to rotate but not unlatch; (3) *Locked* – connection site geometry held fixed.

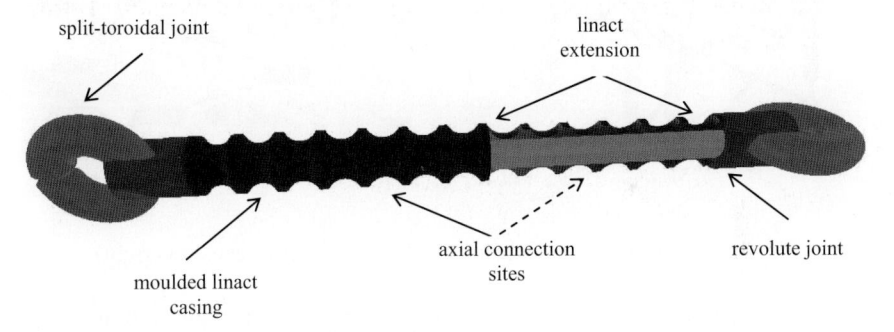

Fig. 2. Core *ORTHO-BOT* Module

By following the heterogeneous design principle, specialist modules and end-effectors (e.g. cameras, power units, manipulator arms) are conceived having compatible grooved handle attachments for the toroidal joint to attach to, in similar fashion as to axial site connections.

Thus far the design is considered neglecting detailed attention to integration and production issues such as: command and control architecture; modular interfacing; and suitable materials choices for component hardware.

5 Configurations investigated

Based on core *ORTHO-BOT* modules, the key capability of locomotion is considered. One regime previously investigated that would be suitable for this type of design is based on inducing rolling motions by shifting robot Centre of Mass (CoM) [10]. Within a closed system of linacts connected into a faceted "wheel" shape, the structure can be tipped over onto adjoining facets through correctly sequenced actuation.

For unpredictable environments a legged configuration may be desirable for obstacle negotiation. To this end, a walking capability has been demonstrated in *Nastran* simulation of a hexapod system. A minimum of sixteen modules are utilised to form the stable structure depicted in Fig. 3a (some linacts reduced to lines for clarity).

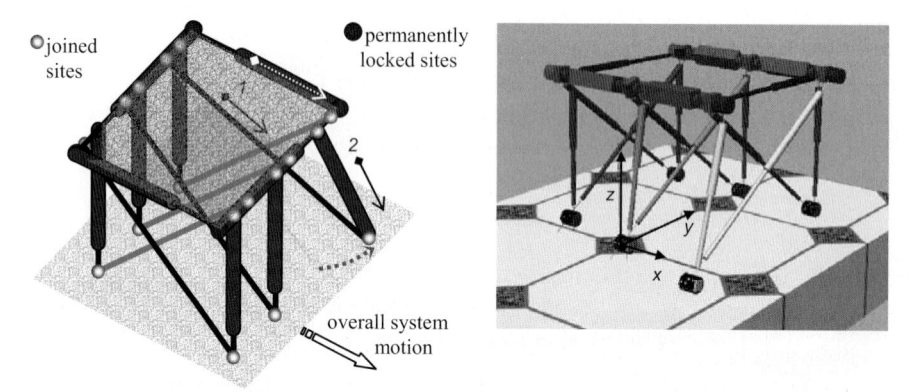

Fig. 3. (a) Hexapod concept (left); **(b)** Simulation screenshot (right)

Forward system motion is achieved through the following actuation sequence: (1) rear leg extension to swing foot forward; (2) front leg extension to place foot down a short distance ahead; *[repeated for each of the 3 front legs to step out]* (3) side linacts extend (white arrows) to square up body shape at front; (4) rear legs move forward in turn; (5) side linacts retract returning system to original pose; *[Sequence restarts]*.

For simulation simplicity the toroidal geometry was replaced with revolute joint combinations to produce representative interactions (Fig. 3b).

Fig. 4. opposite indicates the response of one of the feet stepping out under simple PID length control within the simulation, with each foot following the same profile within the sequence of moves described above (Cartesian axes as indicated in Fig. 3b).

Such a walking configuration can be perceived in a transportation role where the linact connection sites could allow the carriage of additional modules on the robot's "back" (upper plane) for future reconfiguration (given appropriate arrangement or the use of specialist manipulator arm type modules).

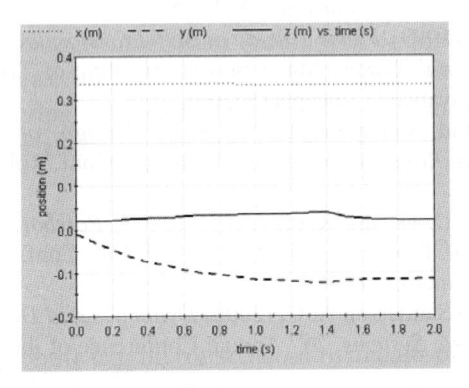

Fig. 4. Stepping forward foot profile

6 Conclusion

A modular robotic design has been produced based on a novel split-toroidal joint geometry. Though capable of exhibiting a maximum of only 2-DOF rotational connection flexibility, configurations have been conceived to address the key capability of locomotion needed for planetary industrial type tasks. In particular, an initial walking capability has been successfully demonstrated in simulation as proof of concept.

With self-reconfiguration being an ultimate goal, the connection geometries described here are to be further developed to assess design feasibility. This will include greater consideration of how to physically enable the interlocking modes required.

Future work will also investigate configuration planning and distributed control schemes.

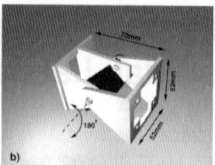

Fig. 1. a) The three modular configurations constructed, composed of two and three Y1 modules: Pitch-Pitch (PP), Pitch-Yaw-Pitch (PYP) and three-modules star. **b)** A cad rendering of the Y1 modules

but they are planned to be used in space applications[7] and urban search and rescue[6].

The amount of different configurations growth exponentially with the number of modules and there is no geometrical limitation to the total number of modules. In this paper we focuses on the minimum number of modules needed to achieve locomotion and to perform motions like lateral rolling[8] and lateral shift. Also, the study of motion of these minimal configurations is developed for a better understanding of the locomotion's properties of more complex configurations.

Three modular robots using one-degree-of-freedom modules are presented (**Fig.1a**). The simplest one has only two modules and it is capable of moving forward and backward. Adding just one more module, three new types of locomotion appears: 2D sinusoidal locomotion, lateral rolling and lateral shift. The control of the movement is based on sinusoidal waves and the coordination is achieve by changing both amplitude and phase.

2 Construction of the modular configurations

The three configurations developed are based on the Y1 modules (**Fig.1b**), designed for the Cube worm-like robot[9]. These modules have only one degree of freedom actuated by a servomotor. All the electronic and power supply are located outside. They were designed for rapid modular robot prototyping and testing. The rotation range is 180^0 and the dimensions are 72x52x52 mm, as shown in **Fig.1b**. These modules are inspired in Polybot G1, designed by Mark Yim at PARC.

Y1 modules can be connected one to each other maintaining the same orientation, so that, they can only rotate on the pith axis. This connection is used for the construction of a chain of modules that rotates in the plane perpendicular to the ground, like in Cube worm-like robot. There is another kind of connection. One module can be rotated 90 in the roll axis and connected to another Y1 module. This configuration has pitch and yaw axis.

3 Configuration 1: Pitch-Pitch (PP)

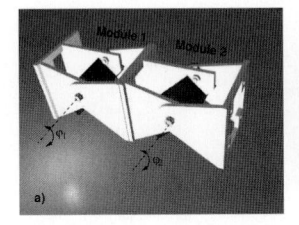

Fig. 2. Pitch-Pitch (PP) configuration, composed of two Y1 modules connected in the same orientation.

This configuration is constructed attaching two Y1 modules as shown in **Fig.2**. Experiments show that this configuration can move on a straight line, backward and forward. Also, the velocity can be controlled. Therefore, this is the minimal possible configuration for locomotion, using this modules.

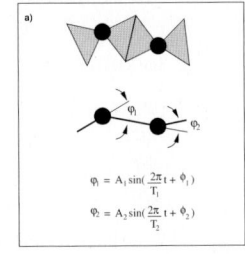

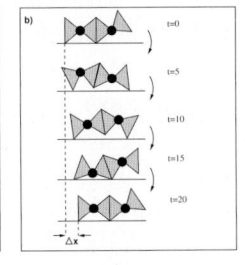

Fig. 3. a) PP configuration parameters and control. b) Locomotion of the PP configuration when $A = 40^0$, $\triangle \phi = 120$ and $T = 20$.

Fig.3a shows the robot parameters. φ_1 and φ_2 are the rotation angles of the modules 1 and 2 respectively. The locomotion is achieved by applying a sinusoidal function to the rotation angles:

$$\varphi_i = A_i \sin\left(\frac{2\pi}{T_i}t + \phi_i\right) \tag{1}$$

where $i \in \{1, 2\}$. The values of the parameters: A_i, T_i and ϕ_i determines the properties of the movement.

In order to simplify the experiments, the following restrictions have been applied: $A_1 = A_2 = A$, $T_1 = T_2 = T$, therefore, φ_1 and φ_2 are the same sinusoidal function with a different phase ($\triangle \phi = \varphi_2 - \varphi_1$). The period has been fixed to 20 unit of time.

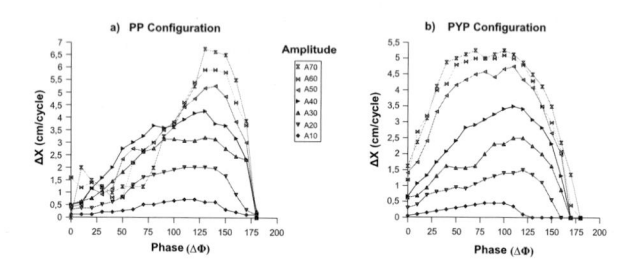

Fig. 4. The distance per cycle roved ($\triangle x$) as a function of the phase and amplitude. **a)** Pitch-Pitch configuration. **b)** Pitch-Yaw-Pitch configuration with $\varphi_2 = 0$

The motion is cyclical, with a period of T. After T unit of time, the movement is repeated. The space per cycle roved by the robot is $\triangle x$. **Fig.4a** shows the relation between $\triangle x$ and the phase ($\triangle \phi$) and amplitude (A) of the waves applied. As can be seen, $\triangle x$ increases with the increment of amplitude. Therefore, the speed of the locomotion can be controlled by the amplitude of the wave.

The difference in phase determines the coordination between the two articulations. If the modules rotates in phase($\triangle \phi = 0$), no locomotion is achieved. The same happens when $\triangle \phi = 180^0$. The best coordination is obtained when $\triangle \phi \in [110, 150]$. For negative values ($\triangle \phi \in [0, -180]$), the locomotion is done in the opposite way.

Fig.3b shows the position of the articulations at five instants, when $A = 40^0$, $\triangle \phi = 120$ and $T = 20$.

4 Configuration 2: Pitch-Yaw-Pitch (PYP)

Three Y1 modules are employed in this configuration. The outermost modules rotate in the pitch axis and the one at the center in the yaw

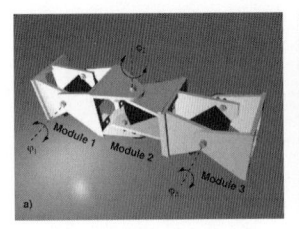

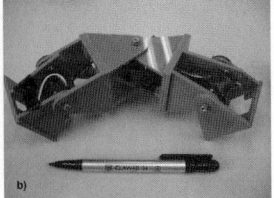

Fig. 5. Pitch-Yaw-Pitch (PYP) Configuration. **a)** A cad rendering showing the three modules and its rotation angle ranges. The module angles φ_1, φ_2 and φ_3 are set to 0. **b)** A picture of the robot

axis (**Fig.5**). Only one more module is added, but three new kind of gaits can be realized: 2D sinusoidal movement, lateral rolling and lateral shift. The same sinusoidal function is applied (equation 1) but in this case $i \in \{1, 2, 3\}$.

4.1 1D Sinusoidal motion

When φ_2 is fixed to 0, this configuration has the same shape as in **Fig.5a** and therefore, it is very similar to PP configuration. It only can move on a straight line, forward and backward. The experimental results are shown in **Fig.4b**. The velocity of the movement increases with the amplitude and there is a phase window in which the coordination is better. For the same amplitude, the space roved is less than in PP configuration but the phase window is wider. As the distance between the outermost modules is greater than in PP configuration, it is most difficult for this two modules to carry out the locomotion. Therefore the $\triangle x$ is smaller.

4.2 2D sinusoidal motion

If φ_2 is between 0° and 40°, the locomotion has the same characteristics than in the previous case, but the movement is not a straight line: The robot trajectory is an arc. The constraints used in this movement are: $A_1 = A_3 = A$, $T_1 = T_3 = T$, $\varphi_2 \in [0, 40]$.

4.3 Lateral Shift

PYP configuration can move using a lateral shift gait. It moves parallel to itself, as shown in **Fig.6**. Three sinusoidal waves are applied to all the modules with the following restrictions: $A_1 = A_2 = A_3 \in (0, 50)$,

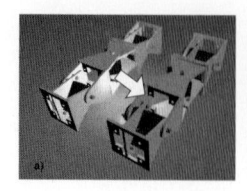

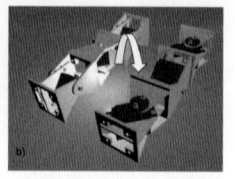

Fig. 6. a) Lateral shift gait in PYP configuration. b) Lateral rolling gait in PYP configuration

$T_1 = T_2 = T_3$, $\phi_1 = \phi_3 = 0$, $\phi_2 = 90$. The amplitude of the waves are the same, with a value greater than 0 and smaller than 50.

4.4 Lateral Rolling

PYP also can perform a lateral rolling gait (**Fig.6**b). The restrictions are the same as in lateral shift but the amplitude of the waves are greater or equal than 60: $A_1 = A_2 = A_3 >= 60$.

When PYP rolls 90° it is converted into a YPY (yaw-pitch-yaw) configuration. Now, it only has one module on the pitch axis. Therefore, it cannot move forward or backward. But lateral shift or lateral rolling can still be achieved. When lateral rolling is performed, configurations PYP and YPY appears alternatively.

4.5 PYP motion summary

Table1 summarizes the conditions needed to perform the different gaits. A sinusoidal wave is used in all the cases except in 2D sinusoidal gait, in which the rotation angle φ_2 has a constant value.

Table 1. Conditions needed to perform the different gaits on a PYP configuration.

Gaits	Amplitude	Phase	Angles
2D Sinusoidal	$A_1 = A_3$	**Fig.4b**	φ_1, φ_3 sin; $\varphi_2 \in [0, 40]$
Lateral rolling	$A \in [60, 90]$	$\phi_1 = \phi_3 = 90; \phi_2 = 0$	$\varphi_1, \varphi_2, \varphi_3$ sin
Lateral Shift	$A \in [0, 40]$	$\phi_1 = \phi_3 = 90; \phi_2 = 0$	$\varphi_1, \varphi_2, \varphi_3$ sin

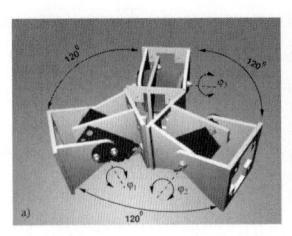

Fig. 7. The three-modules star configuration.

5 Configuration 3: Three-modules star

The last configuration tested was a three-modules star, shown in **Fig.7**. The modules form a star of three points with an angular distance of 120^0.

The locomotion is achieved by means of sinusoidal waves. It can move on a 2D surface, in three directions, as well as performing rotations in the yaw axis. If two adjacent modules are in phase and the opposite has $\triangle \phi \in [100, 150]$, it moves on a straight in the direction of the module out of phase. However, this movement is very surface-dependant.

When the increment of phase between the three modules is 120°, for example, $\phi_1 = 0$, $\phi_2 = 120^o$ and $\phi_3 = 240^o$, the robot performs a slow rotation in the yaw axis.

6 Conclusion and further work

Three different minimal modular configurations has been tested. Using only two modules, sinusoidal locomotion in straight line can be achieved. The difference of phase ($\triangle \phi$) between the two signal determines the coordination. A value between 100° and 150° degrees exhibit the best movement. The sense of the motion can be set by changing the sign of $\triangle \phi$. The speed is controlled modifying the amplitude of the sinusoidal wave.

When adding one more module in the yaw axis, three new gaits can be performed: 2D sinusoidal motion, lateral rolling and lateral shift. All of them are realized using sinusoidal waves, changing the amplitude and phase.

Lateral shift and rolling differs only on the amplitude range. When all the amplitudes are below 40, a lateral shift motion is performed. If a value greater or equal to 60 is applied, lateral rolling is achieved.

By attaching modules in different ways, other configurations can be created, as the three-modules star, which is capable of moving in three directions and rotating parallel to the ground. These motions are very surface-dependant, but shows the versatility of the modular robots. Not only modular snakes can be constructed, but another more complex systems.

In future works, new configuration will be constructed and tested, looking for simple coordination methods. Also, genetic algorithm will be used for the calculation of the optimal parameters (amplitude, phase) of these minimal configurations.

References

1. Mark Yim, Zhang & David Duff, Xerox Palo Alto Research Center (PARC), "Modular Robots". IEEE Spectrum Magazine. Febrero 2002.
2. D. Duff, M. Yim, K. Roufas,"Evolution of PolyBot: A Modular Reconfigurable Robot", Proc. of the Harmonic Drive Intl. Symposium, Nagano, Japan, Nov. 2001, and Proc. of COE/Super-Mechano-Systems Workshop, Tokyo, Japan, Nov. 2001.
3. Butler Z., Murata S., Rus D., (2002) "Distributed Replication Algorithms for Self-Reconfiguring Modular Robots, Distributed Autonomous Robotic Systems 5, pp. 37-48.
4. Marbach, D. and Ijspeert, A.J.: Co-evolution of Configuration and Control for Homogenous Modular Robots. In Proceedings of the Eighth Conference on Intelligent Autonomous Systems (IAS8), F. Groen et al. (Eds.), IOS Press, 2004, pp 712-719.
5. S. Murata, E. Yoshida, A. Kamimura, H. Kurokawa, K. Tomita, and S. Kokaji, "M-TRAN: Selfreconfigurable modular robotic system," Ieee- Asme Transactions on Mechatronics, vol. 7, pp. 431-441, 2002.
6. M. Yim, D. Duff, K.Roufas, "Modular Reconfigurable Robots, An Aproach to Urban Search and Rescue," Proc. of 1st Intl. Workshop on Human-friendly welfare Robotic Systems (HWRS2000) Taejon, Korea, pp.69-76, Jan. 2000.
7. M. Yim, K. Roufas, D. Duff, Y. Zhang, C. Eldershaw, "Modular Reconfigurable Robots in Space Applications", Autonomous Robot Journal, special issue for Robots in Space, Springer Verlag, 2003.
8. M. Mori, S. Hirose, "Three-dimensional serpentine motion and lateral rolling by Active Cord Mechanism ACM-R3". Proceedings of the 2002 IEEE/RSJ. Intl. Conference on Intelligent Robots and Systems. EPFL, Lausanne, Switzerland. pp. 829-834, Oct 2002.
9. J. Gonzez-Gez, E. Aguayo, E. Boemo, "Locomotion of a Modular Worm-like Robot using a FPGA-based embedded MicroBlaze Soft-processor". Proceedings of the 7th International Conference on Climbing and Walking Robots, CLAWAR 2004. Madrid, Spain, Sep. 2004.

Modularity and System Architecture

The Modular Walking Machine, Platform for Technological Equipments

I. Ion[1], I. Simionescu[2], A. Curaj[3], L. Dulgheru[1] and A. Vasile[3]

[1]Department of Technology of Manufacturing, [2]Department of Mechanism and Robot Theory, [3]Department of Automatic Control and System Engineering, "POLITEHNICA" University of Bucharest, Bucharest, Romania
ioni51@yahoo.com

Abstract

Walking robots represent a special category of robot, characterized by having the power source and technological equipment on-board the platform. A walking robot can traverse most natural terrains. The advantages of a legged system for off-road use have been gradually recognized. One of the most important advantages is mobility.

The weight of the power source and technological equipment is an important part of the total load that the walking machine can transport. That is the reason why the walking system must be designed so that the mechanical work required for displacement, and the highest power necessary for actuation, should be minimal.

Keywords: Modular Walking Robot, Static Balancing, Walking machine, Robot control, Shifting Systems.

1 Introduction

The realization of an autonomous walking robot, equipped with capabilities such as object handling, shift, perception, navigation, learning, reasoning, data storing and intelligent checking, enabling them to carry out missions such as changing a multitude of parts in a dynamic world, is a target focusing the activities of many scientific research teams in several countries, worldwide.

The *MERO* modular walking robot made by the authors is a multi-functional mechatronics system designed to carry out planned movements aimed at accomplishing several scheduled tasks. The walking robot operates and completes tasks by permanently interacting with the environment where there are known or unknown physical objects and obstacles. Its environmental interactions may be technological (by mechanical effort or contact) or contextual ones (route identification, obstacle avoidance, etc)

The successful fulfillment of the mission depends both on the knowledge the robot, through its control system, has of the initial configuration of the working place and of the configuration obtained during its movement.

The experimental pattern of the MERO modular walking robot (Fig 1) has been achieved at the "Merotechnica" Laboratory belonging to the Polytechnica University of Bucharest, over 1999-2004, and it was built in two versions, namely as a hexapod and a four-legged walking robot.

Fig. 1 Experimental model of MERO modular walking robot

The MERO modular walking robot is made up of the following parts:
a) the mechanical system made up of one, two or three modules articulated and shaped according to the requirements of the movement on an uneven ground. The robot's shift system is built such that it may accomplish many toes' trajectories, which can alter by each step;
b) the actuating system of feet have a hydraulic drive;
c) the distribution system is controlled by 12 or 18 servo-valves, according to the robot's configuration;
d) the energy feeding system;

e) the system of data acquisition on the shift, the system's configuration and the environment;

f) the control panel processing signals received from the driving and the acquisition systems.

Walking robots may have a lower or higher autonomy degree. This autonomy has in view the power source's supply capability but also orientation and perception capabilities as regards the terrain configuration the robot is running upon, its decision making and the motion manner towards a target. Walking robots having a high autonomy concerning the decision - making should benefit from appropriate driving programs and obviously from high-speed computers.

To control the walking robot shift in structured or less structured environments, we need the following specific functions:
- environmental perception and shaping using a multi-sensor system for acquiring data;
- data collecting and defining the field configuration;
- movement planning ;
- analysis of the scenes;

The control system of the walking robot is modular. The basic modules are the parts named "leg regulation blocks" and the aggregate consisting of an IBM PC-AT 486 DX computer and three conversion modules (two AX 5210 type interfaces and an AX 5212 type exit interface).

The control programs fulfill the following functions:
-edit and memorize the movement and adjustment parameters in the form of real time files;
- operate the robot's movement, launch or stop some movement sequences by the keyboard;
- create "robot files" to record primary movement sequences, whose mixing up result in the legs' walking movements. These files can be edited, recorded, loaded and then operated;
- main parameters can be visualized in real time, as they are required to elucidate, test and adjust the system as a whole;
- calculate and check the movement control, regulate the position, its force sensors detect the steps, automatically generate the movement laws (space, speed, acceleration), real time interpolation, automatic estimation of the offset servo-valves, convert direct and indirect coordinates among the spherical, cylindrical and Cartesian coordinates systems, attached to the walking robot's elements.

2 Static balancing elastic systems

Walking machines represent a special category of robots, characterized by having the power source and technological equipment on the platform. This weight of the source and technological equipments is an important part of the total load that the walking machine can transport. That is the reason why the walking system must be designed so that the mechanical work necessary for displacement, or the highest power necessary to actuate it, should be minimal. The major power consumption of a walking machine is divided into three different categories:
- the energy consumed for generating forces required to sustain the body in gravitational field; in other words, this is the energy consumed to compensate the potential energy variation;
- the energy consumed by leg mechanism actuators, for the walking robot displacement in acceleration and deceleration phases;
- the energy lost by friction forces and moments in kinematics pairs.

The magnitude of reaction forces in kinematics pairs and the actuators forces depend on the load distribution on the legs. For slow speed, joint gravitational loads are significantly larger than inertial loads; by eliminating gravitational loads, the dynamic performances are improved.

The energy consumption to sustain the walking machine body in a gravitational field can be reduced by:
- using the statically balancing elastic;
- optimum design of the leg mechanisms.

The potential energy of a walking machine is constant or has little variation, if the static balance is achieved. The balancing elastic system is formed by rigid and linear elastic elements.

3 The synthesis of static balancing elastic systems

The more usual construction of a leg mechanism has three degrees of freedom (Fig. 2). The proper leg mechanism is a plane one and has two degrees of freedom. This mechanism is articulated to the body and it may be rotated around a vertical axis. To reduce power consumption by the robot driving system, it is necessary to use two balancing elastic systems. One must be between links **2** and **3**, and the other - between links **3** and **4**. Because link **3** is not fixed, the second balancing elastic system cannot be set. Therefore, the leg mechanism scheme in fig. 2 can be balanced partially only.

It is well known and demonstrated that the weight force of an element which rotates around a horizontal fixed axis can be exactly balanced by the elastic force of a linear helical spring [2],[6],[8]. The spring is jointed between a point belonging to the rotating element and a fixed one. The major disadvantage of this simple solution is that the spring has a zero undeformed length. In practice, the zero free length is very difficult to achieve. The opposite assertions are theoretical conjectures only. A zero free length elastic device is given by a compression helical spring. In the construction of this device, some difficulties arise, because the compression spring, corresponding to the calculated feature, must be prevented from buckling. A very easy constructive solution, in which the above mentioned disadvantage is removed, consists of an assembly of two parallel helical springs, as shown in Fig. 2. The equilibrium of forces which act on the link **3** is expressed by the following equation:

$$(m\ BC\ \cos\varphi_{3i} - m_{7F}\ X_F - m_{8I}\ X_I - m_2\ X_{G2})g - F_{s7}\ BF\ \sin(\varphi_{3i} - \psi_{1i} + \alpha_1)$$

$$- F_{s8}\ BI\ \sin(\varphi_{3i} - \psi_{2i} + \alpha_2) = 0,\ i = 1,..,12 \qquad (1)$$

where:
- m is the mass of distributed load on leg in the support phase;
- x_{G2} is the gravity center abscise of the link **2**.
- m_{7F} and m_{8I} are the masses of springs **7** and **8**, concentrated at the points F and I respectively;

$$\psi_{1i} = \arctan\frac{Y_{F_i} - Y_H}{X_{F_i} - X_H};\ \psi_{2i} = \arctan\frac{Y_{I_i} - Y_J}{X_{I_i} - X_J};$$

$$X_{F_i} = BF\cos(\varphi_i + \alpha_1);\ Y_{F_i} = BF\sin(\varphi_i + \alpha_1);$$

$$X_{I_i} = BI\cos(\varphi_i + \alpha_2);\ Y_{I_i} = BI\sin(\varphi_i + \alpha_2);$$

$$F_{s7} = F_{07} + k_7(HF_i - l_{07});\ F_{s8} = F_{08} + k_8(JI_i - l_{08});$$

$$\alpha_1 = \arctan\frac{y_{3F}}{x_{3F}};\ \alpha_2 = \arctan\frac{y_{3I}}{x_{3I}};$$

$$HF_i = \sqrt{(X_H - X_{F_i})^2 + (Y_H - Y_{F_i})^2};$$

$$JI_i = \sqrt{(X_J - X_{I_i})^2 + (Y_J - Y_{I_i})^2}.$$

The Equations (1), which are written for twelve distinct values of the position angles φ_{3i}, are solved with respect to following unknowns: x_{3F}, y_{3F}, x_{3I}, y_{3I}, X_H, Y_H, X_J, Y_J, F_{07}, F_{08}, l_{07}, l_{08}. The un-deformed lengths l_{07} and l_{08} of the springs given with acceptable values from constructional point of view.

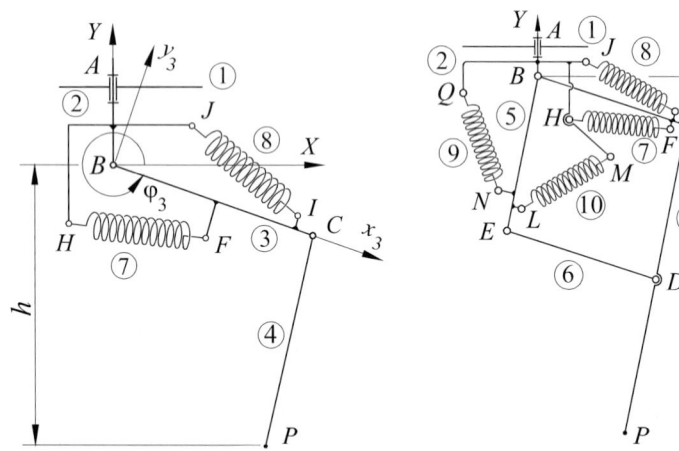

Fig. 2 Elastic system for the discrete partial static of the leg mechanism

Fig. 3 Elastic system for the discrete total static balancing of the leg mechanism balancing

The masses m, m_1, m_2, m_7, m_8 of elements and springs, and the position of the gravity center G_2 are assumed as known. In fact, the problem is solved in an iterative manner, because at the start of the design, the masses of springs are unknowns.

The limits $\varphi_{3\,min}^s$ and $\varphi_{3\,max}^s$ between the position angle φ_3 is compressed in the support phase are calculated as function on the lengths BC and CP on the ground clearance h on the step length. Also as function on the maximum height of the obstacles the limits $\varphi_{3\,min}^t$ and $\varphi_{3\,max}^t$ between the angle φ_3 is compressed in the transfer phase are calculated.

The imposed values φ_{3i}, φ_{3i}, $i = \overline{1,12}$, are chosen in a convenient mode, for example half in the domain $[\varphi_{3\,min}^s, \varphi_{3\,max}^s]$ and remain in the domain $[\varphi_{3\,min}^t, \varphi_{3\,max}^t]$. In the position corresponding to the transfer phase, the mass of distributed load $m = 0$.

If total static balancing is desired, it is necessary to use a more complicated leg structure. In the mechanism leg schema in Fig.3, two active pairs are superposed in B. The second balancing elastic system is set between the elements **2** and **5**.

The equilibrium equations of forces that act on the elements **3** and **5** are respectively:

$$BC(R_{34Y}\cos\varphi_{3i} - R_{34X}\sin\varphi_{3i}) + (m_{7F}X_F + m_{8I}X_I + m_3 X_{G3})g + F_{s7} BF \sin(\varphi_{3i} - \varphi_{7i} + \alpha_1) + F_{s8} BI \sin(\varphi_{3i} - \varphi_{8i} + \alpha_2) = 0;$$

$$BE(R_{56Y}\cos\varphi_{5i} - R_{56X}\sin\varphi_{5i}) + (m_{9N}X_N + m_{10L}X_L + m_5 X_{G5})g + F_{s9} BN\sin(\varphi_{5i} - \varphi_{9i} + \alpha_3) + F_{s10} BL \sin(\varphi_{5i} - \varphi_{10i} + \alpha_4) = 0, i = \overline{1,12},$$

where: $\alpha_3 = \arctan\dfrac{y_{5N}}{x_{5N}}$; $\alpha_4 = \arctan\dfrac{y_{5L}}{x_{5L}}$;

$$F_{s9} = F_{09} + k_9(ML_i - l_{09}); \quad F_{s10} = F_{0,10} + k_{10}(QN_i - l_{0,10});$$

$$R_{34X} = \frac{U(X_D - X_E) - V(X_C - X_D)}{W};$$

$$R_{34Y} = \frac{V(Y_D - Y_C) - U(Y_E - Y_D)}{W};$$

$$U = g[m_4(X_{G4} - X_D) - m(X_P - X_D)]; \quad V = g[m_6(X_{G6} - X_D) - (m_4 - m_6 - m)(X_E - X_D)];$$

$$W = Y_D(X_C - X_E) - Y_C(X_D - X_E) - Y_E(X_C - X_D); \quad R_{56X} = -R_{34X}; \quad R_{56Y} = (m_4 + m_6 - m)g - R_{34Y}.$$

The magnitude of angles φ_{3i} and φ_{5i} are calculated as functions on the position of the point *P*. The variation fields of these, in support and return phase, must not be intersected. In the support phase, the point *P* of leg is on the ground. In the return phase, the leg is not on the ground, and the distributed load on the leg is zero. The not intersecting condition can be easy realised for the variation fields of angle φ_3. If the working positions of link **4** are choose in proximity of vertical line, the driving force or moment in pair *C* is much less than the driving force from pair *B*. This is workable by adequate motion planning. In this manner, using the partial balancing of leg mechanism only, we may reduce energy consumed for walking machine displacement.

The static balancing is exactly theoretically realized in twelve positions of the link **3**, accordingly to angle values $\varphi_{3i}, i = 1,..,12$, only. Due to continuity reasons, the unbalancing magnitude between these positions is negligible. In order to realize the theoretical exact static balancing of leg mechanism, for all positions throughout the work field, it is necessary to use cam mechanisms.

Conclusions

Reduction of power consumed by the driving devices of walking robot leg mechanisms can be achieved by static balancing of displacement systems. The total static balancing of these systems is a difficult task, because the movement of the legs has two phases: support phase and return phase. Static balancing must be done in the support phase only, because the distributed load on a leg in the return phase is zero. In fact, the optimum solution is characterized by:
- reduction of driving force/moment by a partial static balancing;
- construction of a simple static balancing system and with high reliability.

References

1. Kumar V., Waldron K.J., (1988) Force Distribution in Walking Vehicle, Proc. 20th ASME Mechanism Conference, Orlando, Florida, Vol. DE-3, pp. 473-480.
2. Ion I., Stefanescu D.M., (1999) *Force Distribution in the MERO Four-Legged Walking Robot*, ISMCR'99 - Topical Workshop on Virtual Reality and Advanced Human-Robot Systems, vol. X, Tokyo, Japan, June10-11.
3. Ion,I.,Simionescu,I.,Curaj,A.,(2002)*Mobil Mechatronic System With Applications in Agriculture and Sylviculture*. The 2th IFAC International Conference on Mechatronic Systems, December 8-12, -Berkeley –USA
4. McGhee R.B., Frank A.A., (1968) *On the Stability Properties of Quadruped Creeping Gait*, Mathematical Biosciences, Vol. 3, No.2, pp. 331-351.
5. Simionescu I., Ion I., *Static Balancing of Walking Machines*, POLITEHNICA University of Bucharest, Sci. Bull., Series D, vol. 63 no. 1, 2001, pp.15 – 22.
6. Simionescu I., Ciupitu L., *The static balancing of the industrial robot arms, Part. I: Discrete balancing*, Mechanism and Machine Theory, 35 (2000), pp. 1287 – 1298
7. Song S.M., Waldron K.J., (1989) *Machines that Walk,* Massachusetts Institute of Technology Press.
8. Strait D.A., Shin E., *Journal of Mechanical Design* 115(1993) 604-611
9. Waldron, K.J., (1996) Modeling and Simulation of Human and Walking Robots Locomotion, Advanced School, Udine, Italy.
10. http://www.walking-machines.org/

YaMoR and Bluemove – An Autonomous Modular Robot with Bluetooth Interface for Exploring Adaptive Locomotion

R. Moeckel, C. Jaquier, K. Drapel, E. Dittrich, A. Upegui, A. Ijspeert

Ecole Polytechnique Fédérale de Lausanne (EPFL), Logic System Laboratory (LSL), CH 1015 Lausanne, Switzerland
moeckel@ini.phys.ethz.ch, (andres.upegui, auke.ijspeert)@epfl.ch

Abstract

Modular robots offer a robust and flexible framework for exploring adaptive locomotion control. They allow assembling robots of different types e.g. snakelike robots, robots with limbs, and many other different shapes. In this paper we present a new cheap modular robot called YaMoR (for "Yet another Modular Robot"). Each YaMoR module contains an FPGA and a microcontroller supporting a wide range of control strategies and high computational power. The Bluetooth interface included in each YaMoR module allows wireless communication between the modules and controlling the robot from a PC. With the help of our control software called Bluemove, we tested different configurations of our YaMoR robots like a wheel, caterpillar or configurations with limbs and their capabilities for locomotion.

Keywords: Modular robot, Bluetooth, Locomotion, FPGA

1 Introduction

Locomotion with modular robots constitutes a great potential and at the same time a very difficult challenge [1-4]. In comparison to conventional monolithic robots, modular robots present the advantage of supporting a fast reconfiguration of their structure. To build a robot of the desired form a completely new robot does not need to be constructed but it can be reassembled by simply disconnecting and reconnecting modules. Furthermore modular robots constitute a challenging framework for exploring distributed control when each module contains its own controller and sensors.

YaMoR is designed to act as a cheap platform for (1) testing different control algorithm for locomotion and their implementation in both software and hardware, (2) exploring the capabilities for locomotion of a large variety of different robot configurations and shapes as well as for (3) finding new applications for wireless networks.

The main characteristics of our modular robots are: (1) each module contains a Bluetooth interface for inter-module communication as well as for communication between the modules and a base station like a PC — most modular robots use direct electrical connections, which are less flexible — and (2) each module comprises an FPGA for reconfigurable computation — most modular robots use traditional microcontrollers, but see [5] for an example of a robot using a single FPGA for controlling all modules.

We designed and implemented a control software called Bluemove that allows to control the YaMoR modules from a PC via Bluetooth. Bluemove offers an easy way for exploring the capabilities for locomotion of different configurations of modules.

In section 2 we give an overview about the mechanics and electronics of our modular robot. Section 3 describes the Bluetooth interface each robot modules contains. Section 4 gives an introduction to Bluemove, the control software that we use for exploring locomotion. In section 5 we describe first examples of locomotion illustrating the capabilities of YaMoR. Finally section 6 concludes and gives an outlook about future work.

2 YaMoR – mechanics and electronics

YaMoR consists of mechanically homogeneous modules. One of the key features of YaMoR is its low cost: in contrast to the majority of modular robots, YaMoR is constructed with off-the-shelf components. Each module contains a powerful one degree of freedom servo motor (with a 73Ncm maximal torque). Its casing consists of cheap printed circuit boards (PCB) that can also serve as support for printed circuits (see **Fig. 1.**).

The casing of each module is covered with strong velcros. Velcros offers the advantage to connect robot modules together with no restriction on angles between the surfaces of the modules. Unfortunately, it does support self-reconfiguration and the modules can only be connected together by hand.

The modules are autonomous. They are powered by on-board Li-Ion batteries and include the necessary electronics for power management, motor control, communication and running algorithm. To achieve more flexibility and modularity in terms of control each YaMoR module contains three separated control boards: (1) one board including a Bluetooth-ARM microcontroller combination, (2) one board carrying a Spartan-3 FPGA, and (3) a service board containing power supply and battery management.

YaMoR was constructed as a framework for a variety of different projects. For instance, a user may choose between using a microcontroller, an FPGA or a combination of both for implementing the desired control algorithm. Configuring the FPGA to contain a MicroBlaze soft-processor [6], allows exploiting the hardware-software codesign capabilities offered by the platform, taking also advantage of the flexibility provided by the FPGAs partial reconfiguration feature [7].

The YaMoR architecture with distributed electronic components gives a flexible solution for connecting the electronic boards: the FPGA board can be left out if it is not needed to save energy; or it can be replaced by a board with specific sensors if useful. The new sensor board can still take advantage of the electronics mounted on the remaining boards. So a designer for an additional sensor board does not have to worry about power supply or battery management.

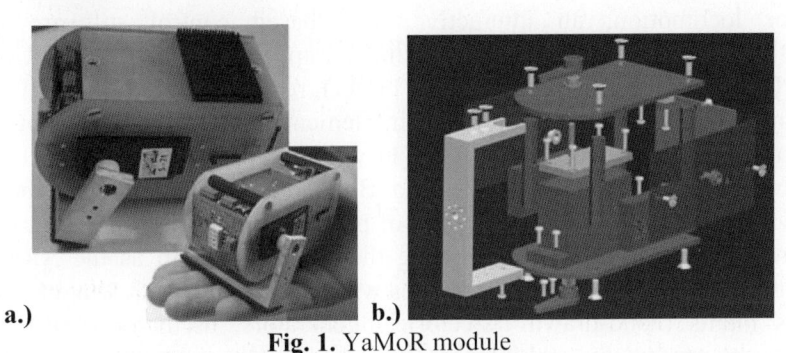

Fig. 1. YaMoR module
a.) Module closed and with open casing. b.) Assembly of YaMoR mechanics.

3 Bluetooth – the wireless interface to YaMoR

We chose Bluetooth for wireless communication given its flexibility and energy efficiency. Wireless communication between modules allows creating a new robot configuration by simply disconnecting and reconnecting the mechanical modules without the need for reconnecting cables or changing the control infrastructure. In comparison to a communication based on wires, Bluetooth has the constraint that a module normally does not know its physically connected neighbours just by communicating with them. However, this constraint can be overcome with the addition of touch or distance sensors.

The ARM on the Bluetooth board is running both a real time operating system and the embedded Bluetooth stack. It can also be used for customized software e.g. a control algorithm or for reconfiguring the FPGA via Bluetooth. The ARM program code and FPGA configuration bitstream are stored inside a Flash memory on the microcontroller board.

The Bluetooth-ARM board was designed to provide a wireless interface that can be easily controlled. The embedded Bluetooth stack allows taking advantage of wireless communication by sending simple commands via UART. For instance, a researcher concentrating on FPGA based algorithms may implement a simple UART module on the FPGA and is able to communicate wirelessly with a PC or other modules.

4 Bluemove – controlling YaMoR via Bluetooth

For easily exploring new configurations of modules and their capabilities for locomotion, an interactive Java based control software called Bluemove has been developed. Using a graphical user interface (GUI) on a PC, a user can quickly start a new project, register all modules used for the current robot configuration and implement a controller. To control the modules Bluemove allows both (1) writing trajectories that can be continuously sent to the modules via Bluetooth and interactively modified without any resetting, (2) the use of plugins for a controlling the modules from a PC, and (3) programming the FPGAs as well as the ARMs for autonomous control in the modules without needing a PC. Plugins can act as inputs (hand-drawn trajectories, generators, oscillators, etc.), filters (signal processors, multiplexers, etc.) and outputs (data sent to the modules, files, streams, etc.). Plugins support the generation of controllers with feedback from sensors. The whole project including the trajectories and plugins can be saved in XML. The main parts of the graphical user interface are:
1. The Module Manager serves to manage all modules belonging to the current robot configuration including module names and Bluetooth addresses of the modules.
2. The Timelines Manager (see **Fig. 2.**) allows generating trajectories for each actuator of the modules registered in the Module Manager by set-

ting key points on the GUI with a mouse. Linear and spline interpolation can be chosen to draw the trajectories and connect the key points set before. Trajectories even can be changed "online" while transmitted to the modules e.g. by changing the position of the key points.
3. The Real-Time Module (see **Fig. 3.**) supports an easy use of Bluemove with the help of plugins. New plugins can be created with a script editor. The relations between different plugins are visualised with the help of a graph.

Bluemove is implemented in Java, taking advantage of its standard and consistent interface for Bluetooth. Given the popularity of Bluetooth and Java, it would be possible to create Bluetooth applications for the modular robot on mobile phones, PDAs or other small systems that support Java and Bluetooth.

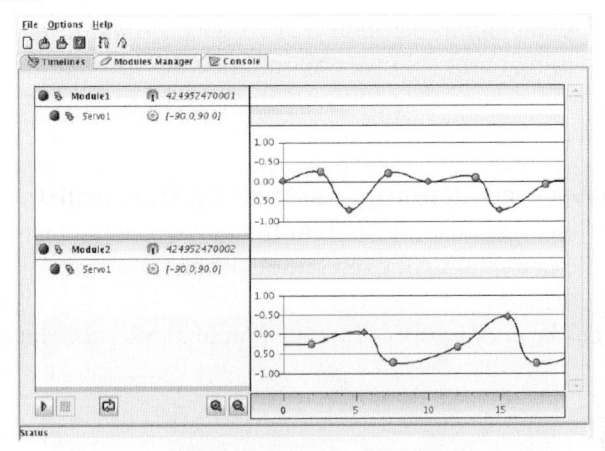

Fig. 2. Timelines Manager

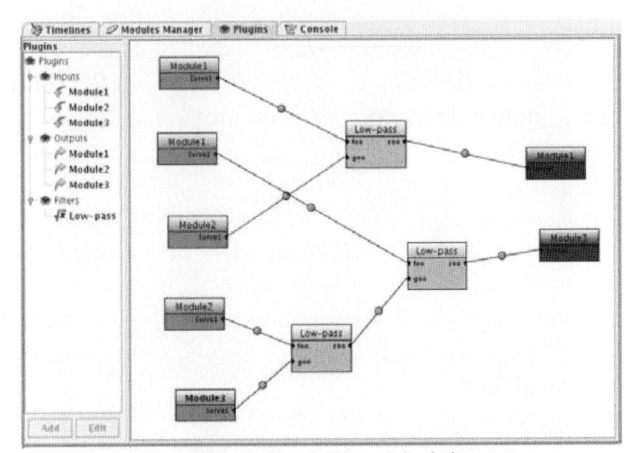

Fig. 3. Real-Time Modules

5 Exploring locomotion

We explored the locomotion capabilities of different YaMoR configurations with up to six modules, using Bluemove to generate the joint angle trajectories for the servo motors. By trial and error, interesting gaits could be generated for a variety of robot structures, such as travelling waves for worm and "wheel" structures (see Fig. 4. for four snapshots of a moving wheel), crawling gaits for limbed structures, and other peculiar modes of locomotion (see **Fig. 5.** for different examples of configurations of YaMoR modules). For videos, the reader is kindly requested to visit the project website [8].

These gaits are only a first step towards adaptive locomotion (see next section), but already represent in our opinion an interesting example of control with a "human in the loop". The user can indeed interactively adjust the gaits in real time, in order to optimize the speed of locomotion for instance, as well as modulate the gaits in terms of speed and direction, by modifying frequency and amplitude parameters.

For exploring more complex shapes of a robot, tests with more than six modules are needed. For example, most of the configurations tested so far do not allow changing the direction of the movement, and would therefore not be capable of avoiding or overcoming obstacles.

Fig. 4. Rolling wheel.

Fig. 5. Different configurations of YaMoR modules

6 Future work

To achieve adaptive locomotion, "human in the loop" control that we were using for the first experiments is clearly not sufficient. In a complex environment a robot has to react and adapt in real time. That is why we designed Bluemove to also support feedback signals from sensors with the help of plugins. Distributed algorithms can be implemented in the ARM and FPGA of each YaMoR module. Both the parameter of these algorithms and the shape of the modular robot can be optimized under different constraints like energy efficiency or speed.

We are currently extending the presented work along two main axes: (1) the design of the next generation of YaMoR modules with sensors capabilities — e.g. IR sensors, inertial sensors, and load sensors — and (2) the implementation of distributed locomotion controllers based on central pattern generators. Central pattern generators (CPGs) are biological neural networks capable of producing coordinated patterns of rhythmic activity while being initiated and modulated by simple input signals. We have extensively used models of CPGs for the control of locomotion in other projects [9, 10]. In particular, simulation experiments have demonstrated they are ideally suited for implementing distributed control of locomotion in simulated YaMoR units [11].

Acknowledgements

We would like to acknowledge André Badertscher for the technical support, Alessandro Crespi, Fabien Vannel and René Beuchat for many useful discussions and hardware debugging. We acknowledge also the Swiss National Science Foundation for the Young Professorship Award granted to Auke Ijspeert.

References

1. Duff D., Yim M., Roufas K. (2001) Evolution of PolyBot: A Modular Reconfigurable Robot. *Proc. of the Harmonic Drive Intl. Symposium, Nagano*, Japan, November.
2. Yim M., Zhang Y., Roufas K., Duff D., and Eldershaw C. (2002) Connecting and disconnecting for chain self-reconfiguration with polybot. *IEEE/ASME Transactions on Mechatronics*, vol. 7, no. 4, pp. 442–451.
3. Kurokawa H., Kamimura A., Murata S., Yoshida E., Tomita K., and Kokaji S. (2003) M-tran II: Metamorphosis from a four-legged walker to a caterpillar. *In Proceedings of the Conference on Intelligent Robots and Systems (IROS 2003)*, vol. 3, pp. 2454–2459, October.
4. Castano A., Behar A., and Will P. (2002) The conro modules for reconfigurable robots. *IEEE Transactions on Mechatronics*, vol. 20, pp.100–106.
5. González-Gómez J., Aguayo E., and Boemo E. (2004) Locomotion of a Modular Worm-like Robot using a FPGA-based embedded MicroBlaze Soft-processor, *Proceedings of International Conference on Climbing and Walking Robots (CLAWAR 2004)*.
6. Xilinx Corp., MicroBlaze™, URL http://www.xilinx.com [June 2005]
7. Upegui A., Moeckel R., Dittrich E., Ijspeert A., and Sanchez E. (2005) An FPGA dynamically reconfigurable framework for modular robotics. *Workshop Proceedings of the 18th International Conference on Architecture of Computing Systems 2005 (ARCS 2005), VDE Verlag*, Berlin, Germany, pp. 83-89.
8. YaMoR project website, URL http://birg2.epfl.ch/yamor [June 2005]
9. Ijspeert A.J. (2001) A connectionist central pattern generator for the aquatic and terrestrial gaits of a simulated salamander. *Biological Cybernetics*, vol. 84, no 5, pp. 331-348.
10. Crespi A., Badertscher A., Guignard A., and Ijspeert A.J. (2005) AmphiBot I: An amphibious snake-like robot. *Robotics and Autonomous Systems*, vol. 50, no. 4, pp.163-175.
11. Marbach D. and Ijspeert A.J. (2005) Online Optimization of Modular Robot Locomotion, *Proceedings of the IEEE International Conference on Mechatronics and Automation (ICMA2005), to appear*.

On the Development of a Modular External-pipe Crawling Omni-directional Mobile Robot

P. Chatzakos, Y. P. Markopoulos, K. Hrissagis[1] and A. Khalid[2]

[1]ZENON S.A, 5 Kanari Str, Glyka Nera Attikis, 15354 Athens, GREECE;
[2]TWI Ltd, Granta Park, Great Abington, Cambridge CB1 6AL, UK

Abstract

In this paper, the development of a novel omni-directional inspection robot is presented, which is capable of delivering NDT sensors to surfaces on straight pipe, pipe bends and branch connections, overcoming the limitation that a test area over a pipe bend or past a branch or other obstruction raise. The lightweight crawler is attached on the outside of the pipe to the thin metal strip that holds the insulation in place without deforming the insulation through the application of a force controlled clamping mechanism while performing longitudinal, circumferential and arbitrary movements. In order to be able to cope with a range of pipe, materials and coverings, to allow for future modifications and to be able to incorporate a wide range of NDT inspection equipment, a modular approach was considered for the design of the mobile robot. Either two different inspection sensors may be mechanically incorporated into the chassis of the crawler and deployed at the same time or just a double-sided acting sensor (e.g. X-Ray).

Keywords: mobile robotics, omni-directional robot, external-pipe crawler, automated inspection, NDT scanner.

1 Introduction

Over 10 million kilometers of pipelines in Europe carry hazardous fluids (1,2). These pipelines are subjected to corrosion by the environment and contents. Other defects in pipes are caused by mechanical fatigue. The inspection of a great majority of the pipes is mandatory to ascertain their structural integrity as failure to inspect can result in leakage of hazardous material into the environment and even explosion. Up to 90% of these pipelines are inaccessible for inspection by current methods (1) because these are covered with coatings, such as paint and insulation. Pipeline spills of hazardous fluids into the environment outnumber all other source, e.g., tanker spills in oceans, combined. In Europe, up to 4 million gallons of oil are leaked into the environment per year (3-9). In Brussels, European Union officials have urged member governments to begin applying new inspection rules and technology to stem this pollution (10).

In recent years, many mechanised inspection techniques, sensors and systems for finding defects and corrosion in pipe have been developed. On the other hand, current inspection methods have major drawbacks because these are based on the following regime: "intelligent pigs" that travel along with the fluid inside the pipe, which, except in transmission lines, is rarely possible because "pig traps" are not present to insert the "pig" and retrieve it. For pipelines covered with coatings, inspections involve removing the coatings, inspecting the pipeline under the coating and then reinstating the coating. These inspection methods add extra costs to the inspection process. Thus, development of mechanised inspection techniques, sensors and systems for finding defects and corrosion in pipe that is inaccessible to the operator, covered with coatings, such as thick paint or insulation, without the need to "dig up" or remove coatings out, is the sine qua non of nowadays non-destructive inspection.

Today, there are no commercially available current inspection techniques that can accurately detect significant corrosion or other types of defects in pipework under thick coatings. Another limitation is that current inspection techniques can only be applied manually by highly trained operators. Recent PANI trials (11), carried out to assess the effectiveness of manual inspections have shown that operators detect only 50% of defects. Commercial scanners have been developed for scanning pipe girth welds and lengths of straight pipe with inspection sensors (12). These are primarily ultrasonic sensors and the scanning is in simple X-Y routines. These scanners either move around the pipe on tracks or along the pipe on magnetic wheels. However these cannot work on curved surfaces around pipe bends and in the vicinity of valves, branches and other features in the pipe. Unluckily, these are areas where corrosion is most likely to occur.

2 Design Scheme

The design concept of the robotic pipe crawler encompasses the following features: longitudinal and circumferential motion and scans of pipe surface, manipulation of pipe bends, elbows, pipe supports and other obstructions, mechanical incorporation and deployment of a wide variety of NDT sensors, modular construction and easily scalable chassis, lightweight, holding on the thin metal band around insulation of pipe without deforming it, limited height and compact so as to cope with restricted access between pipes, execution of inspection routines in arduous on-site conditions and cost-effective design.

In **Fig. 1** the final design and the main subsystems of the crawler are presented. The major brakethrought was to optimise the integration of drives and power transmission components, the structural framing, sensors, and the omni-wheels in quite tight space. Additionally an opening at the frame (see **Fig. 1**) is necessary in order that a stationary obstacle, such as pipe supports, branches, etc., can be avoided. A second function of this opening is to allow for easy and quick placement of the scanner on and off the pipe.

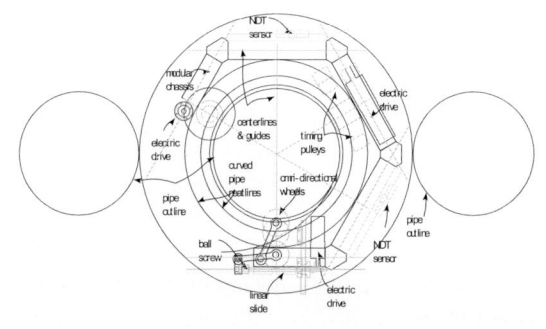

Fig. 1. Schematic sketch of the crawler on a pipe intersection.

One may also notice that various neat-lines are drawn, inside which the mobile system is to be limited, so that manipulation of curved pipe segments is possible. These lines are drawn with respect to the external dimensions of scanner's components that lie around and above the pipe and the values from **Fig. 2** that represent the bounding dimensions of the chassis to facilitate the crawler travelling along pipe bends. In particular, **Fig. 2** demonstrates the effective bounding dimensions of the chassis for three different curvatures of a pipe bend. All the points bellow these curves correspond to a specific set of dimensions, i.e., length and width of the mobile vehicle, which ensure that the system is able to crawl along the curved pipe without bumping into it. Typical values for the curvature of a pipe bend are 1-3.5 times the standard diameter of pipe.

The outside diameter of the pipe is considered to be from 270 to 285 mm, which is analyzed as follows: 6" inch inner pipe diameter, 50 mm thick insulation and 2 mm thick cladding. By picking up the dimensions of the chassis and arranging the placement of various components of the system on the crawler accordingly, i.e., after consulting the nomograph of

Fig. 2, successful manipulation of curved pipe is intrinsically obtained by design. In **Fig. 3** a well executed turn of the developed robotic crawler around a 285 mm outer diameter pipe bend is simulated.

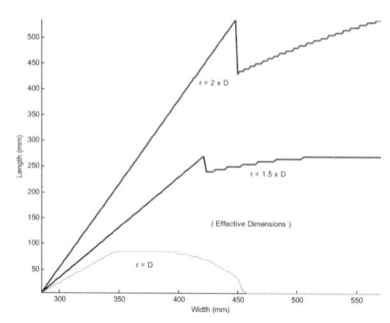

Fig. 2. Bounding dimensions of chassis in order to overcome a pipe bend.

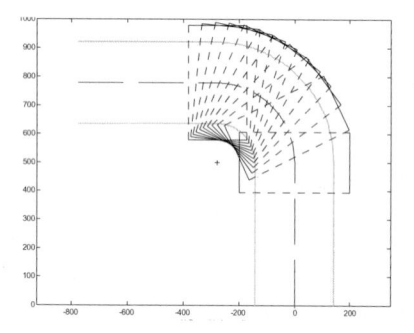

Fig. 3. Simulated turn of the crawler represented by a bounding box.

As seen in **Fig. 1**, three sets of omni-directional wheels are symmetrically placed at 120°. The first set of wheels is responsible for the longitudinal movement of the crawler. Three rows of twin omni-wheels are evenly distributed along the total length of the crawler. The main axes of rotation lie in a direction perpendicular to the symmetry axis of pipe, while the rotation axes of the rollers, of which each omni-wheel consists, are parallel to the pipe so that the circumferential movement of the crawler is not restrained. A second assembly of a twin omni-wheel set and a DC servo motor drives the scanner around the periphery of the pipe.

The last set of wheels consists of two passive, free-rotating omni-wheels and completes the locomotion scheme. This is part of the clamping mechanism, which exerts a controllable force to the pipe. A precision servo drive, a ball-screw and two linear sliders comprise the clamping subsystem. Strain gauges are placed in appropriate location on the scanner and measure variation in strain of the aluminum chassis due to clamping and therefore provide a feedback for the magnitude of the clamping force. This feedback closes the control loop with the torque-controlled DC servo motor. A static analysis of the system, after a constant clamping force has been applied, is presented in **Fig. 4**.

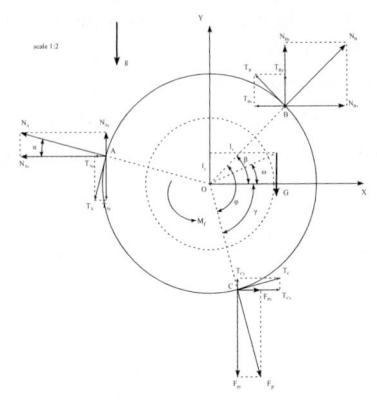

Fig. 4. Reaction forces on the chassis after clamping onto pipe.

The center of mass of the scanner lies inside the outline of the pipe. Ideally, it should coincide with the center of pipe. This deviation causes a moment due to gravitational force to be exerted on the system and tends to rotate the crawler clock- or anticlock-wise according to the orientation of the system with respect to horizontal each time. Thus, for an unbalanced chassis, the circumferential drive should always provide an amount of torque to the omni-wheels so that the crawler is not uncontrollably rotating around the pipe under the gravitational load.

In order to avoid a constantly working motor and because the chassis is not yet passively balanced a suitable combination for the material of the rollers and the magnitude of clamping force should be selected and would lead to a static system. The rollers of omni-wheels are usually made of nylon or polyurethane. Nylon exhibits low sliding and rolling friction coefficients on steel, whereas rollers made of polyurethane provide better grab. The latter were finally selected for the crawler. With a minimum clamping force that was estimated after exhaustive trials and omni-wheels with polyurethane rollers the static friction that is generated is adequate so that the crawler always stays onto the pipe without slipping even when the motors are not providing any torque.

Changes in the magnitude of the reaction forces versus the orientation of the crawler with respect to horizontal are presented in **Fig. 5**. The reaction force on the passive wheels is constant and in fact equals the clamping force. This force determines the reaction on the driving wheels that

masses will result from exhaustive simulation on the CAD model of the crawler. Design and manufacture of custom-made omni-wheels exclusively for use with the proposed clawer is also included in the scope of future work. Finally, a sophisticated control scheme for special, uncommon and fully automated inspection routines will be developed.

Acknowledgements

Support of this work by the European Community (EC CRAFT Project No 508614) is acknowledged.

References

1. Lyons D., *Western European cross-country oil pipelines*, 25-years performance statistics, CONCAWE Oil Pipelines Management Group.
2. Davis P.M., et al. (2000), *Performance of cross-country oil pipelines in Western Europe*, statistical summary of reported spillages, CONCAWE Oil Pipelines Management Group.
3. United Kingdom Offshore Operators Association, *Emissions and discharges*, http://www.oilandgas.org.uk.
4. Environment Canada, *Oil pollution and birds*, http://www.cws-scf.ec.ca.
5. Australian Maritime Safety Authority, *Major oil spills in Australia-Al Qurain*, http://www.amsa.gov.au.
6. Friends of the Earth, *The big waste - adding up the barrels*.
7. Sue Haile (2000), *Oil Pollution*, University of Newcastle on Tyne.
8. Stanislav Patin, *Oil Pollution of the Sea-Environmental Impact of the Offshore Oil and Gas Industry*, www.researchandmarkets.ac.uk.
9. Health & Safety Executive, *Offshore injury and incident statistics (provisional data), 2001-2002*, Hazardous Installations Directorate, Offshore Division, UK, http://www.hse.gov.uk.
10. Frohlich (1998), *Inspection systems and safety report*, European Commission Seveso Directive Seminar, Rome, 23-25.
11. McGrath B (1999), *The effectiveness and performance of normal industrial NDT techniques and procedures*, PANI, Health and Safety Executive (USE) UK, AEA Technology pic report IVC99/56.
12. Heckhauser H., Schulz S. (1995), *Advanced technology in automatic weld inspection of pipeline girth welds*, Insight, Vol.37, No.6, pp.440-445.

Modularity and Component Reuse at the Shadow Robot Company

The Shadow Robot Company[1]

[1]251 Liverpool Road, London, UK. contact@shadow.org.uk

Abstract

A series of robots have been developed by overlapping project teams at the Shadow Robot Company. The development history is recounted highlighting modularity and reuse. Lessons learned for modularity are recounted.

Keywords: modularity, reuse, biped, hand.

1 Introduction

In modern engineering, design efficency is produced by a pattern of re-use of components. Modularity is seen as a technique that can provide significant benefits in robotics. [1]. Experience from software shows reuse is harder than expected if the wrong level of granularity is chosen. We recount some areas of robotic engineering at Shadow where we have and haven't re-used previous components and sub-systems, and see what this can tell us about re-use in general in robotics. As a bonus, we provide a short catalog of re-usable components we have generated.

2 Reuse in other fields

In software engineering, it costs at least 3x as much to develop a reusable component as a good module. [2] Since in software, development cost is usually the most important cost, this leads to little re-use. Where re-use happens, it tends to be at the library and application levels: few organisations develop libraries that see heavy re-use outside the original organisation, although the NAG numerical library is a good example of the reverse [3].

In other areas of engineering, the concept of re-use varies from the trivial to the profound. In civil engineering, design re-use is extensive, but reuse of product minimal. In electronic engineering, component-level re-use abounds, and the development of large-scale programmable logic components has led to an IP core industry. In rocketry, designs are heavily tested, and remain in use and are incrementally modified and cannibalised to a very high degree, with rare failures [4].

3 Project History reuse

The reuse at Shadow seems largely to come from the fact that all members of the company know the project histories of the company to a very high degree. This solves one of the classic problems of reuse, namely "how do I find it?", as well as preventing "Not-Invented-Here" syndrome. This deep knowledge also allows the design of other components to be influenced by the well-explored parameter space of the component being re-used.

Original Shadow Biped

This Air Muscle powered humanoid biped was built with multiple valve driver boards, and ADC multiplex boards using a proprietary interface with long-forgotten computers. Some structural reuse occurred at a construction level. We were unable to reuse anything from it.

Liberator v1

This was a mobile base with combined Air Muscle and motor-powered arm. The I^2C bus [5] with modular boards was used for motor drivers, bump sensors, arm motor and valve drivers. However, the bus was too slow and the host computer (an embedded Forth system) too weird. We were able to re-use the mechanism.

Dextrous Hand A

This was the first prototype Hand system, constructed in maple with a skeletal palm. Three major evolutions of the Hand system have been produced. It used ad-hoc electronics, wired directly to computer interface ports, for valves and later for sensors. None of these electronics were useful after the demise of the interface hardware provider (Same problem with the Biped). This experience led us to want to isolate the electronics from the computer interface.

Dextrous Hand B

We moved to distributed local electronics using CAN [6] as a field-bus. This insulated the system from the host computer, which by this time was a Linux box. This also made software reuse more practical, as Free Software platforms have a habit of remaining around.

Electronics for this hand were all built around PIC microcontrollers. The Hand incorporated two identical valve driver boards with 56 outputs each, and on the palm hosted a CAN node with separate plug-in ADC boards providing 8 8-bit channels per board. During the evolution of the arm, other nodes were constructed reusing the PIC firmware and basic PCB design to provide single-sensor electronics modules for some shoulder joints and later integrated valve driver and ADC boards for local control of valves.

Prototype Insects

The integrated valve+ADC board design from Hand B was then used to make a series of prototype robots, including 6 and 8 legged walkers. This proved an efficient way of building these prototypes. The electronics assemblies were known good, and could be quickly swapped out for testing. Small design updates for differing requirements were made. The host PC software was continuously developed.

Small biped

A toy company from a Scandinavian country asked us to produce a small biped in a short time. We re-used our ADC boards using the same PIC code and circuit, just with a new layout to fit. New DAC/Valve boards were built for driving proprtional valves. Then we made the mistake of re-using the CAN connectors for sensor connectors: One Reuse Too Far.

Liberator upgrade

Switching to a Linux-based host PC prompted us to overhaul the other electronics. The motor drivers were rebuilt as one-board-per-motor with motor + shaft encoder + sonar + bump sensor interfacing, using CAN and PICs. This allowed connection to our standard hardware and RTIO code, which let us do the rework in a very short timescale. The re-use was primarily in PIC code, and host PC.

Hand C

This redevelopment of the Dextrous Hand [7] was a fully-integrated design. Newer smaller valves were embedded in the system, requiring new valve driver module layouts, built using the same software and circuit design as the Hand B valve/adc modules. Integrated pressure sensing on the Hand required additional electronics: most of the design was recycled from the fingertip ADC module. Other sensing was done with the usual ADC module reworked for size.

A tactile sensing unit was developed for the fingertip. The first prototype was synthesized from a valve/ADC module in a day; the final design was completely from scratch in order to cope with extreme space constraints.

Dinosaur

A manual-valve puppet was computerised in 1 hour on-site during a performance workshop. Most of the overhead was the time to connect new valves to plugs. This reused directly the normal valve driver boards and standard host PC software. About 10 minutes was required to convert a keyboard control program to provide a useful set of movements for the dinosaur.

Zephyrus

This six-legged walker was first built with a proprietary microcontroller system, which largely served to persuade us not to use such things. We rebuilt it as one of our first trial PIC systems, and the only reuse that has happened has been conceptual

Zephyrus 2

This is (currently) an 8-legged robot with a modular segmented design. Body modules can be repeated ad nauseam – we used 2 pairs for the implementation. One standard Valve/ADC boards was used per 2 modules. A keyboard node was built for the control of movements by recycling another CAN node.

Chair

We built a series of VR chairs for a client. An RS-232 connected sensor/valve driver board was built for this based on our standard valve/ADC board without CAN. This serial module has since seen quite a lot of reuse for prototyping and test work.

Test Leg

We were asked to build a leg for testing neural control. We used standard valve/adc boards, with the same code as Hand C. The standard RTIO code was requested by the customer once examples of use were shown. We have added new front-end panels and so on.

4 Pattern of Reuse

At this point in time, we had a fairly stable base of firmware for the PIC microcontroller. This code (mostly in the form of complex macros) was then our starting point for each new design. Engineers were encouraged to develop new functionality and find flaws in the old functionality; upgrades of "ancient" boards then happened when they were re-programmed. The existence of this codebase encouraged us to re-use the PIC designs, because we had working hardware, layouts and software as well as host PC interfaces.

5 Issues and Lessons we found in doing this

Lock-in by a hardware vendor can really hurt! There is usually a good reason at the time for using the vendor-specific interfaces – typically, some ease of communication, or perhaps some significant speed optimisation. However, if the vendor changes their business model, your investment rapidly becomes worthless.

The reuse of robot components comes at a price. If no further engineering can be done on the component, then it will usually be the wrong form factor, have the wrong number of I/O connections, and so on. However, evolution of the physical form factor normally is possible. This, of course, requires full access to the design, which means if your component is supplied externally, it had better be Open Hardware.

Some things should not be re-used! And you should have a method for working out whether or not to do reuse beforehand.

When we have used components without changing their physical form, the code on-board has tended to evolve. This evolution is normally fed forward into future iterations of the components. The valve drivers reflect this - the code base for that board now supports timed operation of valves, PID control of valves based on sensor data, and external SPI-connected sensor modules.

The significant advantages of this component-based architecture come when a wholly new robot is to be produced, and it is possible to improvise the first version within hours. Use of the standard components permits us to make use of all software ever developed to talk to any robot – typically, only user-visible names need changing - almost immediately the system can be wired up.

5 Classification of levels of reuse

We have found there are four distinct types of reuse occurring.

5.1 Whole RTIO code

In this case, the Linux code-base is shared across all robot systems under development. Utilities produced during development on one robot can often be used for similar or novel purposes on others. Also, programmer experience on one robot can be transferred to another robot with no retraining!

5.2 Electronics modules

This is the commonly-perceived re-use case, where a physical board is used in more than one robot, reducing parts counts and easing design. This is pretty frequent at the prototyping stage.

5.3 PIC code

Since we only use one microcontroller family to date, we have accumulated a library of well-tested code for it that can be used to quickly produce the code for a novel board design. In particular, this means we can expect all boards to talk the same interface protocol!

5.4 Circuit designs

When a component requires significant optimisation, the point of re-use tends to be the schematic. The layout will be novel, as form-factor is usually the over-riding constraint. Usually, there will be changes made to the schematic, but this is most commonly to reduce generality by removing connectors, extra driver chips, ADCs and so forth.

6 Some Reusable Components

Table 1 Catalogue of reusable components

Item	*Function*	*Availability*	*IO*
Valve+ADC board	Multiple power outputs and analogue inputs	Board, schematic	CAN
Tactile sensor	Multiple tactile readings in small size	Product	SPI
RTIO	Software architectire	Source	various
Motor driver	Motor, encoder, sensor interface	Board,schematic	CAN
Hand	Dextrous robot Hand	Product	CAN
Valve+ADC board	Multiple power outputs and analogue inputs	Board, schematic	RS232

Item	Function	Availability	IO
Sensor board	Multiple ADCs reading local sensors	Board, schematic	CAN

7 Conclusion

For a small project team, reuse is a default mode of operation. Avoiding proprietary solutions may seem more difficult initially, but is worthwhile. Standardising on certain components e.g. microcontrollers promotes reuse, as does the availability of working examples of previous versions.

Acknowledgements

Work described in this paper has been funded by the UK SMART scheme, the National Endowment for Science, Technology and the Arts, Adrenalin Systems, and the Performance Robotics Research Group.

Without the contributions of numerous Shadow people, it would not have happened.

References

1. Virk, G.S. (2003) CLAWAR Modularity for Robotic Systems
   The International Journal of Robotics Research. vol. 22 pp. 265-277.
2. Butler Lampson, "How Software Components Grew Up And Conquered The World",
   http://research.microsoft.com/Lampson/Slides/ReusableComponentsAbstract.htm [2005-06-30]
3. Numerical Algorithms Group. http://www.nag.co.uk/
4. "ARIANE 5 Flight 501 Failure Report by the Inquiry Board", chair J.L.Lions, specifically 3.1(m)-(p)
5. http://www.semiconductors.philips.com/markets/mms/protocols/i2c/ [2005-06-30]
6. http://www.can.bosch.com. [2005-06-30] Also ISO 11898
7. Shadow Robot Company (2003) Design of a Dextrous Hand for advanced CLAWAR applications. *CLAWAR 2003* Catania

CLAWAR Design Tools to Support Modular Robot Design

Gurvinder S. Virk

School of Mechanical Engineering, University of Leeds, LS2 9JT, Leeds, UK. Tel: +44 113 343 2156; Fax: +44 113 343 2150; Email: g.s.virk@leeds.ac.uk

Abstract

Robot component modularity is discussed and the latest CLAWAR work on modularity to develop generic workable concepts that are acceptable to all the stakeholders involved in robot system design. The paper presents an overview of the design tools that are needed to support the modular robot design cycle from starting with the generic user requirements to formulating formal specifications and then designing and testing of prototypes followed by production, using and recycling activities.

Keywords: modularity, robot components, design tools, standards

1 Introduction

Robot component modularity is widely accepted as being extremely useful for encouraging the widespread adoption of robotised technologies in new applications where highly specialised designs need to be easily and quickly realised. The paper presents the latest work carried out by the Members of the CLAWAR Thematic Network in supporting modular robot system design within the various sectors of the robotic community. From its start in 1998, CLAWAR has stressed the importance of modularity and the development of open standards for robot components so they may be integrated to form application specific machines in an easy and flexible manner. The existence of such components would allow technologies to become "reusable and the current problems of "reinventing-the-wheel" scenarios could be reduced. This would assist researchers as well and machine manufacturers and a viable sustainable and vibrant robotics sector would be created. CLAWAR has taken a holistic view by including all the stakeholders in the planning phases to produce a generic modular design philosophy that sub-divides the overall robot design process into a modular format where the individual components can link up to other modules to form the overall system using an "interaction space highway" type of data bus. This involves determining how the modules need to link up. After considerable investigations and discussions it has been established that six interaction variables are needed for this inter-connectivity, namely, (1) power, (2) computer data bus, (3) mechanical linkages, (4) analogue signals, (5) digital signals and (6) working environment (see previous publications by the author concerning CLAWAR modularity [1]-[7].

The focus in year 2 has been to look at design tools that are needed to support this overall modular approach. The design tools need to cover the entire process from starting the requirement formulation phase of the designs needed to delivering the final robot, using it during its working life and recycling. The important stages are felt to be as follows:

1. Gathering the information in the specifying requirement stage;
2. Creating a simulation environment (which includes the environments, the tasks to be carried out, assessing the performance of designs, including standards in different sectors, etc) design;
3. Creation of design concepts for possible solutions; this needs to be supported by the use of expert knowledge;
4. Virtual prototyping and testing of the designs;
5. Assembly of the modules to give sub-systems, super-modules and even the full system; the tool should allow the various integrated sub-systems to be fully tested within appropriate environments and via specific test procedures as required;
6. Engineering design and selection of the modules (via the use of existing methods and tools);
7. Physical prototyping and testing the modules at super module level and at system level is needed.
8. Service and maintenance procedures needed to keep the robot in good working order during its working life.
9. Recycling of the materials at the end of its working life.

The paper presents the details on some of these aspects and specifies the design tools needed. In particular the early sections 1-4 above are covered here. In order to support this entire process, it has been agreed to produce the following design tools:

- Tools to specify the design requirements. These have been broken down into different environments and application sectors. What tasks need to be carried out; how the designs are assessed; how the designs are operated; what standards to use. These tools should allow static and dynamic analysis and simulation within the various environments and the assessment of designs against different metrics. These include cost, specific task needs, speed, reliability, ease of use, etc.
- Tools to help create design concepts (via expert knowledge)
- Tools for virtual and rapid prototyping
- Testing and analysis tools
- Creation of engineering design concepts module library; this should allow specific machines to be tailored to individual requirements
- Physical prototyping and testing

Of course, the question of standards in modular design is important and it is interesting to note that following the development of the new CLAWAR concepts, ISO has put forward a Resolution 276 taken by correspondence to setup an ISO Advisory Group on "Standards for mobile service robots", with Professor Virk as Chairman. This resolution has been approved by Canada, Czech Republic, Finland, France, Germany, Italy, Japan, Republic of Korea, Portugal, Sweden, Switzerland, United Kingdom, USA indicating the high level of interest in this area to develop workable solutions for the new robotised systems needed. An Advisory Group has been setup in June 2005 with ≈30 nominated experts who will deliberate and report their finding to the ISO Plenary meeting in June'06.

2 Operational environments

All of the design aspects have been broken down into applications sectors that individual partners can contribute to. Focussing on the various environments where a CLAWAR machine is likely to be needed, the individual expertise within the group is as follows:

1. Urban environments (Cybernetix)
2. Domestic environments(Shadow Robot Company)
3. Industrial environments(Caterpillar)
4. Underground environments (University of Genova)
5. Quarry environments (University of Genova)
6. Construction environments (Univ Carlos III, Madrid and Shadow)
7. Space environments (Space Applications Services)
8. Pipe and duct environments (ISQ and Fraunhofer-IFF)

9. Land environments (QinetiQ)
10. Sea environments (Helsinki University of Technology)
11. Air environments (BAE Systems)
12. Underwater environments (CSIC-IAI)
13. Archaeological site environments (Politecnico di Torino)
14. Mining environments (Örebro University)
15. Nuclear environments (CEA)
16. Fire fighting environments (University of Genova)
17. Facade cleaning environments (Fraunhofer-IPA)
18. Humanitarian environments (Royal Military Academy)

Detailed descriptions of these environments, the tasks needing to be carried out and the performances levels needed, how to assess the machines, etc for each of the sectors are presented in the Year 2 WP2 Report. In addition several of the other partners have assisted in the specification of the design tools needed.

3 Creation of design concepts

When mobile robots are developed, it is usually assumed that the design principles are known and the main problem is one of applying them. In reality, it is fair to say that there is no rigorous design methodology for realizing the robots. In other words, given a formal description of the terrain that must be crossed, there are no general methods for specifying appropriate machine geometry, body articulation, suspension characteristics, traction mechanisms (wheels, tracks or legs), surface adherence mechanisms and actuation systems. This is a particular problem in the design of high-agility mobile robots for complex terrains. Most prototypes have either been the result of trial-and-error design methods or, in the case of legged machines, have been copied from nature with no scientific justification.

The most common approach has been the informal application of the designer's experience, pre-conceived ideas and preferences, in other words, a non-scientific approach. In many cases, more formal, but still subjective, design methods have been applied. These are usually based on some form of design matrix (e.g. showing design options versus performance metrics) and a subjective scoring system. A good example of this approach, drawn from the CLAWAR partners, is the methodology used to select the locomotion principle in the EU-IST ROBOVOLC project (see www.robovolc.diees.unict.it). They used a design matrix showing locomotion options versus locomotion performance metrics. All of the consortium partners applied a subjective scoring system to complete their matrix and, then, the individual partner matrices were combined to get overall scores for the different locomotion options. This more formal approach reduces the chance of neglecting important aspects of the design problem, and the use of multiple opinions can increase the confidence in the result. However, the subjective scores still reflect the designers' pre-conceived ideas and preferences. A more scientific approach is required if mobile robot design is to move forward, leading to better designs, more able to achieve the tasks required of them and, of course design tools to support this more objective approach are then also needed.

To illustrate the complexity of an optimisation approach to mobile robot design, consider the legged robot case. Obtaining a solution would involve at least three nested optimisation problems covering design, gait, and foot-force control respectively. For each design instance, a nested gait optimisation problem must be solved for a predefined route over a model terrain representing the machine's agility requirements. At each point along that route, a nested control optimisation problem must be solved to establish the instantaneous foot-force distribution. The two outer optimisation loops (design and gait) would be driven by the integral of some performance measure over the route traversed (giving average efficiency for example). The foot-force optimisation loop would be driven by the instantaneous performance (instantaneous efficiency for example).

The gait optimisation problem alone has been the subject of much research and has only been solved for particular well defined and rather simplistic artificial terrains. Real machines need to deal with a range of difficult terrain features (surface transitions, walls, ditches, steps etc.). Gait patterns for these specific terrain features have been determined by trial and error, not by optimisation. For complex terrains, the constraints on mobile robot design and control are likely to be severe, and finding optimisation seeds that are within the feasible design space is a non-trivial problem. Hence, although an optimisation approach may be possible, it will require good seeds for both robot design and control. Obtaining these seeds (concept generation) will itself require predetermined design rules and design data.

The details of the tools needed to support the overall modular design process is beyond the scope of this paper but it is sufficient for our purposes to state that the design process can be broken down into the following areas (and support tools for each created):
- specifying design requirements;
- design concept generation using expert knowledge;
- virtual prototyping;

- and engineering design.

4 Rules and regulations

The rules and guidelines for industrial robotics are a very wide field which is covered by hundreds of documents, depending very much on the final application and the environment where the tool, device or system is applicable. Nowadays in Europe, Directives are the starting point when considering the development of a standardized object. There are many Directives available, but for the robotic industry, it seems that the most important are the Machine Directive and the Low Voltage Directive. Other Directives could also be applied according to the scope of the application like in the case of applications for hazardous explosive atmospheres, Directive ATEX 94/9/EC needs to be taken on-board. For toys there is the 88/378/EEC Directive. Of course health and safety are of high importance and are also covered by the Directive 89/655/CEE. One more example is the Directive for Active implantable medical devices, 90/385/EEC. Standards are currently produced from several viewpoints, namely, 1) mechanical; 2) electrical/electronic; 3) safety and environment; 4) economical; and 5) control/computational. Regarding the area of mobile robotics or mobile service robotics, there are no real standards that apply and these need to be formulated and agreed for the robot systems of the future. Modularity could be important in the respect as well. These aspects will be explored by the ISO Advisory Group in its work.

5 Analysis tools

Tools are needed to analyse different aspects of a robot system and this can be a complicated process. For example, we can consider the locomotion design aspects to illustrate the important aspects that need to be taken into account. When designing the locomotion system of a mobile robot it is necessary to consider several aspects to decide on the final details. This can be carried out by using software tools to model and simulate the different aspects by various performance indices. For locomotion in mobile robots we can use the following metrics:
- Trafficability, which is a robot's ability to traverse soft soils or hard ground without loss of traction

- Manoeuvrability, which addresses a robot's ability to navigate through an environment using various forms of techniques to change the robot's heading (via steering and/or other turning mechanisms)
- Terrainability, which captures a robot's ability to negotiate terrain irregularities while maintaining a margin of safety (both statically and dynamically).

To model these it is necessary to consider all the mechanical and control aspects to study the dynamical performances using different scenarios and determine the best theoretical results. There are many software platforms that are available for doing this. CLAWAR has investigated the list of software packages that are available and formed an opinion on what are the most important for helping the designer analyse the results of his ideas before committing to hardware build. Some of the best software packages that are available include the following:

- For mechatronics aspects we have CONFIG, DynaMechs, Dynawiz XMR, ODE, Visual Nastran, JUICE and ADAMS.
- For control aspects we have ORCCAD, ESTEREL and Matlab.
- General software platforms for robot system analysis include Player/Stage/Gazebo, CLARAty, SynDEx, 4D/RCS, RCS and EyeSim Simulator.

Tools are also needed for virtual prototyping (eg, ADAMS, Pro-Engineer, SolidWorks, etc);

6 Conclusions

The paper has presented the latest work on modularity carried out within the CLAWAR project. The work focuses on design tools needed to support the modularity ideas that CLAWAR has been developing. These design tools cover the entire process from starting the requirement formulation phase of the design to delivering the final robot, supporting it during its useful working life and then finally recycling the materials at the end of its lifetime. In particular the CLAWAR WP2 members have defined a wide range of environments that CLAWAR machines need to operate in and the tasks that have to be carried out. The contributions have been built on expertise of the Partners in the particular application sectors and form a valuable contribution to knowledge in the area. The wide range of environments and tasks clearly demonstrate the enormity of the problem and how difficult it is to develop techniques that can work across the spectrum. In fact this is almost impossible and hence the motivation for modularity is reinforced so that the enormous task can be partitioned into smaller more manageable sub-sections. A common approach must be adopted so that the various components can be linked together. The report has described the importance of rules and regulations in modular developments. The need for software tools to support the designer are also described and these will be followed up next year when benchmarking of the designs is the focus.

Acknowledgements

The author would like to acknowledge the support of the European Commission for funding the CLAWAR Thematic Network under contract G1RT-CT-2002-05080. He would also like to express his gratitude to the WP2 Modularity task members for their excellent contributions.

References

1. Virk, G. S., CLAWAR Modularity: The Guiding Principles, Proc 6th Int Conf CLAWAR'03, Catania, Italy, Professional Engineering Publishing, pp 1025-1031, 17-19 Sept 2003.
2. Virk, G. S., CLAWAR Technical Reports on Modularity, Tasks 1, 6, 11 and 16, EC Contract BRRT-CT97-5030, Univ of Portsmouth, 1999-2002.
3. Virk, G. S., Modularity for CLAWAR Machines – Specs and possible solutions, Proc 2nd Int Conf CLAWAR'99, Portsmouth, UK, pp 737-747, PEP, 14-15 Sept 1999.
4. Virk, G. S., Modularity of CLAWAR machines – Practical solutions, Proc 3rd Int Conf CLAWAR'2000, Madrid, pp 149-155, PEP, 2-4 Oct 2000.
5. Virk, G. S., Functionality modules – specifications and details, Proc 4th Int Conf CLAWAR'2001, Karlsruhe, Germany, pp 275-282, Professional Engineering Publishing, 24-26 September 2001.
6. Virk, G.S., CLAWAR – robot component modularity, Proc 5th Int Conf CLAWAR'2002, Paris, pp 875- 880, Professional Engineering Publishing, 25-27 September 2002.
7. Virk, G.S., CLAWAR modularity for robotic systems, International Journal of Robotic Research, Special Issue on CLAWAR'01, vol 22, No 3-4, pp 265-277, 2003.

Powering, Actuation, Efficiency

Pneumatic Actuators for Serpentine Robot

G. Granosik[1] and J. Borenstein[2]

[1]Institute of Automatic Control, Technical University of Łódź, Łódź, POLAND, granosik@p.lodz.pl, [2]Dept. of Mechanical Engineering, The University of Michigan[*], Ann Arbor, MI, USA, johannb@umich.edu

Abstract

This paper presents analytical and experimental results of an investigation of joint actuators and their suitability for a particular class of complex mechanisms: serpentine robots. In practice, serpentine robots usually comprise of multiple segments connected by joints. Some serpentine robots provide legged, wheeled, or tracked propulsion, and, in addition, actuation for the joints. We compare different types of joint actuators, paying special attention to those properties that are uniquely important to serpentine robots and propose an effective solution based on pneumatic bellows.

Keywords: Pneumatic bellows, joint actuators, serpentine robots.

1 Introduction

Many mechanical systems exist, in which two members are linked by a joint that allows one, two, or more Degrees-of-Freedom (DOF) of motion between the members. Application areas for such systems are robotics in general and, more specifically, so-called "serpentine" mobile robots.

Our own two serpentine robots called *"OmniPede"*

Fig. 1. The "OmniPede," developed at our lab, uses legs for propulsion and pneumatic cylinders for actuation of articulated 2 DOF joints.

---

[*] This research was conducted at the Univeristy of Michigan where Dr. Granosik worked as a post doctoral researcher

and "*OmniTread,*" are shown in Fig. 1 and 2, respectively. Insights gained from the earlier work with the legged OmniPede helped us design the OmniTread. This tracked robot is about 5-10 times more energy efficient than the OmniPede. The OmniTread has also a one order of magnitude larger "Propulsion Ratio" – the dimensionless property that we define as the surface area that provides propulsion, A_p, divided by the inert surface area of the body, A_i

Fig. 2. The OmniTread serpentine robot. All segments are linked by 2-DOF pneumatically actuated joints, which are the focus of this paper.

$$P_r = A_p/A_i \qquad (1)$$

The value of the P_r can vary between 0 and 1; a larger P_r provides superior performance on very rugged terrain (see [3]). In order to increase the P_r, the space taken up by the joints should be as small as possible because it typically has only inert surface areas.

By definition, serpentine robots are relatively long compared to their diameter, so that their lead segments can reach up and over a high obstacle while still being able to fit through small openings. From this geometric constraint, as well as from other unique operational characteristics of serpentine robots, important requirements for joint actuators can be derived:

1. The energy consumption and weight of the actuators should be minimal, because energy is a limited resource in an untethered mobile robot.
2. Serpentine robots should conform to the terrain *compliantly*, so that as many driving segments as possible are in contact with the ground at all times to provide effective propulsion.
3. At other times it is necessary to increase the stiffness of a joint, for example, for crossing a gap or reaching over an obstacle. Serpentine robots must thus be capable of adjusting the stiffness of every degree of freedom individually and proportionally.
4. Joint angles in serpentine robots should be controllable proportionally, to provide full 3-dimensional mobility.
5. Joint actuators should be capable of developing sufficient force to lift at least two lead segments to the edge of a step, in order to climb over it.
6. As discussed earlier, joint actuators should take up as little space as possible, to reduce the size of Joint Space.

In order to identify the most suitable joint actuator for serpentine robots, we analyzed available actuation methods methodically as presented in the next sections.

2 Review of candidate joint actuators

There are many different ways of actuating joints in a mechanical structure. However, only a few of them can provide the range of motion and force required for actuating the joints of a serpentine robot. Table 1 lists some key parameters for candidate joint actuators.

In order to find the best-suited actuator for joints in serpentine robots we performed a detailed analysis. Our analysis was mostly based on the comparison of performance indices of mechanical actuators introduced by Huber et al. [4] and complemented by our own investigations. We focused our attention on the graphs that plot *actuation stress* and *specific actuation stress* versus *actuation strain*. We reproduced the first graph and also part of the second one, with some modifications (explained below), in Fig. 3.

The original paper by Huber et al. did not include electric motors, and it included only select types of pneumatic actuators. To supplement this data, we calculated the performance indices for some electric motors with a ball screw transmission mechanism that produces reasonable linear speed and force. We also calculated the performance indices for a few pneumatic bellows and artificial pneumatic muscles (see Section 3) and added those results in Fig. 3.

Actuators that are closest to the top right corner of Fig. 3 are naturally suited to lifting weights and propelling masses in the orders of magnitude required for serpentine robots. As is apparent from the right part of Fig. 3, the superior characteristics of hydraulics (compared to pneumatics) are

Table 1. Key parameters of different actuators (reproduced from [5])

Drive type Performance compared	Electric	Hydraulic	Pneumatic
Efficiency [%]	(<1) 50-55 (>90)	30-35	15-25
Power to weight ratio [W/kg]	25-150	650	300
Force to cross section area [N/cm^2]	0.3-1.5	2000	100
Durability [cycles]	5-9·105	6·106	>107
Stiffness [kN/mm]	10-120	30	1
Overload ratio [%]	25	50	50-150
Linear movements ranges [m]	0.3 – 5	0.02 – 2	0.05 – 3
Positioning precision [mm]	0.005	0.1 – 0.05	0.1
Reliability (relative)	Normal	Worse	Better

diminished once actuation stress is related to the actuator's density. Also, hydraulics also becomes less desirable over electric motors once efficiency is considered, as was shown in Table 1. One should note that Huber's analysis considers the actuator only, without the volume (or weight) of the compressor, manifolds, valves, fittings, and pipes. In general, these weight factors work in favor of electric systems. This is one of the reasons why electric actuation is usually chosen for freely moving robots while hydraulic or pneumatic actuation is mostly used for tethered robots.

However, the actuation strain of most cylinder-type actuators is limited to 1.0 and only pneumatic bellows produce the largest value (reaching 4) without any external mechanisms. Furthermore, the overload ratio of pneumatic actuators is significantly higher then competitors.

There is also another consideration, which, in our opinion, is of primary importance: natural compliance. We believe that natural compliance is critical for robots, whose propulsion depends on optimal traction between its propulsion elements (i.e., legs, wheels, or treads) and arbitrarily shaped environments, such as the rubble of a collapsed building or the rugged floor of a cave.

As explained in [4], the lines of slope +1 in Fig. 3 are related to the stiffness of the actuators. Hydraulic systems provide several orders of magnitude greater stiffness than pneumatic systems and electric motors without closed loop position control. However, electric motors do require closed-loop control and have to be considered in this configuration. That means that the working stiffness of electric motors depends on parameters

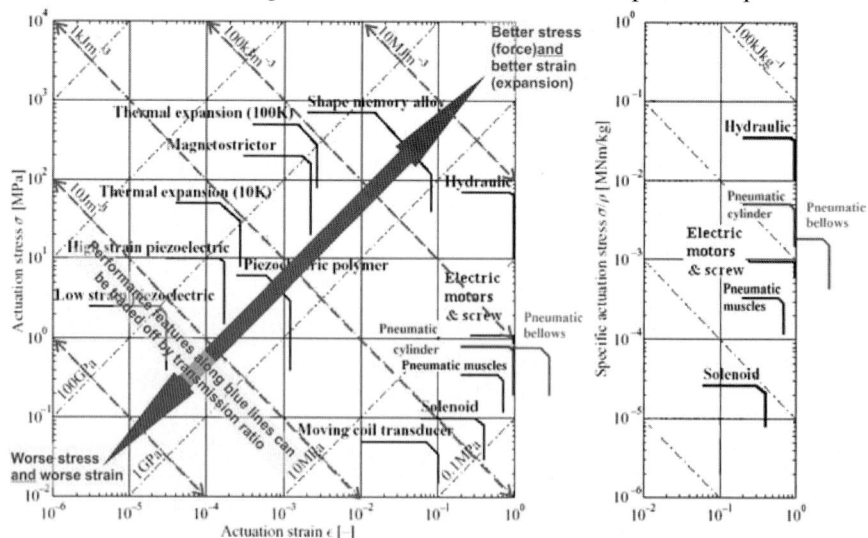

Fig. 3. Actuation stress and specific actuation stress versus actuation strain for various actuators (reproduced from [4] and augmented with our own data.)

of the control loop. Moreover, if gearboxes or transmissions are added, then the elasticity of the actuator is eliminated. This makes electric drives ideal for accurate position control, but not for joint actuation of serpentine robots, which must be able to conform to the terrain compliantly.

Robinson [7] offered a work-around for this inherent limitation. He modified elasticity of an inherently stiff actuator by adding a soft spring in series with an electric motor with ball screw transmission or to a hydraulic cylinder. Special control algorithms allowed his system to produce a controllable force. However, Robinson's approach substantially reduces the actuation strain and increases the weight of the actuator, which is then no longer suitable for serpentine robots.

We therefore conclude that pneumatic actuators are the only devices that provide *natural compliance*. In practice, pneumatic actuators behave as natural air springs, and, when used in closed-loop systems, can work as position-force actuators. Moreover, changes in working pressure can control the stiffness of pneumatic actuators from very limp (compliant) to very stiff. It is this fundamentally important property that makes pneumatic actuation the preferred choice for serpentine robots.

3 Pneumatic actuators

There are three "mainstream" types of pneumatic actuators: cylinders, bellows, and artificial pneumatic muscles. Cylinders and bellows develop force in quadratic proportion to their diameter d. In pneumatic muscles

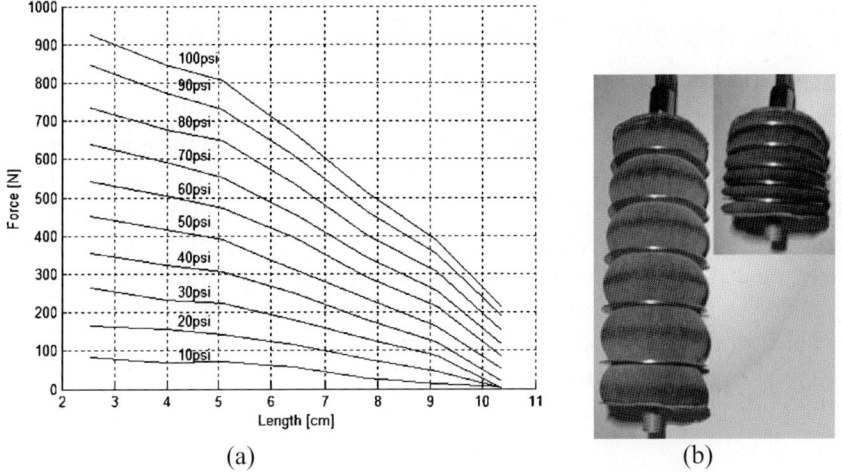

(a)   (b)

Fig. 4. Pneumatic bellows developed at the University of Michigan (a). Static characteristics, (b). Pneumatic bellows: extended and compressed.

force is related to diameter and length, and the actuation force can be much larger than the force generated by a cylinder with the same diameter. Unfortunately, the force drops very quickly with contraction. Characteristics of different types of pneumatic muscles can be found in [2, 8]. The actuation force of bellows (see Fig. 4) also drops with expansion, but not nearly as dramatically as that of artificial muscles.

One serpentine robot with pneumatic actuation is MOIRA [6], which uses cylinder-type actuators in the space of the joints. However, in doing so joints take up even more space than segments. We believe that this is a less advantageous design, because it increases the robot's inert surface area A_i and thus reduces the propulsion ratio P_r.

To avoid this situation, cylinders or pneumatic muscles would have to be placed within a segment to actuate the joints. These actuators would take up much or most of the available space within a segment and dramatically limit the space available for other components.

In contrast to cylinders and artificial muscles, we believe that pneumatic bellows are an ideal solution because they allow the integration of four large-diameter pneumatic actuators in so-called "Joint Space" (i.e., the space occupied by the joint – see Fig. 5). As shown in Fig. 6, bellows have the very suitable property of taking up minimal space when deflated, and maximal space when inflated. They can thus be placed in Joint Space, without taking up any segment space. The location of the pneumatic actuators in Joint Space also allows for larger actuator diameters than what would be possible if the actuators had to be placed in Segment Space, where space is shared with all other onboard components.

4 The Integrated Joint Actuator for serpentine robots

Based on the discussion thus far, we chose pneumatic bellows as the best-suited actuator for serpentine robots. In accordance with that choice we designed our "Integrated Joint Actuator" (IJA) for serpentine robots. Fig. 5 shows a cross-section of the IJA. The design assumes that there is a 2-DOF universal joint in the center, connecting any two adjacent segments. An arrangement of four equally spaced bellows is used to actuate the two degrees of freedom of each joint. Each closed end of a bellows is rigidly fastened to the front or rear "firewall" of a segment. Compressed air can be pumped into the bellows or exhausted from the bellows via an appropriate hole in the firewall. The maximum bending angle in our IJA is up to 25° in each direction.

Fig. 6 shows the case of the OmniTread lifting its two lead segments (according to requirement #5), each of weight W. To accomplish this task, the IJA of Joint B inflates bellows B1 and B2 and exhausts bellows B3 and B4. This creates a lifting torque τ_p that must overcome the reactive moment from the weight of the two segments, $M_{react} = L_1 W + L_2 W$.

One must further keep in mind that a fully symmetric serpentine robot can roll on any side and may even move on one of its four edges. In such an extreme case, only one single bellows would be able to contribute to the lifting torque τ_p and the lever arm for producing this lifting torque has length D, as shown in Fig. 5. During experiments we measured the minimum value of the pressure difference $p_A - p_B = 63$ psi needed for generating a torque $\tau_p = 25$ Nm, which is sufficient to lift up the two lead- or tail-segments of the OmniTread.

In the nominal case of Fig. 6 (OmniTread lying on a side, not an edge), not just one but two bellows-pairs provide the lifting torque, albeit at a reduced moment lever $D/\sqrt{2}$. The available lifting torque in that case is larger than in the case of the OmniTread laying on its edge and can be generated by an even smaller pressure difference. In this case two front segments can be lifted up by the pressure difference $p_A - p_B = 47$ psi generating a torque $\tau_p = 27$ Nm.

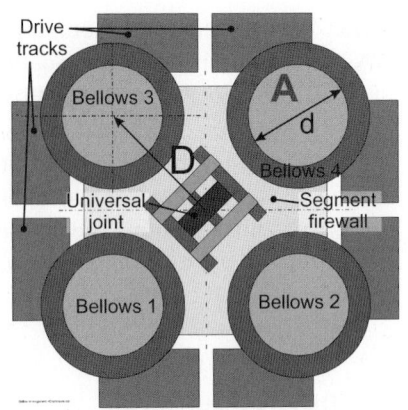

Fig. 5. Cross-section of the integrated joint actuator.

5 Conclusions

This paper focuses on the problem of joint actuation in serpentine robots. Based on our experience with the design of such robots, we defined in this paper the unique requirements for joint actuation in serpentine robots.

Our paper then introduces our solution for this problem, a system

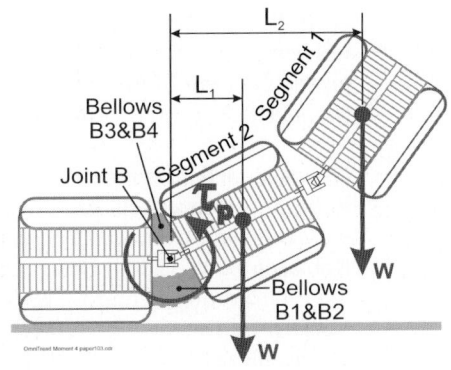

Fig. 6. Serpentine robot lifting up its first two segments to reach up the edge of a step.

called "Integrated Joint Actuator" (IJA) for serpentine mobile robots. The IJA uses specially designed pneumatic bellows as actuators in combination with our unique Proportional Position and Stiffness (PPS) control system described in details in [1].

While the combination of requirements defined in this paper may be unique to serpentine robots, subsets of these requirements are typically found in many existing mechanical structures. Our IJA should thus also appeal to researchers outside of the small community of serpentine robot developers.

Acknowledgements

This work was funded by the U.S. Department of Energy under Award No. DE-FG04-86NE3796.

References

1. Borenstein J., Granosik G., (2005) Integrated, Proportionally Controlled, and Naturally Compliant Universal Joint Actuator with Controllable Stiffness. *U.S. patent #6,870,343* issued March 22, (Rights Assigned to the University of Michigan).
2. FESTO, http://www.festo.com. [25.06.2005]
3. Granosik G., Borenstein J., (2005) Integrated Joint Actuator for Serpentine Robots, *To appear in October issue of the IEEE/ASME Transactions on Mechatronics.*
4. Huber J. E., Fleck, N. A. and Ashby, M.F. (1997) The selection of mechanical actuators based on performance indices, *Proc. of the Royal Society of London. Series A.* 453, pp. 2185-2205, UK.
5. Olszewski M., Janiszowski K., (1994) Solving of the main problems of position control of pneumatic drives of machinery and industrial robots, (in polish) *Proc. of 2nd Conf. on Mechatronics*, pp.69-74, Warsaw, 22-23 Sept.
6. Osuka K, and Kitajima, H., (2003) Development of Mobile Inspection Robot for Rescue Activities: MOIRA." *Proc. 2003 IEEE/RSJ Int'l Conf. on Intelligent Robots and Systems*, pp. 3373-3377, Las Vegas, NV, Oct.
7. Robinson D.W., (2000) Design and analysis of series elasticity in closed-loop actuator force control," *PhD thesis*, MIT.
8. Verrelst B., Daerden F., Lefaber D., Van Ham R., Fabri T., (2000) Introducing pleated pneumatic artificial muscles for the actuation of legged robots: a one-dimensional set-up, *Proc. of 3rd Int'l Conf. on Climbing and Walking Robots*, Madrid, 2-4 Oct.

Nontraditional Drives for Walking Robots

Teodor Akinfiev, Roemi Fernandez and Manuel Armada

Industrial Automation Institute, Spanish Council for Scientific Research. Department of Automatic Control. La Poveda 28500, Arganda del Rey, Madrid, Spain. teodor@iai.csic.es

Abstract

The results of a series of researches on the new nontraditional drives are presented. These drives with changing transmission ratio and dual properties have been developed for walking machines, whose working elements make movements with stop and stoppage. Analytical calculations, simulations and experiments with prototypes confirmed a high effectiveness of the designed drives.

Keywords: Start-stop movement, drive, effectiveness.

1 Introduction

It is well known that traditional drives in walking robots are characterized by extremely low speed and high energy expenses [1]. Usually, the increase of speed of machines can be achieved by an increase of capacity of a drive. Walking robots perform a movement with a stop and stoppage and are characterized by alternation of acceleration and braking. In such machines the most part of energy expenses during acceleration is being expended on overcoming of inertia forces, later all this energy being lost during the braking process. From the theory of oscillations it is known that acceleration and braking in resonance oscillatory systems are performed at the expense of passive spring elements, the engine serving only for making up for friction losses. It has been shown previously that the use of resonance drives allows to increase equipment's speed substantially and simultaneously to lower energy expenses more than 10 times [2, 3].

Resonance drives have a very big efficiency only in case, when friction forces are not too great. In case of friction forces increasing, the efficiency of resonance drive diminishes considerably. Additionally, resonance drive design is more sophisticated than the design of traditional drives.

Alternatively, conventional drives have low effectiveness in start-stop movement because during initial and last stages of motion it is necessary to have a reduction gear with high transmission ratio for intensive acceleration (braking) of mobile link so that the motor accelerates up to the highest speed very fast. Its torque falls down extremely fast. However, working element' speed is low during the middle stage of trajectory so that the quickness of the system is not very high. On the other hand, the use of reduction gear with small transmission ratio allows getting only low acceleration of mobile element and system's quickness is not high again.

Appropriate solution can be achieved by using a reduction gear with changing transmission ratio (Fig. 1). Such kind of gears provides high transmission ratio in initial and last parts of trajectory and low transmission ratio in the middle part of trajectory. It has been shown [4, 5, 7, 8, 11, 13, 16] that the use of such reduction gears increases quickness and diminishes energy expenses because in this case electric motor works with high efficiency during the whole trajectory. In this work, various drives with continuously changing transmission ratio are considered.

Fig. 1. Examples of different types changing transmission ratios.

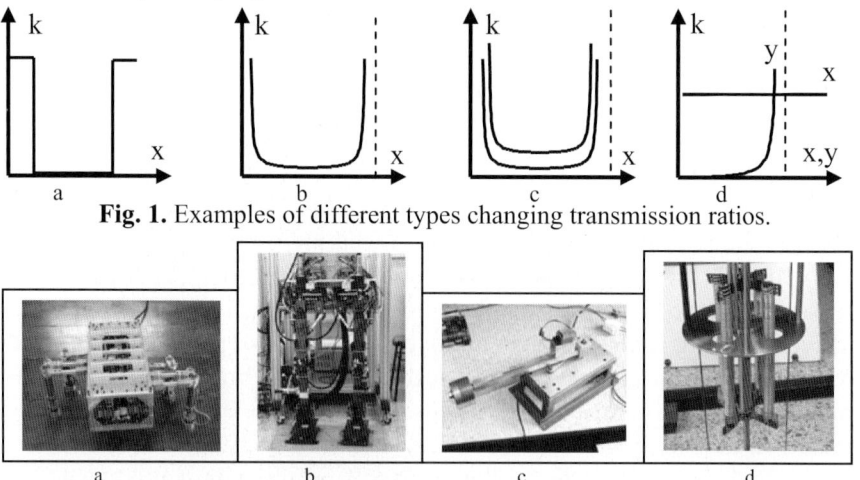

Fig. 2. Experiments with prototypes and robots

2 Design of drives with changing transmission ratio

2.1 Simplest quasi-resonance drive

Simplest quasi-resonance drive [4] is intended for progressive movement of mobile link from initial to end position (and back) with stoppage in extreme positions. On Fig. 3 a cinematic circuit of offered drive is presented.

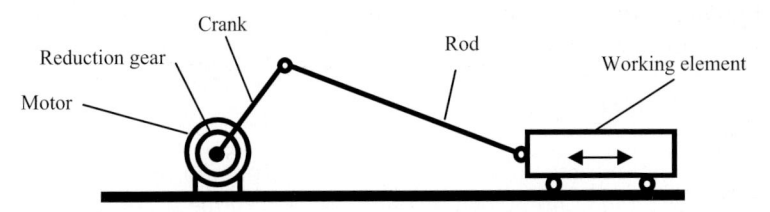

Fig. 3. Simplest quasi-resonance drive

On Fig. 1b a transmission ratio between motor' speed and working element' speed as a function of motor angular coordinate is shown. It is clear that close to extreme positions, this ratio is increasing limitlessly, and in the middle position it has minimal magnitude. It was shown theoretically that transmission ratio with these properties is most favorable from the point of view of energy expenses and time of movement [4, 5].

2.2 SMART drive

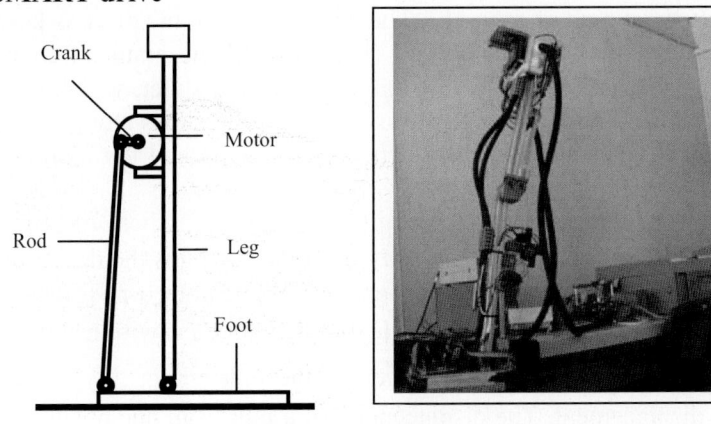

Fig. 4. SMART drive

A different type of quasi-resonance drive was elaborated [7] for angular movement of working element - Special Mechatronic Actuator for Robot joinTs (SMART). SMART is kind of quadric-crank mechanism (Fig. 4) and has a relation between transmission ratio and angular coordinate of motor similar to the one shown on Fig. 1a. SMART is effective even in quasi-static regime and can be used for biped robots (Fig. 2b). The matter is that SMART drive changes transmission ratio and this reduces torque of motor, especially in the end points, where transmission ratio is equal to infinite and system is self-braking. So, at the end points torque of motor is equal to zero.

2.3 Dual SMART drive

Usually, in walking machines a drive of a leg works in two essentially different regimes. One regime occurs when any leg of the robot rests on a surface so that the drive of this leg is utilized for the moving of a heavy (in relation to a fixed leg) body of the robot (normally, quasi-static condition). The second regime of the same drive occurs when the robot rests on alternate leg (or alternate legs): the same drive is utilized for the movement of the lightweight (in relation to its fixed body) robot's leg. Clearly, these two regimes are rather different. The drive, tuned on one of the regimes (usually, on the first one, as the most unfavorable) in optimal mode, appears to be inefficient in the second regime because, generally, the movement of a leg is progressing too slow, despite of high power of a drive motor.

To solve the problem that arises because the same drive is to be used for realization of two essentially different types of motion, there has been designed a special type of drive [8]. The construction of the drive is based on the use of crank-slider mechanism (Fig. 1c, 5). The crank is connected with motor. The crank causes the movement of the slider, which turns the working element (leg of robot), and simultaneously slides along this working element. The reduction ratio between the crank and the working element is presented on Fig. 1c.

The reduction ratio tends to infinity in end points, where the angle between a crank and a link measures $\pi/2$. In these points the deviation of a link from its medium position is maximal (Fig. 5). Two positions of the crank – one, within limits of an angle α_1, and another, within limits of an angle α_2 correspond to each position of the link within the limits of its possible displacement. The displacement of a link from one position to another can be carried out in two ways – by displacement of a crank within the limits of the angle α_1 or within the limits of the angle α_2. At displacement within the limits of the angle α_1 the absolute magnitude of the reduction ratio will be greater than at displacement within the limits of the angle α_2. This enables to use the movements of a crank within the limits of one angle, when the loading is small (and to gain high speeds of displacement), or within the limits of another angle, when the drive loading is big (with smaller velocities of displacement). The changeover from one working angle to the other is carried out with the help of displacement of the crank through one of critical points. This drive can be tuned on two different regimes of driving. It can switch from one regime to another immediately during operation.

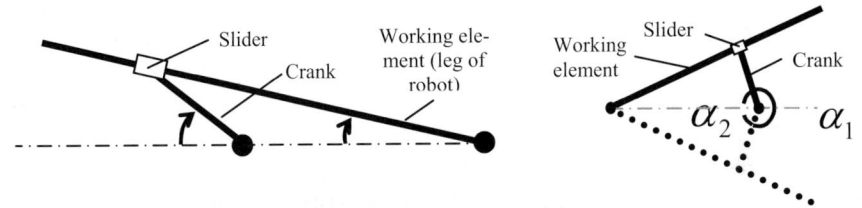

Fig. 5. Dual SMART drive

2.4 Dual drive for walking robot

If a technological operation (welding, painting etc.) is performed when robot's body moves [10], there is no point in the use of resonance drive for robot speed increasing. The matter is that while using resonance drive, the law of motion is completely determined by resilient elements and hardly can be corrected by a motor. For realization of technological process, a totally different law of motion may be required. The requirements of a technological process superimpose limitations on the law of motion of the robot's body, but do not superimpose any limitations on the law of motion of legs of the robot in a phase of their motion. So, to increase productivity the legs have to move as rapid as possible.

In this case it is desirable to have a special drive, which would possess double properties: while a body of the robot is moving, this drive would work as a conventional drive, and while legs of the robot are moving, it would work as a resonance drive. It is necessary to underline that the same drive will be utilized for the movement of both the body of the robot (when robot's legs rest on a base) and its legs (when the corresponding leg does not contact with a base). The proposed solution [11] is illustrated by the example of quadruped walking robot.

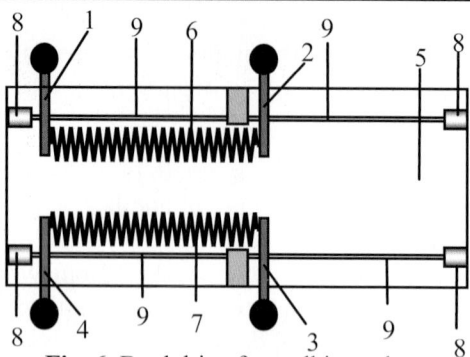

Fig. 6. Dual drive for walking robot

The quadruped walking robot is shown on Fig. 6. On the robot's body 5 all legs are fixed. The robot has two resilient elements. One extreme point of the first resilient element 6 is connected with the leg 1, and another extreme point of the first resilient element 6 is connected with the leg 2. One extreme point of the second resilient element 7 is connected with the leg 4, and another extreme point of the second resilient element 7 is connected with the leg 3. The electric motors 8 are connected to the legs through a screw-nut transmission 9.

Each of drive motors provides two rather different modes of movement. The movement of robot's body is carried out in a regime of a conventional drive. While this moving, the springs do not affect the movement of the body, because the distance between legs does not change so that the law of motion of the body is determined only by force of motors.

The second mode of movement (a movement of any leg) is performed with both spring and corresponding motor effecting upon a leg simultaneously. During this movement the spring helps to get high acceleration of a leg on the first part of a trajectory and heavy braking on the second part of a trajectory, the motor serving only for compensation of friction losses during this movement in the same way as it occurs in resonance drives. Thus, the designed drive has double properties - while moving of robot's body it works in a traditional drive mode, and at moving of each of legs it works in a resonance drive mode.

Similar ideas were used for the robot (Fig 2a) with different transmission ratio in direction X and Y (Fig 1d).

3 Experiments and simulations

In the process of the comparative experiments two different control algorithms have been used. In the cases when it was only necessary to displace working element from on position to another reaching the end point with zero velocity (the law of the motion was not important in that case), the universal algorithm of adaptive control have been used, which has been elaborated earlier for resonance drives [17]. In the cases when the law of the motion was important, a backstepping tracking algorithm has been employed [14, 15]. The experiments have shown that both algorithms provide a reliable work of the drives [6].

Experiments with biped robot's leg have shown that power consumption of SMART drive is less then 2 W. The nonlinear SMART drive and a classical drive were compared experimentally. The SMART actuator demonstrated an average saving of energy 48% in comparison with the classical drive [12]. Also compared were the dual SMART drive and a classical drive. The dual SMART drive proved to reduce motion time 69% and energy consumption 72% during the first regime, and to reduce the motion time 40% and the energy consumption 41% during the second regime [6]. Even when regimes of operating of the dual SMART drive are rather different, effectiveness of motor is about 90% that is unusual for start-stop regime while using traditional drives [9].

Dual SMART drive, compared with SMART drive, can reduce not only the time of motion 54% but, simultaneously, the energy consumption 65%.

These results show high effectiveness of the use of nontraditional drives for walking robots.

4 Conclusions

The design of nontraditional drives including drives with dual properties for mobile robots is considered. It is shown that effectiveness of these drives is similar to that of resonance drives having, at the same time, simpler design and wider fields of applications. The results obtained were confirmed by analytical calculations, simulations and experimentally.

Acknowledgments

The authors would like to acknowledge the financial support from the Spanish Ministry of Education and Science: Project "Theory of optimal dual drives for automation and robotics" (Akinfiev) and Fellowship F.P.U. (Fernandez).

3 Measures for performance evaluation

Two global measures of the overall mechanism performance in an average sense are established [9]. The first, the mean absolute density of energy per travelled distance (E_{av}), is obtained by averaging the mechanical absolute energy delivered over the travelled distance d:

$$E_{av} = \frac{1}{d}\sum_{i=1}^{n}\sum_{j=1}^{m}\int_{0}^{T}\left|\tau_{ijm}(t)\dot{\theta}_{ij}(t)\right|dt \quad \left[\text{Jm}^{-1}\right] \tag{3}$$

The other, based on the hip trajectory tracking errors (ε_{xyH}), is defined as:

$$\varepsilon_{xyH} = \sum_{i=1}^{n}\sqrt{\frac{1}{N_S}\sum_{k=1}^{N_S}\left(\Delta_{ixH}^{2} + \Delta_{iyH}^{2}\right)}, \Delta_{i\eta H} = \eta_{iHd}(k) - \eta_{iH}(k), \eta = \{x, y\} \quad [\text{m}] \tag{4}$$

where N_s is the total number of samples for averaging purposes.

The performance optimization requires the minimization of each index.

4 Simulation results

To illustrate the use of the preceding concepts, in this section we develop a set of simulation experiments to estimate the influence of parameters L_S and H_B, when adopting periodic gaits [8]. We consider three walking gaits (Walk, Chelonian Walk and Amble), two symmetrical running gaits (Trot and Pace) and five asymmetrical running gaits (Canter, Transverse Gallop, Rotary Gallop, Half-Bound and Bound). These gaits are usually adopted by animals moving at low, moderate and high speed, respectively.

In a first phase, the robot is simulated in order to analyse the evolution of the locomotion parameters L_S and H_B with V_F, being the controller tuned for each gait while the robot is walking with $V_F = 1$ ms^{-1}.

In a second phase, the controller is tuned for each particular gait and locomotion velocity and the quadruped robot is then simulated in order to compare the performance of the different gaits *versus* V_F.

For the system simulation we consider the robot body parameters, the locomotion parameters and the ground parameters presented in Table 1. Moreover, we assume high performance joint actuators with a maximum torque of $\tau_{ijMax} = 400$ Nm. To tune the controller we adopt a systematic method, testing and evaluating a grid of several possible combinations of controller parameters, while minimising E_{av} (3).

Table 2. Quadruped controller parameters

Gait	ϕ_1	ϕ_2	ϕ_3	ϕ_4	β	Kp_1	Kd_1	Kp_2	Kd_2
Walk	0	0.5	0.75	0.25	0.65	1000	40	2000	40
Chelonian Walk	0	0.5	0.5	0	0.8	5000	200	2500	20
Amble	0	0.5	0.75	0.25	0.45	1000	20	1000	60
Trot	0	0.5	0.5	0	0.4	1000	140	2000	20
Pace	0	0.5	0	0.5	0.4	1000	60	500	40
Canter	0	0.3	0.7	0	0.4	1000	0	1500	20
Transverse Gallop	0	0.2	0.6	0.8	0.3	6000	40	1000	40
Rotary Gallop	0	0.1	0.6	0.5	0.3	4000	0	500	80
Half-Bound	0.7	0.6	0	0	0.2	4000	0	3000	20
Bound	0	0	0.5	0.5	0.2	2000	0	500	20

4.1 Locomotion parameters *versus* body forward velocity

In order to analyse the evolution of the locomotion parameters L_S and H_B with V_F, we test the forward straight line quadruped robot locomotion, as a function of V_F, when adopting different gaits often observed in several quadruped animals while they walk / run at variable speeds [8].

With this purpose, the robot controller is tuned for each gait, considering the forward velocity $V_F = 1.0$ ms^{-1}, resulting the possible controller parameters presented in Table 2.

After completing the controller tuning, the robot forward straight line locomotion is simulated for different gaits, while varying the body velocity on the range $0.2 \leq V_F \leq 5.0$ ms^{-1}. For each gait and body velocity, the set of locomotion parameters (L_S, H_B) that minimises the performance index E_{av} is determined.

The chart presented in Fig. 2 depicts the minimum value of the index E_{av}, on the range of V_F under consideration, for three different robot gaits. It is possible to conclude that the minimum values of the index E_{av} increase with V_F, independently of the adopted locomotion gait. Although not presented here, due to space limitations, the behaviour of the charts min[$E_{av}(V_F)$], for all other gaits present similar shapes.

Next we analyse how the locomotion parameters vary with V_F. Figure 3 (left) shows, for three locomotion gaits, that the optimal value of L_S must increase with V_F when considering the performance index E_{av}. The same figure (right) shows that H_B must decrease with V_F from the viewpoint of the same performance index.

For the other periodic walking gaits considered on this study, the evolution of the optimization index E_{av} and the locomotion parameters (L_S, H_B) with V_F follows the same pattern.

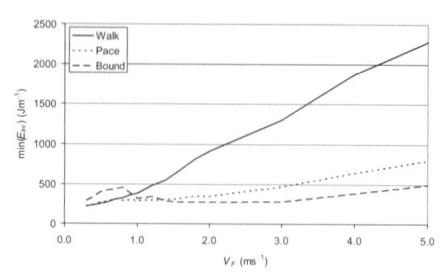

Fig. 2. $\min[E_{av}(V_F)]$ for $F_C = 0.1$ m

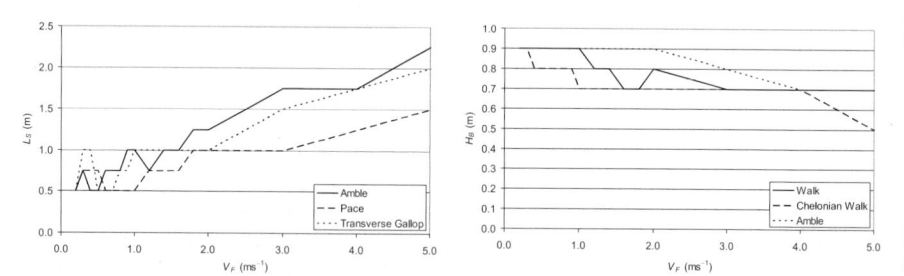

Fig. 3. $L_S(V_F)$ (left) and $H_B(V_F)$ (right) for $\min(E_{av})$, with $F_C = 0.1$ m

Therefore, we conclude that the locomotion parameters should be adapted to the walking velocity in order to optimize the robot performance. As V_F increases, the value of H_B should be decreased and the value of L_S increased. These results seem to agree with the observations of the living quadruped creatures [11].

4.2 Gait selection *versus* body forward velocity

In a second phase we determine the best locomotion gait, from the viewpoint of energy efficiency, at each forward robot velocity on the range $0.1 \leq V_F \leq 10.0$ ms^{-1}. For this phase of the study, the controller is tuned for each particular locomotion velocity, while minimizing the index E_{av}, and adopting the locomotion parameters $L_S = 1.0$ m and $H_B = 0.9$ m.

Figure 4 presents the charts of $\min[E_{av}(V_F)]$ and the corresponding value of ε_{xyH} for the different gaits. The index E_{av} suggests that the locomotion should be Amble, Bound and Half-Bound as the speed increases. The other gaits under consideration present values of $\min[E_{av}(V_F)]$ higher than these ones, on all range of V_F under consideration. In particular, the gaits Walk and Chelonian Walk present the higher values of this performance measure.

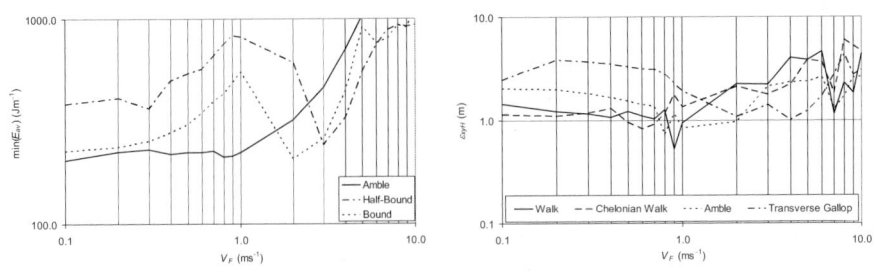

Fig. 4. min[$E_{av}(V_F)$] and the corresponding value of ε_{xyH}, for $F_C = 0.1$ m

Analysing the locomotion though the index ε_{xyH}, we verify that for low values of V_F ($V_F < 1\text{ms}^{-1}$), the gaits Walk and Chelonian Walk allow the lower oscillations of the hips. For increasing values of the locomotion velocity the Amble and Transverse Gallop gaits present the lower values of ε_{xyH}.

From these results, we can conclude that, from the viewpoint of each proposed optimising index, the robot gait should change with the desired forward body velocity. These results seem to agree with the observations of the living quadruped creatures [11].

In conclusion, the locomotion gait and the parameters L_S and H_B should be chosen according to the intended robot forward velocity in order to optimize the energy efficiency or the oscillation of the hips trajectories.

5 Conclusions

In this paper we have compared several aspects of periodic quadruped locomotion gaits. By implementing different motion patterns, we estimated how the robot responds to the locomotion parameters step length and body height and to the forward speed. For analyzing the system performance two quantitative measures were defined based on the system energy consumption and the trajectory errors. A set of experiments determined the best set of gait and locomotion variables, as a function of the forward velocity V_F.

The results show that the locomotion parameters should be adapted to the walking velocity in order to optimize the robot performance. As the forward velocity increases, the value of H_B should be decreased and the value of L_S increased. Furthermore, for the case of a quadruped robot, we concluded that the gait should be adapted to V_F.

While our focus has been on a dynamic analysis in periodic gaits, certain aspects of locomotion are not necessarily captured by the proposed measures. Consequently, future work in this area will address the refinement of our models to incorporate more unstructured terrains, namely with distinct trajectory planning concepts. The effect of distinct values of the robot intra-body compliance parameters will also be studied, since animals use their body compliance to store energy at high velocities.

References

1. Poulakakis I., Smith J. A. and Buehler M. (2004) Experimentally Validated Bounding Models for the Scout II Quadruped Robot. *Proc. of the 2004 IEEE Int. Conf. on Rob. & Aut. (ICRA'2004)*, New Orleans, USA, 26 April – 1 May.
2. Zhang Z. G., Fukuoka Y. and Kimura H. (2004) Stable Quadrupedal Running Based on a Spring-Loaded Two-Segmented Legged Model. *Proc. of the 2004 IEEE Int. Conf. on Rob. & Aut. (ICRA'2004)*, New Orleans, USA, 26 April – 1 May.
3. Iida F. and Pfeifer R. (2004) "Cheap" Rapid Locomotion of a Quadruped Robot: Self-Stabilization of Bounding Gait. *Proc. of the 8th Conf. on Intelligent Autonomous Systems (IAS'2004)*, Amsterdam, The Netherlands, 10-13 March.
4. Kohl N. and Stone P. (2004) Policy Gradient Reinforcement Learning for Fast Quadrupedal Locomotion. *Proc. of the 2004 IEEE Int. Conf. on Rob. & Aut. (ICRA'2004)*, New Orleans, USA, 26 April – 1 May.
5. Palmer L. R., Orin D. E., Marhefka D. W., Schmiedeler J. P. and Waldron K. J. (2003) Intelligent Control of an Experimental Articulated Leg for a Galloping Machine. *Proc. of the 2003 IEEE Int. Conf. on Rob. & Aut. (ICRA'2003)*, Taipei, Taiwan, 14-19 September.
6. Hardt M. and von Stryk O (2000) Towards Optimal Hybrid Control Solutions for Gait Patterns of a Quadruped. *Proc. 3rd Int. Conf. on Climbing and Walking Robots (CLAWAR 2000)*, Madrid, Spain, 2 – 4 October.
7. Silva M. F., Machado J. A. T. and Lopes A. M. (2005) Modeling and Simulation of Artificial Locomotion Systems. *ROBOTICA* (accepted for publication).
8. URL: http://www.biology.leeds.ac.uk/teaching/3rdyear/Blgy3120/Jmvr/ /Loco/Gaits/GAITS.htm **[2005-06-30]**
9. Silva M. F., Machado J. A. T., Lopes A. M. and Tar J. K. (2004) Gait Selection for Quadruped and Hexapod Walking Systems. *Proc. of the 2004 IEEE Int. Conf. on Computational Cybernetics (ICCC'2004)*, Vienna, Austria, 30 August – 1 September.
10. Silva M. F., Machado J. A. T. and Lopes A. M. (2003) Position / Force Control of a Walking Robot. *Machine Intelligence and Robotic Control*, vol. 5, no. 2, pp. 33–44.
11. Alexander R. McN. (1984) The Gaits of Bipedal and Quadrupedal Animal. *The Int. J. of Robotics Research*, vol. 3, no. 2, pp. 49–59.

Bellows Driven, Muscle Steered Caterpillar Robot

G. Granosik and M. Kaczmarski

Institute of Automatic Control, Technical University of Łódź, POLAND,
granosik@p.lodz.pl

Abstract

This paper presents an investigation into the design of a caterpillar-like robot Catty that utilizes only pneumatic actuators both for propulsion and steering. The proposed approach offers a robot of fully compliant behavior with many advantages. The aim of this research is to verify the mobility of the proposed design and applicability of the caterpillar robot for inspection purposes. **Fig. 1** shows a single segment prototype system of the proposed robot, which is able to perform 2-D motion on the smooth non-vertical surfaces. It can also bend the body almost 90 deg. in vertical plane and provides an opportunity for concave transitions.

Keywords: Caterpillar-like robot, pneumatic bellows, pneumatic muscles.

1 Introduction

The Robotics Group from Technical University of Lodz has been successfully applying various pneumatic drives into robotic applications for several years now [7, 4, 3, 5, 8]. The main advantages of pneumatic actuators are large power to weight ratio and low (and controllable) stiffness. Moreover, overload ratio is much higher than in electric or hydraulic drives. This feature is especially important in mobile robotics where contact with

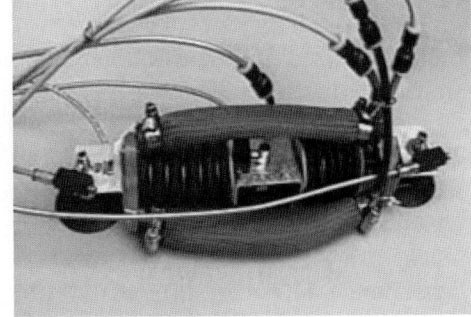

Fig. 1. Prototype of caterpillar robot Catty

environment is permanent and unexpected shocks can be really significant. The detailed, qualitative and quantitative analysis of pneumatic actuators for some class of mobile robots can be found in [6]. The main disadvantage of pneumatics is low efficiency and problematic position control. These problems were addressed in many papers and in different ways; the review of many methods and derivation of position-force control for pneumatic cylinders can be found in [5]. We realize that employing pneumatic drives in autonomous mobile robots is very difficult but we believe that in tethered robots this choice is the best in many cases, including the one described here.

Imitation of the nature is one of the most exploited methods used in design and control of robots. We also used this approach in some of the projects mentioned earlier and this is a continuation of that. Caterpillar and inchworm locomotion drew attention of robot's designers due to its simplicity and effectiveness in very constrained spaces. These motion patters can usually be described using finite state models and thus make gait generation very easy [2].

The planar inchworm robot designed by Chen's group uses the unique mechanical arrangement of the actuators to allow for quick change in travel direction and to permit rotational movement [12]. Pneumatic cylinders are used to achieve the motions instead of rotary motors. Four actuators are linked together using pivot joints to form a square loop. At the four pivoting joints, grippers are used to attach the pivot joints to the travel surface. A combination of actuator extensions and retractions together with synchronized gripping action allow the execution of direct forward motion and rotational turning motion on a horizontal surface. The concept retains the original inchworm gait of attaching at least one segment of the body to the travel surface before propelling the other segments forward. Authors showed that the proposed planar inchworm robot is able to move from a starting point to a destination within specified positional accuracy based on simple gait planning algorithm.

Another walking strategy is presented by Kotay and Rus in their Inchworm robot for inspection of the 3-D steel structures [9]. The Inchworm is a biologically-inspired mobile robot whose movement is similar to that of an inchworm caterpillar. It consists of a three degree-of-freedom main body which can flex and extend to propel itself, along with electromagnets at each end of the body which provide the anchoring force for the motion. A fourth degree of freedom is provided by a pivot joint which allows the body of the inchworm to rotate relative to the attachment mechanism of the rear foot. This allows the robot to turn. The Inchworm can operate on steel surfaces of arbitrary orientation, i.e. it can climb vertical steel walls and crawl across a steel ceiling using the electromagnets to attach itself to the surface. The Inchworm can also make transitions between surfaces, allowing it to navigate autonomously in unknown environments. Robot is driven by simple servomotors used in model aircrafts.

The same driving mechanism is employed in CRAWLY – a robotic caterpillar [10]. The locomotion mathematical model of this robot is based on a sinusoidal type of formula that makes it capable to scrawl on a variety of different types of terrain. CRAWLY robot is constructed by six small servos and is controlled through an SSC II serial servo controller. The robot's 0.5kg body is divided into seven segments and is 60cm long. Modifications in CRAWLY's locomotion mathematical model can make its chassis capable to travel inside pipes.

Another simple INCHWORM ROBOT is also fashioned after animal and utilizes four RC servo motors. It is designed mainly for edutainment purposes and is commercially available [11].

Our approach combines imitation of the nature with our experience with pneumatic actuators. We try to model a multi-directional planar robot based directly on the original structure of caterpillar body and by taking advantage of natural elasticity of pneumatic bellows and muscles.

2 Robot design

Catty is the walking and climbing robot schematically shown in **Fig. 2**. Modeled robot consists of two bellows mounted in series on triangular chassis, three suction cups for gripping to surface and three pneumatic muscles mounted to the tips of walls W1 and W2. Simple linear movements can be obtained by pumping bellows B1 and exhausting

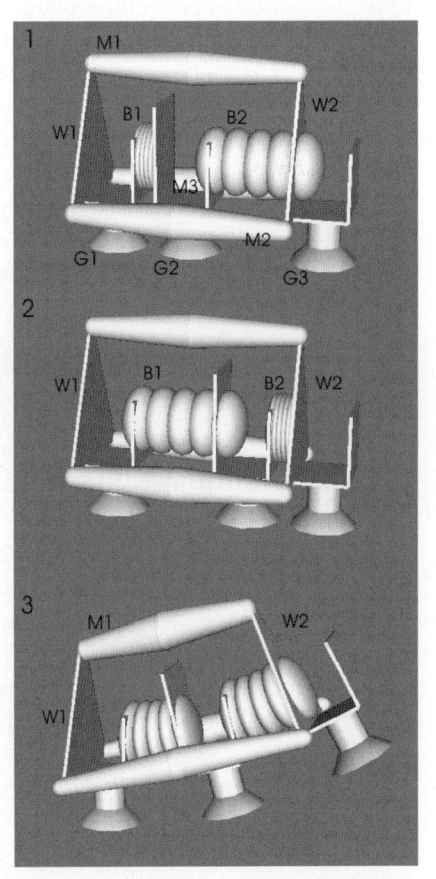

Fig. 2. Model of our caterpillar robot with associated symbols. Linear movement is shown in picture 1 and 2, while picture 3 shows lifting up of the front segment

bellows B2 (transition from picture 1 to 2 in **Fig. 2**) synchronized with appropriate activating of suction cups G1-G3, which provide the anchoring force for the motion. During this reciprocal movement firewalls W1 and W2 remain parallel and at the same distance one from another. Such motion can be produced on a flat surface due to suction cups properties and the fact that suction cup G2 is sliding on the surface. Muscles M1-M3 are released and stretched during linear operation.

Changing of the orientation requires activating of some muscles: usually shortening one muscle (M1, M2, or M3) while releasing others. Lower part of **Fig. 2** shows lifting up of the front segment of our robot. The longer version of the robot (containing at least two segments) would be able to make transitions between perpendicular surfaces.

The idea of reciprocal movements in order to propel robot forward and backward is somehow similar to bridled bellows introduced by Aoki and Hirose in Slim Slime Robot [1]. However, in our design we use different technology – rubber bellows – specially designed for joint actuation in serpentine robots [6]. These bellows have larger elasticity than any metal construction and, more important, large actuation strain (or simply speaking relative elongation). We also use muscles as steering actuators instead of motorized strings.

During the simulation stage we developed simple gait patterns, which were experimentally verified on our test bed, as presented in the next section.

3 Experimental results

Fig. 1 shows photograph of the current version of Catty – our caterpillar robot. Dimensions of robot and actuators are collected in **Table 1**. Robot is currently operated from PC computer (equipped with Lab. card) and off-board solenoid valves. Two different values of pressures are used in proposed pneumatic system. Pneumatic bellows are supplied from 2 bars, while muscles and suction cups require 4 bars.

Table 1. Specification of the caterpillar robot Catty

Height	80 mm
Width	115 mm
Maximal length	290 mm
Step length	60 mm
Bellows diameter, length	43 mm, 25 – 100 mm
Muscle diameter, length	10 mm, 170 – 200 mm

We performed several tests to verify possibility of basic movements of our caterpillar robot. These include straight locomotion, turning in spot and lifting up of the front part of the robot. All tests were performed on smooth and horizontal surface.

The sequence of actions, which propel robot forward, is presented in **Fig. 3** and states of actuators in each phase of movement are shown in Table 2. B1 and B2 state for bellows actuators, while G1 and G2 are rear and central suction grippers, respectively. All other actuators remained released during this test. The anchoring point is indicated by vertical yellow line in **Fig. 3**. Horizontal line shows the moment of transition.

As we expected from modeling and theoretical analysis, evenly distributed muscles and natural stiffness of pumped bellows prevented robot from straying from correct course. Performance was the best when muscles were stretched by expanded bellows. To further verify this natural stiffness property we mounted spherical joint between bellows B2 and wall W2. Even with these extra degrees of freedom Catty held constant direction. At this preliminary stage, our robot is controlled in open loop and we used activation and delay times long enough to ensure proper actions of actuators. As the consequence, the time of single step (total time of phases 1-3) is 16s.

Table 2. States of actuators during linear motion.

Phase	State of actuators: B1, B2, G1, G2
1	1, 0, 1, 0
2	0, 0, 0, 1
3	0, 1, 0, 1
4	1, 0, 1, 0
5	0, 0, 0, 1
6	0, 1, 0, 1

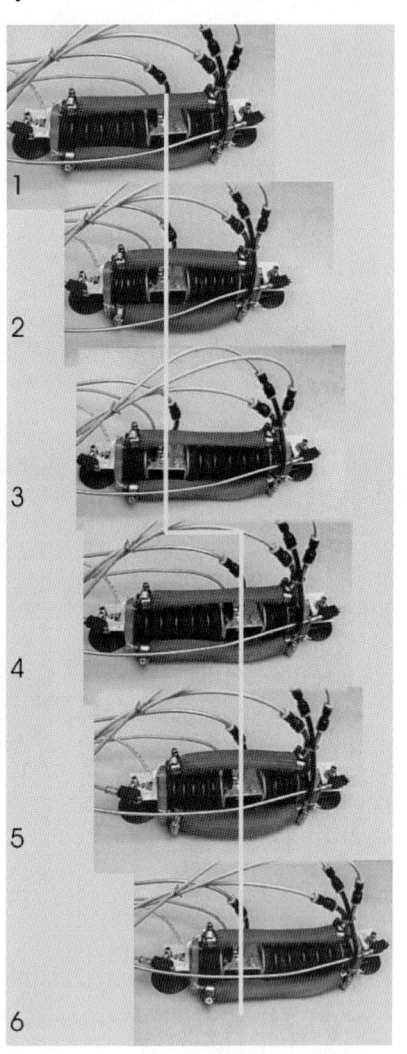

Fig. 3. Six snapshots taken from a basic step. Yellow line indicates position of the central suction cup.

The next experiment confirmed the ability of turning in spot. Similarly to previous test, we designed the minimal sequence of actuations in order to turn robot around vertical axis located somewhere in the middle of the robot's body. Twelve snapshots from rotating the robot counter-clockwise

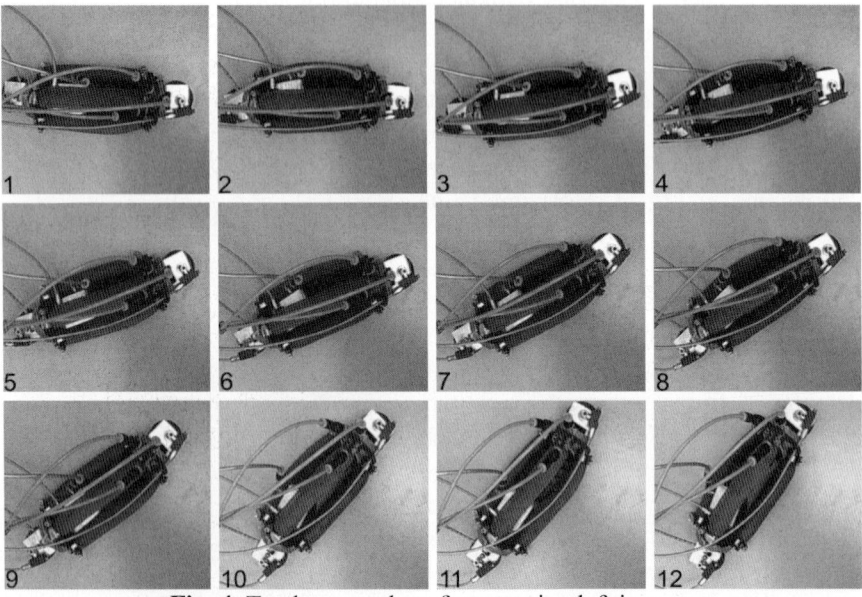

Fig. 4. Twelve snapshots from turning left in spot.

of about 50 deg are presented in **Fig. 4**; in addition, states of actuators used in this experiment are collected in **Table 3**.

The turning procedure is simply based on activating one of the lower muscles M2 or M3 together with one of the distal suction grippers G1 or G3. The whole experiment, being a six times repetition of phases 1 and 2 lasted 60s. In the **Fig. 5** we presented a closer view of the robot during turning in the spot.

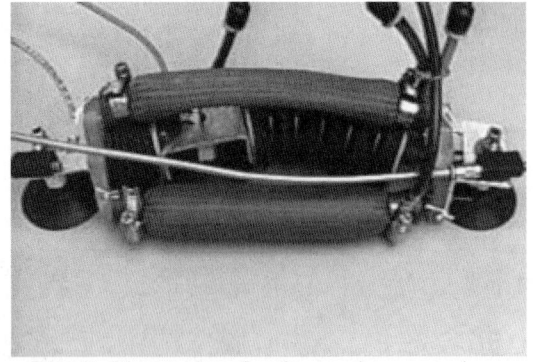

Fig. 5. Closer view for the turning procedure, muscle M2 and bellows B2 are activated

Table 3. States of actuators during turning in spot

Phase	State of actuators: B2, G1, G3, M2, M3
1, 3, 5, 7, 9, 11	1, 1, 0, 1, 0
2, 4, 6, 8, 10, 12	1, 0, 1, 0, 1

Contracted muscle M2 together with expended bellows B2 produce bending angle of about 5 deg.

The last experiment shows the lifting up of the front section of Catty and releasing it back to horizontal orientation, as presented in **Fig. 6**. States of actuators are collected in **Table 4**. As shown in picture 7 in **Fig. 6**, robot can bend its body even up to 90 deg, which gives the possibility of transition from horizontal to vertical plane. However, further movement on vertical plane is impossible at this stage of design. This would require more segments in Catty's body and will be tested in the future.

4 Conclusion

This paper has presented an investigation into the design and development of a caterpillar-like robot Catty. It is worth noting that the experimental results demonstrated the expected outcome of the mobility but maneuverability is slightly limited in case of turning. Nonetheless, robot performs well on horizontal surface and provides a certain degree of ability for climbing, which will be explored in the near future.

Table 4. States of actuators during lifting up and releasing front suction cup of Catty.

Phase	State of actuators: B1, B2, G1, M1, M2, M3
1	0, 0, 1, 0, 0, 0
2	0, 0, 1, 1, 0, 0
3	1, 0, 1, 1, 0, 0
4	1, 1, 1, 1, 0, 0
5	1, 1, 1, 1, 0, 0
6	1, 1, 1, 0, 0, 0
7	1, 1, 1, 0, 0, 0
8	1, 0, 1, 0, 0, 0
9	0, 0, 1, 0, 0, 0
10	0, 0, 1, 0, 0, 0
11	0, 0, 1, 0, 1, 1

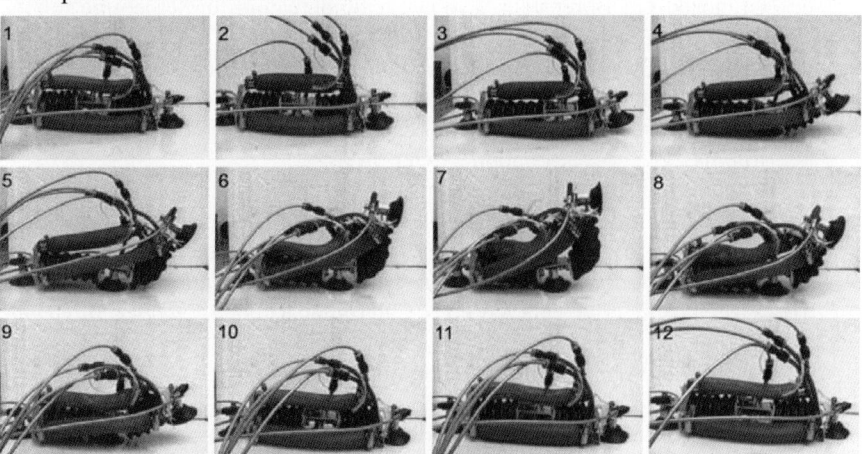

Fig. 6. Snapshots from lifting up and releasing of the front suction cup.

The originality of our design lies in application of three kinds of pneumatic actuators in propulsion mechanism of robot. They are pneumatic bellows, muscles and suction cups. Bellows actuators and muscles used for Catty were built in our Laboratory with off-the-shelf components, which allowed us to model their properties and exactly fit our needs. In this initial stage of project robot is teleoperated using off-board valves and PC computer. An investigation is currently underway to design a cheap pneumatic valve to mount on the robot.

References

1. Aoki, T.; Ohno, H.; Hirose, S.; Design of Slim Slime Robot II (SSR-II) with Bridle Bellows, *Proc. IEEE/RSJ International Conference on Intelligent Robots and System*, pp. 835-840, vol.1, Oct. 2002
2. Chen, I.-M., Yeo, S. H., Gao, Y., "Locomotive Gait Generation for Inchworm-Like Robots Using Finite State Approach," *Robotica*, Vol. 19, No.5, pp535-542, 2001
3. Dąbrowski T., Feja K., Granosik G., Biologically inspired control strategy of a pneumatically driven walking robot, *Proc. Of the 4th Int. Conference on Climbing and Walking Robots CLAWAR 2001*, pp 687-694, Karlsruhe 2001
4. Granosik G., Jezierski E., Application of a maximum stiffness rule for pneumatically driven legs of walking robot, *Proc. of 2nd Int. Conference on Climbing and Walking Robots*, pp 213-218, Portsmouth 1999
5. Granosik G., An adaptive position/force control of a pneumatically driven manipulator, PhD dissertation (in polish), Technical University of Łódź 2000
6. Granosik, G., Borenstein, J., Pneumatic actuators for serpentine robots, *To appear in Proc. of CLAWAR 2005*
7. Jezierski, E.; Mianowski, K.; Collie, A.A.; Granosik, G.; Zarychta, D. Design and control of a manipulator arm for a walking robot, *IEE Colloquium on Information Technology for Climbing and Walking Robots*, pp. 2/1-2/3, London 29 Oct. 1996
8. Kaczmarski, M. "Vision system for supervision of artificial hand", MSc dissertation (in polish), Technical University of Łódź, 2004
9. Kotay, K., Rus, D., Navigating 3d Steel Web Structures with an Inchworm Robot, *Proc. of the Conference on Intelligent Robot Systems*, 1996
10. Vastianos George, "CRAWLY - Caterpillar Robot", *Encoder Magazine, Seattle Robotics Society*, Issue Number 46, July 2002
11. www.microrobotna.com/Inchworm.htm
12. Yeo, S.H., Chen, I.-M., Senanayake, R. S., Wong P. S. "Design and Development of a Planar Inchworm Robot," *17th IAARC Int. Symp. Automation Robotics in Construction*, Taipei, 2000.

On the Application of Impedance Control to a Non-linear Actuator

H Montes, M Armada and T Akinfiev

Automatic Control Department, Industrial Automation Institute – CSIC, Madrid, Spain   hmontes@iai.csic.es

Abstract

Non-linear actuators are employed to drive some joints of the SILO2 humanoid robot. Saving energy along locomotion cycle is the main objective of this nonlinear actuator. In order to increase knowledge about the dynamic characteristics of this four-bar mechanism, called SMART, one force sensor has been implemented on one bar of this actuator. Moreover, taking advantage from that extra sensitivity it has been possible to implement force feedback control algorithms. In this paper, an impedance controller for the robot joints actuated by the non-linear mechanism is presented. Experimental evaluation, including potential energy analysis of the controller, are explained.

Keywords: Impedance control, nonlinear actuator, humanoid robot, compliance control, terrain adaptation.

1 Introduction

Force control applied to robots has been approached by many relevant researchers in the last three decades. The first realizations in this field have been highlighted in a first state of the art in force control by [1], and other researchers have evidenced the subsequent progress in force control in the robotic field [2-5]. Also, very detailed monographs in the area of force control have been published [6-7]. Presently, the majority of the researchers consider that the force control strategies can be grouped in two categories: a) those that carry out an indirect force control; and, b) those strategies that implement a direct force control [7]. The principal difference between the two categories is that in the first one the force control is carried out by means of position control (without the explicit closing of a force feedback loop), while the second category is characterized by controlling (directly) the contact force to the desired value due to the closing of an explicit force feedback loop. To the first category belongs the stiffness control [8], [9], and the impedance control [10-12], where the position error is related with the contact force through the mechanical stiffness or an impedance of adjustable parameters (that can be simply stiffness in its particular case). The impedance can be attributed to a mechanical system characterized by a matrix of mass, a matrix of damping and a matrix of stiffness, which will make possible the dynamic desired behaviour that the robot should have in its interaction with the environment.

2 Brief description of SMART nonlinear actuator

SMART (Special Mechatronic Actuator for Robot joinTs) has been developed in order to improve the performance of humanoids robots and has been implemented in some joints of the SILO2 biped robot [13]. This actuator is characterized by the change in the reduction value from any minimal value in the half zone of the trajectory until the infinite in the extreme positions (see **Fig. 1(a)**). It has been demonstrated, that after applying this nonlinear actuator it improves considerably the efficiency in the robot locomotion [14].

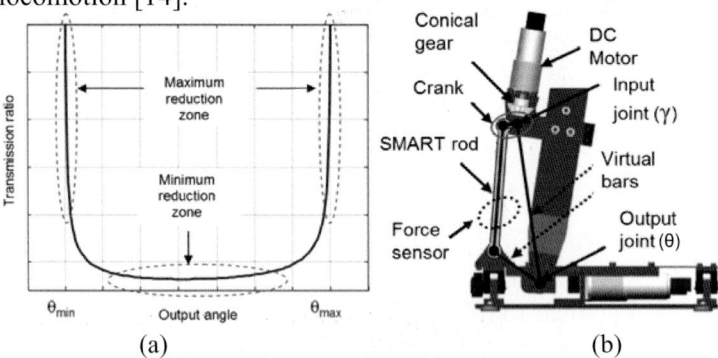

Fig. 1. (a) Transmission ratio function; and (b) Ankle joint: SMART nonlinear actuator

When the nonlinear actuator is used in the zone of minimum reduction, the external dynamic effects are significant and they influence directly the joint control. On the other hand, in the zones of maximum reduction it carries out a quasi-braking phenomenon, and so the external dynamic effects are of much less influence. The non-linear mechanical transmission shown in **Fig. 1(a)** can be modelled by an approximated function of the inverse transmission ratio, dependent of the output angle (θ). This equation is described as follows,

$$R_T^{-1} = \cos(\theta - \bar{\theta})[\theta_m^2 - (\theta - \bar{\theta})^2 f(\theta)]^{-1/2} \quad (1)$$

where,

$$\theta_{min} \leq \theta \leq \theta_{max}; \quad \theta_m = \frac{\theta_{max} - \theta_{min}}{2}; \quad \bar{\theta} = \frac{1}{n}\sum_{i=1}^{n}\theta_i; \quad f(\theta) = \left(\frac{\theta_m}{\sin(\theta_m)} \cdot \frac{\sin(\theta - \bar{\theta})}{(\theta - \bar{\theta})}\right)^2$$

This approximate function is very useful to make calculations, analysis and estimations of preliminary designs with any type of used non-linear actuator in the SILO2. The SMART actuator has been implemented by using a four bar linkage mechanism, where the individual links are of different lengths [1]. This mechanism is made up of two real bars (rod and crank) and two virtual bars [15] (see **Fig. 1(b)**).

3 Force control system

In order to achieve a compliance degree and/or an improvement on stability for the SILO2 using the SMART actuator, control strategies that include force feedback are proposed. These strategies can be stiffness control, impedance control or force/position control [1], [8], [10].

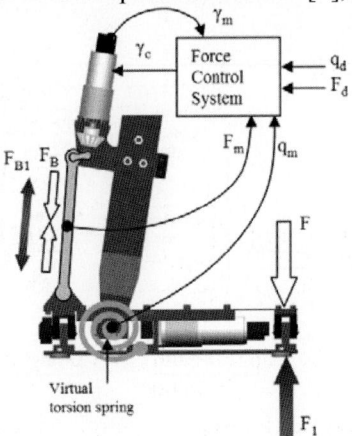

Fig. 2. Conceptual diagram of the force control effect on a joint

Fig. 2 shows the effect that the force control can cause on a joint that is driven by the SMART mechanism. It is assumed that the SMART mechanism is equipped with a force sensor. Therefore, the joint will behave as a torsion spring under the influence of external force, when being controlled with some strategy of force control. The force control system will act in the servomotor so that the displacement angle of the joint is proportional to the magnitude and the direction of the external force applied.

The force control strategies that are proposed in this paper have a position internal loop of high stiffness. In this case a microcontroller PID carries out the position control. The feedback of the position internal loop is made by means of incremental optical encoder placed on the servomotor. On the other hand, these control strategies have, in addition, two external loops, one of position and another one for the force feedback. The external loop of position is provided by the measurement of the output angle, obtained of the absolute sensor of rotary angle placed on each output axis of the SMART mechanism. The external loop of force is composed of the deformation measurement in each SMART rod, carried out by the strain gages placed on that bar.

4 Impedance controller

In order to implement an impedance control it is necessary to specify the desired dynamic behaviour of the system with the environment, by means of the mechanical impedance of the robot. **Fig. 3** shows a block diagram that represents the impedance control implemented in the non-linear actuator of SILO2.

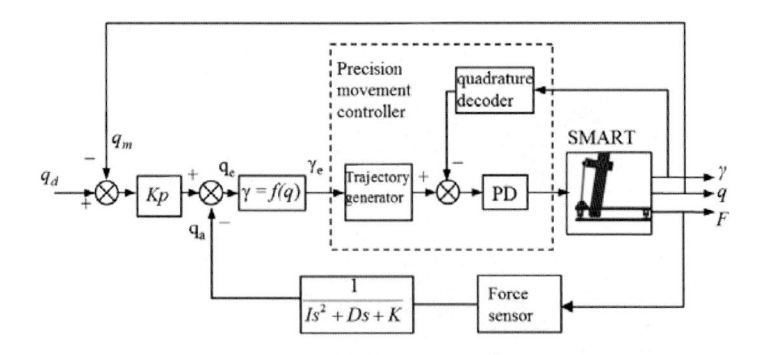

Fig. 3. Impedance control scheme

The impedance control is a more general approach than the stiffness control, since it introduces a dynamic reference to the control objective.

In this control system all their parameters can be experimentally tuned. The output position (q_e) is calculated by means of the following law,

$$q_e = K_p(q_d - q_m) - F\left(\frac{1}{Is^2 + Ds + K}\right) \quad (2)$$

Afterward, q_e is converted to values of input angles through of the inverse kinematics function, given by the following equation,

$$\gamma = 2\arctan\left(\frac{a_1 + \sqrt{a_1^2 - b_1^2 + c_1^2}}{b_1 + c_1}\right) - \beta \quad (3)$$

where,

$$\begin{aligned}a_1 &= k_3 \cos\beta + \text{sen}(q + \delta)\\ b_1 &= k_1 - k_2(\cos\beta \,\text{sen}(q + \delta) - \text{sen}\,\beta \cos(q + \delta))\\ c_1 &= -k_3 \,\text{sen}\,\beta + \cos(q + \delta)\\ \delta &= \pi - \alpha\end{aligned} \quad (4)$$

being, k_1, k_2, and k_3 constants related with the SMART linkage dimension, and α and β constants angles related with this mechanism.

It is necessary to make this transformation, because the angle γ is the input command of the precision movement controller (see **Fig. 3**).

5 Experimental evaluation

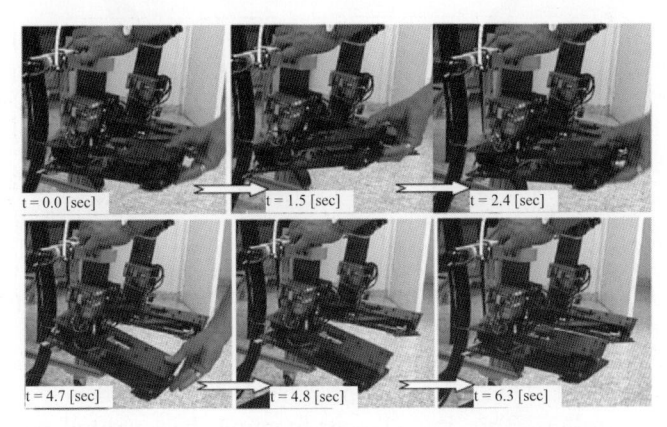

Fig. 4. Graphic succession of the experiment

A photographic sequence of the accomplishment of the experiment is shown in **Fig. 4**.

In this experiment, the foot was moved by hand up to the ankle joint ends positions [15], [16]. The force applied by the hand causes tension and compression forces on the SMART rod. Tension force on the SMART rod the foot means it was moved up, and with the compression force, it was moved down (see **Fig. 5**).

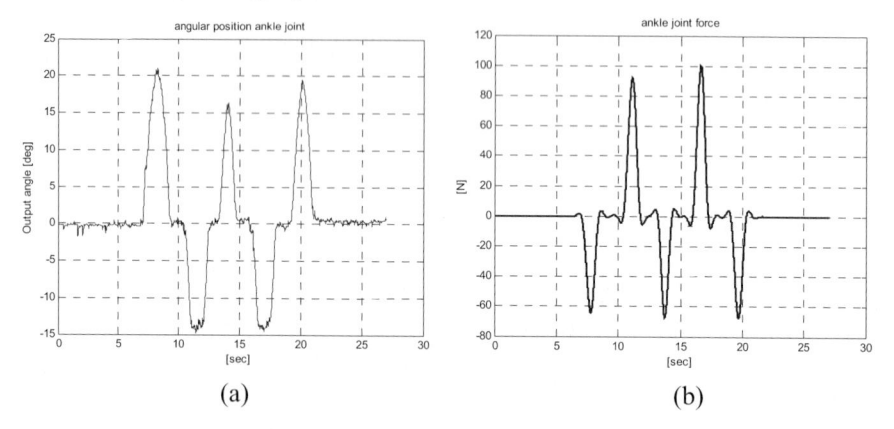

Fig. 5. (a) Output angle of the ankle joint; and (b) Force measurement on the SMART rod under impedance control

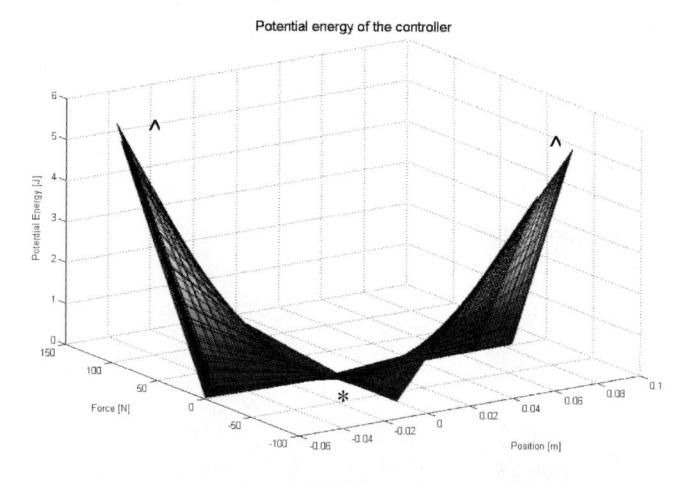

Fig. 6. Potential energy of the impedance controller during the experiment

The angular displacement of the ankle joint is related to the force exerted on the foot. This relationship is the desired impedance implemented in the controller. The impedance function has natural frequency of $\omega_n = 0.628\ rad/s$ and damping coefficient of $\varsigma = 0.7$; with these characteristics the system is under damped, and it is possible to carry out the foot movement by the hand (see **Fig. 4**). In practice, the effect of the stiffness of the impedance controller can be explained using the concept of a potential energy field with reference to the foot movement (see **Fig. 6**). In this figure (the potential energy is in absolutes values), the starting position is marked with (*) and the end positions are marked with (^). The desired impedance of the controller can be modelled accordingly to the required characteristics of interaction with the environment. This is possible by modifying the constants ω_n and ζ or varying the matrix values of *I*, *D*, and *K* of the impedance function (see Equation (2)).

6 Conclusions

In previous works [15], [16], [17], force sensor was implemented in the SMART nonlinear actuator to know the intrinsic characteristic of this mechanism. It has been possible to implement force control algorithm because the relationship between the force measurement on the SMART rod and the output angle of this mechanism is known. As this relationship is linear it has been possible to carry out impedance controller in the SMART actuator. In these experiments, the behaviour of the ankle joint was like a torsion spring. It has been demonstrated that the mechanical impedance (or stiffness in particular case) is depending of ω_n and ζ, or I, D, and K. The potential energy achieved of the impedance controller verifies the performance of the torsion spring of the ankle joint.

Acknowledgements

Authors would like to acknowledge the help provided by the mechanical and electronic shops of the IAI-CSIC.

References

1. Whitney, D. E. (1987) Historical Perspective and State of the Art in Robot force Control. *The International Journal of Robotics Research*, Vol. 6, No. 1, Massachusetts Institute of Technology, pp. 3-14.
2. Carelli, R. and Mut, V. (1993) Adaptive motion-force control of robots with uncertain constrains. *Robotics & Computer Integrated Manufacturing*, Vol. 10, No. 6, pp. 393-399.

3. Grieco, J. C., Armada, M., Fernandez, G., and González de Santos, P. (1994) A Review on Force Control of Robot Manipulators. *Journal of Informatics and Control*, vol. 3, no 2-3, pp. 241-252.
4. Montano, L. y Sagüés, C. (1997) Control de esfuerzos en un robot industrial. *Informática y Automática*, vol. 30, no. 2, pp. 35-50.
5. De Schutter, J., Bruyninckx, H., Zhu, W.H., and Spong, M.W. (1998) Force control: A bird's eye view. In *Control Problems in Robotics and Automation*, Springer-Verlag, London, pp. 1-17.
6. Gorinevsky, D.M., Formalsky, A.M., and Schneider, A.Y. (1997) *Force Control of Robotics Systems*. CRC Press LLC, Boca Raton, FL.
7. Siciliano, B. and Villani, L. (1999) *Robot Force Control*. Kluwer Academy Publishers, Norwell, Massachusetts.
8. Salisbury, J.K. (1980) Active stiffness control of a manipulator in Cartesian coordinates. In: *Proc. of 19th IEEE Conference on Decision and Control*, Albuquerque, NM, pp. 95-100.
9. Klein, C.A. and Briggs, R.L. (1980) Use of Active Compliance in the Control of Legged Vehicles. *Journal of IEEE Transactions on Systems, Man, and Cybernetics*, vol. SMC-10, no. 7, pp. 393-400.
10. Hogan, N. (1985) Impedance Control: An Approach to Manipulation. *Journal of Dynamic Systems, Measurement, and Control*, vol. 107, pp. 1-24.
11. Surdilovic, D. (1998) Synthesis of Impedance Control Laws at Higher Control Levels: Algorithms and Experiments. In *Proc. of the IEEE International Conference on Robotics & Automation*, Leuven, Belgium, pp. 213-218
12. Ferretti, G., Magnani, J.A., Rocco, P., Cecconello, F., and Rosetti, G. (2000) Impedance Control for Industrial Robots. In *Proc. of the IEEE International Conference on Robotics & Automation*, San Francisco, CA, pp. 4028-4033.
13. Akinfiev, T., Armada, M., y Caballero, R. (2000) Actuador para las piernas de un robot caminante. Patente de Invención ES2166735A1.
14. Caballero R., Akinfiev T., Montes H., and Armada M. (2001) On the modelling of SMART nonlinear actuator for walking robots. In *Proc. of Int. Conf. on Climbing and Walking Robots*, Sept. 24-26, pp. 159-166.
15. Montes, H., Pedraza, L., Armada. M., Akinfiev, T., and Caballero, R. (2004a) Adding extra sensitivity to the SMART nonlinear actuator using sensor fusion. *Industrial Robot: An Intl. Journal*, vol. 31, no. 2, pp. 179-188.
16. Montes, H., Pedraza, L., Armada, M., and Akinfiev, T. (2004b) Force Feedback Control Implementation for SMART nonlinear Actuator. *In Proc. of Int. Conf. on Climbing and Walking Robots*, Sept. 22-24, Madrid, Spain.
17. Montes, H., Pedraza, L., Armada, M., Caballero, R., and Akinfiev, T. (2003) Force sensor implementation for enhanced responsiveness of SMART nonlinear actuators. In *Proc. of Int. Conference on Climbing and Walking Robots*, Sept. 17-19, Catania, Italy, pp. 887-894.

MACCEPA: the Actuator with Adaptable Compliance for Dynamic Walking Bipeds

Ronald Van Ham, Bram Vanderborght, Michaël Van Damme, Björn Verrelst & Dirk Lefeber

Vrije Universiteit Brussel, Department of Mechanical Engineering, Pleinlaan 2, 1050 Brussel, Belgium

Abstract

Walking robots can be divided into two categories: on one hand the fully actuated robots that don't use passive dynamics, and on the other hand the energy efficient passive walkers. For autonomous robots the energy storage is a problem, forecasting a bright future for passive walkers. At this moment the passive walkers are restricted to one walking speed due to the eigenfrequency, which is fixed by the mechanical constructions. Several actuators with adaptable compliant have been designed, but due to size, complexity or controllability these are difficult to implement in bipeds.

Another application of the use of adaptable compliance is safe robot-human interaction. Sometimes a robot has to be stiff, e.g. for pick and place operations, but when moving between humans, a robot is preferably compliant. Also for exoskeletons or rehabilitation devices this compliance can improve ergonomics and speed up the rehabilitation process.

The MACCEPA is a straightforward and easy to construct rotational actuator, of which the compliance can be controlled separately from the equilibrium position. The generated torque is a linear function of the compliance and of the angle between equilibrium position and actual position. This makes this actuator perfectly suitable for dynamic walking, human-robotic interfaces and robotic rehabilitation devices.

Keywords: Adjustable Compliance, Equilibrium Position, Actuators, Compliance control, Spring

1 Introduction

Humans, like most walking animals, are walking efficiently by using the kinetical energy and the potential energy of the lower limbs [1,2]. Human joints are actuated by at least 2 muscle groups, giving the possibility to change the stiffness of a joint and to control the equilibrium position. By controlling both the compliance and equilibrium position a variety of natural motions is possible, requiring a minimal energy input to the system.

One of the first realizations of a compliant actuator was the MIT Series Elastic Actuator [3], which has an inherent, but fixed compliance. For shock absorbance this is useful, but in order to use natural dynamics this approach is limited to one eigenfrequency since the spring constant is fixed. This is comparable to passive walkers [4], which are able to walk energy efficiently, although restricted to a single walking speed.

Different designs with adaptable compliance have been made: at Carnegie Mellon University the AMASC (Actuator with Mechanically Adjustable Series Compliance) [5], at the Vrije Universiteit Brussel the Robotics and Multibody Mechanics research group has developed the PPAM (Pleated Pneumatic Artificial Muscle) [6] used in the biped Lucy [7], at the University of Pisa, Italy, [8] the Variable Stiffness Actuator (VIA) is developed, at Georgia Institute of Technology, USA, a Biologically Inspired Joint Stiffness Control [9] is made. All these design work on the same principle: two antagonistic coupled non linear springs.

Adaptable compliance can be used in walking bipeds, but also in a number of other disciplines. The ultimate (lower)leg prostheses is an actuated system, which moves naturally as a human body would do. The knowledge acquired by developing walking bipeds is applicable in the field of leg prostheses, and so is the use of an actuator with adaptable compliance. Other possible applications are rehabilitation robots. Such devices are imposing gait-like motion patterns to, for instance, the legs of a patient. In the beginning of the rehabilitation process it is preferred to have a relatively high stiffness, which could be gradually lowered when a patient has regained a certain level of control over his/her legs. While classical industrial robots are built to be as stiff as possible, resulting in unsafe devices for humans, the new trend is to incorporate compliance to make it possible to have a safe human-robot interaction. For example [10], describing a soft robot arm which will assist the user to carry the load, while the operator only has to push gently on the load.

As is shown in the above examples a straightforward, easy to control actuator with adaptable compliance has a bright future.

2 Working principle

In **Fig. 1** the essential parts of a Maccepa (Mechanically Adjustable Compliance and Controllable Equilibrium Position Actuator) are drawn. As can be seen there are 3 bodies pivoting around a common rotation axis. To visualize the concept, the left body in **Fig. 1** can be seen as an upper leg, the right body as the lower leg and the rotation axis, which goes through the knee joint. Around this rotation axis, a lever arm is pivoting, depicted as a smaller body in **Fig. 1**. A spring is attached between a fixed point on the lever arm and a cable guided by a fixed point on the right body to a pretension mechanism.

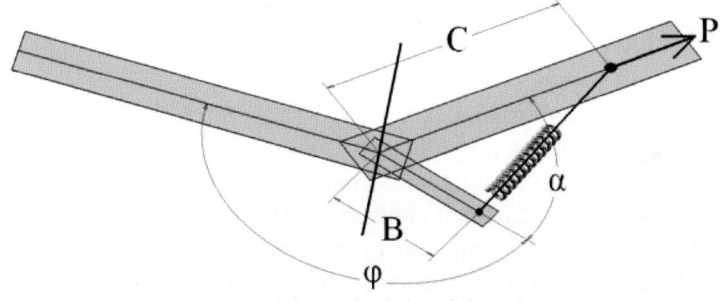

Fig. 1. Working principle of the Maccepa

The angle φ between the lever arm and the left body, is set by a classical actuator. When α, the angle between the lever arm and the right body, differs from zero, the force due to the elongation of the spring will generate a torque, which will try to line up the right body with the lever arm. When the angle α is zero—this is the equilibrium position—the spring will not generate any torque. The actuator, determining the angle φ actually sets the equilibrium position. A second actuator, which pulls on the cable connected to the spring, will set the pretension of the spring. This pretension will vary the torque for a certain angle α, thus controlling the spring constant of an equivalent torsion spring.

3 Calculation of the torque

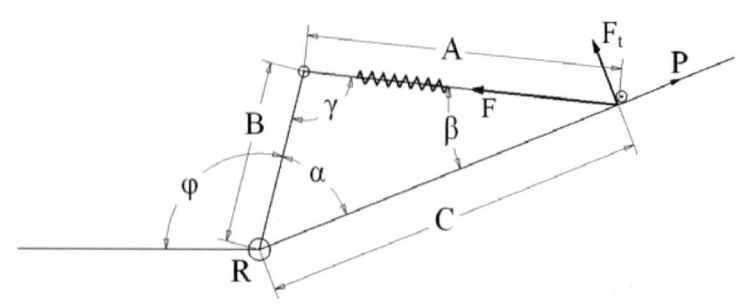

Fig. 2. Scheme of the Maccepa

R = Rotation point
T = Torque applied by Maccepa
F = Force due to extension of the spring
k = Spring constant, assume a linear spring
B = Lever arm motor, which controls equilibrium position
C = Distance between joint and spring tension mechanism
L = Length of the cable + restlenght of the spring – position of the pretensioner
P = Pretension of the spring, function of the position of the second actuator
α = Angle between lever arm and right body
φ = Angle between left body and lever arm, equilibrium position

$$T = k.B.C \sin \alpha \left(1 + \frac{P - L}{\sqrt{B^2 + C^2 - 2BC \cos \alpha}} \right) \tag{1}$$

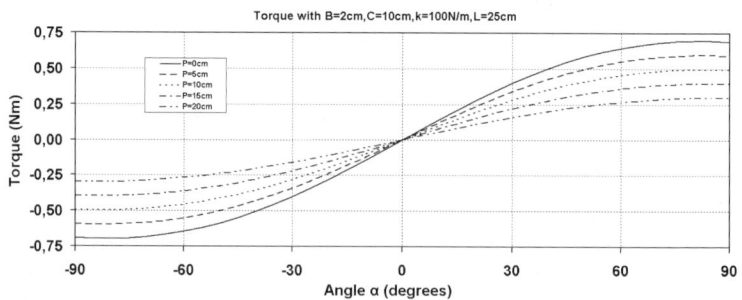

Fig. 3. Torque as a function on angle α when pretension (P) is altered

MACCEPA: the Actuator with adaptable compliance for walking bipeds. 763

Fig. 3 shows the torque generated by the Maccepa is symmetrical around the equilibrium position. The torque (see formula 1) is also independent from the angle φ, which means the compliance and equilibrium position can be controlled independently. For the linearity one can see that around the equilibrium position the plot is rather linear, but for larger angles the plots are not linear anymore.

Before talking about linearity, one should define the working range. As useful working range we assume -45° to 45°. It is worth mentioning, the range of the joint is not limited to 90° with this choice, since this range of -45° to 45° means the angle between actual position and the equilibrium position. The equilibrium position can vary over a range of 360° and even more. When using the joints for bipedal walking—either robots or prostheses—the choice of -45° to 45° is perfectly justifiable. So when looking at the linearity only the range between 0° and 45° will be studied because of the symmetrical torque characteristics.

4 Influence of design variables

The variables k, B and C are chosen during design and are fixed during normal operation. In this chapter the influence of these four variables will be shown in detail.

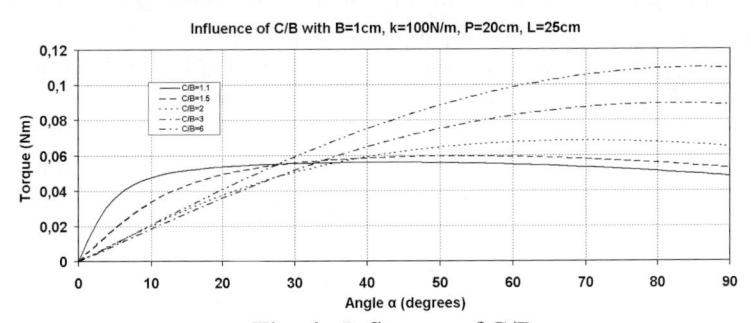

Fig. 4. Influence of C/B

In **Fig. 4** one can see that the C/B ratio determines the non-linearity of the curves. One can see the bigger the ratio C/B the more linear the curves. Note that looking at the formula of the torque one can see that B and C can be exchanged without changing the result, so C/B should be big enough or B/C should be big enough. Simulations showed that from a B/C or C/B ratio of a little above 5 the correlation coefficient is 0.99. This can be used as a guideline during design, when working with a range of -45° to 45°.

In **Fig. 5** the influence of the length of the lever arm, B is depicted. If the length of the lever arm is doubled, the torque is also approximately doubled. Since B and C can be exchanged, the influence of C is analogous.

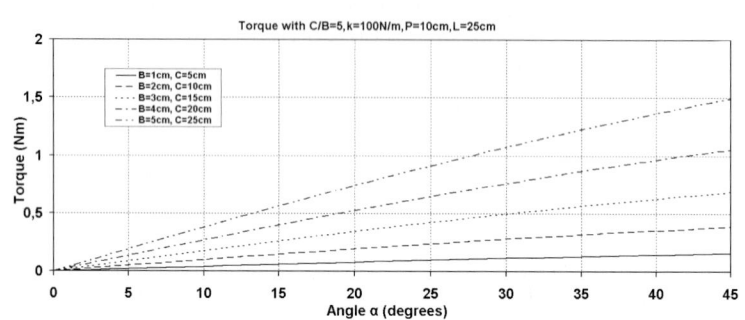

Fig. 5. Influence of the lever arm B (or C) on the torque with C/B = 5

The Influence of k is linear, looking at formula 1. As shown in **Fig. 3** the torque can be adjusted by controlling the pretension.

5 Experimental Setup

A first prototype was built. In **Fig. 6** a CAD drawing of the setup is depicted. On the left side the body with the pretension mechanism is shown, on the right side the body with the actuator, which controls the equilibrium position. The rotation axis between these bodies is equipped with 2 roller bearings. A potentiometer is placed on this axis to measure the angle between the two bodies, the actual angle of the joint. Actuation is done by servomotors, which are position controlled actuators with integrated position measurement and position controller.

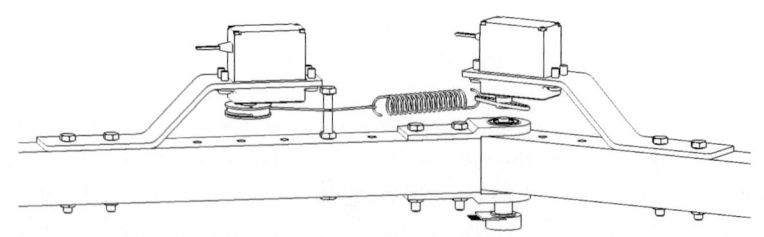

Fig. 6. CAD drawing of the first Maccepa prototype

In a first experiment the difference in natural movements with altered compliance was witnessed. The setup is placed so the rotation axis is vertical, as such gravity does not influence our experiment. During the experiment the equilibrium position is set to 0 degrees, defined as the position where both arms are aligned. In **Fig. 7** the joint, which is made stiff, is pulled manually out of the equilibrium position and released. As a result the joint starts oscillating around the equilibrium position with a certain frequency. After the joint stops oscillating (around 2 sec), the joint is made more compliant, and pulled again out of the equilibrium position (2.5 sec). Releasing the joint will result in a lower eigenfrequency.

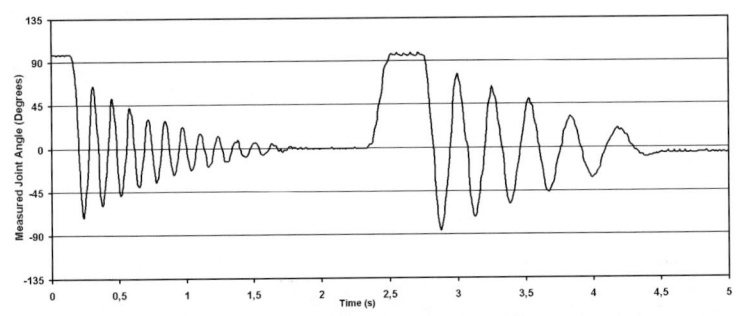

Fig. 7. Variation of the natural frequency for different settings of the compliance.

As can be seen from the experiment, the compliance can be controlled, by only changing the position of one of the actuators, which is not the case in other designs with adaptable compliance.

6 Conclusions

The Maccepa (Mechanically Adjustable Compliance and Controllable Equilibrium Position Actuator) is presented in detail. The design variables, limited in number, are explained and their influence on the torque characteristics is shown. Compared to other compliant mechanisms the Maccepa is straightforward and relatively inexpensive. The control of the equilibrium position and compliance is completely independent. Thus to control the one parameter it requires only the action of one of the two actuators, e.g. one servo motor. These advantages make it the ideal actuator for use in applications where adaptable compliance is required or useful, e.g. dynamic walking or any robotic application interacting with humans.

Acknowledgements

Bram Vanderborght is supported by the Fund for Scientific Research Flanders (Belgium)

References

1. Thomas A. McMahon and George C. Cheng. The mechanics of running: How does stiffness couple with speed? Journal of Biomechanics, 23:65–78, 1990.
2. T. A. McMahon. The role of compliance in mammalian running gaits. Journal of Experimental Biology, 115:263–282, 1985.
3. Pratt, G. A. and Williamson, M. M. [1995]. Series elastic actuators, Proceedings IEEE-IROS Conference, Pittsburg, USA, pp. 399-406.
4. Wisse, M. [2004]. Essentials of Dynamic Walking : Analysis and Design of Two-Legged Robots, PhD thesis, Technische Universiteit Delft.
5. Hurst, W., Chestnutt, J., E. and Rizzi, A., A. [2004]. An actuator with physically variable stiffness for highly dynamic legged locomotion, In Proceedings of the IEEE International Conference on Robotics and Automation, New Orleans, USA, pp. 4662-4667.
6. Daerden F. & Lefeber D., The concept and design of pleated pneumatic artificial muscles, International Journal of Fluid Power, 2(3):41–50, 2001.
7. http://lucy.vub.ac.be/
8. Giovanni Tonietti, Riccardo Schiavi and Antonio Bicchi [2005]. Design and Control of a Variable Stiffness Actuator, Proceedings of the 2005 IEEE ICRA International Conference on Robotics and Automation, Barcelona, Spain, April 2005
9. Shane A. Migliore, Edgar A. Brown, and Stephen P. DeWeerth [2005]. Biologically Inspired Joint Stiffness Control, Proceedings of the 2005 IEEE ICRA International Conference on Robotics and Automation, Barcelona, Spain, April 2005
10. http://mech.vub.ac.be/softarm

A Design of a Walking Robot with Hybrid Actuation System

Katsuhiko Inagaki[1] and Hideyuki Mitsuhashi[2]

[1]Department of Applied Computer Engineering, TOKAI University, JAPAN inagaki@tokai.ac.jp; [2]Graduated School of Engineering, TOKAI University, 4aeem037@keyaki.cc.u-tokai.ac.jp

Abstract

This paper describes a mechanical design of a walking robot that has hybrid actuation system. In our design, engine actuator is applied as power source of the walking robot to realize practical cruising range. To perform that, two actuation mechanisms are applied: *(i)* electrical motor based mechanism and *(ii)* is powder clutch joint driving mechanism. This actuation system offers an opportunity to distribute engine power to many joints of walking robot, and to realize separate motion of every joint.

Keywords: Walking robot, Hybrid actuation, Leg mechanism

1 Introduction

The aim of our investigation is to improve the cruising range of walking robot by using a hybrid actuation system. Improvement of cruising range is one of the noteworthy problems to build up a practical walking robot. Typically, an engine actuator and conventional actuator are used to solve the problem. However, it is very difficult to apply engine to walking robot. In case of wheeled machine such as automobile, it is easy to apply engine, because required degree of freedom (DOF) is only two. On the contrary, in case of walking machine, it is very difficult to apply engine, because general walking robot consists of multiple DOF. One of the effective methods to use engine is to apply hydraulic actuation system [1], [2], but many equipments are required to use this actuation system.

On this point of view, we have already proposed a new mechanism that is named powder clutch joint driving mechanism [3]. By use of the mechanism, we can distribute power of an engine to multi-DOF and can control their motion separately. The merit of the proposed mechanism is simplicity of the mechanism and circuit to drive it. However, as a demerit of the mechanism, we can point that the mechanism does not have any ability of energy storage such as battery, spring etc. To improve energy efficiency, it is very effective way to make energy consumption flat, during walking period. Energy storage equipment is very important factor to realize the smoothing.

In this investigation, we tried to avoid the demerit of the joint driving mechanism by use of hybrid actuation system. In our system, some part of DOF is driven by the joint driving mechanism, and the other DOF are driven by electric motor. Electrical energy to drive the motors is generated by generator, and that is driven by engine. Additionally, a storage battery is used to level energy consumption of the engine. This means that a part of engine power is used for the joint driving mechanism, and the other part is used for electric motors. **Fig.1** shows system diagram of our hybrid actuation system for walking robot.

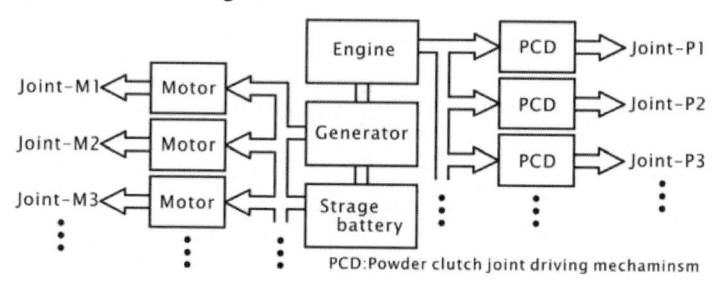

Fig.1 System diagram of hybrid actuation system

2 Powder clutch joint driving mechanism

Here, we describe the mechanism of the powder clutch joint driving mechanism. **Fig. 2** shows overview of the joint driving mechanism. A pair of electromagnetic powder clutch and an electromagnetic disc clutch is used to drive one DOF. The input shaft connected to the disc clutch is connected to engine. Usually, the disc clutch is engaged to transmit the power of engine to the powder clutches. Then, Input side of the powder clutches is connected by timing belt, and output side of the clutches is connected by spur gears. If clutch-A is engaged, the output shaft rotates in the opposite direction to the input shaft. On the contrary, if clutch-B is engaged, rotation of the output shaft may be in the same direction to the input shaft. Therefore, even if input shaft is rotated into the same direction, we can choose the rotational direction of the output shaft by selecting powder clutch to be engaged.

The proposed mechanism has several kinds of merits as follows.

1) The driving mechanism can distribute power of a single engine and make independent motion for all joints.

2) For each rotational direction, different reduction ratio can be applied.

3) Brake motion to lock the output can easily be realized.

4) Size and weight of driving circuit for the mechanism is very compact.

First of all, the reason of why we have employed powder clutches is as follows. Engaging torque of the powder clutch can be adjusted by changing the exciting current. Therefore, the mechanism has a merit that we can provide separate motions to many number of DOF, even if we use common power source for the joint driving mechanisms. This advantage is very suitable to use engine actuator.

Additionally, we can set different reduction ratio for each rotational direction. In **Fig. 2**, reduction ratio of the timing belt side is 1:1, but reduction ratio of the spur gear side is 2:1. Thus, reduction ratio during clutch-A is engaged is twice of the other situation (clutch-B is engaged). Specification of this merit is discussed in the next section.

It is worth noting that, if both the powder clutches are engaged, the mechanism can realize the brake motion. In this mode, we need to pay attention that the disc clutch must be released to avoid engine stall.

Finally, energy consumption to transmit power of 300[W] that is rating capacity of our joint driving mechanism is only 18[W]. Thus, the size and weight of driving circuit to drive the mechanism is quite small and light.

3 Joint configuration

In our hybrid actuation system, it is important factor to consider which actuation is suitable for each joint, electrical motor or powder clutch joint driving mechanism. On this point of view, design of joint configuration is important, because joint configuration affect to purpose and characteristic of joint. **Fig. 3** shows overview of our leg mechanism. This leg has three DOF; two rotational joints and one prismatic joint. The two rotational joints make horizontal 2D motion and the prismatic joint makes vertical linear motion. For convenience, we call the column part of the leg as *vertical leg*, and call the part of closed linkage as *horizontal leg*.

In this joint configuration, only the vertical joints are affected by gravitation. This design is based on GDA (Gravitationally Decoupled Actuation) [4]. Under the GDA, joints are used either to propel the body, or to support body weight. Thus, we can easily choice suitable actuation for each joint, electrical motor or powder clutch joint driving. Specifically, powder clutch joint driving mechanism is applied to the vertical joint, and electrical motors are applied to the horizontal joints. The reason is explained in following subsections.

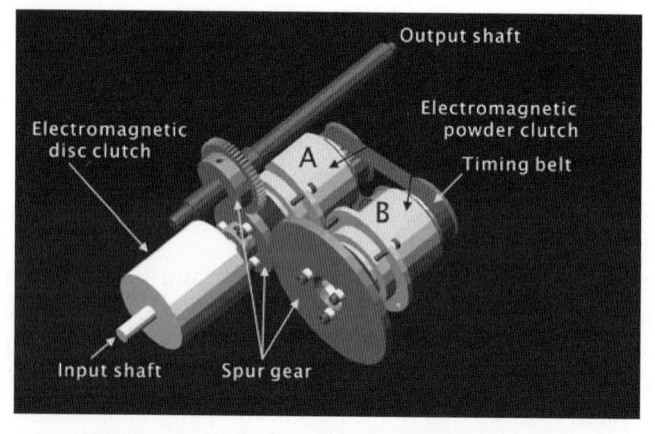

Fig. 2. Powder clutch joint driving mechanism

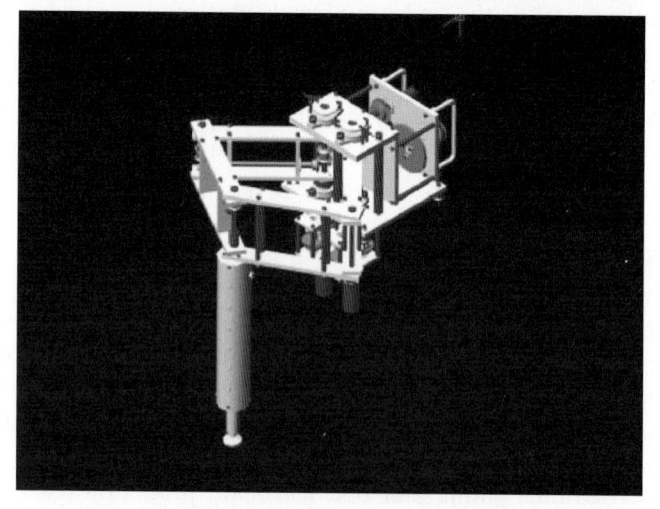

Fig. 3. Overview of leg mechanism

3.1 Vertical joint

One of the most important factors for vertical joint is to adopt gap of required characteristic for standing phase and swing phase. In the standing phase, vertical joint needs enough torque to support the body weight. On the contrary, enough speed is required to return to the standing phase so quickly. As a method to adopt these needs, applying of two states variable transmission is effective. As mentioned before, the powder clutch joint driving mechanism can easily realize the transmission. That is a main reason why we applied the driving mechanism.

Additionally, there is one more reason why we apply the joint driving mechanism to the vertical joint. That is the mechanism has bake motion mode. This mode is very suitable for standing phase, because vertical motion of leg should be locked during flat terrain waking.

As mentioned before, vertical joint makes vertical prismatic motion. To do that, a ball screw is placed in the column shown in **Fig. 3**. Then, the joint driving mechanism is placed on the base part of the leg. The output power of the driving mechanism is transmitted by several shafts that are in the horizontal leg.

To confirm the validity of our vertical joint, we have developed a test bench shown in **Fig. 4**. This equipment has a vertical leg in the bottom part. Then, the joint driving mechanism is in the middle of the equipment, and an engine is placed on the top. These main parts are supported by for shafts that are connected with liner bearing. Thus, main parts can make vertical motion along with the shafts. However, since each shaft has stopper in the middle position, vertical motion is limited at this point. Therefore, we can realize two phases that correspond to standing phase and swing phase.

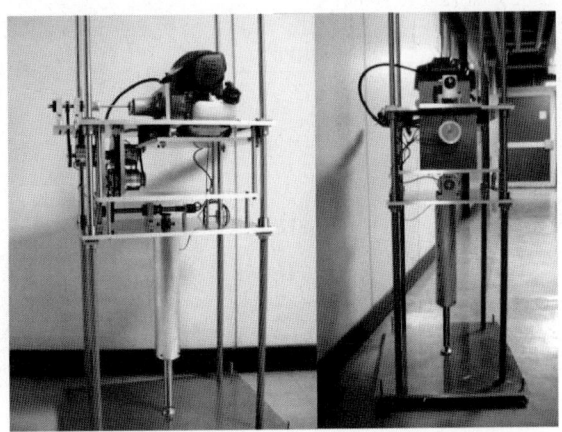

Fig. 4. Overview of test bench for vertical joint

We have done a simple experiment for the test bench. In the experiment, vertical leg makes stroke motion with computer control. **Fig. 5** shows the result of experiment. By this experiment, we could confirm that the joint has enough power to drive the body part.

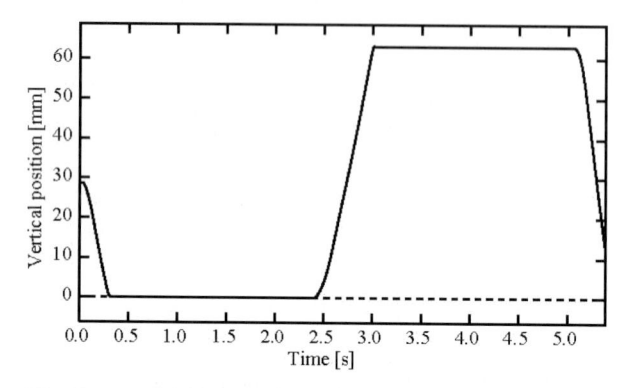

Fig. 5. Result of experiment with simple stroke motion

3.2 Horizontal joints

As mentioned before, vertical leg has two rotational joints to make horizontal 2D motion. The 2D motion is realized by use of closed link mechanism show in **Fig. 6**. This mechanism consists of 5 rotational joints (2 active joints placed in bottom side and 3 passive joints). Inverse kinematics of the horizontal joints is as follows.

$$\theta_{L2} = \cos^{-1} \frac{P_x^2 + P_y^2 - 2l^2}{2l^2} \quad (1)$$

$$\theta_{L1} = \tan^{-1} \frac{-\sin\theta_{L2}}{\cos\theta_{L2} + 1} - \tan^{-1} \frac{P_x}{P_y} \quad (2)$$

$$\theta_{R2} = \cos^{-1} \frac{(P_x^2 - d) + P_y^2 - 2l^2}{2l^2} \quad (3)$$

$$\theta_1 = \tan^{-1} \frac{-\sin\theta_{r2}}{\cos\theta_{r2} + 1} - \tan^{-1} \frac{P_x - d}{P_y} \quad (4)$$

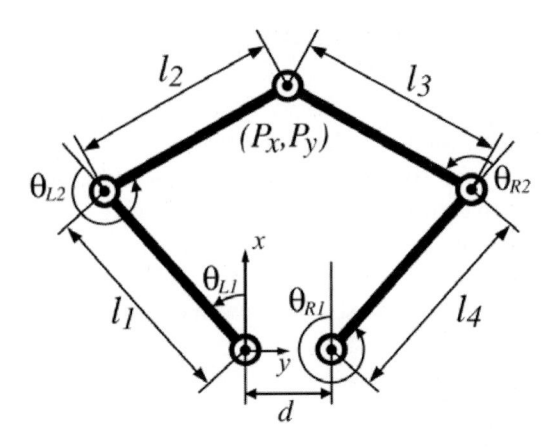

Fig. 6. Configuration of horizontal joints

For the two active joints, we have applied electrical motors due to the following reasons. First, it is impossible to make energy consumption flat by use of powder clutch driving mechanism. To overcome this problem, combination of generator and storage battery is effective. Moreover, high accuracy is required for these joints, because the positioning accuracy affect to the walking trajectory. In case of the multi-legged waking robot, since standing legs make a closed linkage through the ground, every standing leg must keep synchronized motion for each other. Other wise, serious problem may be occurred such as slipping of legs or vibration of the body.

In our study, to improve the accuracy of these joints, newly developed gear box is applied. Backlash of gear box mainly have bad influence on accuracy of the joint positioning. To solve this problem, our gear box uses a pair of gear sets as shown in **Fig.7**. By use of this method, backlash is absolutely reduced, because gear in output shaft is pressurized.

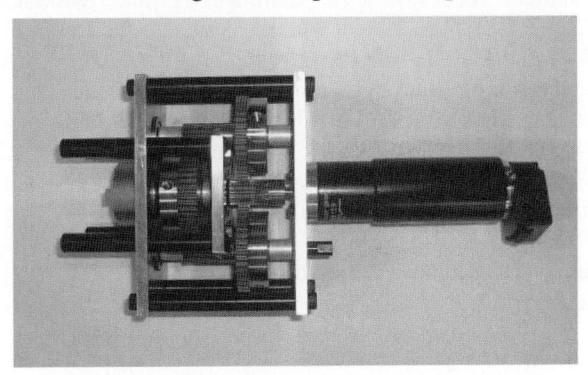

Fig. 7. Overview of backlash-less gear box

4 Conclusions and future work

Based on the proposed methods, we have developed a leg module as shown in **Fig. 8**. The total weight of the leg module is about 30[Kg]. Two 90[W] DC motors, two 10[Nm] powder clutches and a 10[Nm] disc clutch are mounted. An investigation is currently underway to develop a quadruped walking machine, by use of these modules. An engine whose displacement is 25[cc] is carried on the back side of the module in Fig.8. However, the engine may be used as a common power source for all modules in the quadruped robot.

Fig. 8. Overview of the prototype of leg module

References

1. Hartikainen, K., Halme, A., Lehtinen, H. and Koskinen, K. (1992) MECANT I: A Six Legged Walking Machine for Research Purposes in Outdoor Environment. *IEEE International Conference on Robotics and Automation.* Los Alamitos, pp.157-163.
2. Hodoshima, R., Doi, T., Fukuda, Y., Hirose, S., Okamoto, T., Mori, J. (2004) Development of TITAN XI:a Quadruped Walking Robot to Work on Slopes Design of system and mechanism, *Proceedings of International Conference on Robots and Sysrems(IROS2004)* Sendai, JAPAN, pp.792-797
3. Inagaki, K., Nishizawa, H., and Kawasumi, S (1999) A leg mechanism for self contained multi-legged walking robot, 2nd *International Conference on Climbing and Walking Robots (CLAWAR99)* Portsmouth, UK, pp.641-648
4. Hirose, S., (1984) A study of design and control of a quadruped walking vehicle. *Int. J. of Robotics Research*, 3(2), pp.113–133

Manipulators Driven by Pneumatic Muscles

K. Feja, M. Kaczmarski[1] and P. Riabcew

[1]Institute of Automatic Control, Technical University of Lodz, POLAND
mkaczmar@wpk.p.lodz.pl

Abstract

In recent years more and more publications reporting alternative actuators, including artificial muscles, for climbing and walking robots have appeared in the literature. This paper considers pneumatic artificial muscles, better known as McKibben muscles and fluidic muscles, the so called FESTO actuators. The dynamic and static characteristics of these muscles, including a comparison between theoretical and experimental characteristics, are first presented and then the use of such actuators in walking machines demonstrated.

Keywords: Artificial muscles, FESTO actuators, McKibben muscles, manipulators, pneumatic muscles.

1 Introduction

Living beings usually have two or three degrees-of-freedom (DOF) joints in their skeleton. Developers of robots are attempting to imitate the nature, although robots very rarely have 2 DOF or 3 DOF joints. It is very difficult to construct such joints, specially using traditional drives such as electric motors. Moreover, these drives belong to the category of "rigid drives" and walking robots have some problems in walking on uneven terrain. On the other hand, adjustable stiffness drives and 3 DOF joints are commonly encountered in the natural world. As an example, natural muscles have small stiffness during marching and allow balance the body, in contrary to a manipulation task, where the same muscles are very tight. Therefore, current designs of robots include 3-DOF joints driven by "compliant drives", called artificial muscles, which easily allow constructing a manipulator with 2- or 3-DOF. However, there are difficulties in control and position measurement of links [1], [6] with such muscles, that are yet to be resolved.

2 Biological inspiration

The literature in recent years has seen an increasing number of publications dealing with analysis of movement of humans and animals. Many of these take an energy aspect into consideration. It was shown, that oscillatory movement of legged living creatures is very efficient from the energy point of view [7]. Proper endings of bones at every joint can cooperate effectively with muscles, which put tension on bones preventing displacements. This is one of the leading properties of mammal joints. Moreover, there are no position sensors in animal joints.

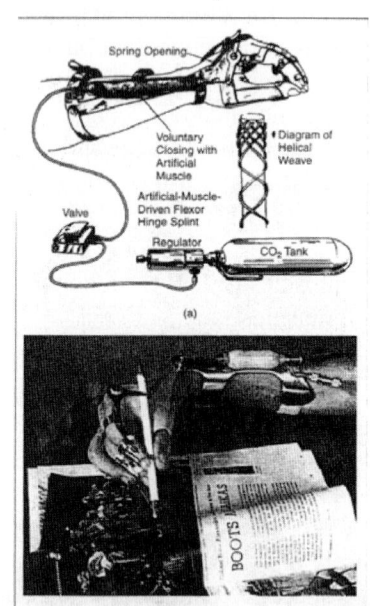

Fig. 1. Orthopedics arm to assist opening and closing fingers of handicapped hand driven by McKibben artificial muscles.

Artificial pneumatic muscles, which are similar to natural muscles have been known for years. S.Garasiev designed the first actuator of this kind in 1930 and was called "artificial muscle". Morin in 1947 took up another attempt in this way, using a kind of bladder filled with compressed air. In the 1950s physician J. L. McKibben invented an artificial muscle to motorize pneumatic arm orthotics to help control handicapped hands, see **Fig. 1**. The artificial muscle was composed of a rubber inner tube covered with a shell braided according to helical weaving [2]. Nowadays this kind of artificial muscle is known as "McKibben muscles" or "McKibben actuators". One of the leaders in producing very powerful pneumatic actuators used in construction of walking machines and in another industrial applications is FESTO [5].

A further type of actuator, very similar to natural muscles, is known as "electro-active polymers" or "EAP". Such an actuator is activated by electric pulses.

3 Properties of pneumatic muscles – McKibben muscles

A classical McKibben artificial muscle cosists of an inner rubber tube covered by a braided mesh and closed with two fittings, one being the air input and the other one the force attachment point [4], see **Fig. 2**.

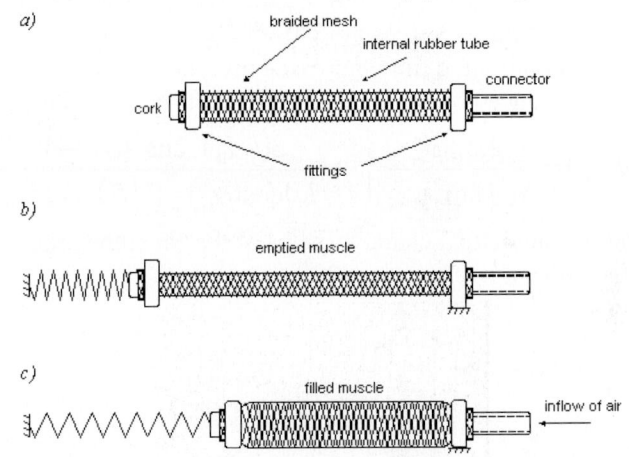

Fig. 2. Structure and principle of working of McKibben muscle.

An initial tension of muscle by the spring increases the length of the muscle, because of the elasticity of inner rubber tube. This increase in length is proportional to the external force, for small forces. For large external forces, there is a restriction on length of the muscle due to limitations in length of threads in the braid. When the muscle fills up with pressurized air, this increases its volume and radial dimension, and at the same time decreases its axial length and produces force. Contraction in McKibben muscles equals 30% of initial length of the muscle. The authors have identified static and dynamic characteristics of three kind of pneumatic muscles, namely fluidic actuator FESTO, pneumatic muscle of Shadow Robot Company Ltd. and pneumatic muscles constructed in their laboratory on the basis of McKibben muscle design.

Theoretically the relation between force *f* produced by McKibben muscle, pressure in the muscle *p* and relative length of muscle l/l_0 is given as:

$$f(p, l/l_0) = \frac{\pi r_0^2 p}{\sin^2(\alpha_0)} \left(3(l/l_0)^2 \cos^2(\alpha_0) - 1\right) \qquad (1)$$

where l_0 – initial length of McKibben muscle, r_0 – initial radius of inner rubber tube of muscle, α_0 – initial angle of braid covering muscle. These parameters describe a pneumatic muscle in static state, shown on Fig. 3 (graph a). McKibben muscle produces a force depending on:
 1) pressure of air in the muscle *p*,
 2) nonlinear, relative length of muscle l/l_0,
 3) area cross section πr_0^2

When external force equals zero, the muscle is pressurized and has maximum volume. In such conditions, internal pressure in the muscle is p_m. If the inlet and outlet valves are closed, external force causes decrease in volume, and at the same time increased pressure in the muscle. In such a case the force produced by a pneumatic muscle is described as:

$$f(p, l/l_0) = \frac{2\pi r_0^2 p_m}{3\sqrt{3} \sin^2(\alpha_0)} \left(\frac{3(l/l_0)^2 \cos^2(\alpha_0) - 1}{(l/l_0)\cos(\alpha_0) - (l/l_0)^3 \cos^3(\alpha_0)} \right) \qquad (2)$$

Fig. 3 shows theoretical characteristics of the Mckibben muscle with parameters: *l_0=300mm, r_0=7mm, α_0=20⁰*.

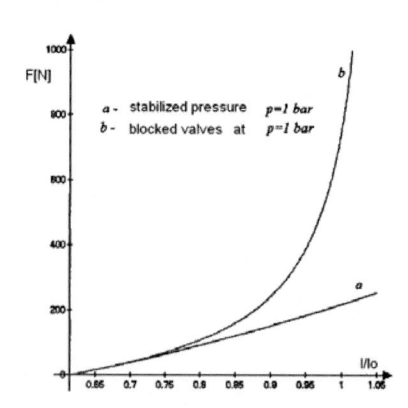

Fig. 3. Theoretical characteristic of McKibben muscle.

At low pressure the rubber tube doesn't stick to braided shell, because of elasticity of rubber. Therefore, the force produced by pneumatic muscle is expressed as:
$$f(p,\varepsilon) = p\frac{\pi r_0^2}{\sin^2(\alpha_0)}\left(3(1-\gamma\varepsilon)^2 \cos^2(\alpha_0) - 1\right) \quad (3)$$

where γ – corrective coefficient equals 1.25-1.35 and depends on pressure according to $\gamma = a_\gamma e^{-p} + b_\gamma$, $\varepsilon = (l/l_0)/l_0$, a_γ and b_γ – coefficients obtained empirically. A plot of equation (3) is shown in **Fig. 4**.

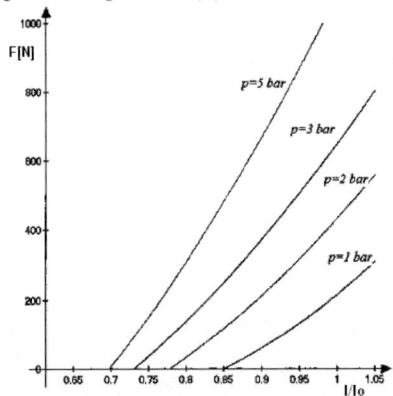

Fig. 4. Theoretical characteristic of pneumatic muscles according to equation (3), for $a_\gamma = 3.55$ and $b_\gamma = 1.25$ [1].

It was noticed that there were differences between theoretical and experimental results. The origins of the discrepancies were identified as:
1. deformed shape of the muscle at both ends, non-cylindrical shape
2. elasticity of internal rubber tube,
3. static friction in the muscle

For these reasons there is a histeresis effect in the pneumatic muscle, as noted in **Figs. 5 – 7**.

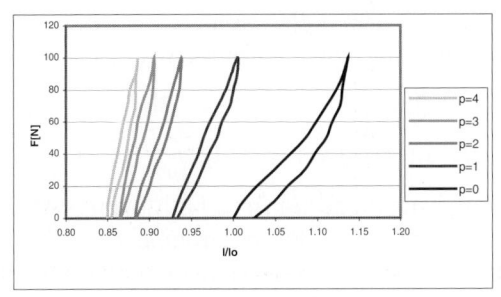

Fig. 5. Real characteristic of pneumatic muscle Shadow Robot Company Ltd. (Diameter=20 mm, Length=210mm)

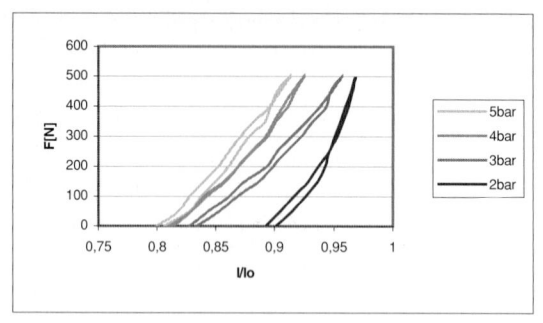

Fig. 6. Real characteristic of fluidic actuator FESTO.
(Diameter=20 mm, Length=200mm)

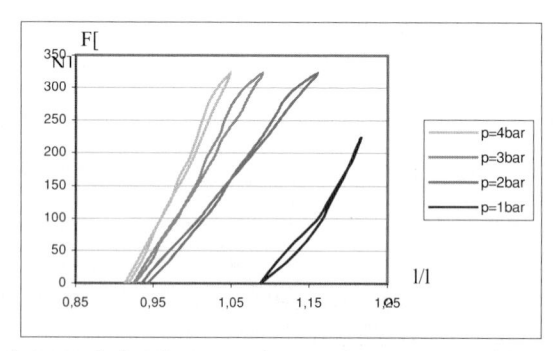

Fig 7. Real characteristic of pneumatic muscles constructed in our laboratory.
(Diameter=30 mm, Length=500mm)

The main features of pneumatic muscles are:
1. high maximum force to own weight ratio (1kW/kg),
2. high force generated from area cross section: 300N/cm^2,
3. energy efficiency: 30% - 40%,
4. maximum pressure: 6 bar,
5. range of operating temperatures: 0-50^{0}C,
6. compliance in varying directions.

Pneumatic muscles have some stiffness, which can be described as:

$$k = \frac{df}{dl} = \frac{2\pi v_0^2 p_m}{3\sqrt{3} l_0 \sin^2(\alpha_0)\cos(\alpha_0)} \left(\frac{3(l/l_0)^4 \cos^4(\alpha_0)+1}{(l/l_0)^2 \left(1-(l/l_0)^2 \cos^2(\alpha_0)\right)^2} \right) \quad (4)$$

Experimental investigations confirm, both manipulators driven by pneumatic muscles in contact with the environment and some components of environment may be protect from damages by adjustable stiffness of pneumatic muscles [3]. They also have natural damping effect (see **Fig. 8**)

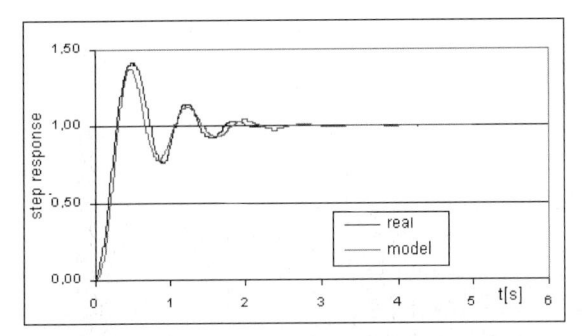

Fig 8. Step response of pneumatic muscle.

Such muscles are described by:

$$G(s) = \frac{l(s)}{p(s)} = \frac{k}{(T_1 s + 1)(s^2 + 2\zeta\omega s + \omega^2)} \qquad (5)$$

where T_1 – time of inertia, ζ – coefficient of damping, ω – natural frequency of vibration and l – length of muscle, p – step input of pressure.

4 Applications

Laboratory tests for identification of air muscles were an introduction to the realization of the structure of the manipulator, based on the human hand. This project consist of two main parts. The first is construction of an artificial palm (see Fig 9). This has 9 DOF, which is enough to provide most of the typical human grasps. Each rotary joint is actuate by an antagonist pair of the McKibben muscles.

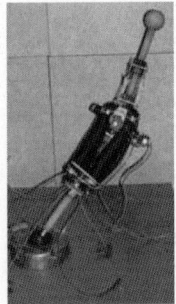

Fig. 9. The view of two arms manipulator with 4 DOF and an artificial hand

The second part of the project is the construction of the manipulator, which can be used to change position and orientation of the artificial hand (see **Fig 9**). It consists of two arms connected with a joint with more than 1 DOF. The first joint is a type of ball joint. The second joint is the Cardan type, and is driven by four pneumatic muscles.

5 Conclusions

Pneumatic muscles are lightweight and efficient actuators which can compete with pneumatic cylinders and electric motors. They are easy to assemble and repair. They are used in walking robots [5] and in rehabilitation robots, especially orthotics artificial limbs, because of their properties, which are similar to human muscles [6]. Pneumatic muscles have adjustable stiffness and therefore can protect manipulators from damage while in contact with the environment. Susceptibility of actuators such as McKibben muscles allows to use them to drive manipulator joints with more than 1 DOF.

References

1. E. Jezierski, *Dynamika robotów*, WNT, Warszawa, 2005
2. B. Tondu, P. Lopez, *Modeling and Control of McKibben Artificial Muscle Robot Actuators*, IEEE Control Systems Magazine
3. B. Tondu, P. Lopez, *The McKibben muscle and its use in actuating robot-arm showing similarities human arm behavior*, Industrial Robot, Volume 24, Number 6/1997.
4. B. Tondu, P. Lopez, *Theory of an artificial pneumatic muscle and application to the modeling of McKibben artificial muscle*, C.R. Acad. Sci. Paris, t. 320, Serie IIb, p.105-114, 1995.
5. T. Kerscher, J. Albiez, K. Berns, *Joint control of the Six-Legged Robot AirBug Driven by Fluidic Muscles*, CLAWAR, 24 – 26 September, 2001.
6. K. Feja, E. Jezierski, P. Riabcew, *Napęd mięśniowy przegubu o dwóch stopniach swobody*, VIII Krajowa Konferencja Robotyki, 23-25 czerwca, Polanica – Zdrój 2004.
7. H. Witte, R. Hackert, *Transfer of biological principles into the constructions of quadruped walking machines*, Proc. of the second Workshop on Robot Motion and Control, 254-249, Bukowy Dworek, 2001.

Sensing and Sensor Fusion

New Advances on Speckle-velocimeter for Robotized Vehicles

A. Aliverdiev[***], M. Caponero[**], C. Moriconi[*],
P.A. Fichera[*], and G. Sagratella[*]

[*]Robotics and Informatics division, ENEA Casaccia, V. Anguillarese 301, 00060 Roma (Italy), moriconi@casaccia.enea.it
[**]Applied Physics division, ENEA Frascati, V. E. Fermi 45, 00044 Frascati (Italy), caponero@frascati.enea.it
[***] Institute of Physics, Daghestan Scientific Center of Russian Academy of the Sciences, Yaragskogo, 94, 367003, Makhachkala (Russia), aliverdi@frascati.enea.it

Abstract

The paper points out the progress in the exploitation of high, fast and precise non-contact velocimetry for robot applications. A technique for the precise measurement of the speed between two sliding surfaces has been developed during the research project for the realisation of an autonomous robot. The robot is devoted to the scouting of dangerous sites and to the execution of measurements in these places for the exploration of an extreme Antarctica environment (RAS project). This technique is based on the precise calculation of the common movement of a laser speckle field. This approach allows the realisation of a velocimeter suitable for use in extreme conditions. A description of the adopted methodology and the obtained results are the main topic of our work.

Keywords: Speckle velocimetry, Robotics.

1. Introduction

It is well known that the problem of very accurate dynamic control of a dynamic system in low-medium friction conditions is not an easy task. Because of the very poor precision of the odometry signal (caused by sliding) and of the GPS data (that is at a low data rate) an additional, more reliable, data source is required. The exploitation of a laser velocimetry technique could allow us to solve this problem because of the potentially very high precision and quick response. To achieve this aim, a complex theoretical and experimental work has been undertaken, whose preliminary results already look rather satisfactory.

Our method consists of (i) to create a speckle picture using a high coherent light source, (ii) to record it with a high-speed camera, and (iii) to analyze it and to calculate the velocity. Using this method it is possible to characterize the relative motion on even very smooth surfaces to achieve a result, comparable with the ENEA current realisations in the robotics field.

The creation and recording of a speckle pattern field, points (i) and (ii), is a solvable technical problem, so we will not focus on these.

Concerning the velocity calculation, point (iii), several approaches that rely on both time and space statistics have been suggested. Differential [1] and integral [2] (with respect to time) intensity functions have been used. Our initial approach was aimed at modifying the analysis of an integrated statistics. Namely, instead of a set of averaging times, we take the average for one value of the averaging time T, but for a set of angles in the spatial-temporal plane [3,4]. The result is the transformation of the intensity function $I(x,t)$, just like a direct Radon-transformation, but with fixed integral limits: $g(s,\phi) = \int_{-T \cdot v_0}^{T \cdot v_0} I(s \cdot \cos(\phi) - p \cdot \sin(\phi), \frac{s}{v_0} \cdot \sin(\phi) + \frac{p}{v_0} \cdot \cos(\phi)) dp$. The maximum value of the r.m.s. spatial deviation ($\sigma_s'^2$) corresponds to the angle, which is a representation of the real velocity. Considering that in case of non-zero value of the orthogonal speed, traces drawn by speckles quickly abandon the researched area, we suggest to modify the first equation, namely, to add the integration on the orthogonal direction, that allows us to neglect this effect.

Moreover it is necessary to remark that for the calculation of velocity we need to make some primarily mathematical development of the input function I to plane the Gauss profile of the laser spot, and some optimisations in the procedure of the maximum contrast search.

For control of the precision of our method, we have developed the "angle-resolved velocimetry", that is the velocimetry for the set of the angles between the X-axis and the direction of the velocity measurement. The typical result is presented in fig. 1. A local coordinate system (x', y') is rotated for the angle α with reference to the coordinate system (x, y) and the velocity calculation is then applied in the direction of X'-axis. Although in real-time operation this algorithm is exceedingly computationally heavy (we calculate only two orthogonal components v_x and v_y) this scheme allows us to estimate of the absolute and relative error of our calculations. For good conditions, the error of measurement can be reached up to 1%.

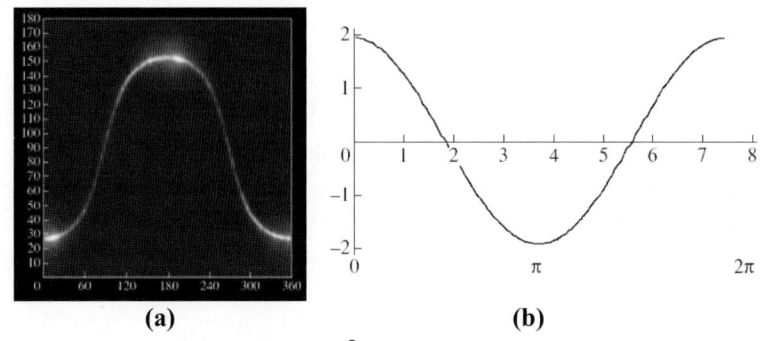

Fig. 1. The dependence of the $\sigma_s'^2$ from ϕ (a vertical axis) and from the angle of a turn α (a horizontal axis) (a), and the dependence of the velocity from α (b). Velocity is in relative units (pixel/frame)

But turning to the common characterization of our method, the creation of an adequate speckle field, point (i), really is not a trivial problem, because real speckle pattern statistics depend strongly on the scattering properties of a surface, and the optical system. Therefore we put this task to experimental analysis to find this dependence, and to make a conclusion about the real application.

2 The experimental set-up

To carry out the demonstration tests we have built an experimental breadboard. A beam generated by a He-Ne laser source is projected through a system of lenses on to a flat white surface. The light scattered by the white surface is collected, through the system of lenses, by the high-rate camera "Dalsa" (Model DS-41-065K0955). The acquisition equipment is composed by a very fast PC frame grabber and a dual processor XEON 2.4 GHz PC. The described optical equipment (camera and laser) is mounted on two degrees of freedom Cartesian robot, to simulate the vehicle motion. The axes motion is achieved by means of a couple of DC brushless motors, equipped with both differential encoders and tachometers to increase the response characteristics. The robot control system is based on a VME bus controller composed of a CPU board (Motorola MVME162) for the calculations inherent to the control law and to the trajectories definition, a piggy-back module (mounted on the CPU) for the encoder reading and the low-level control of trajectories, a digital input board for the reading of the proximity switches limiting the axes excursion, and an analogue input VME board dedicated to the management of the X axis movement.

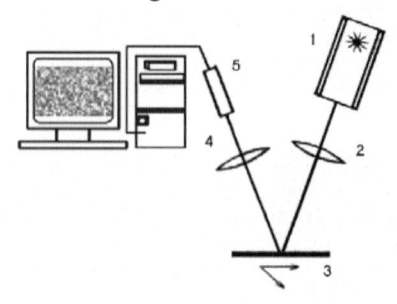

Fig.2. The sketch of experimental set up

3 Experiment results and discussion

We have tested our velocimeter with different kinds of surfaces, and different trajectories of motion.

To analyse calibration of the system from the focalisation of the camera, we have provided a number of experiments with the fixed by the robot control system velocity, and different focal parameters. The typical results are present in the figures 3-4. These figures present: (a) the time-dependence of the speckle field along axis-X (the vertical axis), integrated by the axis Y; (b) the time-dependence of r.m.s. spatial deviation (σ'^2_s) from velocity-corresponded angle (vertical axis, see fig. 1 (a)); and (c) the time-dependence of the measured velocity (in the relative units pixel/frame). The real velocity in the experiments, presented in the figures 3-5, was 20 cm/sec. The time (t) is in relative units (frames). The camera rate was 400 frames/sec.

These 3 images give us the possibility to clearly see the performance of our algorithm. The velocity is determined by the angle of inclination of the speckle traces in fig. 3(a), it results in the vertical position of the maximum of σ'^2_s (fig. 3 (b)), and finally it gives the searched velocity (fig. 3(c)).

For good focalisation (fig. 3) we can see the very good precision of the measurement. The r.m.s. deviation is less than 1%, which is the possible error of the real velocity variation. We can write the relation between real velocity v (in cm/sec) and calculated velocity V (in pixel/frame) as

$$v = V \cdot C \tag{1}$$

where C – the constant coefficient, which depend from the optical system and the camera rate.

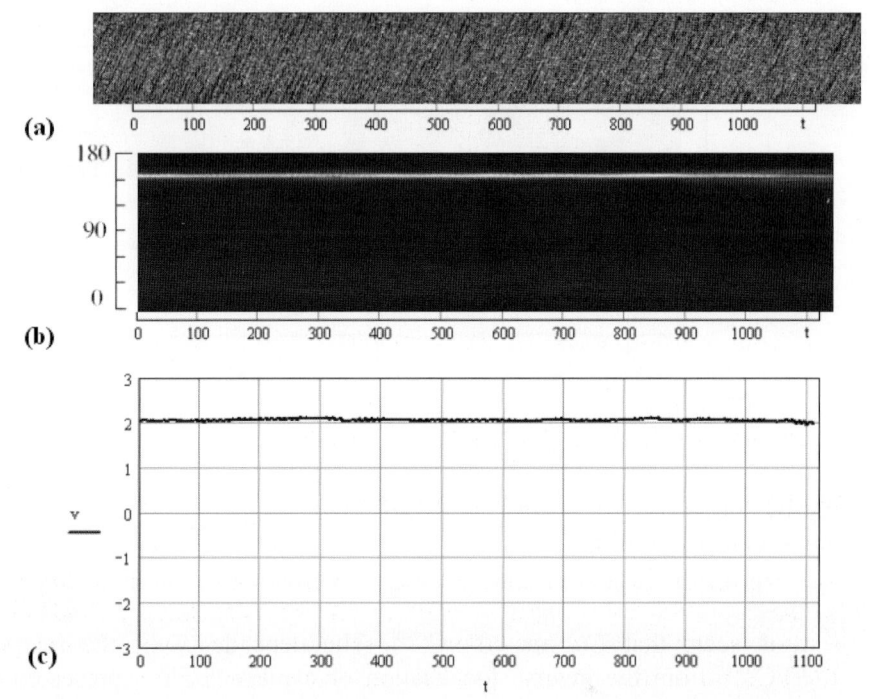

Fig. 3 The velocity measurement for the maximal focalisation of camera. (a) the time-dependence of the speckle filed along axis-X (the vertical axis), integrated by the axis Y; (b) the time-dependence of r.m.s. spatial deviation from velocity-corresponded angle; and (c) the time-dependence of the measured velocity (in the relative unites pixel/frame).

Figure 4 demonstrates the result of the measurement of the same velocity but with the focalisation of camera on 2 meters (that is much more than the real distance between the camera and the moving surface). We can see clear speckle traces, good precision of the velocity calculation, but the value of the calculated velocity is different.

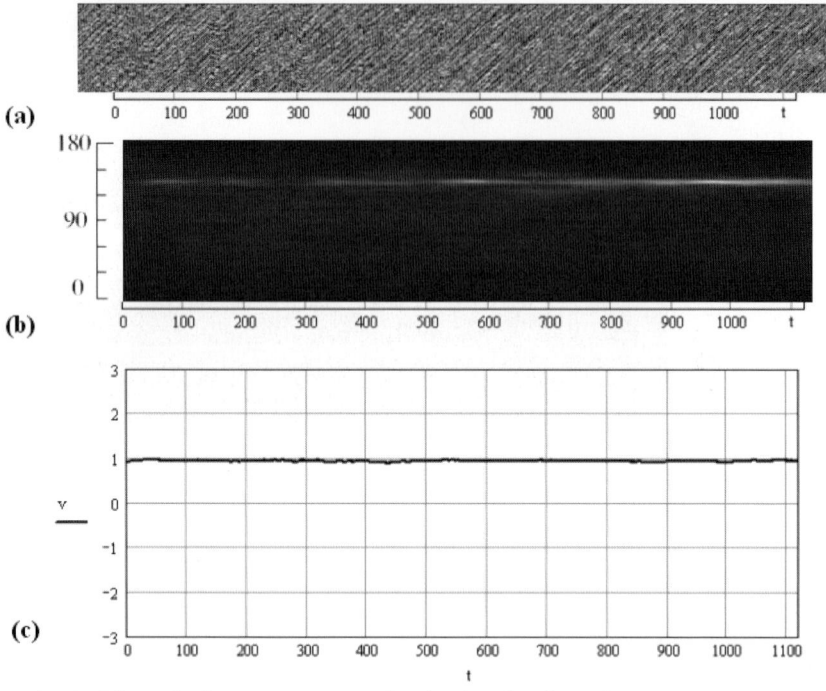

Fig. 4. The velocity measurement for the focalisation of camera on 2 m.

It is seen, that $C(2\ meters) < C^{focal}$. The dependence of the relation $C(F)/C^{focal}$ from the inverse focalisation of camera ($1/F$) is presented in figure (5).

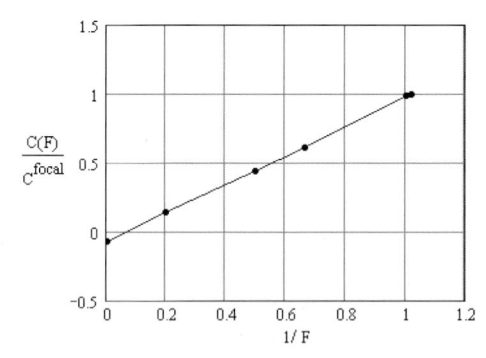

Fig. 5. The dependence of the relation $C(F)/C^{focal}$ from the inverse focalisation of camera ($1/F$) is presented in the figure (5).

We can see, that it is close to being linear, and for the focalisation on infinity it is negative. For an illustration of this phenomenon see fig. 6. We can see, that speckle traces are not as good as in cases considered previously, and they really have the opposite inclination. Nevertheless, with some additional optimisation we can calculate the corresponding velocity.

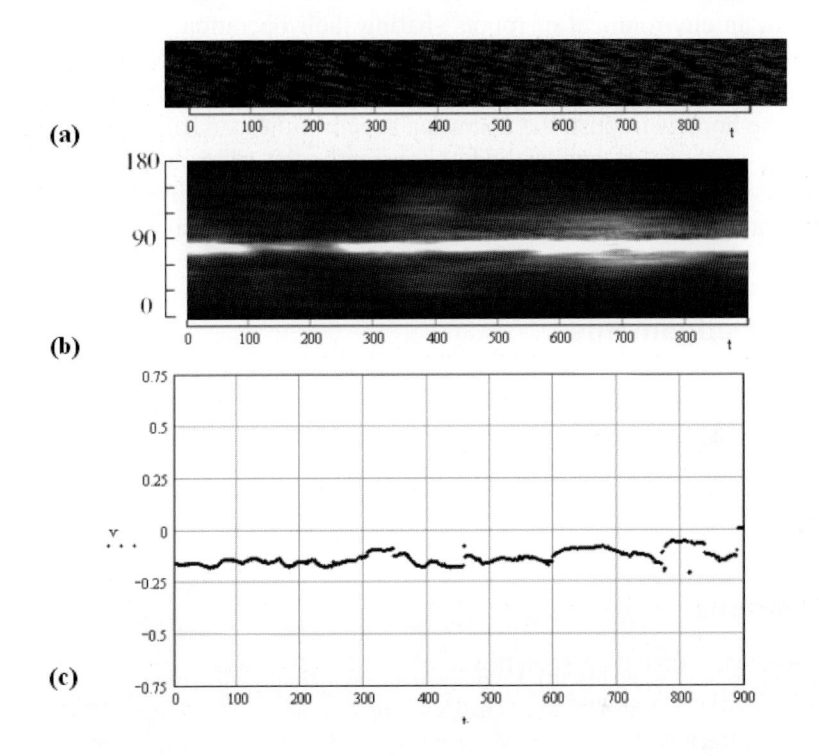

Fig. 6. The velocity measurement for the focalisation of camera on 2 m.

The precision of the velocity measurements strongly depends on the absolute velocity value. For a value of about 0.15 (as in the last case) it cannot be good, because of discrete limitation. But for larger velocities it is better.

Usually the optimisation of the system with unfocused parameters is not considered, but it seems to be quite interesting, and there is the possibility to apply it in some specific applications. Unfortunately, the velocity calculation in these conditions is less stable than in the case of a good focalisation, and it is more sensible to the nature of surface. We are now continuing the tests of our set-up for various other surfaces.

Conclusion

The system can also be easily applied to different environments, for example in an environment of robots sharing their operation space with humans giving precise data related to differential motion of different parts of the robot body (e.g. to the complexities of motion of humanoid-like systems); the hope is to push the operation speed of these machines up to levels close to those of human bodies (currently the technology is still far from this result). This is a very important feature, often neglected or underestimated, to succeed in co-operation with humans in real life.

Acknowledgements

The authors would like to thank ENEA, the CNR-NATO program, and the Ministry of Education and Science of Russia for the program "The development of the potential of Higher School" which made this study possible (34054).

References

1. Fercher A. F. (1980) Velocity measurement by first-order statistics of time-differentiated laser speckles. *Opt. Commun*, vol. 33, pp. 129-135.
2. Fercher A. F., Peukert M. and Roth E. (1986) Visualization and measurement of retinal blood flow by means of laser speckle photography. *Opt. Eng.*, vol. 25, pp. 731-735
3. Aliverdiev A., Caponero M., Moriconi C. (2002) Speckle Velocimeter for a Self-Powered Vehicle. *Technical Physics*, vol. 47, no. 8, pp. 1044-1048.
4. Aliverdiev A., Caponero M., Moriconi C. (2003) Speckle-velocimeter for robotized vehicles. *Proceedings of the SPIE. ALT'02 International Conference on Advanced Laser Technologies. Edited by Weber, Heinz P.; Konov, Vitali I.; Graf, Thomas.*, vol. 5147, pp. 140-147
5. Aliverdiev A., Caponero M., Moriconi C. (2003) Some Issues Concerning the Development of a Speckle Velocimeter. *Technical Physics*, vol. 48, no. 11, pp. 1460–1463.

Information Processing in Reactive Navigation and Fault Detection of Walking Robot

A. Vitko, M. Šavel, D. Kameniar[1] and L. Jurišica[2]

[1]Faculty of Mechanical Engineering, Slovak University of Technology Bratislava, Slovakia; [2]Faculty of Electrical Engineering and Information Technology, Slovak University of Technology Bratislava, Slovakia
vitko@kam.vm.stuba.sk

Abstract

The paper first summarizes main approaches to the sensor integration for the purposes of reactive navigation and the fault detection. Then the topic is focused on the results obtained by the originally developed fusing and classification algorithm for detection and classification of abnormal behaviorus appearing in a walking robot

Keywords: sensor integration, navigation, fault detection and classification, robot

1 Introduction

Conversion of the sensor data into a more information-rich structure that would be able to reflect a both complex state of a robot and its environment is a subject of intensive study. Augmentation of the robot with ability to integrate information coming from multiple sensors drastically expands its autonomy. An autonomously operating robot is required to respond to instantaneous incentives coming form the surrounding environment. In this view autonomously operating robot is a particular example of an intelligent system. The control community is familiar with the term of "intelligent control", connoting the abilities that the conventional control system cannot attain, like making complex decisions, adapt to new conditions, self-organizing, planning and the more. It is important to note

preserves main features of already accepted input patterns. The ART1 (see Fig.1) consists of two fully connected layers F_1 and F_2. So called pattern-representation layer F_1 accepts an instantaneous input pattern and sends it through the button-up weighted connections b_{ij} (initially set to unit values) to cluster-representation layer F_2. (A binary coded input pattern "I" is a concatenation of particular sensor outputs and possibly flags of the communication errors). Due to the competition based on the *winner-takes-all* paradigm, which is supported ensured by lateral negative feedback (-ε), the neuron in the F2 layer what receives the highest button-up activity is declared a winner. Its output is set to unit value and subsequently projected back to F_1 through the top-down weights t_{ij} and a chosen measure similarity between the projected winner and the input pattern is evaluated.

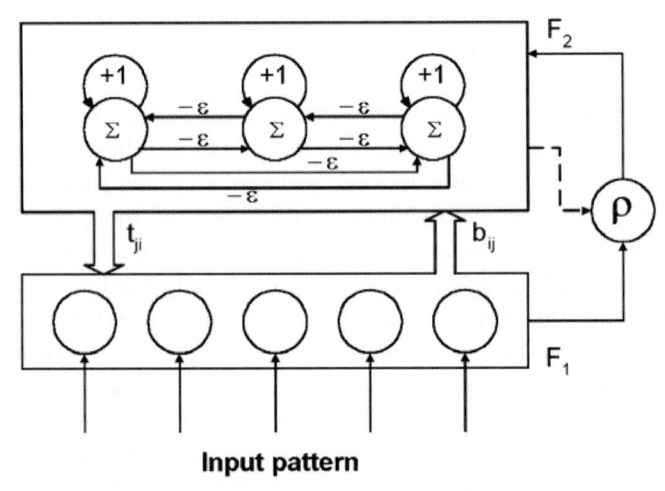

Fig. 1. Block scheme of the ART-1 network

If the similarity reaches greater value then the user given value of the *vigilance* ρ, so called *resonance* occurs, and vectors of the weights t_{ij}, b_{ij} are (by different way) moved closer to the input pattern (learning stage). If the resonance does not occur, the winner is disqualified and the process searches for the second best matching neuron, which is again submitted to the vigilance test. Searching repeats until either vigilance test passes or no more neurons are available for testing. It is just the value of vigilance what calibrates how much novelty the neural network can tolerate before it clusters and classifies an input pattern into particular class. Experiments have shown that the ART-1 is fully justified for being used as a means for the novelty detection and subsequent classification of behaviour patterns.

3 Experimental results

Faults may be caused by increased friction in bearings, slipping or dragging clutches, lack of lubrication or a partial loss of energy delivered to a particular joint. Besides, there could also be faults that were caused by incorrect coordination of legs due to improper timing (fall out of phase or fall out of step and the like). Malfunctions may remain hidden for long time and may gradually lead to fatal failures. Such faults may jeopardize the robot's walking stability or even cause robot overthrowing. Faults are commonly manifested through abnormal trajectories of the joint torques or forces. Therefore, the learning ART classifier was designed just for the task of detection and classification of any abnormal joint torques

The leg can be either in a stance state, when it supports the robot body or in a swing state, when it moves in air to the position where it can begin a new stance. A time-course of the normal (faultless) torque exerted in a femur joint is shown in Fig. 2. One complete step cycle is performed in three phases, each lasting one second. As seen from the figure, these three phases can be easily observed from torque-time dependence. A particular phase is supplemented with a simple imbedded sub-figure depicting the leg configuration and corresponds to the phase. During the first phase the leg remains in a flexed configuration in the stance. The femur joint exerts the torque value about 30 Nm, which maintains an attitude of the robot body. The second phase starts at one second. The leg is uncoupled from the ground and starts its swing movement in a direction of walking. While the torque exerted in the femur joint causes raising the leg, the coxa joint is rotating the leg about the vertical axis and the tibia joint is extending the leg. When reaching the highest position and maximum extension the leg ends its second phase. At this time instant the femur joint exerts maximum torque of about 70 Nm. Just after the third second the femur torque slightly decreases so as to make the foot go down until it reaches the ground. At this moment (at about the fourth second) the leg is entering into its stance state again, and supports the robot body.

During learning, the neural network was first taught to learn the above-described normal torque. As a result, the neural network appointed the normal torque course as the centre of a receptive field of the cluster of all "approximately normal" torque courses (torque patterns). This was done by adaptation of the bottom-up weights leading to most left neuron in the layer F_2. From this time on the unit value of this neuron indicated that the current input belonged to the cluster of "approximately normal" torque courses and this cluster represented a class of normal torque courses. Then a training list, i.e. a series of faulty torque patterns, generated by Sim-mechanics Toolbox, was repeatedly presented. The experimental results have shown that the learning task could be considered as accomplished (the weights reached their steady values), after presentation about 5 or 6 epochs. After learning stage the neural network became able to classify successfully any other set of faulty courses.

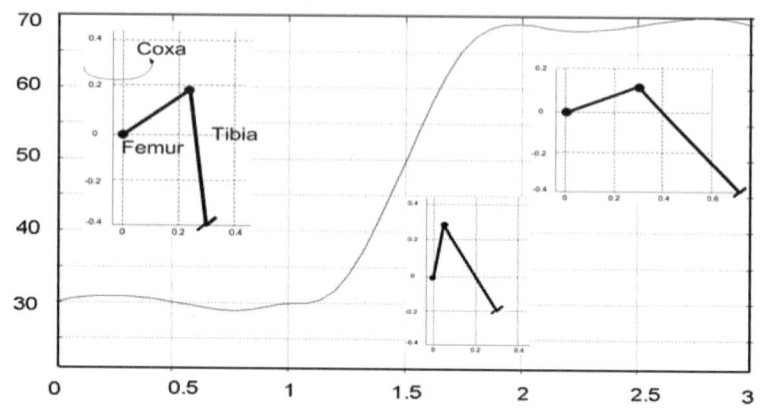

Fig. 2. Normal torque in the femur joint

Results of classification are briefly summarized in Figs. 3 Except for the normal torque, faulty torque patterns were presented to already learned ART network. Each of them corresponds to a particular fault as indicates the text under figures.

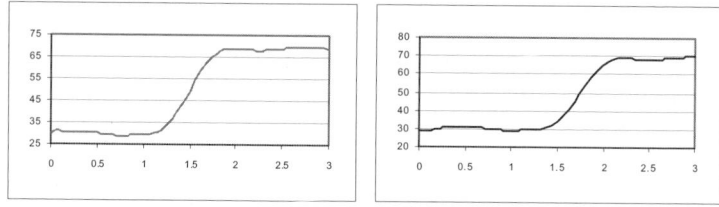

Class 1: Normal torque for slightly dragging bearing

Normal torque delayed by 0.24 s

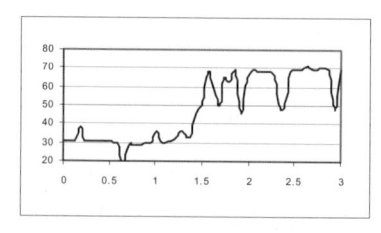

Normal torque with imposed strong noise caused by dragging and slipping clutch

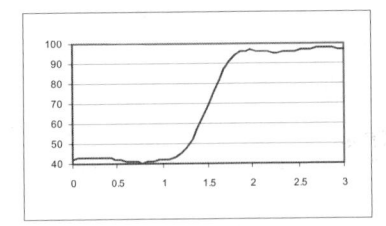

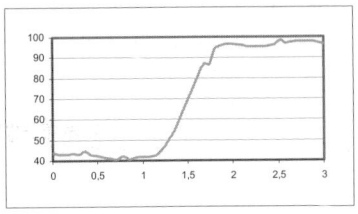

Class 2: Increased torque due to increased both friction in bearing and dragging

Increased and noisy torque due to in friction in bearing ; clutch is slightly load

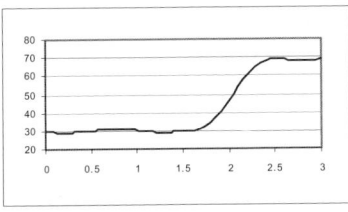

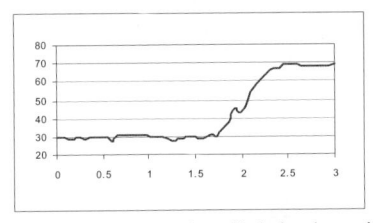

Class 3: Normal torque delayed by 0.54 s

Normal torque for slightly dragging bearing and delayed by 0.54 s

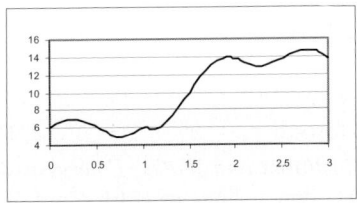

Class 4: Torque for the foot sinking the into a soft ground (almost zero load)

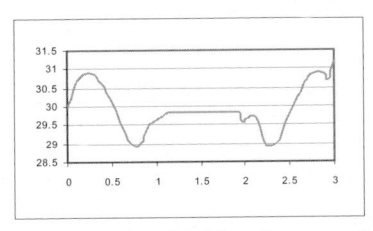

Class 5: The foot was raised and laid to the ground without rotation in the coxa jointwhat caused a plateau in the middle

Fig. 3. Results of classification for the vigilance value 0.8

For the value of vigilance "ρ" set to 0.8, the input patterns were classified into five distinctive classes. Raising the vigilance to 0.9 meant that the system was no longer willing to tolerate so much novelty (dissimilarity to the normal pattern) as before.

Conclusion

The ART1 neural network is very flexible and reliable means for detection of any dissimilarity (i.e. novelty) to the normal behaviours of the walking robot and classifies them into distinctive classes. Contrary to other neural or statistical approaches, there is no need to specify number of classes in advance. Based on a chosen value of the vigilance ρ, the network classifies input patterns into so many classes, how many is required for separation of those input patterns, which are more dissimilar to each other then it is allowed by the vigilance. In the more complex cases the system can detect and classified even contexts i.e. a composite state of the robot together with its environment.

Acknowledgment

The work has been supported by the Slovak Ministry of Education grant VEGA No. 1/0153/03

References

1. Brooks R. A. (1989) *A Robot that Walks: Emergent Behaviours from a Carefully Evolved Network.* Proceedings of IEEE Conf.on Robotics and Automation, pp. 692-696, 1989.
2. Fu K.S. (1971) Learning Control Systems and Intelligent Control Systems: An Intersection of Artificial Intelligence and Automatic Control. *IEEE Trans. AC. vol. 12, no 2, p.p 17-21*
3. Jang J.S.R., Sun E.C.T*., Mituzami E.,(1997) Neuro-fuzzy and Soft Computing.* 1st edition, Prentice Hall, Inc. Upper Saddle River
4. Zadeh L.A. (2001) A New Direction in AI. Toward a Computational Theory of Perception. *AI Magazine,* pp 73-84
5. Carpenter. G.A, Grossberg. S. A. (1987) Massively parallel Architecture for a Self-organizing Neural Pattern Recognition Machine, *Comp. Vision, Graphics and Image Proc.* vol. 37, pp 54-115
6. Baraldi, A., Alpaydin,(2002) Constructive feedforward ART Clustering Networks. *IEEE Trans on Neural Networks,* vol. 13, pp. 645-661

Intelligent Sensor System and Flexible Gripper for Security Robots

R.D. Schraft; K. Wegener; F. Simons; K. Pfeiffer

Fraunhofer Institute for Manufacturing Engineering and Automation (IPA)
Nobelstr.12 70569 Stuttgart / Germany

Abstract

During the last years, mobile service robots more and more entered a new field of application: Security & surveillance. Mobile robots do not become tired or bored, are incorruptible and reliable and can be equipped with all kinds of sensors to detect hazards and intruders where humans can not. It adds that mobile robots can easily be replaced in case of damage or destruction and no one complains about it except the purchase department. However, known security robots are just passive alarm systems, which interact only with optical and acoustical signals with their environment. It might be necessary to take samples for further inspection, e.g. in case of chemical contamination or to pick up objects dropped to the ground. Further these robots could be able to enlarge their detection zone by distributing small intelligent sensor probes. This paper introduces an autonomous security robot with a flexible gripper to manipulate its environment. Further the concept of an intelligent sensor system will be described. This system consists of a number of probes that can be distributed in the environment by a special probe repository and the robots gripper.

1. Security and surveillance robots

The market of mobile security and surveillance robots is not very big yet. Most of the available models are for indoor use only. **Table 1** shows a representative extract of the security robots available on the market.

Table 1: Security robots on the market

Robot	Manufacturer	Applic. Area	Sensors	Interaction
Banryu	Tmsuk	Outdoor, home only	Camera, micro, odor	Acoustic
Ofro	Robowatch	Outdoor, industrial	Thermo graphic, DGPS, opt. gas & chemical detector	Optic & acoustic
Mosro Mini	Robowatch	Indoor, home or office use only	Camera, heat sensor, microphone, opt. gas sensor	Optic & acoustic
Mosro	Robowatch	Indoor, office buildings & industry	Camera, radar, gas & smoke detector, motion detector, opt. microphone	Optic & acoustic
Secur-O-bot	Fraunhofer IPA & Neobotix	Indoor, office buildings & industry	Camera, radar, motion detector, microphone, opt. gas & fire detector	Optic & acoustic
Robot X	Secom	Indoor, office buildings	360° Camera	Optic, acoustic & smoke generator
SMIS	Robosoft	Indoor, office buildings	Camera, microphone	Optic & acoustic
Alsok	Sohgo Security Services	Indoor, office buildings	Camera, microphone, motion detector, fire sensor	Optic & acoustic

All these robots are just passive alarm systems, which interact only with optical and acoustical signals with their environment. An interesting idea of a little bit more active system was realised in the Robot X from Secom, which has got a smoke emitter to confuse and repel intruders.

Even if this is the most interactive system within a security robot, there is still no physical interaction possible, which might be desired, because most of the introduced robots can be teleoperated to have a closer look at the origin of the alarm. It could be helpful to take samples for further inspection, e.g. in case of chemical contamination or to pick up something an intruder dropped to the ground.

1.1 Secur-O-bot

Fraunhofer IPA and Neobotix developed an innovative and autonomous security robot, which is characterised through its robust mechanics and chassis on the one hand and through its powerful navigation system which based on high-end laser scanners and other sensor systems: Secur-O-bot. This mobile platform is equipped with a stereo camera on a 360° pan unit, an IR motion detector, a microphone and a speaker. It has differential drive kinematics and is powered by two 12V lead batteries which enable about 10 hours of continuous operation. An explanation of the robots hardware is given in **Fig. 1**.

Intelligent Sensor System and Flexible Gripper for Security Robots   803

The payload of the robot is more than 30 kg and therefore optional sensors like radar sensors, thermo graphic cameras, gas detectors or fire and smoke detectors can be added depending on the application.

Fig. 1. Explanation of Secur-O-bot

1.2 Vision of future security robots

To eliminate the above described disadvantages of state of the art security robots, future systems should be able to enlarge their detection zone and should be able to manipulate their environment. This vision of future security robots is depicted in the following **Fig**.

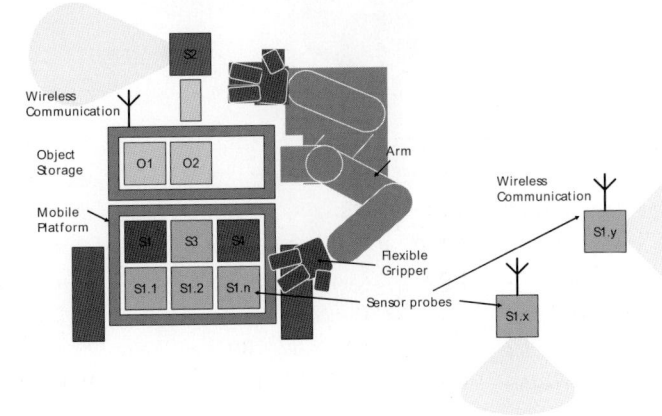

Fig. 2. Robot with sensor probes, a robot arm, and a flexible gripper

Key systems of these types of robots are intelligent sensor systems and flexible grippers. Two new approaches are described in the following two chapters.

2. Intelligent sensor system for security robots

One disadvantage of security robots is the small area that can be covered simultaneously by the sensors of one robot and the fact, that this area can only be enlarged by increasing the range of the sensors or the number of robots in use. In addition the whole protection area of a security robot is basically limited by the range of the communication method and can be increased mostly by changing the infrastructure.

One way to handle both problems are sensor probes, which are defined in [1]. A mobile robot equipped with sensor probes can easily adapt its surveillance space to the environment and the application. In addition the robot can build up its own communication infrastructure.

The first prototype built up at Fraunhofer IPA is a sensor probe with a standard VGA camera, a WLAN access point and a battery pack with voltage regulation [1]. A 3D drawing of this sensor probe is shown in **Fig. 3**. For integrating the probes into a mobile robot it needs to have a storage device, which is able to place the probes and pick them up again. A first design of a probe repository integrated to Secur-O-bot is shown in **Fig. 4** and works with only two actuators and holds six sensor probes.

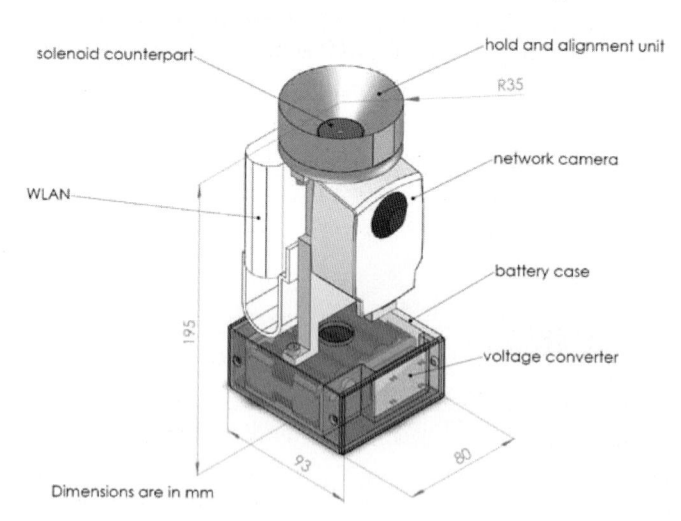

Fig. 3. Prototype of sensor probe with camera and mechanical interface

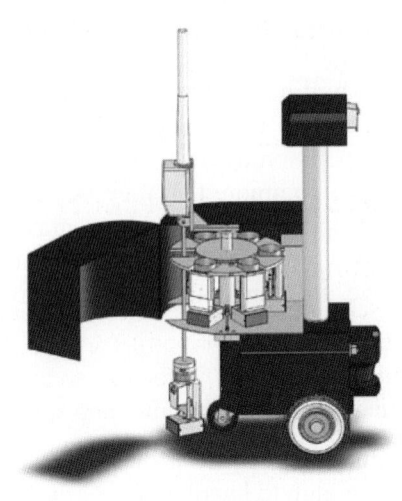

Fig. 4. Illustration of robot with probe repository and picking operation

For placing a probe will be positioned over the entry of the probe repository and a linear drive connects itself to the probe with an electro magnet and performs a pick and place operation to place the probe right behind the robot. The picking works vice versa.

If a probe is equipped with a directional sensor like a camera, it is important to assure the correct orientation of the probe when placing it. The robot and/or the sensor probe are most likely not in the same orientation during the pick up process as they where during the placement. Thus either the correct orientation has to be regained inside the storage or before returning it in the storage. The mechanical design of the first prototype assures an inside-reorientation. An outside-reorientation could simplify mechanical design of the storage and be realized by an additional flexible gripper. This security robot could place sensor probes all around and not only behind itself and the flexible gripper could handle different types of probes with no common mechanical interface.

3. Flexible Gripper

Based on the above defined requirements flexible grippers are needed which are able to handle unknown objects and which are cheap enough to be integrated in future products.

3.1 Requirements for a flexible gripper used in security robots

A flexible gripper which will be used in security applications has to grasp sensor probes on the one hand and unknown objects, like pieces of evidence, door handles or operation devices, on the other hand. The class of unknown objects has all in common, that they were designed for humans. Therefore the flexible gripper should be able to grasp objects with a size equal or less than objects which are designed for a human hand. When we analyse those objects we will find most of them have some features in common:

- Symmetric Design
- The basic geometries are cylinders, cubes and spheres

An analysis how human beings grasp these objects shows that almost three type of grasps are enough to handle most of them:

- Two-finger-grasp,
- three-finger-parallel-grasp and
- three-finger-centric-grasp.

From the economical view, the complexity of the system and due to that the cost has to be reduced compared to flexible devices like artificial hands.

3.2 State of the Art of Flexible Grippers

Grippers for industrial applications have a low degree of flexibility against varying object geometries, but they are cheap in purchase [2]. Artificial hands [3] which are mainly provided in the field of service robots have a high degree of flexibility, but they are too complex to find an economic application.

The task of an artificial hand is beside of grasping the internal manipulation of objects. Therefore these hands need more than nine degrees of freedom with a minimum of three fingers [4].

At least neither industrial grippers nor artificial hands are meeting the requirements in total. Grippers which are cheap in price and flexible to manipulate unknown objects without internal manipulation has to be developed.

3.3 The IPA-Hand

At Fraunhofer IPA a cinematic-concept based on just two electrical motors has been developed [5]. The gripper consists of one thumb and two fingers. The thumb can be switched between two fixed positions (Ref. **Fig. 5**).

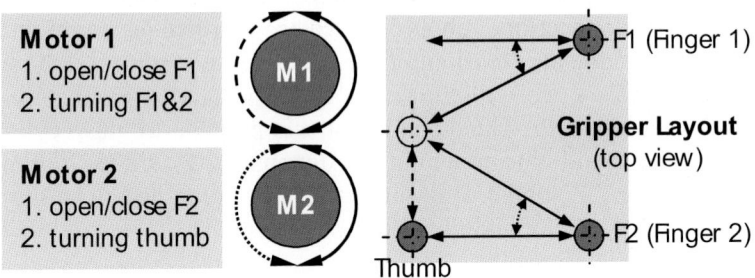

Fig. 5. Concept of the IPA-Hand

Each motor has two functions. The first is to open and close its dedicated finger; the second is to switch to another configuration. This is possible with a special differential gear and a mechanical break which is active in the back position of each finger.

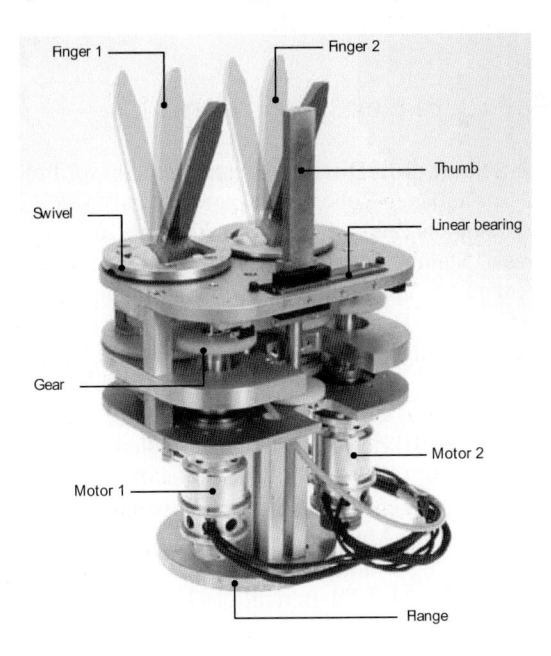

Fig. 6. Realization of the IPA-Hand

Fig. 6 shows the realization of the IPA-Hand. The IPA-Hand is controlled by an embedded PC. The so called IPC@Chip from Beck Inc. The motors are interfaced via CAN-Bus. Process sensors like force or distance sensors can be adopted via an integrated AD-Converter. The commands, e.g. for switching the configuration or a gripping command, are provided via Ethernet. The Ethernet interface enables users to integrate the IPA-Hand into different security robots very easily. Further on an integrated web-server allows the user to monitor the states of the system and the hand integrated sensors.

First tests with the IPA-Hand show that it is possible to grasp cylindrical, cubical and round objects within the dimensions 20 x 20 x 20 mm^3 to 80 x 80 x 300 mm^3. Potential of optimization lies in the design of the fingers and the integration of additional sensor systems.

4 Conclusions

Both, the intelligent sensor system as well as the flexible gripper are new approaches in the field of robots for security and surveillance. The concepts and the first experimental results show the high potential of these systems for use in future security robots.

References

1. Pfeiffer, Kai; Schraft, Rolf Dieter: Decentralized Sensor Probe Positioning for Security and Surveillance Robots. In: Tagungsband "SSRR 2004: IEEE International Workshop on Safety, Security and Rescue Robotics", Bonn 2004
2. Wolf, Andreas; Steinmann, Ralf: Greifer in Bewegung : Faszination der Automatisierung von Handhabungsaufgaben. München; Wien : Hanser, 2004
3. Lotti, F.; Vassura, G.: Design aspects for advanced Robots Hands: Mechanical design. In: Tagungsband "Towards Intelligent Robotic Manipulation", IROS 2002.
4. Butterfaß, Jörg: Eine hochintegrierte multisensorielle Vier-Finger-Hand für Anwendungen in der Servicerobotik. Aachen : Shaker, 2000. Darmstadt, Diss. 1999
5. Schraft, Rolf Dieter; Wegener, Kai: Greifen geometrievarianter Objekte. In: VDI/VDE-Gesellschaft Meß- und Automatisierungstechnik (GMA) u.a.: Robotik 2004: Tagung München, 17. und 18. Juni 2004. Düsseldorf : VDI Verlag, 2004, S. 117-124 (VDI-Berichte 1841).

Search Performance of a Multi-robot Team in Odour Source Localisation

Chris Lytridis[1], Gurvinder S. Virk[2], Endre E. Kadar[3]

[1] Department of Electronic & Computer Engineering, University of Portsmouth, UK; [2] School of Mechanical Engineering, University of Leeds, UK; [3] Department of Psychology, University of Portsmouth, UK

Abstract

In this paper, the latest developments from research carried out by the Portsmouth/Leeds group on odour source localisation using multiple robots are presented. Previous studies have demonstrated the capability of individual robots to find an odour source using three biologically inspired navigational strategies, namely chemotaxis, biased random walk (BRW) and combined chemo-BRW strategy. This work has been extended to multi-robot searching scenarios and it has been shown that co-operation leads to improvements in search efficiency and robustness of the search strategies. The present paper investigates and discusses the effect of team size on the search performance of a multi-agent system in a searching task. The results show that interactions between robots during a search affect the performance significantly and can be quantified, even though the performance of multi-robot searches clearly improves compared to single robot searches.

Keywords: navigation, multi-robot teams, chemical fields, searching

1 Introduction

The potential applications of gas-sensitive robots have attracted much interest in recent years. Such applications include; location of hazardous chemicals, demining, etc. Research focuses on the development of appropriate odour sensors and effective navigational algorithms. Odour based navigation at present faces two main limitations; (a) the current odour sensors have limited sensitivity and slow response and recovery characteristics and (b) the navigational strategies cannot fully cope with the unpredictability of real chemical fields. One of the first works in this area was carried out by Russell with a trail following robot [1]. Further studies have followed (for instance see [2], [3]) which demonstrated successful target localisation. In some applications, it is beneficial to use multiple robots. Such an approach offers advantages such as averaging of measurement errors, tolerance to unpredictable failures and the possibility of replacing a single complex (and potentially expensive) robot with many low-cost simple robots [4]. Some researchers have explored the potential of using multiple robots in odour-based searching with promising results [5], [6].

The initial studies carried out by the authors were based on simulations and the relative merits of the chemotaxis and BRW strategies have been demonstrated [7]. These strategies have been tested in a variety of simulated fields [8], [9], [10]. It was found that chemotaxis is efficient in stable fields, but unreliable in noisy fields. In contrast, BRW was robust even in turbulent fields but its efficiency is worse than that of chemotaxis in stable fields. A combined chemo-BRW strategy has been developed, and the initial results were promising. Simulations have also shown the potential of a multi-agent approach and a cooperation strategy has been proposed [10]. The theoretical work has been confirmed in experiments using BIRAW robots [11], [12].

The present paper discusses how the interactions between robots in a multi-robot search scenario affect the search performance. The factors that limit the benefits of using multiple robots are identified and the partial influence of each factor is investigated. The paper is laid out as follows: in Section 2 the navigational strategies that were used are reviewed. Section 3 describes the experimental setup and in Section 4 the experimental results are presented. Finally, in Section 5 the results are discussed and recommendations for future work are made.

2 Navigational Strategies

The strategies that are used have been discussed in detail in previous studies [7], [10]. With chemotaxis, the robot turns by a constant angle towards the sensor with the strongest reading and moves forward by a constant length. Chemotaxis produces smooth paths but is unreliable in high levels of noise. On the other hand, the BRW strategy is essentially a uni-sensory strategy that uses Poisson statistics to determine a random turning angle and the length of the next step. Longer step lengths are generated when the gradient in the direction of movement is positive and, over time, this causes the searching agent to drift towards the source. In simulations as well as in experiments with the BIRAW robots, the BRW strategy has been shown to be robust even in high levels of noise. However, the method is much less efficient even in smooth fields since the random turns often produce more meandering paths. Chemotaxis and BRW have been merged to produce a combined chemo-BRW strategy which uses bilateral sensor comparison for steering and random step length generation. It has been shown that this strategy is reliable as well as efficient.

The experiments consist of two parts: a) Single robot searches and b) non-cooperative multi-robot searches (with teams of two and three robots). In the non-cooperative multi-robot searches, the robots use the single robot strategies together with a collision avoidance strategy.

3 Experimental Setup

The odour source is a dish of methylated spirits. Since diffusion is slow, a fan was used to speed up the spreading of the chemical. For the concentric field the fan was placed on top of the dish and a downward flow was established. For the plume the fan was positioned such that lateral flow is created. Figure 1 shows the concentration distributions for the two field types.

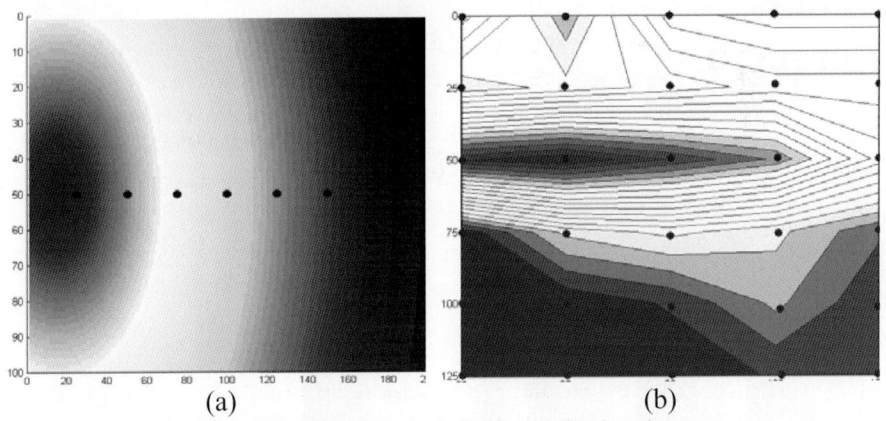

Fig. 1. (a) The concentric field and (b) the plume

In this paper, non-cooperative search data are selected for the analysis in order to isolate the factors that affect the performance and that are related exclusively to the use of multiple robots, from cooperation. Figure 2 shows a sample search with 3 robots in the plume using chemotaxis. The robot paths are superimposed on the sensor readings recorded during the trial.

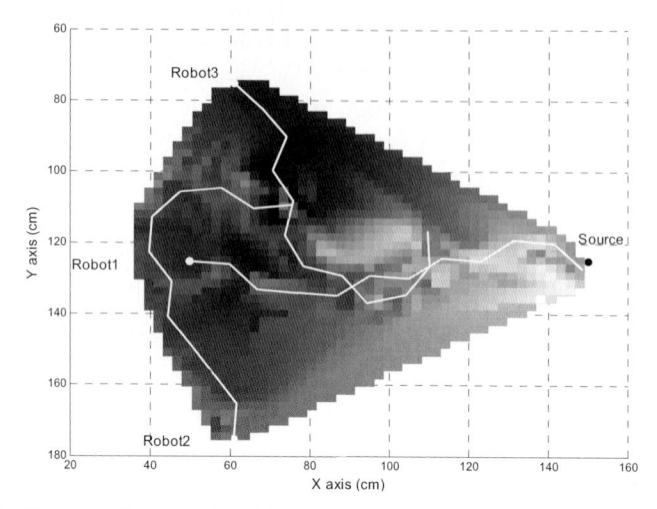

Fig. 2. Co-operative search with three agents using the chemotaxis strategy

In all experiments, the initial distance from the odour source is 1.25 m and the initial orientations of the robots were randomly selected. The initial distances between robots was 60 cm when two robots were used and 50 cm when three robots were used. These distances were chosen based on the plume width so that the robots start their search inside the plume.

4 RESULTS

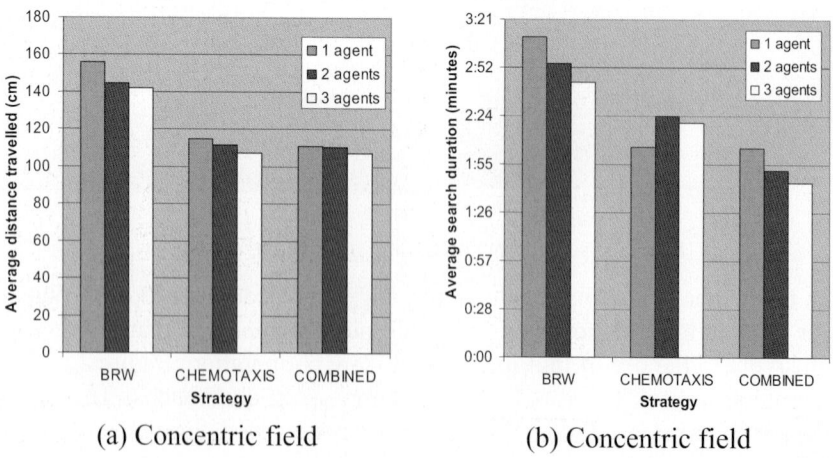

Figure 3 shows the performances for single and multi-robot odour source localisation experiments for all field conditions.

(a) Concentric field   (b) Concentric field

Search performance of a multi-robot team in odour source localisation 813

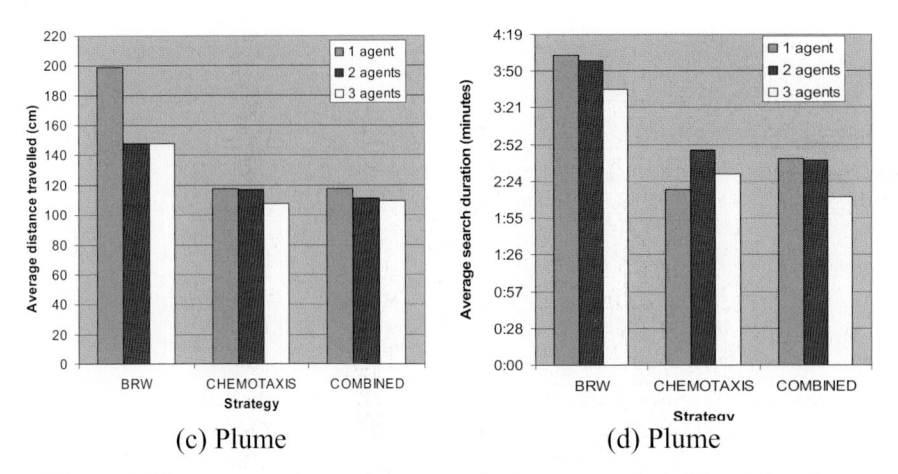

(c) Plume (d) Plume

Figure 3. Non-cooperative performances in the concentric field and the plume

Empirical observations suggest that the avoidance manoeuvres between agents hindered the search and reduced the efficiency, especially for chemotaxis trials where the search duration does not improve when more robots are used. This increase in the search duration occurs due to the clustering of the robots near the source. Specifically, obstacle avoidance causes the robots to turn away from the source to avoid collisions with each other. After avoidance, the robots resume the search behaviour. Such interactions between robots are observed more frequently with chemotaxis since the robots move towards the source in a deterministic fashion and at a constant rate, as opposed to BRW and the combined method in which randomness causes greater spatial separation.

Increased interactions between robots are observed in the plume. Based on the sensors' narrow detection range, the plume is very narrow compared with the size of the arena and the size of the robots. In this field, the robots first move towards the centreline of the plume and then ascend the gradient along the longitudinal axis towards the source. Therefore, in many cases the plume's centreline becomes cluttered early in the search.

To investigate the extent of this effect in the two field types, the performance of non-cooperative searches without interactions (i.e. independent searches) had to be assessed and compared to the experimentally measured performance. To create the hypothetical independent searches, the data from the single robot trials were used. The performances of single robot trials were grouped randomly in sets of two and three trials and the best performance of the grouped set was used as the performance of the hypothetical multi-robot search. Figure 4 shows the predicted performances of the independent searches.

(a) Concentric field

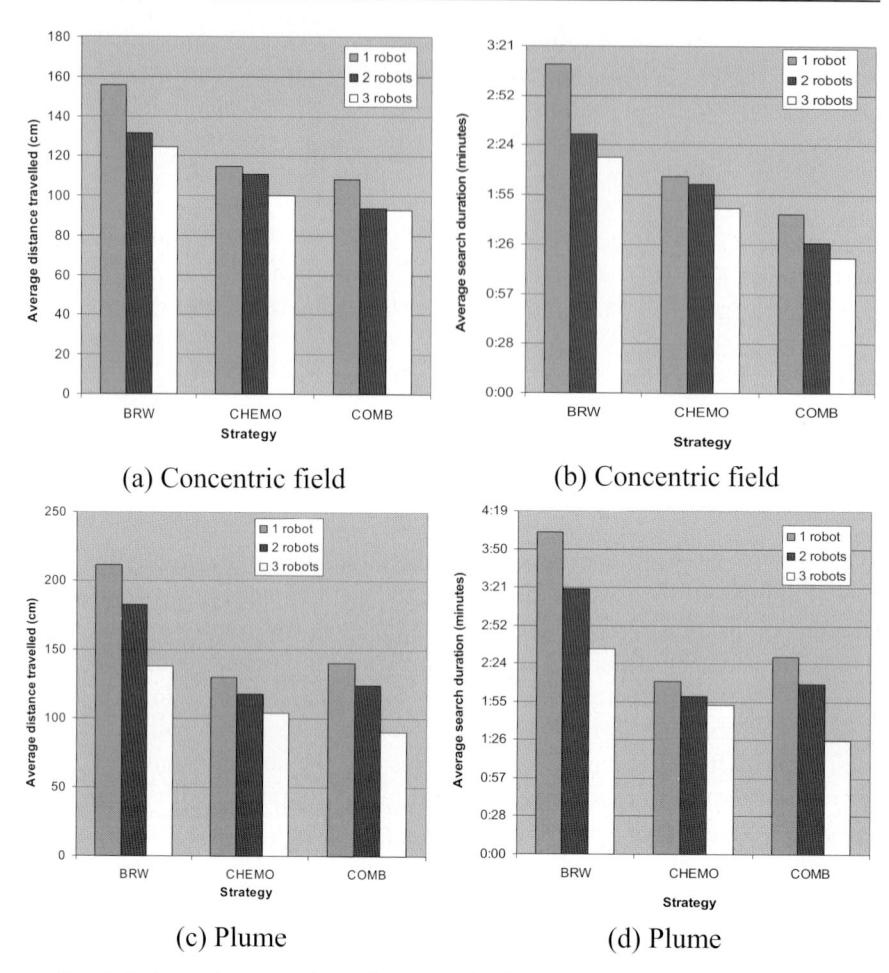

(b) Concentric field

(c) Plume

(d) Plume

Fig. 4. Independent search performance in the concentric field and the plume

Figure 4 shows that the improvements are more pronounced in all cases. Table 1 shows the quantified interactions, produced by calculating the percentage difference between the experimental data and the baseline data.

Table 1. Interactions effect (a) the concentric field and (b) plume

		BRW		CHEMOTAXIS		COMBINED	
		Dist.	Dur.	Dist.	Dur.	Dist.	Dur.
(a)	2 agents	9%	14%	1%	16%	15%	22%
	3 agents	13%	17%	7%	23%	14%	25%
(b)	2 agents	-10%	16%	2%	29%	-4%	20%
	3 agents	7%	28%	3%	25%	20%	35%

Comparison of mean values is only an indication of their differences, since the results may be misleading due to large variance and small sample size (number of trials). In this case, an analysis of variance (ANOVA) can be carried out in order to identify the significance of the results. In our case, to verify the significance of the results, a four-way ANOVA on the distance travelled was performed with the team size (one to three robots), the strategy (BRW, chemotaxis and the combined method) and the type of performance data (experimental vs. baseline) for both field types (concentric field and plume). The main effect of the team size is highly significant (F=10.48, $p<0.001$), which proves that increasing the number of robots, increases the interactions between the robots. The main effect of the strategy (F=56.86, $p<0.001$) is also highly significant, which means that the amount of spatial interactions depends on the chosen navigational strategy. The same is true for the field type since the main effect is found to be highly significant (F=8.88, $p<0.01$). However, the main effect of the type of performance data (F=1.827, $p = 0.18$) is non-significant and this suggests that collision avoidance manoeuvres do not have a significant effect on the distance travelled. It was found that there were no interaction effects between variables. The corresponding four-way ANOVA on the search duration reveals there is a highly significant main effect of the number of robots (F=21.41, $p<0.001$), the strategy (F=101.95, $p<0.001$) and the chemical field type. In this case however, main effect of the type of data is highly significant (F=69.39, $p<0.001$), meaning that collision avoidance manoeuvres significantly affect the search duration. As before the main effect of the interactions between the variables is non-significant.

5 Conclusions

The effect of spatial interactions between robots during a search has been investigated using the BIRAW robots in odour source localisation experiments. The procedure that was described in this paper is a way to estimate the effect of spatial interactions in a multi-robot team, using single-robot searching experiments as baseline data. Statistical analysis has shown which factors affect the performance more significantly and can help determine - given the experimental circumstances - whether the use of multiple robots is beneficial. The results have shown that several factors affect the performance of the multi-robot team, including the number of robots, the chemical field type and the navigational strategy. The interactions between the robots mainly affect the duration of the search and not so much the distance travelled. Current research focuses on the development of algorithms which minimise the effect of interactions between robots in such multi-robot search tasks. A solution currently being investigated is the use of the existing communication capabilities of the BIRAW robots in order to alter the robot paths when the spatial arrangement suggests clustering.

References

[1] Russell, R. A. (1995). Laying and sensing odour markings as a strategy for assisting mobile robot navigation tasks. *IEEE Robotics & Automation Magazine*, 2(3), 3-9.
[2] Ishida, H., Hayashi, K., Takakusaki, M., Nakamoto, T., Moriizumi, T., & Kanzaki, R. (1995). Odour-source localization system mimicking behaviour of silkworm moth. *Sensors and Actuators A-Physical*, 51(2-3), 225-230.
[3] Duckett, T., Axelsson, M., & Saffiotti, A. (2001). Learning to locate an odour source with a mobile robot. *IEEE International Conference on Robotics and Automation* (pp. 4017-4022).
[4] Cao, Y. U., Fukunaga, A. S., & Kahng, A. B. (1997). Cooperative mobile robotics: antecedents and directions. *Autonomous Robots*, 4(1), 7-27.
[5] Sandini, G., Lucarini, G., & Varoli, M. (1993). Gradient driven self-organizing systems. *IEEE/RSJ International Conference on Intelligent Robots and Systems*, Vol 1-3 (pp. 429-432).
[6] Hayes, A. T., Martinoli, A., & Goodman R. M. (2003). Swarm robotic odour Localization: Off-line optimization and validation with real robots. *Robotica*, 21(4), 427-441.
[7] Kadar, E. E., & Virk, G. S. (1998). Field theory based navigation for autonomous mobile machines, *Proceedings of the IFAC Workshop on Intelligent Components for Vehicles (ICV '98)*, Seville, Spain, 23-24 March.
[8] Virk, G. S., & Kadar, E. E. (2000). Trail following navigational strategies. *Proceedings of the 3rd International Conference on Climbing and Walking Robots (CLAWAR 2000)* (pp. 605-613). Madrid, Spain, 2-4 October.
[9] Kadar, E. E., & Virk, G. S. (1998). Field theory based navigation towards a moving target. *Advanced Robotics: Beyond 2000: 29th International Symposium on Robotics,* Birmingham, England, 27 April - 1 May.
[10] Lytridis, C., Virk, G. S., Rebour, Y., & Kadar, E. E. (2002). Odour-based navigational strategies for mobile agents. *Adaptive Behaviour*, 9(3-4), 171-187.
[11] Lytridis, C., Fisher, P., Virk, G. S., & Kadar, E. E. (2003). Odour source localization using co-operating mobile robots. *Proceedings of the 6th International Conference on Climbing and Walking Robots (CLAWAR 2003)*, Catania, Italy, 17-19 September.
[12] Lytridis, C., Virk, G. S., & Kadar, E. E. (2004). Cooperative smell-based navigation for mobile robots. *Proceedings of the 7th International Conference on Climbing and Walking Robots (CLAWAR 2004)*, Madrid, Spain, 22-24 September.

A "T-shirt Based" Image Recognition System

P.Staroverov, C.Chicharro, D.Kaynov, M.Arbulú, L.Cabas and C.Balaguer

Robotics Lab, Department of Systems Engineering and Automation
University Carlos III of Madrid
Avda. Universidad 30, Leganés 28911 (MADRID) Spain
{pstarove, dkaynov, marbulu, lcabas, balaguer}@ing.uc3m.es
cristina.chicharro@alumnos.uc3m.es

Abstract

This paper presents a simple image recognition system for the Rh-0 humanoid robot that has been developed in the University Carlos III of Madrid, Spain. This system extracts some basic features from the image that may be used in the robot's navigation and path planning. The simplicity of this system is based on the assumption that the person interacting with the robot is wearing a red T-shirt as well as on some other assumptions. In the paper we describe the vision system's structure and the steps followed in the development of the image recognition system. Experimental results and discussion complete the description of the system.

Keywords: Humanoid Robot, Vision System, Image Recognition

1 Introduction

The humanoid robot Rh-0 has been developed in the University Carlos III of Madrid. It has 21 degrees of freedom (DOF) in its articulations and 2 more DOFs in the camera attachment that permit pan and tilt movements. The concept of a humanoid robot suggests that the robot will have to interact with the humans, because in the opposite case it might be more effective to create a robot in another form-factor that may have lower complexity and higher effectiveness than a humanoid one. It is the similarity that makes the humanoid robots look friendlier than other types of robots, but this friendliness only counts when the person interacting with the robot can see the robot itself or at least an image of the robot. Similarly, it is very important for the robot to be aware of the person's location. The objective is to establish a very simple, rapid and robust interaction between the humanoid robot and the person. That is why we have chosen a person, wearing a red T-shirt as the search target, who may be compared to an information desk person in a large supermarket. When we talk about the human-robot interaction and even about performing some cooperative tasks, the robot's awareness of the person's location is crucial. The most evident way to obtain someone's location is using sensory information, and the first sensory system that comes to mind is the vision system. If the information provided by the vision system is not sufficient to detect the person's position due to some reason (for example, the person is not in the field of view of the camera, or it is too dark), then the information from other sensors (such as microphones, laser rangefinders etc.) may be used, if there are any. At the moment, the only sensor type installed on the Rh-0 robot is the camera, though other sensors will be installed shortly. Therefore, our task is to obtain the person's location using visual information.

This paper is organized as follows. In section 2 some related works are discussed, in section 3 the development of our recognition system is presented in detail along with the experimental results and section 4 concludes the paper.

2 Related works

Recently, autonomous humanoid robots have progressed significantly, with several successful projects developed in Japan (HONDA ASIMO, KAWADA HRP-2P, SONY QRIO), Germany (JOHNNY), Russia (ARNE and ARNEA) and other countries. There is no doubt that the Japanese humanoid robots have the longest history and the research teams from Japan have the most experience in the development of humanoid robots. Let us make a brief overview of the vision systems of the most well-known humanoid robots.

ASIMO robot developed by HONDA has 2 colour CCD cameras connected to a PCI-bus framegrabber that provides images to a Mobile Pentium III 1.2GHz based PC running on Windows NT 4.0 [1]. ASIMO's vision software is rather sophisticated and may perform moving objects recognition, 3D objects recognition, gesture and posture recognition, face detection and recognition etc.

HRP-2P robot by KAWADA Industries has three cameras connected to a PCI-bus framegrabber feeding images to a 1 GHz Pentium III PC running on ALT-Linux [2,3]. It uses proprietary VVV System software that is capable of reconstructing 3D shape models, detecting objects, measuring position, tracking objects, generating a 3D terrain map of the worksite, detecting ground topology etc.

SONY QRIO robot uses 2 colour CCD cameras, an FPGA module controlled by an 8-bit microcontroller for stereo processing that is connected to a 64-bit RISC CPU via OPEN-R bus [4]. It uses an APERIOS operating system and vision software that can compute distance to objects, extract the floor plane etc. Furthermore, ball and landmark detection using colour segmentation as well as face recognition helps QRIO to understand the outer world.

3 Development of the "T-shirt based" image recognition system

3.1 Hardware and software

It may be clearly seen from the examples above that many if not the majority of the researchers tend to use several cameras for stereo vision connected via a framegrabber to a robust PC system. However, for our Rh-0 robot shown in **Fig.1** below, we have chosen to use a single camera connected to the PC via TCP/IP.

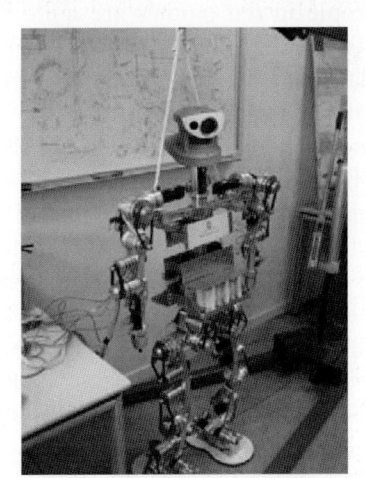

Fig. 1. Rh-0 humanoid robot

The camera model is AXIS Communications 213PTZ, it is a network camera that has remote control over its pan/tilt/zoom and other functions. The pan movement is limited to 170° in either direction that will give us the possibility in the future to work with 360° view images. The images are transferred via network to the Advanced Digital Logic MSM855 PC/104 board based on Intel Pentium-M 750 CPU. The operating system is Windows XP and the image processing program is created using MathWorks MATLAB [5].

3.2 Formulation of the problem

Given that this is our first experience in this field, we wanted to keep everything as simple as possible yet effective. There was no sense in trying to develop a complex vision system without making some easy tests first. Hence, it was decided to develop a program that searches for a person in the image without thinking about obstacles, face recognition and other more advanced things. To simplify our task, some assumptions had to be made. First of all, we suggested that the person interacting with the robot wear a red T-shirt. In this case the search for a person in the image is limited to the search of the red T-shirt that is a simpler task. This is quite a feasible assumption due to the fact that during the presentations and similar events the robot's operators wear some type of corporate clothes. Next, we assumed that the T-shirt is the largest red object in the image. Having in mind that the robot is making its first steps in the corridors of our department, where the only red objects you will normally see are fire-extinguishers, and people in red clothes are quite rare, this one may be a viable assumption, too. Further on, we decided that the T-shirt in question has to be not just a plain red one, but must have a small picture on the front side, that would permit us to distinguish whether the person is facing the robot or not. Lastly, it was decided to work in the RGB colour space due to the fact that the light intensity in the lab and the passage is nearly constant.

Now, let us say a few words about the features that we want to extract from the image. First of all, we need to obtain the estimated distance d from the robot to the person that may be calculated using the relationship between the real height of the T-shirt in cm (which is known to us) and the height of the T-shirt in the image in pixels. The second parameter we want to get is the angle α of the person's position with respect to the robot's sagital plane (where $\alpha=0°$ means that the person is in the center of the robot's view), this one can be easily calculated using the position of the T-shirt's center line in the image. The last parameter that we would like to measure is the angle β of the person's rotation ($\beta=0°$ means that the person is facing the robot), which can be calculated using the relationship between the width and the height of the T-shirt, having in mind the presence or the absence of the picture in the T-shirt to determine the orientation. **Fig.2** gives a better idea of the parameters that we would like to measure.

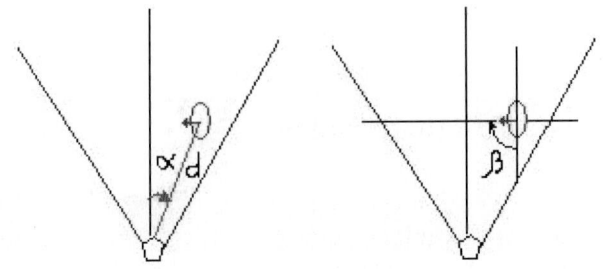

Fig. 2. The parameters to be estimated (top view; the robot is at the bottom of the picture and the three lines showing the field of view and the center line).

Quite a few sample photos were taken, where all the parameters (d, α and β) were known to us and increased gradually in small increments. The photos were then analyzed in order to get an identification of the functions that relate the parameters in question to some known values, such as the height of the T-shirt etc.

The image recognition process can be divided into two stages in our case; during the first stage the T-shirt should be found in the image and during the second stage the values of the parameters should be estimated.

3.3 First stage: image recognition

In order to find the T-shirt in the image the following steps were followed:
- The original colour image (**Fig.3**(a)) was divided into three images containing its R, G and B components.
- Noise was filtered in every component image separately ((**Fig.3**(b,c,d,) showing R, G, B components, respectively).
- The sum of the green and blue component images (**Fig.3**(e)) was subtracted from the red component image (**Fig.3**(b)). As a result of this operation (**Fig.3**(f)), only the pixels that had a high red value and low green and blue values, received a high resulting value.
- A threshold function was applied to the resulting image. The threshold value was calculated according to Otsu's method.
- The image was converted to B&W using the obtained threshold value (**Fig.3**(g)).
- The white regions (corresponding to red objects) in the image were labeled

- The areas of the white regions were compared, it was assumed that the region with the highest area corresponded to the T-shirt. As a result, an image containing only the highest area region was constructed (**Fig.3**(h))

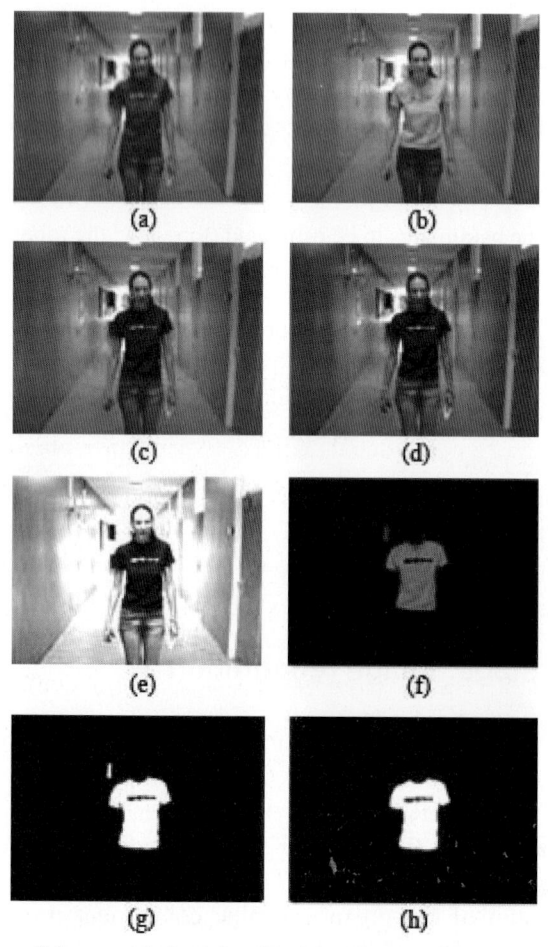

Fig. 3. (a) One of the sample images, (b), (c), (d), (e), (f), (g), (h) – various stages of image processing, explained in the text above

3.4 Second stage: calculating the parameters' values

The recognized images were then used to find functions that approximated the relationship between the variables d, α and β and the known values, such as the real T-shirt width and the recognized images' features. The resulting function for d is presented below (Equation (1)):

$$d = 7.048 \cdot \left(\frac{h_i}{h_r}\right)^{-1.007} \quad (1),$$

where h_i is the T-shirt height in the image in pixels and h_r is its real height in *cm*. If the resulting distance is bigger than some reasonably high value, for example, 30 m, it is supposed that there is no T-shirt in the image, because other red objects may be responsible for that result.

The approximating function for α is as follows (Equation (2)):

$$\alpha = \frac{180°}{\pi} \tan^{-1}\left(\frac{1.2\left(C_x - \frac{1}{2}S_x\right)}{8.3d^3 - 113.07d^2 + 404.82d}\right) \quad (2),$$

where C_x and S_x are properties of the recognized image, namely the centroid and the resolution in the *x* axis.

In order to get a good approximation for the angle *β*, a 3D model of the T-shirt had to be constructed because the function obtained using the real T-shirt images had a big error due to the lack of uniformity. The function obtained using a 3D-model of the T-shirt can be seen above (Equation (3)), this function was created without taking into account the orientation:

$$\beta_* = -1165.6\left(\frac{w_i}{h_i}\right)^4 + 1746\left(\frac{w_i}{h_i}\right)^3 - 66.38\left(\frac{w_i}{h_i}\right)^2 - 934.6\left(\frac{w_i}{h_i}\right) + 413.9 \quad (3),$$

where w_i is the T-shirt width in the image. After calculating β_* the proper orientation should be set using the image properties *Orientation* and *EulerNumber*. If *Orientation*>=0 then $\beta = -\beta_*$, and if *EulerNumber=1* (in our case it means that the picture on the T-shirt cannot be seen, because the person is not facing the robot) then $\beta = \beta_* + 180°$. Otherwise $\beta = \beta_*$. But even after getting a better approximation using a 3D model, the resulting error was a bit large, and it was decided to round the resulting angle *β* to a 45° precision that seems usable for our type of application.

3.5 Experimental results and discussion

Some new photos were taken, which were different from the ones used for creating the approximating functions. These photos were then fed to our recognition program and the results are presented in the **Table 1**.

Table 1.

Photo #	Real d,m	Est. d, m	Real α	Est. α	Real β	Est. β
1	2.5	2.25	0°	2.08°	0°	0°
2	2.5	2.51	-21°	-24.85°	0°	0°
3	2.5	2.38	-21°	-25.1°	-90°	-90°
4	5.5	5.82	0°	0.8°	0°	0°
5	5.5	6.13	20°	-23.17°	90°	90°

It may be stated that the results are satisfactory for our type of application that proves the validity of the model.

4 Conclusions

A simple "T-shirt based" image recognition system for the humanoid robot Rh-0 is presented in this paper. While this system may not be very important in the future development of humanoid robots by itself, it is a good example of simple yet effective systems that work, and as a first step in the development of our vision system it proved to be quite adequate.

Future work includes the addition of a module that will permit calibration of the model for several types of T-shirts, not necessarily red ones, and to associate every T-shirt with a specific person for use in further interaction. Also it has been decided to develop a more sophisticated vision system that could take obstacles into account along with navigation and path planning.

References

1. Sakagami Y. et al. (2002) *"The intelligent ASIMO: System overview and integration"* Proceedings of the 2002 IEEE/RSJ Intl. Conference on Intelligent Robots and Systems, EPFL, Lausanne, Switzerland, pp.2478-83
2. Kaneko K. et al. (2002) *"Design of Prototype Humanoid Robotics Platform for HRP"* Proceedings of the 2002 IEEE/RSJ Intl. Conference on Int. Robots and Systems, EPFL, Lausanne, Switzerland, pp.2431-36
3. Yokoyama K. et al. (2003) *"Cooperative Works by a Human and a Humanoid robot"* Proceedings of the 2003 IEEE International Conference on Robotics and Automation, Taipei, Taiwan
4. Sabe K. et al. (2004) *"Obstacle Avoidance and Path Planning for Humanoid Robots using Stereo Vision"* Proceedings of the International Conference on Robotics and Automation (ICRA 2004), New Orleans
5. González R.C. *"Digital Image Processing Using Matlab"* Pearson/Prentice Hall, 2004

Object Shape Characterisation using a Haptic Probe

O.P. Odiase*, R.C Richardson

School of Computer Science, University of Manchester, Oxford road, Manchester, M13 9PL, *corresponding author: O.P.Odiase@cs.man.ac.uk

Abstract

One of the fundamental challenges in robotics is object detection in unstructured environments where objects' properties are not known a priori. In these environments sensing is susceptible to error. Conventional sensors, for example sonar sensors, video cameras and infrared sensors, have limitations that make them inadequate for successful completion of outdoor tasks where vision is partially or totally impaired. For instance, in harsh atmospheric conditions, the poor image quality that video cameras suffer and the low spatial resolution sonar sensors exhibit make them unsuitable for object detection. Therefore, we propose an alternative approach for object characterisation in unstructured environments using active touch. A simulated anthropomorphic haptic probe system with 3 DOF and a force sensor was used in this approach. The system measures and analyses forces at contact points, and data useful for object characterisation were realised. Two shapes, a sphere and a cuboid, were modelled and used in the object detection simulation and satisfactory results were obtained.

Keywords: Robots, active touch, anthropomorphic probe, object detection, sensors.

1 Introduction

For robots to achieve a substantial degree of autonomy, it is crucial that they recognise and characterise objects they encounter [1, 2]. Outdoor applications, such as remote planetary explorations, search and rescue operations, and pipe and duct inspections involve recognition of objects with undefined features [3]. These environments require robots to adapt to the prevalent atmospheric conditions and to the surfaces of objects they encounter [4, 5].

Conventional sensors, for example sonar sensors, video cameras, infrared sensors and laser range finders, have limitations that make them better suited for characterising recognisable objects in structured environments where information on objects present and/or the atmospheric condition is available [6]. These limitations include the poor image quality that video cameras suffer in harsh atmospheric conditions [7], the low spatial resolution sonar sensors exhibit and the non-linear amplitude response of infrared sensors [8, 9]. In addition, these sensors are unable to determine object compliance, that is, the tendency of an object to move due to the effects of forces applied to it, for instance bending and twisting[10].

Multiple sensing methods can be used in a system [7]. The Haptic Vision system [6] combines active touch and video cameras but significantly relies on the video camera to extract and model the object's geometric properties. Consequently, the object detection process is prone to errors in unfavourable conditions. Another alternative is to integrate the Spatial-Filtering tactile sensor [11], but its delicate detector, which needs to be protected from physical damage, shock and chemical contamination, is unsuitable for use in unstructured environments.

As a solution, we propose a method for object perimeter and compliance detection using active touch. At the heart of this work is a simulated anthropomorphic haptic probe system shown in **Fig. 1.** Its main features are the three links each with a revolute joint, the 3 DOF and the force sensor.

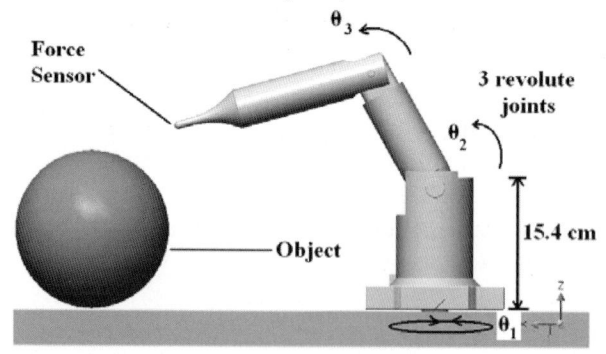

Fig. 1. Anthropomorphic Haptic Probe

MSC.visualNastran 4D [12], a software tool that automatically computes forces, torques, friction, velocity and acceleration was used for the simulation. It simulates contacts, collisions, and friction, and calculates loads and stresses throughout the assembly taking into consideration systems lags, gravity and environmental constraints.

2 Description of the haptic probe system

Fig. 2. shows a block diagram of the haptic probe system. It comprises modules for trajectory planning, 3D mapping, and position and interaction control. It applies the forward and inverse kinematics algorithms for the translation of joint angles to global positions and from global positions to joint angles respectively. The inputs to the system are the global *x y z* coordinates of the initial and final points of the intended path. These inputs are then used to generate the probe's trajectory.

When in operation, the system detects objects in its path and the force sensor measures the contact force which is resolved into its *x y z* components f_x f_y f_z. The position controller facilitates smooth motion along the path while the interaction controller compensates for the high forces and instability the system experiences during contact.

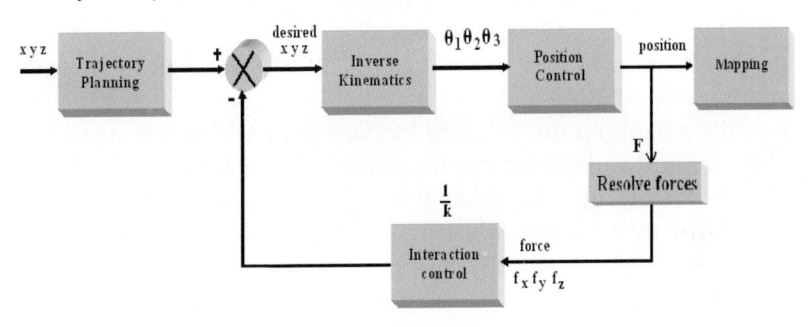

Fig. 2. Block diagram of the haptic probe system

The object detection process is a recurring cycle of trajectory planning, data acquisition and data analysis over a set period. The probe's trajectory in joint space and time is described in terms of joint angles, θ_1 θ_2 θ_3, derived from the *x y z* inputs which represent the orientations of the three links, the base, upper arm and lower arm respectively.

For a smooth point to point motion, the probe has to move from an initial to a final joint configuration in a given time *t*, thus generating four constraints on $\theta_i(t)$: the initial and final positions and velocities. These constraints are satisfied by a cubic polynomial shown in Equation (1) and each joint's velocity and acceleration are derived using Equations (2) and (3) respectively [13].

$$\theta_i(t) = a_0 + a_1 t + a_2 t^2 + a_3 t^3 \qquad (1)$$
$$\dot{\theta}_i(t) = a_1 + 2a_2 t + 3a_3 t^2 \qquad (2)$$

$i = 1,2,3$
$$\ddot{\theta}_i(t) = 2a_2 t + 6a_3 t \qquad (3)$$

By solving for the coefficients a_0, a_1, a_2 and a_3 the trajectory connecting any initial joint angle to a desired final angle in time t is realised. A typical trajectory profile of a joint moving from 45° to 135° in 0.4 seconds is shown in **Fig. 3**.

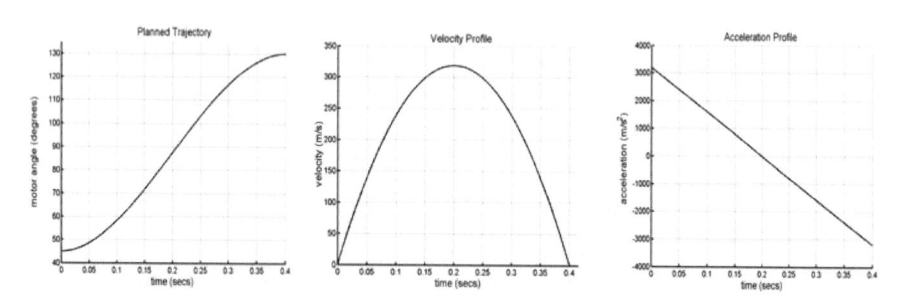

Fig. 3. Typical trajectory profiles of the haptic probe

The joint's angle moves smoothly and half way through its transition it attains its peak velocity and zero acceleration. Contact at this point will produce the maximum force sensed by the sensor which is at its maximum momentum mv. To minimise this force, the distance between the probe and the object needs to be adjusted so that contact is detected at the earlier or later stages of its transition when its velocity is minimal. This interaction between the probe and the object is controlled by the position based impedance controller which allows the probe's position and force relationship to be specified using Equation (4) [14].

$$x_d = (x - x_c), \quad x_c = \frac{F_x}{K_p} \qquad (4)$$

where x_c is the required change in position in response to the force F_x in the x axis, and K_p is the stiffness of the probe (see **Fig. 4.**). Thus a relationship between the probe's stiffness and that of the object can be derived.

Information on the object's material property can be inferred from the effect the contact force has on it thereby enabling it to be categorised by its mechanical behaviour [15]. One such category is tensile materials, which tend to resist being pulled upon or pushed. This behaviour can be investigated using stress-strain tests where stress σ is the force per unit area at a given cross section while strain ε is the proportional change in the dimensions of the material as a result of the applied force [16].

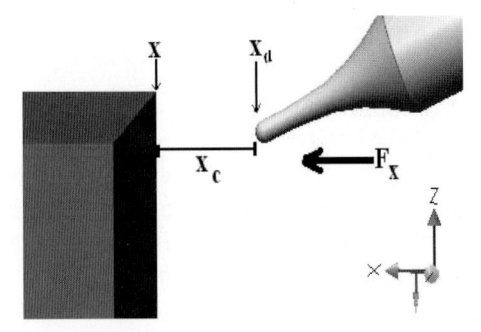

Fig. 4. Position-based impedance control scheme

The amount of internal stress in a material is proportional to its strain and is given by:

$$E = \frac{\sigma}{\varepsilon}, \text{ where } E = \text{Young's Modulus of Elasticity} \quad (5)$$

At yield stress, the material deforms permanently; ductile materials will stretch and brittle materials will either crack or break. Measuring and analysing the stress and strain experienced during contact will provide information on the objects material properties.

3 Object detection simulation

The ultimate objective of this simulation is to determine the identity and location of an object primarily by using active touch with the minimal number of measurements. The haptic probe, in this case stationary, begins its search with no more information other than there is an object in the environment and is one of the two shapes modelled; a sphere or a cuboid. The operating area was restricted to reduce the probe's blind search time.

Points in space were chosen at random (spatially sampled) which were either sampled coarsely where the points chosen are far apart or finely with points much closer. The finer the points, the better the information obtained about the object shape and material structure. The disadvantage in fine sampling is that it takes a longer time to explore the object's surface and hence requires more energy.

The system implemented the blind search by initially sampling points coarsely until contact has been made at which point sampling becomes finer. Mapping at this stage was achieved by plotting points where contact was detected.

4 Simulation Results

Typical force values at point where contact was detected on the sphere and cuboid are shown in **Table 1.** and **Table 2** respectively. **Fig. 5.** shows the plots of the x and z coordinates (side view) of the contact points on the sphere and cuboid. The measure of the forces detected is an indication of the stiffness of the object and hence useful for determining the objects material property. The force experienced by the sphere was distribution on the $x\,y\,z$ axes while that of the cuboid was predominantly on the x axis and seldom on the z axis.

Table 1. Typical contact force distribution on the sphere

	Coordinates (m)			Forces (N)		
	x	y	z	F_x	F_y	F_z
1	0.3228	-0.0135	0.1581	-0.51	-0.09	0.39
2	0.3306	0.0743	0.1091	-1.27	1.35	0.16
3	0.3531	-0.0831	0.1401	-0.46	-0.81	0.39
4	0.3305	0.0747	0.1094	-0.88	0.94	0.12
5	0.3606	-0.0806	0.1525	-0.03	-0.06	0.04

Table 2. Typical contact force distribution on the cuboid

	Coordinates (m)			Forces (N)		
	x	y	z	F_x	F_y	F_z
1	0.35.55	-0.0816	0.1722	3.96	0.0	0.00
2	0.35.25	0.0597	0.1009	2.19	0.0	0.00
3	0.34.85	-0.0611	0.1199	0.35	0.0	0.00
4	0.35.81	-0.0684	0.1974	0.54	0.0	-1.47
5	0.38.82	-0.0281	0.1981	0.0	0.0	-1.42

Other methods for point sampling include approaching the object from a random direction or sliding along the object or preferable moving outside the operating area and approaching the object in a direction orthogonal to the preceding measured contact. In this case, the directions of the probe's approaches were limited by its stationary nature.

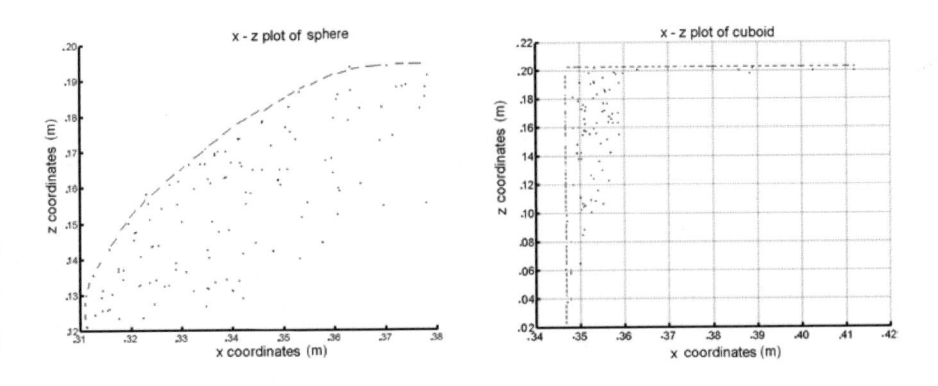

The dashes illustrate the perimeter of the object

Fig. 5. Plots of x and z coordinated of points where contact was detected

5 Conclusion

Conventional sensor technology can at best accurately detect an object's perimeter in structured environments. These sensors provide little or no information about the object's material properties or compliance and do not miniaturise well. An alternative approach has been demonstrated which is know as active touch.

Models of a sphere and a cuboid were placed in the operating area of the haptic probe and used in the object detection simulation. Initial results were satisfactory and showed that the sphere's perimeter was readily detected when compared with the cuboid.

Although perimeter detection was achieved, the fixed nature of the probe restricted its versatility in approaching the object from different directions. Further work will seek to demonstrate its use when mounted on a mobile platform. An adaptive trajectory generation algorithm, based on the available data, would also be developed to minimise the probing effort required to build highly diagnostic 3D map that includes compliance information. Furthermore, work would be done in measuring and analysing the stress and strain experienced during contact with a view to establishing a relationship between the probe's stiffness and the stiffness of the object.

References

1. H. R. Choi, J. H. Kim, and S. R. Oh, "Determination of Three Dimensional Curvature of Convex Object via Active Touch," in *Proc. IEEE Intl. Conf. on Intelligent Robots and Systems*. Victoria, B.C., Canada, 1998, pp. 1652-1657.
2. H. Maekawa, K. Tanie, K. Komoriya, M. Kaneko, c. Horiguchi, and T. Sugawara, "Development of a Finger-Shaped Tactile Sensor and its Evaluation by Active Touch," in *Proc. IEEE Intl. Conf. on Robotics and Automation*. Nice, France, 1992, pp. 1327-1334.
3. R. D. Howe, "Tactile Sensing and Control of Robotic Manipulation," *Journal of Advanced Robotics*, 1994.
4. A. M. Okamura, M. L. Turner, and M. R. Cutkosky, "Haptic Exploration of Objects with Rolling and Sliding."
5. J. L. Schneiter, "An Objective Tactile Sensing Strategy for Object Recognition and Localisation." Cambridge, Massachusetts, 1986, pp. 1262-1267.
6. H. T. Tanaka and K. Kushihama, "Haptic Vision: Vision-Based Haptic Exploration," in *IEEE*, 2002, pp. 852-855.
7. G. Dudek, P. Freedman, and I. M. Rekleitis, "Just-in-time Sensing: Efficiently Combining Sonar and Laser Range Data," in *Proc. of IEEE Intl. Conf in robotics and Automation*, vol. 1, 1996, pp. 667-671.
8. G. Benet, F. Blanes, J. E. Simo, and P. Perez, "Using Infrared Sensors for Distance Measurement in Mobile Robots," *Robotics and Autonomous Systems*, pp. 1-12, 2002.
9. F. Blanes, G. Benet, P. Perez, and J. E. Simo, "Map Building in an Autonomous robot using Infrared Sensors," 2002.
10. P. J. McKerrow, *Introduction to Robotics*: Addison-Wesley Publishing Company Inc., 1995.
11. M. Shimojo and M. Ishikawa, "An Active Touch Sensing Method Using a Spatial Filtering Tactile Sensor," 1993.
12. MSC.Software, "MSC.visualNastran.4D," 7.2.1 ed, 2004.
13. J. J. Craig, *Introduction to Robotics - Mechanics and Control*, third ed: Pearson Prentice Hall, 2005.
14. R. C. Richardson, "Actuation and Control for Robotic Physiotherapy," in *School of Mechanical engineering*. Leeds: University of Leeds, 2001, pp. 139-142.
15. M. Beals, L. Gross, and S. Harrell, "Spider Silk: Stress-Strain Curves and Young's Modulus," 1999.
16. P. J. McKerrow, *Introduction to Robotics*: Addison-Wesley Publishing Company Inc., 1995.

Detection of Landmines Using Nuclear Quadrupole Resonance (NQR): Signal Processing to Aid Classification

S. D. Somasundaram[1], K. Althoefer[1], J. A. S. Smith[2], and
L. D. Seneviratne[1]

[1] Dept. of Mechanical Engineering, King's College, The Strand,
London, WC2R 2LS. samuel.somasundaram@kcl.ac.uk
[2] Dept. of Chemistry, King's College, The Strand, London, WC2R 2LS.
john.smith@kcl.ac.uk

Abstract

Nuclear quadrupole resonance (NQR) is a sensor technology that measures a signature unique to the explosive contained in the mine, thus providing a means of efficiently detecting landmines. Unfortunately, the measured signals are inherently weak and therefore detection times are currently too long (especially for TNT-based landmines) to implement in a man-portable detection system. However, the NQR hardware is light enough to be integrated into a robot based system. This paper investigates several power spectrum estimation algorithms applied to NQR signals in order to distinguish between data containing signals from explosive and data that does not.

Keywords: NQR, Explosives Detection, Power Spectrum Estimation.

1 Introduction

1.1 The Problem

According to the "International Campaign to Ban Landmines", there may be more than 100 million landmines buried in more than 80 countries and more are still being laid. They estimate that if the current rate of demining remains, it will take over 500 years and US$33billion to clear them. This has led to an increase in demining research, especially in the areas of robotics and sensors in an attempt to automate the mine detection and removal process.

1.2 Why NQR?

Many current landmine detector technologies, such as metal detectors and ground penetrating radar (GPR), suffer from a high false alarm rate as they detect secondary features not specific to landmines [1, 2]. However, NQR detectors measure signals that are unique to the explosive contained within the landmine, thus providing a way to reduce false alarm rates. Further, current methods of detecting landmines have serious disadvantages; metal detectors, for example, have difficulties in magnetic soils and with mines of low metal content, and GPR in clay or wet and conducting soils and with mines very close to the ground surface. Although mine detection using NQR faces problems with interference and very low signal-to-noise ratio (SNR), recent reports indicate that the unique NQR signature offers an exceptionally high probability of detection [2–5].

1.3 NQR

NQR is a radio frequency (RF) technique which may be used to detect the presence of the ^{14}N nucleus. The majority of explosives found within landmines are nitrogen-rich (e.g., RDX, TNT and PETN). The ^{14}N nucleus has spin quantum number, $I = 1$, and thus behaves as though it has a non-spherical charge distribution. This results in the nucleus possessing an electric quadrupole moment, which when placed in a non-zero electric field gradient (EFG) means some nuclear orientations are more energetically favourable than others. The EFG at the nucleus is due to neighbouring charges, electrons and protons, and is therefore directly related to the structure of the compound. Applying radiation to the sample at the correct frequency drives transitions between the resultant quadrupolar energy levels and once the applied radiation is turned off, the signal produced by the sample is monitored. Two types of signal are often measured when NQR is used to detect explosive compounds. The first, called the free induction decay (FID) is the signal monitored immediately following a pulse. The second, is an echo train, which is produced by subjecting the sample to a series of pulses. Unfortunately, after an FID or echo train has been acquired a wait time (up to 30 seconds for TNT) must be adhered to in order to allow the ^{14}N spins to "relax" and return to equilibrium. Furthermore, the measured signal is weak and corrupted by noise and interference, so many measurements must be averaged before one can say whether explosive is present or not. The need to average many measurements combined with the restriction to "wait" in between measurements ultimately means that with current technology, implementation of a man-portable detector is not feasible. Implementation on a robotic platform is the long-term aim of the research at King's College. The envisaged robot system will place the antenna over the area to be investigated until enough data is collected to decide whether a mine is present or not. Once a decision has been made, the robot could mark the area (physically or electronically) before moving to the next area to be tested. An intelligent enough robot may even be able to disarm the mine. Such a robot is expected to survey a minefield without human intervention.

At present, there are no commercially available NQR landmine detectors and therefore, the equipment required is relatively expensive compared to, for example, metal detectors. The cost of a laboratory spectrometer and the associated hardware is currently around US$30,000. However, this cost could be reduced if one were to build a dedicated NQR landmine detector which need only detect certain materials.

In this paper, a Discrete Fourier Transform (DFT) based algorithm [3], the periodogram [6,7], an autoregressive (AR) spectral estimation based algorithm, using four different approaches to estimate the AR parameters (Yule-Walker, covariance, Burg and modified covariance) [6–9] and an eigendecomposition based spectrum estimator called "MUltiple SIgnal Characteristion" (MUSIC) [6–8] are applied to shielded NQR data to show how the classification of NQR data (into a set that contains explosive and a set that does not) can be facilitated. We note that, in the field, RFI becomes a limiting factor, especially for the detection of TNT, as the NQR signal lies in the AM radio band. There are two main approaches to RFI cancellation, passive and active. Passive techniques use specially designed antenna called gradiometers to cancel farfield RFI [2,5,10], but with the disadvantage that the SNR of the NQR signal is reduced as compared to using a simple antenna. Active techniques use a second set of reference antenna to measure the non-stationary RFI and apply adaptive noise cancellation techniques to mitigate the RFI [7,11].

2 Detection Algorithms based on Power Spectrum Estimation

2.1 Data model

An FID can be well modelled as a sum of damped sinusoids [12],

$$y(t) = \sum_{k=1}^{d} \alpha_k e^{(iw_k - \beta_k)t} + w(t), \qquad (1)$$

where α_k, w_k and β_k represent the (complex) amplitude, frequency and damping of the kth sinusoidal component, and $w(t)$ represents an additive coloured noise. The number of sinusoidal components, d, is equal to the number of resonant NQR lines and is dependent upon the excitation frequency, excitation bandwidth, the pulse width and the impulse response of the receiver hardware. We have developed a model of an echo, which may be viewed as a back-to-back FID, and is defined as

$$y(t) = \sum_{k=1}^{d} \alpha_k e^{iw_k t - \beta_k |t - t_0|} + w(t), \qquad (2)$$

where t_0 is the echo centre, which may be determined from knowledge of the multiple pulse-sequence parameters.

Clearly, one could estimate the power spectrum, to find the distribution of power over frequency, and use this as the basis for a detector. Power spectra can be estimated in a number of ways and the topic has received tremendous attention in recent decades [6]. There are two main categories, non-parametric and parametric. Non-parametric techniques, such as the periodogram, make no assumption about the underlying process, whilst parametric techniques, such as the AR techniques and MUSIC, allow us to incorporate information about the process in the spectrum estimation algorithm, in the hope of obtaining a higher resolution spectrum. The AR techniques model the NQR signal as the output of a causal, linear time-invariant filter driven by white noise. This is a reasonable choice as the AR equation may model spectra with narrow peaks [6]. Once the model has been constrained to be an AR model, the user must choose the order of the AR model and choose the technique to estimate the AR parameters. The estimated parameters are then inserted into the parametric form of the spectrum.

MUSIC is a subspace technique that assumes the NQR signal can be modelled as a sum of complex exponentials in white noise, and is based upon eigendecomposition of the autocorrelation matrix into two subspaces, the *noise* subspace and the *signal* subspace. Once these subspaces have been determined, a frequency estimation function is used to extract estimates of the frequencies [8].

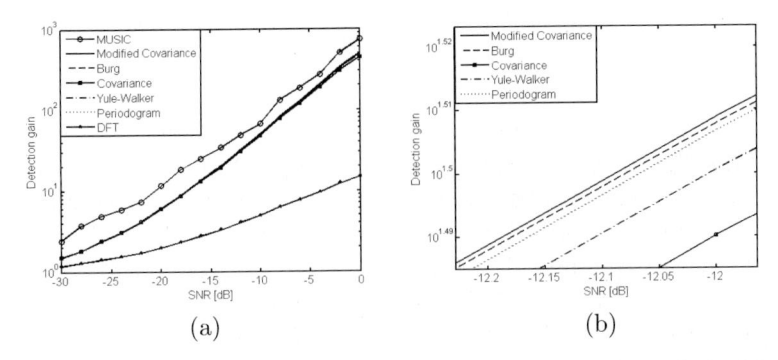

Fig. 1. Detection gain versus SNR for, a) all the detectors, and b) the Modified Covariance, Burg, Covariance, Yule-Walker and Periodogram detectors (zoomed portion).

It is worth noting, that MUSIC and the chosen AR techniques can be evaluated using computationally efficient algorithms such that the computational burden of computing the power spectra is not a limiting factor.

Once the power spectrum has been obtained, the detection variable is chosen as the integral, over a predetermined interval, of the spectrum. The interval is selected to include only the areas of the spectrum where one would expect to get power due to the NQR signal.

3 Numerical examples

Two sets of data, one containing signals from TNT and another not, were collected in a shielded environment. The mass of the TNT sample used was 180g, which is similar to the masses found in anti-personnel (AP) mines. The sample was placed within a solenoidal coil within a Faraday shield. Each data file consists of 4 summed echo trains, which would take around one minute to acquire. The algorithms were evaluated using both real and simulated data.

Fig. 1 shows the detection gain of the detectors (expressed as a ratio between the detection thresholds for a sample containing TNT and for one without TNT) as a function of SNR for simulated data using 100 Monte-Carlo simulations. The figure shows that the detector with the highest gain is the MUSIC-based detector, whilst the DFT-based detector has the lowest gain. The rest of the detectors have similar detection gains to each other, with the modified covariance based detector having a slightly larger gain. From the figure we would expect the MUSIC-based detector

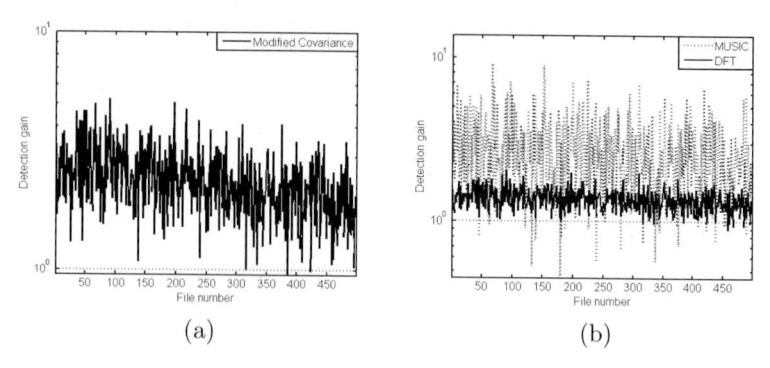

Fig. 2. Detection gain as a function of file number for a) the Modified covariance based detector, and b) the Music and DFT based detectors.

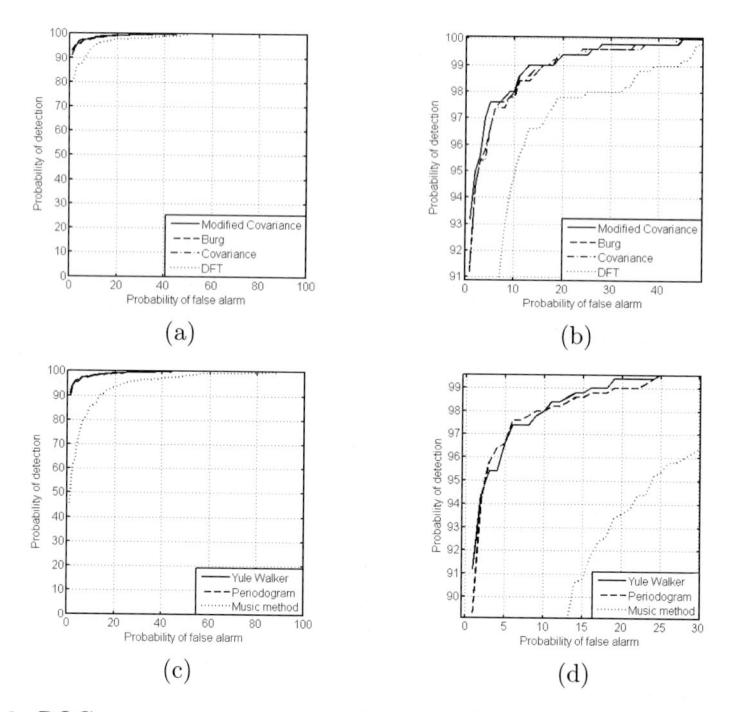

Fig. 3. ROC curves using real data for, a),b) the Modified covariance, Burg, Covariance and DFT detectors, and c),d) the Yule-Walker, Periodogram and Music detectors.

to perform the best. However, the figure only shows the average gain for each SNR; the variance of this gain is also important.

Fig. 2 illustrates the detection gain of the modified covariance, MUSIC and DFT based detectors, as a function of file number for 500 files each containing 4-summed echo trains. The other AR based detectors and the periodogram based detector have similar plots to **Fig. 2**a. The MUSIC detector has the largest average gain, but also has the largest variance.

As we are primarily concerned with detection, we now investigate receiver operator characteristic (ROC) curves for the detectors based on real data, illustrated in **Fig. 3**. The figure shows that, with the exception of the MUSIC-based detector, the various power spectral techniques have similar performances, being significantly better than the DFT-based method. It appears that the modified covariance method shows a slight improvement as compared to the other methods; this result should be expected given the preferable quality of the modified covariance method AR parameter estimates. The MUSIC-based detector, whilst having the largest average gain, is let down by the large variance of that gain.

4 Conclusions

In this paper, we evaluated the performance of average power detectors based on several power spectrum estimation techniques. Our results show that the modified covariance based detector performed best out of those compared. Future work includes incorporating techniques to cope with RFI, as discussed in Section 1.3.

5 Acknowledgement

The authors would like to thank Dr. Mike Rowe, Dr. Jamie Barras and Miss Elizabeth Balchin for their help in the NQR laboratory.

References

1. Somasundaram, S. D., Smith, J. A. S., Althoefer, K., and Seneviratne, L. D. (2004) Detection of Landmines Using Nuclear Quadrupole Resonance (NQR): An Overview. *HUDEM Conference paper.*
2. Garroway, A. N., Buess, M. L., Miller, J. B., Suits, B. H., Hibbs, A. D., Barrall, A. G., Matthews, R., and Burnett, L. J. (June 2001) Remote Sensing by Nuclear Quadrupole Resonance. *IEEE Trans. Geoscience and Remote Sensing*, vol. 39, no. 6, pp. 1108–1118.

3. Smith, J. A. S. (1995) Nitrogen-14 Quadrupole Resonance Detection of RDX and HMX Based Explosives. *European Convention on Security and Detection*, vol. 408, pp. 288–292.
4. Rowe, M. D. and Smith, J. A. S. (1996) Mine Detection by Nuclear Quadrupole Resonance. *The Detection of Abandoned Landmines (IEE) Eurel*, vol. 43, pp. 62–66.
5. Deas, R. M., Burch, I. A., and Port, D. M. (2002) The Detection of RDX and TNT Mine like Targets by Nuclear Quadruple Resonance. In *Detection and Remediation Technologies for Mines and Minelike Targets, Proc. of SPIE*, vol. 4742, pp. 482–489.
6. Stoica, P. and Moses, R. (1997) *Introduction to Spectral Analysis*. Prentice Hall, New Jersey.
7. Tan, Y., Tantum, S. L., and Collins, L. M. (2002) Landmine Detection with Nuclear Quadrupole Resonance. *Geoscience and Remote Sensing Symposium*, vol. 3, pp. 1575–1578.
8. Hayes, M. H. (1996) *Statistical Digital Signal Processing and Modeling*. John Wiley and Sons, Georgia.
9. Cervantes, H. R. and Rabban, S. R. (1999) Application of autoregressive spectral estimator in 2D NQR nutation spectroscopy. *Solid State Communications*, vol. 110, pp. 215–220.
10. Suits, B. H. and Garroway, A. N. (2003) Optimising surface coils and the self-shielded gradiometer. *Journal of Applied Physics*, vol. 94, pp. 4170–4178.
11. Jiang, Y., Stoica, P., and Li, J. (2004) Array Signal Processing in the Known Waveform and Steering Vector Case. *IEEE Trans. Signal Processing*, vol. 52, no. 1, pp. 23–35.
12. Malcolm-Lawes, D. J., Mallion, S., Rowe, M. D., and Smith, J. A. S. Time-Domain Data Analysis of NQR Response. *Patent application number 9915842.0.*

Software and Computer-aided Environments

Simulator for Locomotion Control of the Alicia³ Climbing Robot

D. Longo, G. Muscato, S. Sessa

Università degli Studi di Catania - D.I.E.E.S.
V.le A. Doria 6 95125 Catania - ITALY
dlongo@diees.unict.it, gmuscato@diees.unict.it, ssessa@diees.unict.it

Abstract

The climbing robot Alicia³, is mainly made of three identical modules linked together by two arms actuated by two air pistons.

Each module consists of a cup, an aspirator, two wheels that are actuated by two DC motors with encoders and gearboxes, and four passive wheels to guarantee plain contact of the system on the wall.

The cups can slide over a wall by mean of a special sealing maintaining the vacuum inside the cup and at the same time creates the right amount of friction according to the system weight and the target wall kind.

This paper focuses on description of Alicia³ simulator, implemented in Simulink®, which allows testing locomotion and finding an adequate control parameters setting.

A viewer, developed in Delphi™ 7 using Mex-function to integrate it in Simulink®, is used in order to visualize Alicia³ in 3D and the sequence of operations executed to follow the reference trajectory.

Keywords: Climbing robot, maintenance and inspection, locomotion control, robot simulator.

1 Introduction – Description of Alicia³ Robot

Climbing robots are useful devices that can be adopted in a variety of applications such as maintenance, building, inspection and safety in the process and construction industries. Possible applications of these systems are inspection of external/internal surface of aboveground/underground petrochemical storage tanks [1], concrete walls and metallic structures [2] [3] [4] [5].

The structure of the Alicia II module currently comprises three concentric PVC rings held together by an aluminium disc. The bigger ring and the aluminium disc have a diameter of 30 cm. The sealing system is allocated in the first two external rings. The third ring (the smallest one) is responsible for semi-rigid contact between the robot and the wall, using a special kind of spherical ball bearing.

An aspirator is used to depressurize the cup formed by the rings and the sealing, so the whole robot can adhere to the wall like a standard suction cup. While the system has to move over the target surface, the cup must not adhere with a high degree of friction, so a particular kind of sealing is required between the wall and the robot. The sealing must guarantee negative internal pressure and should allow the robot to pass over small obstacles like screws or welding traces. This configuration allows the robot to climb a vertical surface with a minimum curvature radius of 1.8 m.

The total weight of the module is 7 kg. The DC motors/gearboxes used in each module are able to move a mass of up to 15 kg in a vertical direction at a maximum speed of 2 m/s.

Fig. 1. The Alicia^3 robot

The basic idea for the Alicia^3 robot (**Fig. 1.**) is to use three of these modules to allow the whole system to better deal with obstacles on the target surface. When obstacles higher than 1 cm are encountered, the base module fails to pass over them (this limitation is due to the height of the cup sealing that cannot be higher than a few centimetres). In this case, the Alicia^3 robot can pass over the obstacle by detaching the three modules one by one, although it does so at a lower speed. The structure has been designed in such away that only two modules at a time can support the weight of the whole robot. The two links between the three modules are actuated with two pneumatic pistons [6].

2 Alicia³ Simulator – global description

The Alicia3 simulator developed in Simulink® allows testing locomotion strategies, but can be also used to tune parameters for locomotion control for the entire robot, as well as for the base system module Alicia II (Fig. 2.).

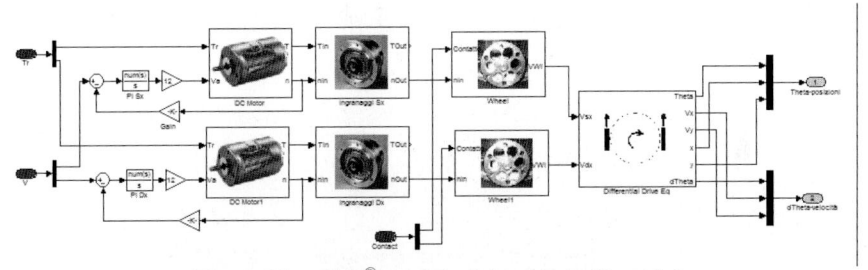

Fig. 2. Simulink® model of the Alicia II module

Fig. 3. Graphical 3D viewer

A graphical 3D viewer, developed in Delphi™ 7 using the GL-Scene free package and a Mex-function to integrate it in Simulink®, visualize Alicia³ sequence of operations (**Fig. 3.**). Basic blocks that represent the behaviour of the main system components have been implemented. Connecting together all these blocks, a realistic model of the Alicia³ robot behaviour has been obtained. The implementation of these basic blocks will be described afterwards. For each block, a set of parameters has been defined and can be adjusted according to system components. In the Simulink® diagram there is the possibility to modify all these parameters. The control algorithms implemented in the simulator are the same of that of the real robot.

2.1 DC motor and Gearboxes blocks

The first block uses the well-known permanent magnets DC motor equations:

$$\Omega = \frac{\frac{K}{K_v}}{1+s\frac{R_a I_m}{K_t K_v}} V'_c - \frac{\frac{R_a}{K_v K_t}}{1+s\frac{R_a I_m}{K_t K_v}} \tau_r \quad (1)$$

with hypothesis that:

$$T_e << T_m$$

where:
- Ω : Laplace transform of angular speed;
- V'_c : Laplace transform of coil voltage;
- τ_r : Laplace transform of torque;
- K : Transformation constant;
- K_v : Voltage constant;
- K_t : Torque constant;
- R_a : Coil resistance;
- L_a : Coil inductance;
- I_m : Inertial moment;
- T_e : Time constant of the electrical sub-system;
- T_m : Time constant of the mechanical sub-system.

This dynamical model uses a saturation at $\pm 12\,V$ on coil voltage to consider physical limits.

The second block represents the gearboxes module and uses two fundamental parameters:
- K_r : transmission ratio;
- η : transmission efficiency;

2.2 Wheel block

The wheel block examines the effect of diameter of the wheel on the robot movement and models the contact between wheel and wall.
In particular, the hypothesis used in contact model is that:

$$v = K_w n \quad \text{if there is a contact}$$
$$v = 0 \quad \text{if there is no contact} \tag{2}$$

where:
n : turns per second of the motor after gearboxes;
v : speed of the wheel;
$K_w = \pi \cdot d / 60$: wheel constant;
d : diameter of the wheel.

This model is very simple even if not perfectly adhering to reality, but allows understanding the control behaviour in the presence of no perfect contact between wheel and wall.

2.3 Differential drive block

Each robot module has two wheels in a differential drive configuration that is a non-holonomic constrain. The equations, in state form, that models this constrain are [7]:

$$\begin{cases} \dot{x}_c(t) = V \cos(\theta(t)) = \dfrac{v_d + v_s}{2} \cos(\theta(t)) \\ \dot{y}_c(t) = V \sin(\theta(t)) = \dfrac{v_r + v_l}{2} \sin(\theta(t)) \\ \dot{\theta}(t) = \omega = \dfrac{v_r - v_l}{b} \end{cases} \tag{3}$$

where:
- x_c, y_c : absolute coordinate of the centre of the robot;
- θ : inclination of the robot;
- v_l, v_r : left - right speed;
- b : wheelbase;

3 Control Algorithms for locomotion

A multi-level approach has been used to control the locomotion of the Alicia[3] robot. At the lowest level, there is a classical speed-position PID control of each wheel.

This type of control is inadequate to move the robot, because a disturbance (e.g. no contact wheel-wall) could cause a misalignment between modules that leads to mechanical stress on the structure.

To avoid these problems a cascade control loop that use inclination of the modules has been implemented; the orientation of each module is obtained by reading three inclinometers.

The basic movements that the robot can execute are:
- *Go forward (backward);*
- *Turn;*
- *Pass over an obstacle;*

For *go forward* movement the wheels have the same nominal speed reference, but the principal target is that the modules must be aligned, so that the nominal reference speed is modulated by the inclination of the module:

$$v_{fw} = v_{ref} \pm K_{fw}(\theta - \theta_i) \quad (4)$$

where:

v_{ref} : nominal reference speed;

v_{fw} : reference speed for PID wheel control

θ_i : initial inclination

θ : actual inclination

While the inclinometers used in real robot were affected by noise and offset problems, the use of the simulator permitted us to implement more robust sensor readings by simulating different kind of signal filtering.

The *turn* of the robot is divided into two steps: in the first one, the external modules turn 90° with respect to the central module; in the second one, speed reference for each wheel is determined from kinematic calculus. Every five seconds the first operation is repeated to avoid mechanical stress.

The *pass over an obstacle* is a complex operation that requires various steps:
- Anterior module is raised
- Robot goes forward with centre and posterior modules until the anterior module passes over the obstacle
- Anterior module is lowered

- Centre module is raised and posterior module goes forward
- Anterior and posterior modules go forward until the centre module passes over the obstacle;
- Posterior module passes over the obstacle as described for the anterior.

This operation does not require a particular control, but is important to correctly follow all the steps. (**Fig. 4.**)

Fig. 4. Steps sequence for passing over an obstacle

4 Conclusion

A modular approach has been used to design the simulator of the Alicia^3 robot, each basic block of the robot has been modelled and tested separately. The global system model was then built by putting together these basic blocks.

Every block on the simulator models a specific component; these blocks are then linked together to compose the whole system.

The advantage of this kind of software structure is that the block reuse and readability of the model is improved.

A robot simulator has been developed to allow parameter tuning, system and control performance test and mission trajectory planning. For example it is possible to simulate robot movement in a certain environment before performing the real mission. The virtual environment in 3D can be used during the training phase for technical staff that have to tele-operate the robot (for complex operations e.g. pass over an obstacle).

References

1. American Petroleum Institute (1992), "Tank inspection, repair, alteration and reconstruction", Standard 653, API, Washington, DC.
2. Berns K. and Hillenbrand C. (2003), "Climbing robots for commercial applications – a survey", Proceedings of the 6th International Conference on Climbing and Walking Robots CLAWAR 2003, 17-19 September, Professional Engineering Publishing, Catania, Italy, pp. 771-6.
3. La Rosa G., Messina M., Muscato G. and Sinatra R. (2002), "A low cost lightweight climbing robot for the inspection of vertical surfaces", Mechatronics, Vol. 12, pp. 71-96.
4. Schraft R.D., Simons F., Schafer T., Keil W. and Anderson S. (2003), "Concept of a low-cost, window-cleaning robot", Proceedings of the 6th International Conference on Climbing and Walking Robots CLAWAR 2003, 17-19 September, Professional Engineering Publishing, Catania, Italy, pp. 785-92.
5. Weise F., Kohnen J., Wiggenhauser H., Hillenbrand C. and Berns K. (2001), "Non-destructive sensors for inspection of concrete structures with a climbing robot", Proceedings of the 4th International Conference on Climbing and Walking Robots CLAWAR 2001, 24-26 September, Professional Engineering Publishing, Karlsruhe, Germany, pp. 945-52.
6. D. Longo, G. Muscato (2004), "A modular approach for the design of the Alicia3 climbing robot for industrial inspection", Industrial Robot: An International Journal, Vol. 31, N. 2, pp.148-158.
7. J. Borenstein, H.R. Everett and L. Feng, (1996) "Where am I? Sensors and methods for mobile robot positioning", University of Michigan.

A General Platform for Robot Navigation in Natural Environments

Enric Celaya, Tom Creemers, and José Luis Albarral

Institut de Robòtica i Informàtica Industrial (IRI), UPC-CSIC
Llorens i Artigas 4-6. 08028 Barcelona, Spain. email: celaya@iri.upc.edu

Abstract

A user interface for the remote control of vision-based robot navigation in previously unknown, indoor or outdoor environments has been developed. Visual feedback from the camera(s) of the robot is provided to the user, allowing him to select a visual target to be reached by the robot and launch an autonomous navigation process. Manual control can be taken back by the user at any time. The interface is built as a modular platform, capable of accommodating different types of robots and different algorithms for vision and navigation. There are hardware-dependent modules, whose purpose is to isolate the particularities of each robot from the rest of the control system, and must be implemented for each different robot to be used with the platform. Other modules are hardware-independent, and they implement generic tasks like landmark detection or path planning. Alternative versions of these modules can co-exist in the system, so that the user may select the preferred combination of them for each navigation task.

Keywords: Visual navigation, robot interface, unstructured environments.

1 Introduction

Legged robots excel in their capability to evolve in natural unstructured environments. This capability makes them suitable for non-conventional applications of mobile robots, such as intervention in catastrophic situations and dangerous areas, scientific missions in volcanoes, or planetary exploration. For this kind of applications traditional assumptions about robot navigation do not fit well. For example, the existence of a map of the navigation area will only be available in very exceptional situations, and the possibility of the robot building one itself is excluded since, in most cases, the task will require a single walk towards the goal, so that there is little opportunity to gather information about alternative paths. Also, the availability of global positioning systems can not be assumed, and even the supposition that navigation takes place on a flat surface must be often discarded. Yet the usual distinction between obstacles and free space blurs for a legged robot which is able to climb over many obstacles.

To fully take into account the peculiarities of legged robots we need a more flexible approach to navigation control that differs from what is the norm for conventional applications of wheeled robots. Assuming that no previous knowledge of the environment will be available, it makes little sense to specify the navigation target as a point of given coordinates in a reference system or a position in a map (simply because there is no map at all or, more realistically, because the operator lacks the required information). In this paper, we present a platform for the definition and control of navigation tasks that avoids the usual assumptions that have become a common place in many indoors navigation tasks.

2 Vision-based navigation

Probably, the most natural way to command a robot in an unstructured environment is by means of visual information, so that the user employs his own ability for interpreting the scene to direct the robot towards the places he selects in the environment. With this purpose, the software platform we have developed allows the definition and monitoring of the navigation task by means of the visual information acquired by cameras carried on the robot and made available to the user.

A key feature of the platform is its robot-independent design, so that the user can safely ignore the specific robot that is actually performing the navigation task, the kind of sensors it has, and even the locomotion system it uses (currently, we are using the same interface to control a legged and a wheeled robot). The only requirements the robot must fulfil to be used with this platform is to be provided with a camera, preferably mounted on a pan/tilt mechanism, proximity or contact sensors for obstacle detection, and some form of short term odometry measurement.

A typical navigation process using the visual interface would be as follows: After initiating the interface, the user selects the robot it wants to communicate with, and begins to receive images from its camera(s). He can look around the scene by using the pant/tilt capability of the camera or by manually driving the robot in the desired direction, until he eventually determines a desired target for navigation. Then he uses the mouse to select the desired target on the screen and an autonomous navigation process may be started in which the robot, using a navigation algorithm, will try to reach its goal. During this process, the user can simply watch how the navigation evolves, or suspend it at any time to manually correcting the robot's trajectory or providing a new navigation target.

A major issue in this process is the specification of the navigation target by the user: what for a human can be a perfectly identifiable object, can not be easily recognized by the robot's vision system. To solve this problem, we follow the approach of limiting the possible choices of the user to only image features that the vision system detected as identifiable landmarks. To do this, one or more landmark detection algorithms must be available and put in operation by the user. Detected landmarks are then presented on the screen so that the user can select the desired one with a simple mouse click. According to this, a landmark-based navigation algorithm must be available, able to drive the robot to the target landmark using only the information about the landmarks detected by the robot, the obstacles found on its path, and its odometry measurements.

3 A platform for robot navigation

To allow an easy integration of the different elements of the navigation system, we have designed a modular structure in which independent modules can be connected together with pre-defined communication interfaces between them. The purpose of the platform is not simply that of providing a general tool to facilitate the interface with the robot's sensors and actuators as do other tools, like the Player/Stage project [1]. Instead, our platform constitutes a tool specifically designed to perform vision-based navigation tasks, and defines an adequate interface and a modular structure to solve this problem in a well defined way.

The structure of our robot navigation platform consists of a main process, denoted RobotCtrl, that is under direct management of the user interface (GUI) and that centralizes the communication with all other modules in the system (**Fig. 1**).

We distinguish between hardware-dependent and harward-independent modules.

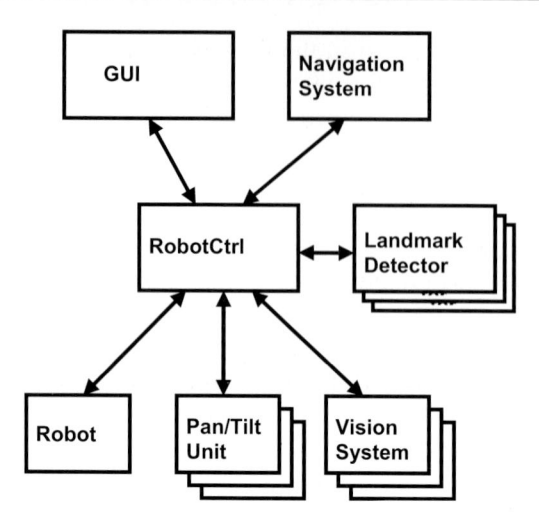

Fig. 1. General structure of the navigation platform.

3.1 Hardware-dependent modules

Hardware-dependent modules must be implemented once for each robot that has to be used with the platform. They provide the required functionalities using the specific sensors and actuators with which the robot is endowed. There are three hardware-dependent modules:

- The *Robot* module: Implements the translation from the generic commands for advance and turning into the appropriate ones required by this particular robot, as well as virtual sensors for collision detection and odometry measurements using the real sensors installed in the robot.

- The *Pan/Tilt Unit* module: It translates the generic commands for pan and tilt to those required by the particular pan/tilt device. One instantiation is required for each pan/tilt device mounted on the robot.

- The *Vision System* module: It translates the generic commands for image acquisition and camera settings to those appropriate for the particular camera. One instantiation is required for each camera.

3.2 Hardware-independent modules

Hardware-independent modules provide specific functionalities defined in the system whose implementation does not depend on a particular robot.

There can be several alternative implementations for each module, and the user will be able to select which of them will actually be used.

There are two main hardware-independent modules:

- The *Landmark Detector* module: It receives a camera image and returns a list of detected landmarks with a unique identification label and a relative position estimation. Several landmark detectors may be active at the same time for the same camera, according to the user choices.

- The *Navigation System* module: It receives as input the information about the landmarks detected by the vision system, the navigation target, and the odometry measurements. Its output is the instantaneous driving command to be executed by the robot.

4 The user interface for vision-based navigation control

The user interface has the aspect shown in **Fig. 2**. On the left part, there is a main window displaying a simulated *world view* representing the robot and its environment, and below it, a text window that displays messages informing about key events recorded by the system.

At the right part of the interface, there is a driving wheel for the control of the advance direction and a slider for the speed of the robot, as well as a security stop button. Below them, a set of option selectors allow the user to specify if the robot is manually driven or if an autonomous navigation process should temporarily take the control of it.

Besides the main interface window, one ore more camera windows (**Fig. 3**) can be used to monitor the images actually captured by the camera(s) carried by the robot.

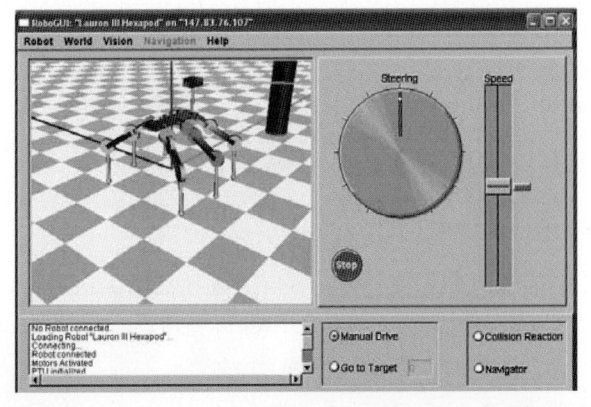

Fig. 2. The user interface for robot navigation con-

4.1 The world view

The purpose of the *world view* is to provide the user with a graphical representation of the current situation as perceived by the robot. The position of the robot in this simulated world corresponds to its estimated position obtained by odometry. Detected landmarks and obstacles are displayed at their estimated positions with respect to the robot. The user can choose what information is displayed in the *world view*: for example, the user may want to visualize only the currently observed landmarks, or all those that have been observed recently. The same applies for detected obstacles. There is also the possibility to introduce user-defined landmarks and obstacles that the robot will regard as if it was actually detecting them.

4.2 The camera view

A camera window (**Fig. 3**) allows the user to control the gaze direction of this camera by operating the controls for pan and tilt.

A landmark detection process can be activated for a camera by selecting it from a list of available landmark detectors. Detected landmarks are indicated with an overlaid mark in the image. The user can use the mouse to select a landmark as the navigation target by clicking on it. A landmark can also be chosen in the camera view as the camera target, in which case the camera starts a visual tracking of it using its pan/tilt capability.

4.3 Speed and direction control

No matter the shape of the terrain on which the robot navigates, it must follow the ground profile, and therefore, its possibilities of movement are restricted to two dimensions. The motion of the robot will be described as composed of trajectories along arcs of circumferences tangent to the x-axis

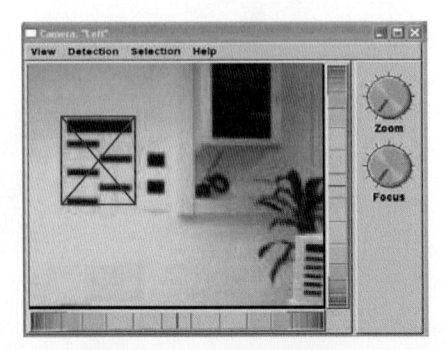

Fig. 3. Camera window

of the robot. Thus, its instantaneous motion is determined by its tangential speed *v*, and its angular speed *w*, which are related with the turning radius *R* by the relationship

$$v = R\,w. \tag{1}$$

A user-intuitive form of control is to use a driving wheel to set the turning radius, and a slider for the speed along the trajectory. A simple-minded approach would determine *R* according to the driving wheel angle α, and *v* according to the slider position *c*. The drawback of this approach is that when the radius *R* goes to 0, corresponding to a turn in place, for any value of $v \neq 0$, the rotation speed *w* grows to infinity, according to (1).

To avoid this problem, we use the following relationships:

$$w = c\,\sin(\alpha)\,/L$$
$$v = c\,(1-\sin(\alpha)),$$

where L is a constant that determines the influence of the steering angle in the turning radius and, intuitively, must be related to the size of the robot.

4.4 Autonomous Navigation.

Once a navigation target has been set, the user can launch an autonomous navigation process by selecting a navigation algorithm. The platform provides a very simple navigation algorithm (the *go to target* option) to drive the robot towards the currently perceived direction of the target. Such reactive strategy, which does not even use a map, should probably not be taken as a real navigation algorithm. Nonetheless, for many easy environments, it is enough to drive the robot to its target.

5 Navigation experiments

The platform has been tested with robots as diverse as a four wheeled Pioneer II endowed with two low resolution b/w cameras, and a six-legged Lauron III robot [2] with a colour camera. Some navigation tests were performed to compare the behaviours of both robots in the same environment in which we distributed a number of artificial landmarks detectable by a specially designed vision algorithm. We disposed some obstacles that could not be traversed by the wheeled robot, but that the legged robot could negotiate walking over them. When both robots were confronted to the same navigation task, which involved reaching a goal behind one of the obstacles, both robots followed very different paths despite the fact that both were driven by the same navigation algorithm [3]. Thus, the wheeled robot detected an untraversable obstacle in its way towards the target and the navigation algorithm devised an alternative path around the obstacle. The legged robot, also reached the obstacle, but, instead of recognizing it as an obstacle, it used its climbing capabilities to walk over it. Since the robot succeeded in traversing the obstacle, it never informed to the navigation system of the presence of an obstacle, and it proceeded directly towards its goal without deviation.

6 Conclusions

We have developed a platform for the control of vision-based navigation in unknown, unstructured environments. The platform can be used with both legged and wheeled robots, allowing the user to ignore what kind of robot is actually performing the task. The addition of autonomous navigation capabilities to the platform allows the user to relinquish direct control of the robot as soon as a visual target has been selected. This facility results in less demanding conditions for the user, who can limit his intervention to eventual re-definitions of the current navigation target.

Acknowledgements

This work has been partially supported by the Spanish Ministerio de Ciencia y Tecnología and FEDER funds, under the project DPI2003-05193-C02-01 of the Plan Nacional de I+D+I.

References

1. B.Gerkey, R. Vaughan, and A. Howard. The Player/Stage Project: Tools for Multi-Robot and Distributed Sensor Systems. Proc Int. Conf. on Advanced Robotics (ICAR 2003), July 2003, Coimbra, Portugal. pp. 317-323.
2. Gaßmann B., Scholl K., Berns K.: Behavior Control of LAURON III for Walking in Unstructured Terrain, *Int. Conf. on Climbing and Walking Robots*, CLAWAR 2001, September 2001, Karlsruhe, Germany, pp. 651-658.
3. D. Busquets, C. Sierra, and R. López de Màntaras. A multi-agent approach to qualitative landmark-based navigation. *Autonomous Robots*, 2003, Vol. 15, pp. 129-153.

Simulations of the Dynamic Behavior of a Bipedal Robot with Torso Subjected to External Eisturbances

C. Zaoui[1], O. Bruneau[2], F.B. Ouezdou[3], A. Maalej[1]

[1]Laboratoire d'Automatique et de Dynamique des Systèmes, University of Sfax, Tunisia, Chiheb.Zaoui@ipeiem.rnu.tn
[2] Laboratoire Vision et Robotique, ENSIB, University of Orléans, France
   olivier.bruneau@ensi-bourges.fr
[3] Laboratoire d'Instrumentation et de Relations Individu Système, University of Versailles, France, ouezdou@liris.uvsq.fr

Abstract

This paper deals with the study of the stabilization of a robot subjected to external disturbances. Initially the stabilization is carried out with a trunk having 4 dofs (degrees of freedom), three translations and one rotation (design of robot ROBIAN). In the second time the stabilization is performed with a system with arms and having 10 dofs. At first, for a vertical posture of the robot, the trunk bodies of ROBIAN are used to compensate the external three-dimensional efforts applied to the robot. The study is based on the *General State Equation* formalism (GSE). During the simulation, this study allows us to determine on-line, the required movements and accelerations of the trunk bodies in order to maintain the robot stability. In the following stage, we study the required movements of a system which is made up of a trunk and two arms in order to ensure the robot stability in presence of disturbances or during a handling of an object. The same formalism is selected to study the dynamics of the new upper part of the robot. This study shows the importance of the arms for the robot stability.

Keywords: Bipedal robot, Simulation of the dynamic behavior, stabilization.

1 Introduction

For several years, many researches have focused on the importance of the upper part of the biped robots and on their stability role during walk or during tasks of handling. The dynamic behavior simulation of the movements of a human has been yet carried out in several studies. Zordan and Hodgins [1] proposed a dynamic controller to reproduce movements of the upper part of a human or the realization of certain tasks of handling, based on recorded data, starting from a system of motion capture. The study of capture was also carried out by Gravez [2] [3] for the dynamic simulation of the virtual mockup having a realistic trunk with 13 dofs. Masaki Oshita [4] presented an approach based on dynamics to control the movement of the virtual mockup. This method enables to control the movements of a virtual mannequin in the presence of external forces or of impulses due to collisions. Paolo Baerlocher and Ronan Boulic [5] have used the inverse kinematics for the control of a virtual mockup. They presented a recursive algorithm forcing an arbitrary number of strict priorities in order to carry out contradictory constraints on the virtual mockup. A study, using dynamic simulation, showed that we could reduce a realistic trunk having 13 dofs [6] to only 4dofs (one rotation and three translations) [7], keeping the same efforts applied to the lower part of the biped during walk. The design of the trunk of ROBIAN robot was based on this observation [8]. The objective of this work is to study the stability of the robot, subjected to an external disturbance and performing tasks of handling. In this study, tests of simulation are made on the trunk of ROBIAN. Initially, the dynamics equations of the trunk are developed in the presence of external disturbing forces. The results of simulation make it possible to determine the movements and accelerations of mobile masses necessary to stabilize the robot. In a following stage, we study a system made up of a trunk with two arms, and calculate necessary motions to ensure the robot stability in the presence of the disturbances or during handling of an object.

2 Stabilization with a simple trunk with 4 DOFs

2.1 Method

The simple trunk is a mechanism having 4 dofs : three translations and one rotation (see **Fig. 1.**).

Simulations of the dynamic behavior of a bipedal robot ... 861

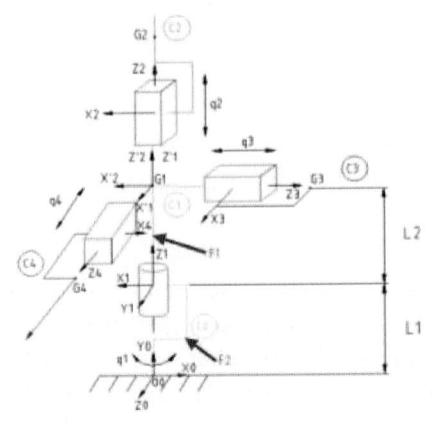

Fig. 1. Trunk with 4 dofs R3P subjected to the external forces F1 and F2

The equations of the dynamics of the upper part of the robot are developed using the formalism of Newton-Euler. The efforts applied to the trunk to located on point O_0 are given by:

$$\begin{cases} \vec{F}_0 = \sum_{i=0}^{4} m_i (\vec{\gamma}_i - \vec{g}) - \sum_{i=1}^{2} \vec{F}_i \\ \vec{M}_0 = \sum_{i=0}^{4} (\vec{h}(i,O_0) - \vec{M}(m_i \vec{g}, O_0)) - \sum_{i=1}^{2} \vec{M}(\vec{F}_i, O_0) \end{cases} \quad (1)$$

- $\vec{h}(i,O_0)$: Derivation of angular momentum of the body C_i calculated at the point O_0

- $\begin{Bmatrix} \vec{F}_0 \\ \vec{M}_0 \end{Bmatrix}$: Efforts exerted by the frame on the body C_0 at point O_0.

- $\vec{F}_1$ and $\vec{F}_2$: External forces applied respectively to the bodies C_1 and C_0.

Direct and inverse kinematics are obtained using modified DH parameters.

The motion equations based on the formalism of Newton-Euler are calculated for each body C_i of the trunk at the rigid support point O_0. They can be written in the following form:

$$\begin{bmatrix} \ddot{q}_1 \\ \cdot \\ \cdot \\ \ddot{q}_4 \end{bmatrix} = \begin{bmatrix} g_1(q_1,...,q_4,\dot{q}_1,...,\dot{q}_4,F_x,F_y,F_z,M_y,P_g,P_i,F_1,F_2) \\ \cdot \\ \cdot \\ g_4(q_1,...,q_4,\dot{q}_1,...,\dot{q}_4,F_x,F_y,F_z,M_y,P_g,P_i,F_1,F_2) \end{bmatrix} \quad (2)$$

- $q_i, \dot{q}_i$ and $\ddot{q}_i$: positions, velocities and accelerations of the ith joint.
- F_x, F_y, F_z et M_y: Components of the measured efforts exerted by the support to the trunk at point O_0. We recall that $M_x = k_1.F_z$ and $M_z = k_2.F_x$
- F_1 and F_2: External forces applied respectively to the bodies C_1 and C_0 with $F_1 = [F_{1x}, F_{1y}, F_{1z}]^T$ et $F_2 = [F_{2x}, F_{2y}, F_{2z}]^T$.
- P_g and P_i: Geometrical and inertial parameters of R3P mechanism.

In this part, the bodies $C_{i\ (i=1..,4)}$ represent the mobile masses which are used to compensate the external disturbances.

The modeling and the simulation of the dynamic behavior of the ROBIAN's trunk subjected to the disturbing forces F_1 and F_2 is carried out using the Adams software. The forces F_1 and F_2 are applied respectively to the body C_1 at point A and to the body C_0 at point B (see **Fig. 2.**).

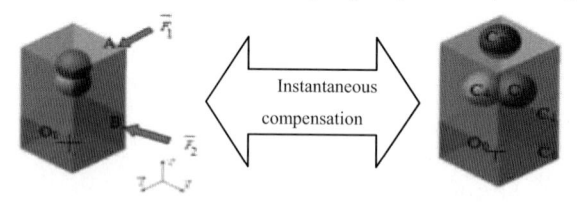

Fig. 2. Instantaneous compensation of the trunk

2.2 Results of simulation of a three-dimensional disturbance

Based on the GSE (General State Equation) the system (2) is solved to calculate the 4 accelerations to be generated to compensate instantaneously the external forces applied to the robot. The GSE is built as follows:

- $\vec{X} = [x_1,......,x_8]^T$: Vector of system state $x_1 = q_1, x_2 = \dot{q}_1,....,x_7 = q_4, x_8 = \dot{q}_4$.
- $\vec{U} = [u_1,......u_{10}]^T$: Input vector of the system whose components are respectively F_x, F_y, F_z, M_y, F_{1x}, F_{1y}, F_{1z}, F_{2x}, F_{2y} and F_{2z}.
- $\vec{Y} = [q_1, q_2, q_3, q_4]^T$: Output vector of the system.
- $\dfrac{d\vec{X}}{dt} = [\dot{q}_i, \ddot{q}_i]^T$ i= (1..6)

For this example the external forces $F_1 = [50N, 50N, 50N]^T$ and F_2 [10N, 10N, 10N]^T are applied during 0.25 second. The first simulation is carried out without compensation: the robot rocks and falls down just after the application of the disturbance (see **Fig. 3.**). The same simulation is carried out with compensation of the R3P trunk: the robot is stabilized (**Fig. 4.**).

Fig. 3. Robot without compensation of disturbance

Fig. 4. Compensation of the three-dimensional disturbances

3 Stabilization with trunk and arms

In this section, a system with trunk and arms is studied. The motions of the arms required for the robot stability are calculated in presence of disturbances. The same formalism is used to study the dynamics of the new upper part of the robot which has 10 dofs (see **Fig. 5.**).

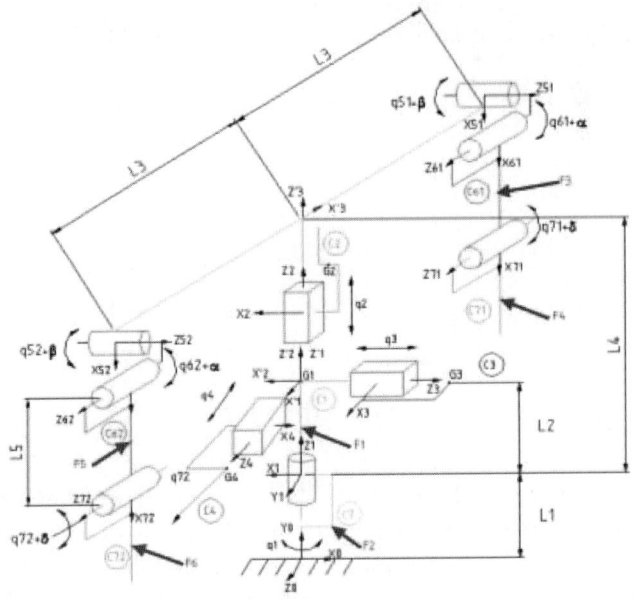

Fig. 5. Torso and arms

Each arm is connected to the body C_1 and has three degrees of freedom, 2 rotary joints on the shoulder and 1 rotary joint on the elbow. The external forces $F_i = [F_{ix}, F_{iy}, F_{iz}]^T_{(i=1..6)}$ are applied to the bodies $C_1, C_2, C_{61}, C_{71}, C_{62}$, and C_{72} respectively. The angles α, β and δ are parameters which enable to choose an initial configuration for the two arms.

3.1 Method

The formalism of Newton-Euler is used to write the equations of the dynamics of the upper part of the robot. The efforts applied by the rigid support to the trunk at point O_0 are written as follows:

$$\begin{cases} \vec{F_0} = \sum_{i=0}^{4} m_i(\vec{\gamma}_i - \vec{g}) + m_{61}(\vec{\gamma}_{G61} - \vec{g}) + m_{71}(\vec{\gamma}_{G71} - \vec{g}) + m_{62}(\vec{\gamma}_{G62} - \vec{g}) + m_{72}(\vec{\gamma}_{G72} - \vec{g}) - \sum_{i=1}^{6} \vec{F_i} \\ \vec{M_0} = \sum_{i=0}^{4}(\vec{h}(i,O_0) - \vec{M}(m_i \vec{g}, O_0)) + (\vec{h}(61, O_0) - \vec{M}(m_{61} \vec{g}, O_0)) + (\vec{h}(71, O_0) - \vec{M}(m_{71} \vec{g}, O_0)) \\ \quad + (\vec{h}(62, O_0) - \vec{M}(m_{62} \vec{g}, O_0)) + (\vec{h}(72, O_0) - \vec{M}(m_{72} \vec{g}, O_0)) - \sum_{i=1}^{6} \vec{M}(\vec{F_i}, O_0) \end{cases} \quad (3)$$

The 6 motion equations of the arms can be written in the following form:

$$\begin{bmatrix} \ddot{q}_{51} \\ \ddot{q}_{61} \\ \ddot{q}_{71} \\ \ddot{q}_{52} \\ \ddot{q}_{62} \\ \ddot{q}_{72} \end{bmatrix} = \begin{bmatrix} g_1(q_i, \dot{q}_i, F_x, F_y, F_z, M_x, M_y, M_z, P_g, P_i, F_1, F_2, F_3, F_4, F_5, F_6) \\ \cdot \\ \cdot \\ \cdot \\ \cdot \\ g_6(q_i, \dot{q}_i, F_x, F_y, F_z, M_x, M_y, M_z, P_g, P_i, F_1, F_2, F_3, F_4, F_5, F_6) \end{bmatrix} \quad (4)$$

The system (4) allows the calculation of the 6 accelerations to be generated for the bodies C_i $_{(i=61, 71, 62, \text{ and } 72)}$ to compensate the external forces applied to the robot instantaneously.

3.2 Results of simulation of a three-dimensional disturbance

The external force F_1 [20N, 20N, 20N]^T is applied on the body C_1. The modeling of the robot and the external force F_1 are carried out. The first simulation is carried out without compensation by the arms; the robot rocks and fall at the time of application of the disturbance (see **Fig. 6.**).

Fig. 6. Robot without compensation of disturbance

The same simulation is carried out with compensation of the external forces using the 6 dofs of the arms to preserve balance (see **Fig. 7.**).

Fig. 7. Compensation of a three-dimensional disturbance with arms

4 Stabilization during handling of an object

4.1 Method

In this part, a new initial configuration is selected for the upper part of the robot. The bodies C_{61} and C_{71} are fixed and carried an object is handled by a simulated external force F4 = [0, 20N, 0]. During the handling, the robot is stabilized using the 3 mobile masses in translation $C_{i\ (i=2,\ 3,\ 4)}$ and of the two bodies $C_{i\ (i=62,\ 72)}$ of the free arm.

4.2 Results

At first, a simulation is carried out without compensation by the arms and masses, the robot rocks and falls down during handling. The same simulation of the complete robot (trunk, arms, and locomotion device) during handling is carried out. The movements required to maintain the stability of the robot are solved by the GSE (see **Fig. 8.**).

Fig. 8. Simulation of the handling of an object

5 Conclusion

The objective of our work was to show, by using the dynamic simulation of the virtual mockup of ROBIAN, the importance of the upper part in stabilization of the robot in the presence of external disturbances. The trunk of robot ROBIAN is a RPPP mechanism having a rotation and three translations. A method based on *General State Equation* formalism, was developed to calculate the movements and the accelerations of the parts of the trunk for the instantaneous compensation of the external forces. The addition of the arms proved to be the most adequate solution to ensure the stability of the robot during disturbances or object handling.

References

1. Zordan V. B. and Hodgins J. K. (1999) Tracking and modifying upper-body human motion data with dynamic simulation. Eurographics Animation Workshop in Computer Animation and Simulation.
2. Gravez F., Bruneau O., Ouezdou F.B., (2001) Capture de mouvement pour la simulation dynamique de mannequin virtuel, $15^{\text{ème}}$ Congrès Français de Mécanique, Nancy.
3. Gravez F., Bruneau O., Ouezdou F.B., (2000) Three-Dimensional Simulation of Walk of Anthropomorphic Biped, 13th CISM-IFToMM Symp. on Theo. and Practice of Robots and Manipulators, Romansy.
4. Masaki O. and Akifumi M. (2001) A Dynamic Motion Control Technique for Human-like Articulated Figures, Eurographics, volume 20, number 3.
5. Baerlocher. P and Boulic. R (2003) An Inverse Kinematic Architecture Enforcing an Arbitrary Number of Strict Priority Levels The Visual Computer, The Visual Computer.
6. Mohamed B., Gravez F., Bruneau O., Ouezdou F.B. (2002) , Four Dof Torso dynamic effects on biped walking gait, 14th CISM-IFToMM Symp. on Theo. and Practice of Robots and Manipulators, Romansy.
7. Gravez. F. , Mohamed B., Ouezdou F.B., (2002) Dynamic Simulation of a Humanoids Robot with Four Dofs Torso IEEE-International Conference on Robotics and Automation (ICRA)", p. 511-516. Washington, D.C., USA URL.
8. Ouezdou F.B., Konno A., Sellaouti R., Gravez F., Mohamed B., Bruneau O. (2002) ROBIAN biped project – a tool for the analysis of the human-being locomotion system", 5th International Conf. on Climbing and Walking Robots, pp 375-382, CLAWAR 2002.

System Analysis, Modelling and Simulation

Analysis of the Direct and Inverse Kinematics of ROMA II Robot

J.C. Resino, A. Jardón, A. Gimenez, C. Balaguer

RoboticsLab, Universidad Carlos III de Madrid, 28911, Leganes, Madrid
{ajardon, agimenez, balaguer}@ing.uc3m.es

Abstract

The development of the climbing robots during the last years has stimulated the scientific research and has brought a great interest in search for robots more and more profitable and optimized for their particular tasks. The project "ROMA II", having in mind these objectives, is dedicated to creation of a climbing robot able to move in three-dimensional (3D) environments to perform tasks of inspection and maintenance. The scientific works developed in this project are focused to four very important areas of robotics: design of robots, software engineering, control engineering and motion planning.

Keywords: Climbing robots, kinematics analysis, path-planning simulation.

1 Introduction

The climbing robot "ROMA II" consists of three integrated systems: mechanical, electrical and pneumatic. The project specifications require that the robot have reduced weight, minimum of degrees of freedom necessary for operation in the working environment, simplicity of control and reception systems and a support system based on suction cups [1].

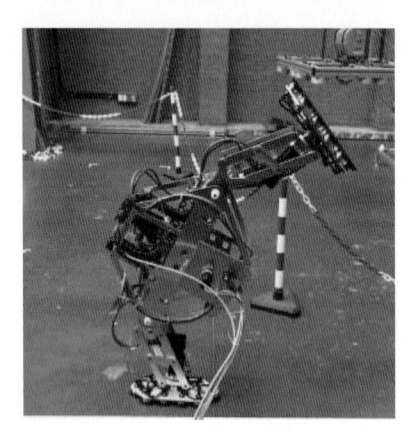

Fig. 1. ROMA II robot

To perform motion planning and navigation of the robot "ROMA II" it is necessary to know the direct and inverse kinematics structure of the ROMA II. The three actuator systems of the robot:

- Three three-phase brushless permanent magnets AC motors. Two of these serve to control the orientation of the claws and one to raise the body.
- Two cylinders to extend and take away the "arms", these can act only if one of the claws is raised.
- Two pneumatic claws equipped with suction cups to provide the necessary surface fixation.

Due to the special locomotion actuator system, the kinematics study is more complex than usual manipulators. It is not appropriate to use the traditional Denavit-Hartenberg algorithm, to obtain a kinematics model. The robot has two pneumatic cylinders with absolute positions, and there are two symmetric joints which move at the same time, with the same absolute value but in opposite direction. This paper describes the special equations to compute the direct and inverse kinematics for this special robot.

2 Direct kinematics

The robot has one actuator that moves two joints at the same time Q1, and two joints that can be moved by two pneumatic cylinders: D1, y D2 that allows two different positions in each leg: D2a, and D2b. All these joints can be seen in figure 2. The data that it is used in the direct kinematics are: two continuous joints (Q2,Q3), one symmetric joint (Q1), and two digital joints (D1, D2).

Table 1 presents the most important geometric parameters related to the calculations of the direct and inverse kinematics of the robot

Table 1. ROMA II robotParameters

	Dim.	Parameter
Axis distance Q1-Q1	555 mm	L2
Maximum length	1122 mm	
Maximum height	637 mm	
Maximum width	195,5 mm	
Arm height	407 mm	L1
Gait	400 mm	
Range angle D2a-D2b	23,5°	
Range of the axis cylinder	130 mm	x1

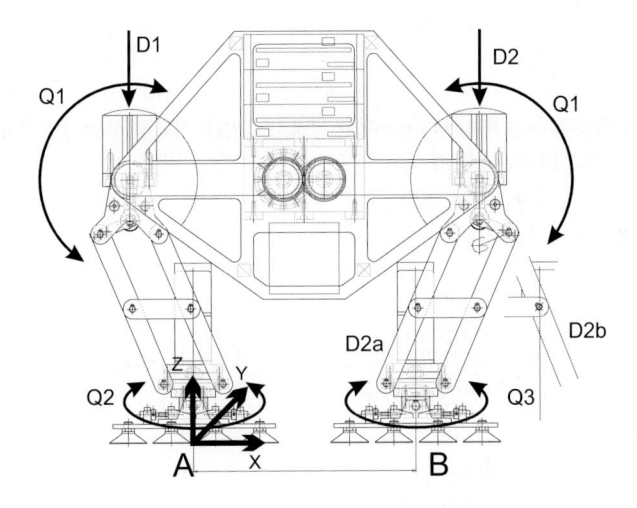

Fig. 2. ROMA II robot kinematics chain

Furthermore, the robot can be fixed in two different edges: A or B. For this reason, two direct kinematics must be known, for edge A or B. The next equation shows this formula for the A edge. The same thing happens with the inverse equations. These equations involve the whole variables, "digital" (D1, D2: values 0 or 1) and "continuous" variables. The direct kinematics expression from point A is:

Position

$$x = [L_1 \cos 90° + (d_1 * x_1) + L_2 \sin(90° - q_1) + L_1 \cos(90° - (2*q_1) - (d_1 * 20°))] \cos(q_2)$$
$$y = [L_1 \cos 90° + (d_1 * x_1) + L_2 \sin(90° - q_1) + L_1 \cos(90° - (2*q_1) - (d_1 * 20°))] \sin(q_2) \quad (1)$$
$$z = [L_1 \sin 90° - (d_1 * x_2) + L_2 \cos(90° - q_1) - L_1 \sin(90° - (2*q_1) - (d_1 * 20°))]$$

Orientation

$$\phi = \beta = q_2$$
$$\theta = \alpha = (2 \cdot q_1) + (d_1 \cdot 20°) \quad (2)$$
$$\psi = \gamma = q_3$$

The position equations show that there is one "discrete" variable, d1. Its value will be 0 or 1. During the sequence motion the two cylinders D1 and D2 are activated or deactivated at the same time, so we only need one variable to control the effect of the pneumatic cylinders. It is important to note that this is one possibility to introduce mechanisms and actuators which work using "discrete" positions, instead of the normal actuators which has a "continuous" position. The variables related to d_1 are: x_1 which is a known measure, the length of the axis cylinder, and x_2, which is obtained by the (3) expression and represents the triangle generated by the length of the arm L_1 and the length of the axis of the cylinder x_1.

$$x_2 = \frac{2L_1 - \sqrt{4L_1^2 - 4x_1^2}}{2} \quad (3)$$

The direct kinematics from point B is the same expressions that appear in (1) and (2). The reason is that the robot has a symmetric kinematics configuration.

3 Inverse kinematics

The inverse kinematics from the A point is, taking into account all the variable joints:

$$\alpha = (2 \cdot q_1) + (d_1 \cdot 20°) \begin{cases} q_1 = \dfrac{\alpha - (d_1 \cdot 20°)}{2} \\ d_1 = \dfrac{\alpha - (2 \cdot q_1)}{20°} \end{cases}$$

$$q_2 = \arctan\left(\pm\dfrac{y}{x}\right) \qquad (4)$$

$$q_3 = \gamma$$

$$z = \left[L_1 \sin 90° - (d_1 \cdot x_2) + L_2 \cos\left(90° - \left(\dfrac{\alpha - (d_1 \cdot 20°)}{2}\right)\right) - L_1 \sin(90° - (\alpha - d_1 \cdot 20°)) - (d_1 \cdot 20°)\right] (5)$$
$$z + L_1(\cos(2\cdot\alpha) - 1) = \left[-(d_1 \cdot x_2) + L_2[\sin(2\cdot\alpha)\cdot\cos(d_1 \cdot 40°) - \sin(d_1 \cdot 40°)\cdot\cos(2\alpha)] \right]$$

In order to check all these expressions and to verify the calculated trajectories to send to the robot, a simulation tool has been developed using MATLAB. This tool allows simulating the different sequences of motions that ROMA II will do during an inspection task. Finally this tool writes a file with the different sequences of the joints of the robot during a motion and can be sent directly to the robot computer. Figure 3 shows the simulation tool, where it is possible to use the direct or inverse kinematics selecting the fixed end, A or B, change the velocity of the movements, and the lengths of the chains.

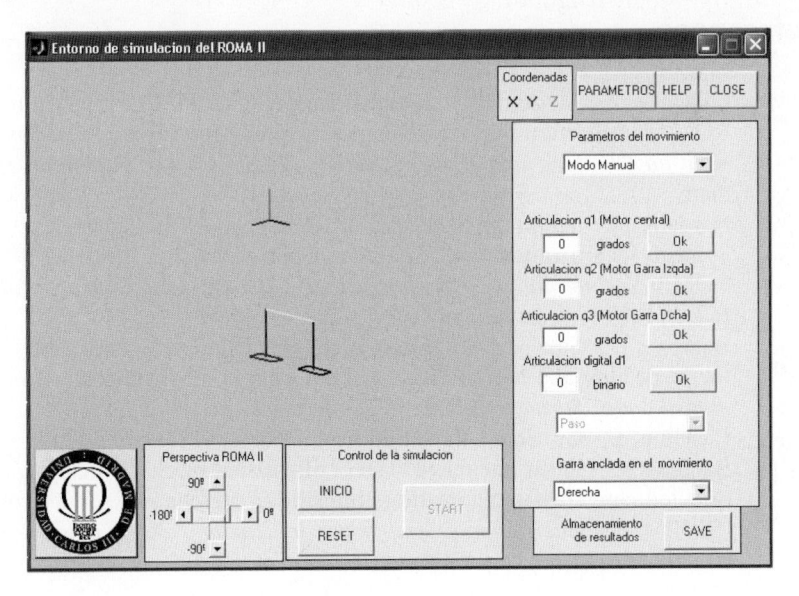

Fig. 3. ROMA II robot simulation tool

This tool can change the view of the camera and study all the different motions of the robot, read the different values of the joints and the expressions that are used in every case, when the robot is connected to the platform A or B.

Acknowledgements

The authors gratefully acknowledge the funds provided by the Spanish Government through the MCYT Project (n° TAP98-0274).

3 Conclusions

In this paper we present a new methodology to obtain the direct an inverse kinematics when the climbing robot has "discrete" joints. This fact is very usual when the robot designers use pneumatic cylinders to move is robot. In this way, is possible to introduce the different position controls that are used in other fields of robotics research, such as robotic manipulators. Using the simulator, it is possible to have a previous study of the future paths that will be done by the ROMA II robot.

References

1. Gimenez, A., M. Abderrahim and Balaguer C. (2001) Lessons from the ROMA1 inspection robot development experience. *International Symposium on Climbing and Walking Robots (CLAWAR'01)*, Karlsruhe, Germany
2. Nardelli, M., Jardon A., Staroverov P., Gimenez A., Balaguer C. (2003). System Identification and Adaptive Control of the Climbing Robot ROMA II. *International Symposium on Climbing and Walking Robots (CLAWAR'03)*, Catania, Italy
3. Balaguer C., Giménez A., Jardón A. (2005) Climbing Robots' Mobility for Inspection and Maintenance of 3D Complex Environments. *Autonomous Robots, Springer Science*, Vol. 18, no. 2.
4. Castejón C, Giménez A., Jardón A., Rubio H., García-Prada J.C., Balaguer C. (2005) Integrated system of assisted mechatronic design for oriented computer to automatic optimising of structure of service robots (SIDEMAR). *International Symposium on Climbing and Walking Robots (CLAWAR'05)*, London, England

Simulation of a Novel Snake-Like Robot

R. Aubin[1], P. Blazevic[1], and J.-P. Guyvarch[2]

[1] Laboratoire de Robotique de Versailles, France.
    [aubin,blazevic]@lrv.uvsq.fr
[2] TDA Armements, France.
    jean-paul.guyvarch@fr.thalesgroup.com

Abstract

This research focuses on simulation of a new snake-like robot. Most of the snake-like robots developed until now are based on the lateral undulation mode of locomotion. Nevertheless, it seems difficult to apply this mode of locomotion in realistic conditions such as non-planar and complex grounds, since most of these robots use wheels. As a result the rectilinear mode of locomotion is used for this snake-like robotic structure. After introducing the biological standpoint and locomotion principles of the robot, the simulation environment is presented. Simulations of a five module version of the snake-like robot performing a simple trajectory tracking with a constant rectilinear propulsive component are shown. Simulation results of two direction control methods are presented. The paper is concluded based on the results and future research prospects are introduced.

Keywords: Snake-like robot, rectilinear, locomotion, simulation.

1 Introduction

The propulsion mode used by most of the snake-like robots developed until now assumes very strong constraints in relation to the environment. For example, passive wheels based robots, and more common structures using the difference between lateral and longitudinal friction forces [1] are not suitable for moving on real grounds. The propulsion of these robots is inspired by the lateral undulation locomotion mode observed in nature

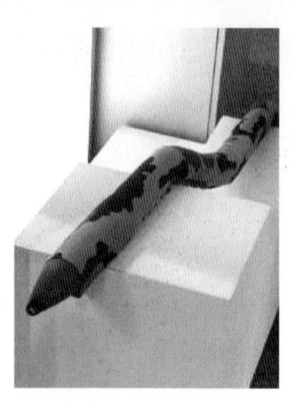

Fig. 1. The TDA's VIPeRe prototype at Eurosatory 2004

[2]. This mode can hardly be used in uneven ground conditions because wheels require homogeneous planar surfaces. Finally, most of these robots are not designed for three-dimensional propulsion because of their planar morphology.

A response to the mobility problem on real ground conditions is the use of tracked vehicles [3, 4] inspired by a holistic approach of the rectilinear mode of locomotion.

The solution proposed in this paper is inspired by the local implementation of the rectilinear mode of locomotion as observed in real snakes [5], **Fig. 2**.

This paper describes the simulation environment and first simulations of a new design of biologically inspired snake-like robot. These simulations are based on a robot developed by the TDA company and supported by European patents [6, 7]. A first prototype was built as shown in **Fig. 1**. The novel aspect of this design is that it uses a mechanism generating periodic movements in order to propel each module of the robot.

Detailed description of the mechanism of the robot is not provided in this Paper, and to date implemented rectilinear movement has been implemented without overall direction control on the robot.

2 Biological Standpoint and Principles of the Robotic Design

The main classes of limbless locomotion are lateral undulation, rectilinear, concertina and sidewinding. Most of the snake-like robots based on the lateral undulation mode use wheels and present dynamic singularities in their configuration (straight line, circular arc), bearing in mind that on

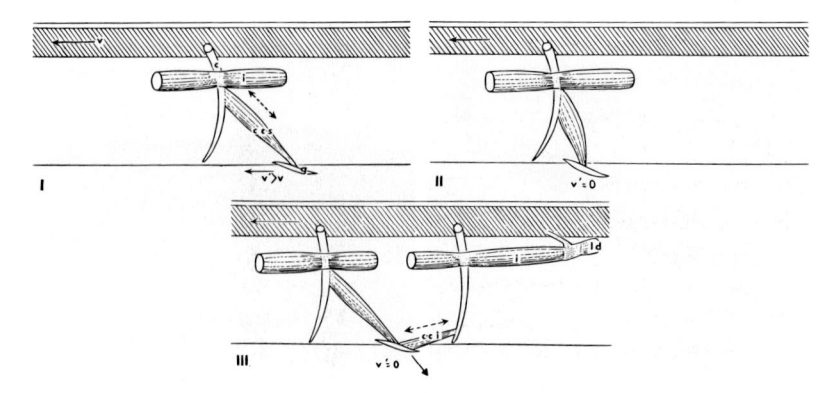

Fig. 2. Rectilinear mode (from left to right displacement) – from [5]

any planar surface, the shortest path between any two points is a straight line.

As a consequence, the rectilinear mode of Locomotion is chosen in this study. Among the real snakes that adopt this mode, it can be observed that propulsion is the result of addition of local muscular activities (**Fig. 2**). The concept of this mode of locomotion relies on an overall continuous movement from local periodic movements. During the rectilinear displacement, the backbone and the ribs are fixed and rigid. Muscular activities between ribs and skin result in local propulsive action.

The design in this study combines lateral and rectilinear locomotion modes. The propulsion is produced by a specific mechanism that locally reproduces the rectilinear locomotion mode as observed in real snakes [5]. Combining rectilinear propulsion with lateral undulation is necessary in order to control the overall direction taken by the robot, and allows to increase structural stability. Lateral undulation implies managing contact points and controlling lateral forces to work properly as a propulsive component. However, lateral undulation is not used as a propulsive component in the frame of this research, but only as a steering funtion. The rectilinear mode is mainly chose to propel the robot because of the relative simplicity of its implementation in comparison with the lateral undulation mode. Furthermore, linear trajectories can be performed in a straight forward manner.

To help the design of prototypes an insight into how to control this robotic structure is needed. As a consequence, a dynamic simulation was conducted in order to evaluate and tune up control algorithms.

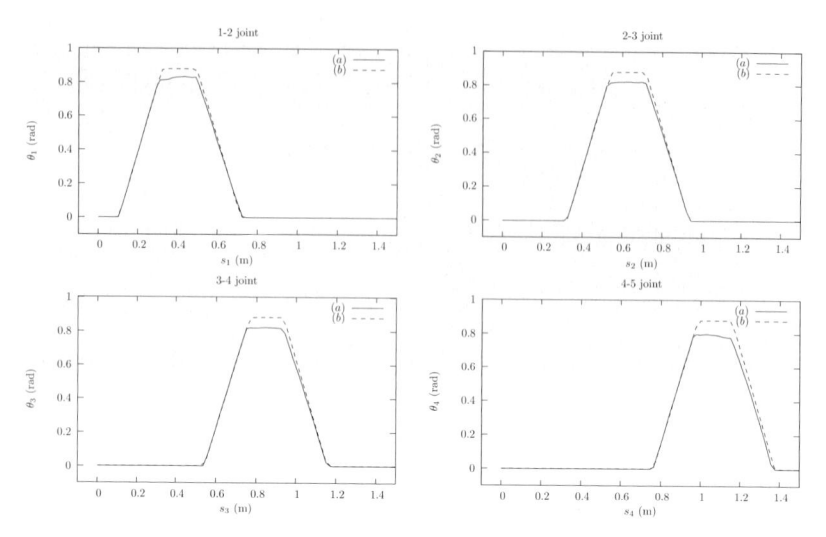

Fig. 5. Angular position of each joint

l be the length of the first segment of the trajectory

This mechanical design involves slippage at some contact points. The simulation allows to approximate this slippage and be taken into account within the control strategy. Results depicted in **Fig. 4** and **Fig. 5** are obtained by controlling the angular positions from angular velocities with a proportional feedback control:

$$\dot{\theta}_i = K\left(\theta_{Mi} - \theta_{Di}\right) \qquad (1)$$

$$\theta_{Di} = \alpha_i - \alpha_{i+1} \qquad (2)$$

$$\alpha_i = \begin{cases} 0 & \text{if } s_i < l + i\,L, \\ \frac{\pi}{2} & \text{if } s_i \geq l + i\,L + \frac{R\pi}{2}, \\ s_i - \frac{l + i\,L}{R} & \text{if } l + i\,L \leq s_i < l + i\,L + \frac{R\pi}{2}. \end{cases} \qquad (3)$$

The different results are obtained with two control approaches:

(a) s_i is calculated by multiplying the constant propulsive linear velocity of the ith module's center of mass with the time step,
(b) s_i is calculated according to the simulated planar displacement of the ith module's center of mass (taking slippage into account).

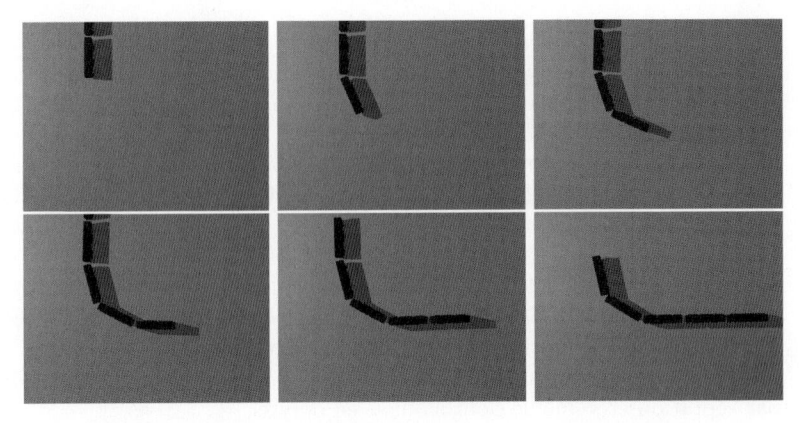

Fig. 6. Evolution of the simulation – (b) control method

Obviously, the trajectory is better followed by method (b) (**Fig. 4**). This result shows that the simulation is relevant to the control system because method (a) is open-loop in position and thus cannot adjust the command to errors induced by slippage. Consequently, it is verified that slippage cannot be neglected in our control strategy.

4 Conclusion and Future Work

The simulation environment and first simulations of the snake-like robot inspired by the rectilinear mode of locomotion have been introduced.

Future simulations will use learning methods in order to estimate the slippage of the robot in a model based control. learning methods (neural networks and artificial evolution) will also be used as proposed in [10] to minimize trajectory tracking errors.

In the long term, it is aimed to implement three-dimensional movements for the robot to achieve wide range ground operability, and resulting control algorithms will be tested and embedded in the prototype.

References

1. Saito, Masashi, Fukaya, Masakazu, and Iwasaki, Tetsuya (2002) Serpentine Locomotion with Robotic Snakes. *IEEE Control Systems Magazine*, pp. 64–81.
2. Hirose, Shigeo (1993) *Biologically Inspired Robots. Snake-like locomotors and manipulators.* Oxford University Press.

3. Arai, Masayuki, Takayama, Toshio, and Hirose, Shigeo (2004) Development of "Souryu-III": Connected Crawler Vehicle for Inspection inside Narrow and Winding Space. In *Proc. of the IEEE/RSJ Int. Conf. on Intelligent Robots and Systems (IROS)*, pp. 52–57.
4. Kamegawa, Tetsushi, Yamasaki, Tatsuhiro, Igarashi, Hiroki, and Matsuno, Fumitoshi (2004) Development of The Snake-like Rescue Robot "KOHGA". In *Proc. of the IEEE Int. Conf. on Robotics & Automation (ICRA)*, pp. 5081–5086.
5. Gasc, Jean-Pierre (1974) *L'interprétation fonctionnelle de l'appareil musculo-squelettique de l'axe vertébral chez les serpents*. vol. 83 of *A*, Mémoires du Muséum National d'Histoire Naturelle, Paris.
6. Guyvarch, Jean-Paul Procédé de propulsion d'un véhicule à déplacement par reptation et véhicule pour sa mise en œuvre. .
7. Guyvarch, Jean-Paul Véhicule à déplacement par reptation propulsé conjointement par ondulations longitudinales et latérales. .
8. The Open Dynamics Engine website. .
9. Ďurikovič, Roman and Numata, Katsuhiro (2004) Human hand model based on rigid body dynamics. In *Proc. of the IEEE Conf. on Information Visualization*, London, UK, pp. 853–857.
10. Chocron, Olivier, Brener, Nicolas, Bidaud, Philippe, and Ben Amar, Faz (2004) Evolutionary Synthesis of Structure and Control for Locomotion Systems. In *Proc. of the Int. Conf. on Climbing and Walking Robots and the Support Technologies for Mobile Machines*, Madrid, SP.

An Actuated Horizontal Plane Model for Insect Locomotion

John Schmitt

Department of Mechanical Engineering, Oregon State University, Corvallis, OR 97331, U.S.A. schmitjo@engr.orst.edu

Abstract

We analyze a simple three degree of freedom model for running in the horizontal plane: the lateral leg spring (LLS) model with an actuated hip. The leg attachment point within the body is actuated to produce a force in the elastic leg that matches experimental force profiles. While the resulting bipedal model accurately reproduces force profiles by construction, we find that a single leg cannot reproduce the moment profiles and yawing variations seen in an insect equipped with a tripod of legs. We find stability is strongly dependent upon foot placement and that the leg attachment point motion within the body differs from that seen in the insect.

Keywords: Legged locomotion, horizontal plane, stability.

1 Introduction

Simple two and three degree of freedom models have been used in both the vertical [1, 2] and horizontal planes [3, 4, 5] to represent locomotion of upright and sprawled posture animals. The horizontal plane LLS models were inspired by the remarkable ability of insects to stably run over rough ground at high speeds, with apparently relatively little neural reflex control. In this work, we extend the LLS model to include actuation inputs that quantitatively reproduce experimental force profiles. While previous LLS models produce stable periodic gaits without feedback that qualitatively match experimental results, quantitative comparisons reveal reversed fore-aft and lateral force magnitudes and moment profiles that remain an order of magnitude too small. Ultimately, we aspire to explain and replicate insects' remarkable stability properties through simple reduced order models. Actively embedding self-stabilizing reduced order models into higher degree of freedom robotic representations may simplify the control required for locomotion over rough terrain.

2 Prescribed actuation model

As illustrated in **Fig. 1**, the actuated pivot model consists of a rigid body of mass m and moment of inertia I moving under the influence of forces produced by an effective elastic leg. The effective leg is attached at a prescribed point in the body and represents the cumulative effects of the tripod of legs acting during locomotion. We develop the equations of motion relative to the inertial axes using force and moment balances in an (x, y, θ) frame. In the inertial frame, let $\mathbf{r} = x\hat{\mathbf{e}}_x + y\hat{\mathbf{e}}_y$ and θ denote the position of the center of mass and the orientation of the body with respect to the inertial frame. Let $\mathbf{r_F}$ denote the position of the active foot placement point and $\mathbf{R}(\theta)$ the usual rotation matrix from the body frame to the inertial frame. Since forces are only applied at the feet, the equations of motion in the inertial frame can be written as:

$$m\ddot{\mathbf{r}} = \mathbf{F}, I\ddot{\theta} = (\mathbf{r_F} - \mathbf{r}) \times \mathbf{F} \tag{1}$$

where

$$\mathbf{F} = \mathbf{R}[\theta(t)]\mathbf{f} \tag{2}$$
$$\mathbf{r_F} = \mathbf{r}(t_n) + \mathbf{R}[\theta(t_n)]\{d_1\hat{\mathbf{e}}_1 + d_2\hat{\mathbf{e}}_2 + \mathbf{R}[(-1)^n\beta]l\hat{\mathbf{e}}_2\} \ . \tag{3}$$

In (2-3),$\mathbf{f}$, β and $(d_1(t), d_2(t))$ represent the leg force vector in the body frame, leg touchdown angle with respect to the body axis, and distance to the leg attachment point from the center of mass in the body frame respectively. The second expression specifies foot placements in the inertial frame using the convention that the left leg is active on even steps ($n = even$) and the right leg is active on odd steps ($n = odd$).

The leg is attached at a pivot point in the body whose movement is actuated such that the force developed in the elastic leg

$$F = k(l - |\mathbf{q}|) \tag{4}$$

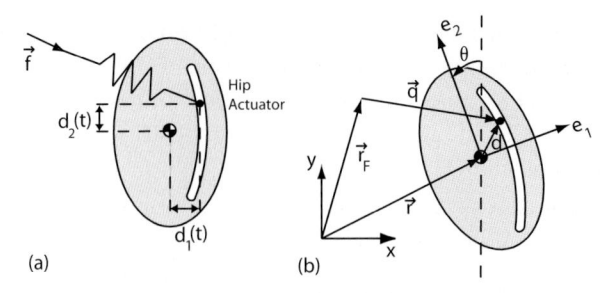

Fig. 1. The actuated hip model

matches force profiles observed experimentally. In (4), l, k, and $\mathbf{q}$ represent the nominal leg length, spring stiffness, and vector leg length in the inertial frame respectively. The forces observed experimentally for each leg in the fore-aft and lateral directions during each stance phase are detailed in [6, 7, 8] and summarized here through the effective leg profile

$$\mathbf{F}_{\text{effective}} = (.0032(-1)^n \sin(\Omega t), -.004 \sin(2\Omega t)) \ . \tag{5}$$

where Ω denotes the leg stride frequency. To reproduce quantitatively correct force profiles of the cockroach *Blaberus discoidalis* [6, 7], the actuation inputs of the model illustrated in **Fig. 1** are derived from equations relating the force developed in the elastic leg to a prescribed force profile for a summed tripod

$$F_x = A_x \sin(\Omega t) = k(l - |\mathbf{q}|) \frac{q_x}{|\mathbf{q}|} \tag{6}$$

$$F_y = A_y \sin(2\Omega t) = k(l - |\mathbf{q}|) \frac{q_y}{|\mathbf{q}|} \tag{7}$$

where (A_x, A_y) denote the summed tripod force magnitudes provided in (5). The pivot position enters the force equations (6-7) through the calculation of the leg length in the inertial frame

$$\begin{bmatrix} q_x \\ q_y \end{bmatrix} = \begin{bmatrix} x(t) - x(t_F) \\ y(t) - y(t_F) \end{bmatrix} + \mathbf{R}(\theta(t)) \begin{bmatrix} d_1(t) \\ d_2(t) \end{bmatrix} - \mathbf{R}(\theta(t_F)) \begin{bmatrix} b_x \\ b_y \end{bmatrix} \tag{8}$$

where b_x and b_y are the foot placements in the body frame and t_F denotes the time of foot placement.

Coupling between the translational and rotational dynamics preclude solving for the body rotation and subsequent actuation inputs directly. However, since the body rotation of *Blaberus discoidalis* remains within

±5° of zero during the stance phase [8], we assume $\theta(t) \equiv 0$ to decouple the dynamics in deriving the hip actuation protocol.

With the center of mass motion available from integration of the effective force profiles, the motion of the pivot position is determined from equations (6-8), with $\theta(t) = 0$ as

$$d_1(t) = b_x - x(t) + \frac{A_x l}{\sqrt{4A_y^2 \cos^2(\Omega t) + A_x^2}} - \frac{A_x \sin(\Omega t)}{k} \qquad (9)$$

$$d_2(t) = b_y - y(t) + \frac{2A_y l \cos(\Omega t)}{\sqrt{A_x^2 + 4A_y^2 \cos^2(\Omega t)}} - \frac{A_y \sin(2\Omega t)}{k}. \qquad (10)$$

With a fixed nominal leg length, equations (9-10) imply that the leg is placed down at a constant angle for each stance phase

$$\beta = \arctan\left(\frac{d_1 - b_x}{b_y - d_2}\right) = \arctan\left(\frac{-A_x}{2A_y}\right) \qquad (11)$$

Additionally, perfectly symmetric strides for straight ahead running require $d_2(t_{end}) = -d_2(t_F)$, which yields $b_y = \frac{v_{des} t_{des}}{2}$, where $t_{des} = 1/2f$ represents the desired stance phase duration based on stride frequency. Therefore, the only free parameters in the simulations that follow include: b_x, l, k, Ω, and v_{des}.

3 Results

While the pivot position is a predefined function of time, the coupled translational-rotational equations of motion are non-integrable. Gait characteristics and stability are therefore determined through numerical simulations, as in previous work [3, 4]. The equations of motion are simulated using the Runge-Kutta integrator ode45 in Matlab. We locate periodic orbits with a Newton-Raphson routine and determine the stability of such gaits by computing the eigenvalues of the left leg to left leg Poincare map linearized about the periodic orbit. The Jacobian used in computing the eigenvalues is computed via central difference approximations.

Using parameters characteristic of the insect, as in [4], result in the periodic gait illustrated in **Fig. 2**. As expected, the force and velocity profiles of periodic gaits match those seen experimentally. Moment and yaw variations, while qualitatively correct for a range of foot placements, are an order of magnitude smaller than experimental results. Movement of

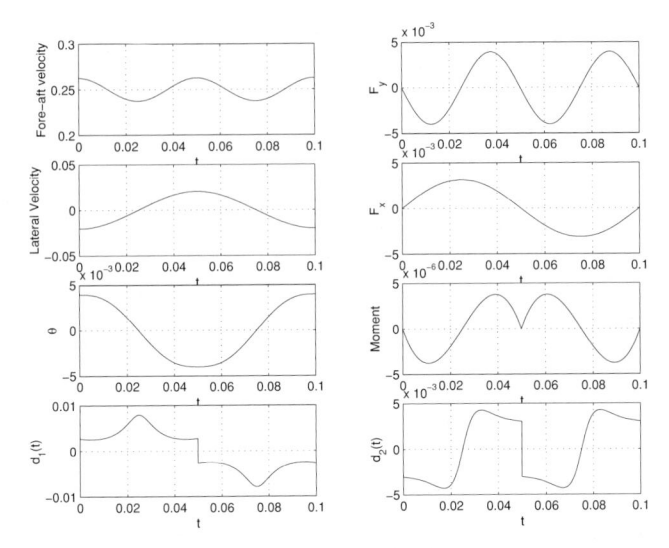

Fig. 2. Gait characteristics with $k = 2.25N/m$, $l = 0.01m$, $b_x = -.001m$, $v_{des} = .25m/s$, $t_{des} = .05s$, $\Omega = 10Hz$, $A_x = \pm.0032N$ and $A_y = -.004N$.

the leg attachment point is exactly opposite to that observed experimentally [9]; the hip pivot moves from rear to front during a stance, while remaining opposite the center of mass from the foot placement point. While force profiles are equivalent in both model and experiment, the leg force line of action remains close to the center of mass during the stance phase, resulting in a moment arm magnitude of 0.004 m versus 0.05 m in experiment [9]. While the pivot position movement differs from experiment, the angle the leg force line of action makes with the body centerline is quantitatively similar. Therefore, the reversed motion of the leg attachment point during stance, while necessary to reproduce the correct force profiles, creates a moment arm that prevents matching moment and yaw oscillations observed experimentally.

For a given desired fore-aft velocity, stride stability depends primarily on the foot placement b_x, as illustrated in **Fig. 3**. As the foot placement point moves further from the body, the intersection point of the line of action of the leg force with the body centerline moves rearward, eventually moving behind the center of mass at the start of the stance. This reversal not only produces inverted moment and yaw profiles, but also destabilizes the stride.

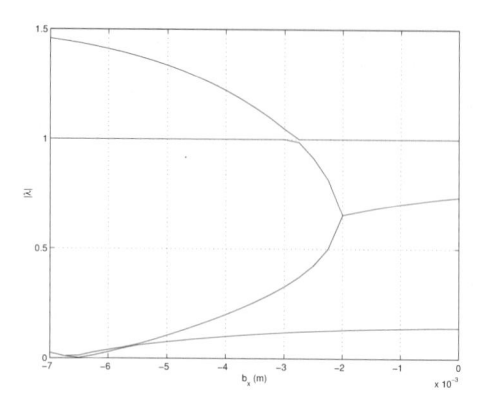

Fig. 3. Eigenvalue variation for foot placement $b_x = 0$ to $b_x = -0.007$

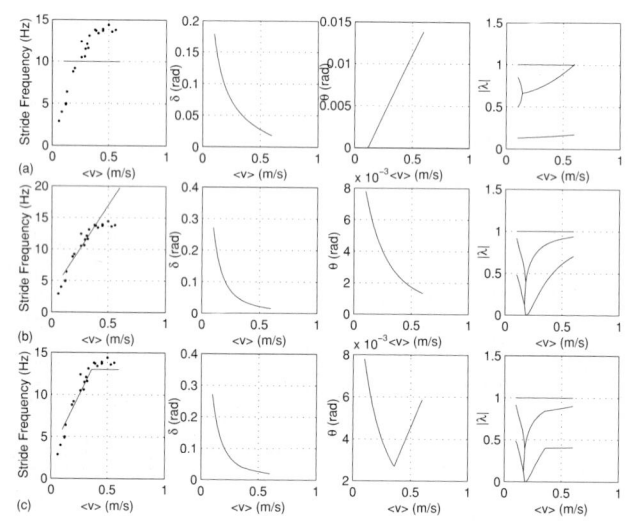

Fig. 4. Gait families for $v_{des} = 0.1 - 0.6$ and (a) $\Omega = 10$ Hz (b) Ω varying linearly from $5.8 - 19.7$ Hz and (c) Ω varying linearly up to $v_{des} = .35$ m/s and constant thereafter at 13 Hz. Panels display the stride frequency, heading angle (δ), body rotation (θ), and eigenvalues ($|\lambda|$).

While *Blaberus discoidalis* naturally uses its preferred running speed of 0.25 m/s, the insect also employs of range of running speeds between .1 − .6 m/s. Periodic gaits are achieved over the entire speed range by varying v_{des} while also varying Ω in three different manners. In the first instance, we hold the leg cycle frequency constant at 10 Hz, producing the gaits illustrated in the first row of **Fig.4**. The second protocol changes the leg cycle frequency linearly over the entire speed range, as illustrated in the second row of **Fig. 4**. This strategy mimics the leg cycle frequency changes that are used by insects at the lower end of the speed range [9]. The last protocol varies Ω linearly until $v_{des} = 0.35$ m/s, at which point it is held constant at 13 Hz. The combination of frequency and stride length changes of this final protocol best represent the stride length and frequency results of [9] considered here.

The resulting gait families and their associated stability are illustrated in **Fig. 4**. For appropriate foot placement, the resulting gaits are stable over the entire speed range, with maximal stability occurring slightly below the preferred fore-aft velocity. This is in contrast to the passive LLS models that experienced a saddle node bifurcation at the lower end of the speed range, where body kinetic energy was not sufficient to compress the elastic leg and move the body past the foot placement. The actuation included in the current model ensures that the body moves past the foot placement at low speeds, thus removing the possibility of a saddle node bifurcation and generating stable periodic gaits at lower speeds.

4 Conclusions

In this paper we have constructed a horizontal plane bipedal model for insect locomotion with an actuated leg attachment point. The actuation protocol is constructed to produce force profiles during normal locomotion that replicate those observed experimentally for an insect running in an alternating tripod gait.

As expected, the actuation protocol produces a family of periodic gaits, dependent on a desired fore-aft velocity v_{des}, that replicate the force and velocity profiles observed experimentally. However, the actuation required to replicate experimental force profiles results in a small moment arm, preventing moment magnitudes from matching experimental results. This is due to the fact that motion of the pivot point is exactly opposite that seen in experiments. While the angle of the line of action of the force developed in the leg is similar, the aft to fore motion of the pivot point during each stance phase serves to minimize its ability to produce moments of the order expected. As a result, it appears that while a single leg can be appropriately actuated to reproduce either the force or moment profiles generated by a tripod of legs, no actuation protocol can be constructed to simultaneously match all profiles in a bipedal model. Evidently, moment couples produced by a tripod of forces are largely additive rather than opposing, thus preventing a single leg from matching both force and moment profiles simultaneously.

The inclusion of actuation in these previously passive models serves to stabilize the body velocity, in addition to velocity heading angle and angular velocity. We find stability of periodic gaits for all protocols is strongly dependent upon the foot placement, b_x. As the foot is placed further away from the body, stability decreases. Evidently, inclusion of passive elements that enable prescribed forces to respond to body motions is important for stability. Changes in the desired fore-aft velocity lead to a family of gaits, whose stability is maximized when stride length and leg cycle frequency values are changed to mimic those seen experimentally. Additionally, the stability of gaits produced by the actuated pivot model are superior in almost all instances to those produced by other actuation protocols considered, such as an actuated leg attachment and nominal leg length protocol. This suggests that gait stability is enhanced when passive elements included in actuated models remain unactuated.

References

1. R. Blickhan. The spring-mass model for running and hopping. *J. Biomechanics*, 11/12:1217–1227, 1989.
2. R. Blickhan and R.J. Full. Similarity in multi-legged locomotion: bouncing like a monopode. *J. Comp. Physiol. A*, 173:509–517, 1993.
3. J. Schmitt and P. Holmes. Mechanical models for insect locomotion: Dynamics and stability in the horizontal plane – Theory. *Biological Cybernetics*, 83(6):501–515, 2000.
4. J. Schmitt and P. Holmes. Mechanical models for insect locomotion: Dynamics and stability in the horizontal plane – Application. *Biological Cybernetics*, 83(6):517–527, 2000.
5. J. Schmitt, M. Garcia, R. Razo, P. Holmes, and R.J. Full. Dynamics and stability of legged locomotion in the horizontal plane: A test case using insects. *Biological Cybernetics*, 86(5):343–353, 2002.
6. R.J. Full, R. Blickhan, and L.H. Ting. Leg design in hexpedal runners. *J. Exp. Biol.*, 158:369–390, 1991.
7. T.M. Kubow and R.J. Full. The role of the mechanical system in control: a hypothesis of self-stabilization in hexapedal runners. *Phil. Trans. Roy Soc. Lond. B*, 354:849–861, 1999.
8. R. Kram, B. Wong, and R.J. Full. Three-dimensional kinematics and limb kinetic energy of running cockroaches. *J. Exp. Biol.*, 200:1919–1929, 1997.
9. L.H. Ting, R. Blickhan, and R.J. Full. Dynamic and static stability in hexapedal runners. *J. Exp. Biol.*, 197:251–269, 1994.

Industrial Applications

Machine Vision Guidance System for a Modular Climbing Robot used in Shipbuilding

J. Sánchez, F. Vázquez, E. Paz

Departamento de Ingeniería de Sistemas y Automática. E.T.S.I. Industriales. Universidad de Vigo. Campus Universitario, 36310 Vigo, Pontevedra, (Spain)   jslopez@aisa.uvigo.es

Abstract

During naval construction, vessels are divided into various blocks or sub-units that are manufactured in different workshops and then carried to a building dock or shipway where they are welded together. This paper describes a versatile lightweight climbing robot that is capable of moving on the hull and a compact machine vision guidance system that allows it to follow grooves on the surface of the hull. This capability is useful for tasks such as inspecting welding seams, reducing the need for scaffolds. The robot is designed as a three wheeled Synchro-Drive vehicle. That allows it to perform movements in any direction without changing its orientation. The force of attraction necessary to stick the robot to the ship's hull is provided by three permanent magnets that are placed at the bottom of the robot. To guide this robot when it operates in automatic mode, a real time machine vision system has been developed. This system is prepared to detect and track the groove that is formed between two blocks of a ship when they are joined for welding. This groove is detected by the vision system using a multiline laser beam and a camera with an appropriate filter. The camera used for this application is a compact vision system that includes image capturing, image processing and communications capacities in a single unit.

Keywords: Machine vision, real time, laser, climbing robot.

1 Introduction

Naval construction is a complex process. Formerly, most construction took place in a building dock, with the ship constructed almost piece by piece from the ground up. However, advances in technology and more detailed planning have made it possible to construct the vessel in sub-units or blocks that have utilities and systems integrated within. These sub-units are manufactured in different workshops and then carried to a building dock or shipway where they are welded together to form the vessel. The work in those workshops has an acceptable degree of automation due to the relatively small size of the block and a closed and controlled work area. In the erection phase the working conditions are very different.

The building dock is an open air area with a great deal of activity and with a large structure to work on. The blocks are placed by huge wheeled cranes, on the ship's structure and at a great height. Scaffolds are raised all around the ship to allow workers to perform many tasks such as welding. Most of the tasks performed on the ship's hull are done manually, which is both expensive and dangerous.

One of the solutions in automating this work is to create a climbing robot that can move over the ship's hull to perform diverse tasks. Many efforts have been made in this direction and many prototypes and some commercial devices have been developed. Some of them use vacuum gripper 'feet' to stick to the wall [1] but for a ferromagnetic wall it is better to use magnetic devices because they can easily generate more than 10 times the adhesive force per unit area. A combination of permanent magnets and electromagnets was used in the six-legged robot called REST [2] developed by CSIC (Centro Superior de Investigaciones Científicas - Spain). This robot was designed to weld vertical seams but it had problems with its high weight (250Kg) and difficulties with continuous movements. Experience shows that legged robots are suitable for irregular surfaces but in flat surfaces wheeled robots can achieve higher velocities and smoother trajectories. An example of this kind of vehicle is a product called M3500 [3] developed by a company called UltraStrip, in collaboration with Carnegie Mellon University, which is used to strip paint from the hull. This robot has powerful permanent magnets close to the hull and performs well. However, the total weight of the robot is also considerable, creating problems with slipping when on wet hulls and making manual manipulation difficult. Other designs use permanent magnet disks as wheels [4]. This has the advantage that the magnet touches the surface, thus improving adhesive force but producing abrasion on the surface.

Our development group has design a semiautonomous modular robot that is capable of climbing ferromagnetic walls. It can carry cameras and sensors to perform inspection tasks such as verifying the welding seam or the hull's thickness. To perform other kinds of operations, such as welding, two or more robots must work together in cooperative mode. This last functionality is still work in progress.

To allow the robot to work in automatic mode, a guidance system is necessary. Some tasks such as welding are performed by following the groove between two blocks. An absolute positioning system is not necessary in these cases and it is enough to have local guidance system capable of detecting and tracking the joint between the two blocks. The problem of detecting the seam between two different metal pieces that are going to be welded has been addressed in various papers before and the solutions can be divided into the non-touch sensors and touch sensors. The touch type sensor detects the seam mechanically [5]. They have a simple structure and low price but they have low accuracy and are not suitable in some applications depending on the metal piece's thickness. The non-touch type, based on machine vision with laser lighting are accurate and flexible and can be used in almost all the applications in seam tracking [6][7]. Although these systems used to be expensive and complex, advances in technology have made it possible to create a complete machine vision system in a compact package at a relatively low-cost.

This paper is divided into two sections. First the modular climbing robot is described and then the machine vision guidance system is presented.

2 Robot description

The robot has been designed as a three wheeled synchro drive vehicle. A scheme of the mechanical structure is shown in figure 1-a.

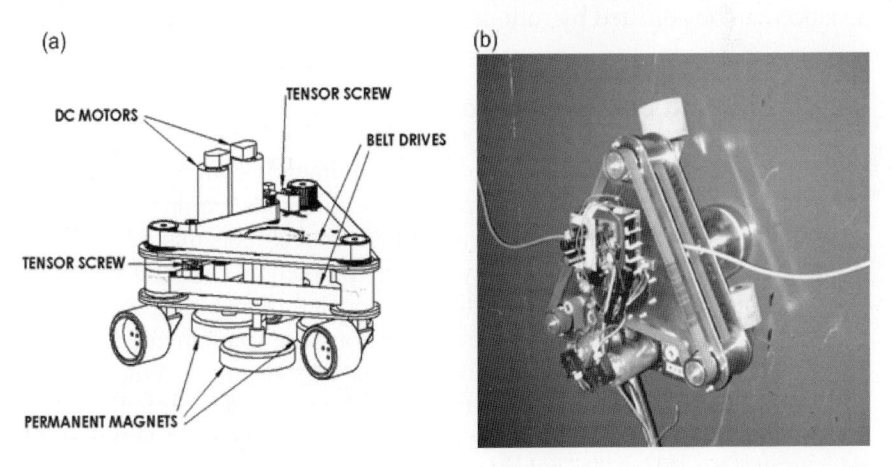

Fig. 1. Pictures of the modular climbing robot. (a) CAD model of the robot. (b) Robot in working position over a vertical steel surface.

The three wheels give the robot three points of support which is suitable for curved surfaces such as the hull of a ship. The synchro drive system makes the robot less sensitive to slippage because all the wheels are tractors, and allows making movements in any direction without changing its orientation and its weight distribution. The robot is prepared to work in the position show in figure 1-b. The weight is concentrated in the bottom part to make this position stable, although small deviations may occur. To overcome this problem, the robot carries an inclinometer that provides actual orientation measures to the control system. The robot is semiautonomous and it needs a power supply of 24 Volts DC to work properly. This kind of power supply is easily found in an industrial facility such as a shipyard.

2.1 Mechanical structure

The body of the robot is formed by two aluminium plates joined by four steel pieces. Two DC motors are placed on the top, one rotates the wheels to produce motion and the other motor turns all the wheels to change direction. The movement of the motors is transferred to the wheels by a belt drive, lighter and cleaner than a chain drive. Three permanent magnets are placed at the button of the robot. The magnetic force is highly dependent on the distance between the magnet and the wall. The closer the distance the higher the force but if they are placed too close to the surface the robot will not be able to pass through small obstacles such as weld bead. This distance can be adjusted by rolling the magnet over its support.

2.2 Control and communications

The robot is designed to obey locomotion commands coming from a manual joystick or the guidance system. These commands, composed by a direction and a velocity, are translated to motor movements that take into account the information provided by the inclinometer. The PID algorithm that controls the motors insures that the acceleration during movements does not exceed limit values. Communications are implemented in CAN bus so that multiple robots can be coordinated to work together. In this multi-robot scenario, each robot acts as a module of a larger robotic system that is capable of carrying a larger payload. Master-slave mode, where all slaves follow the master's movements, is currently supported. All the control functions and the communications are implemented in a control board carried by the robot. The core of the control is a Microchip microcontroller that provides a good performance at a low price.

3 Machine vision guidance system

The guidance system is divided into two subsystems, a machine vision system that detects the position of the groove and a tracking algorithm that controls the movements of the robot.

3.1 Seam detection based on machine vision

This subsystem is based on a laser triangulation technique that is widely used to extract 3D characteristics from 2D image. The principle of function is show on figure 2-a.

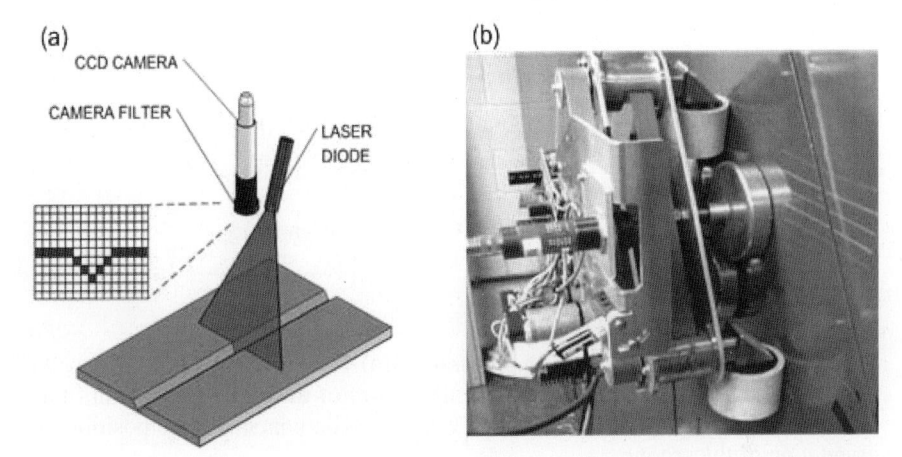

Fig. 2. Figures of the machine vision system. (a) Scheme of the principle of function of laser triangulation. (b) Picture of the actual system.

A laser plane of light is projected at a known angle with respect to a camera installed over the observed surface. The cross-section of the laser plane and the surface produces a bright line. This line is observed by the camera, allowing it to calculate the surface profile.
With only one laser line it is possible to get information about where the groove is but not about the groove's direction. This data is important for a mobile robot because it may not be aligned with the path. To solve this problem, at least two laser lines must be used to get two position points. This way the path is completely defined. In this system we use three laser lines to obtain redundant information and make the solution more robust. A picture of the actual system developed is shown on figure 2-b.

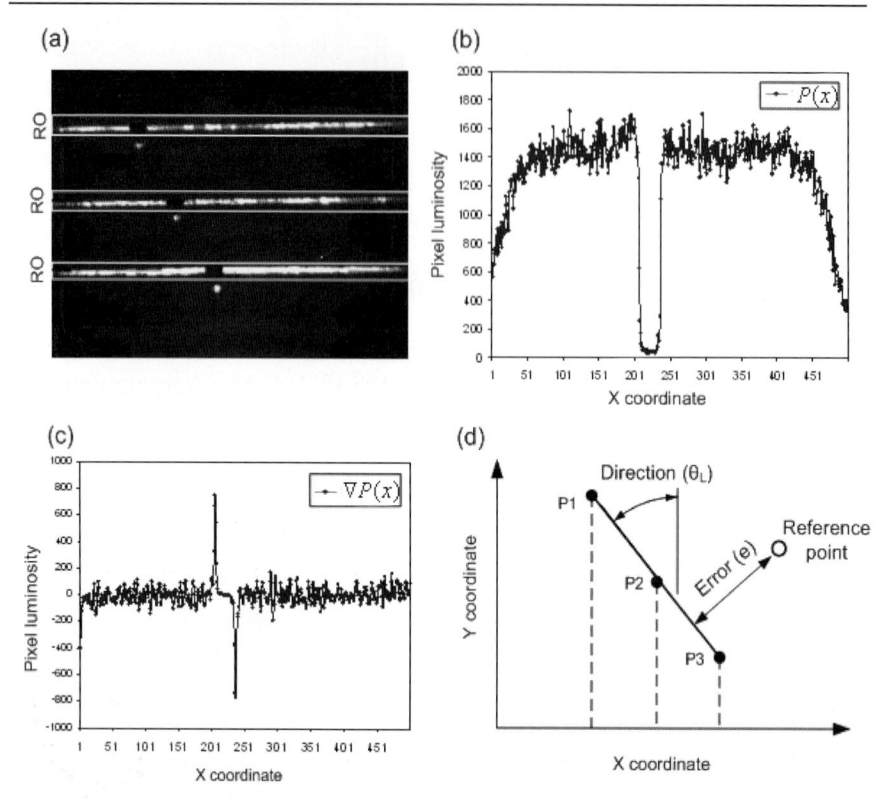

Fig. 3. Different stages in image processing. (a) Placement of the ROI over the captured image. (b) Result of summing the values of the pixels. (c) Result of applying the gradient operation to a ROI vector. (d) Estimation of the position and orientation of the groove.

The three laser plane lights come from a laser diode with a special lens that divides the laser beam. The camera is a compact vision system that includes image capturing, image processing and communications capacities. It is equipped with a band pass filter that is centered on the laser's wavelength, making the solution more immune to changes in ambient light. The laser lines are placed parallel to the rows of the camera sensor in order to simplify image processing. A scheme of the process applied on an actual image is shown in figure 3.

When the system starts, it exams the image row by row, locating where the laser stripes are. When they are located, a region of interest (ROI) is placed over each stripe. That means that only the rows inside a ROI are processed, saving computing time. After that the expression shown on

Equation 1 is applied in each ROI, where $V_{(x,y)}$ is the image pixel value at a given coordinate and the pair (j,k) are the limits of the ROI.

$$P_{(x)} = \sum_{i=k}^{i=j} V(x, y_i) \ / \ \forall x \in (ROI) \tag{1}$$

The result is vector $P_{(x)}$, which contains a proportional measure to the mean value of the lightning inside each ROI (figure 3-b). Once these vectors are calculated for each ROI, a median filter is applied to reduce the high frequency noise and a Prewitt filter is used to find the maximum and minimum gradient points (figure 3-c). These points represent the edges of the groove, therefore the position of the groove can be obtained by computing the point exactly halfway between these two edges. When the process ends we have three estimated groove points which are checked to confirm that they are aligned.

This method is fast and robust, allowing the analysis of three laser stripes with good performance (15 frames per second) and using relatively cheap hardware.

3.2 Tracking algorithm

This algorithm uses the information about the path direction (θ_L) and the error (e) to generate the next movement command. Its calculations are based on a PID algorithm as shown on Equation 2.

$$\theta_R = \theta_L + K_P \cdot e + K_I \cdot \int e \cdot dt + K_D \cdot \frac{d}{dt} e \tag{2}$$

The final orientation of the robot (θ_R) is a contribution between the path angle and the PID controller. Without the PID control, the robot would follow a direction that is parallel to the path. The proportional component (K_P) makes the error smaller and the integral component (K_I) corrects the alignment error between the camera and the robot. The derivative component (K_D) provides a fast response when a wheel slippage occurs.

4 Conclusions

The three-wheeled climbing robot has been tested in the laboratory with some small mechanical problems that can be easily solved and is prepared for the next part of our research, making two robots work together. This robot is lightweight and can be transported and handled by an operator without help or need for cranes. The total weight of the robot is 30 Kg and it can carry an extra payload of 10 Kg.

The machine vision guidance system has been tested with the robot following straight and curved paths. The laser location method has proven to be a robust method for detecting grooves and its accuracy is better than 0.2mm. It could be used in a real shipyard making some small changes such as increasing the laser's power.

Acknowledgements

This work is being performed as part of the activities of the DASA (Diseño y Automatización de Sistemas Avanzados) Research Group, in the Systems Engineering and Automation Department of the University of Vigo. The project is partially financed by the Dirección Xeral de Investigación e Desenvolvemento of the Xunta de Galicia (PGIDIT 02 DPI 30301 PR). Special thanks to J.M. Sotelo for his work in this development.

References

1. White T., Hewer N., Luk B.L. and Hazel J. (1998) The design and operation performance of a climbing robot used for weld inspections. *Proceedings of the IEEE, Int. Conference on Control applications*, Trieste, Italy 1-4 september, Volume 1, pp 451-455
2. Grieco J.C., Prieto M., Armada M. and Gonzalez de Santos P. (1998) A six-legged climbing robot for high payloads. *Proceedings of the IEEE, Int. Conference on Control applications*. Trieste, Italy 1-4 september, Volume 1, pp. 446 – 450
3. Ross B., Bares J. and Fromme C., (2003) A Semi-Autonomous Robot for Stripping Paint from Large Vessels. *The international journal of robotics research*. Vol. 22, No. 7-8, pp. 617-626
4. Hirose S. and Tsutsumitake H., (1992) Disk Rover: A Wall-Climbing Robot Using Permanent Magnet Disks. *Proceedings of the IEEE/RSJ Intern Conf on Intelligent Robots and Systems Raleigh*. NC July, Volume 3, pp. 2074 - 2079
5. Byoung-Oh K. Yang-Bae J. and Sang-Bong K. (2001) Motion control of two-wheeled welding mobile robot with seam tracking sensor. *International Symposium on Industrial Electronics*. Volume 2, pp. 851 - 856
6. Haug K. and Pritschow G. (1998) Robust Laser stripe Sensor for Automated Weld seam tracking in Shipbuilding Industry. *Proceedings of IECON'98*. Aachen. vol.2, pp. 1236-1241
7. Wilson M., (2002) The role of seam tracking in robotic welding and bonding. *Industrial Robot*. Volume 29, No 2, pp. 132-137

A Locomotion Robot for Heavy Load Transportation

Hidenori Ishihara and Kiyoshi Kuroi

Kagawa University, Japan
ishihara@eng.kagawa-u.ac.jp, s05g511@stmail.eng.kagawa-u.ac.jp

Abstract

This paper deals with a new type of 4-leg locomotion mechanism with jack-type legs. This research aims at developing the locomotion robot for heavy load transportation. The jack-type leg can hold heavy load without active control by actuator and lift up it by small power. The proposed mechanism achieved the locomotion by crawl step that forwards a leg by twisting the waist. As the experimental results, the developed prototype walked at 0.001m/s and held the load of 45 kg. This paper introduces the basic ideas and mechanism of robot with the proposed jack-type legs and discusses the basic performance of developed prototype.

Keywords: 4-leg locomotion robot, Transportation of heavy load, Jack, Crawl step, Reduction of energy consumption

1 Introduction

Various applications of locomotion robots have been introduced into many fields. Especially, the leg locomotion mechanism at the irregular terrain has been taken notice into, and the rescue activities become the important applications of the locomotion robot.

In such applications, it is important to carry the heavy load such as victim. However, the leg mechanism holding and carrying the load has never been achieved in the practical use. Four legged robot driven by DC motors realizes smooth action like as an animal, but it can only carry it own body. This is caused by shortage of torque generated by motors. Additionally, it consumes the large amount of energy because a motor requires electric current during generating the torque.

The hydraulic actuators can generate larger torque than DC motors. It however requires the large driving equipments such as compressor. Furthermore, it is difficult to design the hydraulic system because it consists of complex combination of hydraulic and electrical system.

We have taken account into the jack that is used to lift up a heavy load such a car. The jack can lift up a several ten-time heavier load than the driving power, and hold it without energy consumption.

This paper proposes the way to introduce the jack system into the mechanism of mobile robot. In the second section, design concept and crawl step that robot uses to move is explained. The third section introduces the prototype designed along the design concept. In the fourth section, the effectiveness of proposed mechanism is discussed through the performance of developed prototype. The finally section is conclusion of this report.

2 Design concept

This robot utilizes the mechanism of jack for the leg structure. The advantages of the jack-type leg are:
1) Lifting up heavy load by small power,
2) Holding the heavy load without energy and active control by the actuator,
3) Easy control with simple mechanism,
4) Keeping the posture even if the power supply is down due to the trouble.

Meanwhile, the jack has the simple mechanism that stretches toward a single direction. Therefore, it is required to design the mechanism to forward the leg as the locomotion mechanism. As a solution of this problem, the crawl step with twisting the waist is introduced as the moving mechanism.

The crawl step is the typical step of reptile such as a crocodile and lizard. The crawl locomotion has the following features.
1) Three legs always are keeping touch with the ground to hold its body.
2) There is no movement to top and bottom.
3) Crawl step can carry a leg forward by twisting the waist even if the leg does not have the mechanism to forward it.

Crawl step is suitable for heavy load transportation because of above point (1) and (2). In point (3), it is possible for the proposed jack-type leg to be applied into the moving mechanism.

3 Design of prototype

The prototype robot was developed in order to confirm the effectiveness of proposed mechanism with the jack-type leg and crawl step. The specifications of first prototype robot were decided as follows.
1) Dimension of prototype is 300 x 300 x 300 mm.
2) The maximum load is 40 kg.

Figure 1 shows the image of prototype robot. This chapter introduces the idea to design the structure of the robot as shown in **Fig. 1**.

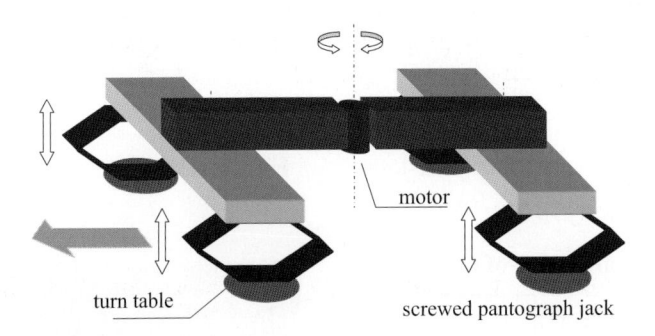

Fig. 1. Image of prototype robot

3.1 Structure of leg with the pantograph jack mechanism

Figure 2 shows the basic structure of leg using the pantograph jack mechanism. The leg has a joint of the pantograph jack and a turn ankle.

Figure 3 is the kinematic model of pantograph jack. In **Fig. 3**, Fa and Ft are stresses at arm, W is load to the system, and q is an angle between the arm and thread. Here, the stress at thread is derived as:

$$F_t = 2F_a \cos\theta = \frac{W}{\tan\theta} \tag{1}$$

The torque to turn the screw is expressed as:

$$T = \frac{d}{2} F_t \tan(\phi + \varphi) \tag{2}$$

Where, d is the diameter of thread, ϕ is the angle of friction, and φ is the lead angle.

Finally, the specifications of the jack-type leg were decided as shown in **Table 1** by adding the consideration of strength of each part. **Figure 4** indicates the photographs of the developed leg.

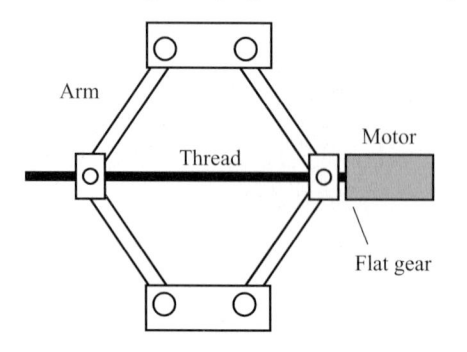

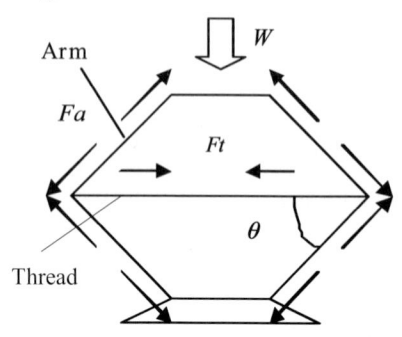

Fig. 2. Basic structure of jack-legs

Fig. 3. Kinematical model of legs

Table 1. Specifications of the leg

Height of leg	160 mm
(Shrunk condition)	154 mm
Width of leg	60 mm
Depth of leg (Including motor)	180 mm
Weight	630 g
Motor	DC Motor (2224 U006SR)
Tread	Trapezoidal

Fig. 4. Jack-type leg

3.2 The body with the structure of twisting the waist

The waist plays the very important role for the proposed robot's locomotion with jack-type leg, crawl step. Its bending angle decides the distance of moving. According to the calculation of torque to the waist and consideration of mechanical strength, the specifications of body were decided as shown in **Table 2**. **Figure 5** shows the photograph of developed body. The designed body consists of the frames, a DC motor and a worm gear set.

Table 2. Specifications of the leg

Length of body	200 mm
Width of body	242 mm
Height of body	242 mm
Weight of body	2.4 kg
Motor	Dc motor (Tamiya 380K75)
Worm Gear Ratio	50:1

Fig. 5. Developed body

3.3 Structure of prototype

Figure 6 shows the photographs of developed prototype. Dimension of prototype is 378 x 242 x 253 mm and its weight is about 4.9 kg. Current prototype does not have the controller inside it.

It does not contain the loading platform. The loading platform is one of the most important parts of transporter, but this prototype was only designed for testing the locomotive ability.

Fig. 6. Prototype robot

4 Experiments

The prototype robot without the loading platform was developed initially. This chapter shows only the basic properties of robot for walking and the capacity of holding the load.

4.1 Basic walking

Figure 7 shows the basic walking action. The robot moves forward along the steps as shown in Fig. 7. In these experiments, the robot was controlled manually.

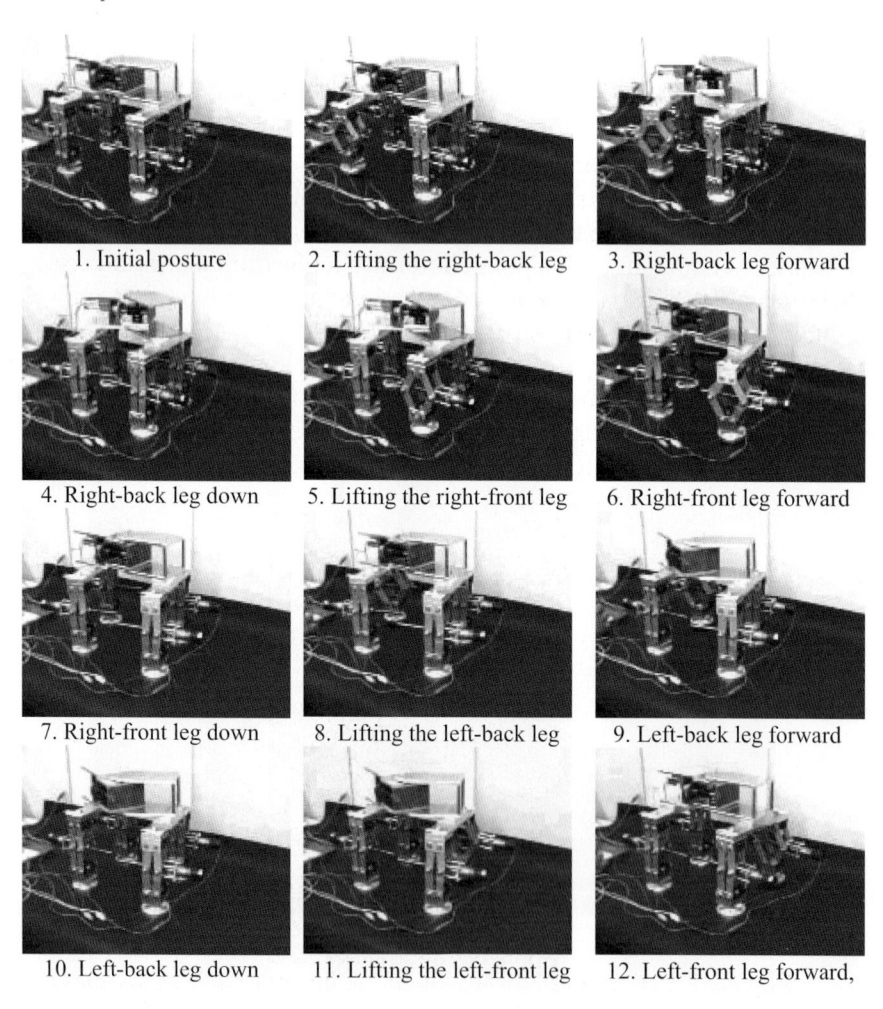

1. Initial posture
2. Lifting the right-back leg
3. Right-back leg forward
4. Right-back leg down
5. Lifting the right-front leg
6. Right-front leg forward
7. Right-front leg down
8. Lifting the left-back leg
9. Left-back leg forward
10. Left-back leg down
11. Lifting the left-front leg
12. Left-front leg forward,

Fig. 7. Walking flow of developed prototype robot

The walking speed was about 0.001m/s. This is related to the speed to bend its waist and the decline of body frame. Especially, the decline of body was bigger than we designed and therefore the leg must be lifted up larger than we expected. Also, the bending speed was slow because the worm gear is used. Furthermore, getting the control automatic, the locomotion speed is expected to be faster.

4.2 Static loading capacity

This research aims at the development of the robot for heavy load transportation. The loading capacity is the most important factor of its function for this robot.

The loading capacity was about 45 kg when the robot was standing. **Figure 8** shows the example of test to hold the heavy load on the robot. We also tested the capacity at walking, but it could not be measured because it is impossible to put the heavy load on the back of the robot stably. When the developed robot is walking, it twists its waist. The heavy load is swung by bending its waist. Therefore, the robot can't keep balance because the center of gravity is moved dynamically.

In other word, the developed robot needs the loading platform that keeps its posture independently of actions of body and legs. During walking, the waist bends repeatedly so that the platform on the top of the robot also twists. Therefore, the platform must keep its posture stable though the robot's body is twisted. It will be installed at the next step to improve the performance of transportation.

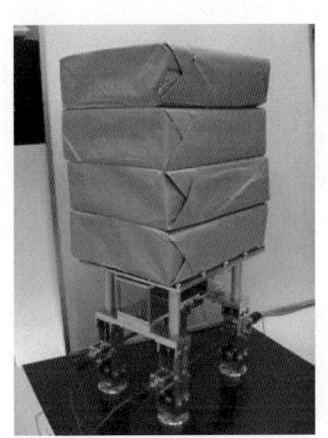

Figure 8 Photograph of test to hold the heavy load: The weigh of load is 27.2 kg.

5 Conclusions

This paper proposed new mechanism of 4-leg robot for heavy load transportation. The prototype robot has the jack-type leg, and it moves by crawl step. The jack-type leg can lift up heavy load without large torque, and keep its posture without any power. Crawl-step realized the locomotion of the prototype robot with the proposed leg mechanism.

It was confirmed that the robot with the jack-type leg can move by the crawl-step. The prototype robot also gave proof that it can lift the heavy load and hold it without power.

Now, we are trying to install the loading platform. Then, we will test the ability of heavy load transportation at walking. Additionally, this robot was developed for the locomotion at the rough terrain, but now it has not been tested. The reason is that this prototype does not contain the sensor which detects the condition of ground. In the next step, some sensors to recognize the ground will be installed and the locomotion on the rough ground will be realized.

References

1. K. Kato and S. Hirose, "Development of Quadruped Walking Robot, TITAN-IX –Mechanical Design Concept and Application for the Humanitarian Demining Robot-" Advanced Robotics, 15, 2, pp. 191-204, 2001
2. H. Kimura and Y. Fukuoka, "Biologically Inspired Adaptive Dynamic Walking in Outdoor Environment Using a Self-contained Quadruped Robot: 'Tekken2'", Proc. of IEEE/RSJ Int. Conf. on Intelligent Robots and Systems (IROS2004), CD-ROM, 2004
3. Y. Sugahara, T. Endo, H.O. Lim and A. Takanishi, "Design of a Battery-powered Multi-purpose Bipedal Locomotor with Parallel Mechanism," Proc. of the 2002 IEEE/RSJ International Conference on Intelligent Robots and Systems (IROS02), pp. 2658-2663, 2002.
4. K. Kurashige, T. Fukuda, and H. Hoshino, "Motion Planning Based on Hierarchical Knowledge for Six Legged Locomotion Robot," Proc. of IEEE Int'l Conf. on Systems, Man, and Cybernetics (SMC'99), Vol. VI, pp. VI924-929, 1999
5. M.Lasa, M.Buehler, "Dynamic compliant quadruped walking," Proc. Of IEEE Conf. on Robotics and Automation (ICRA01), CD-ROM, 2001
6. Ren C. Luo, S. H. Henry Phang, and Kuo L. Su, "Multilevel Multisensor Based Decision Fusion for Intelligent Animal Robot," Proc. Of IEEE Conf. on Robotics and Automation (ICRA01), CD-ROM, 2001
7. K. Takita, T. Katayama, and S. Hirose, "The Efficacy of the Neck and Tail of Miniature Dinosaur-like Robot TITRUS-3," Proc. Of IEEE/RSJ Conf. on Intelligent Robots and Systems (IROS02), CD-ROM, 2002

Using Signs for Configuring Work Tasks of Service Robots

Mikko Heikkilä, Sami Terho, Minna Hirsi, Aarne Halme, and Pekka Forsman

Automation Technology Laboratory, Helsinki University of Technology, PL 5500, 02015 TKK, Finland, E-mail mikko.heikkila@tkk.fi

Abstract

This paper describes how to use signs as a part of the work task scenarios with service robots. The signs are introduced as an alternative or complementary tool for passing information of the task plan of a work task from the human operator to the service robot. The signs can be used for pointing a direction, bounding an area, marking a route, or defining location of a target. There are two kinds of signs, passive and active. The passive signs have all the information in their appearance. In active signs the information content can be modified, for example with radio communication. Moreover, active signs can be equipped with sensors such as GPS-receiver and compass to measure their location on the map. One of the advantages of using signs to control work tasks is usability. The signs are just carried to the working area and no other actions are needed. Tests in real outdoor and non-structured environments with the WorkPartner robot are realized and experimental results are shown.

Keywords: service robot, task planning, active sign, passive sign

1 Introduction

Signs are widely used with robots for navigation. The locations of signs are typically fixed and the robot knows them. In navigation they are usually called beacons or artificial landmarks. For example automatic guided vehicles follow stripes on the floor, or navigate by using reflective stripes as fixed landmarks.

Other approach is to use movable landmarks to provide relative instead of absolute location information. The appearance and position of the sign may also give more information. Movable and variable-looking signs are suitable for defining tasks for service robots. The signs can be used for pointing a direction, bounding an area, marking a route, or defining location of a target. Also usability of these kinds of signs is in high level. Configuring the tasks is easy because the signs are just carried to the working area and no other actions are needed.

The signs can be divided to two main categories. Passive signs contain all the information in their appearance. For example normal road signs belong to this category. In active signs the information content can be modified, for example with radio communication. Moreover, active signs can be equipped with sensors such as GPS-receiver and compass to measure their location on the map.

This paper presents a concept for using signs as a part of service robots' task definition and execution. The robot used in implementation and tests of the concept is WorkPartner service robot developed in Automation Technology Laboratory in Helsinki University of Technology (TKK, formerly known as HUT). The robot is designed for outdoor environments. The WorkPartner project has been reported in seven previous CLAWAR conferences [1], [2], [3], [4], [5], [6] and [7].

2 Previous Research

The concept of using active markers, placed in the operation environment, for exchanging information between robots and the environment was introduced by Asama [8]. The markers, called Intelligent Data Carriers (IDCs), are portable electronic data carrier devices. Wireless data exchange between the IDCs and the robots is realized through RF-ID weak radio communication. The main application area planned for the IDC-devices was related to cooperative organization of multiple robots and their operation environment. For example, information about the existing free routes in an unknown environment exchanged between the members of a robot society by means of IDCs. When a robot member detects that a route is leading to a dead-end, it can write the information to the IDC, placed at the beginning of the route, when it comes back from the dead-end.

In what follows, we will present results of using IDCs or, as they are called here, active signs for configuring work tasks for service robots. In addition to active signs also more simple passive signs are introduced as an alternative or complementary tool for passing information of the task plan of a work task from the human operator to the service robot.

3 Experimental results

Two work tasks were used to test the concept. In the first task the robot ploughed an area from the snow, and in the second task the robot brushed the ground. In both tasks signs were used to indicate the working area of the robot. Both active and passive signs are used in the research.

3.1 Case study: Passive Signs

In the first task the passive signs were used. The four signs were corner points of the working area. The signs showed direction to the next sign. Fig. 1 shows a sign. There are two spheres in the passive sign; the bigger orange one and the smaller yellow one.

Fig. 1. A passive sign placed in one corner of the working area. The sign is pointing in to the direction of the next corner point. The position vector pointing from the center of the big ball to the center of the small ball defines the direction.

Passive signs were found by using computer vision. The spheres appeared as circles in the image. The image was segmented by color. The orange and yellow circles were found from the threshold image using circle-fitting algorithm (Fig. 2). The distance to the sign was determined by measuring the size of the circle on the image plane. 3D pose of the sign was calculated based on the measured distance and the camera orientation. [9]

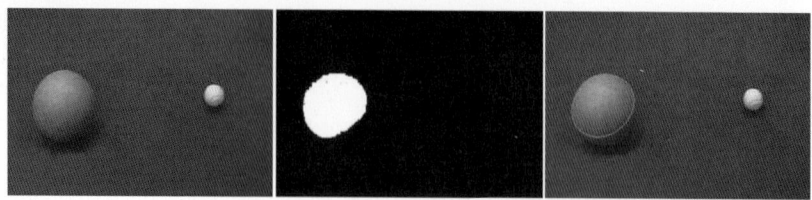

Fig. 2. The original image, image thresholded with orange color and the found orange circle [9]

In the task the working area was determined based on the information acquired from the signs. During the task execution the robot wandered a route inside the working area and moved the tool, the plough, up and down. More specially, the robot wandered to one end of the working area, lowered the tool, and started to clean the area. After reaching the other end of the area, the robot raised the tool again, and moved back to the other end. After a few rounds the whole area was cleaned.

3.1.1 Results

During the task execution the accuracy of the image handling method was not measured. The accuracy was measured before using signs in the work tasks [9]. Based on the results it was easier to develop the control system for the work tasks. To find out the accuracy of the image handling method, the distances of the spheres were measured also with a laser pointer [9]. The laser pointer is the second eye of WorkPartner and the accuracy of the pointer is about 1-2 mm. In the experiments the distances measured by the vision system and the laser pointer were compared [9].

The distance from WorkPartner's head (both from camera and from laser pointer) to the sign was transformed to the location relative to the base coordinate frame of the robot (Fig. 3.) by calculating needed kinematics transforms. The figures 4 and 5 show the errors of the locations. The errors are same as the caps between calculated locations of the sign measured by the laser pointer and the camera. In the experiments the sign was pointing about 90 degrees to the right at the front of the robot. [9]

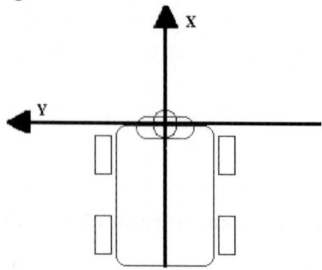

Fig. 3. The base coordinate system of the robot [9]

The results shown in figures 4 and 5 demonstrate that it is possible to localize the passive signs properly when the measuring distance is from 2 to 4 meter. There was more error in the X-coordinate because the measured distance from the camera is more related to the image handling than the angle of the pan-tilt-unit of the robot's head. The heading of the sign was calculated based on the position of its spheres. The cap between correct angle of the sign and calculated was from -10 to 10 degrees. The error was quite huge but small enough in this case. [9]

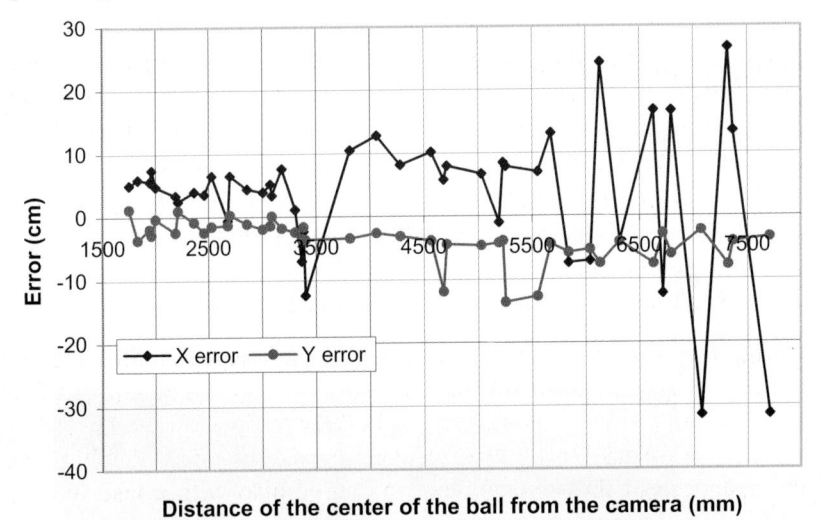

Fig. 4. The accuracy of the localization (orange ball) [9]

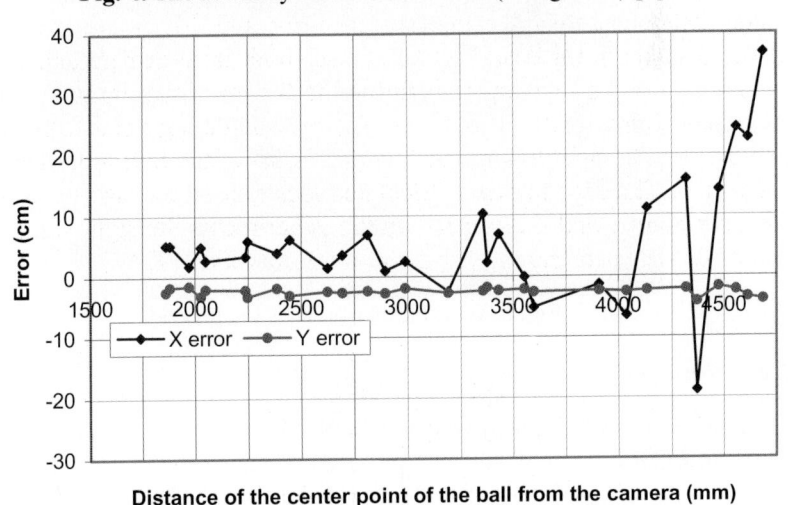

Fig. 5. The accuracy of the localization (yellow ball) [9]

In the experiment the accuracy of the measures was good enough because it was not needed to know exactly positions and headings of the signs. The idea was to say to the robot "go to that direction" or "I am somewhere here". This includes same kind of information as between humans. The image handling caused most errors when measuring locations of the signs. Changes in lighting conditions were the most effective aspect in image handling. Task planner should take account the limitations of the accuracy when planning the movements and routes for the robot.

3.2 Case study: Active Signs

The active sign was used in the brushing task (Fig.6.). The active sign includes microcontroller, Bluetooth and GPS modules. The sign was located in the center of the working area. Before starting to clean the area the WorkPartner communicated with the sign via Bluetooth. Received data from the sign included the location of the sign and the radius of the working area. The robot planned the brushing task based on the received information in same way as in a case of passive signs.

Fig. 6. WorkPartner brushes the ground.

3.2.1 Results

The active sign was located with the GPS. The accuracy of measured location was suitable enough for our purposes, because the nature of the work tasks and the adaptability of the robot do not need the exact location. At the more difficult tasks the robot needs to locate and identify the objects for manipulation, hence the ordinary GPS is precise enough to identify the general work area.

The usability of the active sign is also affected by the range of Bluetooth, because it is used for the communication of the robot and the sign. In our case the range of the Bluetooth has not been an issue. Robot usually works in areas where distances are easily inside the range of the Bluetooth and there have been no obstacles to block the Bluetooth signal, such as walls, between the robot and the sign.

The location of the active sign in the middle of the working area is a problem. The robot should be able to remove the sign from its location while brushing. The better choice is to set the sign to the corner of the working area but then for example a compass or other means of determining the direction of the sign is needed. This is one part of the future work.

4 Conclusions

The results of the tasks indicated that using signs is a very effective way to define the working area of a service robot. In the experiment two different kinds of signs, active and passive, were tested in two tasks. In the first task the passive signs were used. The four signs were corner points of the working area. The signs showed direction to the next sign. The active sign was used in the second task. The sign was located in the center of the working area and the sign included microcontroller, Bluetooth and GPS modules. The passive sign were located with the camera of the robot and the active sign via Bluetooth.

In the research the signs were just carried to the working area in the beginning of the tasks. During the tasks execution the robot localized signs and found all the needed information from the signs. The usability of using the signs was very high.

In the tasks the signs bounded the work area. Other approaches to use the signs in the future are pointing a direction, marking a route, or defining location of a target etc. Both active and passive signs are suitable for these purposes. Other potential development area in the future is to develop a "sign language", a common reference for using of signs with the work tasks of the service robots. A very typical example of the sign language is the road signs. They are looking same all over the world.

References

1. Leppänen I, Salmi S, Halme A (1998) WorkPartner – HUT-Automations new hybrid walking machine. 1st International Conference on Climbing and Walking Robots, Brussels
2. Halme A, Leppänen I, Salmi S (1999) Development of WorkPartner-robot – design of actuating and motion control system. 2nd International Conference on Climbing and Walking Robots, Portsmouth
3. Halme A, Leppänen I, Salmi S, Ylönen S (2000) Hybrid locomotion of a wheel-legged machine. 3rd International Conference on Climbing and Walking Robots, Madrid
4. Halme A, Leppänen I, Montonen M, Ylönen S (2001) Robot motion by simultaneous wheel and leg propulsion. 4th International Conference on Climbing and Walking Robots, Karlsruhe
5. Ylönen S, Halme A (2002) Further development and testing of the hybrid locomotion of WorkPartner robot. 5th International Conference on Climbing and Walking Robots, Paris
6. Luksch T, Ylönen S, Halme A (2003) Combined Motion Control of the Platform and the Manipulator of WorkPartner Robot, 6th International Conference on Climbing and Walking Robots, Catania
7. Ylönen S, Heikkilä M, Virekoski P (2004) Interaction between Human and Robot – Case Study: WorkPartner-robot in the ISR 2004 Exhibition, 7th International Conference on Climbing and Walking Robots, Madrid
8. Asama H (2001) Towards Emergence in Distributed Autonomous Robotic Systems – Intelligent Data Carrier for Cooperative Organization of Mobile Robots and Their Operating Environment. 3rd International Conference in Field and Service Robotics, Helsinki
9. Suontama H-L (2003) Use of Signs for Configuring Tasks for Service Robots. Master's Thesis (in Finnish), Helsinki University of Technology, Espoo

System for Monitoring and Controlling a Climbing and Walking Robot for Landslide Consolidation

Leif Steinicke, David Dal Zot and Tanguy Fautré

Space Applications Services, Leuvensesteenweg 325, B-1932 Zaventem, Belgium leif.steinicke@spaceapplications.com,

Abstract

This paper describes the application of space robotics control technology applied to the slope/landslide consolidation sector. The description is made from the point of view of monitoring and controlling a walking/climbing machine used in the Roboclimber project.

1 Introduction

Fig. 1. Early conceptual prototype of the Roboclimber

Figure 1 shows an early conceptual prototype of the Roboclimber, used to test the concept, operating on a near vertical rock face. Stabilising a risky slope initiates with a geological survey, after which a series of holes are made to consolidate the wall. The objective of a tele-operated climbing robotic system for the maintenance and consolidation of mountain slopes is to drill the deep holes for the insertion of 20-metre long rods. This saves time and operating costs and is less dangerous. Roboclimber is made by a consortium of European SMEs and uses the same expertise as that used in ESA robotic missions and instruments used to control the attitude of spinning satellites.

2 CLAWAR Technology Used for Slope/Landslide Consolidation

2.1 Slope consolidation scenario

This section describes a typical use of monitoring and control technology in the field of slope consolidation. Figure 2 shows Roboclimber on a slope to be consolidated..

Fig. 2. Roboclimber in action in the mountains of North-East Italy

The following describes the different stages involved in consolidation of a slope, using a climbing and walking robot. The general approach is that the robot climbs down from an anchoring point at the top of the slope, and then is controlled to walk to a number of positions on the slope where it will drill rods into the slope.

The typical steps are thus:
- setting the anchor on the top of the slope;
- climbing down the slope from the anchoring point
- positioning the robot on the slope at a drilling location;
- drilling a rod (or several rods) into the slope at drilling location;
- move to new drilling location and repeat drilling, until all reachable drilling locations, from the current anchor point, have been consolidated with rods;
- moving the robot back up the slope and walk it to the next anchoring position;
- iterate the process until the entire slope is consolidated with rods.

During the whole process, at all times the monitoring and control system will keep the operator aware of the situation through a Human-Machine Interface (HMI) providing him with necessary information about the robot, such as its position, its speed, leg position, and various relevant telemetry for each particular task.

3 Requirements on Robot Technology in Slope/Landslide Consolidation

From the typical slope or landslide consolidation scenario described in the previous section, a number of requirements can be derived for the robot technology applied in this sector, and the associated monitoring and control.

3.1 Requirements on the robot

3.1.1 Moving and positioning

A tele-operated climbing robot shall be able to move vertically, laterally and horizontally during slope consolidation.

The robot structure should be provided with an ad hoc base plate to act as a skid support for quick vertical movement and as a protection against protruding rocks.

The robot shall be able to move laterally and vertically by a combination of the ropes by which it is suspended from the top of the slope and its own legs.

The positioning of the robot shall be able to be provided by the robot/robe system.

Due the working conditions all on-board equipment should be well-anchored with the robot frame and work independently from the position on the slope.

3.1.2 Power supply

A climbing robot for slope consolidation is likely to require three types of power sources:
- pneumatic: compressed air (at e.g. 12-20 bar) for a drilling unit (for the operations of drilling and flushing debris from the drilling);
- hydraulic: oil (at e.g. 200 bar) for a drilling unit (to rotate and advance), for the leg movements and other services;
- electric: for sensors, control system, cameras and lighting.

3.1.3 On-board equipment

A robot for slope consolidation is likely to require the following types of on-board equipment:
- drilling rig;
- drilling rods;
- drilling bit;
- 1 camera with pan/tilt/zoom) for monitoring leg positions;
- 1 fixed camera for monitoring the drilling area;
- electronics unit for performing the on-board control.

3.1.4 Other system features

The robot shall incorporate features that actively and passively ensures that the robot will not become disconnected from the ropes that support it from the top of the slope.

3.2 Requirements on the robot HMI

The HMI shall allow the operator to move the robot as a whole or component by component depending on the operational context (climbing, pre-drilling positioning, walking to anchor points, etc.).

The HMI shall incorporate a simple to use and easily recognizable emergency stop button.

The HMI shall allow the operator to manage and monitor the drilling, using high level macros or low level primitives.

The HMI shall be able to operate in a harsh environment: wide temperature range, humidity, precipitation, dust, etc.

The HMI shall be able to operate in a bright daylight environment, leading to special needs of screens to provide high contrast, and provide features to reduce reflections of light.

The HMI shall be ruggedized to sustain being dropped and/or treated roughly.

The HMI shall be user friendly and accessible to non computer-literate users. This requirement translates into employing devices such as joysticks and buttons, rather than keyboard and mouse.

The HMI should follow a layered approach allowing the operator to access high level or low level commands depending on the situation.

The HMI should support different modes, each mode being tailored and adapted to a particular phase of the overall operation. Such modes include:

- Prestart mode
  Used to specify terrain-specific data.
- Positioning mode
  - Moving/Climbing/Anchoring: the operator uses the joystick to indicate the robot where to move (high level control).
  - Finetuning: for a given leg, the operator will send directives such as "move forward 0.5 meter". Alternatively, the Operator can adjust its rotation, extension and elevation.
- Drilling mode
  - High level operations: the operator specifies a depth and the HMI automatically has the robot drill a hole of the specified depth (high level command).
  - Medium level operations: the operator controls the robot on a rod by rod basis and sends commands such as "insert rod", "remove rod" and so on.
  - Low level operations: likely to be directly accessed only for testing and debugging. At this level, the operator would control the drilling subsystem on a very low level and would send commands such as "rotate joint x by n degrees".

3.3 Robot HMI design

3.3.1 General considerations

The command console will be PC-based. It must be light so that it can be carried around and easily strapped around the neck. As the robot has to be operated outdoors, this command console has to be harsh condition resistant (rain, dust) and the screen must have outdoor-displaying capabilities.

Finally, the console should ideally be used as a remote control, to prevent the user from being too close from slopes and the display should be large enough for displaying camera views.

The camera system is network based, TCP/IP cameras typically broadcast video in the form of MJPEG, MPEG-1 or MPEG-4 streams.

3.3.2 Tablet PC as a command console

The hardware solution that has been chosen is a Fujitsu-Siemens tablet-PC.

Fig. 3. Fujitsu-Siemens Tablet-PC

The 4121 model has an outdoor screen and a harsh environment case that enables manipulation even with dust or rain, and the screen is large enough to hold all the relevant information for controlling the robot. It also has an embedded 802.11b wireless Ethernet card.

3.3.3 HMI Operator Mode

When running in Operator Mode, Roboclimber will be controlled using one joystick and a set of buttons (provided by the joystick). These buttons are used to change the operational modes. All the commands are sent via a TCP/IP connection. Figures 4 and 5 show examples of what the operator sees when he is controlling the robot in "high level mode". With a single button, he can switch to the desired sub-system he wants to control.

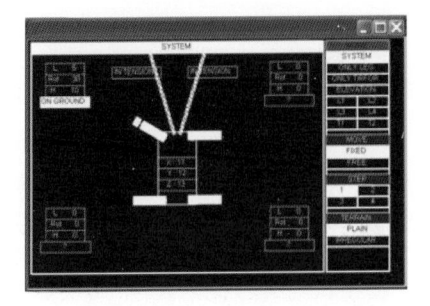

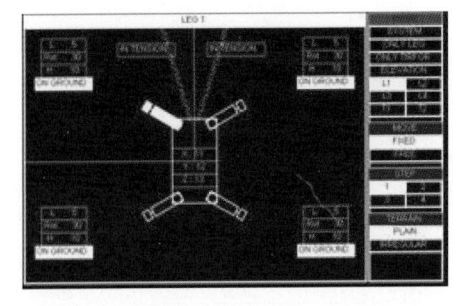

System for monitoring and controlling a climbing and walking robot for... 923

Fig. 4. System Level Control View & **Fig. 5.** Leg Level Control View

Operating the robot is done by selecting the desired move and setting the value or increment to reach. Figures 6 and 7 shows how the user fine-tunes the rotational angle of the upper left leg, by setting the angle to 42 degrees. An arrow indicates the direction of the leg after having confirmed the command and real-time display is provided during the move.

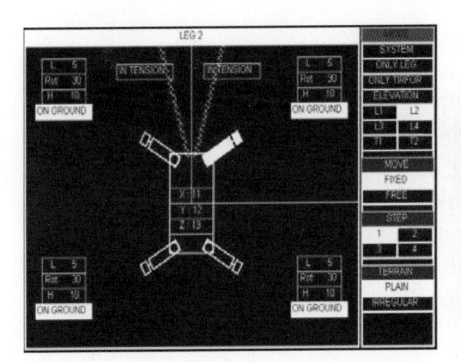

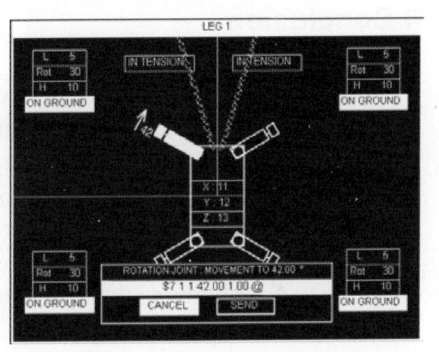

Fig. 6. Leg Level Control View & **Fig.7.** Command Confirmation Dialog

It is possible to send emergency stop signals during a move. Also, for security reason, the server running on the robot ping the client regularly and it stops immediately any move if the client does not answer (for instance, if the console runs out of battery or if it breaks down).

3.3.4 HMI Superuser Mode

Superuser mode provides access to low-level functionalities for testing or for resolving contingencies.

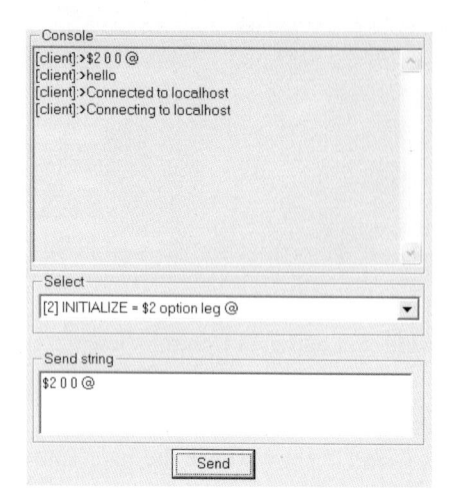

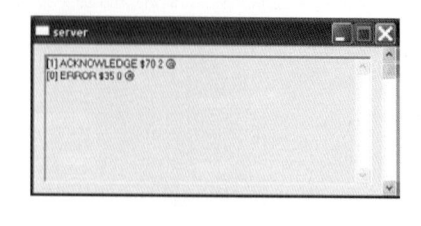

Fig. 9 & 10. Low-Level Control Dialog Boxes

4 Conclusion

Stabilizing risky slopes is a dangerous job. Today it is done manually by workers who have to build and climb tall scaffolding and remain exposed to siliceous dust and loud penetrating noise. Maybe the most important aspect of the Roboclimber is that it makes risky jobs safer. Thanks to automated interventions and the use of remote control, accidents related to operating on high scaffolding can be eliminated: any sudden soil movement will endanger nobody as the operators are working from a safe distance.

References

1. URL http://www.esa.int/esaCP/SEM9R03AR2E_index_0.html
2. Anthoine P., Armada M., Carosio S., Comacchio P., Cepolina F., Klopf T., Martin F., Michelini R.C., Molfino R.M., Nabulsa S., Razzoli R.P., Rizzi E., Steinicke L., Zannini R., and Zoppi M. (2004) A Four-Legged Climbing Robot for Rocky Slope Consolidation and Monitoring. *Int. World Automation Congress WAC2004*, Seville, Spain, ISBN 1-889335-20-7. CD proceedings.
3. Nabulsi S., Montes H. and Armada M. (2004) ROBOCLIMBER: Control System architecture. *Proc. Int. Conf. of Climbing and Walking Robots, CLAWAR 2004*, Madrid, Spain, 22-24 September.

Non-destructive Testing Applications

Small Inspection Vehicles for Non-Destructive Testing Applications

M. Friedrich, L. Gatzoulis, G. Hayward, W. Galbraith

Centre for Ultrasonic Engineering, University of Strathclyde, 204 George Street, Glasgow G1 XW, UK, markus.friedrich@eee.strath.ac.uk

Abstract

Miniature autonomous vehicles offer high potential for inspection tasks in areas where access is difficult or that are hazardous for human intervention. This paper focuses on the design of small mobile robots for non-destructive testing (NDT) of ferromagnetic materials. The major challenge in this connection is miniaturisation, as climbing skills are required in order to cope with various types of inclined or curved surfaces and ceilings. The proposed compact design involves permanent magnets that provide both the holding force and the source of the magnetic field for surface inspection using Hall Elements. Experiments show that artificial defects up to 1 mm width can be detected reliably with this passive inspection method.

Keywords: MFL-Inspection, climbing robot, Hall Sensor, magnetic wheels

1 Introduction

Miniature and high performance mobile robots are well suited for remote inspection tasks in areas that are only accessible through narrow passageways or present a health and safety hazard for human operators. In particular, infrastructures that consist of large numbers of components with small structural dimensions motivate the application of autonomous micro systems for inspection and on-site manipulation. The major requirements for these vehicles are mobility and complete autonomy in terms of energy requirements. Furthermore, in distributed robotic systems which incorporate a large number of collaborating miniature or micro-scale vehicles, the individual robots can be much smaller, lighter and be less expensive.

Non-destructive testing (NDT) plays a key role in the prevention of structural failure. An initial inspection of new structures, in order to detect manufacturing flaws, is usually followed by repeated inspections during service to locate fatigue cracks, corrosion or accidental damage. General requirements for NDT are simplicity of application, sensitivity to small anomalies and reliability. In this context, inspection of diversely orientated ferromagnetic surfaces constitutes an application area for robotic vehicles that provides holding force by means of magnetic adhesion.

Using magnetic wheels is a common way of attaching mobile climbing robots to ferromagnetic surfaces [1, 2, 3]. However, utilising their field for NDT applications in order to reduce overall vehicle weight and size is a novel extension and affords considerable potential. Application areas of magnetic flux leakage (MFL) based inspection, performed by a team of heterogeneous autonomous climbing robots, are: offshore platforms, pipe work in the oil and gas industry and nuclear power plants, railway lines, hulls of vessels and storage tanks.

2 Locomotion and Climbing Mechanism

As the mechanical structure of the vehicle is of major importance with regard to final performance, the suitability of general locomotion and climbing techniques has been investigated firstly with regard to miniaturisation and NDT applications. Compared with crawling, sliding, tracks, running and walking [4, 5] the active powered wheel was considered as being the most suitable locomotion mechanism for the inspection vehicles. It is the most energy-efficient motion mechanism, taking into account the limited on-board energy supplies of autonomous vehicles. The compact design, due to mechanical simplicity offers the highest potential for miniaturisation. Moreover, a wheeled design is appropriate for structural condition monitoring, where predominantly hard and flat surfaces are to be expected and stable balance is needed.

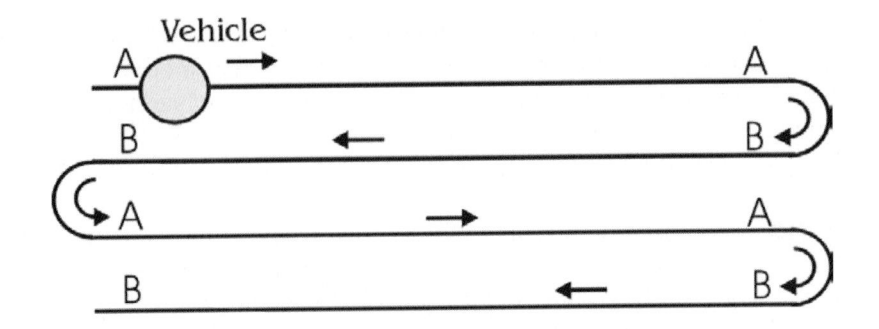

Fig. 1. Scanning Pattern for non-destructive inspection

Autonomous non-destructive testing is not restricted to purely horizontal surfaces, hence the vehicle requires climbing skills in order to cope with various types of inclined and curved surfaces and ceilings. To highlight the benefits of different climbing techniques, four different categories are set apart, based on the physical principle of adhesive force: magnetic adhesion, dry adhesion, adhesion based on vortex regenerative air movement and suction adhesion. Overall robot size and complexity increases in that listing, and since miniaturisation is regarded as being crucial, a vehicle design based on magnetic wheels offers the highest potential.

The different vehicle kinematics that result from different wheel designs and wheel arrangements have been evaluated for NDT applications by using the concept of *holonomy* and determining the *degrees of freedom* (DOF) and *differential degrees of freedom* (DDOF) [1]. While the DOF is equal to the number of independent coordinates that a robot can achieve on a two-dimensional surface, the DDOF of a mobile robot is equal to the number of degrees of freedom that can be immediately manipulated by changes in wheel velocity. As visualised in the trajectory shown in **Fig. 1,** the scanning vehicle requires high directional stability between the points *AA* and *BB* and high maneuverability for the close turns between *AB* and *BA*. Therefore DDOF has to be two. An immediate position change orthogonal to the current moving direction is supposed to be impossible due to the wheel constraints. A nonholomic configuration with DDOF=2 and DOF=3 therefore provides the required amount of maneuverability for accurate straight line movements on the one hand and turnarounds on one spot on the other hand. Hence, the most compact configuration of that kind consisting of two independently driven fixed wheels supported by an omni directional wheel was selected for the vehicle. Furthermore, the combined drive and steering mechanism minimises steering errors. Odometric pose estimates for dead reckoning are expected to be highly precise as the intense attraction force between magnetic wheels and the surface prevents slipping.

3 NDT method based on MFL inspection

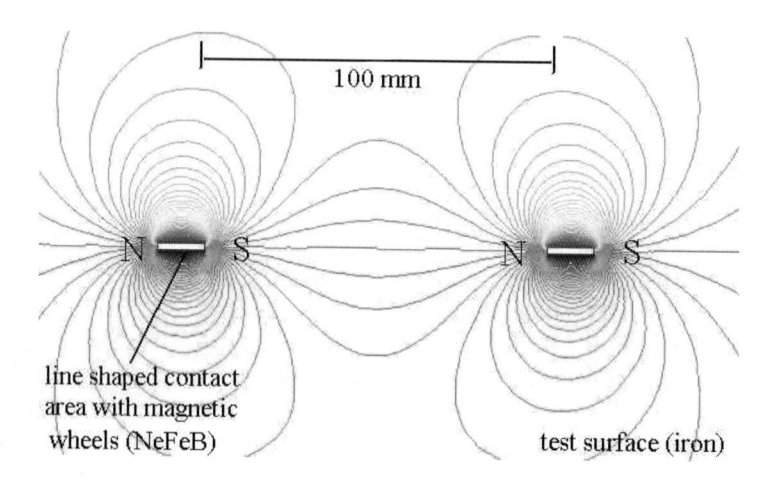

Fig. 2. Qualitative distribution of flux lines on surface of ferromagnetic material

The six primary methods for non-destructive testing of engineering structures are visual inspection, dye penetrant inspection, radiography, magnetic inspection, ultrasonic inspection, and eddy current inspection [6, 7]. Despite being restricted to ferromagnetic surfaces, a method based on magnetic flux leakage (MFL) inspection has been selected, as it is straightforward to implement with the existing vehicle design: there is no need for additional devices for magnetising the test material as the magnetic wheels already create a field on the test surface: **Fig. 2** illustrates the distribution of the flux lines of that field between the linear contact area of the two driven magnetic wheels and the test material and shows the general applicability of MFL inspection. Information about dimension and extent of the expected flux leakage is obtained by using the software package Femlab (Comsol). One result based on the wheel arrangement presented in **Fig. 2** is shown in **Fig. 4**. The simulated change of the magnetic field strength above a 1 mm wide surface breaking flaw is 0.3 mT. Simulations with alternative software packages QuickField and Maxwell also served to substantiate this result.

4 Experimental Results

The variation in flux was detected by a linear Hall Effect sensor (Honeywell SS94A2) mounted underneath the vehicle, which offers the advantages of extremely small size and low power consumption, qualities needed for small autonomous inspection robots. The wheels were machined out of 38 mm x 4 mm NeFeB disc magnets with a surface field strength of 1T. Experiments show, that the resulting holding force allows for additional payload and battery weight up to 4 kg for upside-down operations.

While the vehicle is moving along the surface of a test specimen the hall sensor passes above sets of artificial flaws varying in depth and diameter from 1 mm to 5 mm. These are detected reliably by the change of magnetic field strength directly above the imperfections, as shown in Fig. 5. It was also found that the Hall sensor output exhibits high sensitivity to changes in vehicle speed, sensor lift-off and variations of permeability of the test specimen. However, it is possible to measure a decrease of magnetic field strength above a 1mm-flaw of 0.25 mT, which verifies the theoretical simulations.

In comparison, stationary large scale MFL instruments with Hall elements can detect through-hole defects as small as 0.34 mm in diameter [8, 9]. However, those inspection instruments are far less flexible and of multiple size and weight dimension of a small inspection vehicle as they require complex magnetic circuits and extensive hardware like plotters and PC. Furthermore, those systems are not affected by the energy restrictions of autonomous robots.

Fig. 3. Test vehicle and sets of artificial flaws

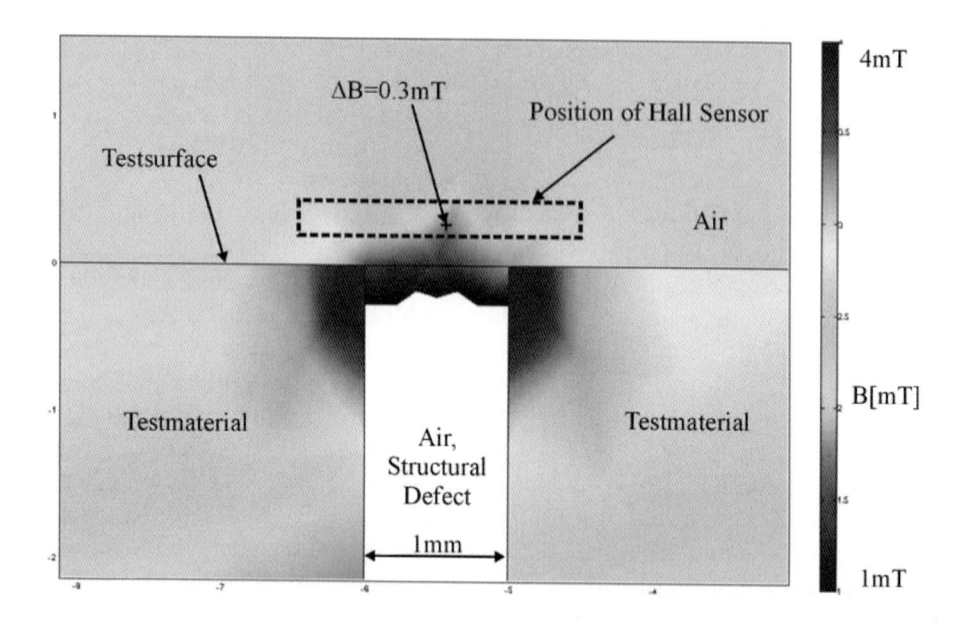

Fig. 4. Magnetic field above material with flaw

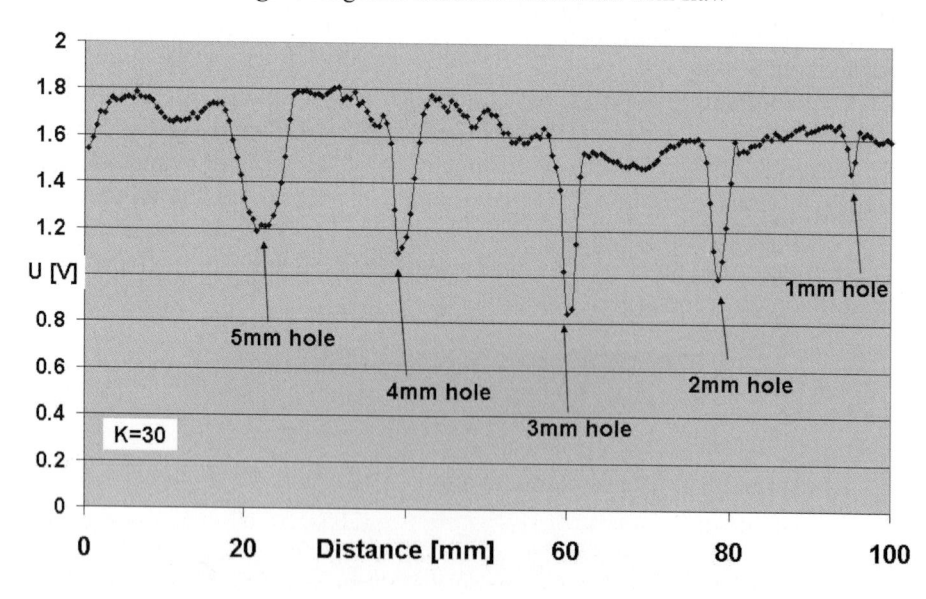

Fig. 5. Detection of artificial flaws with a Hall Sensor

5 Summary and further work

Artificial surface breaking cracks and flaws in ferromagnetic materials as small as 1 mm can be detected reliably by means of a passive method based on MFL inspection which uses the field created by the magnetic wheels of an autonomous inspection vehicle. In this way overall vehicle weight and size can be reduced and energy consumption minimized.

In order to increase sensitivity, reliability, and ability of discrimination between actual flaws and noise future experiments will be based on measurements taken from an array of Hall elements. Micro-machined Hall arrays, that cover areas as small as 1mm^2 with an active area per probe of 6 μm^2 [10], offer high potential for coping with significantly worse surface conditions as they are found in real world. Furthermore, the decrease of magnetic field strength has to be investigated in more detail, since the main principle MFL inspection is based on detecting an increase due to the leakage field that occurs at cracks and flaws.

Acknowledgements

This research received funding from the Engineering and Physical Sciences Research Council (EPSRC).

References

1. Shuliang L. Yanzheng Z. Xueshan G. Dianguo X. Yan W. (2000) A wall climbing robot with magnetic crawlers for sand-blasting. Spray-Painting and Measurement, *High Technology Letters*, vol. 10, pp. 86-88
2. URL http://www.ais4ndt.com/scanners_crawler.html, **[24. 6. 2005]**
3. Iborra A. Alvarez B. Ortiz F. Marin F. Fernandez C. Fernandez-Merono J.M. (2001) Service robot for hull-blasting, Dpto. Tecnologia Electronica, Univ. Politecnica de Cartagena, Spain, in *27th Annual Conference of the IEEE Industrial Electronics Society (IECON)*, vol. 3, pp. 2178–2183
4. Siegwart R. Nourbakhsh I. (2004) *Introduction to Autonomous Mobile Robots*, Cambridge, MIT Press
5. Takahashi H. Morisawa M. Ohnishi K. , (2004) Mobility of a mobile robot, Dept. of Syst. Design Eng. Keio Univ. Tokyo Japan in *8th IEEE International Workshop on Advanced Motion Control (AMC '04)*, pp. 253–257

6. Grandt A. F. (2004) *Fundamentals of structural integrity: damage tolerant design and nondestructive evaluation*, John Wiley, Hoboken, New Jersey
7. Halmshaw R. (1991) *Non-Destructive Testing*, 2nd edition, Edward Arnold, London
8. Connor S.O. Clapham L. Wild P. (2002) Magnetic flux leakage inspection of tailor welded blanks *Measurement Science and Technology*, vol. 13, pp. 157–162
9. Song X., Wu X., Kang Y. (2004) An inspection robot for boiler tube using magnetic flux leakage and ultrasonic methods, *Insight*, No. 46, pp. 275–277
10. Aytur T. Beatty P.R. Boser B. Anwar M. Ishikawa T. (2002) An Immunoassay Platform Based on CMOS Hall Sensors, *Solid-State Sensor, Actuator and Microsystems Workshop,* Hilton Head Island, South Carolina, pp.126–129

Automated NDT of Floating Production Storage Oil Tanks with a Swimming and Climbing Robot

T.P. Sattar, H.E. Leon Rodriguez, J. Shang and B. Bridge

Department of Electrical, Computer and Communications Engineering,
London South Bank University, 103 Borough Road, London SE1 0AA,UK
sattartp@Lsbu.ac.uk

Abstract

This paper describes the design and development of a prototype swimming and wall-climbing robot that gains access to internal tank wall and floor surfaces on Floating Production Storage Oil (FPSO) tanks for the purposes of carrying out Non-Destructive Testing (NDT) of welds while the tank is in-service and full of oil. A brief description is given of the inspection environment and the three NDT techniques (ultrasonic phased arrays, eddy current arrays, and Alternating Current Field Measurement ACFM arrays) that will be used to detect weld cracks and floor corrosion and pitting.

Keywords: FPSO NDT, Swimming & Climbing Robot

1 Introduction

FPSO tanks [1] are found on ships moored near off-shore oil platforms. They are used to store oil before transportation to the mainland. For structural safety and environmental reasons, it is necessary to test the welds periodically.

The main inspection task is to test the integrity of welds on plates used to strengthen the walls and floor of the tank. Currently they are inspected manually by emptying the tank of product after thoroughly cleaning it. Human operators then enter the tank to perform ultrasonic NDT. There is a large cost associated with the cleaning and inspection tasks. A pair of tanks are emptied, cleaned and inspected in 3-4 weeks with 60-70 man-days work and costs between £25-30k. A pair of FPSO tanks and ballast tanks

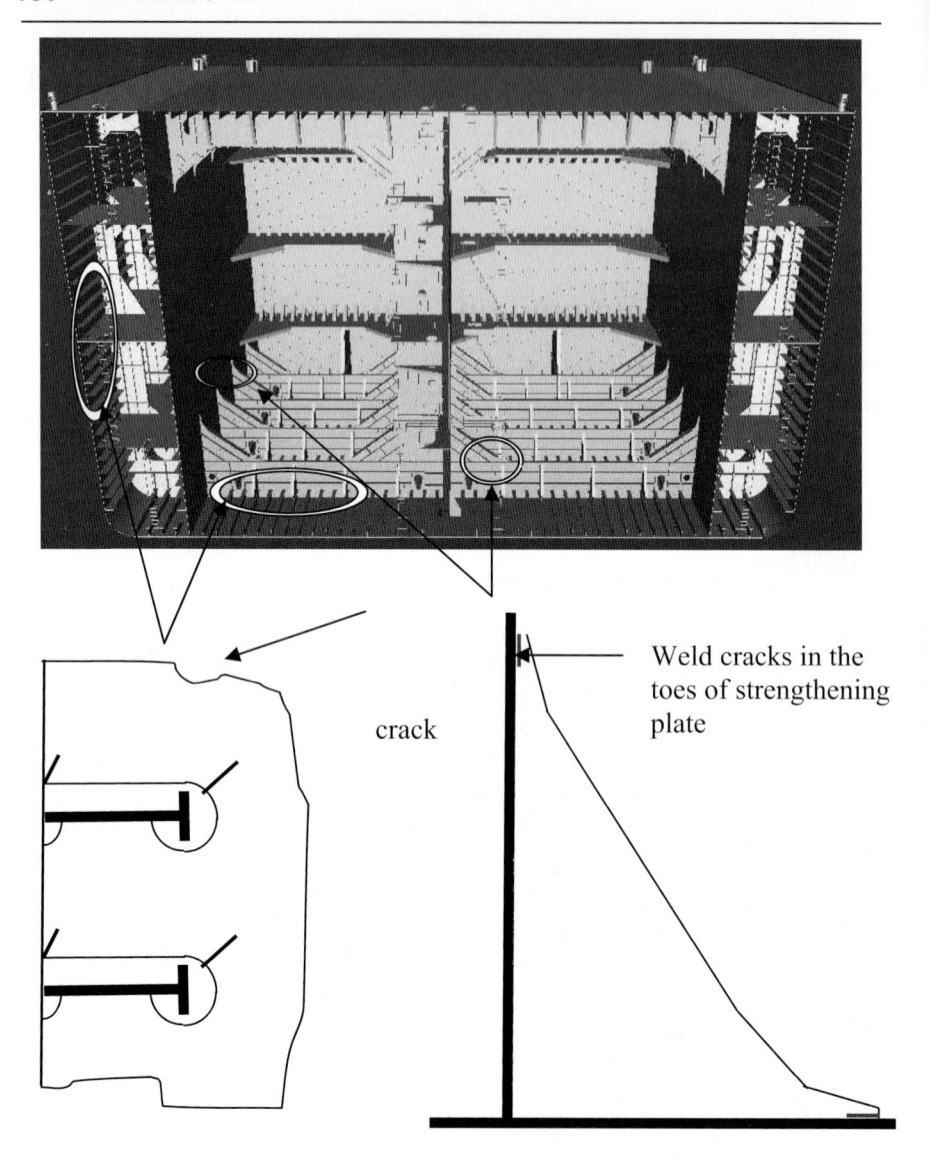

Fig. 1. Cross-section of ship showing two FPSO tanks in the middle and two ballast tanks on the sides.

inspected in the first five years costs £60-70k. This cost rises to £150-200k to inspect 3 pairs of cargo tanks and 3-4 pairs of ballast tanks after ten years. These costs can be reduced substantially by sending a robot into the tank without first emptying it thereby saving the cost of cleaning and emptying the tanks.

Weld cracks are caused by fatigue and are of two types. Low-Cycle fatigue is driven by panel deflection when filling and emptying tanks causes cracks at the toe of a bracket, generally in the secondary material. The drawing on the bottom right in figure 1 shows a bracket with cracks at its toe ends and the location of the brackets in the FPSO tank. High-Cycle fatigue is driven by wave pressure on the side and bottom shell of the tanks. It causes cracks at cut-outs where shell longitudinal strengthening plates connect to cut outs in the frames. Figure 1 is a drawing (bottom left) of cracks at cut outs and the location of longitudinal strengthening plates.

It is also required to test for corrosion caused by coating breakdown on the tank bottom. Pits can develop at the rate of 2-3 mm/year and even faster at the rate of 5mm/year if more corrosive crude is present. The bottom plate is usually 18-25mm thick.

2 The inspection environment

Obtaining access to welds on strengthening plates on the walls and the floors of the tank is not easy. The environment is very cluttered so that a very large walking and climbing robot would be required to step over plates. However, the robot is required to be compact, mass approximately 20kg, so that it can be inserted through a manhole of minimum diameter 600mm. In FPSO's owned by BP, the manholes are two elliptical hatches into each cargo tank approx. 900x600mm in size. FPSO's operated by Petrobras have approx. 600x800mm openings. The robot should be transportable by one or at most two operators, and should be able to operate between two adjacent longitudinal strengthening plates separated by a distance of 900 mm with the transverse frames separated by a distance of 4.5m. Both the walls and the floor are cluttered with strengthening plates so that unhindered robot motion on the walls or the floor by a small robot is not possible.

Access to welds can be obtained by swimming over the plates from one section of the tank to another and then landing on a wall or floor between the plates. The NDT inspection requirement is to inspect vertical welds as well as horizontal welds.

3 Access to welds with a swimming and climbing robot

FPSO Tanks in the North Sea are cleaned first with pressurised crude oil to agitate wax and sludge, and then cleaned with hot sea water. The water is removed prior to inspection by human operators, normally only a few centimetres are left on the bottom. The water in the tank tends to be fairly clean, though when disturbed can mix with the oil residues. Therefore, an opportunity exists to change the inspection method by filling the tank with water and then gaining access to welds by swimming to a test site. The robot can adhere to the wall and climb to a suitable location to scan a weld with NDT probes deployed by an arm.

Brazilian off-shore FPSO's are not cleaned with hot water because they are operated in higher temperatures. Therefore, they are cleaned with pressurized oil only thus eliminating the cost of establishing a process that cleans the water before returning it to the sea. Here, a swimming robot would have to operate in crude oil and would therefore have to meet intrinsic safety requirements for operation in explosive environments. Provided the cleaning systems are working, there should be very little residue on the floor, though there is always a waxy film. In places there will be a 2-3mm of wax, like shoe polish on the bottom of the tank but not on the side walls.

4 NDT requirements and proposed techniques

Work in tanks is always potentially hazardous and thus a system which can minimise the need for personnel entry is obviously beneficial. The development of a remote tool will only improve structural integrity, and thus potentially reduce leakage, if it can deliver a higher level of inspection than is currently achieved.

The deck and bottom plating is normally 20-25mm thick. Bottom stiffeners are typically T-shaped with the web 650 x15mm, and the flange 250x25mm. Side shell stiffeners are generally bulb bars, 200-400wide and 12-15mm thick. Apart from butt joints joining plates together, all connections are fillet weld 6-10mm throat thickness depending on the section. Welding size is variable and depends on the design. Coating is provided by a paint layer 300-500 microns thick. The NDT is required to identify through thickness cracks that are 5-10mm long (these are currently generally found by visual inspection). If smaller cracks can be detected that would be bonus. It is also important to identify coating defects or pitting on the bottom plates.

The project aims to develop an NDT Sensor Payload that is suitable for robotic deployment and that obtains better NDT data in a hazardous environment than possible with manual inspection. Towards this end, array probes are being developed to use the following NDT techniques: Ultrasonic phased array technique, Eddy current technique, and the ACFM (Alternating Current Field Measurement) technique.

Phased Arrays: Phased array systems offer the possibility of performing inspections with ultrasonic beams of various angles and focal lengths using a single array of transducers [2]. Software control over beam angle and focusing is achieved by application of precisely controlled delays to both the emission pulse and received signal for each element in an array of transducers, hence the term "Phased Array".

Signals from multi-element transducers in the array produce A-Scan responses that are comparable with those obtained using a fixed angle probe with a conventional pulse-echo imaging system. Therefore, imaging and image interpretation remain the same as for a conventional pulse-echo system. The data can be processed to provide top, side and end view images of the inspected volume of material. In addition to standard imaging, a focused beam is created using a few of the many elements contained in a Phased Array probe (up to 128). The beam is then multiplexed to the other elements to allow a high speed scan of the component with no transducer movement along that axis. Flaw detection and sizing ability is only limited by the beam width which, in theory, can be less than 1mm (0.04"), depending on the excitation frequency. Phased-array probes are being designed to be used to scan for defects in fillet welds. They have a stainless steel housing with dimensions of L 60 mm x W 20 mm with cable output from the centre on top, with Wedge material being Rexolite (v = 2350 +/- 30 m/s) and the angle is 34 degrees.

The Eddy Current Array technique will be used to inspect the tank floor for corrosion pitting in the presence of sludge and wax. The robot will carry a set of array probes and its motion will result in a surface scan. A feasibility study with a conventional eddy current system shows that the conventional eddy current may work to pick up corrosion type of defect, if all the parameters are optimised. Therefore a system is being developed that consists of a conventional eddy current probe array.

ACFM (Alternating Current Field Measurement) techniques: ACFM is an electromagnetic inspection technique that provides one pass inspection. It has a high tolerance to lift off and requires no electrical contact so that it can be used to detect through coatings. It provides crack detection and sizing and is suitable for weld inspection. To cover the weld cap, toes and HAZ of a welded joint, it is necessary to either use a simple single sensor probe and scan several times or use a multisensor array probe and cover the required area in a single pass. Work is progressing to develop an underwater ACFM array probe that can be deployed on the robotic vehicle. The probe is to be used to detect and size surface breaking cracks at or in close proximity to the welds to be inspected. Modifications will be needed to make existing instrumentation intrinsically-safe. The objective is to scan a fillet weld in one pass, accommodate different cap widths, and minimise weight. The current probe design is estimated to weigh approximately 1.2kg in air. The maximum dimensions are 87mm wide x 117mm long x 117mm high. The materials have been chosen to be mainly stainless steel for structural parts and PEEK for parts that will not be subject to great loads.

5 Robot Design

The robot design selected for construction is shown in figure 2 with the robot travelling on the floor. In swimming mode the bottom side is the one fitted with the scanning arm while the wheels point in the direction of forward motion. This design is a further development of a wall climbing robot called RobTank that has been developed earlier for in-service inspection of oil and chemical storage tanks [3]. The wall-climbing ability of this robot has been tested and demonstrated in water tanks. It can make transitions from a floor to a wall and vice-versa. Further development of this design

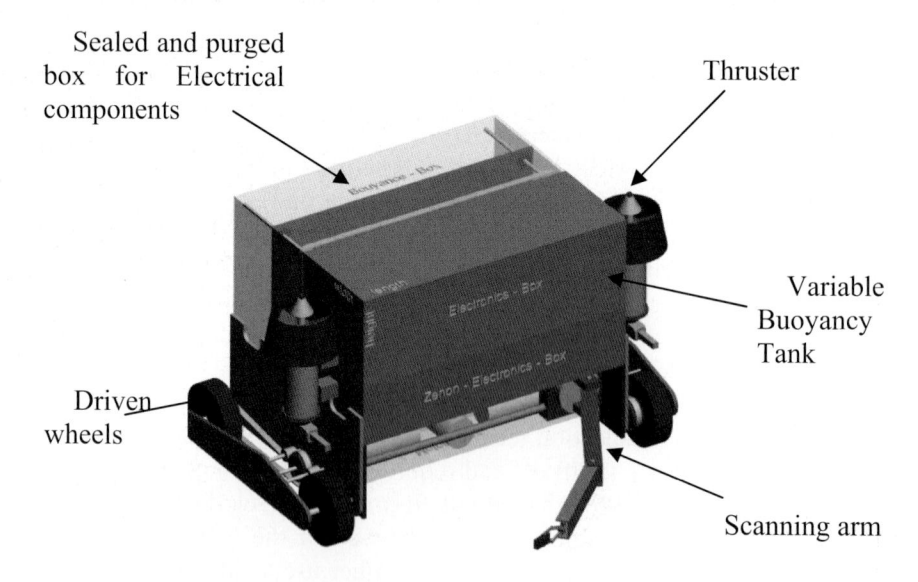

Fig. 2. Conceptual design of FPSO swimming and wall climbing robot

has added a variable buoyancy tank that can quickly and accurately control buoyancy around the neutral buoyancy of the robot.

All motors, motion controllers, navigation electronics, NDT flaw detectors are housed in an air pressurized central chamber to prevent the ingress of water through any leaks at the rotating shafts emerging from the central chamber and through NDT sensor probe cables. The reason for placing most hardware systems onboard the robot is to reduce the size of the umbilical cord so that cable management becomes easier.

The outer dimensions of the robot are (mm): 540L x 300W x 300H. Its mass in air is 12 kg and it can carry a payload of 8kg. On-board embedded servo controllers with encoder feedback control the speed and position of the robot. High level control is from an operators console via RS 485 twin pair communications with on-board controllers.

Both depth and horizontal motion is controlled simultaneously to swim the robot to a test site on a wall. After contact with a wall, thrust forces obtain adhesion to the wall while actuated wheels move the robot on the wall. The robot manoeuvres freely on the wall and can be driven down from a wall to the floor of the tank and back on to it.

After insertion of the robot through a manhole in the top deck, positive or negative buoyancy control is used to swim the robot vertically to a specified depth and to maintain that depth with neutral buoyancy. Two independent, speed controlled thrusters move the robot in a horizontal plane in the forward and reverse direction or rotate it to face in any direction.

Figure 3 shows the prototype swimming and wall climbing robot. Figure 4 shows the robot climbing on the glass wall of a water tank. Adhesion to the wall is obtained via forces generated by two thrusters. The robot is able to change surfaces (floor to wall and back) and climb on the wall without the aid of neutral buoyancy. A variable buoyancy tank is currently being developed to change buoyancy around neutral by obtaining volume change. The tank should enable the robot to swim to a given depth and to be parked on the floor with negative buoyancy when inspecting the floor.

Future work: A scanning arm will be mounted on the robot to deploy ultrasonic phased array probes to perform tests on the toe of fillet welds. Vision systems and range finders will be used to detect tank walls in low light levels and poor visibility. These sensors will be made an integral part of the control system to provide feedback to maintain straight-line motion when moving towards a wall and to avoid collisions.

Fig.3. Prototype swimming and wall-climbing robot

Fig.4. Wall-climbing robot climbing on wall without buoyancy tank

Aknowledgements

This work is funded by the European Community through the CRAFT project FPSO INSPECT (COOP-2004-508599) [4]with the following partnership: NDT Consultant (UK), TSC Inspection Systems (UK), Isotest Engineering (Italy), Tecnitest Ingenieros (Spain), Spree Engineering Ltd (UK), Miltech Hellas S.A. (Greece), ZENON (Greece), Kingston Computer Consultancy (UK), BP (UK), Petrobras (Brazil) and LondonSouth Bank University (UK). The Project is coordinated and managed by TWI (UK)

References

1. Shimamura Y. (2002) *FPSO/FSO: State of the art*, J. Mat. Sci. Technol. 2002, pp 60-70
2. American Society for Nondestructive Testing Handbook (2002), Vol. 7 – Ultrasonic Testing.
3. Sattar T.P., Zhao Z., Feng J., Bridge B., Mondal S., Chen S., (2002) *Internal In-service Inspection of the floor and walls of Oil, Petroleum and Chemical Storage Tanks with a Mobile Robot*, Proc. Of 5th International Conference on Climbing and Walking Robots and the Support Technologies for Mobile Machines, Edited by Philipe Bidaud and Faiz Ben Amar, ISBN 1 86058 380 6, 2002, pp 947-954, Professional Engineering Publishing Ltd. UK.
4. *Non-Intrusive In-service Inspection Robotic System for Condition Monitoring of Welds inside Floating Production Storage and Offloading (FPSO) Vessels* (2004), EU 6th Framework Programme, Co-operative Research Project, COOP-CT-2004-508599, December 2004.

7-axis Arm for NDT of Surfaces with Complex & Unknown Geometry

T. P. Sattar and A. A. Brenner

Department of Electrical, Computer and Communications Engineering, London South Bank University, 103 Borough Road, London SE1 0AA, UK.   Sattartp@Lsbu.ac.uk

Abstract

The aim of the work described here is to design a dexterous and intelligent robotic arm to automate the non-destructive testing and evaluation of geometrically complex industrial products. The robot is required to scan contoured surfaces with ultrasonic and eddy current sensors without the need to position the test pieces precisely, thereby reducing inspection times. In addition, the arm should be lightweight so that it can be carried by wall-climbing and walking robots but at the same time it should be rugged enough to be operated in industrial environments where the tool (sensor probe) comes into contact with rigid and unknown test surfaces. To realise these aims, a seven-axis lightweight and rugged robot arm has been specially developed for non-destructive testing (NDT) and is reported here. The arm weighs 22 kg and is designed for portability, dexterity, ruggedness in an industrial environment and 1 mm repeatability. Its NDT payload is 4 kg. Current work is directed at developing a control system for the arm with a proven operating system with further work required to make the arm adapt to uncertain surfaces.

Keywords: 7-Axis Arm, Scanning Arm, NDT Robot Arm

1 Introduction

Progress in the automation of Non-Destructive Testing (NDT) for both rapid inspection of complex pieces and for remote inspection in hazardous environments requires the development of robotic scanning systems that are dexterous, lightweight, easily transportable from site to site and that can deploy sensor probes on uncertain surfaces with the required probe stand-off or contact force.

The above characteristics provide the flexibility to use the same scanning arm to provide NDT in various industrial inspection applications thereby reducing instrumentation and tool costs. A lightweight and rugged arm that can be mounted on different types of mobile robots including walking and wall-climbing robots provides the flexibility to scan vertical and large surfaces that are located remotely e.g. ship hulls, petrochemical storage tanks, bridges and tall buildings. Applications of 6-axis arms for inspection invariably use rails to move the along weld lines e.g. the NOMAD project [1] For greater flexibility it would be necessary to use mobile robots to carry the scanning arm. Our arm mounted on different wall climbers would give inspection capability e.g. weld inspection on hulls of cargo container ship [2]. A scanning arm with dexterity and flexibility, approaching that of a human arm, and the ability to adaptively scan complex surfaces that are not precisely known will open up areas of applications in various hazardous or dangerous environments such as nuclear power plant, off-shore oil platforms, large bridges, hulls of ships, oil and chemical storage tanks where the work is difficult for the human operator.

The remote location of a test site where the robotic system will operate, also usually means that the location and geometry of a test surface is not known precisely with the resultant danger that the scanning arm could apply too much or too little surface contact force resulting respectively in damage to the arm/surface/NDT probe or loss of measurement data. There are different kinds of approaches for the recognition of the surface. The problem is the need first to measure the surface and then to control the tool to follow the surface outline. There are several pure spatial approaches.[3] to measure the surface. Our approach is to perform adaptive force control by using a force sensor.

2 The robotic system

The robotic system comprises of three main components; a mechanical arm and its AC drive motors, 7 servo amplifiers and a controller. The scanning arm is called the ANSALDO arm. It is interfaced with a Adept control system. The robot arm has 7 revolute joints and a force sensor mounted in its wrist to enable the adaptive control of probe contact forces. The ANSALDO robot arm is controlled by the 9-axis MV-10A Adept motion controller, which provides the possibility to use the Denavit-Hartenberg algorithm kinematics model to manipulate the robot. The controller and the operator's PC/Laptop have a server-client relationship. The operator can access and control the Adept controller via the Adept Windows software and delegate the tasks to the robot easily. The motion controllers are interfaced via three interface panels to the servo drives. These provide the connections for the motor brake signals, motor shaft encoder signals, over travel and drive enable and drive error signals from the 7 motors and their corresponding servo drives.

The robot arm

The main difficulty in designing the seven axis arm was to get the required accuracy of spatial position repeatability and payload of the end effecter while trying to minimise overall mass of the arm.

Also, unlike pick and place robots designed to operate with their base on a flat bench top or floor the arm dynamics must be such as to allow accurate operation with the base in any orientation, as will arise when the arm is transported by the climbing robot vehicles.

The arm is also required to have the dexterity of the human operator with respect to the manipulation of NDT sensors.

Technical specifications of the arm are summarised in Table 1.

- The dexterity of the human operator in manually scanning ultrasonic sensors stems largely from elbow and wrist movements. With 2 elbow joints and 3 wrist joints the robot arm simulates all the key axes of rotation in the human elbow and wrist i.e. all the elbow and joint axes having a large rotation range. However the range of rotation angle about these 5 axes is up to 3 times greater than is possible with the human arm.
- The remaining 2 perpendicular axes (turret and shoulder) are equivalent to one of the waist rotational axes and one of the shoulder axes in the human operator but again with a several fold increase in the range of rotation.
- The arm is comparable to the human arm in the lengths of the upper and lower arm and shorter in waist to shoulder distance. So it is suitable for operation in confined spaces.
- The arm is mounted on the payload platform via 4 bolts equally spaced round the turret base-plate and an adapter plate within a few minutes.

Most of the weight of the arm arises from 7 electric motors. All non-load bearing components such as external covers are made in high grade plastic whilst the arm body is manufactured in lightweight high strength anodised aluminium alloy. Internal gears are made from titanium alloy grade 5.

All joints are controlled by an incremental encoder built into each motor. The arm control system thus records the current position of each joint in a non volatile memory. All cables for power, sensor transmission and reception signals and ultrasonic couplant feed are routed through the inside of the arm itself and out through the base of the turret (waist). The arm is made rugged and weatherproofed for a severe environment with anodised aluminium surfaces and sealed and lubricated for life bearings for all joints.

Table 1. Summary of technical specifications of 7-axis arm

Joint #	Joint Name	Joint Speed	Joint Workspace
1	Turret	30° / sec	± 170°
2	Shoulder	30° / sec	± 170°
3	Elbow Roll	45° / sec	± 180°
4	Elbow Pitch	45° / sec	± 150°
5	Wrist Roll	90° / sec	± 180°
6	Wrist Pitch	90° / sec	± 120°
7	Hand Roll	90° / sec	± 180°

Robot Mass = 22 kg, *Payload* = 40 N, *Arm Length* = 300 mm, *Forearm Length* = 300 mm, *Accuracy* = 1 mm

The axis of rotation of joint 1, 3, 5 & 7 are perpendicular to the axis of rotation of joint 2, 4, 6 & 8. The four characteristic parameters of the robot (two joint and two link parameters) are shown in Table 2. The joint distance d, the link length a and the link twist angle α are fixed parameters.

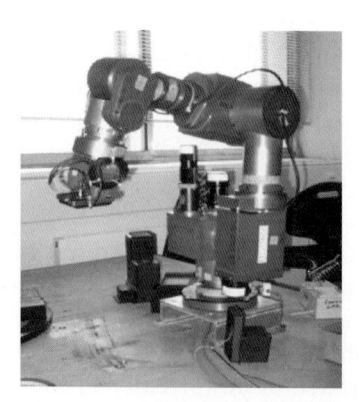

Fig. 1. 7-axis revolute joint ANSALDO arm

In operation the ANSALDO robot has a repeatability of 1 mm which is sufficient for ultrasonic and eddy currents probes to find any cracks. The robots servo control is sufficient and provides a good system response, e.g. to a step, as can be seen in figure 2.

3 The advantages of the compilation

The ANSALDO arm provides the dexterity and flexibility approaching that of a human arm whereas the Adept controller enables assembly operations to be performed at higher speeds than are possible with other force sensing units. This is integrated by the use of the AdeptForce package that allows the robot to react to sensed forces and moments. The Adept system has also the advantage to be easily extendable into an automation system with vision guidance and inspection, by use of the AdeptVision Module and software. The high level language V+ implements extensive vision tools for vision-related operations like image capture, enhancement, and analysis. It is a programming language that provides a solution to programming needs in a robotic work cell, including safety, robot motion, vision operations, force sensing and I/O.

The efficient use of system memory and reduction in overall system complexity of the Adept system combined with the ANSALDO robot arm enables a real-time scanning system which permits complex motions to be executed quickly, using the V+ continuous trajectory computation. The system allows on-line program generation and modification and can concurrently interact with an operator. Besides a real-time operating system, the Adept controller also offers advanced servo-control features and the Kinematics Device Modules which result in a powerful control system. We have used the "6/7 Axis Rotary Device Module"[15] which allows V+ to understand the geometry and specifications of a particular robotic mechanism. This software defines the characteristics of a particular 7 axis robot. It is a separate file stored in the system file which implements all the characteristics of the particular robot. Figure 3 shows the ANSALDO robot in its zero position, all the angles are set to zero. It is the graphical representation of the data or knowledge the Adept controller has about the robot it is controlling. This software enables a more structured view and enables the operator to use instructions which can move the robot in a less complicated way. The robot device module defines a robot with seven axes where the third axis is the redundant one.

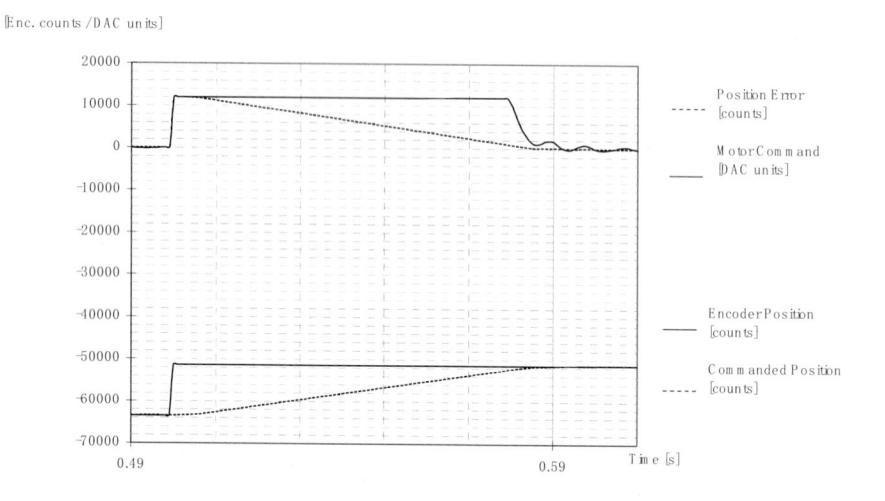

Fig. 2. Step response and error for one joint of tuned system

Table 2. The resulting characteristic table for the revolute ANSALDO arm

Axis (Joint)	Joint angle θ [°] variable	Joint distance d [mm] - fixed -	Link length a [mm] - fixed -	Link twist angle α - fixed -
1 (Base)	θ_1	185	0	$-\pi/2$
2 (Shoulder)	θ_2	150	0	$\pi/2$
3 (Redundant waist)	θ_3	310	0	$\pi/2$
4 (Elbow)	θ_4	80	0	$-\pi/2$
5 (Yaw)	θ_5	300	0	$-\pi/2$
6 (Pitch)	θ_6	0	0	$\pi/2$
7 (Roll)	θ_7	260	0	0

4 Contact force

The main difficulty in automating NDT is to establish an accurate adaptation to the test surface (remotely located or of complex shape such as aircraft turbine blades that have a complex surface geometry) to obtain high quality data from possible defects. It is then necessary to respond to surface changes while going over the surface and at the same time to also orientate the NDT probe in the right way and position to get the best results..

A force sensor from JR3 is already mounted on the robot's wrist. This will provide surface adaptation in real time thus allowing task-based control strategies to be employed. Although the most common method to profile a surface is to use vision sensors, force control has advantages when pushing a sensor probe against a surface to maintain a given contact force. Since this is how ultrasound inspection is performed the force sensor for adapting to the unknown surface is a very good option. Other research shows the possibility of a multi-sensor based robot which would further expand this project by the implementation of a visual sensor, like a camera [4]. The Adept Controller provides the possibility to add a force control module to control the contact force in two different modes, guarded and protect mode [5].

The RJ3 sensor (Model: 90M31A) has onboard electronics to provide force and torque data at very high bandwidths and with very low noise. Data for all six axes is returned at a rate of 8 kHz. The sensor has a load capacity of minimum 25,50 lbs up to 68,58 lbs [6].

5. Conclusion

The ANSALDO robot arm with a weight of 22 kg is a reasonably lightweight robotic scanning system and therefore easily transportable. It is designed for portability, dexterity, ruggedness in an industrial environment. It is built with rugged gears to withstand possible collisions when working in probe contact mode. A great deal of flexibility is provided to reach a given point inside its workspace and to reach around possible obstacles. The combination of the Adept controller and the 7 axis ANSALDO scanning arm has resulted in a fast and dexterous robot that has a powerful control system which with further additions of force control and vision will allow rapid inspection of complex pieces and inspection on uncertain surfaces located in remote and hazardous environments. In operation the ANSALDO robot has a repeatability of 1 mm which is sufficient for ultrasonic and eddy currents probes to find most defects.

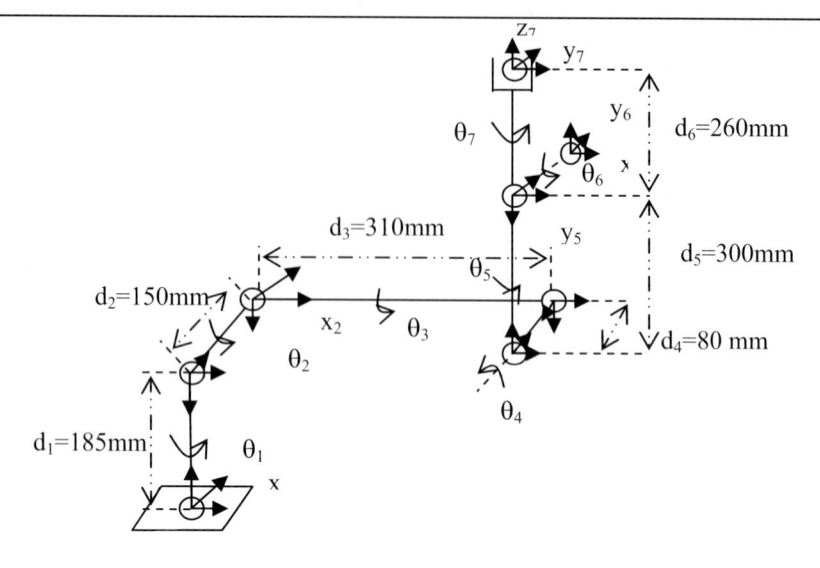

Fig.3. The robot arm (zero position) configuration applied to device module.

Acknowledgements

The authors gratefully thank the EPSRC (UK) and TWI Ltd (UK) for their funding of Ms Alina-Alexandra Brenner's CASE Award.

References

1. URL http://www.automation.com/store/pid9244.php [Jun 2005]
2. Sattar T.P., Alaoui M., Chen S., Bridge B. (2001) ROBHULL: *A magnetically adhering Wall Climbing Robot to Perform Continuous Welding of Long Seams and Non Destructively Test the Welds on the Hull of a Container Ship*, 8th IEEE Conference on Mechatronics and Machine Vision in Practice, M2VIP2001, pp. 408-414, ISBN 962-442-191-9, Paper No. 129.
3. Smith M L (2001): *Surface Inspection Techniques*, Edmundsbury Press Limited, Suffolk, UK.
4. Xiao D., Ghosh B. K., Xi N, and Tarn T. J (1999): *Sensor-based hybrid position/force control of a robot manipulator in an uncalibrated environment*. IEEE transactions on control systems technology, Vol.8, No.4, Jul 2000.
5. Adept Technology, Inc.: Adept Force VME User's Guide (1995), USA.
6. URL http://www.jr3.com/dsp.html [Jun 2005].

Personal Assistance Applications

Elderly People Sit to Stand Transfer Experimental Analysis

P. Médéric, V. Pasqui, F. Plumet and P. Bidaud

Laboratoire de Robotique de Paris (LRP)
CNRS FRE 2507 - Université Pierre et Marie Curie, Paris 6
18, route du Panorama, BP 61, 92265 Fontenay-aux-Roses, France

Abstract

This paper describe an assistive device for the elderly providing support during the sit to stand transfer. A postural stabilization criteria and an assisting force in case of instability are presented.

Keywords: assisting device, transfer assistance, stability criteria

1 Introduction

The elderly population is growing likewise numbers of resident in assisted living facilities. Medical staff are more busy with daily living tasks such as escorting the patient or assist them during transfer. Due to the disparity between the member of the medical staff and the elderly, a lot of patient are not assisted and stayed in an inactivity leading to the loss of the walking function and mortality [1].
Moreover many nursing home workers are suffering of low back injuries caused by the high frequency of patient lifting [2],[3]. On the one hand conventional walker have been developed to provide support during the walk on the other hand patient lifter have been use concerning the transfer of the patient. However these kind of device are not totally useful for all patient, pathologies and medical staff [4], various institutes started the development of intelligent robot.
The Pam-Aid [5] and the Pamm projects with the SmartCane [6] and the smart walker [7] provide support during the walk and allow obstacles avoidance and path navigation. The Care-O-Bot [8],[9], and the Nursebot

[10], are assistive devices for the daily life. The Care-O-Bot is equipped with a manipulator arm to catch object, the movement of its platform is deal with sensors on the handles and a path navigation system, it also provide information to the user like taking medicine.

The Nursebot, which do not afford a physical support, is studied for the users having failing memory and disorientation. The goal of this project is to augment the human interaction by reminding people about important routine activities and guiding them with a path navigation system.

All the previous works raise the problems of path navigation with obstacle avoidance and shared control. Relating to the transfer assistance of the elderly a power-assistance device has been developed by the Ritsumeikan university [11], it ensure a physical support during the walk and the sit to stand transfer. The entire system have to be in a defined room, the trajectory and the supporting forces do not evolve in accordance with the configuration of the patient and this kind of assistive device can not be use outside a defined room. An assistive device (AD) have been developed in our laboratory. We emphasis on a device combined the transfer and the walk. This device provide support during the walk and also during the sit to stand transfer, it is primarily intended for elderly patients who have falling background [12]. We focus on postural stability especially during the verticalisation, postural observation and estimation is implemented on the assistive device in order to monitor the patient and modify its posture.

This paper presents experimental results done with elderly and a postural stabilization criteria during the transfer.

2 System overview

The assistive device (AD) is a two degree of freedom mechanism mounted on an active mobile platform (see **Fig. 1**). During the transfer assistance the handles of the AD pull slowly the patient to the up position. This is obtained by using two parallel and independent mechanisms combined in a serial way.

The trajectory of the handles relay on a trajectory generator, it allow the patient to get off its retropulsion configuration to an antepulsion configuration. Force sensors are added on each handle to measure force interaction between the user and the device. Details on the design the AD and its trajectory generator can be found in [12].

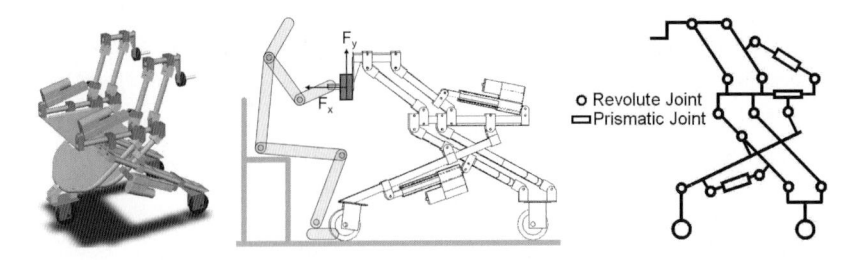

Fig. 1. Description of the assistive device

3 Transfer assistance for elderly

In order to demonstrate the efficiency of our AD during the transfer, several experimental standing-up and siting-down trials were executed. Measurement series has been done on a set of 19 elderly patients in Charles-Foix Hospital. The patients are affected with different pathologies disturbing the transfer. The device is used to provide support during the transfer, gerontologists choose, according to the patient, the position of the initial and final points of the assistive trajectory. During the movement the patient hold the handles of the prototype, force measurement are done on each handles. On the 19 patients only 2 could not stand up with the help of the AD, these two patients have serious impairment (acute hemiplegia) that did not allow us to stand up without the assistance of member of the medical staff.

Obviously, patient who could not stand up with an external help could not use the device to stand up. Patient who could stand up without a help could stand up thanks to the AD.

Due to their physical condition and age the patient could not produce many trial, nevertheless concerning the force measurement we can observe a repeatability of the measurement on each user. The global shape is the same for each trials of the same patient but it is not the same between the patients.

Concerning the sit to stand transfer the effort along the **y** axis (**Fig. 1**) do not evolve in the same way, see **Fig. 2**. For the patient 1 its seems that the patient lift up the right handle (**Fy** > 0) whereas for the patient 2 (if we do not consider the first trial **Fy** ≈ 0) the force interaction **Fy** is still evolve in the negative value meaning that the patient is still pushing during the transfer with a maximum force **Fy** = −140 N around 19 % of his weight. The opposite direction of the effort **Fy** between the left and the right handle indicate that the patient 1 is tipping over on his left side.

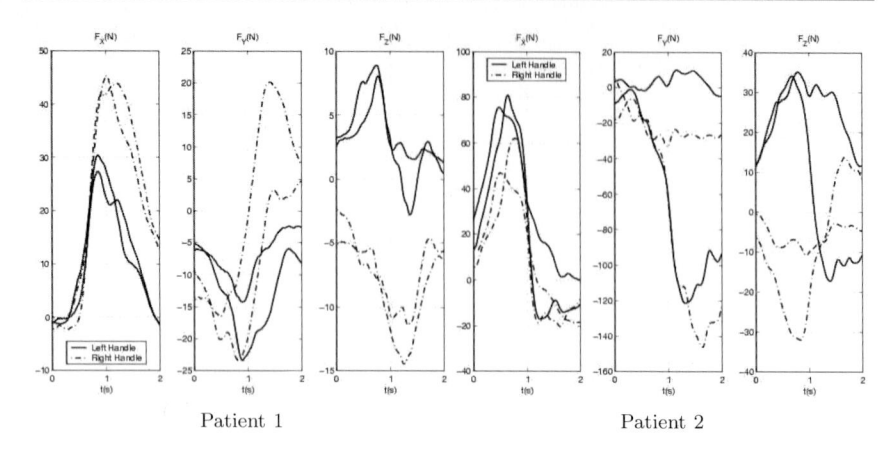

Fig. 2. Force interaction during sit to stand transfer

During the stand to sit transfer we observe the same phenomena of opposite effort (**Fy**) for the patient 1, see **Fig. 3**. The analysis of the stand to sit transfer is more difficult because most of the patient let their body fall back to the chair. The interpretation of the effort is getting more complex, the measurement of the force interaction between the patient and the chair could be useful for a complete identification of the transfer.

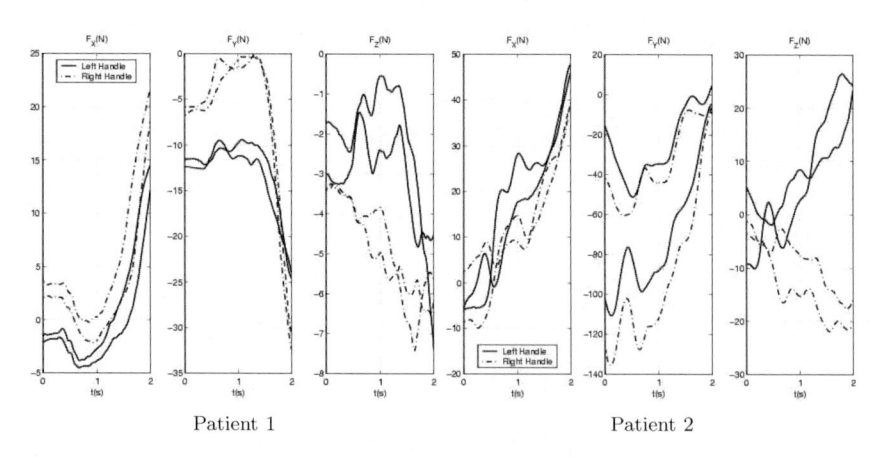

Fig. 3. Force interaction during stand to sit transfer

Theses previous interpretations do not allowed us to identify the posture of the patient during the transfer. Observations of the force interaction between the patient, the chair and also the floor added with measurement of motion of the body segment could give an interesting observation of the patient's posture (see **Fig. 4**).

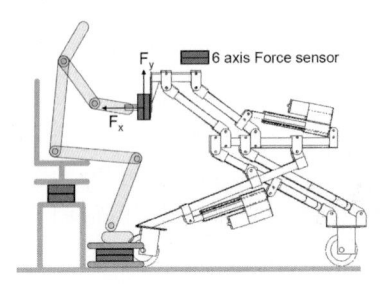

Fig. 4. Future test bench

In order to compensate the fall during the transfer we design a criteria of stability, next section will present this criteria.

4 Stability criteria

Compensation of the fall during the transfer relay on a stability criteria, the AD have to produce the trajectory and force compensation helpful for the user in case of instability. In the following sub section we will introduce a stability criteria of the patient.

4.1 Design of a stability criteria

This stability criteria is based on the establishment of the Zero Moment Point (ZMP) [13] on a simplified model of the patient.
The patient is a 7 bar mechanism studied in the sagittal plane (x_o, y_o) (see **Fig. 5**), m_i is the mass of the ith segment, M is the total mass of the patient, X_{zmp} is the position of the ZMP along the $\mathbf{x_o}$ axis in the frame R_o.

Using the Newton-Euler formulation at the ZMP we get the equation of the position of the ZMP during the transfer:

$$X_{zmp} = \frac{H_Z + M(\ddot{Y}_G X_G - \ddot{X}_G Y_G) + \sum_{i=1}^{i=n} m_i g X_{Gi}}{M(\ddot{Y}_G + g)} \quad (1)$$

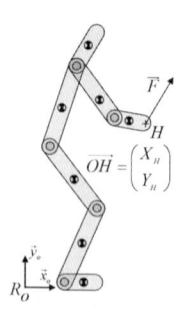

Fig. 5. Simplified model

By considering the effect of inertia of each segment lower in front of the inertia of the trunk, the expression of the ZMP become:

$$X_{zmp} = \frac{\ddot{Y}_G X_G - \ddot{X}_G Y_G + g X_G}{\ddot{Y}_G + g} \quad (2)$$

The stability criteria of the patient is done by using the ZMP equation, the patient posture is defined stable if X_{zmp} is included in the foot's area. In case of instability the AD had to produce the assisting force to remain the ZMP into the foot's area. In the next sub section will present the assisting force model.

4.2 Assisting force model

The goal of the assisting force is to ensure the postural stability of the patient. The force interaction **F** between the patient and the AD is defined into the hand position of the patient at the point H (see **Fig. 5**). By included **F** into the Newton Euler formulation (eq. 1) we get an other expression of the X_{zmp}:

$$X_{zmp} = \frac{M(\ddot{Y}_G + g)X_G - M\ddot{X}_G Y_G - F_X Y_H + F_Y X_H}{M(g + \ddot{Y}_G) - F_Y} \quad (3)$$

With the stability criteria :

$$a \leq X_{zmp} \leq b \quad (4)$$

According to eqs. 3 and 4 if the patient is in an instable posture the AD had to provide the assisting force to remains the patient into a safe posture. To solve this problem we express the force interaction **F** under a non linear problem under constrained, by finding the minimal effort **F** to the following formulation:

Minimize $F = \sqrt{F_X{}^2 + F_Y{}^2}$ subject to:

$$\begin{bmatrix} Y_H & -(X_H + a) \\ -Y_H & (b + X_H) \end{bmatrix} \begin{bmatrix} F_X \\ F_Y \end{bmatrix} \leq \begin{bmatrix} M((g + \ddot{Y}_G)(X_G - a) - \ddot{X}_G Y_G) \\ M((g + \ddot{Y}_G)(b - X_G) + \ddot{X}_G Y_G) \end{bmatrix} \quad (5)$$

With this model of assisting force we need kinetics information of the patient. A new test platform added with motion sensors on the patient will permit us to validate this model.

5 Conclusion

An assisting device for elderly patient has been developed. First experiments with elderly have been done on the transfer function. These first experiments showed the efficiency of the assisting device, only 2 out of 19 patients could not stand up with it. With its sensors the assisting device can provide information on the posture of the patient and also providing support to the user in case of instability. This device can also be a tool for the kinesitherapist, it can give a powerful programmable device for the monitoring and rehabilitation of the lower limbs. A stability criteria and an assisting force models have been done to observe posture and to stabilize the patient during the transfer. Experimentation and validation of these models will be done soon.

Acknowledgments

This work is partly supported by the French RNTS (Réseau National des Technologies de la Santé) program under grant N° 02B0414 (Monimad Project).

References

1. Hirvensalo, M., Rantanen, T., and Heikkinen, E. (2000) Mobility difficulties and physical activity as predictors of mortality and loss of independence in the community-living older population. *Journal of the American Geriatric Society*, vol. 48, pp. 493–498.

2. Stobbe, T.J., Plummer, R.W., Jensen, R.C., and Attfield, M.D. (1988) Incidence of low back injuries among nursing personnel as a function of patient lifting frequency. *Journal of Safety Research*, vol. 19, no. 1, pp. 21–28.

3. Zhuang, Z., Stobbe, T. J., Hsiao, H., Collins, J. W., and Hobbs, G. R. (1999) Biomechanical evaluation of assistive devices for transferring residents. *Applied Ergonomics*, vol. 30, no. 4, pp. 285–294.

4. Ruszala, S. and Musa, I. (2005) An evaluation of equipment to assist patient sit-to-stand activities in physiotherapy. *Physiotherapy*, vol. 91, no. 1, pp. 35–41.

5. Lacey, G., Namara, S. Mac, and Dawson-Howe, K. M. (1998) Personal adaptive mobility aid for the infirm and elderly blind. *Lecture Notes in Computer Science*, vol. 1458, pp. 211–220.

6. Dubowsky, S., Genot, F., Godding, S., Kozono, H., Skwersky, A., Yu, H., and Yu, L.S. (2000) Pamm - a robotic aid to the elderly for mobility assistance and monitoring: A "helping-hand" for the elderly. In *IEEE International Conference on Robotics and Automation*, San Francisco, USA, vol. 1, pp. 570–576.

7. Yu, H., Spenko, M., and Dubowsky, S. (2003) An adaptive shared control system for an intelligent mobility aid for the elderly. *Autonomous Robots*, vol. 15, pp. 53–66.

8. Schraft, R.D., Schaeffer, C., and May, T. (1998) Care-O-bot™: The concept of a system for assisting elderly or disabled persons in home environments. In *Proceedings of the 24th Annual Conference of the IEEE Industrial Electronics Society: IECON 98*, Aachen, Germany, vol. 14, pp. 2476–2481.

9. Graf, B., Hans, M., and Rolf, D. S. (2004) Care-O-bot II - development of a next generation robotic home assistant. *Autonomous Robots*, vol. 16, pp. 193–205.

10. Baltus, G. et al. (2000) Towards personal service robots for the elderly. In *Proceeding of the 2000 Workshop on Interactive Robotics and Entertainment (WIRE-2000)*, Pittsburgh,USA.

11. Nagai, K., Nakanishi, I., and Hanafusa, H. (2003) Assistance of self-transfer of patients using a power-assisting device. In *IEEE International Conference on Robotics and Automation*, Taipei, Taiwan, pp. 4008–4015.

12. Médéric, P., Pasqui, V., Plumet, F., Bidaud, P., and Guinot, J.C. (2004) Design of a walking-aid and sit-to-stand transfer assisting device for elderly people. In *7th Int. Conference on Climbing on Walking Robots (CLAWAR'04)*, Madrid, Spain.

13. Vukobratovic, V. and Borovac, B. (2004) Zero moment point thirthy five years of its life. *International Journal of Humanoid Robotics*, vol. 1, no. 1, pp. 157–173.

A Portable Light-weight Climbing Robot for Personal Assistance Applications

A. Gimenez, A. Jardon, R. Correal, R. Cabas, and C. Balaguer

RoboticsLab, Universidad Carlos III de Madrid
{agimenez, ajardon, rcorreal, rcabas, balaguer}@ing.uc3m.es

Abstract

Human care and service demands will need innovative robotic solutions to make easier the everyday of elderly and disable people in home and workplace environments. The main objective of this work is to develop a new concept of climbing robot for this type of service applications. ASIBOT is a 5 DOF self-containing manipulator, that includes on-board all the control system. The main advantage of this robot is its light weight, about 11 kg with 1.3m reach. The robot is totally autonomous and needs only power supply to be operated. The robot is a symmetrical arm able to move between different points (Docking Stations) of the rooms and, if it is necessary, "jump" to (or from) the environment to the wheelchair. In this way the ASIBOT robot could became a home companion and assistance for numerous people.

Keywords: Rehabilitation robotics, light weight robot, Service robots, hard-soft architectures

1 Introduction

During the last years the rehabilitation technology has developed towards more flexible and adaptable robotic systems. These robots aim at supporting disable and elderly people with special needs in their home environment. Furthermore, most advanced countries are becoming to be aged societies, and the percentage of people with special needs is already significant and due to grow. There have been very interesting developments in this field, such as PARO [1], a robot which provide psychological and social effects to human beings, or Dexter [2] and MANUS [3], which are mounted on a wheelchair and helps in welfare tasks to the disabled people. Another interesting robot is HANDY 1 [4] a feeding helper and RAID [5] a robotic workstation for vocational purposes. The Korean developments in rehabilitation robots KARES I (wheelchair mounted) and KARES II (mobile platform mounted) [6] are also very interesting. Only the MANUS and HANDY 1 are a commercially available robotic system capable of assisting the most severely disabled people in self-feeding, and personal hygiene tasks. The modular aids system concept [7] represents the meeting point between two lines of work, robotics and domotic devices, both integrated in the "smart home" environment [8] [9], thanks to a shared home network. The integral assistance systems are robotic modules and technological aids in general for personal assistance, such as robots, mobile bases, electric wheelchairs, aids for standing up and walking, etc, all together integrated under the same network that manage and supervise the devices in the same way as common home appliances like lighting system, doors and windows, intruder detection or air conditioning.

2 The portable aid robot concept

The European Union MATS project (IST-2001-32080), with the participation, among others, of the University of Staffordshire (UK), Scuola Superiore Sant'Ana in Pisa (Italy), University of Lund (Sweden) and University Carlos III of Madrid, has the objective to develop the robotized system that joint both, the static and moving aid systems into one climbing robot [10], [11],[12]. The other partners are not involved in the development of the robotics system, but they are the "end-users". The robot is able to be attached to the wheelchair and helps the disabled person in his/her life domestic tasks. But at the same time the robot is able to "jump" from/to the wheelchair to/from the domestic environment and vice versa.

This robot presents a new concept in the rehabilitation robotics [13]. The main advantage of the ASIBOT robot concept is the combination of both, static and moving systems, into one climbing robot. The robot is able to be fixed to the wheelchair and helps the disabled person in his/her life domestic tasks. At the same time the robot is able to work in the day-life environment climbing the walls, tables and other surfaces.

2.1 Climbing ability

The climbing ability of this 5 DOF robot is due the successive fixation and release at Docking Stations (DS) [14]. There are three different kinds of DS:

- Fixed DS. This kind of mechanisms is fixed to the walls and others places of the house where is needed for any special task such as on the table to place the plates into the dish-washer.
- Mobile DS. When the robot needs to move a large distance between two DS it is better to move in high velocity. This is possible if the DS is fitted to a mobile trolley that can move in a rail located into the wall.
- DS on the wheelchair. It is a special DS, which is located on the wheelchair. There is a special DS in the room which allow the transition between the room DS and the wheelchair.

The robot presents an innovative grasping method based on special connectors and a bayonet fitting, that allows the robot to move itself along normally unused home spaces, like walls and ceilings. In each tip of the robot there are special conical connectors that are also a grip. Those parts play three functionalities:

1. The tip docked works as a fixation part to the DS.
2. In the tip of the conical part there are electrical contacts for the power supply to the robot.
3. The free tip of the robot is able to manipulate things thanks to the retractile fingers that work as a hand. The electrical contacts on the top of the cone are hidden to avoid electrical risks.

In this way the ASIBOT robot extends the human abilities and is able to perform a big variety of domestic operations: house-keeping, assistance, entertainment, etc.

User requirements have driven the research and development processes in the project with continual user evaluation and peer review of the results obtained at every stage. Every relevant aspect of the lives and environments of potential users was explored in detail by acknowledged experts in their field. Physicians, therapists and psychologists has contributed to the process of eliciting and evaluating the views and expectations of the end users who can benefit most from the application of the MATS system [14]. Among these requirements the user demands a technological solution that wouldn't waste the limited spaces in their homes. So they rejected assistance solutions based on mobile platforms or permanently fixed to wheelchair.

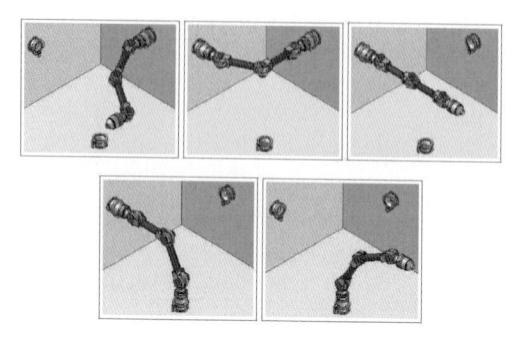

Fig. 1. The robot is able to climb using the environment connectors.

The results of this user requirements study was used to generate functional and performance specifications for the system's components, which was then be designed and manufactured to satisfy the users needs. Mainly, the ideas of portable aid core and minimum waste of space are consequences of the users needs.

Fig. 2. The robot is able to transfer from the environment to the wheelchair.

2.2 Modular and portable system

ASIBOT is designed to be modular and capable of fitting into any adapted environment. This means that all the control system is on-board. The only device needed to operate the robot is a PDA based *HMI* that send commands to perform the tasks. This degree of flexibility has significant implications for the care of the disabled and elderly people. A great deal of functional flexibility and versatility will be derived from the use of software and the integration of the system into "smart home" environments [12].

Fig. 3. ASIBOT is very easy to transport. The robot is inside the bag.

The modularity of the system makes possible for the system to grow as the level of disability of the user changes [13]. This robot stands out because the capability to be transported by one person and carried to another place in an easy way. Furthermore, it is only necessary a Docking station and 24 V - 600W to connect it in any place. So its possible to use the same robot in several places. The robot is transported by the caregiver inside a bag like **Fig. 3**, or by the own user attached to the wheelchair. To reach this objective, a core aid technology has been developed taking into account two ideas, the portability requirement of the system, and the modularity to integrate the robot in the environment.

According to the social inclusion idea, the handicapped people like others need to travel daily from home to the work place and vice versa, and their robot assistance may travel with them. The modular aid system concept [7] the robot assistance is part of the structure that provide additional actuation capability. The question is to define which components of the smart home adapted would travel with the user. When the user arrives to their office would need the assistant to perform several task so its necessary to adapt the office. Using the portable aid core is not necessary duplicate expensive components in the adapted environments. The main functionalities are located in the portable aid core formed by the arm to perform extended manipulation capability and by the interface, to perform interaction with the robot and the environment. **Figure 4**, shows the main components of the assistance system, the components inside the box constitute the portable aid core.

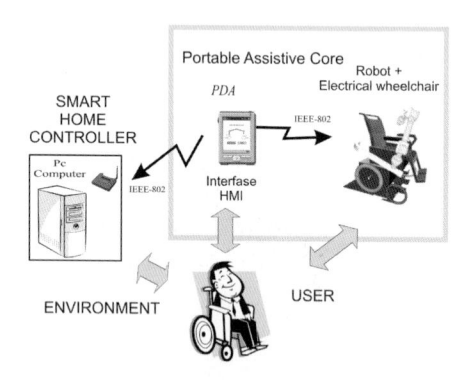

Fig. 4. Schematic diagram of the ASIBOT system.

When the user arrives to the office the user itself can connect the aid core to the local smart controller. The on board control and communication architecture allows its operation from a simple HMI, running on a PDA. The modularity of the software architecture allows operating the robot from multiple wifi-based interfaces, such PDA, joystick, smart-scan devices, chin-control, etc [5]. Only the selected HMI has the direct operation over the robot at the same time. An hospital or geriatric environment is the maximum exponent of the practical approach that the minimum portable aid core offers.

2.3 Interface

A good user interface is necessary for the acceptance of service robots in rehabilitation; it will be only effective if the underlying system has a certain degree of intelligence. There are two ways to interact with the HMI. Like a normal application, clicking buttons, writing text and values using the stick or even the finger because the screen is tactile.Another way, in order to allow the use of this robot by the most severely disabled people, is via voice.The user can use a set of defined commands to manage the interface and can perform exactly the same tasks permitted using the tactile screen. Also, the HMI has a voice synthesis module to generate voice message to the user when he/she is controlling the robot using the voice.

The communications architecture allows the HMI to use the services present in the environment. In this way is possible to integrate the aid core on to the WIFI infrastructure of the domestic environment obtaining control of home appliances (video, TV, lighting equipment, heating-cooling systems, access to internet, etc); in the same way when the user arrives to the office or to a friend adapted home can connect it self to the same services. The modularity architecture allows to incorporate more functionalities to the system without modifying the robot, simply introducing more sensors and different interfaces HMI depending on the requirements of the end-user.Also is possible to choose as a HMI the device that make the user feel more comfortable like a traditional joystick This device connects directly with the PDA and generates the commands for direct control of the arm.

3 Conclusions

An excellent ratio weight/number of DOF/length is presented by the ASIBOT robot. Moreover, given that all the control system is on-board, the robot can be very easily transported from one environment to another. Extremely easy adaptation of the home environment to the ASIBOT robot, by introducing low cost DSs and a unique power supply. The robot is designed to be modular and capable of fitting into any adapted environment. The robot can move accurately and reliably between rooms and up or down stairs, and can transfer from being wheelchair- mounted to floor, or wall-mounted. This degree of flexibility will have significant implications for the care of the disabled and elderly people with special needs. ASIBOT satisfied the users needs related to portability and minimum waste of space. The robot is currently finished and is under testing. The actual tests demonstrate that with the tolerances of the DS location in order of some mm and some degrees, it is possible to perform the docking process in an automatic way with any type of compliance.

References

1. K. Wada, T. Shibata, T. Saito, K. Tanie (2002) Analysis of factors that bring mental effects to elderly people in robot assisted activity. International

Conference on Intelligent Robot and Systems, IEEE/RSJ
2. L. Zollo, C. Laschi, G.Teti, B. Siciliano, P. Dario (2001) Functional compliance in the control of a personal robot. International Conference on Intelligent Robot and Systems, IEEE/RSJ
3. H. Kwee (1997) Integrated control of MANUS and wheelchair. International Conference on Rehabilitation Robotics ICORR'97, Bath UK
4. M. Topping (2002) An overview of the development of Handy 1, a rehabilitation robot to assist the severely disabled. Journal of intelligent and robotic systems, Vol. 34 pp. 253-263
5. Dallaway JL, Jackson RD (1993) The RAID workstation for office environments. Proceedings of the RESNA 93 Annual Conference. 504-506
6. Pyung Hun Chang, Hyung-Soon Park (2002) Development of a Robotic Arm for Handicapped People: A Task-Oriented Design Approach. Department of Mechanical Engineering, KAIST, 373-1, Guseong-dong, Yusong-gu, Daejeon 305-701, Korea
7. C. Laschi, E. Guglielmelli, G. Teti, P. Dario (1999) A modular approach to rehabilitation robotics. 2nd EUREL Workshop on Medical Robotics, September 23-24, pp. 85-89, Pisa, Italy
8. Ad van Berlo (1998) A "smart" model house as research and demonstration tool for telematics development. Proceedings of the Technology Initiative for the integration of Disabled and Elderly people, TIDE Conference, July
9. Hammond J., Sharkey P., Foster G. (1996) Integrating augmented reality with home systems. Proceedings of 1st International Conference on Disability, Virtual Reality and Associated Technologies, pp 57- 66
10. C. Balaguer, A. Gimenez, M. Abderrahim (2000) A climbing autonomous robot for inspection applications in 3D complex environment. Robotica, vol. 18, September
11. C. Balaguer, A. Gimenez, A. Jardon (2003) MATS: An assistive robotic climbing system for personal care & service applications. 1st Workshop on Advanced in Service Robotics ASER'03, Bardolino Italy
12. C. Balaguer, A. Gimenez, A. Jardon (2005) "Climbing Robots" Mobility for Inspection and Maintenance of 3D Complex Environments. Autonomous Robots, Springer Science, Vol. 18, N 2
13. K. Kawamura, S. Bagchi ,M. Iskarous, M. Bishay (1995) "Intelligent robotic systems in service of the disabled". IEEE Transactions on rehabilitation engineering, vol. 3, no. 1
14. A. Gimenez, C. Balaguer, A. Sabatini, V. Genovese (2003) The MATS system to assist disabled people in their home environments. IEEE/RSJ International Conference on Intelligent Robots and Systems 2003, Las Vegas USA

Modeling and Control of Upright Lifting Wheelchair

S. C. Gharooni, B. Awada and M. O. Tokhi

Department of Automatic Control and Systems Engineering, The University of Sheffield, UK;   c.s.gharoni@sheffield.ac.uk

Abstract

This study is concerned with lifting and stabilization of a wheelchair, in which torque is applied to the back motors of the wheelchair, to force enough power to lift it upwards and then stabilize itself in the upright position. To achieve this; a wheelchair model is built in the Visual Nastran software environment. The development of fuzzy control strategies for control and stabilization of the wheelchair in the upright position is investigated and verified within simulation studies. The potential benefit of fuzzy control in this context is demonstrated.

Keywords: Fuzzy control, inverted pendulum, upright wheelchair.

1 Introduction

Electronic wheelchair users would often like to be lifted upwards to talk to people directly or reach high places to pick up or put things down. This operation can be performed by lifting the wheelchair upwards and then stabilizing it as an inverted pendulum. Since the manual wheelchair came into production and became more widely used, engineers from different disciplines attempted to improve the manual wheelchair in terms of reliability, practicality size, weight, performance and automation. They searched for ways in designing more comfortable wheelchairs that will not exhaust the user, and therefore automated wheelchairs came into production [1]. With simple command-action motor-driven wheelchairs were soon developed into more sophisticated control systems that can climb over pavements, go downstairs/upstairs, run over rough material such as wood and sand, lift themselves upwards, and run steadily on two wheels while going forwards, backwards or even sideways. The iBot wheelchair available from 'Johnson & Johnson' [2] is capable of performing all the above mentioned tasks.

This study is concerned with the design of a controller that will be able to lift and stabilize a wheelchair while a weight is placed on the chair. This task requires a high control precision with reliable and robust operation, offering the following advantages [3]:

- Allows the user to speak to a walking person while looking at them directly.
- Gives the user the ability to reach high places to pick up or put things down.
- Gives the user a better feeling when moving around in the city or at home, as they are at the same height as the others.

2 Methods

A three-dimensional wheelchair model resembling the real one was developed for this study. The model and associated simulations were implemented in Visual Nastran (VN) [4]. A fuzzy logic (FL) [5] controller was developed in Matlab/Simulink [6], connected to Visual Nastran for simulating the wheelchair motion. The performance of the system was assessed with random disturbances.

2.1 Modeling the wheelchair

The wheelchair model built in VN has most of the physical properties of a wheelchair in terms of dimensions and weight, although it represents the wheelchair in its simplest form. While building the model some design compromises have to be made, as for this stage of study not every aspect of the wheelchair would be modeled in VN. Below is a point wise comment on the model differences between the V.N. wheelchair and the real one:

1. The weight and size of motors were not accounted for in the model. This is due to the fact that the amount of torque required to lift the wheelchair is still not known and therefore choosing a motor weight, size or power would not be appropriate at this time.
2. The weight of wheels was modelled in VN as 5 kg although in the real one the normal wheel would not weigh more than 2 kg. This is due to the following two reasons:

- To stabilize the wheelchair because light wheels would make the wheelchair unstable, and difficult to keep on the ground.
- To account for the un-modelled weight of the motors, as eventually the weight of motors will have to be added. The motors will be connected to the axis and the wheel, which means that any extra motor weight will directly influence the wheels' stability and grip to the ground.

To account for the weight of the human body on the wheelchair, a humanoid model in full weight scale was put on the chair at the centre of the seat, so that its weight is acting directly downwards. Fig. 1 shows the wheelchair model in VN in with the actual humanoid model.

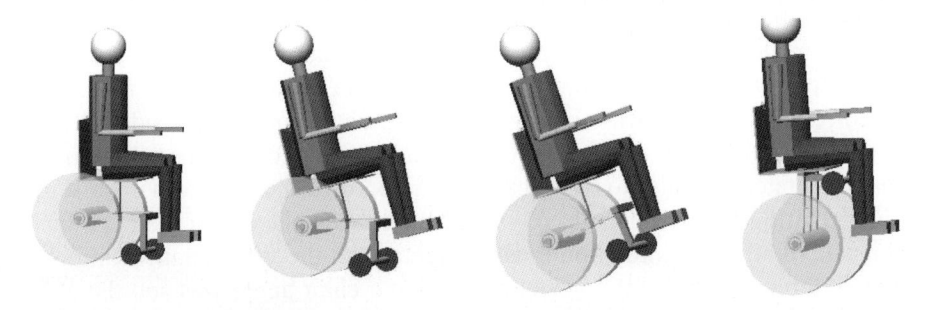

Fig.1. Model of wheelchaie in VN in various posisions.

2.2 The Control system

The initial aim of the controller is to lift the wheelchair, and then stabilize it in its position. The wheelchair must be able to carry a range of weights (50-100 kg). The two back wheels of the wheelchair are fitted with torque motors, which should apply enough torque on the back wheels to lift the whole wheelchair upwards while moving a very small distance (forwards or backwards) from the original position. The torque must not be too high otherwise the wheelchair will flip over, or the wheels will slip, or it will lift up very quickly resulting in instability. The torque must not be too low, or else the wheelchair will move forward and not lift up. At the start of the lifting stage, a large torque results in a high angular velocity. The velocity should decrease as the wheelchair comes closer to the upright position, and once the wheelchair is upright the velocity must be zero. The better the velocity regulation is, the smoother the wheelchair lifting would be.

The angle the wheelchair makes with the horizontal (ground) is 140°, so by testing and with basic mathematical calculation the angle at which the wheelchair is upright was found to be 50° (140° – 90° = 50°). This is an important value because it is the controller's set point or target. The amount of initial torque required to be applied was found out to be around –100 N to –150 N depending on the weight on the chair. This value was found by experimental testing on the wheelchair model, where a manual slider control was connected to the rear torque motors and different torques were applied until a torque was found that was able to lift the wheelchair appropriately i.e. not too fast and not too slow.

2.3 Designing the Chair Rotation Control

Rotating the chair depends on the wheelchair orientation angle. Therefore, the controller input was chosen to be the wheelchair angle and the controller output was chosen to be the chair orientation angle. To rotate the chair an orientation motor was fixed between the chair connecting rod and the axis connecting rod. In the first 0.5 seconds of the control the wheelchair has to rotate gradually to the back by 20°, and then the control will shift to start rotating the wheelchair up. Once lifting takes place then the chair has to start rotating This rotation must coincide with the wheelchair lifting, so for different values of the wheelchair angle a chair angle was set, the values are shown in Fig. 2.

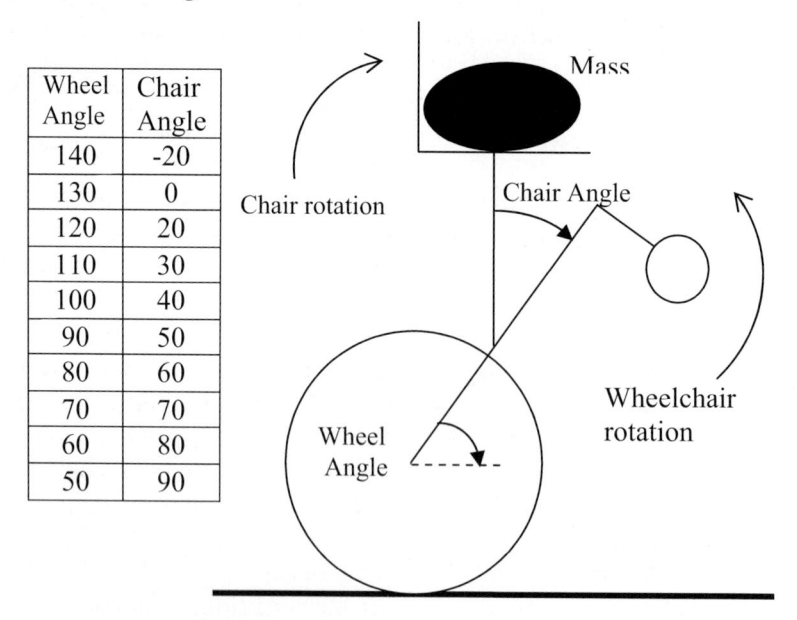

Wheel Angle	Chair Angle
140	-20
130	0
120	20
110	30
100	40
90	50
80	60
70	70
60	80
50	90

Fig. 2. Wheelchair and chair rotation during lifting up process.

2.4 Designing fuzzy logic controller

The idea of fuzzy control is to build a fuzzy model of a human control expert who is capable of controlling the plant without thinking in terms of a mathematical model. The main advantage of the FLC approach is the possibility of implementing rules of thumb, experience, intuition, prediction and heuristics without the need for a mathematical model [7]. To design the fuzzy controller, membership functions (MFs) for both inputs and output must be drawn. The variables that are required by the controller as input are error and change of error of the wheelchair orientation. The change of error MFs were designed by observing the system and recording the maximum and minimum change of errors reached so that MFs can be designed. The MFs for error has a minimum of –50° and a maximum of 90°. This is due to the reason that at the start the angle is 140°, which means that the maximum positive error is 90°. The torque requirement was found to be between –150 Nm and +150 Nm.

2.4.1 Fuzzy rule base

A set of fuzzy if-then rules is contained in the fuzzy rule base, and the fuzzy inference engine uses these rules to determine a mapping from fuzzy sets in the input to fuzzy sets in the output universe based on fuzzy logic principles. It was found that the easiest way to design a rule-base was by deciding on where the 'OK' has to be for the system to become stable or to reach the required target.

Table 1. The Error and change of error rules table

$\dot{e}$ / e	NVL	NL	NM	NS	NO	OK	PO	PS
NM	PL	PL	PL	PL	PL	PM	PS	OK
NS	PL	PL	PL	PL	PM	PS	OK	NS
OK	PL	PL	PL	PM	PS	OK	NS	NM
PS	PL	PL	PM	PS	OK	NS	NM	NL
PM	PL	PM	PS	OK	NS	NM	NL	NL
PL	PM	PS	OK	NS	NM	NL	NL	NL
PVL	PS	OK	NS	NM	NL	NL	NL	NL

The first 'OK' was placed in the zero error, zero change of error cell (the dark shaded cell in Table 1). The other OKs were then placed diagonally above and below. As the 'OK's are in their position, the next decision that had to be taken was; to which side of the 'OK' shall a negative or a positive torque be placed. The continuity rule was used to fill up the remaining cells. This means that after the 'PS', an 'OK' will come then an 'NS', followed by an 'NM' and etc., i.e. no jumps of torque output between the neighboring cells, ensuring that the control is smooth and stable.

2.4.2 Upright stability

Although the above controller was designed for the lifting stage it can be used for the stabilization stage too. The rules designed for both controllers are the same, so the rule-base does not have to be changed. The same MFs can be used, and the only change would be to the input signal, i.e. when the final stabilization starts to take place, the error would be very small. Therefore a gain can be placed to step it up, so that the controller can give a finer and more precise output.

2.4.3 Implementing controller

Following the controller design, Simulink design, and tuning of the parameters (output torque gain, the change of error multiplication graph, and the error gain), the controller was tested with a 70 kg weight, and for some disturbance forces applied to the wheelchair. The final Simulink block diagram incorporating the controllers, the plant, the chair rotation control, and the controller switching action, is shown in Fig. 3.

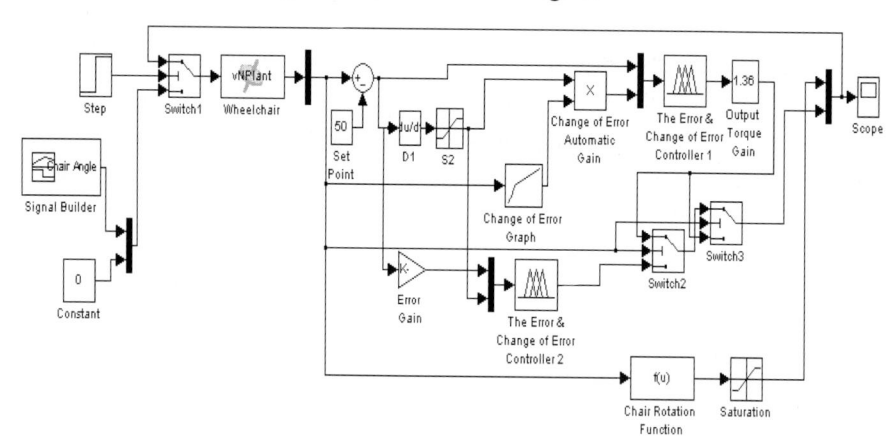

Fig. 3. The error/change of error control in simulink.

3 Results

To test the controller a force was placed on the seat in the y-direction i.e. as if someone is pushing the chair from the back or from the front. This experiment was performed to test if the wheelchair is pushed by mistake from the most dangerous area (the top of the chair) and whether the wheelchair will stay upright and stable.

The position where the disturbance force was applied and wheelchair trajectory transient to upright are shown is shown in Fig. 4. It shall be noted that although the figure shows that the forces are in the same direction, they are not in practice, as this is the total modulus of force. The first impulse is positive, the second is negative, and the third is positive. All the forces were applied for a period of 0.12 s.

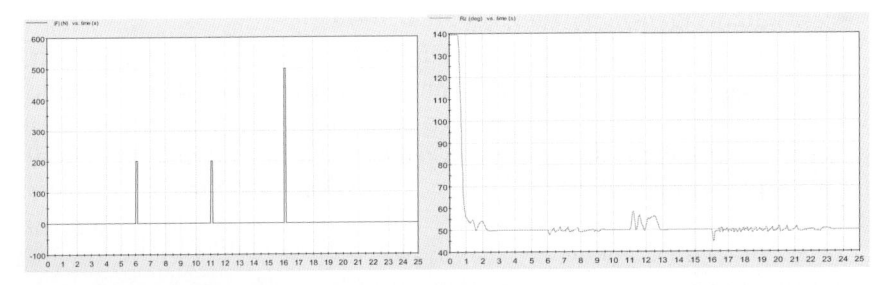

Fig. 4. The disturbance impulse force and the wheelchair angle.

Fig. 5 (left) shows the angular velocity of the wheelchair. As noted, the velocity at the start increases and then decreases with few oscillations in between, hence a smooth lifting control. The wheelchair lost its speed and stopped oscillating when a 200 N or a 500 N force was applied.

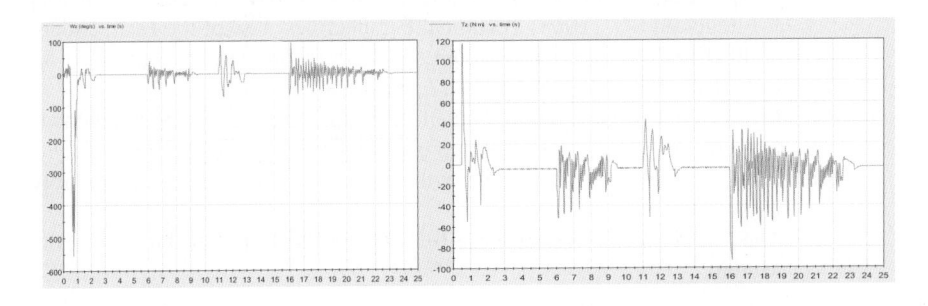

Fig. 5. The wheelchair angular velocity and torque (y-axis) against time (x-axis).

It can also be noticed from the torque graph (Fig. 5 right), that when the controller reaches the set-point, the torque oscillates slowly between –4N and –5N, i.e. the oscillation frequency is not high, which means that the torque moves smoothly from 4N to 5N and vice versa just to keep the wheelchair upright, which is normal and causes no vibration.

4 Conclusion

The controller was tested for the following functions:
- Lifting the wheelchair upwards in a stable smooth manner
- Stabilizing the wheelchair in an upright position and stopping it moving.
- Lifting weights in the range of 50 to 100 kg.
- Disturbance adaptation, i.e. if a force is applied on the wheelchair, it will not deviate significantly from the set point, and it will be able to re-stabilize itself.

As demonstrated, the controller reduces the angle from 140° until it reaches the set point (50°).

References

1. Wellman P, Krovi W, Kuma V, Harwin W, (1995) Design of a Wheelchair .with Legs for People with Motor Disabilities, *IEEE Transaction on Rehabilitation Engineering*, Vol. 3, pp. 343-353.
2. Johnson and Johnson (2004) website: http://www.independencenow-europe.com/uk/main.html (Date of last update: March 2004)
3. Cooper RA, Boninger ML, Cooper R, Dobson AR, Schmeler M, Kessler J, Fitzgerald SG. (2003) Technical Perspectives: Use of the Independence 3000 IBOT Transporter at Home and in the Community, *Journal of Spinal Cord Medicine*, Vol. 26, No. 1, pp. 79-85.
4. MSC SOFTWARE. Stunt Man, free Visual Nastran demo downloaded from website http://www.krev.com/vn4d/simulations.html.
5. Ross, T.J., 1995, Fuzzy Logic with Engineering Applications, (McGraw-Hill, New York).
6. The Math works, Version 6.5.0.1890913a Release 13, June 19, 2002, Matlab.
7. Lee, C. C. (1990a). Fuzzy Logic in control systems: Fuzzy Logic Controller Part I. IEEE Transcations on Systems, Man and Cybernetics, 20(2), pp. 404-18.

A Humanoid Head for Assistance Robots

K. Berns and T. Braun

Robotics Research Lab, University of Kaiserslautern, Germany

Abstract

For the development of a humanoid robot able to assist humans in their daily life, adequate interaction is a key feature. If one considers that in human communication more than 60% is non verbal (communication with facial expressions and gestures) an important research topic is how should a robot head look like and what skills must such a head have to be able to interact with humans. In the following first the simulation system is introduce which was used to test mimic and the expression of emotion. Based on this simulatiuon results the mechatronical head of the University of Kaiserslautern is designed. Finaly a real-time capable method for image based face detection is presented, which is a basic ability needed for interaction with humans.

Keywords: Humanoid robot head, expression of emotion, simulation mechanical design, face detection.

1 Introduction

Worldwide, several research projects focused on the development of humanoid robots. Especially for the head design there is a strong discussion if it should look like a human head or if a more technical optimized head construction [1], [2] should be developed. The advantage of a technical head is that there is no restriction according to the design parameters like head size or shape. On the other hand if realistic facial expressions should be used to support communication between a robot and a person human likeness could increase the performance of the system. In **Fig.1** some existing robot heads are classified according to their human likeness and technical complexity[1].

At the University of Kaiserslautern a human like head construction is under development. The aim of the project is to build a head able to observe humans and its environment and to communicate with humans verbally and non-verbally based on facial expressions.

In the following, the design criteria of our head are introduced. Starting from Ekmans *Facial Action Coding System* [3], adequate action units are selected for the emotions that shall be expressible by the head. Then, our simulation system is described which was implemented to test the behavior of the skin when activating the action units. The results are used to for the construction of the robot head. In parallel to these experiments, software was developed to detect the head of a human being based on colour information and to extract features. This work, which is a precondition for the interaction of the robot head with a human, is described at the end of the paper.

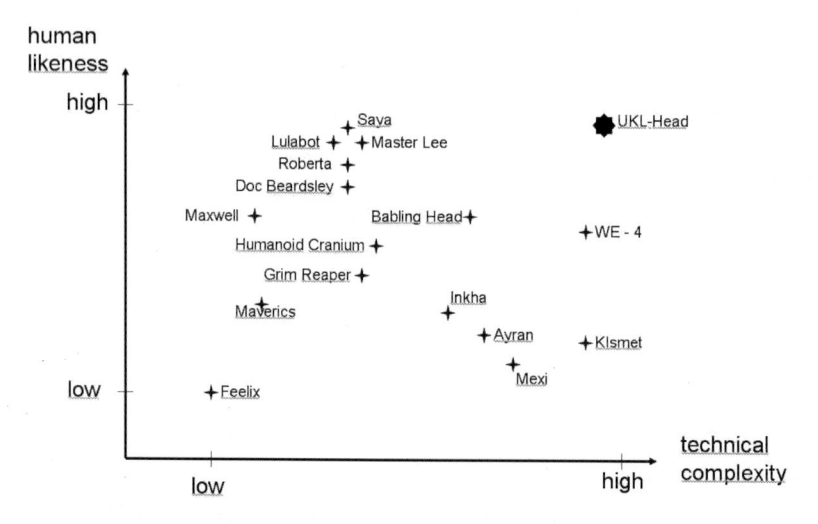

Fig. 1. Classification of humanoid robot head according to human likeness and technical complexity. The technical complexity is meassured by the number of sensors/actuators and the interaction skills with humans.

---

[1] Also see http://www.androidworld.com/prod04.htm

2 Expression of emotion

In [3] a set of action unit are introduced which are used mainly for the expression of emotions by means of facial expression. Each of these units consists of one or several muscles which are connected to specific parts of the skin. Ekman has shown that an activation of the muscles of a specific set of action units lead to an expression of the basic emotions like joy, fear, sadness, surprise or disgust. E.g. disgust is expressed if the action units 'brow lowerer', 'lip corner depressor' and 'chin raiser' are activated at the same time. If each of the action units is implemented by one motor and only the minimum number of units is selected for the expression of emotions, at least 30 motors must be installed in a robot head to control the skin in a humanlike way. One reason for the big number of action units is that muscles are only able to contract and therefore only able to move the skin in one direction. On the other side to move a bigger area of the skin it is necessary to fix several muscles on different contact points. To transfer the design principles of a human head to robot head able to express emotion like humans, it is necessary to answer the following questions:

- Which areas of an artificial skin must be moved and in which direction to simulate the Ekmans action units?
- How important is the velocity when moving from one emotional state to another one?
- Which trajectories in the emotional space look natural when switching between emotion?
- Is it possible to express an emotion like fear more or less strong through different activations of the corresponding action units?

3 Simulation

To answer the questions raised above, a simulation of the artificial skin was implemented together with the Computer Graphics Group (Prof. Hagen) of the University of Kaiserslautern. Starting with a 3D laser scan (using the Minolta-Digitizer Vi900) of the artificial head covered with a silicon skin, a triangular mesh with about 1,6 Mio triangles was generated and reduced by standard mesh simplification algorithms to about 100000 triangles. The simplification was done in such a way that smooth areas are represented by only a few meshes while areas with a lot of skin retain a high number of triangles. Then, the resulting mesh was modelled as a spring/damper system, which describes the behavior of the silicon skin when external forces are applied. Based on this simulation of the artificial skin, several tests were performed to find out where and how the artifical skin can be moved in order to simulate an action unit. Also the influence of the magnitude of the force was evaluated. After the realisation of the Ekmans action units in the simulation system, tests were performed to see if the basic emotions can be expressed adequately. Therefore, 60 students have been interview to classify the simulated faces according to the basic emotions. As result, about 75% of the expressed emotions are correctly classified. In **Fig.2** the simulated head is shown with the silicon skin attached.

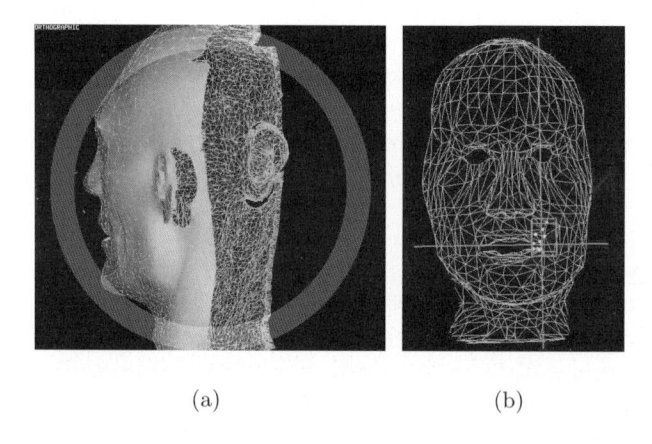

(a) (b)

Fig. 2. In (a) the simulation of the head with the artificial skin is shown. Figure (b) presents some mesh points which are selected to be affected by drawn force vector. Based on the results of the simulation it was possible to select an adequate set of mesh points and the direction, in which these points should be moved when the corresponding action unit was activated.

4 Mechatronics

The mechatronical system of the head is still under development. Due to the simulation results the selected mesh points, which belong to an action unit, are connected with small metal plates. These plates are glued on the silicon skin. Wires are connected to these plates which allow their movements in direction of the related action unit. As actuators, 10 servos are used to pull and push the wires. Additionally, a servomotor is used to raise and lower the lower jaw. An inertial system which is able to determine the pose of the head is installed. The integration of microphones and loudspeaker in the head is in preparation. Also the mechanical eye construction with 3 DOF each and a neck construction with 3 DOF is under development. The control of the servo motors as well as the determination of the pose from the inertial system is done with a DSP (Motorola 56F803) connected to a CPLD (Altera EPM 70 128). The calculation of movements of the action units according to the emotions which should be expressed is done on a Linux-PC. The DSPs are connected via CanBus to the PC. In **Fig.3** the head and the carrier is shown.

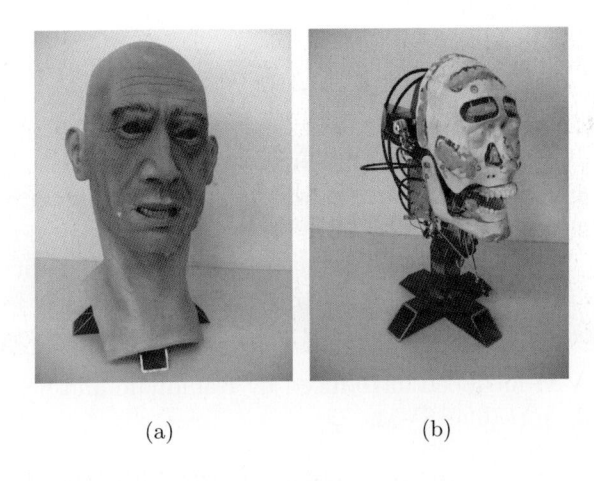

(a) (b)

Fig. 3. Mechanical construction of the humanoid robot head with and without the silicon skin. (The silicon mask is designed by the company Clostermann Design Ettlingen, Germany.)

5 Face detection for interaction with a human operator

One of the next experiments planned for the humanoid robot head is the expression of emotion based on the facial expressions and gestures of a human operator. For example, the robot should be able to 'mirror' the emotions expressed by the human. A necessary precondition for this is the detection of humans and especially their faces. For this, a detection algorithm based on the camera images that will be obtained by the artificial eyes has been implemented and will now be described.

To allow a reliable detection of faces in the expected dynamic environment, a pattern recognition method is needed that can cope well with variable lighting and differing facial poses or expressions. The Support Vector Machine (SVM) has been shown to yield excellent face detection performance even in difficult conditions and thus seems to be a good choice in this context. However, the computational demands of a SVM-based face detection system are high compared to haar-based classifiers or neural-network approaches.

In order to reduce the computational complexity of a SVM-based face classifier and be able to exploit its high classification ability, we have developed an approach that speeds up the face detection task using three complementing techniques:

1. A special, highly efficient *Sequential Reduced SVM (SRSVM)* is used for classification of image patches instead of a regular one.
2. Prior to the application of the SRSVM, it's search space is reduced to image parts containing *face candidates*. These candidates are determined quickly using skin-color filtering and geometrical constraints.
3. The search space is further reduced by fusing range information with the captured image. This yields distance information for each face candidate and is used to *restrict the scale* at which to look for faces.

The **SRSVM** has been introduced by Romdhani and Schkopf [4] as a way to speed up regular SVMs.

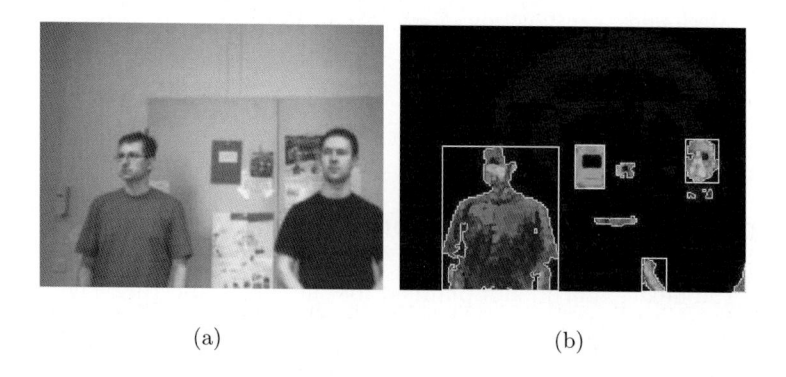

(a)   (b)

Fig. 4. Original image (left) and Skin-filtered image (right). Detected outlines are marked, valid face candidates are indicated by boxes.

To confine the application of the SRSVM to image areas that are likely to contain faces, **face candidate** areas are extracted from the input image prior to classification. These candidate areas are determined by applying a skin-color filter [5] to the color input image. Further processing by morphological smoothing, binarization and contour extraction yields outlines of potential faces. These outlines are then tested using several geometrical validation rules: 1. The contour bounding box (BB) must have a minimum size of 19x19 pixels. 2. The width of the BB must lie between it's height and half of it's height. 3. The relative amount of filled pixels in the BB after binarization must be above 70 percent. **Fig.4** shows an example for the face candidate generation.

Since the employed SVM can only classify image patches with a fixed size[2], an input image containing faces of unknown sizes must be examined at several scales. To avoid this time consuming multi-scale analysis, the distance between the robotic head and the depicted human is calculated with stereo-image processing. At the moment a testbed (see **Fig.5**)for the stereo-camera system (equipped with 2 dragonfly cameras), which will be integrated later on as eyes of the humanoid head, is used to directly determine the appropriate scale for the face detection.

Fig. 5. A testbed for the stereo-camera system, which will be included as eyes in the humanoid robot head.

To actually detect faces in realtime using the collected information, the extracted face candidates are scaled according to their corresponding distance measures. Then, the area occupied by the candidate BB is split into 19x19 subwindows and classified by the SRSVM into faces and non-faces. The basic SVM needed for the face detection system has been trained on over 20000 face examples and 102000 non-face examples which have been collected from various image databases. The initial training

---
[2]The image window used in our implementation is 19x19 pixels wide.

using a gaussian kernel with a width of $\sigma = 0.04$ resulted in a SVM with 27267 support vectors. This set was subseqently reduced to 100 support vectors that formed the basis of the SRSVM. We have found that an image patch can be classified by the SRSVM with only 4 SVs on average, taking $12\mu s$ per patch on a 1.6 GHz P4. Compared to a normal SVM, the resulting speed gain is immense. Nevertheless, the detection performance is only slightly degraded. In total, the face detection part is able to process 3 fps while leaving enough processing time for the motion and safety systems.

6 Conclusion

In this paper a humanoid head construction is introduced, which will be used to interact with humans. Main focus of the present research is how the facial expressions of a human being can be transferred to a robot head. From our point of view the complexity of this problem will be reduced, if the robot head is human-like. Based on a simulation of the silicon skin of the head the implementation of Ekmans action units for the head design was performed. Additionally, face detection software was implemented to detect possible interaction partners. The next steps will consist of the completion of the mechatronical design and the implementation of a behaviour based control concept for the interaction with humans.

References

1. Atsuo Takanishi, Hiroyasu Miwa, Hideaki Takanobu (2002) Development of human-like head robots for modeling human mind and emotional human-robot interaction. *IARP International workshop on Humanoid and human Friendly Robotics*, pp. 104–109.
2. Breazeal, Cynthia (2003) Emotion and sociable humanoid robots. *Int. J. Hum.-Comput. Stud.*, vol. 59, no. 1-2, pp. 119–155.
3. Ekman, P. and Friesen, W. (1978) *Facial Action Coding System*. Consulting psychologist Press, Inc.
4. Romdhani, S., Torr, P., Schölkopf, B., and Blake, A. (2001) Computationally Efficient Face Detection. *Proceedings of the 8th International Conference on Computer Vision*, vol. 2, pp. 695–700.
5. Jones, M. and Rehg, J. (1999) Statistical Color Models with Application to Skin Detection. *IEEE Conference on Computer Vision and Pattern Recognition*, vol. 1, pp. 274–280.

An Application of the AIGM Algorithm to Hand-Posture Recognition in Manipulation

Daniel García, Miguel Pinzolas, J.L. Coronado and Pablo Martínez

Universidad Politécnica de Cartagena, Dpto. Ingeniería de Sistemas y Automática. Paseo Muralla del Mar s/n. 30202. Cartagena (Murcia). Spain
{ daniel.garcia, miguel.pinzolas, jl.coronado, pablo.martinez }@upct.es

Abstract

In this paper a visual module is proposed as a part of a visuo-motor system, implemented on an anthropomorphic robotic platform. This platform is composed by a robotic arm with a human-like hand, and a robotic stereo head to provide visual information. The platform is able to reproduce a human hand-posture in the robotic human-like hand. The vision algorithm provides information about positioning, orientation and posture to the robotic hand. This is done in non-structured environments, in which illumination and workspace distribution can arbitrarily change. The recognition module is inspired in the Elastic Graph Matching algorithm. The main improvement of the proposed application is the ability to cope with affine transformations of the image, that is, with rotated, translated and scaled hand-postures, without the extensive use of comparisons.

Keywords: Computer Vision, Pattern Recognition, Graph Matching, Affine-Invariant

1 Introduction

One of the major problems in robotics about manipulation is the need to get some information concerning to the workspace in which the robotic platform is involved. A robust vision system is required to get a correct identification of each element (object, robotic arm, robotic hand, human hand) in order to achieve the manipulation task on the right way. The vision algorithm must be able to reproduce the position, orientation and posture of the human hand in the robotic human-like hand. This paper is focused on this issue.

The hand recognition problem is a labeling problem based on models of known hand-postures. A model of a hand comprises a set of characteristic features that unequivocally distinct that precise hand from the rest in the model base. It is usual that models employ local features to describe significant characteristics of the hand, and relational features that describe positional relations among the local ones so as to preserve its geometry. These kinds of representations are commonly called *model graphs*, and the process of comparison between models in the model base and hands in the image is named *graph matching*.

This is the case of the Elastic Graph Matching algorithm [1,2] on which the algorithm proposed in this paper is based. This EGM algorithm lies in an elastic model with nodes that carry local information extracted from several locations of the model hand, and edges representing spatial relations (distance and relative orientation) between the nodes. The obtained model is a graph containing a model description with both local and relational features. In the recognition process, a model graph is matched onto the input image. At the beginning of the matching process, nodes are placed on the surface of the image in the shape of not deformed graph. After that, they move and slightly deform the graph to precisely fit it in successive iterations.

The matching index is measured by a cost function that involves the similarity of node features between the model nodes and the image ones, and the deformation of the edges suffered in the fitting process [2]. In the EGM, nodes of the graph are labeled with a local image description, provided by a Gabor or Mallat filter [1,2].

The EGM algorithm described presents several drawbacks. First, if the hand size is different from the model one, the coarse positioning usually fails because some nodes are no longer inside of the mask zone, wherefore the method is not very scale-independent. Second, the recognizer is not invariant against hand rotation in the image plane. Third, the use of Gabor filters to determine the local features that describes the nodes, makes the algorithm not invariant against background changes when those nodes belong to the hand fringes.

As a choice, the AIGM algorithm [3], which is able to deal with rotation, illumination and scale changes without an intensive use of models, is used instead. In this algorithm, the chosen nodes are selected among borders detected by a Canny algorithm [4]. Each node is characterized by two local features: a rotation-dependent feature (curvature) and a rotation-independent one (gradient). Also edges are established between nodes to have into account the relational features (orientation and length) which will assure that the geometry of the hand is maintained. To take into account the topological restrictions, in the matching procedure a reinforcement method is proposed which favors those possible node locations in the input images whose relative position coincides with the corresponding edge in the model. Consequently, if an image with a rotated or scaled hand is supplied, the algorithm returns the best matching model in the model base, the angle of rotation respect the model, and the scale change. The AIGM (Affine-Invariant Graph Matching) algorithm is explained in the following sections.

2 Algorithm

As declared above, nodes are selected amongst borders detected by a Canny algorithm, and they are characterized by two local features: gradient $G(N_k)$, which is rotation-dependent, and curvature $C(N_k)$, which is rotation-independent. Only the gradient direction at node N_k is considered, because the gradient modulus is highly dependant on illumination, and it can give false results with darker or brighter backgrounds.

The relational features are represented by edges linking pairs of nodes (N_k, N_j) in the model graph. These features are the edge orientation $\mathrm{dir}\left(\vec{E}(N_k, N_j)\right)$ (rotation-dependent) and its length $\left|\vec{E}(N_k, N_j)\right|$ (rotation-independent).

All the local and relational features are represented in Figure 1.

Local features:
- Gradient $G(N_k)$
- Curvature $C(N_k)$

Relational features:
- Distance $\left|\vec{E}(N_k, N_j)\right|$
- Orientation $\mathrm{dir}(\vec{E}(N_k, N_j))$

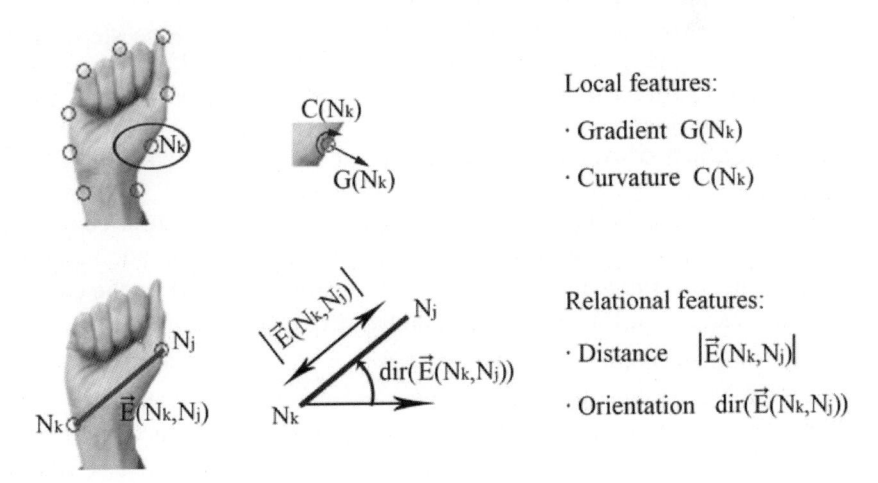

Fig. 1. Local and relational features in a model graph

The steps followed by the matching process are the following:

1. Detection of borders in the image using a Canny algorithm.
2. Extraction of local features $G(B_m)$ and $C(B_m)$ for each border point B_m detected in the image.
3. For each node N_k in the model, nodes B_m that comply with $C(B_m) \approx C(N_k)$ are found. These nodes are possible candidates $C_k^m(\alpha_m)$ to node N_k if model graph is rotated α_m. Angle α_m is deduced by $\text{dir}(G(B_m)) - \text{dir}(G(N_k))$.
4. Every node candidate $C_k^m(\alpha_m)$ votes for the candidates $C_l^p(\alpha_m)$ that have similar $|\vec{E}(N_k, N_j)| \approx |\vec{E}(C_k^m, C_l^p)|$ with respect to the model and that are placed in the correct direction $\text{dir}(\vec{E}(N_k, N_j)) + \alpha_m \approx \text{dir}(\vec{E}(C_k^m, C_l^p))$.
5. Extraction of the scale as the average value of the ratio between all the edge modulus in the image and all the edge modulus in the model.
6. Selection of the three most voted candidates to node N_k and execution of a second voting stage, using the scale calculated in Step 5.
7. Extraction of the matching index as the average voting value of the selected set of nodes. Definition of the rotation angle α_r as the mean of the difference between the orientation in model nodes and the candidates in the image.

In the following a detailed explanation of some of these steps is performed.

2.1 Selection of node candidates

After getting the borders detected by the Canny algorithm, a comparison between each border point B_m and the model node N_k is made in order to select suitable candidates to a node. This analogy is based on the curvature feature, calculated for each node as shown in Equation 1.

$$C = \frac{f_{xx}f_y^2 + f_{yy}f_x^2 - 2f_{xy}f_xf_y}{(f_x^2 + f_y^2)^{3/2}} \qquad (1)$$

Functions f_x and f_y are first-grade derivatives of the intensity function in the image $f(x,y)$ centered in the pixel (x,y), and f_{xx}, f_{xy} and f_{yy} functions are second-grade derivatives. The kernels used for the convolutions are the Prewitt approximations to the X and Y components of the gradient.

Later on, the difference among the gradient direction at the selected nodes in the image and the model ones is calculated. This is the supposed rotation angle α_m of the model in the image.

After this stage, a set of points have been selected amongst the set of border points, to be candidates to a node N_k. These candidates are labeled as $C_n^i(\alpha_k)$, meaning the i-th candidate of node n if rotation α_k is supposed.

2.2 Selection of the three most voted candidates

Once the first voting stage is completed, a list of the three most voted candidates to each node N_k is created. Only these tree nodes for each node N_k are taken into account in the second voting stage. The scale calculated in the first voting is used for modifying the model size so as to fit correctly in the input image.

2.3 Reinforcement of adequate node candidates

Once the node candidates are localized for each node in the graph, topological requirements must be taken into account to select the most possible node location. One of them is the direction of each edge established between connected nodes, and the other one is their relative distance.

Let us define the neighborhood H_n of a node n in a model graph as all the nodes belonging to the graph that are connected to n by an edge. Let be $m \in H_n$ a neighbor of n. The edge joining the two nodes in the model graph can be represented by the oriented segment $\vec{E}_{nm} = \{E_{nm}, \alpha_{nm}\}$, being E_{nm} the module and α_{nm} the direction of the segment. Let $C_n^i(\alpha_k)$ be the i-th candidate of node n if rotation α_k is supposed, and $C_m^j(\alpha_k)$ the j-th candidate of node $m \in H_n$ for the same α_k. The edge between these two node candidates is represented by $\vec{E}_{C_n^i C_m^j} = \{E_{C_n^i C_m^j}, \alpha_{C_n^i C_m^j}\}$.

Then, the voting result for candidate $C_n^i(\alpha_k)$ is updated as:

$$V_{C_n^i}(\alpha_k) = \max\left(V_{C_n^i}(\alpha_k), V_{NEW}\right) \quad (2)$$

$$V_{NEW} = \left[\operatorname{abs}\left(\cos(\alpha_{C_n^i C_m^j} - \alpha_{nm})\right) \; 2 \operatorname{logsig}\left(-0.25 \; \operatorname{abs}(E_{C_n^i C_m^j} - E_{nm})\right)\right]$$

2.4 Extraction of the real scale

The most probable position of graph node n in the image is given by $D_n(\alpha) = \max_i \left(V_{C_n^i}(\alpha_r) \right)$. The probable scale can be determined once $D_n(\alpha_r)$ has been fixed for every n, by comparing the average ratio of the lengths between edges linking nodes $D_n(\alpha_r)$ and the corresponding edges in the model graph, as shown in Equation 3.

$$S = \text{mean}\left(\frac{E_{nm}}{E_{D_n D_m}} \right) \qquad (3)$$

2.5 Selection of the rotation angle

The rotation angle of the image can be obtained by $\alpha_r = \text{mean}\left(\alpha_{D_n} - \alpha_n \right)$.

2.6 Calculating the matching index

The total graph similarity value is computed as:

$$V = \frac{1}{N_G} \sum_{n=1}^{N_G} D_n(\alpha_r) \qquad (4)$$

The number of nodes in the model graph is determined by N_G.

Once all the model graphs have been matched on the image, the chosen model graph is the one having the maximum value of V.

3 Experimental Results

In order to test the behavior of the proposed method, a database [5] of hand-posture images has been used. The images included in this database are scaled, translated and with illumination changes. Some of them have been rotated in order to prove the efficiency of the algorithm, obtaining the results shown in Table 1.

A representation of how the correct model nodes are fit in the border points of the input image is shown in Figure 2. In this figure, there are several examples of hand-posture images scaled, rotated, translated and with a background, in which the matching process is achieved successfully.

An application of the AIGM algorithm to hand-posture recognition .. 991

Table 1. Results obtained with 210 images from the database.

	Correct Ident.	Incorrect Ident.
Rotated Images (50)	94,73 %	5,27 %
Scaled Images (105)	92,57 %	7,43 %
With Illumination changes (105)	90,84 %	9,16 %

There are two main reasons of failure for the algorithm: first, if there is another model in the model base that can be fit inside the correct one, as shown in Figure 3; second, if the border detection fails and consequently, there are false nodes in the input image.

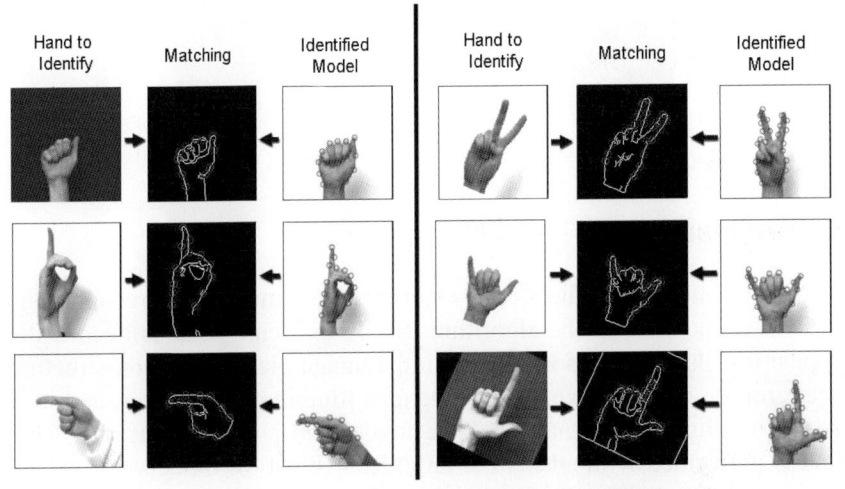

Fig. 2. Matching of hand-postures. Triesch PETS2002 Datasets

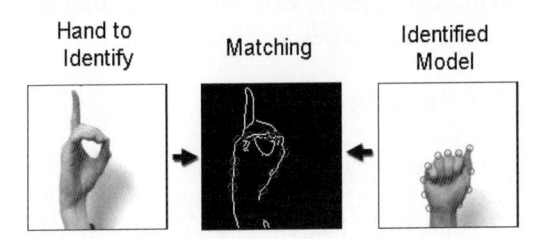

Fig. 3. A case of an erroneous matching.

The proposed application has also been proved with an own hand-posture images database for grasping and manipulation, obtaining the results shown in Table 2. A representation of the matching process is also shown in Figure 4.

Table 2. Results obtained with 50 images from the own database.

	Correct Ident.	Incorrect Ident.
Rotated Images (20)	91,15 %	8,85 %
Scaled Images (20)	87,56 %	12,44 %
With Illumination changes (10)	89,02 %	10,98 %

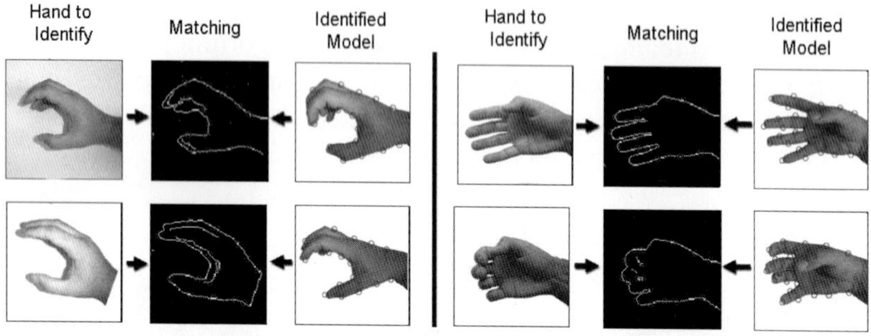

Fig. 4. Matching of hand-postures in grasping and manipulation

4 Conclusions

An new application of the AIGM algorithm for hand-posture recognition is presented and tested. The algorithm is able to identify hands even if they appear translated, scaled or rotated in the image plane, or in non-structured backgrounds. It is also very robust against illumination changes, and it can deal with slight deformations of the model and partial occlusions. All of this is performed without neither an extensive use of models nor exhaustive comparisons. The results show that the algorithm behaves adequately, correctly identifying the hand-posture and calculating the exact scale and rotation angle in most of the cases.

References

1. M.Lades, J.C.Vorbrüggen, J.Buhmann, J.Lance, C.Malsburg, R.P. Würtz, and W. Konen, Distortion invariant object recognition in the dynamic link architecture. *IEEE Transactions on Computers,* 42, 1993, 300-311.
2. J. Triesch, *Vision-based robotic gesture recognition,* (Doctoral Thesis, Bochum University, 1999).
3. M.Pinzolas,R.Chumilla,JL.Coronado, *Affine-Invariant Graph Matching,* 3rd IASTED International Conference on Visualization, Imaging and Image Processing, 2003, Benalmádena, España. Vol:1, pp: 1002-1008.
4. J. F. Canny, A computational approach to edge detecttion. *IEEE Transactions on Pattern Analysis and Machine Intelligence,* 8(6):679-698, 1986.
5. PETS2002 datasets, The University of Reading, UK. Website: http://pets2002.visualsurveillance.org

Security and Surveillance Applications

AirEOD: a Robot for On-board Airplanes Security

S. Costo, F. Cepolina, M. Zoppi, R. Molfino

PMAR laboratory of Design and Measurement for Automation and Robotics, Dept. of mMechanics and Machine Design, via All'Opera pia 15A - 16145 Genova, Italy.   {costo, cepolina, zoppi, molfino}@dimec.unige.it

Abstract

Terrorism is a very actual problem. Although terrorist attacks came in succession from the early seventies and eighties, recent events have raised the technological scale and the dimension of the actions. High level of security is required and bomb-disposal experts are frequently exposed to risks in mission. The effects on the society are blameworthy. Robotic systems can represent an effective mean to provide a high level of security without any risk for humans but several classes of applications exist for which no robot available on the market can be used. One of these application fields is the disposal of bombs inside airplanes. The paper describes the design and development of AirEOD, a mobile robot for rescue of a potential bomb in an airplane. The paper focuses on the mechatronic design of the robot and on its control system.

Keywords: Security robots, bomb disposal, airplane.

1 Introduction

Security has become a critical area with an always growing importance. Robots for security tasks deserve today a high attention and in the following years we can foresee that they will be increasingly required and used in place of man, as much as effective robotic systems will be designed and made available.

Robots show the best instrument for security tasks since they enable continuous operating, provide constant working capability and inspection quality, reduce operation costs compared to traditional methods.

An important security task is the defusing or disposal of bombs or devices supposed explosive. The defusing usually consists in destroying the potential bomb by means of water cannons or micro-charges (if the environment allows). The disposal consists of grasping the potential bomb and carrying it afar from the place where it had been placed, in a safe area where any eventual explosion would not create damages and where further intervention is easier.

In both cases, of course the risk for the human operators (artificers) is strongly reduced if they can remotely control a robot moving in the dangerous scenario instead of doing the job themselves. Moreover, places exist that the artificers cannot reach equipped with the cumbersome and heavy safety suits; if no robots are available, they have to carry out the mission with a minimal protection.

The robots used for the disposal of bombs are name EOD (explosive ordnance disposal) robots. An EOD robot is supposed to be able to move agile in its working environment, it should reach every place where the bomb might have been placed and its manipulation capabilities should allow a delicate grasping and handling of the object.

Tasks and environments exist where no general purpose or specific robots available today can be used to perform EOD missions. The present paper is addressed to a new robot specifically designed for one of these environments: the inside of a passengers airplane. Here only disposal is allowed since every defusing technique (water cannons, etc.) would damage the airplane. The cost of stopped-airplane is very high and the disposal has to be carried out in the minimal time possible.

An EOD robot operating on-board an airplane should have two main characteristics:

- small mass and size in order to be carried on-board the airplane manually by two people and to move in the narrow spaces available on-board;
- an arm long and agile to reach every place inside the airplane, the place under the window seats as well as the hat-boxes and the catering boxes.

The EOD robots available on the market are usually derived from military robots and robots for the treatment of dangerous substances; they cannot operate in narrow spaces such as planes and busses. Robots specially suitable for EOD have been designed only later.

The only commercially available system equipped with a really dexterous arm that could be employed, in principle, in an airplane is *Vanguard*, by *EOD Performance*, but its size and mass are not compatible with the space on-board an airplane; the other EOD robots, suffer lack of arm dexterity, and generally have a cart too large and heavy.

Since none of the available EOD robots could be used, a new dexterous one named AirEOD has been specifically designed in collaboration with the Italian manufacturer *Ansaldo Ricerche*. The performance requirements and design guidelines of the new robot have been defined together with the end-users, the bomb squads of the airports of Genoa and Malpensa 2000.

In the following the design of *AirEOD* is presented and the control issues are outlined.

2 AirEOD mechatronic design

The main performance requirements are summarized hereafter with reference to the main subsystems of AirEOD: the mobile platform and the dexterous arm (**Fig. 1.**).

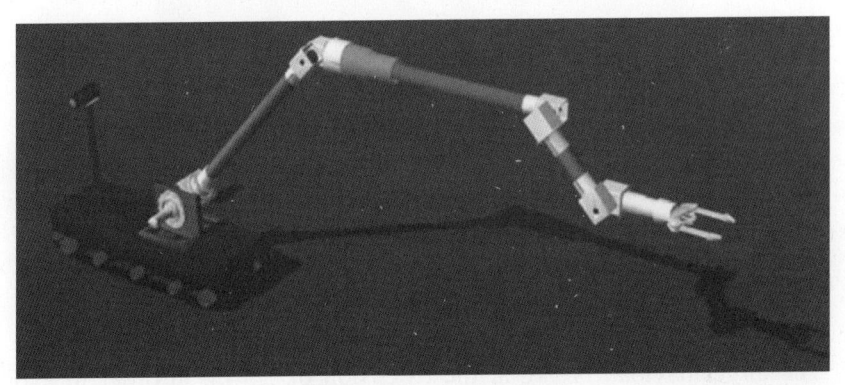

Fig. 1. The AirEOD robot

The mobile platform subsystem includes power supply, locomotion system, batteries, structural components, on board control and communication units: it should be able to operate at very slow speed within close range of suspicious items, avoiding jerking motions and vibrations; the travel and turn maneuvers have to be done reliably in very tight spaces, overcoming, if necessary, ropes and cables laying on the floor; the platform, whose mass has to be limited within 80-90 kilograms, should be endowed with carrying handles to enable two men lifting and transportation.

The manipulator subsystem should have adequate reach and degrees of freedom to easily reach suspicious items put down the seats or inside the hatboxes; the wrist dexterity and the gripper architecture have to allow to "rope and hook" procedures to firmly handle and robustly move packages to less dangerous locations without slippage, the manipulator should be able to position cameras and inspection tools, to lift foreseen (weight and dimensions) loads without causing the tilting and instability of the base, to withstand the forces generated by improper operations.

One of the most problematic issue is the arm geometry and dexterity. The robot needs to work both under the seats and in the hatboxes. A seven mobilities robotic arm has been selected; redundancy is required to improve dexterity in order to easily move around obstacles such as the frames of the seats. Speed is not a critical issue: time-to-mission is limited by economic reasons but the operations on the suspicious payload have to be carried out slowly. A compact rest position of the arm is required to simplify the transport of the system inside the airplane (**Fig. 2.**).

Fig. 2. AirEOD robot in rest position

The target payload is 4kg at the extension of 1.5 m; in order to avoid overturning (with a fixed platform counterweight), the arm must be as lighter as possible. For this reason, the actuators of the first two links are relocated on the mobile platform. The actuator of the second link, the shoulder, is relocated through a shaft concentric to the first link. These two actuators have a double reduction system: an Harmonic Drive unit coupled with a *Minimotor* unit compound by a brushless motor and a planetary reducer. This mechanical solution provides the desired actuation torque with a very compact and light motor.

The arm geometry and dexterity has been checked by simulation in a 3D virtual environment representing the reference airplane section.

Simulations of typical missions of the robot, inspection and/or extraction of suspicious objects differently located, allowed to heuristically optimize the robotic system geometry and cameras positioning in order to guarantee a good view of the target object while avoiding overlapping with robotic arm and airplane structures. Missions are performed by coordinating the platform movement and the arm trajectory. Two kinds of tasks are, at present, foreseen: inspection tasks and "rope and hook" tasks. During inspection tasks the platform goes on the corridor very slowly while the arm examines the prescribed area (on bottom, under seats, or on top, inside previously opened hatboxes) by swinging the cameras and, if a suspicious object is found, it stops, the object is carefully analyzed and, in the case it is considered dangerous, a rope and hook task is programmed. During "rope and hook" the robotic unit reaches a convenient position in the corridor and then, with the aid of the visual sensors approaches the object and operates it.

The robot is equipped with three cameras: the first, with pan and tilt movements, is placed on the base; the second and the third ones are fixed to the arm for local viewing. In order to improve the operator evaluation of the distance of the gripper from the target an ultrasound sensor is also provided. It is very useful for correctly interacting with the worked object, since every involuntary collision could trig the bomb.

The design process of the AirEOD availed of fully implemented digital mock-ups and extended virtual test campaigns.

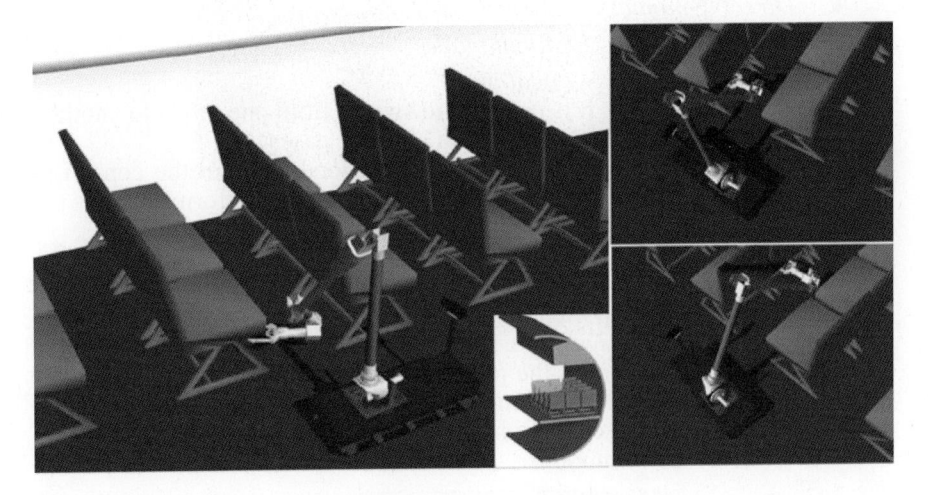

Fig. 3. Cockpit model

A 3D mock-up of an airplane was prepared to simulate the robot in action (**Fig. 3.**).

The robot frame is made of aluminum and composite hollow pipes that envelope the cables, the actuators and the gears. The overall weight is low compared to the payload.

The clamp (**Fig. 4.**) has a compact design and its architecture and geometry are selected to grasp both big and small objects, from 4 mm to 100 mm diameter. The fingers use four-bar mechanisms assuring the parallelism of the tips and a linear force/displacement characteristic.

Fig. 4. The clamp (left), dynamics modeling (center) and placement on the arm (right)

3 AirEOD control system and interface

The user can select one of three control schemes developed for the AirEOD robot:
- *Direct dynamics*
- *Inverse dynamics - Global mode*
- *Inverse dynamics - Local mode*

During the simulation, it is possible to switch from one mode to another. PID controllers are used for all joints in all three cases.

The control of the mobile platform is independent from the control of the arm, thus the pose of the arm gripper refers to the platform and it is not possible to keep the gripper fixed while the platform moves.

The initial control imposed torque thresholds to the motors together with a torque control scheme, but this led to instability. Then the velocity control scheme natively supported by ODE has been adopted.

In the case of *Direct dynamics* scheme (**Fig. 5.**), each joint is controlled separately from the others in the joint space. This scheme is used for large arm displacements while reaching the workspace region containing the target object to be disposed, e.g., in the bottom-front of a seat.

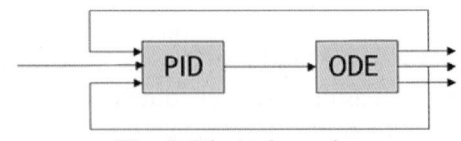

Fig. 5. Direct dynamics

The *Inverse dynamics* scheme (**Fig. 6.**), allows the control of the robot in the end-effector space. In *Global mode* the pose of the gripper is controlled with respect to the mobile platform reference frame, while in *Local mode* the gripper is controlled with respect to its own reference frame.

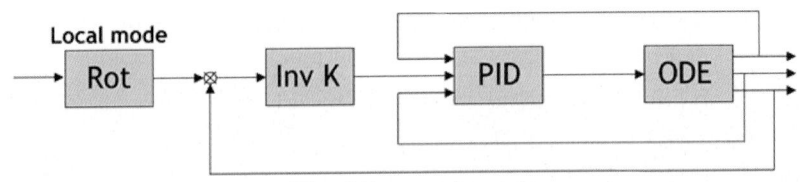

Fig. 6. Inverse dynamics

The normal robot operation sequence is:
- *Direct dynamics* for fast approach of the workspace region where the target object relies
- *Inverse dynamics – Global mode* to make the gripper closer to the target
- *Inverse dynamics – Local mode* to reach the target

The user can control real-time the robot, using the keyboard. The AirEOD moves inside a virtual airplane. The simulation, based on Open Dynamics Engine (ODE), supports dynamics and collision detection. The graphic engine Open Scene Graph (OSG) enhances the reality of the environment. The system can also be used for training of bomb-disposal experts that will maneuver explosive ordnance disposal robots.

Figure 7 shows some frames of the robot during a rescue operation in virtual reality.

Fig. 7. Graphic representation of the robot

4 Conclusions

Robotics deserves an increasing interest in security. Robots can be used to carry out dangerous missions with no direct risks for human operators, who can maneuver them from safe places.

Tasks and environment exist where no available robotic system can be used. The paper addresses the application to bomb-disposal inside airplanes.

A new robot, named AirEOD, is presented. It is composed of a small and low-mass mobile platform and a long and dexterous arm reaching every difficult place on-board, including the hatboxes, the catering boxes, the under-seat of the window seats. The effectiveness of the design has been validated by virtual tests on the virtual mock-up of the robot.

An architecture for the control system is proposed and discussed.

Acknowledgements

The bomb squads of the airports of Malpensa and Genova are hereby kindly acknowledged for their contribution in the definition of the requirements for AirEOD and for the fruitful discussions.

References

1. Anitescu M., Cremer J., Potra F.A. (1995) Formulating 3D Contact Dynamics Problems, *Mechanics of Structures and Machines*, 22/4: 405-437
2. Anitescu M., Potra F.A., Stewart D.E. (1998) Time-Stepping for Three-Dimensional Rigid Body Dynamics, *Comp. Methods Appl. Mech. Eng.*, 177/3-4: 183-197
3. Anitescu M., Potra F.A. (2001) *A Time-Stepping Method for Stiff Multibody Dynamics with Contact and Friction*, Argonne National Laboratory, Argonne, IL
4. Baraff D. (1994) Fast Contact Force Computation for Non-Penetrating Rigid Bodies, *Computer Graphics Proceedings, Annual Conference Series*, Orlando: 23-34
5. Battista U., Cepolina F., Costo S., Zoppi M. (2004) *Videogame for safe flights*. Int. Workshop on Robots in Entertainment and Leisure and Hobby ELH04, Wien, Austria, December 1-2.
6. Bowen I.G., Fletcher E.R., Richmond D.R. (1968) *Estimate of Man's Tolerance to Direct Effects of Air Blast*, Technical Progress Report, DASA-2113, Defense Atomic Support Agency, Department of Defense, Washington, D.C.
7. Craig J. (1989) *Introduction to Robotics: Mechanics and Control*, Addison-Wesley, Reading, MA
8. Featherstone R. (1987) *Robot Dynamics Algorithm*, Kluwer Academic Publishers, Norwell
9. Haug E.J. (1989) *Computer Aided Kinematics and Dynamics of Mechanical Systems*, Allyn and Bacon, Needham Heights
10. Mirtich B.V., Canny J. (1995) *Impulse-Based Dynamic Simulation of Rigid Body Systems*, PhD Thesis, University of California, Berkeley

AIMEE: A Four-Legged Robot for RoboCup Rescue

Martin Albrecht, Till Backhaus, Steffen Planthaber, Henning Stöpler, Dirk Spenneberg, and Frank Kirchner

University of Bremen, Robotics Group, Faculty of Mathematics and Computer Science, Bibliotheksstr. 1, D-28359 Bremen, Germany
malb@informatik.uni-bremen.de

This paper presents a servo-based four-legged robot – named AIMEE – for the RoboCup Rescue competition.
The robot is described with regard to the mechanics which are based on a very modular construction kit, the electronics based on a custom-made MPC 565 microcontroller board, and a software concept using a new behavior-based microkernel. The bio-inspired control approach we use on AIMEE is based on Central Pattern Generators (CPG), posture-control primitives, and reflex-models.

Keywords: RoboCup Rescue, Four-Legged Robot, Bio-Inspired Design.

1 Introduction

The AIMEE robot (see **Fig.1**) is a four-legged walking robot which is planned to be capable to operate in extremely difficult environments, e.g. the roughest areas of the RoboCup Rescue Competition which include steep ramps of up to 40 degrees and clustered terrains ("Random Stepping Fields" presented in **Fig.1**) with height differences of up to 15 cm.

Our robot first participated in this competition at the German Open 2005 in Paderborn. By achieving the third place, AIMEE has shown that a four-legged robot programmed with a bio-mimetic control approach can compete with the existing wheeled and tracked vehicles. In this paper, we briefly present our bio-inspired concept (section 4) for the robots locomotion (section 5), the mechanics & electronics (section 2), and software-concepts (section 3). The later are not discussed in depth but rather are presented as far as they support the bio-inspired control concept. The RoboCup Rescue Competition is currently dominated by wheeled and tracked systems. However, in the broader field of autonomous robotics, several walking robots exist, e.g., the four-legged TEKKEN [1] or the eight-legged SCORPION [2] robot. Humanoid robots like the JOHNNIE [3] often make use of inverse-kinematics and need excessive computational power which barely fits into a small system like our own. Our system, on the contrary, is based on a bio-mimetic control approach – so are the TEKKEN and the SCORPION robot – utilizing weak computational power. On TEKKEN, two bio-inspired control concepts, CPGs and reflexes, were combined for the first time to achieve mammal-like walking. It does not include posture control primitives in its control approach and is limited to certain terrain for operation.

AIMEE, however, was designed to operate in highly uncertain terrain, i.e. mechanics are not only optimized for forward walking but have four active degrees of freedom per leg for very high flexibility in motion. As The AIMEE robot is based on concepts developed in the SCORPION project, they share a lot of internal logic therefore. AIMEE and SCORPION are both based on Central Pattern Generators, reflexes for stability, and posture control primitives. As the roughest so-called "red" areas in the RoboCup Rescue scenarios have seen the introduction of "Random Stepping Fields" (see **Fig.1**) in 2005, we believe that the importance of climbing and walking machines will increase in that area.

Fig. 1. AIMEE on a "Random Stepping Field"

2 The Mechanics & Electronics

The robot is about 30 cm in length and depending on the position of the legs between 15 and 30 cm in height. The weight of AIMEE is 3.5 kg. The corpus of the robot consists of a aluminum strut which serves as backbone. In the center of the backbone is a plastic box which contains the electronics. The neck of the robot is an extension of the backbone and consists of two small joints. They move the head which bears a camera and an infrared headlight in horizontal as well as vertical direction.

AIMEE's four legs are all equal in design. Each leg consists of four identical servo actuators which are connected with a custom-built Robot Assembly Kit developed for education and prototyping by the Robotics Lab of the University of Bremen (see **Fig.2**). The "Bremen Robot Assembly Kit" consists of prefabricated aluminum parts. This concept allows fast assembly and disassembly of our system for rapid testing of different configurations.

Each servo has a maximum torque of up to 3,5 Nm. We believe that the strength of those motors allows to develop a trotting behavior in the future. The whole angle range of the joints is 270 degrees.

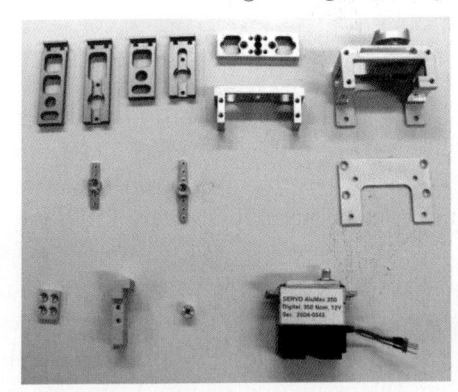

Fig. 2. Bremen Robot Assembly Toolkit

A shoulder joint is situated next to two shoulder plates attached to the backbone. Its motion axle is orthogonal to the backbone. A second shoulder joint is attached below the first one but its motion axle is turned around 90 degrees so that the joint can move the leg sidewards. The last two joints represent the knee and the ankle. They are connected to each other. Their motion axle is the same as that of the first shoulder joint. The feet of the robot have the form of a half cylinder which stands orthogonal to the backbone and points downward. To reduces contact shocks, the feet are equipped with springs. For the implementation of mammal-like walking (which is described in section 5) only the three joints facing forwards are used.

The CPU is a MPC 565 microcontroller running at 40 Mhz. It is mounted on a custom developed mainboard with 2 MB flash memory and 8 MB RAM. It offers 3 time processing units (TPUs) with 32 ports for, e.g., servo control. The board has 64 analogous ports for servo positions and servo current[1] handled by an AD converter, furthermore, 16 additional A/D ports can be used for sensors. Communication is realized via the reliable but low bandwidth DECT standard at 19200 baud.

---
[1] This feature is not yet used.

Additional used sensors are a gyroscope to measure the walking direction, a tilt sensor to measure the pitch, and several micro switches used to detect ground contact and to trigger reflexes as described in section 4. Blocking of servo motors is detected by comparing the written outputs with the real values provided by servos.

For operation in the RoboCup Rescue Competition a camera and an infra-red sensor at the front are used. Two rotating ultrasonic sensors (Devantech SRF10) on top of the electronics box are used for generating a map of the explored area.

3 The Microkernel

Behavior-based programming is a well-known concept for programming mobile robots and is especially suited to program walking robots because of the already integrated principle of parallelism of processes. Many behavior-based programming architectures, e.g. the Process Description Language [4], emulate parallelism by using a cyclic executive for behavior processes and writing their influences on the actual hardware all at once, e.g., at the end of the loop.

But this simple concept has some problems. Single-Loop systems cannot automatically handle the case when behavior process execution takes longer than the estimated fixed loop period. To avoid loop overruns, the loops are typically much longer than the estimated required time for executing all processes. This allows later extension and prevents loop overruns. But this leads to idle CPU time which could be used more effectively. Additionally, most behavior programming models do not provide an easy mechanism to allow behavior processes to run on different frequencies which is desirable if they, e.g., deal with hardware of different reactivity. Finally, out-of-order execution of certain processes as reflexes is not foreseen in these concepts.

Real-time operating systems on the other hand offer high reactivity and satisfy real-time needs but are not specially designed to support behavior programming, consequently they offer a lot of features not needed for our special purpose. Therefore we developed a microkernel labeled M.O.N.S.T.E.R.[2] [5] which features hard- and soft-periodic processes[3], preemptive reflexes, behavior processes, and hardware drivers on different execution frequencies, online-adaptation of these execution frequen-

[2] **M**icrokernel for **s**cabrous **t**errain **e**xploring **r**obots
[3] I.e. processes with hard real-time constraints and those without them

cies, competition for hardware resources between the behavior processes, and a background process.

In fact, the alterations to a behavior programming architecture which are necessary in order to gain these features are quite few. To achieve the desired tolerance for the case when behavior process execution takes longer than the estimated loop period, a mechanism to simulate the ideal case – i.e. when all processes could be executed on time – is required. Let q_i be a variable behavior processes are writing to, let further $inf_{b,i}$ be the influence of behavior process b on that quantity and $w_{b,i}$ the weight[4] assigned to that influence, then the new value $q_i(t)$ at time step t is computed as follows:

$$q_i(t) = \frac{\sum_{b=0}^{n} w_{b,i}(t) \cdot inf_{b,i}(t)}{\sum_{b=0}^{n} w_{b,i}(t)}$$

$$w_{b,i}(t) = \begin{cases} w_{b,i}^{set}(t), & \text{if } b \text{ active in step } t \\ w_{b,i}(t-1) \cdot dec(b,i), & \text{else} \end{cases}$$

If a process b has not been executed in timestep t, its weight is decayed ($dec(b,i)$) otherwise the weight assigned to its influence will be used. Furthermore, a background process has been introduced which will be executed whenever there is spare CPU time, i.e., when process execution takes less time than the fixed loop cycle.

This concept comes with its costs. Behavior processes have to be written in a fashion allowing them to be delayed. This is done by taking the actual timeslice into acount when calculating new values which passed since the last execution of a behavior process. It is used as an estimation of the time till the process gets executed next. But these adaptations arise quite naturally when one understands behavior processes as difference equations. As an example our CPG process uses Bezier Splines to describe the joint movements and in dependendance on the number of passed timeslices it will move along these Splines further or shorter.

4 The Bio-Inspired Control Concept

Our bio-inspired locomotion control is inspired by concepts first used in the eight-legged walking robot SCORPION [2]. In this approach, rhyth-

---

[4] used to control the strength of the taken influence in competition with other behavior processes

mic motion produced by models of Central Pattern Generators (CPGs) is combined with posture control primitives. In addition the control approach features a reflex model which allows to deal with sudden disturbances from the environment. The CPGs functionality is modeled by using third-order Bezier Splines which allow very smooth and adaptive motion patterns. These splines can be controlled by three parameters: the phase, the amplitude, and the frequency. An example of a four-legged walking pattern described with these splines is given in section 5.
More details on this approach can be found in [6].

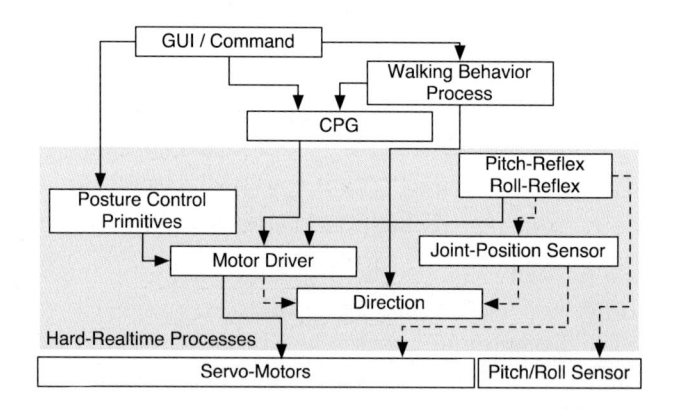

Fig. 3. Control concept as implemented on our robot

5 Mammal-Like Walking

The bio-inspired control concept has been used to implement mammal-like walking. A simplified hierarchy of processes competing for hardware resources for that task can be found in **Fig.3**. Boxes are processes, and arrows indicate that a process writes to (continuous line) or reads from (dashed line) another process or hardware resource.

CPG splines are an adaption of the rhythmic patterns found in cats [7] to our robot's mechanics. Considering AIMEE's hardware limitations compared to a cat, some alterations had to be made, e.g., AIMEE has a stiff backbone which a cat does not, and AIMEE has four equal legs whereas a cat has different front and rear legs. The resulting pattern can be found in **Fig.4**. The phase displacement between the legs are the same as in a cat's walk. Turning is due to the stiff backbone implemented as a behavior which is not found in cats by using amplitude modulation of the CPG splines and sidewards movements of the legs.

There are three reflexes currently implemented on our robot: One is a stumbling control reflex which compares desired angle values with real angle values as reported from the joints. If they differ more than a defined tolerance threshold, a fixed movement is triggered to lift the leg in order to overcome obstacles. The second is the hole reflex which monitors the foot pressure sensors (micro switches) and represses the CPG movement if a foot is not found to be on the ground in a stance phase. Posture control is both possible as a reflex by monitoring the pitch/roll sensor aswell as as a conscious decision used, e.g., on ramps. Reflexes, the posture control process, and the CPGs compete with each other for hardware resources. If one takes more influence on the joints the others will take less influence.

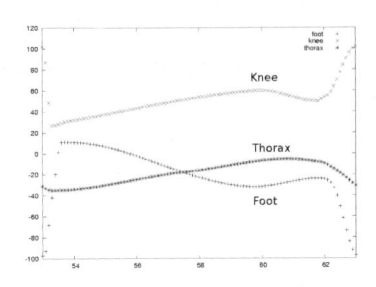

Fig. 4. Plot of the three relevant joints in a single-leg when walking forward

6 Conclusion / Outlook

We have described the concepts of our walking robot which makes it capable of operating in uncertain terrain. So far, experiments have shown that our approach leads to a stable and reliable gait with a speed of 15 cm per second. Even without evaluation of pitch/roll or gyro sensors, the robot is able to walk through plain terrain and to overcome obstacles up to 7 cm. We believe that in the further development the speed of walking can be increased, especially by using reinforcement learning algorithms to optimize the CPG patterns.

In July 2005, our "Bremen Rescue Walkers"-Team will participate in the RoboCup 2005 in Osaka, Japan [8]. Our next goal in order to compete with the non-walking teams at that competition is mastering the steep ramps mentioned in section 1. Currently AIMEE can master a maximum of 25 degrees of steepness. Finally, we are intending to implement stair climbing in the near future.

Acknowledgements

The authors would like to thank Larbi Abdenebaoui, Gerrit Alves, Patrick Amberger, Sebastian Bartsch, Roozbeh Bayat, Uwe Bellmann, Robert Borchers, Friedrich Boye, Justus Brueckel, Heiko Diesing, Simeon Djoko Dzoukou, Jan Evers, Arne Garbade, Stefan Haase, Markus Hagen, Jan Hardel, Wolfgang Kohnen, Jens Kleinwechter, Daniel Khn, Istvan Lovas, Arne Martens, Paul Niechwiedowicz, Malte Roemmermann, Felix Schlick, Alexander Schmidt, Jie Tang, Tobias Quintern, Michael Rohn, Martin Rohlfs, Joachim Voelpel, and Gerrit Wollter for their ongoing commitment to the "Projekt Laufroboter" which results are presented here. Furthermore, we like to thank all members of the Robotics Lab, especially Jens Hilljegerdes and Stefan Bosse, for all their help on the mechanics and electronics.

References

1. Kimura, Hiroshi, Fukuoka, Yasushrioo, and Cohen, Avis H. (2004) Biologically inspired adaptive dynamic walking of a quadruped robot. In *Proc. of 8th Int. Conf. on the Simulation of Adaptive Behavior*, pp. 201–210.
2. Spenneberg, Dirk, McCullough, Kevin, and Kirchner, Frank (2004) Stability of walking in a multilegged robot suffering leg loss. In *Proc. of ICRA 04*, vol. 3, pp. 2159–2164.
3. Löffler, K. and Gienger, M. (2000) Control of a biped jogging robot. In *Proc. of the 6th International Workshop on Advanced Motion Control*, Nagoya, Japan, pp. 307–323.
4. Steels, Luc, Stuer, Peter, and Vereertbrugghen, Danny (1996) Issues in the physical realisation of autonomous robotic agents. In *From Animals To Animats 4*.
5. Spenneberg, Dirk, Albrecht, Martin, and Backhaus, Till (2005) M.o.n.s.t.e.r.: A new behavior-based microkernel for mobile robots. In *Proc. of the 2nd European Conference on Mobile Robots*.
6. Spenneberg, Dirk (2005) A hybrid locomotion control approach. In *Proc. of CLAWAR 2005*.
7. Orlowsky, G.N., Deliagina, T.G., and Grillner, S. (1999) *Neuronal Control of Locomotion*. Oxford University Press, Oxford, GB, 1 edition.
8. (2005) Robocup 2005 - official website. .

Modular Situational Awareness for CLAWAR Robots

Yiannis Gatsoulis[1] and Gurvinder S. Virk[2]

Intelligent Systems Group, School of Mechanical Engineering, University of Leeds, Woodhouse Lane, Leeds, West Yorkshire, U.K. LS2 9JT
(`menig`[1], `g.s.virk`[2]`@leeds.ac.uk`

Abstract

Field operational results have shown that two of the most important features needed by search and rescue robots and so that to be effectively used in such missions are modularity and situational awareness (SA); both are currently missing. Modular aspects are needed so that different branches of the rescue teams can easily and rapidly re–configure the robots for their own mission goals as well as to expand the use of the robot. The SA is needed as this will significantly improve the quality and effectiveness of the operational time. Field operation studies have shown that more than 50% of the deployment time is spent in the operator trying to identify the state of the robot [1, 2]. This paper focuses on identifying potential issues and inspirations for implementing an adaptive SA module able to cope with the characteristics of the operator as well as the robot and the initiation of a formal methodology for SA assessment. The first results of the authors' research in this area are presented.

Keywords: Situational awareness, human–robot interaction, search and rescue robots, semi–autonomous robots, modularity, SAGAT.

1 Introduction

The domain of search and rescue operations is an extremely risky, difficult and time constrained one. Mobile robots could be used for search and rescue (SAR) missions in which the risk is too high for a rescuer or to go into inaccessible areas. However, the current developed systems are still far from being used in real scenarios and their research and development is still an open field. Operational results [3, 1, 2] have identified some fundamental drawbacks in the efficient use of SAR robots. Two of the most important ones are *modularity* and *situational awareness* (SA).

- Modularity is widely accepted as a desirable design concept because of its well known benefits of compatibility, re–usability, local scope of problem resolving, etc. The methodology has been proposed by the EU CLAWAR 1 and 2 Thematic Networks (www.clawar.com). Detailed discussions appear in previous work [4, 5].
- SA of a system in a broad context refers to the ability of determining the status and role of the "system" and its environment with respect to each other, and based on this understanding along with previous understandings and knowledge, make predictions about future actions and events.

This paper focuses on the latter drawback, that of SA, as it was shown that that more than 50% of the mission time is wasted in obtaining this [1, 2]. Related work and the concept of SA as well as methodologies for assessing it are discussed in detail in the rest of the paper.

2 Related work

A broad definition to SA of a system is *the ability of determining the status and role of each entity of the holistic system, and based on this understanding along with previous understandings (SA) and knowledge, make predictions about future events and actions.*

A well structured model of SA developed primarily for aircraft navigation and air traffic control operators but generic enough to be applicable to a number of domains is presented in [6]. According to this model, SA consists of three levels, namely perception, comprehension and projection. In the perception level (Level 1), the necessary cues are identified and perceived from the large volume of perceiving data. The comprehension level (Level 2) is concerned with the comprehension of the meaning of the Level 1 data. This is usually achieved through the formation of a mental model of the system, its environment and its current goals. In the projection level (Level 3) the system tries to predict the future status of the Level 2 elements. Based on this decomposition of SA a generic design process has been proposed. It consists of three phases; requirement analysis of the task, design on how the information should be presented in a meaningful and consistent way and assessment of the system according to the SA. One such assessment method is the Situational Awareness Global Assessment Technique (SAGAT), proved to be successfully used [7] in the area of avionics. In frequent random intervals during the task execution, the task is paused and all display informations are blanked. The user is then asked to answer some questions related to what he/she is currently doing.

To the authors' knowledge there is only one study published which is focused on SA itself for search and rescue robots [3]. It was conducted during an urban search and rescue disaster response drill with teleoperated robots in Miami, US. The methods used included observation and videotaping of five robot operators, encoding their actions and verbal communications with other team members, analysing the data and assessing their SA according to independent raters. The conclusions from this study illustrate the poor SA of the operators as they spent more than half of the operational time trying to build it or maintain it. However, this study is operator–centric, ie the SA of only the operator is under investigation or in our context the holistic system consists of only the human operator. Our case though is both operator–centric as well as robot–centric. Although, the primary requirement is the operator's SA we believe that this cannot be improved without investigating the robot's SA too.

3 Methods

The long term main objectives are to improve the SA of the holistic system and to reduce the time spent during an operation on determining this. The design process is inspired by Endsley's SA decomposition into three levels [6], the perception level, the comprehension level and the prediction level. For that to be realised we should firstly identify the mission goals and develop mental models for the operator(s), the robot(s) and the environment if appropriate. The high level mission goals are searching, identification of victims or clues leading to victims and safety of the system. Searching is primarily concerned with navigation of the robot in a confined area and consists of elements such as obstacle avoidance, localisation and mapping and percentage of area covered. Identification of victims includes determining if they are alive, their position and the route leading to them. Safety includes avoidance of hazards that can damage the robot as well as avoidance of trigerring a further disaster. Mental models realising the mission goals should be developed for the operator and the robot. It is expected that these mental models should be different from each other as the perception and comprehension levels for each entity are very different. The operator primarily relies on visual information from the camera to build a SA while for the robot this is a very difficult process. On the other hand the rest of the robot sensors can give it a SA which the operator may not be able to build. For example, a robot can perceive a potential hazard in front of it, such as a gap, which is out of the field of view of the camera and hence undetectable to the user.

In a sense there are individual SA for each of the entities. However, these are only parts of the overall picture. By communicating with each other they build the overall SA of the system. This raises two questions. *How* they are presented and *when*. The answers to these questions depend on the mental models of the entities.

In our case we are looking at the SA of the holistic system (operator, robot and environment). However, in these early experiments we are focusing on the operator's SA; more precisely we are investigating the:

- The quality of the SA of the operators.
- The efficiency of the operators when presented with unprocessed data.

The variables under investigation in the experiments are the time needed for an operator to cover a specified area, errors determined by independent observations in the operator's actions due to incomplete SA, analysing the statements of the operator during the trial and afterwards, analysing the operator's SA during the operation.

The analysis of these variables will provide a first indication for the above benchmarks to help us better understand how to design and implement the mental models of the entities, how to design and implement the presentation level, how to improve the current system and how to formulate the unprocessed data to meaningful information.

At these first stages the experiments are carried out on a simulator, and these are presented in this paper. The material is described in more detail in Section 4. An experimental trial consists of an initial briefing explaining the environment, the mission goals, the robot and its capabilities, the control interface and its capabilities and the procedure of the SAGAT method. The operators are videotaped and are informed that if they want to they can think loud, although they are not encouraged any further as this may distract them from the mission. After the mission is over the operators are asked to write down what according to their opinions might be helpful in improving their performance. This with conjunction of the videotapes can provide some suggestions for further work, but by no means is it used as a metric as they are only subjective opinions.

The possible major goals which an operator can have in mind are searching an area, identifying a victim and evaluating his/her condition, rescuing a victim, concerns about the safety of the robot and the victims and returning to base (exit point).

Due to the robot setup and the missions goals, only searching, identification and return to base are realistic and these are the ones under investigation. For each one of them there is a set of questions. The questions regarding the searching of an area can be grouped in the analysis into localisation awareness (position and trajectory evaluation), time awareness (estimated time elapsed or estimated time needed to complete the mission), coverage awareness (how much area has been covered or can be covered in the remaining time), spatial awareness (if there are any obstacles), cue awareness (finding survivors).

The identification of a victim goal can be analysed into victim recognition awareness (that a cue signify a victim), localisation awareness (position of the victim), and status awareness (health condition).

Lastly, the return to base goal is concerned with localisation awareness (position of the robot and planned trajectory to exit point evaluation), distance and time awareness (time needed to reach the base).

Some of these can be measured independently from the rest, for example spatial awareness in searching goal, while some are affected by the others. For example the (victim) localisation awareness in identification is affected by the (robot) localisation awareness of searching. Furthermore, some of them are used using averaging of all questions falling into the particular subcategory and some are measured in a Delphi evaluation where an expert is assessing them. The same procedure applies for the questions themselves. The overall operator awareness cannot be directly assessed as a sum of the different types of sub–awareness. This is wrong and deceptive because these types and their questions are not all necessarily measured using the same metrices. So each one had to analysed individually.

4 Materials

Experiments were carried out in simulation, using the Player/Gazebo package as it provides a powerful 3D robot–environment simulator [8, 9]. The environment was the entrance hall of the School of Mechanical

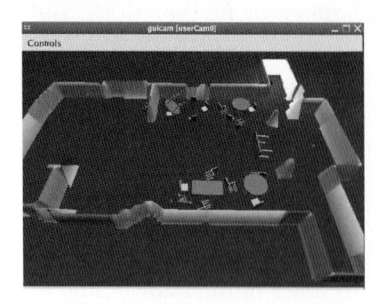

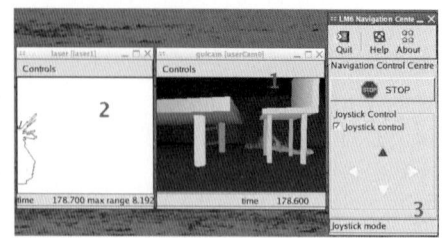

Fig. 1. The virtual environment and the control interface

Engineering building at the University of Leeds. This was chosen as it provides a realistic office environment where a disaster can occur and can be modified according to the experiments requirements to match real world characteristics. During the first stages no ambient intelligence is assumed. A picture of the virtual environment lighted so that it is visible is shown in **Fig. 1**. In the experiments the range of view of the operator is about 4–5 meters. In the arena there are 2 victims both alive, but with one of them lying on the floor being unconscious. The simulated robot is an Activmedia Pioneer 2DX which moves forward, backward, turn on spot right and left with a constant speed of 0.2 m/sec. It carries a camera and a SICK LMS200 laser range finder. These characteristics were simulated because the load of the simulated robot is close to the "teleoperated camera on wheels" [1] used in the World Trade Centre robot SAR operations, with the significant difference that there the operators had no clue of the searched area and a map was also not available. **Fig. 1** shows the camera window with the timer (1), the range finder display (2) and the control interface (3). A simple paper drawing of the area served as a map to the user.

5 Results

A sample of 19 trials have been conducted. A trial consists of the actual simulation time and the SAGAT questions sessions time. **Table 1** shows the mean times for the overall trials, the simulator runs and the SAGAT stops. It is worth noting that the SAGAT stops, in which between 5 and 8 questions were asked, occupied approximately one third of the total time and it was found not to distract the operators, which is one of the

Table 1. Mean durations of the trials, simulator runs and the SAGAT stops

	Mean		Mean
Experiment	33:01 mins	Simulator	20:45
Successful runs	19:59 mins	SAGAT Stops	3 stops
SAGAT total	10:34 mins	SAGAT each	3:16

primary requirements for the success of the SAGAT method. As expected since most of the operational time is spent on searching, 52% of the stops involved questions regarding searching and navigating the area. 28% of the stops involved questions regarding the identification of a victim, 16% involved the return to base goal. The rest 4% were irrelevant.

The results have to be analysed individually according to the three goals under investigation. The quality of SA when the searching for victims goal is active is measured according to localisation, time, coverage, spatial and cue awareness.

Localisation awareness is assessed primarily with the accuracy of the position and the trajectory the robot has followed so far to the true ones using a 0–1 scale, with 1 being same as the true ones. The corresponding values for position and trajectory accuracy were 0.57 and 0.62. The spatial distribution was wide which means that when the operators were lost their approximations were far from the real ones. This was also the main reason for 75% of mission failures. Another factor that can influence the localisation awareness of the operator is the physical landmarks used. In twelve people that they were asked to identify their physical landmarks 3 of them used fixed landmarks such as the walls of the room, 5 used objects of the enviroment such as chairs and tables, while 4 used both types.

The time awareness was assessed by comparing the estimate time given from the operator to complete the mission with the actual time needed for finish. In 7% of the cases the operator could not even estimate the approximate time, 39% of the replies were significant understimates, 26% were quite close to the real time and 28% of the cases were significant overestimates. The reasons for this low time awareness were poor localisation awareness and inadequate perception and comprehension of the existing data. For example in some cases the user only realised very late that he/she was running out of time.

Coverage awareness is evaluated through how much area has been searched so far. 16 out of the 19 people were asked and most of them more than once. In 19% of the responses the operators did not know how much area they had covered, 27% of the responses were underestimates of the true searched area, 19% were overestimates and the rest 35% were approximately accurate. In total 65% were incorrect answers which is a very high percentage since the operators still had the paper version map in front of them helping them to calculate their estimate.

Spatial awareness was evaluated through questions about if there are obstacles within 3 metres and within 3–8 meters, which is approximately the range of the laser range finder. For close by obstacles the side which the obstacle lies was also asked. In the range of 0-3 metres 81% of the answers were correct with the rest 19% being wrong. In the range of 3-6 metres the wrong answers are doubled to 38% with 62% correct ones. The main reasons are inproper or not at all use of the laser range finder, which can be further explained due to unfamiliarity with the sensor as well as the non user friendliness of the current display. For the range of 3–6 metres another reason that explains the increase of wrong answers is that the operators form zones of interest or priotity with the highest being near the robot and slowly decaying the further away something lies. Although, this approach is common and helps in reacting to immediate situations it does not help to plan ahead. About spatial awareness, two operators out of the nineteen expressed ambiguouty if they could fit through a gap. According to the evaluator's observations most of the operators had collisions with objects in particular on the sides of the robot. Two of the robot losses occured due to a collision with an object leaving the robot incapabable of moving.

Cue awareness was pretty straightforward as the only source of information for identifying if there is a victim nearby or not was the camera. However, in two different cases the operators missed to see the survivor from a distance of approximately 3 metres.

The goal of identifying a victim or if there is a cue that signifies a victim was also straightforward and simple since the victims were clearly "look like victims" and moreover there was only one sensor, the camera, capable of being used for this goal. For the victim localisation awareness out of the possible 38 victims in the 19 runs, 36 were discovered from which 19 were marked in their true positions. Obviously, victim localisation awareness is directly correlated with robot localisation awareness, and these results are expected to agree with the above ones presented. In the victims health status awareness the results for the lying on the floor unconscious victim are interesting since the expected answer is they cannot tell as their status cannot be evaluated with just a camera. However,

42% expressed an opinion on whether being dead or alive while the rest 58% suggested they need more information to answer.

In the return to base goal the results are also directly correlated to the robot localisation awareness. However, there is no indication of distance measurement and this is reflected on the answers to assessing the distance and time awareness, as 23% had no clue of estimating the distance since they did not know where they were, another 23% understimated the distance to the exit point, while 38% of the replies were overestimates. Only an 8% was within the 4 metres difference limit. Similar results were obtained for the time awareness since most of the operators did not take into account the speed of the robot.

As presented in the beginning of the section the SAGAT method proved did not interfere with the subjects' awareness when resuming their simulator run. The only drawback was that the operators were helped after the first couple of stops in seeking particular information. However, the not so frequent and unexpected times of stop in conjunction with the randomness of questions was a counterbalance to this.

6 Conclusions

In this paper the results of some early experiments regarding the situational awareness (SA) of a teleoperator and its operator were presented. By using a virtual environment and a simulated robot with an overhead camera and laser range finder the operator is asked to search for survivors in an office environment area. This initial robot setup was chosen because this was approximately used in the search and rescue operations in the World Trade Centre. Although our aim is to evaluate the SA of the holistic system in these experiments only the operator's SA is under investigation. For this assessment the SAGAT method was used and proved to be a reliable in our case. The results have shown that there are significant problems in the operators' SA, mainly regarding localisation of the robot, efficiency of the search strategy, interpretation and correlation of the data provided, even in this simple case. These resulted in robot loss in many of our experiments.

Based on the conclusions and the experience gained from these tests we are aiming to improve the human–robot interaction interface and investigate if these additions in one dimension will increase the complexity of the system or can improve the SA and up to what extent. Furthermore, there are a few remarks on the current SAGAT methodology which need to be improved as well as adapt to the new requirements of the new system. Our intention is to carry out these experiments both in simulation as well as a real robot platform in a more dynamic disaster environment.

References

1. Casper, J. (2002) Human-robot interactions during the robot-assisted urban search and rescue response at the World Trade Center. M.S. thesis, Department of Computer Science and Engineering, University of South Florida.
2. Micire, M. (2002) Analysis of the robotic–assisted search and rescue response to the World Trade Centre disaster. M.S. thesis, Department of Computer Science and Engineering, University of South Florida.
3. Burke, J.J., Murphy, R.R., Coovert, M., and Riddle, D. (2004) Moonlight in miami: A field study of human-robot interaction in the context of an urban search and rescue disaster response training exercise. *Human–Computer Interaction, special issue on Human–Robot Interaction*, vol. 19, no. 1–2, pp. 85–116.
4. Virk, G.S. (2003) CLAWAR modularity for robotic systems. *The International Journal of Robotics Research*, vol. 22, no. 3–4, pp. 265–277.
5. Gatsoulis, Y., Chochlidakis, I., and Virk, G.S. (2004) A software framework for the design and support of mass market CLAWAR machines. In *Proc. of IEEE International Conference on Mechatronics and Robotics (MECHROB'04)*, Aachen, Germany.
6. Endsley, M.R., Bolte, B., and Jones, D.G. (2003) *Designing for Situation Awareness: An Approach to User–Centered Design*. Taylor and Francis.
7. Endsley, M.R., Sollenberger, R., and Stein, E. (2000) Situation awareness: a comparison of measures. In *Proc. of the Human Performance, Situation Awareness and Automation: User Centered Design for the New Millenium Conference*.
8. Gatsoulis, Y., Chochlidakis, I., and Virk, G.S. (2004) Design toolset for realising robotic systems. In *Proc. of CLAWAR 2004, 7th International Conference on Climbing and Walking Robots and the Support Technologies for Mobile Machines*. Springer–Verlag.
9. Gerkey, B.P., Vaughan, R.T., and Howard, A. (2003) The player/stage project: tools for multi–robot and distributed sensor systems. In *Proc. of the International Conference on Advanced Robotics ICAR 2003*, pp. 317–323.

Space Applications

Design Drivers for Robotics Systems in Space

L. Steinicke

Space Applications Services, Leuvensesteenweg 325, B-1932 Zaventem, Belgium.
leif.steinicke@spaceapplications.com

Abstract

This paper provides a summary overview of use of fixed and mobile robotics in the space domain, grouped by a number of typical applications domains. Also, some drivers for designing space robotics are described.

1 Space Applications of Robotics

1.1 In-orbit assembly of large space structures.

The prime example of robotics for in-orbit assembly are those on the International Space Station (ISS), either in place already or planned for future installation. The large Canadian robotic arm, Canadarm 2, which was launched to the ISS early in the ISS build-up, is used extensively for mounting arriving modules and elements to the existing in-orbit ISS infrastructure. This robot arm is operated by astronauts inside the Space Station, with no possibility for remote control from the ground yet. However, the Canadian Space Agency is studying ground-based control of Canadarm. Space Applications Services are involved in a project to demonstrate ground-based control of the European Robotic Arm, under the auspices of the European Space Agency (see below).

The European Robotic Arm, ERA, is planned to be launched to the ISS and mounted on the Russian segment, to perform a number of operations. One intended use of ERA is to assembly a huge mirror for a European X-ray astronomy satellite called XEUS.

The X-ray Evolving Universe Spectroscopy mission is the follow-on to ESA's highly successful XMM-Newton mission. XEUS will consist of two spacecraft – one carrying the mirror, the other carrying the detectors –

formation flying 50 meters apart. It is designed to search for the first giant black holes that formed in the Universe, over 10 billion years ago.

Fig. 1. The International Space Station. The main external robot is clearly visible.

Being an enormous free-flying X-ray telescope, XEUS will orbit much lower than previous ESA satellites for X-ray astronomy. But by orbiting close to the International Space Station, XEUS can be brought alongside the ISS every few years and have new mirrors and features added. By being serviceable by robotics at the ISS, XEUS can last up to 25 years.

The XEUS mirror is far too large to be launched in its assembled state by any existing launch vehicle. Instead, the Space Shuttle will bring a total of 8 mirror segments to the ISS, where the ERA (or alternatively the Canadarm) will be used to assemble the mirror segments into a circular mirror, like petals of a flower.

ERA is also designed for on-board control by astronauts inside or outside the ISS, but the activities mentioned above for determining the feasibility of ground-based control of ERA, could change this control concept, given the extreme scarcity of crew time on-board the ISS.

Fig. 2. Hand-over of large object from Space Shuttle to Space Station robot.

1.2 Inspection and repair of spacecraft.

Various concepts have been studied over many years for inspection and repair robots in space, but the main challenge remains safety considerations. The Inspector system, a small spacecraft flying around the Space Station to inspect for damage to the hull, has been proposed, but this is considered to be too much of a risk of collision to be viable.

1.3 Salvaging/rendering harmless expended satellites

The use of robotics on specifically designed salvage spacecraft is gaining economical interest, and may become reality within the next 5 to 10 years. This is driven by the increasing problem of failed communications satellites, or simply satellites that have reached the end of their operational life time, occupying lucrative slots in geostationary orbit (GEO). As these slots are far and few between, it becomes economically viable for telecommunications operators to pay for the service of a salvage spacecraft to remove a "dead" satellite into a safe so-called "graveyard" orbit, freeing a slot for a new communications satellite to take its place.

Europe is studying the ROGER concept (RObotic GEostationary orbit Restorer) for a satellite equipped robotics means, to "grab" or "catch" a defunct satellite and bring it to a safe orbit away from the GEO orbit.

1.4 Planetary exploration

This domain of use of robotics in planetary exploration can be sub-divided into a number of categories.

1.4.1 Rovers

These are the mobile robots that bring scientific instruments around the terrain of a planetary body (or in the Apollo days, astronauts around on the Moon). The most famous examples include the Pathfinder, and more recently, Spirit and Opportunity Mars rovers. The Americans have opted for 6-wheeled rover designs, whereas Europe's planetary rover concepts have been mainly tracked as a result of locomotion trade-offs. However, based on their proven success, Europe may also turn to wheeled concepts for future Mars rovers, such as the ExoMars rover.

Fig. 3. The Nanokhod micro-rover for planetary exploration, developed by the European Space Agency

Fig. 4. NASA's most recent Mars rover, Opportunity.

1.4.2 Arms, mechanisms, automation, special devices

In terms of pure robotics, the most common devices used for planetary exploration are different types of robotic arms. These are used to apply sensors to rock and soil, mounted on the lander spacecraft, such as on the VIKING lander, or the ill-fated Beagle 2 lander.

The same Beagle 2 lander also include a novel robotics device, called a "mole": an over-size pencil shaped device that, tethered to the lander, would have hammered itself into the terrain in the vicinity of the lander, to make sub-surface measurements.

Additionally, a vast and diverse range of automated devices and mechanism are employed in space missions in general, and for planetary exploration in particular, to deploy supports, antenna masts, cameras, rover deployment slides, etc.

1.4.3 Flying robots, aka Aerobots

Planets or their moons that feature an atmosphere, such as Mars, Venus, and Titan, can be explored by flying robots, so-called AErial ROBOTic platforms, or Aerobots, for short. No Aerobot has yet been deployed in any space exploration mission, but Aerobot concepts and the associated technology for navigation, etc., is being developed on both sides of the Atlantic. An advantage of aerobots is the inherent potential of being capable of exploring larger areas than a ground-based rover.

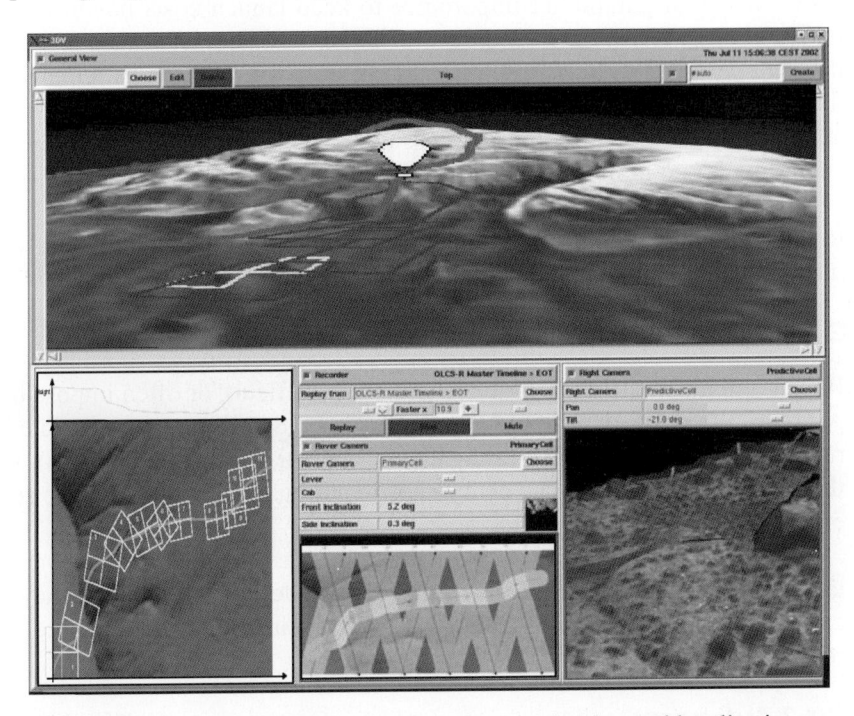

Fig. 5. Snapshot of a Man-Machine Interface for a navigation and localization system for a planetary aerobot

2 Design Drivers for Robotics Systems in Space

Modularity in space robotics systems is primarily found in software components for ground-based monitoring and control such as the FAMOUS suite, and in the robot controller on the robot itself that accepts higher-level commands from the ground control and transforms these into low-level commands controlling the motors and gears of the joints of the robot arm.

The robotic systems that are on the space segment, I terms of hardware, is rather more unique, designed to be optimal for the specific nature and objectives of the actual mission. However, some design drivers are generically applied due to the specificities of the space environment, in the largest sense.

2.1 Lightweight

Low mass is of paramount importance to keep launch costs down, essentially. Though NASA are sending ever larger and heavier rovers to Mars, they are designed to be as low-mass as possible within the requirements of its mission. Robotics arms, for instance on the International Space Station, are also designed to be as light as possible, using, e.g. lightweight alloys. It should be borne in mind that though matter has no "weight" in the weightlessness of space, matter does of course have inertia, so space robots have to have the necessary strength, stiffness, force, etc., to be able to deal with moving objects or astronauts at a given speed from one location to another.

2.1.1 Robustness

Repair and maintenance is space is obviously difficult, or often impossible. A failure in an external robot arm on the Space Station can in principle be repaired by astronauts performing Extra-Vehicular Activities (EVAs), but this is extremely costly. The chances of repairing a rover on Mars are self-evident.

Consequently, the robustness of the robot design is of crucial importance. Robustness translates into simplicity of design, into redundancy of critical systems, and into modularity where repair is at all possible, i.e. for robotics on manned missions.

2.1.2 Autonomy

Where robotics on Earth may – depending on the application – be designed to be very directly controlled by human operators, the design driver for space robotics, specifically on unmanned missions, is for a high level of robot autonomy, in order to minimize the interaction needed from Earth, or because "joy-stick" type robot control is simple not feasible. This is typically the case for robots used in planetary exploration: the time for a radio signal to travel from Earth to Mars, and to get a response back, is in the best case some 10 minutes. And, continuous radio contact is out of the question also. Consequently, a Mars rover has to feature a high level of autonomy, to navigate through unknown terrain, to reach target destinations, to avoid or overcome obstacles, and to "recognize" a situation it can not deal with by the on-board autonomy and stop and wait for ground control intervention.

2.1.3 Controllable at high-level of abstraction

The guiding factors here are in line with those above, namely to be able to control a space robot within unavoidable constraints of limited communications bandwidths and long delay times in the end-to-end control chain. Also important as a driver is the situation that the operator controlling the robot in space, whether the operator is in a control centre on Earth or in space, close to the robot, as on the ISS, the operator may not necessarily be a robot expert, but rather a scientist or an astronaut chosen for possessing completely different skills and expertise. Such operators do not know, and do not want to know about inverse kinematics, etc.; they need to be able to control the robot at the level of tasks, destinations, etc.

Fig. 6. Example of operator control through a high-level abstraction user interface.

2.1.4 Optimized locomotion

For space rovers, the means of locomotion must be adapted and optimized for the terrain to ensure that the rover can traverse the terrain in order to meet its mission objectives, and not get stuck, tip over, etc. Many forms of locomotion has been studied and tested in labs, testbeds, and Earth terrains similar to those of landing sites on the Moon and Mars, including wheeled, tracked, legged, hopping, rolling. Whereas Europe has opted for a tracked concept inn its planetary exploration programme up till now, the fact that NASA has had undisputed success with (six-)wheeled rovers on Mars may lead to a revision of the way forward in terms of future European planetary exploration rover locomotion.

Walking robots has been abandoned for planetary exploration, due to a too high mass-to-payload ratio associated with this form of locomotion, coupled with a relatively higher complexity of the associated technology which violates the design driver of robustness.

2.1.5 Emphasis on collision avoidance

Robot systems must be designed such as to avoid collisions by all means. For a robot arm on the Space Station, this means primarily not colliding with the hull or any appendage of the Station itself which could have catastrophic consequences. This is achieved by a combination of soft and hard stops to prevent the operator (who as stated above cannot be assumed to be a robot expert) from driving the robot arm or its end effector into a collision situation.

For a planetary rover, the emphasis is of course on not colliding with obstacles that can damage the rover, or get it irrecoverably stuck.

In both cases, a specific aspect is that space robots are deliberately designed to move at relatively low speeds and rates of acceleration, as an inherent contribution to allowing collision avoidance measures to be effective, under all possible circumstances.

Credits: Images are courtesy of ESA and NASA.

3 Conclusion

A number of typical applications of robotics in space have been presented. Requirements and constraints that characterize the space environment have been identified that dictate the design of space robotics and their operations. A number of the ensuing drivers for the design of robotics for space applications have been introduced.

References

1. URL http://www.esa.int/techresources/ESTEC-Article-fullArticle_item_selected-21_3_00_par-48_1090331583509.html
2. URL http://marsrovers.jpl.nasa.gov/home/
3. URL http://mpfwww.jpl.nasa.gov/MPF/mpf/rover.html
4. URL http://robotics.jpl.nasa.gov/
5. URL http://www.roboticsonline.com/

A Robotics Task Scheduler - TAPAS

G. Focant, B. Fontaine, L. Steinicke[1] and L. Joudrier[2]

[1]Space Applications Services, Leuvensesteenweg 325, B-1932 Zaventem, Belgium; [2]ESA/ESTEC, Keplerlaan 1, NL-2201 Noordwijk ZH, The Netherlands

Abstract

Within the frame of the ESA project TAPAS (TAsk Planner for Automation & robotics in Space), a generic scheduling tool has been developed. Its prime objective is to interactively assist a robot operator in planning and scheduling the activities of a robot-tended payload facility concurrently used by several (scientific) users, taking into account resources and various constraints. The system is potentially reusable in other space automation and robotics scenarios where this decentralised payload operations concept is applicable. One of the facilities analysed to establish the requirements was ERA, a robotic arm foreseen to be attached to the International Space Station. This abstract mainly gives an overview of the scheduling engine which is at the core of the software and constitutes its intelligence.

Keywords: Space, robotics, scheduling, resources, constraints.

1 Background

Over the past few years, there have been a number of attempts by ESA to develop generic mission planning and scheduling tools for satellite and robotics operations, which would support users in scheduling operation plans, taking into account constraints, resources, events and conflict detection and resolution, while providing a user-friendly graphical interface.

2 Description

The tool is able to handle the scheduling of predefined plans under a number of user-defined constraints. It does not concern itself with low-level issues such as path planning, collision avoidance, etc. but rather concentrates on resource handling, constraint solving and schedule verification.

For each user request, the input data consists of the existing schedule (initially empty) and a scheduling problem; the output is an updated schedule where the plan or event submitted in the request has been assigned a start or occurrence time that satisfies the various timing, resource and world state constraints. During schedule execution, requests for starting individual tasks are sent to the robot monitoring and control station (RMC), which in turn drives the robot controller (see **Fig. 1.**).

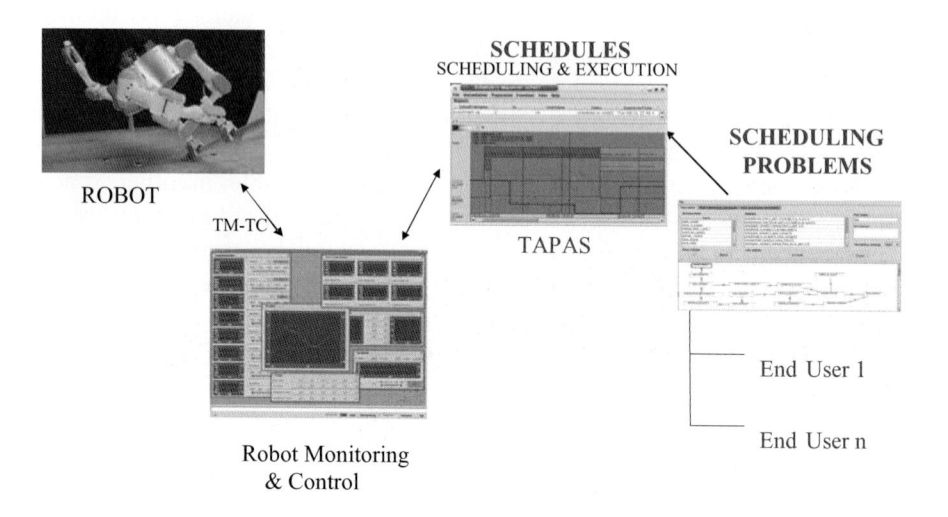

Fig. 1. End-to-end process

A typical scheduling problem consists of a plan, i.e. a set of (possibly partially) ordered tasks, and a number of scheduling constraints (see **Fig. 2.**). Each task has so-called pre-conditions and termination effects, which respectively describe the requirements on the world state for the task to be able to start, and the modifications of the world state caused by the successful termination of the task (see **Fig. 3.**). The world state is maintained for all the significant points of the schedule.

A robotics task scheduler - TAPAS   1033

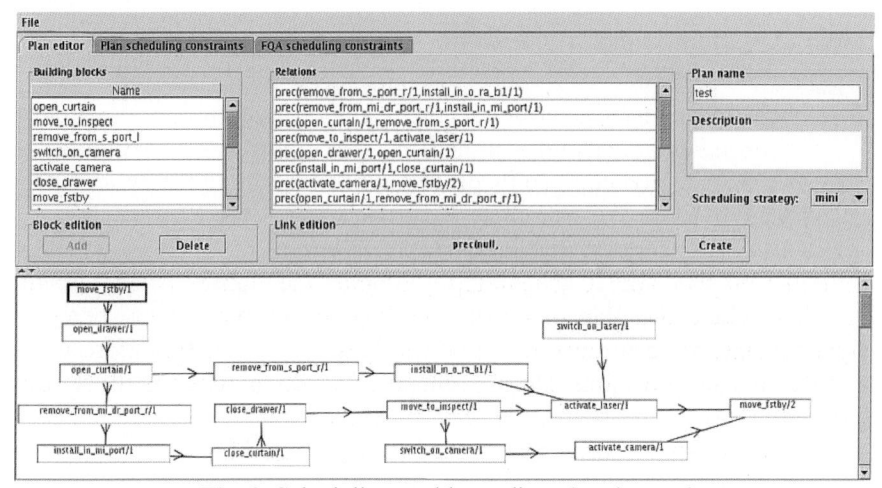

Fig. 2. Scheduling problem editor showing a plan

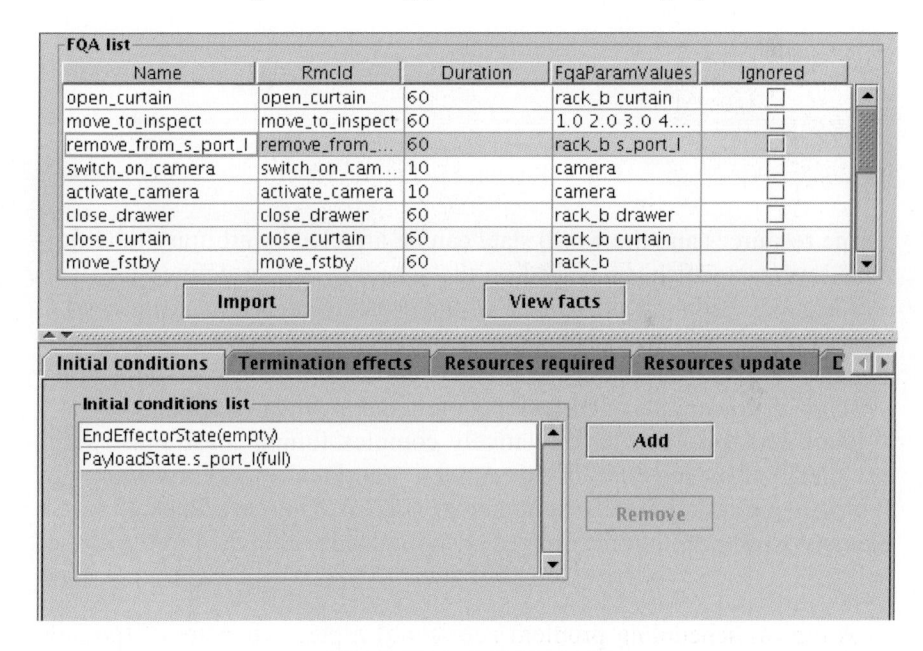

Fig. 3. Task definition editor

Tasks also have a number of resource requirements. Resources can be shareable or not, depletable or non-depletable, and have an availability profile, which is a boolean function of time (see **Fig. 4.**).

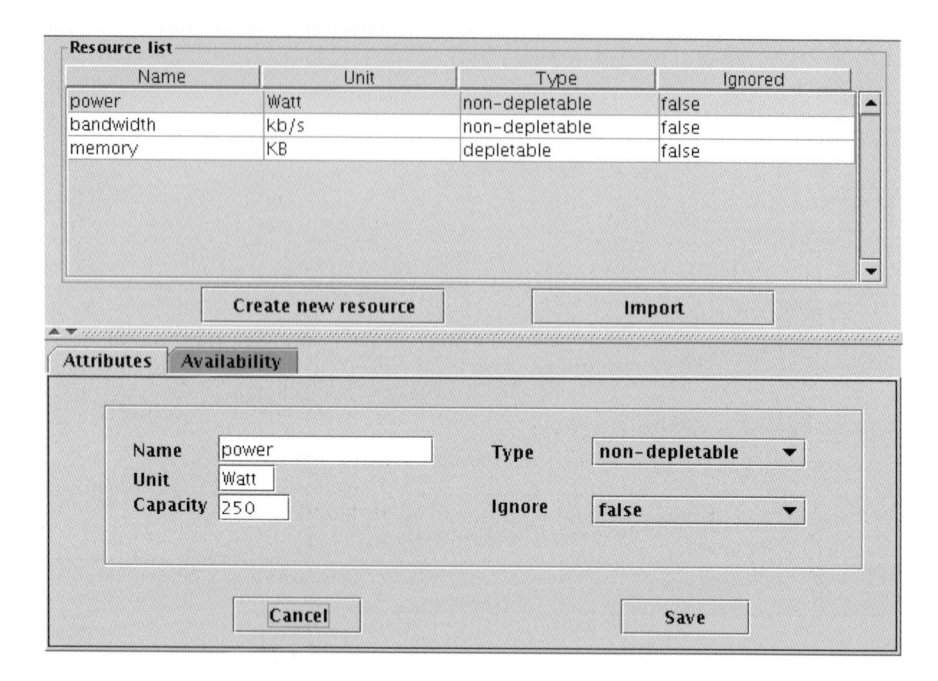

Fig. 4. Resource definition editor

The resources and the world state can be affected by arbitrary modifications (events), which are placed on the schedule at user-defined moments (see **Fig. 5.**). Such events, as well as the possible cancellation (unscheduling) of plans, can cause damage to a schedule; this damage is detected and signaled to allow the repair the invalidated tasks. This can be done either by relaxing constraints or by re-scheduling the invalidated tasks.

Plans and tasks can have relatively complex timing constraints, which are specified by the user as expressions, translated to FD constraints by Prolog using a definite clause grammar (DCG). The start times of scheduled tasks can be either maximized or minimized within the limits imposed

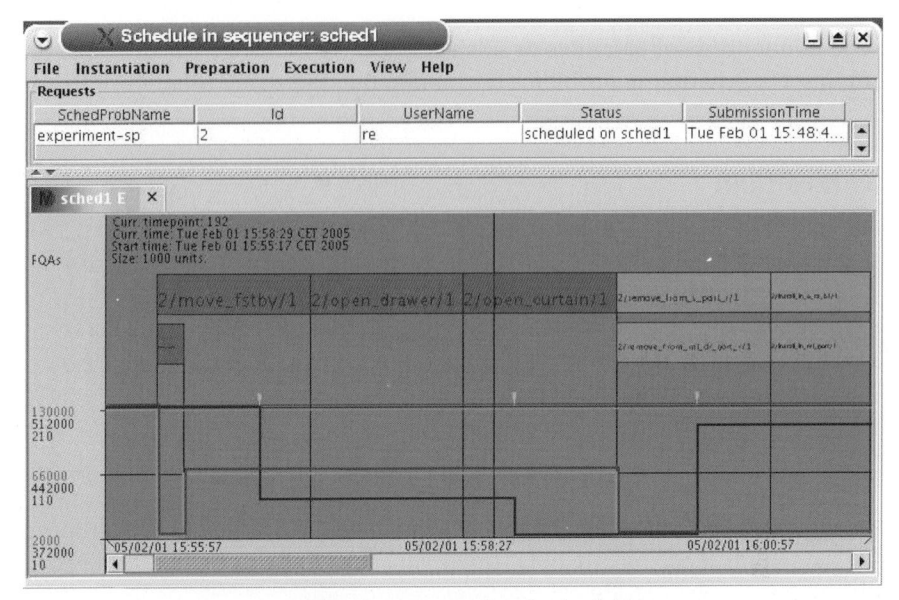

Fig. 5. An example of schedule

3 Techniques

In order to be able to fit the development within the tight budget and timespan and to allow a rapid protoyping approach, the constraint logic programming technique was adopted. The scheduling kernel is based on SICStus Prolog, a logic programming language with powerful constraint-solving capabilities as well as some basic built-in scheduling algorithms. A finite domain time model has been chosen. The scheduling core is integrated with the main code, written in Java, providing a networked multi-user application with an advanced graphical view of the operations (schedules, resources, tasks, etc.) and a user-friendly interface. Bidirectional communication between Java and Prolog allows to follow and control the progress of scheduling, which in some cases may take time. The target platform is Linux, but it can also run on Windows thanks to its portability.

4 Conclusion

Despite the numerous requirements and the limited budget, the result is a generic and extensible scheduler which, although tailored for the specific needs of robotic arm activities, can very easily be modified to handle the scheduling needs of satellites or other robots. We believe it is sufficient for solving the scheduling problems encountered in the space robotics domain.

References

1. Zweben M. and Fox M. (1994) *Intelligent scheduling.* Morgan Kauffman.
2. Sterling L. and Shapiro E. (1994) *The Art of Prolog.* MIT Press.
3. Bratko I. (2001) *Prolog programming for articifial intelligence.* Addison-Wesley, 3rd ed.
4. Smith D.E., Frank J. and Jonnson A.K. (2000) *Bridging the gap between planning and scheduling.* Knowledge Engineering Review, 15(1), 2000.
5. Van Harmelen, F. (Ed). *Proceedings of ECAI 2002* (15th European Conference on Artificial Intelligence).

Mobile Mini-Robots For Space Application

Man-Wook Han

Institute for Mechanics and Mechatronics
Division for Intelligent Handling and Robotics (IHRT)
Vienna University of Technology
Favoritenstr. 9-11, A-1040 Vienna, Austria
Tel.: +43-1-58801 31801, Fax. +43-1-58801 31899
email: han@ihrt.tuwien.ac.at

Abstract

The concept of solar power from space (SPS) was proposed in 1968. The basic idea of this concept is the generation of emission-free solar energy by means of solar cells in outer space and the transmission of the energy to the earth using microwave or laser beams. Because of high launch costs the structure - consisting of solar cells as well as microwave transmitters - should be light weight. Instead of the conventional rigid structure a concept (Furoshiki Concept) of a large membrane or a mesh structure where the corners held by satellites was proposed. The next step to be realized is the transportation of solar panels and microwave transmitters on this mesh structure. This paper deals with the development of low cost mobile mini robots able to move and place the solar cells as well as transmitter on a mesh structure in outer space.

Keywords: solar power from space, crawling robot, space application

1 Solar power from space

The concept of the solar power from space (SPS) is proposed to generate the electricity in high earth orbit and to transmit the energy to the Earth using wireless power transmission, for examples by means of microwave and laser beams. Comparing to a terrestrial solar power plant, the solar power plant in space has always an unobstructed view of the Sun, independent of season, time of day and receives 8 times higher energy intensity. But the construction of such a solar power plant in outer space is difficult and expensive.

In 1968 the SPS concept was introduced. At the moment SPS is not able to compete with conventional energy sources because of the high cost of space transport systems and missing wireless energy transmission technologies. There is an approach to transmit the power to Earth using microwaves from a small antenna on the satellite to a much larger one on the ground - rectenna. Microwaves are used for radar and microwave ovens and for transmitting telephone, facsimile, video and data.

The SPS essentially consists of three parts:
- a huge solar collector
- a microwave antenna on the satellite
- a large antenna on Earth to collect the power

The Earth-based "rectenna" is also a key concept. It consists of a series of short dipole antennas, connected with diodes. Microwaves broadcast from the SPS are received in the dipoles with about 85% efficiency. With a conventional microwave antenna the reception is even better, but the cost and complexity is considerably greater. Rectennas would be about 5 km across, and receive enough microwaves to be a viable concern. Some have suggested locating them offshore, but this presents problems of its own.

For best efficiency, the satellite antenna must be between 1 and 1.5 kilometers in diameter and the ground rectenna around 14 kilometers by 10 kilometers. For the desired microwave intensity this allows transfer of between 5 and 10 Gigawatts of power. To be cost effective it needs to operate at maximum capacity. To collect and convert that much power the satellite needs between 50 and 150 square kilometers of collector area thus leading to huge satellites.[4]

2 Furoshiki Net Concept

The major problem of building a Solar Power Satellite is the immensely high launching cost – for example approximately 10,000 USD per one Kilogram using the Space Shuttle. It is not so realistic to build a large solar paddle and a huge space antenna with a conventional rigid structure. Construction would involve a number of launchers and on-orbit assembly work. Some huge space antennas can be folded, but they require a complicated system for deployment.

The idea of "Furoshiki Satellite" (Fig. 1), a large membrane or a mesh structure - 1km by 1km in size, extends a huge Furoshiki by satellites which hold its corners. This membrane is folded in a very small volume during launch and is deployed and controlled by several satellites at its corners or using centrifugal force generated by rotating the central satellite.

The attitude and the shape of the membrane or mesh structure are controlled by these corner satellites. "Furoshiki" comes from the traditional Japanese light square-shaped lapping cloth, which can be easily folded and deployed by hands. It is expected that such a structure will reduce the weight per area of the space structure and, if the control technology is acquired, it can be efficiently folded to be launched and easily deployed. [3]

The Furoshiki net concept totally differs from conventional rigid space structures. Four small satellites build a large membrane structure that might be used as
- a solar power plant in orbit
- a giant communication antenna capable of a high-speed, large-volume data transmission
- a guard for satellites against space debris
- a recovery net for satellites or
- a huge satellite radiator. [3]

2.1 Mobile Minirobots - Roby-Space

The main purpose of this project is the development of mobile mini robots that bring the solar cells and transmitters on the net structure to build solar power plant based on the Furoshiki net concept. A sounding rocket brings four satellites (one mother satellite and three daughter satellite), robots, net, solar panel and the microwave transmitters in the orbit. Approximately 60 seconds after launch rocket reaches an altitude of 60 km from the earth. There the mother satellite and three daughter-satellites build the Furoshiki net. Robots transport the solar cells and microwave transmitters on the net structure (Fig. 1). In the frame work of the project a feasibility study will be done to verify the performance of the Furoshiki net as well as the crawling robots.

The main technical challenge is the building of a robot that crawls on the mesh. Most crucial considerations are the holding and moving mechanism and knots of the mesh. The dimension of the mesh will be 3 cm by 3cm or 5 cm by 5 cm.

The features of an autonomous mini robot for this purpose are:
- the maximum dimension 10x10x5cm
- light weight (less than 1 kg)
- free-movement on the mesh
- on-board power supply for approximately 10min
- equipped with a camera sending pictures to the earth
- wireless communication with the mother satellite

- mechanical and electronic robustness against vibration and shock [2]

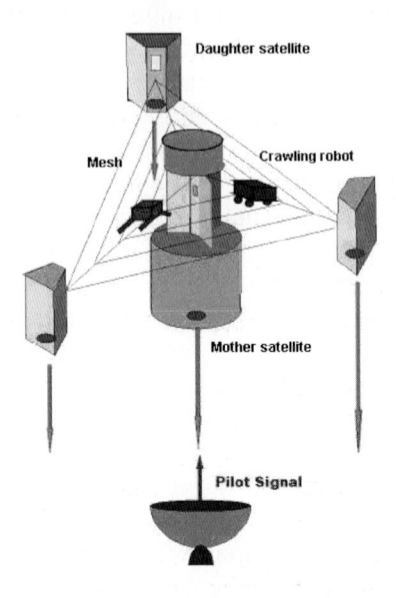

Fig. 1. Conceptual view of solar power plan by means of Furoshiki Satellite[1]

The requirements on the robot are the limited maximum size (10 x 10 x 5 cm), a simple mechanical construction, miniaturized electronics, robustness, "low cost", and independence of the mesh's dimension (from 3 x 3cm to 5 x 5cm). The weight of the robot plays important roles. Even the launching cost per kilogram is very high. Another point to be considered is that in case the robot is too heavy, the satellites can not produce enough tension in the net. For free movement the moving and holding mechanism of the robot should be well designed. Also other difficulties are the vibration and shock by launching of the rocket. The robot should pass the vibration and shock test up to 40 G. Last but not least the working environment of the robot is in outer space – 200 km over the earth. The high/low temperature, the radiation as well as the vacuum and others should be considered in the design phase.

3 Roby-Sandwich

The first prototype, Roby-Sandwich consists of a upper part and lower part. The concept is based on using the free spaces between the wires of the mesh for reference points of the movements and not the wires itself.

The idea behind this is to avoid gripping and following the wires and thus avoiding two critical situations: losing contact to the wires on the one hand and passing the crossing points of the wires on the other hand.

But it is very hard to guarantee accurate movement, because the correction of the position is a hard task to realize and exact positioning not guaranteed.

The whole electronic and locomotion components are placed in the upper part. The two parts are connected by the magnetic force. The electronic part is built up in open architecture. The motion unit controls the motors by a desired trajectory. This desired trajectory (as well as other demand behavior like acceleration and etc.) has to be transferred to the motion unit. For the design of robot the considerations above mean that it consists of a motion unit and a connection via radio to its superior control unit. Remaining to the described system the radio module should be connected to a microcontroller, which selects and processes the incoming information. Afterwards a bus provides the processed information to the motion unit. The electronic part consists of a single board for power electronic, communication and a microcontroller. The task of the microcontroller is to control both DC motors and analyze the radio data. This board is universally useable and in circuit programmable by the serial port. Furthermore it contains a high-speed synchronous serial interface, which gives the possibility to connect several microcontroller boards for different tasks. The electronic part consists of the following components:

- XC167 microcontroller from Infineon with internal RAM (8 kByte) and Flash (128 kByte)
- Voltage supply by switching regulators with high efficiency
- High speed dual full bridge motor driver
- Infrared transmission module
- Bi-directional radio module for the frequencies 433, 869 or 914 MHz
- Status indication by six bright LED's in different colors
- Serial synchronous interface for communication with further modules, e.g. XScale board.
- As the power supply the Lithium-ion (LiIo) rechargeable battery (1700 mAh, 7.2 V) is used.

For the locomotion two PWM controlled DC motors are used. The ppecification of PWM controlled DC Motors is:

- Minimotor Type: 1524 06 SR
- Output power: 1.70 W
- Speed up to: 10000 rpm
- Stall torque: 6.68 mNm
- Encoder resolution: 512 ppr

The robot crawled on the net at the 40 degrees below zero (Fig. 2.). In January 2005 the robot was tested in the micro-gravity environment by means of the parabolic flights in Japan.

Several problems were found. First of all it was not easy to find out the selection of appropriate magnetic force to connect both parts. In case the force is too high the motors have no enough power to crawl. If the force is not enough big, there is a risk to divide the robot in two parts. The tension of the net structure plays an important role. In case the net do not have enough tension also there is a danger that the robot twines the net.

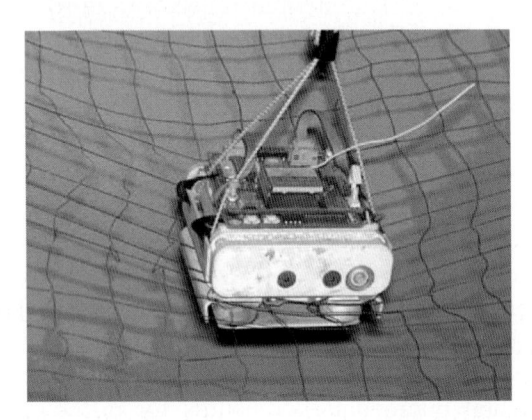

Fig. 2. Roby-Sandwich by the temperature test at the 40 degrees below zero

4 Roby-Insect

In order to avoid several risks as described above, a new robot was built. The major changes were made in the chassis and locomotion. Differing from Roby-Sandwich the robot was built as one part. The height of the robot is the half of the Roby-Sandwich.

The electronic part is the same as the Roby-Sandwich. For the locomotion a specially constructed device is used for the holding of the net and forward movement. The six pin type grippers in each side of the robot (one in right-hand side, another in left-hand side) are responsible for the holding and the movement of the robot on the net. Always at least two devices in each side hold the net. Before the holding devices release the net, the following devices should catch the net and move forwards. Roby-Insect passed the micro gravity tests by means of parabolic flights. During tests the robot moved a distance of 2m within 10 seconds (Fig. 3).

It was found that the tension of the net plays an important role. In case there is not enough tension of the net structure the robot can twine the net. In order to avoid this possible danger a new type of robot is developed.

Fig. 3. Roby-Insect by the micro gravity test

5 Conclusion

In this paper the basic idea of solar power from space and the development of mobile mini-robots were described in detail.

Two robots – Roby-Sandwich and Roby-Insect – were built and tested. Two robots can move on the net well. Roby-Sandwich passed the temperature test (-40°C), but did not pass the micro gravity test – the lower part was released from the upper part. For both robots the tension of the net structure plays an important role.

Because the environmental condition over 200 km over the earth is different from that of the earth – temperature, radiation, vacuum and others, therefore for the selection and assembly of the parts should be carefully investigated. The robot will be sent by a sounding rocket and therefore robot should pass also the vibration and shock tests.

As described above the risks of low net tension and wrapping of the net are still existing. A new type of the robot is now developed. Because of totally new designed mechanical construction and locomotion the risks are decreased substantially in very low level. The electronic part of the robot is not changed. Additionally for the communication Bluetooth module is used. The robot passed already the microgravity tests and shock tests. Also for the behavior of the Furoshiki satellite should investigated in more detail.

Acknowledgement

This project is supported by ESA under ESTEC/Contract No.18178/04/NL/MV- Furoshiki Net Mobility Concept.

References

1. N. Kaya, M. Iwashita, S. Nakasuka, L. Summerer, J. Mankins, Rocket Experiment on Construction of Huge Transmitting Antenna for the SPS using Furoshiki Satellite System with Robots, In Proceedings of The 4th International Conference on Solar Power from Space SPS '04 Together with The 5th International Conference on Wireless Power Transmission WPT 5, 30 June - 2 July 2004, Granada, Spain, pp 231-236
2. P. Kopacek, M. Han, B. Putz, E. Schierer, M. Würzl: "A concept of a high-tech mobile minirobot for outer space"; in: "ESA SP-567 Abstracts of the International Conference on Solar Power from Space SPS'04 together with the International Conference on Wireless Power Transmission WPT5", Granada, Spain; 06-30-2004 - 07-02-2004 (2004), 132 - 133.
3. National Space Development Agency of Japan (NASDA) Report, Feb. 1999 No.80
4. http://www.answers.com/topic/solar-power-satellite

Teleagents for Exploration and Exploitation in Future Human Planetary Missions

G. Genta

Mechanics Dept. and Mechatronics Lab., Politecnico di Torino, Turin, Italy, e-mail: giancarlo.genta@polito.it

Abstract

In view of human Mars exploration and the construction of a Moon Base the main space agencies are planning precursor robotic missions, followed by others for the development of a surface infrastructure needed to sustain the human presence. Astronauts will be accompanied by robots and teleagents cooperating with them in performing demanding and dangerous tasks. The present paper addresses some problems, mainly regarding mobility, energy and autonomy, to be solved to produce efficient robots/teleoperators providing a valid support for the early planetary missions.

Keywords: Space exploration and exploitation, space robotics, Moon, Mars

1 Introduction

A renewed interest in human exploration is flourishing not only among the major spacefaring nations but also in Japan, China and India. Their involvement might even cause the resurgence of strong political motivations for a renewed presence of the traditional spacefaring nations beyond LEO.

Past human lunar missions involved only marginally the use of robots, and mainly in the preparatory phase. However, even at the time of *Apollo* missions, the argument about the alternative between human and robotic exploration started. It went on for about 40 years, and still from time to time resurfaces in papers and conferences. However, both alternatives in a way misfired: if long range human exploration stopped after *Apollo 17*, the promises of Artificial Intelligence (AI), which should have produced by now robotic explorers much more suited than humans to space travel and able to perform the required tasks in a much better way, failed to materialize. This is a generalized situation: e.g. the A.I. embedded in present day motor vehicles is hardly larger than that of a few decades ago, and humans are entrusted practically all high-level control functions. A similar situation can be found in the manufacturing industry where the idea of completely automatic factories are now yielding to more realistic statements that the optimal automation level must be defined in each case.

Strong A.I., with its Von Neumann probes, endowed not only with intelligence but also with consciousness, is strongly challenged on theoretical as well as practical grounds [1, 2]. Without entering this debate, the path leading to intelligent and conscious robots is still long and we cannot count on them for planning realistic future space missions. Intelligent robots swarming in space to explore and send back their findings to the Earth belong to science fiction not less than large nuclear spaceships carrying human crews in the whole solar system. At least we have an idea on how to build the latter, while we do not know even whether the former are theoretically possible.

Even if no (truly) intelligent machine can be built, at least in the foreseeable future, much work can be done in space by *smart* machines, even if nobody knows what are the ultimate limitations of these remote agents and how far it is possible to go along these lines. One of the most important, and also most difficult, tasks of space exploration is the search for extraterrestrial life, at present on Mars and in the future on the satellites of giant planets (Europa, Titan and, who knows, even on the Kuiper Belt bodies [3, 4]). Since we have no definition for life, we cannot program a machine to recognize it and we cannot be sure that even humans can perform this task. Certainly a machine not able to perform following the standards of strong AI has only a limited possibility. Recent experiments performed with a robotic rover in the Atacama Desert in Chile [5] confirmed that this task is very difficult for a machine even for terrestrial life, well known to programmers. The endless debate over the alleged discovery of Martian fossils in the ALH 84001 meteorite, which could not be settled by the best biologists in the better equipped laboratories on Earth, shows that robots will may never be able to successfully search for extraterrestrial life.

Instead of focusing on the alternative between human *or* robotic exploration, the winning strategy is to proceed with a joint human *and* robotic exploration based more on teleagents or telemanipulators than on true robots. However, while teleagents, possibly coupled with virtual reality, would make space exploration much easier and safer, they would require human presence close to the place to be explored. This strategy will then be implemented first on the Moon and then on Mars. Exploration of the outer solar system and beyond following these lines seems to be much more far in the future and perhaps needs some technological breakthrough.

2 Teleagents for Moon exploration

The construction of mobile robots and telemanipulators even for terrestrial use is not an easy task, and we are still far from building machines useful in environments much less hostile than space, with human presence at close distance. Many problems, regarding the mechanical structure, the power sources, the actuators and the control system, are still far from being solved satisfactorily [6]. Military applications are a good example, since are less affected by economical problems than commercial or civil applications: some teleagents have been used recently to help soldiers in the battlefield, particularly for reconnaissance missions but, except for Unmanned Aerial Vehicles (UAV), they are still at the experimental level.

In the case of lunar exploration, the proposed approach follows a multi-step strategy, starting with a series of precursor robotic missions meant to acquire further knowledge of the planet and to select the best landing sites, evolving towards more demanding missions for the development of a surface infrastructure necessary to sustain human presence [7]. When the infrastructure will be in place, astronauts will land and start their tasks, helped by the robots already on site and others that will accompany them.

The technologies involved in building planetary infrastructures range from those typical of space (propulsion, robotics, rendezvous and docking, entry/re-entry, navigation, communications) to technologies that can be borrowed from other fields, like automotive, construction and industrial engineering.

Agents teleoperated from the Earth, an orbiting facility around the Moon, or a Moon Base may be subdivided in two broad categories: exploration and exploitation (or construction) machines. The delay time in the control loop is short enough to allow even fast motion, but must not be overlooked.

2.1 Exploration robots

The main task of robotic explorers is carrying instruments and to sending back to Earth the acquired information. It is doubtful that Lunar Sample Return Missions will be required, since we already have a good supply of lunar specimens gathered in the past, however one of the most interesting scientific goals of future lunar mission is to search for rocks of terrestrial origin thrown into space by a collision of an asteroid or large meteorite with the Earth and then fallen on the Moon. They may contain remains of the oldest Earth lifeforms, supplying essential information on the beginning of life in our biosphere. Since the size and the mass of the instruments are moderate, the rover does not need to be of a large size. We know fairly well the surface characteristics of the Moon and the mobility problems we may encounter and we can exploit the experience of running wheeled man-sized vehicles on the lunar surface for tens of miles in the last *Apollo* Missions, as well as wheeled robots in some Russian missions.

Locomotion can be based on wheels, legs, tracks, unconventional devices or a combination of them. The simplest solution is wheels, which are favored for moving on not too rough terrain at fairly high speeds. Owing to the low mass of the rover and the low lunar gravity, a four wheels configuration seems to be a good starting point. With a human in the loop at a reasonable distance, speeds of several km/h, may be reached and elastic and damped suspensions (automotive type, passive) are much more suited than rocker-boogies. Pneumatic tires are simple, but can decrease reliability; depending on the duration of the mission, non-pneumatic elastic wheels, like the 'twheel' recently introduced by Michelin, may be better suited. Elastic wheels were also initially considered for the Lunar Rover.

The main limitation of mobile teleagents operating on the lunar surface can come from the power source. Depending on size and performance, particularly in terms of gradeability, the power for locomotion can span from 0.1 to 10 kW. Solar cells backed by batteries can be used for low speed daytime operations, particularly if long stops for measurement taking are expected. Fuel cells can be used only for relatively short durations, otherwise complex refueling handled by teleoperators based in a lunar outpost must be planned. Viable options are Radio-Thermal Generators (RTGs) for power and Radioisotope Heating Units (RHU) for thermal control.

More than on the payload, the minimum size of the machine (wheel size and ground clearance) depends on the roughness and morphology of the terrain. A peculiar application is the exploration of the 'lava tubes': solar energy cannot be used and the chance of losing radio contact claims for a higher autonomy. Teleoperated rovers will also prove very useful in exploring difficult and remote locations like canyons, craters and rims. If wheels are used in these lower speed applications, a rocker-bogey configuration is suitable. Alternatives like legged locomotion become interesting and a particularly promising configuration could be twin-rigid frames machines. Perhaps the best solution could prove to be a six-legs twin frames machine with four wheels carried by trailing arms suspensions like the one described in [8], allowing to get to the difficult location at a fairly high speed and then to explore it with better mobility. The power required by a legged rover is small, allowing long operations in the shadow even if the power is supplied by solar panels (with backup battery). For narrow places non-conventional configurations like snakes are worth further study.

2.2 Exploitation and construction teleoperators

To make human exploration more profitable and to overcome the main limit of *Apollo* Missions, the short stays on the Moon, it is worth considering the possibility of preparing a surface infrastructure satisfying critical functions like habitability, mobility, power provision, telecommunication support, and radiation protection. Planetary habitat architectures are highly depending on the site features, the mission scenario, and the available transportation capability. Only for short duration missions the Landing Vehicle might provide all the necessary support to sustain the crew during descent and ascent and during the surface stay, allowing an austere but safe and productive permanence. For longer duration missions, rigid cylindrical modules, derived from the Space Station habitat, are probably the most traditional and proven approach. Positive aspects are the minimum EVA activities required for the habitat activation and modularity allowing reconfiguration and evolution of the Base. The use of inflatable structures could provide substantially more pressurized volume with minimum mass and folded volume. The operation requirements to install such a habitat on a planetary surface would require extensive EVA and telerobotic operations for a proper site preparation including terrain excavation and grading.

The essential tasks of the machines for these operations on the Moon are those typical of construction machines, like cranes, forklifts, graders, dumpers and bulldozers. In particular, the small four-wheeled dumpers with a front blade used in construction sites could be adapted for lunar applications. Lower gravity makes more difficult to transfer forces to the ground, so that bulldozing operations might become problematic, but the lunar surface is covered by layers of regolith characterized by an extremely fine granularity, that can be easily dug and moved around. Since in bulldozing operations inertia forces play a very small role (velocities are too small), it should not be too difficult to simulate on the Earth the lunar working conditions. The weight of the machine can be easily relieved in such a way that the vertical forces on the ground are the same as on the Moon, and a soil with similar characteristics as lunar regolith can be simulated. These simulations will show whether wheeled or tracked bulldozers could perform best on the Moon. It is also possible to increase the vertical forces on the Moon by ballasting the machine using some regolith.

Low gravity makes hoisting operations simpler with no contraindications. The requirements of remote operations make forklifts preferable to cranes. Also here some ballasting can be useful, to prevent the device from tipping over. Since these machines will likely be much lighter than their counterparts on the Earth, the required mobility can be better achieved by using wheels than tracks. However, if heavy ballasting will prove to be needed, alternative configurations can be considered to avoid extremely high energy requirements. A twin rigid-frames legged configuration, for instance, can allow exerting much higher forces, particularly on dusty or sandy terrain, since the feet can sink without slipping.

Since the mobility range of construction teleagents is smaller than that of exploration rovers, the power can be supplied by batteries, periodically charged at a the solar or nuclear power station of the outpost. Nuclear power seems to be the only viable option to sustain the operations of a crewed station on the Moon, where continuous operations must be guarantee over the long day-night lunar cycle. Moreover, substantial amounts of power will be required for in-situ resources utilization, for the production of lunar oxygen, silicon, iron, aluminum and other local resources that could prove very useful for the surface infrastructure development.

Later, when a lunar outpost is operating, it will be possible to operate construction and exploitation teleagents using fuel cells or even standard hydrogen-oxygen internal combustion engines operating on fuel produced by the solar or nuclear power plant of the outpost. Since water can be recovered from the rover exhausts, large water supplies are not required. The technology currently developed for hydrogen motor vehicles will be available when the outpost will be in place.

3 Teleagents for mars exploration

The use of teleagents on Mars is much more problematic than on the Moon, since the distance causes a large control delay. More than teleagents they will be remote agents, designed with a certain degree of decision autonomy (some people think that the strong AI goals will be unattainable even at long term and the abilities of remote agents will remain very limited [1, 2]). The rovers currently operating on Mars incorporate a much higher degree of autonomy than that required for the Moon exploration devices described above. However, this results in very slow motion and reduced capability. Automatic Mars exploration will hopefully continue in the next years at the conceivable pace of two or three missions every launch window opportunity. The objectives of the near future missions will be mostly scientific, mainly in the fields of planetology and exobiology. The latter goals are conceivably at the limit of the 'intelligence' of these machines.

In parallel to the scientific objectives, the near-term missions to Mars will validate critical technologies, like landing of larger masses for the development of a surface infrastructure, and ascent and return capabilities in view of future crewed missions. In sample-return missions, the choice of the samples is at the limit of remote agents' capabilities.

Moreover, the Martian locations with the highest scientific interests for the potential presence of water and hence for the search of pre-biotic and life traces, are difficult to be reached using surface mobility. Alternative solutions based on the use of small robotic planes or balloons have been proposed, but their implementation may prove to be problematic, due to the thin Martian atmosphere. Hybrid wheels-legs configurations may be a suitable option, perhaps combined with rope devices to get into the depth of canyons. Another problem typical of Mars is the available power, since the output of solar cells is much lower than on the Moon. For the current low-speed rovers this has not been much a problem, but if their autonomy will have to grow and higher speed will be required for long-range traverses, power limitation could become the real limiting factor. Again, nuclear energy (RTGs) will become a viable solution.

After the first exploration missions, robotic missions will follow to prepare the outpost site for the first human expedition to Mars. Construction machines and robotic tools will be landed to Mars together with critical elements of the infrastructure like power generators, habitat, laboratory, In-Situ Resources Utilization and telecommunication facilities. Paradoxically, the teleoperation limits might be less severe than with exploration rovers, since the operations needed to develop the outpost could be more easily programmed and then executed autonomously. Again, the power provision criticality might be less severe for construction machines than for rovers. In fact, when the construction machines will operate on the Martian surface, a nuclear power plant will be already in place to satisfy the very demanding power requirements. Note that on Mars the abundance of carbon from the atmosphere makes it possible to produce methane to be used as a fuel, so that the alternative of using internal combustion engines for rovers and above all for construction machines can be considered.

Once the first outpost has been established, either in Mars orbit or on the surface, teleoperators will be controlled and commanded directly by the human explorers and the time delay will be no more a problem.

4 Conclusions

The renewed emphasis on human exploration, first of the Moon and then of Mars, leads to reconsideration of the role of robotic devices in space. The support of robotic capabilities, teleopered and/or autonomous, becomes essential in preparing the way to humans before their arrival, and then in assisting them during dangerous or difficult tasks. Exploration robots must then be complemented by construction and 'exploitation' robots, particularly because the only way to make human exploration cost effective is to 'live off the land'. The degree of autonomy of the robots will be limited, since in the foreseeable future it is not expected that robots will be able to perform very complex tasks completely without human supervision. On the Moon, it will be possible to use teleagents both in the preparatory phase and in the subsequent human missions. On Mars, the preparatory phase will require more autonomous machines with a limited capability of performing complex tasks autonomously.

Besides the degree of autonomy, the applicability of planetary teleagents will depend on suitable improvements in the fields of mechanics, electronics, thermal control, and energy provision. Technology transfer from other fields, like automotive, construction and general robotics will prove to be essential.

References

1   R. Penrose, *The Emperor's New Mind: Concerning Computers, Minds and the Laws of Physics*, Arrow books, London, 1990.
2   F. Lerda, *Intelligenza umana e intelligenza artificiale: est modus in rebus*, Rubbettino, 2002.
3   G. Genta, M. Rycroft, *Space, the Final Frontier?*, Cambridge University Press, 2003R.
4   Forward, *Alien Abodes between Neptune and the Stars*, First IAA Symposium on Realistic Near Term Advanced Space Missions, Torino, June 1996.
5   Mark Peplow, *Roving robot finds desert life*, Nature, March 17, 2005.
6   G. Genta, N. Amati, *Non-Zoomorphic versus Zoomorphic Walking Machines and Robots: a Discussion*, European Journ. Mech. & Env. Eng., Vol. 47, n. 4, 2002, pp. 223-237.
7   F.Ongaro, M.A.Perino et al., *Aurora: The Dawn of European Human and Robotic Planetary Exploration*, 7th ISU Annual International Symp., Strasbourg, France, June 4-7, 2002-05-07.
8   Bona B, Carabelli S et al., *Hybrid wheel-legs locomotion: simplified mechanical and control architectures for realistic, near term, moving robots*, MECHROB 04, Aachen, Sept. 2004.

An Expandable Mechanism for Deployment and Contact Surface Adaptation of Rover Wheels

Philippe Bidaud, Faiz Benamar, and Sébastien Poirier

Laboratoire de Robotique de Paris
Université Paris 6 / CNRS
18, route du Panorama
92265 Fontenay Aux Roses - France
bidaud@robot.jussieu.fr

Abstract

In this paper, an expandable mechanism for unfolding wheels is proposed. This mechanisms combines 2 elementary mechanisms. One allows the deployment of the rim the other one ensuring the contact shape adaptation.

1 Introduction

Mars exploration programs that require surface mobility, such as geology studies, search for past life or infrastructure deployment for future manned missions, make use of wheeled rovers. The total mass of the rover and its volume during launching phase are very constrained, thus resulting in moderate ability to overcome obstacles, despite the efforts in optimizing the kinematics design, due to the limitations on wheel diameter. Meanwhile, the wheel size is one important parameter that determines the rover crossing performances. These limitations can be reduced by introducing mobilities in the system for the deployment of the locomotion mechanism from a stowed configuration or/and using inflating or unfolding wheels that can reach large diameters and thus present high cross-country ability, still being compatible with the launching constraints. The adaptation of the contact surface geometry of the wheels is another important feature which will allow to optimize traction and mobility performances of non-holonomic vehicles with respect to the physical characteristics of the ground. One of the purpose of the INTAS Project untitled Innovative Mars exploration rover using inflatable or unfolding wheels is to study mechanically deployable wheels and compare them with classical rover wheel design, given the same mission objectives and constraints. Many sorts of deployable structures exist, based on a variety of different concepts. Some of these structures form planar or space frames and have already had a variety of applications [3] [5] : space applications for solar arrays, solar sails, etc or architecture applications to support membranes or films. The most impressive example of these deployable structures are certainly retractable roofs within the field of architecture. The paper describes a new type of expandable mechanisms with 2 degrees of freedom (dof). This mechanism combines 2 elementary mechanisms which allows the deployment of the rim the other one ensuring the modification of the rolling tread geometry. To obtain this expandable mechanisms, planar linkages with 1-dof have been design and optimized in their kinematics to obtain the desired expansion ratio for the former one and its rigidity under a radial load for the other one. These planar units are then assemble to form the desired "'solid"'.

2 Deployable mechanisms

Reversible foldable mechanisms are such that their structure remains stress free throughout the deployment. They are capable of expanding from a closed, compact configuration to a pre-determined while modifying their dynamics. They can be classified as follow :

1. structures with rigid 1-D bars connected to each other in various pantographic arrangements of 2-D and 3-D;
2. structures with 2-D panels connected to form various surface structures, basic element being a triangular panel [4];
3. tension structures consisting of cables or membranes, or combination of both, either pre-stressed or pneumatic [5];
4. deployable tensegrity structures composed of rigid rods and cables [6];

The two last one combine are said form-active systems since they allow to create a structure that could autonomously adapt both its shape and its mechanical behavior. This idea will be reused for the design of the "'casing of the tire"' in the deployable wheels proposed in this paper.

Several bar mechanisms have been invented to create reversible foldable structures along its external perimeter or surface. Most of them are made of hinges and pivots and are able to move in a plane, along a cylinder or sphere.

Emilio Pinero pioneered the use of scissor mechanisms to make deployable structures Scissor hinge structures built out of basic units composed of rod elements connected in there center and with each other by hinges. The basic unit (a Scissor-Like Element) is a pantograph. These units can be assembled to form an opened or a closed loop system with a single degree-of-freedom. **Fig.1** considers different planar arrangements of parallel straight or angulated rods. Space frames can be formed by closed

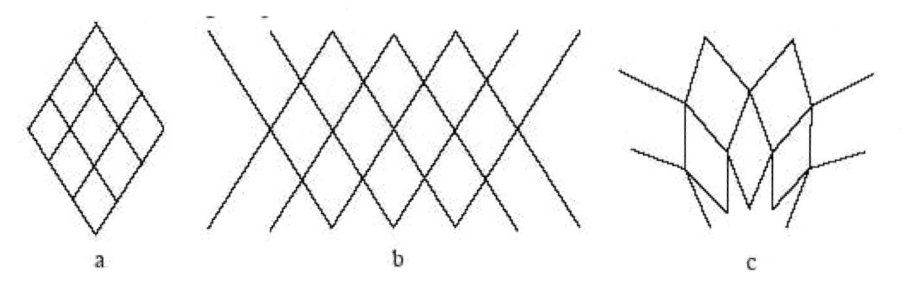

Fig. 1. Multiple scissor-like elements with straight rods or angulated rods

loops of these elements. **Fig.2** shows an example of a mechanism which was adopted in the edge beam of the Hoop-Column Antenna [2] and built with half-scissor mechanisms. This kind of structure must have an even number of equal-length hinged segments with a rotoid joint at each extremity whose axis is orthogonal to the segment. Here, the joints of 2 consecutive segments are not collinear but form an angle in the plane perpendicular to the symmetry axis of the structure. It can be checked that due to its particular geometric properties, the mobility of this mechanism is equal to 1. Hoberman has also created numbers of pantograph mech-

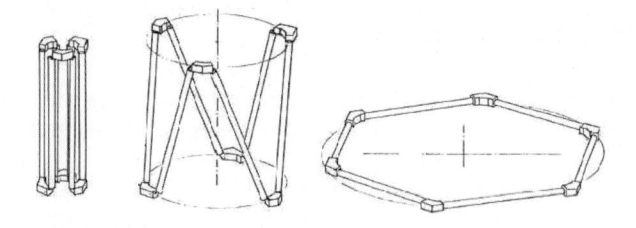

Fig. 2. Half-scissor closed-loop deployable mechanism

anisms that operate in two and three dimensions. A "N-gon" as shown in **Fig.3** can expand out in a symmetric manner by "actuating" only on of its members [7]. When scissors-pairs are connected in closed-loop via angulated elements in such way that the normal line that are perpendicular to the axis of the joined terminal pivot of adjacent scissors-pair intersect, a reversible folding of the structure takes place without internal constraints in the structure.

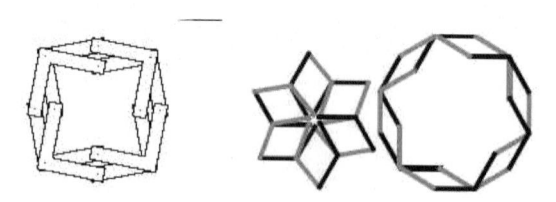

Fig. 3. Expanding "'4-gon"' and "'6-gon"' based on Hoberman design

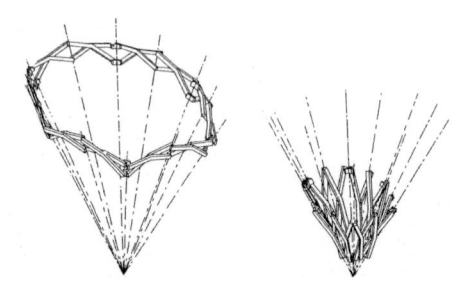

Fig. 4. Configuration where the pivot points of all the scissor-pairs lie on a circle

3 Proposed mechanism design

The mechanism which has been designed for creating the deployable wheel with a contact shape adaptation can be decomposed in 2 subsystems.

1. a mechanism for the radial expansion of the wheel rim,
2. a mechanism for deformation of the contact surface geometry.

Fig.5 shows the deployment of the wheel at 3 different stages.

3.1 The spokes

Its has been designed in the aim to have shape for the wheel similar to the one of the Marshokod robot [1]. The expansion of the rim is based on

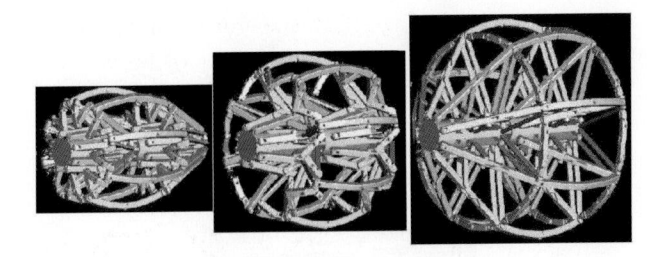

Fig. 5. Folding/unfolding process at 3 different stages. On the left the "'open"' configuration

the use of a set of identical planar mechanisms disposed radially which play the role of spokes. Each of these mechanism can be seen as a parallel mechanism whose limbs are imbricated 4-bars mechanisms (see **Fig.6**). The inputs are the linear motions of points A_1 and A_2 which are coupled to the same linear actuator placed in the hub of the wheel. When these

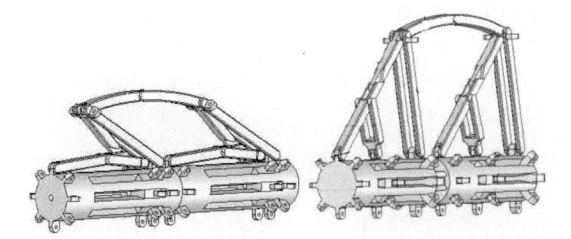

Fig. 6. Mechanism for the expansion of the rim

"'spokes"' are totally deployed, the mechanism is in a stationary configuration. This local singularity happens when the coupling bar is parallel to the upper bar of the 4-bars mechanism. Therefore, the mechanism becomes an immovable structure regardless of the input. Moreover, in this configuration the actuator does not work in any external force applied on the upper 4-bars mechanism.

3.2 The rim

The rim itself is a loop-assembly of rigid 2-bars (a rim sector) joining consecutive spokes, each rim sector being articulated on a spoke by a universal joint. **Fig.7** shows the deployment of a rim sector simultaneously to the deployment of the spokes.

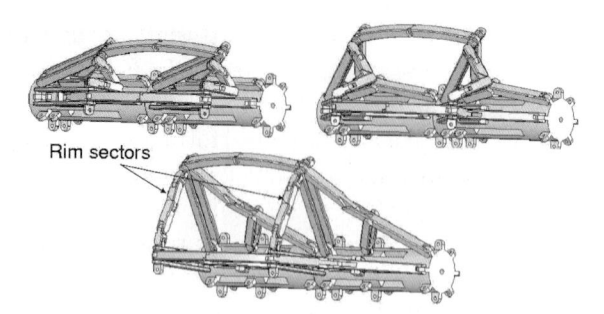

Fig. 7. Deployment of the rim sectors

4 Contact geometry adaptation

The adaptation of the contact geometry has been envisaged by articulating each strut which form the carcass-element (the rib) with a pivot placed at its mid-length. The actuator placed in the hub of the wheel which actuates the deployment will then be used for driving the translation of the 2 parallel rimes with their spokes (see **Fig.8**). So, the angle formed between the two parts of the carcass-element will be modified and consequently the local geometry of the rolling tread as well as the compliance of the contact.

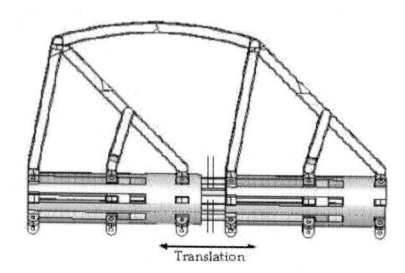

Fig. 8. Mechanism for the control of the carcass-element geometry

4.1 Stiffness analysis

Exposed to a vertical load and a traction force, the carcass will deform. A precise estimation of its deformation has be performed using a finite element linear analysis (FEA) on an isolated radial element. This analysis have been made by using COSMOSWorks design analysis software that is fully integrated in SolidWorks. Schematically, the factors affecting the contact stiffness are the shape (and material) of the carcass-element and its support (the deployment mechanism). In **Fig.9** are displayed the results of the analysis in von Mises stress and represented in a graphical manner. A uniform vertical and ortho-normal load are applied to the carcass-element to emulate a load of 40 daN and a traction of $17 daN$. The maximum stress for a is $1.8\ 10^8 Nm^{-2}$ for aluminum alloy parts of section $10 \times 10\ mm^2$. The possible deformation the carcass-element may achieved with these loading conditions is shown in **Fig.9**. It stays very small (less than $1mm$) and is not enough important to produce a tangential force.

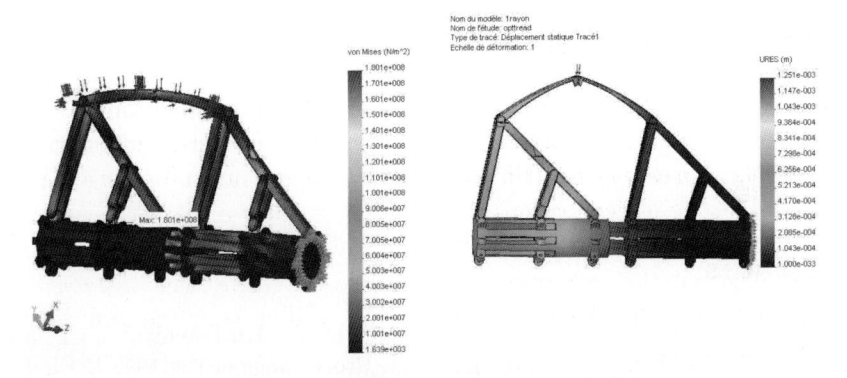

Fig. 9. Display of von Mises stress results (right) and displacement analysis result plot (left)

To increase the carcass deformation, an alternative design for the carcass-element using an arched beam have been explored (see **Fig.10**). A smooth deformation between a geometry of the carcass corresponding to a well-inflated tire and a poorly-inflated tire can be controlled by the contact-variation mechanism described above. The amount of change in carcass shape remains important when using glass-fiber composite materials. Its deformation under a vertical load and its dynamical properties are much more closer to a classical inflated tire.

4.2 Conclusion

This paper has presented an original mechanism designed for creating a deployable wheel with a contact shape adaptation. As far as we know, this is the first time an unfolding wheel integrating a rolling tread adaptation mechanism is proposed.

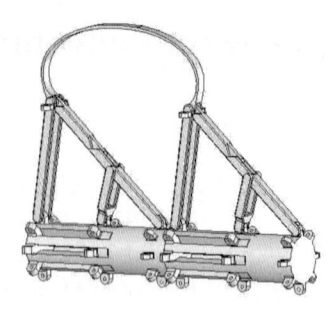

Fig. 10. Carcass-element consisting in an arched beam

Acknowledgments

This research is supported by the International Association for the Promotion of Co-operation with Scientists from the New Independent States (NIS) of the Former Soviet Union (INTAS) under grant number 4063

References

1. A. Kermurdjian, V. Gromov, V. Mishkinyuk, V. Kucherenko, and P. Sologub. Small Marsokhod Configuration. Proceedings of the 1992 IEEE International Conference on Robotics and Automation, May 1992.
2. W.W. Gan, S. Pellegrino, Closed-Loop Deployable Structures, 44th AIAA/ASME/ASCE/AHS/ASC Structures, Structural Dynamics, and Materials Conference, 7-10 April 2003, Norfolk
3. C. Hoberman, Reversibly expandable structures. USA Patent No. 4981732, 1991.
4. S.D. Guest1, F. Kovcs, T. Tarnai and P.W. Fowler, Construction of a Mechanical Model for the Expansion of a Virus, Proc. IASS-2004, 20-24 September 2004, Montpellier, France.
5. Ph. Block, Interactive kinetic structures: Architecture with an organic trait, Joint SMArchS Colloquium / BT Seminar Series, Fall 2003
6. Gunnar Tibert, Deployable Tensegrity Structures for Space Applications, Royal Institute of Technology Department of Mechanics Doctoral Thesis Stockholm, 2002
7. S. K. Agrawal, S. Kumar, M. H. Yim, J.W. Suh Polyhedral single degree of freedom expanding structures. IEEE International Conference on Robotics and Automation (ICRA 2001); 2001 May 21-26; Seoul, Korea. Piscataway, NJ: IEEE; 2001; 4:3338-3343.

A New Traction Control Architecture for Planetary Exploration Robots

Daniele Caltabiano[1], Domenico Longo[2] and Giovanni Muscato[3]

[1]daniele.caltabiano@diees.unict.it, DIEES - Università di Catania.
[2]dlongo@diees.unict.it, DIEES - Università di Catania.
[3]gmuscato@diees.unict.it, DIEES - Università di Catania.

Abstract

Planetary exploration rovers are often redundantly actuated, for mechanical robustness and reliability purposes. This paper describes the Integral Redistribution Traction Control Architecture (IR-TCA) applied to this category of robots. The Traction Control Architecture of mobile robots is typically composed by independent decoupled PID speed controllers, one for each motor. In this paper it will be demonstrated that this solution is not efficient in redundantly actuated robots: in typical situations, the integral part of the controllers introduces an unstable state variable that penalizes the power consumption of the overall system. The IR-TCA is a hierarchical cross-coupled controller, a simple variation to the PID speed controller, easily scalable for more complex systems. The stability of this Traction Control Architecture will also be demonstrated for a two wheels planar robot. The IR-TCA has been implemented on a six wheels skid steering robot called P6W. Some comparative tests will be presented to show the validity of this control architecture.

Keywords: Motion Control, PID, Traction algorithm.

1 Introduction

Electric motors and batteries efficiency is growing more and more, consequently electric vehicles are of great interest for research and for commercial purposes. Moreover the use of internal combustion engines for planetary exploration rovers is not possible due to the different or absent atmosphere of the destination environment.

Skid steering vehicles are very common in outdoor environment for their ability to move and carry weight in very rough terrain. The P6W robot is a skid steering, six independently actuated wheels robot with a 4 D.o.F. articulated chassis; it is a 1:4 small scale prototype of the Robovolc robot: "A robot for volcano exploration" [1], [2]. Planetary exploration rovers have been tested, in the past, in volcanic environment due to the similarities between the terrain morphology and vice versa traction algorithms optimized for volcano exploration can be adopted for planetary exploration. Other examples of skid steering vehicles are M6, Muses, Marsokhod [8] and FIDO [6].

PID controllers are very popular in mobile robotics for their simplicity and scalability, but, it will be demonstrated that, applying this kind of controllers to redundantly actuated robot gives unstable system, with wheels that tends to fight one another. In fact due to the different accumulation of integral errors in the controllers, it happens that some wheels have to increase their speed to compensate this error while the others need to decelerate.

This paper describes the integral redistribution Traction Control Architecture (IR-TCA) that solve the problem found in a decentralized PID architecture, redistributing the integral error between the controllers of the same side of the robot.

2 The IR-TCA on a Planar Robot

For simplicity the problem of fighting wheels will be initially exposed on a two independent wheels planar robot.

It has been supposed that the wheels are in adherence with the terrain; hence, if the front wheel performs a longer trajectory, in respect to the rear one, the integral part of the controller will try to reduce its speed, while the other controller will try to increase the speed of the rear wheel.

Since the derivative part of the controller is not affected by this problem, a simpler PI controller will be considered in the following model.

In order to study the mathematical model of this robot, it has been assumed that the wheels are in adherence with the flat terrain, the mass of the vehicle is concentrated in the two wheels and that the elasticity of the whole system is concentrated in the chassis.

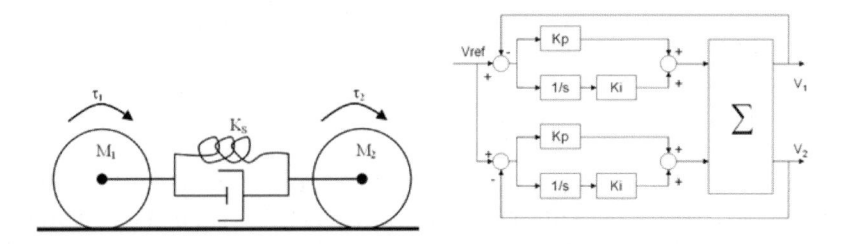

Fig. 1. (a) Simplified model of a two wheels planar robot, (b) scheme of the PI control architecture.

Neglecting the inertia momentum of the wheels and the electrical inertia of the DC motors, the overall system of equations can be written in state space form as:

$$\dot{z} = A z + B V_{ref} \qquad (1)$$

where

$$z = \begin{bmatrix} x_1 - x_2 & \dot{x}_1 & \dot{x}_2 & e_{I_1} & e_{I_2} \end{bmatrix}^T$$

$$A = \begin{bmatrix} 0 & 1 & -1 & 0 & 0 \\ -\dfrac{K_S}{M_1} & \dfrac{-K_m R K_p - K_m^2 - B_B R_m R^2}{R_m R^2 M_1} & \dfrac{B_B}{M_1} & \dfrac{K_m K_I}{R_m R M_1} & 0 \\ \dfrac{K_S}{M_2} & \dfrac{B_B}{M_2} & \dfrac{-K_m R K_p - K_m^2 - B_B R_m R^2}{R_m R^2 M_2} & 0 & \dfrac{K_m K_I}{R_m R M_2} \\ 0 & -1 & 0 & 0 & 0 \\ 0 & 0 & -1 & 0 & 0 \end{bmatrix}$$

$$B = \begin{bmatrix} 0 & \dfrac{K_m K_p}{R_m R M_1} & \dfrac{K_m K_p}{R_m R M_2} & 1 & 1 \end{bmatrix}^T$$

and where x_1 and x_2 are the positions of the two wheels, R is the radius, M_1 and M_2 the two masses, K_S and B_B are respectively the chassis spring and the damper, K_m and R_m are respectively the torque constant and the resistance of the motors, V_{ref} is the reference speed applied to both the controllers, K_p and K_I are the PI coefficients and e_I are the integral speed errors of the controllers.

The eigenvalues of the matrix A shows that there is an unstable variable in the system. In particular, the null determinant of A (rows 1, 4 and 5 are linear dependent) shows that there is an eigenvalue on the origin.

The control low of the IR-TCA is written in eq. (2)

$$\begin{cases} v_k = K_p(V_{ref} - \dot{x}_k) + K_I e_{IR_k} \\ \dot{e}_{IR_k} = V_{ref} - \dot{x} + K_E \left(\dfrac{\sum_{k=1}^{n} e_{IR_k}}{n} - e_{IR_k} \right) \end{cases} \quad k = 1,2 \tag{2}$$

where v is the armature voltage applied to the motor.

Eq. (2) are very similar to those obtained with a PI speed controller: a new factor computes the difference between the average integral error and e_{IR_k} (the integral error of the motor k); this factor is multiplied by a coefficient and is used to update the derivative of e_{IR_k}. In this way all the motors connected in the IR-TCA will converge to the same e_{IR}; thus resulting in an equal distribution of the torques in the wheels at steady state. Hence, in rough terrains, the controllers will give more power to the most stressed motors (like a normal PI controller) but, after a transient, the motor commands will converge again (to a new equilibrium state).

It is easy to demonstrate that the overall system obtained using the new controller is stable.

The K_p and K_I parameters of the IR-TCA do not need to be modified in respect to the PI controller; the K_e coefficient must be chosen according to the desired rigidity of the interaction between the wheels: a larger value will determine a stronger interaction and hence a faster transient; a smaller value imply a little interaction and hence a longer transient. For $K_e=0$ the IR-TCA acts like a decentralized PI speed controller.

The scheme of the IR-TCA is shown in Fig. 2.

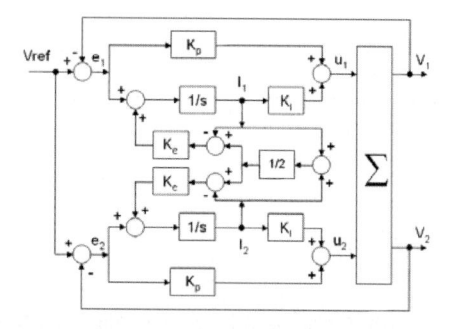

Fig. 2. Scheme of the IR_MCA.

3 Results of Simulations

The simulations have been performed assuming $V_{ref}=0$ and a non zero initial condition on the controllers' integrators.

The model constants have been set as follows: K_s=10000N/m, B_B=100Kg/s, M_1=0.5Kg, M_2=0.5Kg, R=0.055m, K_m=0.59N/m/A, R_m=3.2Ω; while the controller constants were: K_p=100, K_I=50, K_e=1.

The two controllers have been tested using the same PI coefficients giving exactly the same output performances: after a transient (produced by the non zero initial condition on the integrators) the speed error of the two wheels converge to zero. For this reason it has been considered useless to include the images of the speed errors of the wheels in this document.

Fig. 3 shows that using the PI control architecture on the system, after a transient, the equilibrium is maintained applying an equal and opposite voltage to the armature of the two motors, resulting in power loss. While using the IR-TCA, the armature voltages of the two motors converge to zero. The speed of this convergence can be modified tuning the K_e coefficient: caution must be kept choosing this parameter since, as mentioned before, a big value produce a fast transient (power saving), but the controller can not compensate strong local disturbance (bad traction performances), the overall system will perform in a similar manner to the paralleled motors. On the other hand, using K_e=0, the IR-TCA degenerates in the PI control architecture.

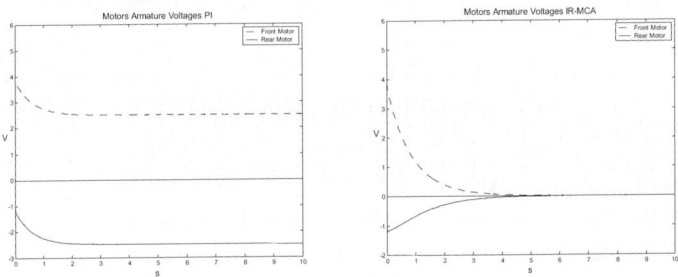

Fig. 3. Motors armature voltages obtained respectively with the PI speed controllers and the IR-TCA.

4 The IR-TCA of P6W

The P6W robot has been realized to easily perform traction control tests of new algorithms in laboratory, in a reproducible manner, prior to implement them on Robovolc.

A powerful 32bit microcontroller, Motorola MPC555, has been adopted to drive this six independent wheels skid steering robot implementing the Integral Redistribution Traction Control Architecture (IR-TCA) and to store data that can be successively post-elaborated (reference and measured currents, measured positions and speeds of the six motors).

Fig. 4 shows the P6W robot and the scheme of the IR-TCA.

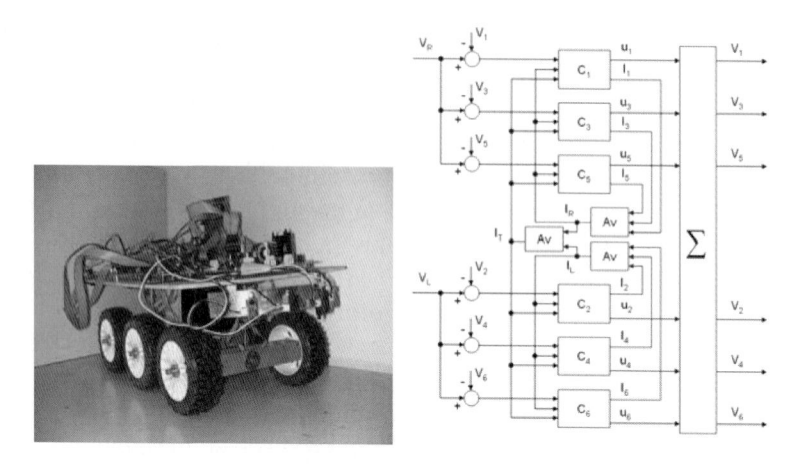

Fig. 4. The robot P6W and a scheme of the IR-TCA.

The extension of the IR-TCA to a six independent wheels skid steering vehicles must be realized into two phases: the integral redistribution algorithm IRA must be applied first between the wheels of the same side (IRA$_R$ for the controllers of the right side wheels and IRA$_L$ for the left side) and then between the two sides (IRA$_T$).

Rotation in skid steering robots is very difficult and, due to the friction (first disengagement) with the terrain, the robot does not rotate for small values of armature voltages (or reference currents).

For this reason in the IR-TCA another IRA (IRA$_T$) is used to negotiate the motor commands between the left and right side of the vehicle.

The C_k blocks are the controller blocks of each wheel, where eq. (2) are implemented.

V_R and V_L are the reference speeds, respectively for the right and left side wheels; V_k is the measured speed of wheel k.

The motor commands (armature voltages or reference currents of the motors) are indicated with u_k.

The integral error accumulated in the motor controller k is indicated with I_k, while the average integral errors, evaluated via the Av blocks of the figure, are indicated with I_R for the right side wheels, I_L for the left ones and I_T for the overall average integral error.

The coefficient of IRA$_T$ has been called K_T and it has been assigned a value of one order lower than K_e.

The tests on the robot have been divided into two phases: in the first experiment it has been demonstrated that the PI controller architecture is not stable on a flat terrain. Following a random trajectory, the wheels tend to fight one another.

For reason of clarity only the results of the right side of the robot are presented. For symmetry the same behavior affects the other side.

Fig. 5 shows the reference current assigned to the right side motors during a random trajectory on a flat terrain, in $t=13$ a zero reference speed is assigned to the PI control architecture. It is evident that, in this case, the front right wheel (wheel number 1) and rear right wheel (wheel number 5) have an opposite reference current while the middle right wheel have a reference current close to zero.

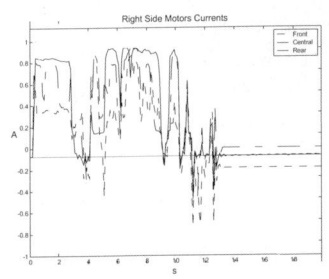

Fig. 5. Example of fighting wheels after a random trajectory.

The second test phase has been performed on a flat terrain, setting a non zero initial condition in the six integrators of the controllers and giving a zero reference speed to the controllers. In order to make clear the behaviors of the controllers another disturb has been added in $t=1s$.

Fig. 6 shows that using the decentralized PI speed controller architecture the motors tend to fight one another, while the currents obtained with the IR-TCA: after a transient, the reference currents of the motors tend to zero.

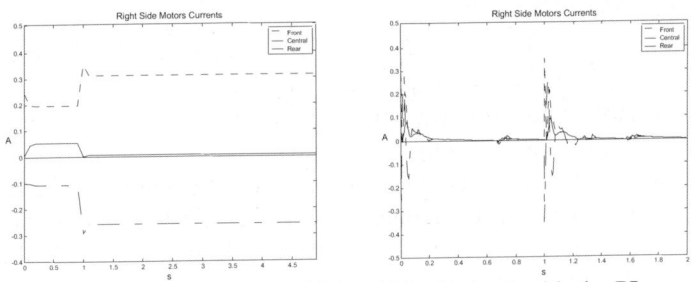

Fig. 6. Motors reference currents (right side) obtained with the PI speed controllers and with the IR-TCA.

5 Conclusion

The power consumption in autonomous electric vehicles is one of the hardest constraints, from the design phase to the applications.

In this paper the Integral Redistribution Traction Control Architecture (IR-TCA) for redundantly actuated electric robots has been presented.

It has been demonstrated that the classical decentralized PID speed controller architecture applied to those robots is not internally stable. This generates excessive power consumption, wasting of electro-mechanic parts, wheel slippage and tendency to side-slip.

The presented simulations and the real tests on a six wheel skid steering vehicle demonstrate that the IR-TCA, being internally stable, avoids useless power consumptions without the needs of external sensors.

The IR-TCA is a variation of the PID architecture and hence inherits all its advantages: simplicity (implementation), scalability (for more complex systems), traction performance.

References

1. The ROBOVOLC project homepage http:\\www.robovolc.dees.unict.it
2. D. Caltabiano, S. Guccione, D. Longo, G. Muscato, M. Coltelli, E. Pecora, A. Cristaldi, G.S. Virk, P. Sim, V. Sacco, P. Briole, A. Semerano, T. White, "ROBOVOLC: A Robot for Volcano Exploration Result of first test campaign", Industrial Robots, Vol. 30, N.3, May 2003.
3. D. Caltabiano, G. Muscato, "A Comparison Between Different Traction Control Methods for a Field Robot", International Conference on Intelligent Robots and Systems (IROS2002), September 30 – October 4, 2002, EPFL, Switzerland.
4. D. Caltabiano, D. Ciancitto, G. Muscato, "Experimental Results on a Traction Control Algorithm for Mobile Robots in Volcano Environment", IEEE International Conference on Robotics and Automation, ICRA 2004, April 26 - May 1, 2004, New Orleans, LA, USA.
5. K. Iagnemma, S. Dubowsky, "Mobile Robots in Rough Terrain - Estimation, Motion Planning, and Control with Application to Planetary Rovers", Springer, March 2004, pp. 81-96.

The Lemur II-Class Robots for Inspection and Maintenance of Orbital Structures: A System Description

Brett Kennedy, Avi Okon, Hrand Aghazarian, Mike Garrett, Terry Huntsberger, Lee Magnone, Matthew Robinson, Julie Townsend

Jet Propulsion Laboratory, California Institute of Technology; 4800 Oak Grove Drive, Pasadena, California 91109; Brett.A.Kennedy@jpl.nasa.gov

Abstract

The assembly, inspection, and maintenance requirements of permanent installations in space demand robots that provide a high level of operational flexibility relative to mass and volume. Such demands point to robots that are dexterous, have significant processing and sensing capabilities, and can be easily reconfigured (both physically and algorithmically). Evolving from **Lemur I**, **Lemur IIa** is an extremely capable system that both explores mechanical design elements and provides an infrastructure for the development of algorithms (such as force control for mobility and manipulation and adaptive visual feedback). The physical layout of the system consists of six, 4-degree-of-freedom limbs arranged axisymmetrically about a hexagonal body platform. These limbs incorporate a "quick-connect" end-effector feature below the distal joint that allows the rapid change-out of any of its tools. The other major subsystem is a stereo camera set that travels along a ring track, allowing omnidirectional vision. The current Lemur IIa platform represents the jumping-off point toward more advanced robotic platforms that will support NASA's *Vision for Space Exploration*, which calls for a sustained presence in space. This paper lays out the mechanical, electrical, and algorithmic elements of **Lemur IIa** and discusses the future directions of development in those areas.

Keywords: limbs, space, assembly, inspection, modular

1 Inspection and Maintenance of Orbital Structures

While the extension of robots' roles in space seems to be a well accepted fact, the priority of development of those roles continues to evolve. The Lemur-class platform shown in **Fig. 1.** was originally developed as a small inspection and maintenance agent for the kilometer-scale Space Solar Power stations under study by NASA in 2001. With the advent of NASA's *Vision for Space Exploration* in early 2004, the roles of a Lemur-class robot have changed to that of an agent on spacecraft (such as the Crew Exploration Vehicle) or as a member of a heterogeneous team on outpost space station construction projects.

Fig. 1. LEMUR IIa uses its limbs for mobility and manipulation

These roles are characterized by the use of the robots as semi-autonomous agents to increase the total work performed or safety thereof without increasing the total number of Extra Vehicular Activities (EVA) required of the astronauts. In particular, these robots will range the exterior of structures and spacecraft performing small-scale assembly tasks. In addition, for the inspection of structures and spacecraft these robots can carry instruments and perform operations unsuitable for free-fliers or robotic arms. Examples of these operations include investigations requiring contact or force exertion and inspection in tight or obscured regions.

2 Other Robots Relevant to the Lemur-class Concept

Of immediate interest are the other robotic systems that have been specifically designed for on-orbit operations. The most heavily used are the robotic arms produced by MD Robotics (Canadarm 1&2). Also highly relevant are some systems that have not flown. The two-armed Ranger system is intended for a Space Shuttle bay experiment. The anthropomorphic

Robonaut is also interesting in that it is modeled after an astronaut in a space suit [6].

Lemur, smaller and less anthropomorphic than the aforementioned robots, is most closely related to the hexapedal MELMANTIS [1] and ASTERSIK [2] robots. Lemur, while sharing these robots' axi-symmetric, limbed layout, differs from these them in that it is self-contained and incorporates a modular end-effector toolset. Also of note is Lemur II's progenitor, Lemur I, an bilaterally symmetric robot [4,5].

3 Lemur Platform Overview

Because the design objectives of orbital robots are so ambitious, extensive prioritization had to be made. JPL decided to focus on a platform for the inspection and maintenance of small scale structures (~1-100m^3) because of the wide applicability of such a platform to space-based structures, large or small. Given that no robot designed by the current effort was intended to be tested in space, the space-environment requirements were given the lowest priority. Conversely, the robot needed to be tested in a 1g environment with as little infrastructure as possible.

Taking some inspiration from nature, it was clear that the primate model of multi-use limbs with tool-using capability could provide a path to the platform complexity desired. However, stepping beyond nature, it was decided that six limbs would be of more use than four. Such a multiplicity of limbs would allow for a stable base of three or more limbs while leaving two or more limbs available for cooperative manipulation operations. The resulting system is a self-contained unit massing less than 10kg.

4 Limbs and Tools

The primary driver for joint design was the independence of each degree of freedom. The second consideration was JPL flight design heritage and path to flight design. These two factors led to the design of two joint types, differentiated only in their packaging (motor parallel to output or at right-angles). Both joint drivetrains have a DC brushed motor coupled to a planetary gearhead and then to a harmonic drive final stage. Feedback sensing is handled by an optical encoder on the motor shaft and a Hall Effect sensor for absolute homing. The drivetrain specifications can be found in Table 2.

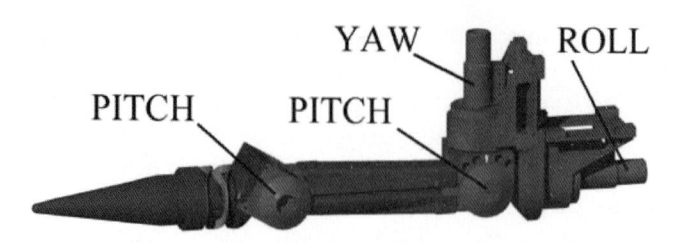

Fig. 2. LEMUR IIa limb layout

The idea that Lemur was to have *limbs*, not arms or legs, dictated the arrangement of the degrees of freedom and the effective range of motion of each. This concept meant that the workspace and dexterity of the limb needed to be the union of those needed for walking and manipulation. Therefore, a 4 DOF limb was designed consisting of a kinematically spherical shoulder and a 1 DOF elbow. The simplifying assumption was made that any initial tool or gripper would be axisymmetric or have passive DOF designed in. The arrangement of the limb can be seen in **Fig. 2.** and the effective range of motion in **Table 1**.

Table 1. LEMUR IIa joint characteristics

1st-stage ratio (planetary)	16.58:1
2nd-stage ratio (harmonic)	100:1
max. torque (Nm)	9.0
max. continuous torque (Nm)	5.0
No-load speed (°/s)	45
Roll travel(°)	360
Yaw min travel (°)	180
Pitch1 min travel (°)	90
Pitch2	180

The final design element of Lemur limbs is the inclusion of a tool quick-release and the tools that mate to it, all of which are shown in **Fig. 3**. The release itself is a socket with a spring-locked ball detent similar to others found throughout industry. To date, four tools have been designed to mate with the quick release. Simplest is the default walking/poking tool. For inspection purposes, a ultra-bright LED task light tool can act alone or in conjunction with a "palm-cam" tool. Finally, a rotary tool with integral reaction torque sensing and its own bit chuck can be used for torqueing fasteners or other rotary operations depending on the bit used. In keeping with the *limb* concept, all of these tools can be used as feet as well as for their manipulation operations.

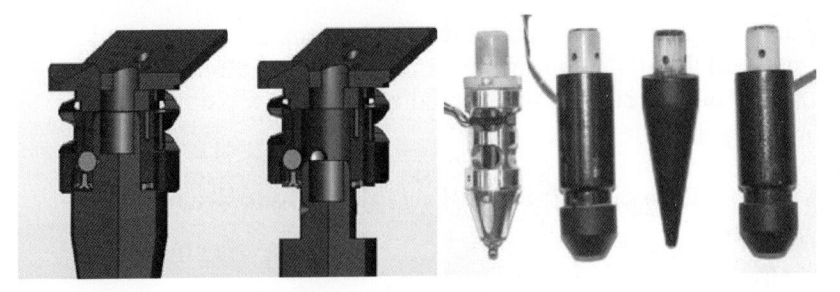

Fig. 3. Quick-release design and the rotary, palm-cam, default, and task-light tools

5 Stereo Camera Assembly

The front hazard avoidance cameras, shown with their track assembly in **Fig. 4.**, are a stereo set of black-and-white cameras. In keeping with the axisymmetry of the rest of the robot, the stereo set is mounted to a carriage that is propelled around the circumference of the body on a circular track. Rather than use a DC brushed motor as is done in the remainder of the robot, an ultrasonic motor was chosen, primarily for its low mass, lack of need of a geartrain, and self-braking capability. Feedback is performed by an encoder mounted to the carriage and by a pair of Hall Effect sensors that read a two-bit magnetic code created by magnets placed 120° apart on the track ring.

Fig. 4. Camera track assembly allows the stereo cameras to rotate through 360°

6 Electronics

Lemur's computing stack is based on the PC104 board standard. Nearly all of LEMUR II PC104 boards are commercial or are now commercially available after JPL sponsored the design. A list of all elements can be found in **Table 2**. A custom actuator driver circuit board was developed for LEMUR II to meet power and mass constraints. The board is capable of driving 12 motors simultaneously under computer control and is built on a standard PC104 footprint. True to flight application development, the board is built around a commercial chip designed specifically for use in robotic spacecraft.

Table 2. LEMUR IIa avionics elements and specifications

Element	Specification
PC104+ Stack	
CPU Board:	Pentium 266MHz, 128 MB RAM
Solid State Disk	64Mb
Ethernet board	
Camera board (2)	8 color channels, NTSC
Digital I/O board	96 channels
Analog input board	16 differential channels, 12 bit resolution
Filter board	2-pole low-pass in hardware, 16 channels
Analog output board	32 channels
Encoder board (2)	30 channels, 16 bit per read
Actuator driver board (3)	12 axes, 6A max per, 3A continuous per
Power System	
Bus	24V
Battery	Rechargeable lithium-ion cells
	Hotswappable tether and battery power

7 Software Architecture

LEMUR II's core control software architecture, written in ANSI C for the VxWorks real-time operating system, is composed of the three layers shown in **Fig. 5.**: *Device Driver*, *Device Abstraction*, and *Applicaton*. The interface with the robot's hardware is provided by the Device Driver Layer. Because of the nature of real-time systems, each device driver execution time was carefully measured, and code was optimized to get maximum performance. Also, to improve reliability and remove any unexpected errors, each device driver code was tested functionally and structurally. Each device driver exports its functionality through a header file to the Device Abstraction layer.

The Device Abstraction Layer combines and abstracts the Device Driver Layer functionality from the rest of the system. Low-level PID motion controller, motor and group queues, coordinate system, limb controller, motion prediction and timeout controller are implemented in this layer.

Each Device Abstraction function exports its functionality through a header file to the Application Layer.

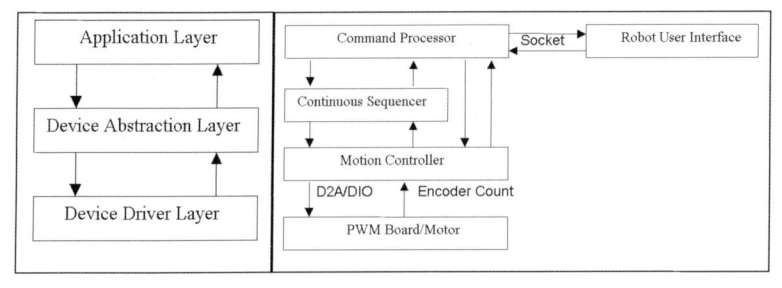

Fig. 5. Division of layers and information flow in Lemur's software architecture

The Application Layer provides interface from high-level control of the robot to the outside world through a set of commands. A user can create and send a sequence of commands to the LEMUR II by using the Robot User Interface, as shown in **Fig. 5**. By using a different set of sequences one can direct LEMUR II to do different tasks.

8 Operational Algorithms

Consistent with the system-level emphasis of the LEMUR program, two algorithms related to navigation and manipulation, namely Hybrid Image Plane Stereo (HIPS) and Barcode Localization, have been developed on the LEMUR I platform and ported to LEMUR II.

HIPS is a vision-based manipulation technique developed to enhance positioning precision beyond the limits of typical calibrated stereo methods. The purpose of Barcode Localization is two fold. First it recognizes a 4x4 grid barcode and reads out the information imbedded on the barcode. Second, when the four corners of the barcode are identified, the algorithm computes the pose of the rover with respect to the template for rover localization. For an in-depth description of these algorithms, please see [3,5].

9 Conclusions and Future Work

The current Lemur IIa robot meets many of the general physical needs for its operations. A major area for development is the inclusion of and use of increased sensing, particularly for force control, both for mobility and manipulation. On delicate structures, effective force control will be paar-mount. Work of a similar nature has been addressed by several researchers for terrestrial applications [7], but we will apply it to space-specific operations.

Upgraded robors will be tested through demonstrations associated with research efforts associated with NASA's *Vision for Space Exploration*. These demonstrations will explore the exact nature of future assembly and inspection operations aboard orbital structures and spacecraft.

Acknowledgements

The research described in this (publication or paper) was carried out at the Jet Propulsion Laboratory, California Institute of Technology, under a contract with the National Aeronautics and Space Administration.

References

1. Koyachi N., Arai T., Adachi H., Asami K., Itoh Y., "Hexapod with Integrated Limb Mechanism of Leg and Arm," *Proc. of IEEE Int. Conf. on Robotics and Automation*, Nagoya, 1995
2. Takahashi Y., Arai T., Mae Y., Inoue K., Koyachi N., "Geometric Design of Hexapod with Integrated Limb Mechanism of Leg and Arm," *Proc. of IEEE Int. Conf. on Intelligent Robots and Systems*, v3, 2000
3. Baumgartner E.T., "Hybrid Image-Plane/Stereo (HIPS) Manipulation". NASA Tech Brief NPO-30492
4. Hickey G., Kennedy B., Ganino T., "Intelligent Mobile Systems for Assembly, Maintenance, and Operations for Space Solar Power," *Proc. of the Space 2000 Conf.*, Albuquerque, NM, 2000
5. Kennedy B, Aghazarian H., Cheng Y., Garrett M., Hickey G., Huntsberger T., Magnone L., Mahoney C., Meyer A., Knight J., "LEMUR: Legged Excursion Mechanical Utility Rover". *Autonomous Robots* vol.11, pp 201-205, Kluwer Press, 2001
6. Rehnmark F., Ambrose R., Goza M., "The Challenges of Extra-Vehicular Robotic Locomotion aboard Orbiting Spacecraft" *Proc. IEEE Int. Conf. on Robotics and Automation*, v 2004, n 2
7. Schmucker, U., Schneider, A., Ihme, T., "Six-legged robot for service operations", *Proc. of the First Euromicro Workshop* 9-11 Oct. 1996 Page(s):135 - 142

Lemur IIb: a Robotic System for Steep Terrain Access

Brett Kennedy, Avi Okon, Hrand Aghazarian, Mircea Badescu, Xiaoqi Bao, Yoseph Bar-Cohen, Zensheu Chang, Borna E. Dabiri, Mike Garrett, Lee Magnone, Stewart Sherrit

Jet Propulsion Laboratory, California Institute of Technology; 4800 Oak Grove Drive, Pasadena, California 91109; brett.a.kennedy@jpl.nasa.gov

Abstract

To extract the full science potential from planetary surface operations, robots must be able to access the entire surface of the planetary body, not just the relatively level areas. Buoyed by the success of *top-to-bottom* strategies employing tethered robots, the stage is set for *bottom-to-top* technologies and techniques to be investigated. To this end, we have designed and built ***Lemur IIb***, a 4-limbed robotic system that is being used to investigate several aspects of climbing system design including the mechanical system (novel end-effectors, kinematics, joint design), sensing (force, attitude, vision), low-level control (force-control for tactile sensing and stability management), and planning (joint trajectories for stability). The technologies developed on this platform will be used to build an advanced system that will climb slopes up to and including vertical faces and overhangs and be able to react forces to maintain stability and do useful work (e.g., sample acquisition/instrument placement). Among the most advanced of these technologies is a new class of Ultrasonic/Sonic Driller/Corer (USDC) end-effectors capable of creating "holds" in rock and soil as well as sampling those substrates. This paper lays out the mechanical, electrical, and algorithmic elements of the ***Lemur IIb*** robot and discusses the future directions of development in those areas.

Keywords: climbing, steep terrain, drilling, autonomy, force control

---

Copyright ©2005 by Springer Verlag. The U.S. Government has a royalty free license to exercise all rights under the copyright claimed herein for Governmental purposes. All other rights are reserved by the copyright owner.

1 Untethered climbing on steep terrain

The creation of a robotic system that will actively cling to vertical or overhanging slopes (i.e., achieve force closure relative to the terrain) requires three areas of development: an end-effector capable of using or creating handholds in various substrates, a robotic platform with sufficient dexterity to properly place those end-effectors, and the force-control and planning algorithms necessary to direct the platform's actuators. In particular, hard-real-time problems of synchronized joint movement are exceptionally acute in limbed systems due to the need for highly coordinated motion within a limb and across multiple limbs (starting, stopping, and smooth motion in between). Moreover, the grouping of the coordinated actuators is constantly changing based on gait requirements. Exacerbating the problem are large numbers of actuators/degrees of freedom (DOFs), intermittent contact, and over-constrained static and dynamic conditions. The problem becomes even more challenging for the operational regime of a free-climbing system in that active force-control for anchoring and stability increases the required level of coordination and synchronization.

Fig. 1. LEMUR IIb on climbing wall and (conceptually) on Endurance Crater

2 Other Robots Relevant to the Lemur IIb Concept

Robots that are comparable to the Lemur IIb (see **Fig. 1.**) fall into three distinct areas: gravity-stabilized legged and wheeled robots, tethered robots, and robots that adhere to the surface. The first category is best represented by MER at JPL and DANTE I (CMU) [1], an 8 legged walking robot for rugged terrain. The second category is characterized by the repelling Cliffbot system at JPL [2] and DANTE II (CMU) [3]. The third category covers robots that use suction cups to adhere to surfaces such as MACS at JPL and Ninja-1 [4], and those that have adhesive end-effectors such as the Gecko robots at iRobot.

The robots in each of these categories cover some portion of the operational envelope of the Lemur IIb. In the first case, the gravity-stabilized robots can traverse over rough terrain. However, these vehicles become unstable on slopes on which gravity becomes a *destabilizing* factor, approximately 45-55 degrees from horizontal. The tethered systems overcome this limitation by counteracting the destabilizing force of gravity with tethers attached to anchors at the top of the slope. This solution itself creates a limitation due to the fact that the traverse must begin at the top of the slope. In addition, the tethered robot may loose contact with the surface (and thus a level of controllability) when an overhang is encountered. An encounter with an overhang may also result in a non-reversible path. The third category relies on active or passive adhesion of the robot to the surface, generally performed by suction cups, magnets, or sticky adhesives. Each of these adhesion methods applies to a very narrow range of substrate properties, primarily smooth, clean, non-friable surfaces (plus ferrous in the case of magnetics). While these techniques are useful for certain scenarios, those scenarios do not as a rule exist for space exploration.

Some of the climbing robots have been developed utilizing bracing contact with the terrain are discussed in references [5-7].

3 Lemur IIb platform overview

The Lemur IIb platform is, with few exceptions, identical to that of Lemur IIa. **Table 1** provides an overview of the robot. The joint design, chassis, electronics, and infrastructure software are all shared. See reference [8] for a more complete description.

Table 1. LEMUR IIb system overview

Mass (kg)	8
Limbs	4
Degrees of Freedom	12
Actuator count (max)	13
Processor speed (MHz)	266

The major divergence lies in the kinematic layout. In an effort to make the challenges of free-climbing more tractable, the decision was made to

restrict initial investigations to the planar problem. In fact, the usual task board is a segment of a gym climbing wall, shown in **Fig. 1**. Given this simplification, the kinematic layout of Lemur IIb was altered from that of Lemur IIa. Kinematic differences include:
- 4 limbs (rather than 6) for decreased system mass and complexity
- 3 degrees-of-freedom per limb (rather than 4) for decreased mass per limb
  - Yaw, Yaw, Pitch (rather than Roll, Yaw, Pitch, Pitch)
- each limb is 25% longer than the limbs of Lemur IIa to increase the reach

These differences result in a platform that is less massive and with a center of gravity closer to the surface of movement, while retaining all of the load carrying capacity, which makes it a more capable platform for climbing inclined planar surfaces.

4 Climbing end-effectors

Three different approaches of differing maturity have been taken with Lemur IIb's end-effectors. The end-effectors that have been used in testing so far are variants on a simple peg. Of greater eventual utility is the Ultrasonic/Sonic Driller/Corer (USDC) that can be used to create handholds. (See **Fig. 2**.) Possibly even more ambitious are end-effectors based on equipment used by human sport climbers (eg. Climbing Cams).

The peg end-effectors come in two configurations: simple peg and self-centering peg. With rubberized contact patches, these pegs provide sufficient friction for testing on inclines up to ~60°. The self-centering peg is a wedge whose apex edge is collinear with the passive wrist axis (normal to surface plane. Contact with a hold causes the wedge to center, but any further self-movement of the robot will not cause any translational errors, unlike a simple round peg.

Due to the advantages of the USDC (*viz.*, simple design, low normal force during operation, dual use for mobility and sampling), this apparatus was chosen as the basis of the drilling end-effector design. In addition, the vibratory nature of the USDC can be used to improve purchase in loose materials.

The USDC is designed for three purposes. First, it has to drill into rock with minimal chance of becoming stuck. Second, it must support the weight of the robot when acting as an anchor. Last, it must be useful in loose scree and other unconsolidated materials. The ability of USDCs to drill into rock has been well demonstrated, but the added desire to minimize the risk of being stuck prompted a redesign that enables the USDC to *hammer out* as well as hammer in. The second requirement drove a new design for the USDC bit that allowed it to act as a cantilevered beam when taking up the weight of the robot. This represents the worst case loading condition. The frictional requirements in unconsolidated materials did not require any change in the USDC actuator design, but did drive the design of the external housing. Through testing, the ability of the USDC to bury itself in loose material was established. The housing, then, was designed to act as a combination "sand anchor" (with longitudinal vanes) and a "ski-pole basket" (with a transverse plate). To date, the climbing-specific USDC has been tested in various materials and the drill rates and breakout loads established. In addition, the penetration depth versus force required for various loose materials has been determined. For greater detail on the design and testing, see reference [9].

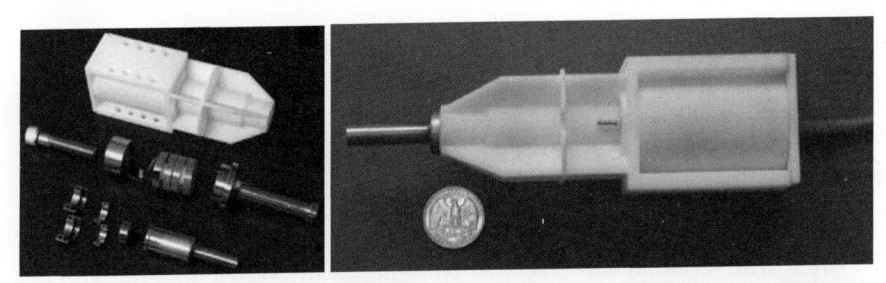

Fig. 2. With the USDC, LEMUR IIb can create its own holds in the terrain.

Emulation of already proven equipment intended for human climbers can provide positive attachment of robots with less operational time and energy than the drilling method. Currently a show-and-tell prototype has been designed and fabricated that incorporates a plain hook and a cam hook (both standard climbing items) into each of three "fingers".

The overall assembly (**Fig. 3.**) can be used for a range of holds:
- Any of the three hooks can be used for horizontal ledges
    - remaining fingers provide in-plane moment support
- The cam hooks work in non-horizontal cracks or slit features
    - Regular hooks are spring loaded, and will move out of the way of the cam hooks
- The linkage driving the fingers can also be made to emulate a "cam", which is a self-locking piece of equipment for pockets or wide cracks

Fig. 3. This end-effector emulates 3 forms of existing climbing equipment.

5 Operational algorithms

Consistent with the task of steep terrain mobility, two algorithms have been implemented on the LEMUR IIb robot.

5.1 Motion planning for robotic climbing

Recent developments in motion planning [10,11] have enabled the LEMUR IIb robot to traverse the climbing wall shown in **Fig. 1**. This planner determines the route through the terrain and the hold-to-hold motions that maintain the robot in static equilibrium.

5.2 Hold characteristic identification using tactile sensing

The above-mentioned planner requires the *a priori* knowledge of three terrain features: contact location, surface normal and coefficient of friction. In application, the robot must acquire the properties of the holds via on-board sensors. Vision-based sensor approaches are inadequate since the surface of interest is often occluded from view. **Fig. 4.** compares the true shape of the hold to an image taken by the robot's left stereo camera.

To this end, we have developed a tactile sensing approach to discern three characteristics of the hold. The kinematics of the LEMUR IIb localizes the hold. The shape of the hold is found by dragging the end-effector along the contour. Lastly, a force-torque sensor measures the contact forces to resolve the coefficient of friction. In operation, a new hold is sensed while three limbs in contact with the terrain support the robot.

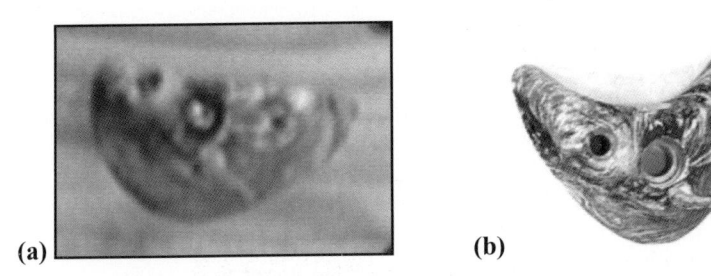

Fig. 4. LEMUR IIb's camera image (a) barely resembles the actual hold shape (b)

Prior to LEMUR IIb sensing the hold, the minimum contact load that disrupts the robot's static equilibrium is determined. This value, less margin, becomes the upper limit for the hold sensing contact force magnitude.

The "free" limb performs the hold sensing via a hybrid force-motion controller. The controller is similar to the work done by Yoshikawa [12]. One distinction is that the manipulator is mounted to a mobile robot and operates quasi-statically. The two technical challenges for this operation are the implementation of force control on manipulator with highly geared, non-back-drivable joints, and contour following with the presence of high surface friction.

Preliminary results show that the LEMUR IIb limb can sense holds while maintaining an average contact force of 3 Newtons and regulating the peak load under 5.1 Newtons. Additionally, the calculated friction coefficients have standard deviations less than 0.09.

6 Conclusions and future work

The LEMUR IIb robot allows the investigation of the technical hurdles associated with free climbing in steep terrain. These include controlling the distribution of contact forces during motion to ensure holds remain intact and to enable mobility through over-hangs. Efforts also can be applied to further in-situ characterization of the terrain, such as testing the strength of the holds and developing models of the individual holds and a terrain map.

Acknowledgements

The work presented in this paper was conducted under contract with the National Aeronautics and Space Administration under the authority of Jet Propulsion Laboratory's Research and Technology Development Program.

References

1. Wettergreen, D., Thorpe, C., and Whittaker, W.L., *Exploring Mount Erebus by Walking Robot*, Robotics and Autonomous Systems, 1993.
2. Huntsberger, T., Sujan. V. A., Dubowsky, S., and Schenker, P. S., *Integrated System for Sensing and Traverse of Cliff Faces*, Proc. Aerosense'03: Unmanned Ground Vehicle Technology V, SPIE Vol. 5083, Orlando, FL, April 22-24, 2003.
3. Bares, J. and Wettergreen, D., *Dante II: Technical Description, Results and Lessons Learned*, Int. Journal of Robotics Research, Vol. 18, No. 7, July, 1999, pp. 621-649.
4. Hirose, S., Nagabuko, A. and Toyama, R., *Machine that can walk and climb on floors, walls, and ceilings*. In 5th Int. Conf. on Advanced Robotics, pages 753–758, Pisa, Italy, 1991
5. Madhani, A. and Dubowsky, S., *Motion planning of mobile multi-limb robotic systems subject to force and friction constraints*. In IEEE Int. Conf. on Robotics and Automation, volume 1, pages 233–239, Nice, France, 1992
6. Neubauer, W., *A spider-like robot that climbs vertically in ducts or pipes*. In IEEE/RSJ/GI Int. Conf. on Intelligent Robots and Systems, pages 1178–1185, Munich, Germany, 1994.
7. Shoval, S., Rimon, E., Shapiro, A., *Design of Spider-Like Robot for Motion with Quasistatic Force Constraints*, In Proc, IEEE Int. Conf. on Robotics and Automation, volume 2, pp. 1377-1383, May 1999
8. Kennedy, B., Okon, A., Aghazarian, H., Garrett, M., Huntsberger, T., Magnone, L., Robinson, M., Townsend, J., *The Lemur II-Class Robots for Inspection and Maintenance of Orbital Structures: A System Description*. To appear in Proc. of the 8th Int. Conf. on Climbing and Walking Robots, London, Sept 2005.
9. Badescu, M., Bao, X., Bar-Cohen, Y., Chang, Z., Dabiri, B. E., Kennedy, B, Sherrit, S., *Adapting the Ultrasonic/Sonic Driller/Corer for Walking/Climbing Robotic Applications*, Proceedings of the SPIE Annual International Symposium on Smart Structures and Materials, San Diego, CA, SPIE Vol. 5762-22, March 6-10, 2005.
10. Bretl, T., Rock, S., Latombe, J. C., Kennedy, B., and Aghazarian, H., *Free-Climbing with a Multi-Use Robot*. In Proc. Int. Symp. on Experimental Robotics (ISER), Singapore, Jun 2004.
11. Bretl, T., Lall, S., Latombe, J.C. and Rock, S., *Multi-Step Motion Planning for Free-Climbing Robots*. In Workshop on the Algorithmic Foundations of Robotics (WAFR), Utrecht/Zeist, The Netherlands, Jul 2004.
12. Yoshikawa, T. and Sudou A., *Dynamic hybrid position/force control of robot manipulators—On-line estimation of unknown constraint*, IEEE Trans. Robotics and Automation, vol. 9, pp. 220–226, Apr. 1993.

__# Tele-operation, Social and Economic Aspects__

Virtual Immersion for Tele-Controlling a Hexapod Robot

Jan Albiez[1], Björn Giesler[2], Jan Lellmann[2], J. Marius Zöllner[1], and Rüdiger Dillmann[1,2]

[1] Interactive Diagnosis- and Servicesystems
   Forschungszentrum Informatik (FZI)
   Haid-und-Neu-Str. 10-14, D-76131 Karlsruhe, Germany
   `albiez@fzi.de`
[2] Institute of Computer Science and Engineering (CSE)
   Chair for Industrial Applications of Informatics and Microsystems (IAIM)
   University of Karlsruhe (TH), Geb.07.21, 76128 Karlsruhe, Germany
   `giesler@ira.uka.de`

1 Introduction

Remote controlled robotic vehicles for different tele-operation tasks have become a common instrument for inspection and search-and-rescue (SAR) applications. These vehicles are usually tracked or wheeled mobile robots controlled by an operator by means of (several) camera links and a joystick system. The benefits of using such a system are clear. Such a vehicle usually has enough payload to accomodate enough cameras to obtain a view on most points of interest needed to control the robot. Data bandwidth poses no problem either since most of the vehicles have a umbilical or a very powerful, and thus heavy, radio link [3]. A problem arising when using such systems in SAR missions is that the weight and the traction system might cause damage to the environment, thus endangering trapped persons.

Legged systems have up to now not been used due to the fact that their complex mechanical system makes remote operation extremely difficult. Recent research in the area of walking machines has therefore mainly concentrated on gaining as much intelligence and autonomy as possible. On inspection and SAR missions, a human operator is currently not replaceable by an artifical intelligence system. While the main reason for using a walking machine in SAR missions is the possibility to reach areas inaccessible for wheeled or tracked robots, the operator must have the possibility to access all the motion control parameters of the legged robot.

Even if the basic motion system is mainly autonomous providing a kind of classical joystick control, the operator has to access all the sensor data and the control parameters to verify and influence the decision-making process of the system. Keeping control of such a system is very difficult for an operator.

The remote control system for a legged system therefore needs to meet the following requirements: Lightweight equipment on the robot and a low data bandwidth is required since the payload of a walking machine is usually very small. The operator interface has to be intuitive and must allow a good and broad overview of the environment surrounding the robot without the need for several camera links.

In this paper we introduce a method for telepresence-based remote operation by using the picture of a panoramic camera as input for tracked video-goggles. In conjunction with a joypad this system gives an operator the feeling of "immersion" into the robot's view of the world. As a test and demonstration plattform, we use a camera mounted on the back of a LAURON III system.

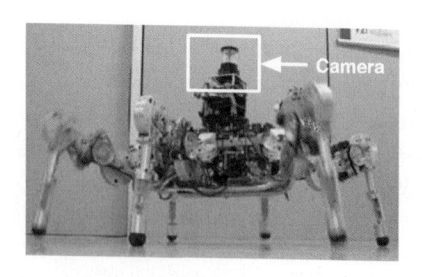

(a) LAURON III

(b) Pan-tilt unit

Fig. 1. The six-legged walking machine LAURON III with the panorama camera mounted on its back and a close up off the pan-tilt unit

2 Test Platform LAURON III

LAURON III (**Fig. 1**, [2]) is a six-legged walking machine built as a universal platform for locomotion in rough terrains. The kinematic structure consists of a main body, a head and six equal legs, each providing three degrees of freedom ($\alpha \pm 80°$, $\beta \pm 100°$, $\gamma \pm 140°$). With all components mounted, the weight of LAURON III is about 18 kg. It has a length and width between the footsteps of about 70 cm and a maximal payload of 10 kg. The maximum distance between body and ground is 36 cm, the maximum velocity is given by $0.3\,\text{ms}^{-1}$. LAURON III matches the requirements for autonomy, whereas the accumulators last for about 45 min (average power consumption 80 Watt).

The sensor suite of LAURON III consist of joint-angle encoders, a 2D inertial system, force sensors mounted in the feet and a pan-tilt unit for mounting other systems like stereo cameras, distance sensors (e.g. point or line laser, ultrasonic sensor) or other sensing equipment like thermal or CO_2 sensors.

In the setup of LAURON III used for the experiments presented in this paper the pan-tilt unit has been equipped with a laser distance sensor (range: 20–500 cm, resolution: 0.5–3 cm).

The system for reactive control and sensor data analysis has been developed over the last years and is very reliable and flexible [1]. Its modular structure, based on the FZI's software framework MCA2 [8], allowed the easy integration of the immersion system.

3 The Immersive Virtual Reality System

The camera images necessary for immersion are captured using an upside-down panoramic camera and transmitted over a wireless link to a host computer. The immersion setup uses a simplified version of the University of Karlsruhe's KARLA augmented reality system (see [6]), which is only used as a virtual reality setup here. A head-mounted display system is used to visualize images acquired from the panoramic camera and undistorted in real-time.

The use of a panoramic camera eliminates the lag induced by the adjustment of a conventional pan-tilt unit on which a conventional camera might be mounted. It also allows free gazing around the last captured scene even if the wireless link has broken down. Thus, diagnostic examination can still be performed even in the case of emergency, which is very useful in search-and-rescue scenarios.

3.1 The Panoramic Camera

The panoramic camera used in our setup consists of a combination of hyperbolic and parabolic mirrors. The mirror constellation is such that image rays from the scene are reflected from the parabolic mirror onto the hyperbolic mirror and from there through a hole in the parabolic mirror onto the image sensor mounted at the bottom of that mirror's base. In the other direction, this setup has the effect of first parallelizing all image rays coming from the sensor and then deflecting them into the scene. Hence, the extensions of all image rays into the parabolic mirror intersect in the paraboloid's focal point (see **Fig. 2(a)**).

(a) Mirror constellation

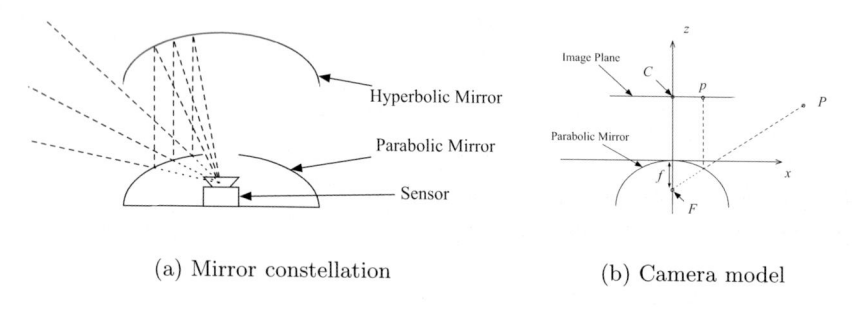

(b) Camera model

Fig. 2. The panoramic camera

This means that the setup can be effectively described by the much simpler one shown in image **Fig. 2(b)** and given by the implicit equation

$$\mathcal{P}: z = f - \frac{1}{4f}(x^2 + y^2) \text{ or} \qquad (1)$$

$$\mathcal{P}: z = ax^2 + ay^2 - \frac{1}{4a}. \qquad (2)$$

Here, f describes the distance from the apex of the paraboloid to its focal point along the z axis, which runs through the paraboloid's axis of symmetry, and $a = -\frac{1}{4f}$ is the opening coefficient derived from f. We use a variant of the method suggested in [5] to calibrate the three parameters a and image center $C = (c_x, c_y)$. It employs the fact that if all image rays intersect in the paraboloid's center of projection, so do the planes created by the images of non-parallel scene lines and the center of projection. The image of a scene line which is not parallel to the paraboloid's axis of symmetry is a circle. Therefore, the camera parameters can be found by finding circles in the image and intersecting the corresponding planes.

After calibration, it is possible to undistort areas of the image, effectively simulating a planar camera whose optical center coincides with the focal point of the paraboloid and whose image plane is tangent to the paraboloid in a single point. This point forms the image center of the virtual planar camera. The undistortion result is smoothed using bilinear interpolation to avoid aliasing effects. Undistorting the image like this takes about 32 ms on our system with PAL resolution of both parabolic and resulting image.

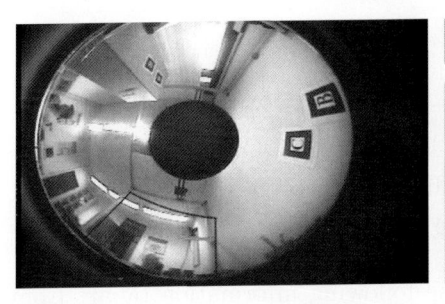

(a) Original image

(b) Undistortion of **Fig.** 3(a)

Fig. 3. Undistorting images from the parabolic camera

3.2 Immersive environment

The undistorted images are presented to the user via a head-mounted display which is fitted with a gyroscopic and inertial sensor. This sensor yields a direction-of-gaze vector which controls both the center of the virtual planar camera and the robot's pan-tilt unit (see **Fig. 4**). The direction of gaze is also used to determine which part of the panoramic image needs to be transferred over the network in the future (namely, a part covering a 180° angle centered around the current direction of gaze). Using this technique together with low-impact JPEG compression, the image data can be transferred even over low-capacity wireless links at about 20fps.

To control the movements of LAURON III the graphical point-and-click control interface has been expanded by the possibility to use a commercial two-stick analogue joypad. Joypads, as they are used in the gaming industry, have the great advantage that they are designed to be used without looking at them (reliefbuttons, hand ergonomic). This eliminates the need of the operator to look at the keyboard or the mouse. The jody-

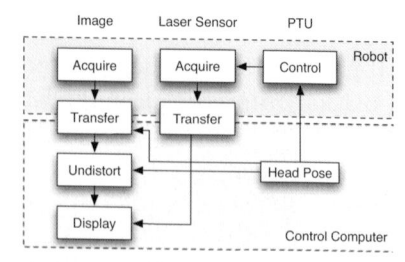

Fig. 4. (left) The HMD with the head pose sensor, (right) flow of image and head pose data.

pad controls the three cartesian walking directions (forward, sideward and turn) and the height of the body (coupled to the step height) and is equipped with a dead-man circuit.

The immersive interface makes the following information (apart from the surround camera view) available to the user: Artificial horizons of both the user (using the HMD inertial sensor) and the robot (using the robot's own inertial sensor), HMD direction of gaze relative to the robot's forward direction, movement of the robot (translation vector, turn angle and body height), foot-ground contact and display and network speed. Additionally, a dot in the user's view marks the current intersection of laser distance sensor and scene and presents the distance obtained from the sensor at that point. See **Fig. 5**.

4 Experiments

The task for the operator(s) during the experiments was to navigate a corridor with various obstacles and walk through a door which is approximately 50 cm wider than LAURON III itself. No direct view of the robot was provided, to make the user rely purely on the immersion system.

The first experiments were been made with the camera mounted topside-up. This allowed a very good overview of the environment around the robot. However, it caused the camera's horizon to lie above the robot; hence it was not possible to look at the legs directly in this setup, making control of the system rather difficult. In the second setup the camera was turned around, looking down on the robot. While this setup reduced the view on the environment around the robot, the operator could look at all

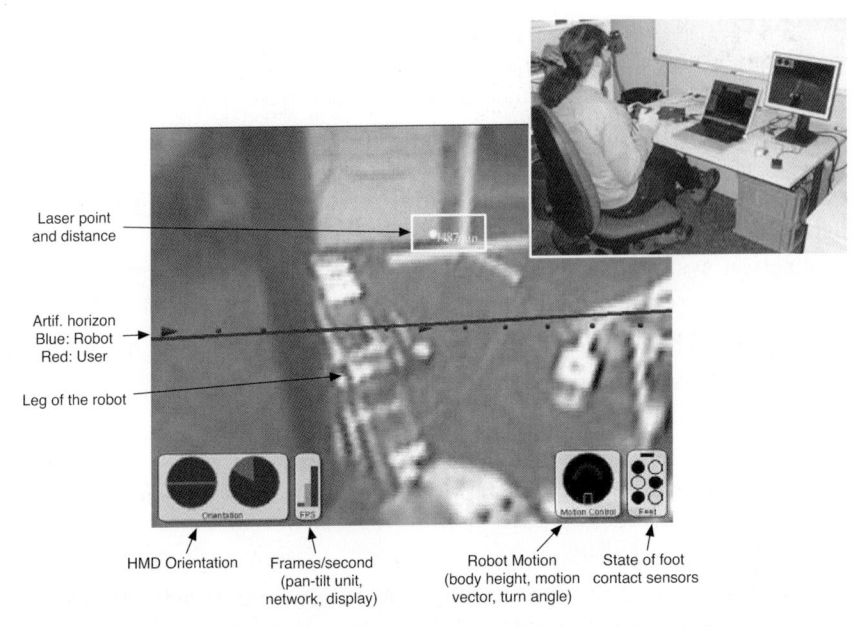

Fig. 5. Immersive interface (user view) and a person using the system

the robot's legs. Even with the reduced overview of the environment this setup allowed moderately good control.

The addition of the pan-tilt-mounted laser range finder improved the results drastically, as it provided an immediate metric estimate of the distance of obstacles, doors and walls simply by looking at them.

The system was evaluated by six people with different degrees of familiarity with the robot ranging from expert to none at all. All of the users reported a very good user experience and an immediate ability to navigate the robot through the immersive interface.

5 Conclusion and Outlook

The possibility to look at each of the legs while they where moving and thus recognizing possible obstacles and avoiding them outranked the better overview of the first setup. Steering LAURON III through narrow spaces proved to be easy and intuitive. Being able to look around the robot by just turning one's own head opened a completly new experience for remote control and tele-operation. This was mainly made possible by the introduction of the panoramic camera and therefore the elimination of lag times entailed by mechanical camera motion rigs.

Future work will concentrate on the remote operation console, replacing the joypad by a tabletop system with more input possibilities and expanding the VR/AR interface by allwoing the inclusion of more external and internal sensor data of the robot. Mid-term goals are the possibility of displaying the data of the 3D world model and the map of LAURON III's localisation system [4] and using a camera mounted on the pan-tilt unit for detailed close-up inspection. The system will be integrated on the dust- and waterproof LAURON IV.

References

1. Albiez J., Dillmann R. (2004). Behaviour Networks for Walking Machines – A Design Method. Proceedings of the *7th International Conference on Climbing and Walking Robots*.
2. Berns K., Cordes St., Ilg W. (1994). Adaptive, neural control architecture for the walking machine lauron. Proceedings of the *International Conference on Intelligent Robots and Systems*, 1172–1177.
3. Cheng-Peng K., Kuu-Young Y. (2003) Challenges in VR-based robot teleoperation Proceedings of the *IEEE International Conference on Robotics and Automation*, Vol. 3, 4392–4397
4. Gassmann B., Frommberger L., Dillmann R., Berns K. (2003). Real-time 3d map building for local navigation of a walking robot in unstructured terrain. Proceedings of the *International Conference on Intelligent Robots and Systems*, Vol. 3, 2185-2190.
5. Christopher Geyer and Kostas Daniilidis. Catadioptric projective geometry. *International Journal of Computer Vision*, 43:223–243, 2001.
6. Björn Giesler, Tobias Salb, Peter Steinhaus, and Rüdiger Dillmann. Using augmented reality to interact with an autonomous mobile platform. In *IEEE Conference on Robotics and Automation (ICRA)*, New Orleans, April 2004.
7. Gu J., Augirre E., Cohen P., 2002 An augmented-reality interface for telerobotic applications Proceedings of the *Sixth IEEE Workshop on Applications of Computer Vision*, Vol. 3, 2719 - 2724
8. Scholl K.-U., Kepplin V., Albiez J., and Dillmann R. (2000). Developing robot prototypes with an expandable modular controller architecture. Proceedings of the *International Conference on Intelligent Autonomous Systems*, S. 67–74.

Economic Prospects for Mobile Robotic Systems, New Modular Components

N. J. Heyes, H. A. Warren

QinetiQ Plc, Farnborough, Hants, GU140LX   njheyes@qinetiq.com

Abstract

The paper for the third year task focuses on new modular components. This year concludes by summarizing the two previous year's by bringing together the results of CLAWAR WP6 and includes other work packages to formulate the specifications for new modular components that can be commercially exploited for multi-role applications.

In year 1, a generic approach was adopted to explore new activities in sectors already exploiting robotic technologies such as the nuclear, construction and automotive industries. Tasks were split into sub-sections with specific activities to assess applications and the risks involved in their exploitation. Levels of economic viability were investigated and recommendations made in order to focus on the future development of new products. We aimed to identify end users at an early stage to maximize the benefits for users and stakeholders.

In year 2, we addressed the key missing ingredients for the absence of a mass market in mobile robotic systems. We took ideas for new market areas and considered the risk and exploitation issues for volume mobile robotics. CLAWAR members gathered information on the current status of mobile robotics markets and attempted to identify areas (from their own expertise) where rules and regulations were required. We encouraged dialogue between different bodies and organizations to aid the development of guidelines. We assessed generic concerns on the introduction of new robotic markets in order to speed up the widespread acceptance of mobile robotic systems.

This year, we have investigated the level of new modular components that are feasible in a selected market. The process is designed to provide an outline formula to define specifications required for new modular components that can be commercially exploited in the future.

1 Introduction

The paper presents the outline CLAWAR focus on new modular components. The work package for 2005 brings together the results of the work in years 1&2 (Tasks WP6.1 and WP6.2) and other work packages. The aim is to outline specifications for new modular components that can be commercially exploited for multi-role activities and applications.

In year 1, a generic approach was adopted to explore new activities in sectors already exploiting robotic technologies such as the nuclear, construction and automotive industries. Tasks were split into sub-sections with specific activities to assess applications and the risks involved in their exploitation. Levels of economic viability were investigated and recommendations made in order to focus on the future development of new products by identifying end users at an early stage thus maximizing the benefits for both users and stakeholders.

In year 2, we addressed the key missing ingredients for the absence of a mass market in mobile robotic systems. We took ideas for new market areas and considered the risk and exploitation issues for volume mobile robotics. CLAWAR members gathered information on the current status of mobile robotics markets and attempted to identify areas from their own expertise, where rules and regulations were needed. We encouraged dialogue between different bodies and organizations to aid the development of guidelines.

This year, 2004/2005 CLAWAR members are viewing modular components and are seeking to outline some generic steps in robotics.

In the full assessment, meetings, workshop and group/individual discussions are reported and address many modularity concept concerns such as industrial & domestic environments, application specifics, moral issues, design limitations, power concerns, human interfaces, safety, performance, communication links, hazards etc. The focus on new modular components is directly related to the results from the work in Tasks 6.1 and 6.2. We are seeking rules and guidelines with WP2 needed for new modular components. We are making technical comparisons for multi-role activities and applications which include further health and safety issues for new markets. Relevant economic data [1] shows the impacts on service robotics worldwide projected sales. Information from the November 2004 robotics workshop is included in the assessment paper. The full paper annexes contain the contract aim, the workshop presentations and relevant information collected by the group.

2 Proposed contents of the full assessment paper due 2005/2006

- Focus on new modular components
- Bringing together the results of the work in Tasks WP6.1 & WP6.2
- Seeking rules and guidelines needed for new modular components
- Multi-role activities and applications
- Health and safety issues for new markets
- Economic impacts from service robotics worldwide sales
- Details of the Robotics Workshop held at Robosoft in November 2004
- Summary of new markets questionnaire findings
- Year 3 summary conclusions and recommendations

2.1 Member information request

Information was sought to obtain ideas and concerns regarding where the modules are required in the new markets and their relationships with existing markets. The following were considered high priority areas.

2.1.1 Manufacturing industry

Current - In-house modularity only, single machine systems, automated assembly tasks carried out, limited flexibility in the systems. Most developed sector but solutions tend to be expensive and not affordable for the vast community. Standards - Machine Directive 98/37/EC, Social directive 95/63/EC. Tend to focus on isolating the machine from the users. Safety is a major issue. Standards for much wider aspects needed (technical, safety, legal, environmental, etc).

Future considerations are safety, open modularity, intelligent cooperative manufactured systems, new non-traditional sectors, pharmaceuticals, bio-tech, food handling.

2.1.2 Maintenance, inspection and construction

Current - Individual solutions for hazardous environments, difficult to access locations, high accuracy requirement, rigorous, repetitiveness, heavy tasks. Standards - Machine directive, ATEX directive, several CENELEC directives and internal guidelines.

Future - Modularity for reducing costs, better reliability, lower maintenance, more generic standards, regulations, easier training as modules are reused.

2.1.3 Hazardous environments

Current - Individual solutions for various hazardous environments, mainly tele-operated. Standards- Nuclear Industry Inspectorate (UK), Machine Directive, ATEX directive, several CENELEC Directives and internal guidelines.

Future - Improved reliability for autonomy, modularity for more cost effectiveness. Re-use must be promoted to make more tasks affordable.

2.1.4 Scientific explorations

Current - No commercial products, R&D prototypes only; some solutions exist for underwater applications. Standards - No specific standards.

Future - Market is niche, always expected to be custom designed prototypes; modularity will help to develop new solutions more quickly and cheaply.

2.1.5 Biomedical and healthcare

Current - Traditional tele-operated systems for specific surgical intervention, limited pre-programmed functionality. Standards - Machine directive, formal approval process but no formal regulatory standards.

Future - Safety, capability/cost balance improved, need specific regulatory standards, address wider needs. Large opportunity exists in this field.

2.1.6 Transportation

Current - Availability of specific off-the-shelf modules (used in CyberCar for example). Standards - Some guidelines exist; CEN EN 1525 (EU), APM standards (USA), BOStrab (Germany) but no widely used standards.

Future - Better cost-effectiveness, need for safety regulations, denser energy sources as modules.

2.1.7 Domestic applications

Current - Vacuum cleaners, lawn mowers. Standards - ISO standards, tech (Bluetooth, RFID, etc).

Future - Safety, 3D navigation systems, usability, capability and cost balance will improve.

2.1.8 Robotic assistance

Current - Rehab robotics for disabled & elderly, limited re-programmed functionality. Standards - ISO standards, tech (Bluetooth, RFID, etc), no specific regulatory standards.

Future - Intelligent adaptable interfaces, capability and cost balance improved, need specific regulatory standards.

2.1.9 Edutainment

Current - In-house modularity only, new mass market in merging education/entertainment. Standards - Rules covering safety, no industry-wide standards.

Future - Reduce costs by open modularity, need for much shorter time to market, use standards from other sectors.

2.1.10 Security and surveillance

Current - Perimeter semi-autonomous bespoke solutions, limited functionality; modular UAVs are well developed.

Standards - Machine directive, in-house guidelines only, DEFSTANs.

Future – Capability and cost balance improves with better integrated technologies (UAV/UGV/UUV), safety standards, wider more generic standards needed.

3 Future applications analysis and summary

In analyzing the application sectors we have highlighted where modules can play their part in forming the next generation of modular robotic devices. It is clear that the current status for innovative robotics is as follows.

3.1 Current status

There are limited sets of solutions available in most sectors. What is available is expensive, not particularly modular and there is little cross fertilization between sectors.

3.2 Standards used

Due to the adoption of a wide range of differing standards, no real generic standards exist. There are few regulations to control robotic activities, e.g. specific rules do operate in the nuclear industry. The demand is there for the technology but there are severe barriers to be overcome for it to be delivered in an effective manner. There are no real standards or regulations that are useful for the manufacturers or for the customers in both system and modular designs. Some specific rules exist but these are not helpful to users or companies. A clear standard for "safety" is needed from module level to user system level for the different sectors.

4 Conclusions

4.1 Future requirements

All the sectors are demanding real time breakthroughs and demand that technology is affordable and realizable. This will only be achievable by cross fertilization, by re-using technologies already developed and by developing modular strategies that incorporate appropriate standards levels.

The modules must be open modules having generic elements of a) input, b) output, c) processing attributes and d) elemental infrastructure for module library inter-linking.

This openness is key to reducing costs, the time to market and will encourage reuse. Fast development of specific modules will require support from an overall modular design methodology process. In addition, specific standards need to be developed and widely accepted. Regulations are also clearly needed so that manufacturers and users know what levels to expect.

There is likely to be a substantial increase in the domestic robot market. Perhaps the more likely area for expansion will be nearer to home – i.e personal robots in and around the house. It is certainly possible that once the idea catches on and when many of the homes in the near future have mobile phone access, then accessible wireless communications solutions will become viable. Phillips at Eindhoven demonstrated this some time ago at their configured house of the future.

We have quizzed many of the "roboteer enthusiasts" and CLAWAR colleagues with >>>>> Have you bought a robot vacuum cleaner or robot lawn mower yet?

Most have said no. – Reasons vary from price, perceived performance, comparison with manual approach and actual needs. It could be said that if we believe in our chosen and dedicated field of robotics – we should be out in the market – promoting – yes **and buying**. We at least then stand accountable by example for the continuation of our robotic engineering profession.

Acknowledgements

Thanks are due to the WP6 contributors who have provided information. Robosoft, Shadow Robot Co, University of Leeds, University of Catania, FZI, Fraunhofer-IPA, IAI-CSIC, RMA-FUB, Poznan University of Technology, QinetiQ, CEA, ISQ, Cybernetix, Politechino di Torino, Kings College London & BAES UK.

References

1. *International Federation of Robotics World 2004 robotics statistical report and analysis*, United Nations publication, ISBN 92-1-101084-5, ISSN 1020-1076.

Any opinions are solely those of the authors or information contributors.

Appendix A: CLAWAR 2005 Organisation

General Chair:

M. Osman Tokhi
Department of Automatic Control and Systems Engineering
The University of Sheffield, UK

International Technical Advisory/Organising Committee

Giovanni Muscato, Italy (Chair)
Manuel Armada, Spain
Yvan Baudoin, Belgium
Karsten Berns, Germany
Philippe Bidaud, France
Krzysztof Kozlowski, Poland
Lakmal Seneviratne, UK

International Programme Committee

Gurvinder S Virk, UK (Chair)
Azad A K M, USA
Arkin R, USA
Balaguer C, Spain
Billingsley J, Australia
Bostater C, USA
Bridge B, UK
Buehler M, Canada
Bugmann G, UK
Chevallerau C, France
Corke P, Australia
Cruse H, Germany
Cubero S, Australia
Davies B, UK
de Almeida A T, Portugal
Dillmann R, Germany
Fortuna L, Italy
Fukuda T, Japan
González de Santos P, Spain
Ge S S, Singapore
Gong Z, China
Gradetsky V, Russia
Halme A, Finland
Hasselvander R, France
Herrera E, Colombia

Kaynak O, Turkey
Kiriazov P, Bulgaria
Kopacek P, Austria
Larin V, Russia
Lefeber D, Belgium
Lopez-Coronado J, Spain
Ma L, China
Molfino R, Italy
M'Sirdi N K, France
Okhotsimsky D E, Russia
Piedbœuf J-C, Canada
Pota H R, Australia
Qiang W-Y, China

Quinn R, USA
Ribeiro M, Portugal
Sa da Costa J, Portugal
Salichs M A, Spain
Shamsudin M, Malaysia
Steinicke L, Belgium
Sy M, Germany
Tanie K, Japan
Vitko A, Slovakia
Waldron K J, USA
Wörn H, Germany
Yigit A, Kuwait
Zomaya A, Australia

National Organising Committee

Tariq P. Sattar (Chair)

Alam M S
Alavi F
Aldebrez F M
Althoefer K
Barnes D
Bell B
Bouazza-Marouf K
Cameron S

Dai J S
Dogramadzi S
Gatsoulis Y
Gharooni S C
Harvey D
Heyes N J
Hossain M A
Howard D

Latif G R
Massoud R
Md Zain M Z
Shaheed M H
Sharma S K
Siddique M N H
Walker R
Warren H A

Climbing Robot Competition Chair

Domenico Longo, Italy

Exhibition Chair

Bill Warren, UK

Appendix B: CLAWAR 2005 Reviewers

Althoefer K, UK
Amati N, Italy
Arkin R, USA
Armada M, Spain
Azad A K M, USA
Baudoin Y, Belgium
Bell B, UK
Berns K, Germany
Bidaud P, France
Billingsley J, Australia
Bouazza-Marouf K, UK
Bridge B, UK
Boxerbaum A, USA
Bugmann G, UK
Cameron S, UK
Chevallerau C, France
Cruse H, Germany
Daltorio K, USA
Dillmann R, Germany
Dunbabin M, Australia
Ge S S, Singapore
Gradetsky V, Russia
Halme A, Finland
Hossain M A, UK

Howard D, UK
Kiriazov P, Bulgaria
Kozlowski K, Poland
Lai X-C, Singapore
Larin V B, Russia
Lefeber D, Belgium
Lopez-Coronado J, Spain
Molfino R, Italy
Muscato G, Italy
Pota H R, Australia
Quinn R, USA
Ribeiro M, Portugal
Rutter B, USA
Sa da Costa J, Portugal
Sattar T, UK
Shaheed M H, UK
Sharma S K, UK
Steinicke L, Belgium
Tokhi M O, UK
Virk G S, UK
Vitko A, Slovakia
Wei T, USA
Yigit A S, Kuwait

Appendix C: CLAWAR 2005 Sponsors & Co-Sponsors

Sponsors

BAE SYSTEMS

CATERPILLAR®

Co-sponsors

The Institution of Electrical Engineers

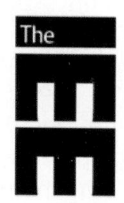

The Institute of Electrical and Electronics Engineers, Robotics and Automation Society
Technical Co-Sponsor

The Institution of Mechanical Engineers, Mechatronics, Informatics and Control Group

International Federation of Robotics

Author Index

Abbate N.	615	Barker A.	89
Abichou A.	221, 253	Basile A.	615
Abourachid A.	319	Beji L.	221, 253
Aghazarian H.	1069, 1077	Benamar F.	475, 1053
Ahmad M. H.	147	Ben Amar F.	533
Akinfiev T.	375, 631, 727, 751	Berns K.	383, 977
Alam M. S.	81, 543, 567, 583, 599	Besseron G.	533
		Bidaud P.	475, 533, 953, 1053
Alarcon P.	425		
Alba D. M.	365, 631	Blanco D.	551
Albarral J. L.	851	Blazevic P.	123, 319, 875
Albiez J.	1087	Boemo E.	667
Albrecht M.	1003	Bonev B.	493
Aldebrez F. M.	543, 599	Borenstein J.	719
Aliverdiev A.	785	Bouazza-Marouf K.	147
Alonso-Puig A.	343	Bourassa P.	277
Althoefer K.	833	Bourgeot J.M.	509
Aoustin Y.	399	Branicky M. S.	65
Arbulú M.	433, 441, 449, 817	Braun T.	977
		Breitwieser H.	97
Armada M.	365, 375, 391, 425, 631, 727, 751	Brenner A.	943
		Bretthauer G.	97
		Bridge B.	139, 935
Asama H.	287	Brockmann W.	107
Aubin R.	875	Brogliato B.	509
Awada B.	969	Bruneau O.	181, 859
Azad A. K. M.	607	Bultstra G.	89
Bacallado G.	365, 631	Buschman H.	89
Backhaus T.	1003	Cabás L.	433, 441, 449, 817
Badescu M.	1077		
Balaguer C.	327, 433, 441, 449, 817, 869, 961	Cabás R.	441, 961
		Caltabiano D.	1061
		Canudas-de-Wit C.	509
Bao X.	1077	Caponero M.	785
Bar-Cohen Y.	1077	Caporaletti G.	245

Carrillo E.	641	Forsman P.	909
Castejón C.	327, 551	Friedrich M.	927
Cazorla M.	493	Ga nik J.	409
Celaya E.	851	Galbraith W.	927
Cepolina F.	995	García D.	985
Chang Z.	1077	Garcia E.	205
Chareyron S.	213	García-Prada J. C.	327
Chatzakos P.	693	Garrett M.	1069, 1077
Chen S.	139	Garzón D.	357
Chicharro C.	817	Gaßmann B.	115
Chugo D.	287	Gatsoulis Y.	1011
Ciardo S.	615	Gatzoulis L.	927
Clark J. E.	261	Genta G.	1045
Coronado J. L.	985	Gerth W.	409
Correal R.	961	Gharooni S. C.	57, 73, 81, 969,
Costo S.	995	Giesler B.	1087
Cotta F.	501	Gimenez A.	327, 869, 961
Creemers T.	851	Gonzalez de Santos P.	205
Curaj A.	677	Gonzalez-Gomez J.	667
Cutkosky M. R.	261	Gorb S.	131
Dabiri B. E.	1077	Grand C.	533
Daerden F.	559	Granosik G.	719, 743
Dal Zot D.	917	Graur A.	229
Daltorio K. A.	131	Guastella C.	615
David A.	181	Guinot J-C.	49
Desbiens A. L.	311	Guyvarch J. P.	875
de Torre S.	441	Hackert R.	319
Dillmann R.	115, 245, 1087	Halme A.	3, 909
Dittrich E.	685	Hamburger V.	383
Drapel K.	685	Han M.-W.	1037
Duhaut D.	641	Harwin W. S.	649
Dulgheru L.	677	Hayward G.	927
Durfee W. K.	19	Heikkilä M.	909
El Kamel M.A.	221	Heller B.	89
Faulisi A.	615	Hennion B.	49
Fauteux P.	277, 311	Hermens H.	89
Fautré T.	917	Heyes N. J.	1095
Feja K.	775	Hilljegerdes J.	335
Fernández R.	375, 727	Hirsi M.	909
Fichera P. A.	785	Hobby J.	89
Focant G.	1031	Höhn O.	409
Fontaine B.	1031	Hongguang W.	173
Fontaine J.-G.	181	Horchler A. D.	131

Hossain M. A.	575, 583	Lefeber D.	189, 559, 759
Howard D.	89	Lellmann J.	1087
Hrissagis K.	693	Leon Rodriguez H.E.	935
Huber M.	115	Lespérance É.	311
Hugel V.	319	Lewinger W. A.	65
Huntsberger T.	1069	Lijin F.	173
Huq M. S.	81	Longo D.	843, 1061
Icardi F.	501	Ludan W.	173
Iida K.	383	Lytridis C.	809
Ijspeert A.	685	Maalej A.	859
Inagaki K.	767	Machado J. A. T.	735
Inoue Y.	157	Macina G.	615
Ion I.	677	Magnone L.	1069, 1077
Ishihara H.	517, 901	Mahalu G.	229
Jaquier C.	685	Mann G.	89
Jardón A.	327, 869, 961	Markopoulos Y. P.	693
Jeziorek P.	41	Martínez H.	493
Joudrier L.	1031	Martínez P.	985
Jurišica L.	793	Maru N.	157
Kaczmarski M	623, 743, 775	Massoud R.	57
Kadar E. E.	809	Matsumoto Y.	485
Kadhim S.H.	551	Matsuyama A.	485
Kaetsu H.	287	Md Zain M. Z.	567, 599
Kameniar D.	793	Médéric P.	953
Kargov A.	97	Mederreg L.	319
Kawabata K.	123, 287	Micheau P.	277
Kaynov D.	433, 441, 449, 817	Mingyang Z.	173
		Minor M. A	525
Kennedy B.	1069, 1077	Mishima T.	287
Kenney L.	89	Mitsuhashi H.	767
Kerr D.	147	Mizoguchi H.	485
Kerscher T.	245	Moeckel R.	685
Khalid A.	693	Mohamed Z.	567
Kirchner F.	335, 1003	Molfino R. M.	501, 995
Kiriazov P.	295	Montes H.	365, 391, 425, 631, 751
Klosek H.	97		
Kobayashi H.	123	Moreno L.	551
Kondaxakis P.	649	Moriconi C.	785
Kozłowski K.	41	Morita K.	517
Krasny D. P.	467	Muscato G.	269, 843, 1061
Kuroi K.	901	Nabulsi S.	391
Lavoie M.A.	311	Nadjar-Gauthier N.	457
Lebastard V.	399	Nunez V.	457

Author Index

Nutter P. 659
Oberle R. 97
Odiase O. P. 825
Ogasawarachi T. 485
Okada T. 197
Okon A. 1069, 1077
Orin D. E. 165, 467
Ouezdou F. B. 859
Palmer III L. R. 165
Pasqui V. 953
Paz E. 893
Peressadko A. 131
Pfeifer R. 383
Pfeiffer K. 801
Phipps C. C. 525
Pinzolas M. 985
Pill J. 49
Planthaber S. 1003
Plestan F. 399
Plumet F. 533, 953
Poirier S. 1053
Ponticelli R. 365, 425, 631
Popa V. 229
Poulain T. 475
Poulsen N. K. 591
Presti M. L. 615
Prieto F. 433
Pylatiuk C. 97
Quinn R. D. 65, 131
Ramchurn V. 659
Ramírez R. 349, 357
Ravn O. 591
Resino J.C. 869
Riabcew P. 775
Richardson R.C. 659, 825
Ritzmann R. E. 131
Roa M. 349, 357
Robinson M. 1069
Rodríguez M.A. 449
Rößler W. 97
Roux M.A. 311
Rubio H. 327
Ruiz V. F. 649

Sagratella G. 785
Sakai T. 197
Sánchez J. 893
Santos V. M. F. 417
Sattar T. P. 139, 935, 943
Sauer P. 41
Šavel M. 793
Schmitt J. 883
Schraft R. D. 801
Schulz S. 97
Seneviratne L. D. 833
Sessa S. 843
Sha N. 89
Shadow Robot Company Ltd
   701
Shaheed M. H. 607
Shang J. 139, 935
Sherrit S. 1077
Shibuya K. 197
Shimizu T. 197
Siddique M. N. H. 575, 583
Silva F. M. T. 417
Silva M. F. 735
Simionescu I. 677
Simons F. 801
Slim R. 253
Slycke P. 89
Smith J. A. S. 833
Somasundaram S. D. 833
Spampinato G. 269
Spenneberg D. 237, 335, 1003
Staroverov P. 433, 441, 449, 817
Steinicke L. 917, 1023, 1031
Stella A. 245
Stöppler H. 1003
Takemura H. 485
Taylor P. 89
Terho S. 909
Testa N. 615
Tokhi M. O. 57, 73, 81, 543, 567, 575, 583, 599, 607, 969

Townsend J.	1069	Wang L.	73
Ueda J.	485	Warren H. A.	1095
Upegui A.	685	Wegener K.	801
Van Damme M.	189, 559, 759	Werner T.	97
Van der Aa N.	89	Wieber P. B.	213
Van Ham R.	189, 559, 759	Yokota S.	123
Vanderborght B.	189, 559, 759	Zaoui C. H.	859
Vázquez F.	893	Zarychta D.	623
Verrelst B.	189, 759, 917	Zhao Y.	73
Vasile A.	677	Zhao Z.	139
Villegas C.	349	Zoellner J. M.	245
Vinayak	303	Zöllner J. M.	115, 1087
Virk G. S.	709, 809, 1011	Zoppi M.	995
Visentin G.	27	Zurlo G. T.	501
Vitko A.	793		
Waliszewski W.	41		